Regions

Miller Projection

SCALE 1:100,000,000

| 0 | 500 | 1,000 | 2,000 | 3,000 | 4,000 | 5,000 | 6,000 | 7,000 | 8,000 |

KILOMETERS

May 1997

Elements of the Nature and Properties of Soils

NYLE C. BRADY
EMERITUS PROFESSOR OF SOIL SCIENCE
CORNELL UNIVERSITY

RAY R. WEIL
PROFESSOR OF SOIL SCIENCE
UNIVERSITY OF MARYLAND AT COLLEGE PARK

PRENTICE HALL
UPPER SADDLE RIVER, NEW JERSEY 07458

Library of Congress Cataloging-in-Publication Data

Brady, Nyle C.
    Elements of nature and properties of soils / Nyle C. Brady, Ray R.
Weil.
            p.   cm.
    Abridged ed. of: The nature and properties of soils. 12th ed.
©1999.
    Includes bibliographical references.
    ISBN 0-13-014497-5
    1. Soil science.   2. Soils.   I. Brady, Nyle C.   II. Weil, Ray R.
Nature and properties of soils.   III. Title.
S591.B792   2000
631.4—dc21   99-25749
                                                              CIP

Publisher: *Charles E. Stewart, Jr.*
Associate Editor: *Kate Linsner*
Production Coordinator: *Adele Kupchik*
Managing Editor: *Mary Carnis*
Manufacturing Buyer: *Marc Bove*
Director of Manufacturing & Production: *Bruce Johnson*
Marketing Manager: *Melissa Bruner*
Formatting/page make-up, production management: *North Market Street Graphics*
Interior illustrations: *Mark Ammerman*
Printer/Binder: *Courier Westford*
Creative Director: *Marianne Frasco*
Cover Designer: *Marianne Frasco*
Cover Illustration: *Connie Hayes*

This text is an abridged edition of *The Nature and Properties of Soils*, 12th Edition,
by Nyle C. Brady and Ray R. Weil.

Printed in the United States of America

10 9 8 7 6 5 4 3 2 1

ISBN 0-13-014497-5

Prentice-Hall International (UK) Limited, *London*
Prentice-Hall of Australia Pty. Limited, *Sydney*
Prentice-Hall Canada Inc., *Toronto*
Prentice-Hall Hispanoamericana, S.A., *Mexico*
Prentice-Hall of India Private Limited, *New Delhi*
Prentice-Hall of Japan, Inc., *Tokyo*
Editora Prentice-Hall do Brasil, Ltda., *Rio de Janeiro*

To all the students and colleagues in soil science who have shared their inspirations, camaraderie, and deep love of the Earth.

CONTENTS

PREFACE   xiii

1 THE SOILS AROUND US   1

1.1	Functions of Soils in Our Ecosystem	2
1.2	Medium for Plant Growth	3
1.3	Regulator of Water Supplies	6
1.4	Recycler of Raw Materials	7
1.5	Habitat for Soil Organisms	7
1.6	Engineering Medium	8
1.7	Soil as a Natural Body	9
1.8	The Soil Profile and Its Layers (Horizons)	11
1.9	Topsoil and Subsoil	13
1.10	Soil: The Interface of Air, Minerals, Water, and Life	16
1.11	Mineral (Inorganic) Constituents of Soils	16
1.12	Soil Organic Matter	18
1.13	Soil Water: A Dynamic Solution	20
1.14	Soil Air: A Changing Mixture of Gases	22
1.15	Interaction of Four Components to Supply Plant Nutrients	22
1.16	Nutrient Uptake by Plant Roots	24
1.17	Soil Quality, Degradation, and Resilience	26
1.18	Conclusion	27
	Study Questions	27
	Reference	28

2 FORMATION OF SOILS FROM PARENT MATERIALS   29

2.1	Weathering of Rocks and Minerals	29
2.2	Physical Weathering (Disintegration)	31
2.3	Chemical Weathering (Decomposition)	33
2.4	Factors Influencing Soil Formation	34
2.5	Parent Materials	35
2.6	Residual Parent Material	35

2.7   Colluvial Debris   36
2.8   Alluvial Stream Deposits   37
2.9   Marine Sediments   38
2.10 Parent Materials Transported by Glacial Ice and Meltwaters   40
2.11 Parent Materials Transported by Wind   42
2.12 Organic Deposits   43
2.13 Climate   45
2.14 Biota: Living Organisms   47
2.15 Topography   50
2.16 Time   51
2.17 Soil Formation in Action   52
2.18 The Soil Profile   55
2.19 Conclusion   57
      Study Questions   58
      References   58

3 SOIL CLASSIFICATION   59

3.1   Concept of Individual Soils   60
3.2   Comprehensive Classification System: *Soil Taxonomy*   60
3.3   Categories and Nomenclature of *Soil Taxonomy*   65
3.4   Soil Orders   66
3.5   Entisols (Recent: Little If Any Profile Development)   70
3.6   Inceptisols (Few Diagnostic Features: Inception of B Horizon)   71
3.7   Andisols (Volcanic Ash Soils)   71
3.8   Gelisols (Permafrost and Frost Churning)   72
3.9   Histosols (Organic Soils without Permafrost)   74
3.10 Aridisols (Dry Soils)   75
3.11 Vertisols (Dark, Swelling and Cracking Clays)   77
3.12 Mollisols (Dark, Soft Soils of Grasslands)   79
3.13 Alfisols (Argillic or Natric Horizon, Medium to High Bases)   81
3.14 Ultisols (Argillic Horizon, Low Bases)   82
3.15 Spodosols (Acid, Sandy, Forest Soils, Low Bases)   83
3.16 Oxisols (Oxic Horizon, Highly Weathered)   84
3.17 Lower-Level Categories in *Soil Taxonomy*   85
3.18 Soil Surveys and Maps   89
3.19 Conclusion   92
      Study Questions   93
      References   94

4 SOIL ARCHITECTURE AND PHYSICAL PROPERTIES   95

4.1   Soil Color   96
4.2   Soil Texture (Size Distribution of Soil Particles)   97
4.3   Soil Textural Classes   100
4.4   Structure of Mineral Soils   104
4.5   Soil Density   106
4.6   Pore Space of Mineral Soils   113
4.7   Formation and Stabilization of Soil Aggregates   115
4.8   Tillage and Structural Management of Soils   121
4.9   Soil Properties Relevant to Engineering Uses   125
4.10 Conclusion   129
      Study Questions   129
      References   130

5 SOIL WATER: CHARACTERISTICS AND BEHAVIOR   131

5.1   Structure and Related Properties of Water   132

5.2 Capillary Fundamentals and Soil Water ... 134
5.3 Soil Water Energy Concepts ... 135
5.4 Soil Moisture Content and Soil Water Potential ... 140
5.5 The Flow of Liquid Water in Soil ... 143
5.6 Infiltration and Percolation ... 146
5.7 Water Vapor Movement in Soils ... 150
5.8 Qualitative Description of Soil Wetness ... 150
5.9 Conclusion ... 155
Study Questions ... 156
References ... 157

6 SOIL AND THE HYDROLOGIC CYCLE   158

6.1 The Global Hydrologic Cycle ... 159
6.2 Fate of Precipitation and Irrigation Water ... 160
6.3 The Soil–Plant–Atmosphere Continuum ... 163
6.4 Efficiency of Water Use ... 167
6.5 Control of Evapotranspiration (ET) ... 169
6.6 Control of Surface Evaporation (E) ... 169
6.7 Percolation-Evaporation Balance ... 171
6.8 Percolation and Groundwaters ... 173
6.9 Enhancing Soil Drainage ... 176
6.10 Septic Tank Drain Fields ... 181
6.11 Irrigation Principles and Practices ... 185
6.12 Conclusion ... 191
Study Questions ... 192
References ... 192

7 SOIL AERATION AND TEMPERATURE   194

7.1 Soil Aeration ... 194
7.2 Means of Characterizing Soil Aeration ... 196
7.3 Oxidation-Reduction (Redox) Potential ... 197
7.4 Factors Affecting Soil Aeration ... 200
7.5 Ecological Effects of Soil Aeration ... 202
7.6 Aeration in Relation to Containerized and Ornamental Plants ... 204
7.7 Wetlands and Their Poorly Aerated Soils ... 207
7.8 Processes Affected by Soil Temperature ... 210
7.9 Absorption and Loss of Solar Energy ... 214
7.10 Thermal Properties of Soils ... 217
7.11 Soil Temperature Control ... 221
7.12 Conclusion ... 223
Study Questions ... 224
References ... 224

8 SOIL COLLOIDS: THEIR NATURE AND PRACTICAL SIGNIFICANCE   226

8.1 Properties of Soil Colloids ... 227
8.2 Adsorbed Cations ... 229
8.3 Fundamentals of Layer Silicate Clay Structure ... 230
8.4 Mineralogical Organization of Silicate Clays ... 232
8.5 Genesis of Soil Colloids ... 236
8.6 Sources of Charges on Soil Colloids ... 238
8.7 Cation Exchange ... 241
8.8 Cation Exchange Capacity ... 241
8.9 Availability of Exchangeable Cations as Plant Nutrients ... 245
8.10 Anion Exchange ... 246

	8.11	Soil Colloids and Environmental Quality	248
	8.12	Conclusion	249
		Study Questions	250
		References	251

9 Soil Acidity, Alkalinity, and Salinity   252

	9.1	Sources of Acidity	253
	9.2	Sources of Alkalinity	256
	9.3	Buffering of Soils	258
	9.4	Variability in Soil pH	260
	9.5	Soil Reaction: Correlations	264
	9.6	Determination of Soil pH	265
	9.7	Methods of Intensifying Soil Acidity	267
	9.8	Decreasing Soil Acidity: Liming Materials	268
	9.9	Reactions of Lime in the Soil	268
	9.10	Lime Requirements: Quantities Needed	269
	9.11	Practical Considerations	271
	9.12	Ameliorating Acidity in Subsoils	273
	9.13	Alkaline and Salt-Affected Soils	273
	9.14	Development of Salt-Affected Soils	274
	9.15	Measuring Salinity and Alkalinity	275
	9.16	Classes of Salt-Affected Soils	276
	9.17	Selective Tolerance of Higher Plants to Saline and Sodic Soils	279
	9.18	Management of Saline and Sodic Soils	281
	9.19	Reclamation of Saline Soils	283
	9.20	Reclamation of Saline-Sodic and Sodic Soils	285
	9.21	Conclusion	287
		Study Questions	288
		References	288

10 Organisms and Ecology of the Soil   290

	10.1	The Diversity of Organisms in the Soil	290
	10.2	Organisms in Action	292
	10.3	Organism Abundance, Biomass, and Metabolic Activity	296
	10.4	Earthworms and Termites	297
	10.5	Soil Microanimals	301
	10.6	Roots of Higher Plants	303
	10.7	Soil Algae	305
	10.8	Soil Fungi	306
	10.9	Soil Actinomycetes	310
	10.10	Soil Bacteria	312
	10.11	Conditions Affecting the Growth of Soil Microorganisms	312
	10.12	Beneficial Effects of Soil Organisms	314
	10.13	Soil Organisms and Damage to Higher Plants	315
	10.14	Ecological Relationships Among Soil Organisms	317
	10.15	Conclusion	320
		Study Questions	320
		References	321

11 Soil Organic Matter   322

	11.1	The Global Carbon Cycle	322
	11.2	The Process of Decomposition in Soils	324
	11.3	Factors Controlling Rates of Decomposition and Mineralization	328
	11.4	Humus: Genesis and Nature	333

11.5	Composts and Composting	335
11.6	Direct Influences of Organic Matter on Plant Growth	338
11.7	Influence of Organic Matter on Soil Properties and the Environment	339
11.8	Management of Amount and Quality of Soil Organic Matter	341
11.9	Carbon Balance in the Soil–Plant–Atmosphere System	343
11.10	Factors and Practices Influencing Soil Organic Matter Levels	344
11.11	Soils and the Greenhouse Effect	349
11.12	Organic Soils (Histosols)	352
11.13	Conclusion	354
	Study Questions	355
	References	355

12 NITROGEN AND SULFUR ECONOMY OF SOILS   357

12.1	Influence of Nitrogen on Plant Growth and Development	358
12.2	Origin and Distribution of Nitrogen	358
12.3	The Nitrogen Cycle	359
12.4	Immobilization and Mineralization	359
12.5	Ammonium Fixation by Clay Minerals	361
12.6	Ammonia Volatilization	361
12.7	Nitrification	362
12.8	The Nitrate Leaching Problem	364
12.9	Gaseous Losses by Denitrification	366
12.10	Biological Nitrogen Fixation	371
12.11	Symbiotic Fixation with Legumes	372
12.12	Symbiotic Fixation with Nonlegumes	375
12.13	Nonsymbiotic Nitrogen Fixation	377
12.14	Addition of Nitrogen to Soil in Precipitation	377
12.15	Reactions of Nitrogen Fertilizers	378
12.16	Practical Management of Soil Nitrogen in Agriculture	378
12.17	Importance of Sulfur	379
12.18	Natural Sources of Sulfur	381
12.19	The Sulfur Cycle	383
12.20	Behavior of Sulfur Compounds in Soils	383
12.21	Sulfur Oxidation and Reduction	385
12.22	Sulfur Retention and Exchange	386
12.23	Sulfur and Soil Fertility Maintenance	387
12.24	Conclusion	389
	Study Questions	389
	References	390

13 SOIL PHOSPHORUS, POTASSIUM, AND MICRONUTRIENTS   391

13.1	Role of Phosphorus in Plant Nutrition and Soil Fertility	392
13.2	Effects of Phosphorus on Environmental Quality	393
13.3	The Phosphorus Cycle	397
13.4	Organic Phosphorus in Soils	400
13.5	Inorganic Phosphorus in Soils	402
13.6	Solubility of Inorganic Phosphorus in Acid Soils	403
13.7	Inorganic Phosphorus Availability at High pH Values	406
13.8	Phosphorus-Fixation Capacity of Soils	406
13.9	Practical Control of Phosphorus Availability	410
13.10	Potassium: Nature and Ecological Roles	411
13.11	Potassium in Plant and Animal Nutrition	411
13.12	The Potassium Cycle	412
13.13	The Potassium Problem in Soil Fertility	415
13.14	Forms and Availability of Potassium in Soils	417

13.15	Factors Affecting Potassium Fixation in Soils	419
13.16	Practical Aspects of Potassium Management	420
13.17	Importance of Micronutrient Elements	421
13.18	Role of the Micronutrients	422
13.19	Source of Micronutrients	422
13.20	Factors Affecting the Availability of Micronutrients	424
13.21	Need for Nutrient Balance	429
13.22	Soil Management and Micronutrient Needs	430
13.23	Conclusion	432
	Study Questions	432
	References	433

14 PRACTICAL NUTRIENT MANAGEMENT   435

14.1	Goals of Nutrient Management	435
14.2	Environmental Quality	438
14.3	Soil–Plant–Atmosphere Nutrient Cycles	441
14.4	Recycling Nutrients through Animal Manures	444
14.5	Storage, Treatment, and Management of Animal Manures	446
14.6	Industrial and Municipal By-Products	448
14.7	Sewage Effluent and Sludge	449
14.8	Practical Utilization of Organic Nutrient Sources	450
14.9	Inorganic Commercial Fertilizers	451
14.10	The Concept of the Limiting Factor	457
14.11	Fertilizer Application Methods	459
14.12	Timing of Fertilizer Application	462
14.13	Diagnostic Tools and Methods	462
14.14	Plant Symptoms and Field Observations	463
14.15	Plant Analysis and Tissue Testing	463
14.16	Soil Analysis	463
14.17	Site-Specific Nutrient Management	468
14.18	Broader Aspects of Fertilizer Practice	469
14.19	Conclusion	470
	Study Questions	471
	References	471

15 SOIL EROSION AND ITS CONTROL   473

15.1	Significance of Soil Erosion and Land Degradation	474
15.2	On-Site and Off-Site Effects of Accelerated Soil Erosion	476
15.3	Mechanics of Water Erosion	480
15.4	Factors Affecting Erosion by Water	482
15.5	Conservation Tillage	489
15.6	Control of Accelerated Erosion on Range and Forest Land	493
15.7	Erosion and Sediment Control on Construction Sites	495
15.8	Wind Erosion and Its Control	499
15.9	Land Capability Classification as a Guide to Conservation	503
15.10	Progress in Soil Conservation	506
15.11	Conclusion	507
	Study Questions	508
	References	509

APPENDIX A U.S. SOIL TAXONOMY SUBORDER MAP AND LEGEND   511

APPENDIX B CANADIAN AND FAO SOIL CLASSIFICATION SYSTEMS   515

APPENDIX C SI UNIT CONVERSION FACTORS AND PERIODIC TABLE OF THE ELEMENTS   519

GLOSSARY   523

INDEX   547

A fundamental knowledge of soil science is a prerequisite to meeting the many natural-resource challenges that will face humanity in the 21st century. But in addition to being of practical importance, we believe that the study of soils can be both fascinating and intellectually satisfying. The soil is an ideal system in which to observe practical applications for basic principles of biology, chemistry, and physics. In turn, these principles can be used to minimize the degradation and destruction of one of our most important natural resources.

The success of the parent text, *The Nature and Properties of Soils, Twelfth Edition,* has helped to define an ecological approach to soil science education around the world. We have been deeply gratified by the positive responses we have received to recent editions of this classic soils textbook. With the widening adoption of *The Nature and Properties of Soils* in introductory soil science courses of all kinds, there has arisen a need for a shorter version that concentrates on the essentials, while maintaining the high standards of rigor and readability that have made the parent text so valuable.

Elements of the Nature and Properties of Soils maintains the ecological approach and general order of topics, but has five fewer chapters and is about 35% shorter than the parent text. We have accomplished this by focusing on the fundamental principles that are the core of soil science, rewriting many sections to make them more concise, and redesigning the art program in a more compact format. In *Elements of the Nature and Properties of Soils* we have paid special attention to the accessibility of the writing and vocabulary. Although we have reduced their number, we have taken care to integrate the many tables and figures into the text and to be sure that each is sufficiently explained by labels and captions to provide maximum educational benefit to our readers.

In this new edition, as in the parent text, our priority is to explain the fundamental principles of soil science in a manner that is relevant to students in many fields of study. Throughout, the text emphasizes the soil as a natural resource and highlights the many interactions between the soil and other components of forest, range, agricultural, wetland, and constructed ecosystems. We recognize that for some readers this book will be their only formal exposure to soil science, while for others it will represent the initial step in a comprehensive soil science education. We have, therefore, sought the advice of professors and students to help us make this book serve both types of readers as an exciting introduction to the fascinating world of soil science, and as a reliable reference in the years to come.

In making the necessary reductions in the length of the text, we have been careful to maintain sections on such topics of current interest in soil science as soil suitability for septic tank drainfields, wetlands and their hydric soils, soil roles in the hydrologic cycle, Gelisols and other changes in *Soil Taxonomy,* environmental uses of clay materials, genetically engineered microorganisms, factors affecting litter quality, and variable-rate agricultural technology. In response to their popularity in the parent text, we have maintained "boxes" that present either fascinating examples and applications or technical details and calculations. Instructors may use these boxes either to *highlight* the material or to make it easy for students to *skip* these materials without losing the logical thread of the text.

Prentice Hall maintains a web site (http://cw.prenhall.com/soils/) for this textbook, giving readers a unique opportunity to enhance their learning with a series of practice quizzes and to broaden their knowledge via a plethora of links to other web sites devoted to topics from every chapter. As is true for the parent text, the artwork in *Elements of the Nature and Properties of Soils* utilizes two colors to make the figures clear, informative, and attractive to the reader. We have found that overhead transparencies can readily be made from these figures by using the photocopier on its lightest setting. To further facilitate the use of the book figures in the classroom, a CD-ROM is available to instructors. This CD-ROM contains most of the line drawings and all 35 color plates from the textbook. An Instructor's Manual contains the CD-ROM as well as chapter summaries. Both the Instructor's Manual and a bank of test items, the PH Custom Test, are available from Prentice Hall.

The devotion and camaraderie we have shared with students and practitioners of soil science is a valued source of inspiration. We are thankful for the many constructive comments we constantly receive by letter, e-mail, and in person from soil scientists, instructors, and students from all parts of the globe. In particular, suggestions from dozens of colleagues were very instrumental in guiding our efforts to produce this condensed version of the textbook.

We would like to give special thanks to the following colleagues who reviewed portions of the text in detail and made extensive suggestions for improvement: Bob Ahrens, Susan Davis, Hari Eswaran, Paul Reich, and Sharon Waltman (USDA/Natural Resources Conservation Service); Kudjo Dzantor, Delvin Fanning, Robert Hill, Bruce James, Margaret Mayers Norton, and Martin Rabenhorst (University of Maryland); Duane Wolf (University of Arkansas); J. Kenneth Torrence (Carleton University); Dan Richter (Duke University); Darrell G. Schulze (Purdue University); Daniel Hillel (University of Massachusetts); Larry King (North Carolina State University); Jimmie Richardson (North Dakota State University); Murray Milford (Texas A & M University); Rattan Lal (Ohio State University); Fred Magdoff and Wendy Sue Harper (University of Vermont); and Mike Swift (U.N. Tropical Biology Program).

Last, but not least, we wish to express our deep appreciation to our wives, Martha and Trish, for their continuing encouragement and understanding as we have devoted much additional time to working on this new condensed textbook. They bear witness to the fact that our efforts to continuously improve *The Nature and Properties of Soils* textbooks is a true labor of love.

N. C. B. and R. R. W.

THE SOILS AROUND US

For in the end we will conserve only what we love.
We will love only what we understand.
And we will understand only what we are taught.
—BABA DIOUM, AFRICAN CONSERVATIONIST

The Earth, our unique home in the vastness of the universe, is in crisis. Depletion of the ozone layer in the upper atmosphere is threatening us with an overload of ultraviolet radiation. Tropical rain forests, and the incredible array of plant and animal species they contain, are disappearing at an unprecedented rate. Groundwater supplies are being contaminated in many areas and depleted in others. In parts of the world, the capacity of soils to produce food is being degraded, even as the number of people needing food is increasing. It will be a great challenge for the current generation to bring the global environment back into balance (Figure 1.1).

FIGURE 1.1   Our planet is unique in that it is covered with life-sustaining water and soil. Great care will be required in preserving the quality of both if our species is to continue to thrive. (Photo courtesy of NASA)

Soils are crucial to life on earth. From ozone depletion to rain forest destruction to water pollution, the global ecosystem is impacted in far-reaching ways by the processes carried out in the soil. To a great degree, soil quality determines the nature of plant ecosystems and the capacity of land to support animal life and society. As human societies become increasingly urbanized, fewer people have intimate contact with the soil, and individuals tend to lose sight of the many ways in which they depend upon soils for their prosperity and survival. The degree to which we are dependent on soils is likely to increase, not decrease, in the future. Of course, soils will continue to supply us with nearly all of our food and much of our fiber. On a hot day, would you rather wear a cotton shirt or one made of polyester? A large percentage of our medicines, including anticancer drugs, will continue to be derived from native and cultivated plants and soil organisms. In addition, biomass grown on soils is likely to become an increasingly important source of energy and industrial feedstocks, as the world's finite supplies of petroleum are depleted over the coming century. The early signs of this trend can be seen in the soybean oil–based inks, the cornstarch plastics, and the wood alcohol fuels that are becoming increasingly important on the market (Figure 1.2).

The art of soil management is as old as civilization. As we move into the 21st century, new understandings and new technologies will be needed to protect the environment and at the same time produce food and biomass to support society. The study of soil science has never been more important to foresters, farmers, engineers, natural resource managers, and ecologists alike.

1.1 FUNCTIONS OF SOILS IN OUR ECOSYSTEM

In any ecosystem, whether your backyard, a farm, a forest, or a regional watershed, soils have five key roles to play (Figure 1.3). *First*, soil supports the growth of higher plants,

FIGURE 1.2   In the future, we will be increasingly dependent on soils to grow renewable resources that can substitute for dwindling supplies of crude oil. Plastics and inks, for example, can be manufactured from soybean oil instead of from petroleum. Soybean oil is edible and far less toxic to the environment. Plastics made from cornstarch, such as packaging foam "peanuts," are readily biodegradable. (Photos courtesy of R. Weil)

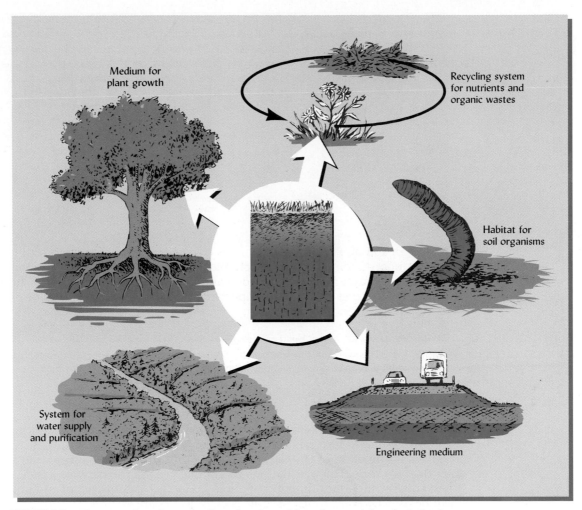

FIGURE 1.3 The many functions of soil can be grouped into five crucial ecological roles.

mainly by providing a medium for plant roots and supplying nutrient elements that are essential to the entire plant. Properties of the soil often determine the nature of the vegetation present and, indirectly, the number and types of animals (including people) that the vegetation can support. *Second*, soil properties are the principal factor controlling the fate of water in the hydrologic system. Water loss, utilization, contamination, and purification are all affected by the soil. *Third*, the soil functions as nature's recycling system. Within the soil, waste products and dead bodies of plants, animals, and people are assimilated, and their basic elements are made available for reuse by the next generation of life. *Fourth*, soils provide habitats for a myriad of living organisms, from small mammals and reptiles to tiny insects to microscopic cells of unimaginable numbers and diversity. *Finally*, in human-built ecosystems, soil plays an important role as an engineering medium. Soil is not only an important building material in the form of earth fill and bricks (baked soil material), but provides the foundation for virtually every road, airport, and house we build.

1.2 MEDIUM FOR PLANT GROWTH

Imagine a growing plant: the aboveground portion may be most familiar, but the portion growing below the soil surface, its root system, may be nearly as large as the portion we see above ground. What things do these plants obtain from the soils in which their roots proliferate? First, the soil mass provides *physical support,* anchoring the root system so that the plant does not fall over. Occasionally, as in Figure 1.4, strong wind or heavy snow does topple a plant whose root system has been restricted by shallow or inhospitable soil conditions.

FIGURE 1.4 This wet, shallow soil failed to allow sufficiently deep roots to develop to prevent this tree from blowing over when snow-laden branches made it top-heavy during a winter storm. (Photo courtesy of R. Weil)

Plant roots depend on the process of respiration to obtain energy. Since root respiration, like our own respiration, produces carbon dioxide (CO_2) and uses oxygen (O_2), an important function of the soil is *ventilation*—allowing CO_2 to escape and fresh O_2 to enter the root zone. This ventilation is accomplished via the network of soil pores.

An equally important function of the soil pores is to absorb water and hold it where it can be used by plant roots. As long as plant leaves are exposed to sunlight, the plant requires a continuous stream of water to use in cooling, nutrient transport, turgor maintenance, and photosynthesis. Since plants use water continuously, but in most places it rains only occasionally, the *water-holding capacity* of soils is essential for plant survival. A deep soil may store enough water to allow plants to survive long periods without rain (see Figure 1.5).

As well as moderating moisture changes in the root environment, the soil also *moderates temperature* fluctuations. Perhaps you can recall digging in garden soil on a summer afternoon and feeling how hot the soil was at the surface and how much cooler just a few centimeters below. The insulating properties of soil protect the deeper portion of the root system from the extremes of hot and cold that often occur at the soil surface.

There are many potential sources of **phytotoxic substances** in soils. These toxins may result from human activity, or they may be produced by plant roots, by microorganisms, or by natural chemical reactions. Many soil managers consider a function of a good soil to be *protection of plants from toxins* by ventilating gases, by decomposing or adsorbing organic toxins, or by suppressing toxin-producing organisms. At the same time, it is true that some microorganisms in soil produce organic, *growth-stimulating compounds*. These substances, when taken up by plants in small amounts, may improve plant vigor.

Soils *supply plants with inorganic, mineral nutrients* in the form of dissolved ions. These mineral nutrients include such metallic elements as potassium, calcium, iron, and copper, as well as such nonmetallic elements as nitrogen, sulfur, phosphorus, and boron. By eating plants, humans and other animals usually obtain the minerals they need (including several elements that plants take up but do not appear to use themselves) indirectly from the soil (see periodic table, Appendix C). Under some circumstances, animals satisfy their craving for minerals by ingesting or licking soil directly (Figure 1.6). The plant takes these elements out of the soil solution and incorporates most of them into the thousands of different organic compounds that constitute plant tissue. A fundamental role of soil in

FIGURE 1.5 A family of African elephants finds welcome shade under the leafy canopy of a huge acacia tree in this East African savanna. The photo was taken in the middle of a long dry season and no rain had fallen for almost five months. The tree roots are still using water from the previous rainy season stored several meters deep in the soil. The light-colored grasses are more shallow-rooted and have either set seed and died or gone into a dried-up dormant condition. (Photo courtesy of R. Weil)

supporting plant growth is to provide a continuing supply of dissolved mineral nutrients in amounts and relative proportions appropriate for plant growth.

Of the 92 naturally occurring chemical elements, only the 18 listed in Table 1.1 have been shown to be **essential elements** without which plants cannot grow and complete their life cycles. The following "sign hanging on a cafe door" may help you remember the

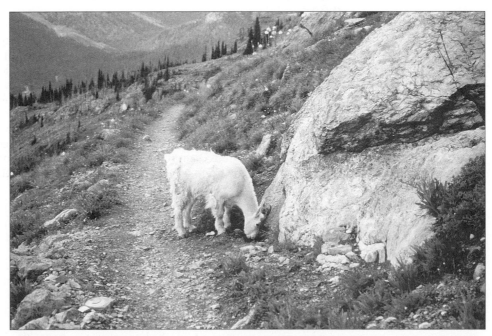

FIGURE 1.6 A mountain goat visits a natural *salt lick* where it ingests needed minerals directly from the soil. Animals normally obtain their dietary minerals indirectly from soils by eating plants. (Photo courtesy of R. Weil)

TABLE I.I   Elements Essential for Plant Growth and Their Sources[a]

The chemical forms most commonly taken in by plants are shown in parentheses, with the chemical symbol for the element in bold type.

Macronutrients: Used in relatively large amounts (>0.1% of dry plant tissue)		Micronutrients: Used in relatively small amounts (<0.1% of dry plant tissue)
Mostly from air and water	*Mostly from soil solids*	*From soil solids*
Carbon (CO_2)	Nitrogen (NO_3^-, NH_4^+)	Iron (Fe^{2+})
Hydrogen (H_2O)	Phosphorus ($H_2PO_4^-$, HPO_4^{2-})	Manganesse (Mn^{2+})
Oxygen (O_2, H_2O)	Potassium (K^+)	Boron (HBO_3)
	Calcium (Ca_2^+)	Zinc (Zn^{2+})
	Magnesium (Mg_2^+)	Copper (Cu^{2+})
	Sulfur (SO_4^{2-})	Chlorine (Cl^-)
		Cobalt (Co^{2+})
		Molybdenum (MoO_4^{2-})
		Nickel (Ni^{2+})

[a] Many other elements are taken up from soils by plants, but are not essential for plant growth. Some of these (such as sodium, silicon, iodine, fluorine, barium, and strontium) do enhance the growth of certain plants, but do not appear to be as universally required for normal growth as are the 18 listed in this table.

18 elements essential for plant growth. Most of the chemical symbols are obvious, but finding those for copper (Cu) and zinc (Zn) may require some imagination.

C.B. HOPKiNS CaFe, Co.
Closed Monday Morning and Night
See You Zoon, the Mg.
C u

Essential elements used by plants in relatively large amounts are called **macronutrients;** those used in smaller amounts are known as **micronutrients.**

In addition to the mineral nutrients just listed, plants may also use minute quantities of organic compounds from soils. However, uptake of these substances is certainly not necessary for normal plant growth. The organic metabolites, enzymes, and structural compounds making up a plant's dry matter consist mainly of carbon, hydrogen, and oxygen, which the plant obtains by photosynthesis from air and water, not from the soil.

It is true that plants can be grown in nutrient solutions without soil (**hydroponics**), but then the plant-support functions of soils must be engineered into the system and maintained at a high cost of time, effort, and management. Although hydroponic production can be feasible on a small scale for a few high-value plants, the production of the world's food and fiber and the maintenance of natural ecosystems will always depend on the use of millions of square kilometers of productive soils.

1.3   REGULATOR OF WATER SUPPLIES

There is much concern about the quality and quantity of the water in our rivers, lakes, and underground aquifers. Governments and citizens everywhere are working to stem the pollution that threatens the value of our waters for fishing, swimming, and drinking. For progress to be made in improving water quality, we must recognize that nearly every drop of water in our rivers, lakes, estuaries, and aquifers has either traveled through the soil or flowed over its surface.[1] Imagine, for example, a heavy rain falling on the hills surrounding the river in Figure 1.7. If the soil allows the rain to soak in, some of the water may be stored in the soil and used by the trees and other plants, while some may seep slowly down through the soil layers to the groundwater, eventually entering the river over a period of months or years as base flow. If the water is contam-

---

[1] This excludes the relatively minor quantity of precipitation that falls directly into bodies of fresh surface water.

FIGURE 1.7 The condition of the soils covering these foothills of the Blue Ridge in West Virginia and Maryland will greatly influence the quality and quantity of water flowing down the Potomac River past Washington, D.C., over 100 kilometers downstream. (Photo courtesy of R. Weil)

inated, as it soaks through the upper layers of soil it is purified and cleansed by soil processes that remove many impurities and kill potential disease organisms.

Contrast the preceding scenario with what would occur if the soil were so shallow or impermeable that most of the rain could not penetrate the soil, but ran off the hillsides on the soil surface, scouring surface soil and debris as it picked up speed, and entering the river rapidly and nearly all at once. The result would be a destructive flash flood of muddy water. Clearly, the nature and management of soils in a watershed will have a major influence on the purity and amount of water finding its way to aquatic systems.

1.4 RECYCLER OF RAW MATERIALS

What would a world be like without the recycling functions performed by soils? Without reuse of nutrients, plants and animals would have run out of nourishment long ago. The world would be covered with a layer, possibly hundreds of meters high, of plant and animal wastes and corpses. Obviously, recycling must be a vital process in ecosystems, whether forests, farms, or cities. The soil system plays a pivotal role in the major geochemical cycles. Soils have the capacity to assimilate great quantities of organic waste, turning it into beneficial **humus**, converting the mineral nutrients in the wastes to forms that can be utilized by plants and animals, and returning the carbon to the atmosphere as carbon dioxide, where it again will become a part of living organisms through plant photosynthesis. Some soils can accumulate large amounts of carbon as soil organic matter, thus having a major impact on such global changes as the much-discussed *greenhouse effect* (see Sections 1.12 and 11.11).

1.5 HABITAT FOR SOIL ORGANISMS

When we speak of protecting ecosystems, most people envision a stand of old-growth forest with its abundant wildlife, or perhaps an estuary such as the Chesapeake Bay with its oyster beds and fisheries. (Perhaps, once you have read this book, you will envision a handful of soil when someone speaks of an ecosystem.) Soil is not a mere pile of broken rock and dead debris. A handful of soil may be home to *billions* of organisms, belonging to thousands of species. In even this small quantity of soil, there are likely to exist predators, prey, producers, consumers, and parasites (Figure 1.8).

How is it possible for such a diversity of organisms to live and interact in such a small space? One explanation is the tremendous range of niches and habitats in even a uniform-appearing soil. Some pores of the soil will be filled with water in which swim organisms such as roundworms, diatoms, and rotifers. Tiny insects and mites may be crawling about in other, larger pores filled with moist air. Micro-zones of good aeration may be only millimeters from areas of **anoxic** conditions. Different areas may be enriched with decaying organic materials; some places may be highly acidic, some more basic. Temperature, too, may vary widely.

Hidden from view in the world's soils are communities of living organisms every bit as complex and intrinsically valuable as their counterparts that roam the savannas, forests, and oceans of the earth. Soils harbor much of the earth's genetic diversity. Soils, like air and water, are important components of the larger ecosystem. Yet only now is soil quality taking its place, with air quality and water quality, in discussions of environmental protection.

1.6   ENGINEERING MEDIUM

"*Terra firma,* solid ground." We usually think of the soil as being firm and solid, a good base on which to build roads and all kinds of structures. Indeed, most structures rest on the soil, and many construction projects require excavation into the soil. Unfortunately, as can be seen in Figure 1.9, some soils are not as stable as others. Reliable construction on soils, and with soil materials, requires knowledge of the diversity of soil properties, as discussed later in this chapter. Designs for roadbeds or building foundations that work well in one location on one type of soil may be inadequate for another location with different soils.

Working with natural soils or excavated soil materials is not like working with concrete or steel. Properties such as bearing strength, compressibility, shear strength, and stability are much more variable and difficult to predict for soils than for manufactured building materials. Chapter 4 provides an introduction to some engineering properties of soils. Many other physical properties discussed will have direct application to engi-

FIGURE 1.8   The soil is home to a wide variety of organisms, both relatively large and very small. Here, a predatory centipede is about to encounter a detritus-eating sowbug. (Photo courtesy of R. Weil)

FIGURE 1.9 Better knowledge of the soils on which this road was built may have allowed its engineers to develop a more stable design, thus avoiding this costly and dangerous situation. (Photo courtesy of R. Weil)

neering uses of soil. For example, Chapter 8 discusses the swelling properties of certain types of clays in soils. The engineer should be aware that when soils with swelling clays are wetted they expand with sufficient force to crack foundations and buckle pavements. Much of the information on soil properties and soil classification discussed in later chapters will be of great value to people planning land uses that involve construction or excavation.

1.7 SOIL AS A NATURAL BODY

You may have noticed that this book sometimes refers to "the soil," sometimes to "a soil," and sometimes to "soils." *The soil* is often said to cover the land as the peel covers an orange. However, while the peel is relatively uniform around the orange, the soil is highly variable from place to place on Earth. In fact, the soil is a collection of individually different soil bodies. One of these individual bodies, *a soil*, is to *the soil* as an individual tree is to the earth's vegetation. Just as one may find sugar maples, oaks, hemlocks, and many other species of trees in a particular forest, so, too, might one find Christiana clay loams, Sunnyside sandy loams, Elkton silt loams, and other kinds of soils in a particular landscape.

A soil is a three-dimensional natural body in the same sense that a mountain, lake, or valley is. By dipping a bucket into a lake you may sample some of its water. In the same way, by digging or augering a hole into a soil, you may retrieve some soil material. Thus you can take a sample of soil material or water into a laboratory and analyze its contents, but you must go out into the field to study a soil or a lake.

In most places, the rock exposed at the earth's surface has crumbled and decayed to produce a layer of unconsolidated debris overlying the hard, unweathered rock. This unconsolidated layer is called the **regolith,** and varies in thickness from virtually nonexistent in some places (i.e., exposed bare rock) to tens of meters in other places. The regolith material, in many instances, has been transported many kilometers from the site of its initial formation and then deposited over the bedrock which it now covers. Thus, all or part of the regolith may or may not be related to the rock now found below it. Where the underlying rock has weathered in place to the degree that it is loose enough to be dug with a spade, the term **saprolite** is used (see Plate 11).

Through their biochemical and physical effects, living organisms such as bacteria, fungi, and plant roots have altered the upper part—and, in many cases, the entire depth—of the regolith. Here, at the interface between the worlds of rock (the lithosphere), air (the atmosphere), water (the hydrosphere), and living things (the biosphere), soil is born. The transformation of inorganic rock and debris into a living soil is one of nature's most fascinating displays. Although generally hidden from everyday view, the soil and regolith can often be seen in road cuts and other excavations (Figure 1.10).

A soil is the product of both destructive and creative (synthetic) processes. Weathering of rock and microbial decay of organic residues are examples of destructive processes, whereas the formation of new minerals, such as certain clays, and of new stable organic compounds are examples of synthesis. Perhaps the most striking result of synthetic processes is the formation of horizontal layers called **soil horizons.** The develop-

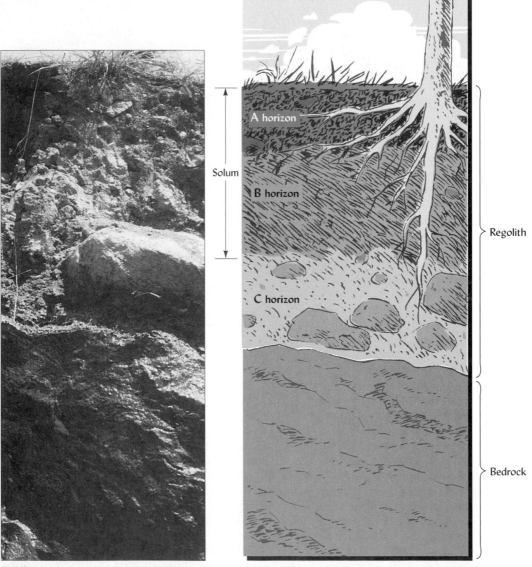

FIGURE 1.10   Relative positions of the regolith, its soil, and the underlying bedrock. Note that the soil is a part of the regolith, and that the A and B horizons are part of the **solum** (from the Latin word *solum,* which means soil or land). The C horizon is the part of the regolith that underlies the solum, but may be slowly changing into soil in its upper parts. Sometimes the regolith is so thin that it has been changed entirely to soil; in such a case, soil rests directly on the bedrock. (Photo courtesy of R. Weil)

ment of these horizons in the upper regolith is a unique characteristic of soil that sets it apart from the deeper regolith materials.

1.8 THE SOIL PROFILE AND ITS LAYERS (HORIZONS)

Soil scientists often dig a large hole, called a *soil pit,* usually several meters deep and about a meter wide, to expose soil horizons for study. The vertical section exposing a set of horizons in the wall of such a pit is termed a ***soil profile.*** Road cuts and other ready-made excavations can expose soil profiles and serve as windows to the soil. Observing how soils exposed in road cuts vary from place to place can add a fascinating new dimension to travel. Once you have learned to interpret the different horizons (see Chapter 2), soil profiles can warn you about potential problems in using the land, as well as tell you much about the environment and history of a region.

Horizons within a soil may vary in thickness and have somewhat irregular boundaries, but generally they parallel the land surface (Figure 1.11). This horizontal alignment is expected since the differentiation of the regolith into distinct horizons is largely the result of influences, such as air, water, solar radiation, and plant material, originating at the soil–atmosphere interface. Since the weathering of the regolith occurs first at the surface and works its way down, the uppermost layers have been changed the most, while the deepest layers are most similar to the original regolith, which is referred to as the soil's *parent material.* In places where the regolith was originally rather uniform in composition, the material below the soil may have a similar composition to the parent material from which the soil formed. In other cases, the regolith material has been transported long distances by wind, water, or glaciers and deposited on top of dissimilar material. In such a case, the regolith material found below a soil may be quite different from the upper layer of regolith in which the soil formed.

A horizons

B horizons

C horizons (parent material)

FIGURE 1.11 This road cut in central Africa reveals soil layers or horizons which parallel the land surface. Taken together, these horizons comprise the profile of this soil, as shown in the enlarged diagram. The upper horizons are designated *A horizons.* They are usually higher in organic matter and darker in color than the lower horizons. Some constituents, such as iron oxides and clays, have been moved downward from the A horizons by percolating rainwater. The lower horizon, called a *B horizon,* is sometimes a zone in which clays and iron oxides have accumulated, and in which distinctive structure has formed. The presence and characteristics of the horizons in this profile distinguish this soil from the thousands of other soils in the world. (Photo courtesy of R. Weil)

Organic matter from decomposed plant leaves and roots tends to accumulate in the uppermost horizons of a soil profile, giving these horizons a darker color than the lower horizons. Also, as weathering tends to be most intense in the upper layers, in many soils these layers have lost some clay or other weathering products by leaching to the horizons below. The organic-matter-enriched horizons nearest the soil surface are called the *A horizons*. In some soils, intensely weathered and leached horizons that have not accumulated organic matter occur in the upper part of the profile, usually just below the A horizon. These horizons may be designated *E horizons* (Figure 1.12).

The layers underlying the A and O horizons contain comparatively less organic matter than the horizons nearer the surface. Varying amounts of silicate clays, iron and aluminum oxides, gypsum, or calcium carbonate may accumulate in the underlying horizons. The accumulated materials may have been washed down from the horizons above, or they may have been formed in place through the weathering process. These underlying layers are referred to as *B horizons* (Figure 1.12).

Plant roots and microorganisms often extend below the B horizon, especially in humid regions, causing chemical changes in the soil water, some biochemical weathering of the regolith, and the formation of *C horizons*. The C horizons are the least weathered part of the soil profile.

In some soil profiles, the component horizons are very distinct in color, with sharp boundaries that can be seen easily by even novice observers. In other soils, the color changes between horizons may be very gradual, and the boundaries more difficult to locate. However, color is only one of many properties by which one horizon may be distinguished from the horizon above or below it (see Figure 1.13). The study of soils in the field is a sensual as well as an intellectual activity. Delineation of the horizons present in a soil profile often requires a careful examination, using all the senses. In addition to seeing the colors in a profile, a soil scientist may feel, smell, and listen to the soil, as well as conduct chemical tests, in order to distinguish the horizons present.

FIGURE 1.12 Horizons begin to differentiate as materials are added to the upper part of the profile and other materials are translocated to deeper zones. Under certain conditions, usually associated with forest vegetation and high rainfall, a leached E horizon forms between organic-matter-rich A and the B horizons. If sufficient rainfall occurs, soluble salts will be carried below the soil profile, perhaps all the way to the groundwater.

FIGURE 1.13 (Left) This soil profile was exposed by digging a pit about 2 meters deep in a well-developed soil (a Hapludalf) in southern Michigan. The top horizon can be easily distinguished because it has a darker color than those below it. However, some of the horizons in this profile are difficult to discern on the basis of color differences, especially in this black-and-white photo. The white string was attached to the profile to clearly demarcate some of the horizon boundaries. Then a trowel full of soil material was removed from each horizon and placed on a board, at right. It is clear from the way the soil either crumbled or held together that soil material from horizons with very similar colors may have very different physical properties. (Photos courtesy of R. Weil)

1.9 TOPSOIL AND SUBSOIL

The organically enriched A horizon at the soil surface is sometimes referred to as *topsoil.* When a soil is plowed and cultivated, the natural state of the upper 12 to 25 centimeters (5 to 10 inches) is modified. In this case, the topsoil may also be called the *plow layer* (or the *furrow slice* in situations such as shown in Figure 4.28). Even where a plow is no longer used, the plow layer will remain evident for many years. For example, if you stroll through a typical New England forest you will see towering trees some 100 years old, but should you dig a shallow pit in the forest floor, you might still be able to see the smooth boundary between the century-old plowed layer and the lighter-colored, undisturbed soil below.

In cultivated soils, the majority of plant roots can be found in the topsoil (Figure 1.14). The topsoil contains a large part of the nutrient and water supplies needed by plants. The chemical properties and nutrient supply of the topsoil may be easily altered by mixing in organic and inorganic amendments, thereby making it possible to improve or maintain the soil's fertility and, to a lesser degree, its productivity. The physical structure of the topsoil, especially the part nearest the surface, is also readily affected by management operations such as tillage and application of organic materials. Maintaining an open structure at the soil surface is especially critical for providing balanced air and water supplies to plant roots and for avoiding excessive losses of soil and water by runoff. Sometimes the plow layer is removed from a soil and sold as topsoil for use at another site. This use of topsoil is especially common to provide a rooting medium suitable for lawns and shrubs around newly constructed buildings, where the original topsoil was removed or buried, and the underlying soil layers were exposed during grading operations (Figure 1.15).

The soil layers that underlie the topsoil are referred to as *subsoil.* The subsoil is not normally seen from the surface, even on plowed cropland, for it lies below the usual depth of plowing. However, the characteristics of the subsoil horizons can greatly influence most land uses. Much of the water needed by plants is stored in the subsoil. Many subsoils also supply important quantities of certain plant nutrients. In some soils there is an abrupt change in properties between the topsoil and the subsoil. In other soils, the change is gradual and the upper part of the subsoil may be quite similar to the topsoil. In most soils the properties of the topsoil are far more conducive to plant growth than those of the subsoil. That is why there is often a good correlation between the productivity of a soil and the thickness of the topsoil in a profile.

FIGURE 1.14 Plant roots respond to the varying conditions they find in the horizons of a soil profile. Roots tend to proliferate in the better-aerated, more fertile, and looser A horizon than in the horizons below. Improvement of soil fertility in the A horizon not only enhances root growth there, but may also increase the vigor and extent of the root system deeper in the profile as well. The roots shown are from crops grown on an Illinois soil (Cisne series) that received no fertilizers or crop residues (left) or that received both fertilizers and crop residues (right). (Photos courtesy of J. B. Fehrenbacher, University of Illinois)

FIGURE 1.15 The large mound of material at this construction site consists of topsoil (A horizon material) carefully separated from the lower horizons and pushed aside during initial grading operations. This stockpile was then seeded with grasses to give it a protective cover. After the construction activities are complete, the stockpiled topsoil will be used in landscaping the grounds around the new building. (Photo courtesy of R. Weil)

Impermeable subsoil layers can impede root penetration, as can very acid subsoils. Poor drainage in the subsoil can result in waterlogged conditions in the topsoil. Because of its relative inaccessibility, it is usually much more difficult and expensive to physically or chemically modify the subsoil. On the other hand, good fertilization of the topsoil can produce vigorous plants whose roots are capable of greater exploration in subsoil layers (Figure 1.14; Box 1.1).

Many of the chemical, biological, and physical processes that characterize topsoils also take place to some degree in the C horizons of soils, which may extend deep into the underlying saprolite or other regolith material. Traditionally, the lower boundary of the soil has been considered to occur at the greatest rooting depth of the natural vegetation, but soil scientists are increasingly studying layers below this in order to understand ecological processes such as groundwater pollution, parent material weathering, and geochemical cycles.

BOX 1.1 USING INFORMATION FROM THE ENTIRE SOIL PROFILE

Soils are three-dimensional bodies that carry out important ecosystem processes at all depths in their profiles. Depending on the particular application, the information needed to make proper land management decisions may come from soil layers as shallow as the upper 1 or 2 centimeters, or as deep as the lowest layers of saprolite (Figure 1.16).

For example, the upper few centimeters of soil often hold the keys to plant growth and biological diversity, as well as to certain hydrologic processes. Here, at the interface between the soil and the atmosphere, living things are most numerous and diverse. Forest trees largely depend for nutrient uptake on a dense mat of fine roots growing in this zone. The physical condition of this thin surface layer may also determine whether rain will soak in or run downhill on the land surface. Certain pollutants, such as lead from highway exhaust, are also concentrated in this zone. For many types of soil investigations it will be necessary to sample the upper few centimeters separately so that important conditions are not overlooked.

On the other hand, it is equally important not to confine one's attention to the easily accessible "topsoil," for many soil properties are to be discovered only in the deeper layers. Plant-growth problems are often related to inhospitable conditions in the B or C horizons that restrict the penetration of roots. Similarly, the great volume of these deeper layers may control the amount of plant-available water held by a soil. For the purposes of recognizing or mapping different types of soils, the properties of the B horizons are often paramount. Not only is this the zone of major accumulations of minerals and clays, but the layers nearer the soil surface are too quickly altered by management and soil erosion to be a reliable source of information for the classification of soils.

In deeply weathered regoliths, the lower C horizons and saprolite play important roles. These layers, generally at depths below 1 or 2 meters, and often as deep as 5 to 10 meters, greatly affect the suitability of soils for most urban uses that involve construction or excavation. The proper functioning of on-site sewage disposal systems and the stability of building foundations are often determined by regolith properties at these depths. Likewise, processes that control the movement of pollutants to groundwater or the weathering of geologic materials may occur at depths of many meters. These deep layers also have major ecological influences because, although the intensity of biological activity and plant rooting may be quite low, the total impact can be great as a result of the enormous volume of soil that may be involved. This is especially true of forest systems in warm climates.

FIGURE 1.16 *Information important to different soil functions and applications is most likely to be obtained by studying different layers of the soil profile.*

1.10 SOIL: THE INTERFACE OF AIR, MINERALS, WATER, AND LIFE

We stated that where the regolith meets the atmosphere, the worlds of air, rock, water, and living things are intermingled. In fact, the four major components of soil are *air, water, mineral matter,* and *organic matter.* The relative proportions of these four components greatly influence the behavior and productivity of soils. In a soil, the four components are mixed in complex patterns; however, the proportion of soil volume occupied by each component can be represented in a simple pie chart (Figure 1.17). Although a handful of soil may at first seem to be a solid thing, it should be noted that only about half the soil volume consists of solid material (mineral and organic); the other half consists of pore spaces filled with air or water. Of the solid material, typically most is mineral matter derived from the rocks of the earth's crust. Only about 5% of the *volume* in this ideal soil consists of organic matter. However, the influence of the organic component on soil properties is often far greater than its small proportion would suggest. Since it is far less dense than mineral matter, the organic matter accounts for only about 2% of the *weight* of this soil.

The spaces between the particles of solid material are just as important to the nature of a soil as are the solids themselves. It is in these pore spaces that air and water circulate, roots grow, and microscopic creatures live. Plant roots need both air and water. In an optimum condition for most plants, the pore space will be divided roughly equally among the two, with 25% of the soil volume consisting of water and 25% consisting of air. If there is much more water than this, the soil will be waterlogged. If much less water is present, plants will suffer from drought. The relative proportions of water and air in a soil typically fluctuate greatly as water is added or lost. Soils with much more than 50% of their volume in solids are likely to be too compacted for good plant growth. Compared to surface soil layers, subsoils tend to contain less organic matter, less total pore space, and a larger proportion of small pores (**micropores**) which tend to be filled with water rather than with air.

1.11 MINERAL (INORGANIC) CONSTITUENTS OF SOILS

Except in the case of organic soils, most of a soil's solid framework consists of **mineral**[2] particles. The larger soil particles, which include stones, gravel, and coarse sands, are

---

[2] The word *mineral* is used in soil science in three ways: (1) as a general adjective to describe inorganic materials derived from rocks; (2) as a specific noun to refer to distinct minerals found in nature, such as quartz and feldspars (see Chapter 2 for detailed discussions of soil-forming minerals and the rocks in which they are found); and (3) as an adjective to describe chemical elements, such as nitrogen and phosphorus, in their inorganic state in contrast to their occurrence as part of organic compounds.

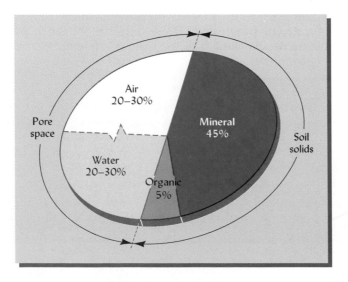

FIGURE 1.17 Volume composition of a loam surface soil when conditions are good for plant growth. The broken line between water and air indicates that the proportions of these two components fluctuate as the soil becomes wetter or drier. Nonetheless, a nearly equal proportion of air and water is generally ideal for plant growth.

generally rock fragments of various kinds. That is, these larger particles are often aggregates of several different minerals. Most smaller particles tend to be made of a single mineral. Thus, any particular soil is made up of particles that vary greatly in both size and composition.

The mineral particles present in soils are extremely variable in size. Excluding, for the moment, the larger rock fragments such as stones and gravel, soil particles range in size over four orders of magnitude: from 2.0 millimeters (mm) to smaller than 0.0002 mm in diameter. **Sand** particles are probably most familiar to us. Individual sand particles are large enough (2.0 to 0.05 mm) to be seen by the naked eye and feel gritty when rubbed between the fingers. Sand particles do not adhere to one another; therefore, sands do not feel sticky. **Silt** particles are somewhat smaller (0.05 to 0.002 mm). Silt particles are too small to see without a microscope or to feel individually, so silt feels smooth but not sticky, even when wet. The smallest class of mineral particles are the **clays** (<0.002 mm), which adhere together to form a sticky mass when wet and hard clods when dry. The smaller particles (<0.001 mm) of clay (and similar-sized organic particles) have **colloidal**[3] properties and can be seen only with the aid of an electron microscope. Because of their extremely small size, colloidal particles possess a tremendous amount of surface area per unit of mass. Since the surfaces of soil colloids (both mineral and organic) exhibit electromagnetic charges that attract positive and negative ions as well as water, this fraction of the soil is the seat of most of the soil's chemical and physical activity.

The proportions of particles in these different size ranges is called **soil texture.** Terms such as *sandy loam, silty clay,* and *clay loam* are used to identify the soil texture. Texture has a profound influence on many soil properties, and it affects the suitability of a soil for most uses (see Section 4.2). To understand the degree to which soil properties can be influenced by texture, imagine dressing in a bathing suit and lying first on a *sandy* beach, and then in a *clayey* mud puddle.

To anticipate the effect of clay on the way a soil will behave, it is not enough to know only the *amount* of clay in a soil. It is also necessary to know the *kinds* of clays present. As home builders and highway engineers know all too well, certain clayey soils, such as those high in smectite clays, make very unstable material on which to build because the clays swell when the soil is wet and shrink when the soil dries. This shrink-and-swell action can easily crack foundations and cause even heavy retaining walls to collapse. These clays also become extremely sticky and difficult to work when they are wet. Other types of clays, formed under different conditions, can be very stable and easy to work with. Learning about the different types of clay minerals will help us understand many of the physical and chemical differences among soils in various parts of the world (see Box 1.2).

Primary and Secondary Minerals

Minerals that have persisted with little change in composition since they were extruded in molten lava (e.g., quartz, micas, and feldspars) are known as ***primary minerals.*** They are prominent in the sand and silt fractions of soils. Other minerals, such as silicate clays and iron oxides, were formed by the breakdown and weathering of less resistant minerals as soil formation progressed. These minerals are called ***secondary minerals*** and tend to dominate the clay and, in some cases, silt fractions.

Soil Structure

Sand, silt, and clay particles can be thought of as the building blocks from which soil is constructed. The manner in which these building blocks are arranged together is called ***soil structure*** (see Section 4.4). The particles may remain relatively independent of each other, but more commonly they are associated together in aggregates of different-size particles. These aggregates may take the form of roundish granules, cubelike blocks, flat

---

[3] Colloidal systems are two-phase systems in which very small particles of one substance are dispersed in a medium of a different substance. Clay and organic soil particles smaller than about 0.001 mm (1 micrometer, μm) in diameter are generally considered to be colloidal in size. Milk and blood are other examples of colloidal systems in which very small solid particles are dispersed in a liquid medium.

BOX 1.2 OBSERVING SOILS IN DAILY LIFE

Your study of soils can be enriched if you make an effort to become aware of the many daily encounters with soils and their influences that go unnoticed by most people. When you dig a hole to plant a tree or set a fence post, note the different layers encountered, and note how the soil from each layer looks and feels different. If you pass a construction site, take a moment to observe the horizons exposed by the excavations. An airplane trip is a great opportunity to observe how soils vary across landscapes and climatic zones. If you are flying during daylight hours, ask for a window seat. Look for the shapes of individual soils in plowed fields if you are flying in spring or fall (Figure 1.18).

Soils can give you clues to understanding the natural processes going on around you. Down by the stream, use a magnifying glass to examine the sand deposited on the banks or bottom. It may contain minerals not found in local rocks and soils, but originating many kilometers upstream. When you wash your car, see if the mud clinging to the tires and fenders is of a different color or consistency than the soils near your home. Does the "dirt" on your car tell you where you have been driving? Forensic investigators have been known to consult with soil scientists to locate crime victims or establish guilt by matching soil clinging to shoes, tires, or tools with the soils at a crime scene.

Other examples of soil hints can be found even closer to home. The next time you bring home celery or leaf lettuce from the supermarket, look carefully for bits of soil clinging to the bottom of the stalk or leaves (Figure 1.19). Rub the soil between your thumb and fingers. Smooth, very black soil may indicate that the lettuce was grown in mucky soils, such as those in New York state or southern Florida. Light brown, smooth-feeling soil with only a very fine grittiness is more typical of California-grown produce, while light-colored, gritty soil is common on produce from the southern Georgia–northern Florida vegetable-growing region. In a bag of dry pinto beans, you may come across a few lumps of soil that escaped removal in the cleaning process because of being the same size as the beans. Often this soil is dark-colored and very sticky, coming from the "thumb" area of Michigan, where a large portion of the U.S. dry bean crop is grown.

Opportunities to observe soils in daily life range from the remote and large-scale to the close-up and intimate. As you learn more about soils, you will undoubtedly be able to see more examples of their influence in your surroundings.

FIGURE 1.18 *The light and dark colored soil bodies, as seen from an airliner flying over central Texas, reflect differences in drainage and topography in the landscape. (Photo courtesy of R. Weil)*

FIGURE 1.19 *Although this Romaine lettuce was purchased in a Virginia grocery store in early Fall, the black, mucky soil clinging to the base of the leaves indicates that it was grown on organic soils, probably in New York state. (Photo courtesy of R. Weil)*

plates, or other shapes. Soil structure (the way particles are arranged together) is just as important as soil texture (the relative amounts of different sizes of particles) in governing how water and air move in soils. Both structure and texture fundamentally influence the suitability of soils for the growth of plant roots.

1.12 SOIL ORGANIC MATTER

Soil organic matter consists of a wide range of organic (carbonaceous) substances, including living organisms (the soil *biomass*), carbonaceous remains of organisms which once occupied the soil, and organic compounds produced by current and past metabolism in the soil. The remains of plants, animals, and microorganisms are continuously broken down in the soil and new substances are synthesized by other microorganisms. Over periods of time ranging from hours to centuries, organic matter is lost from the soil as carbon dioxide produced by microbial respiration. Because of such loss, repeated additions of new plant and/or animal residues are necessary to maintain soil organic matter.

Under conditions that favor plant production more than microbial decay, large quantities of atmospheric carbon dioxide used by plants in photosynthesis are sequestered in the abundant plant tissues which eventually become part of the soil organic matter. Since carbon dioxide is a major cause of the greenhouse effect which is believed to be warming the earth's climate, the balance between accumulation of soil organic matter and its loss through microbial respiration has global implications. In fact, more carbon is stored in the world's soils than in the world's plant biomass and atmosphere combined.

Even so, organic matter comprises only a small fraction of the mass of a typical soil. By weight, typical well-drained mineral surface soils contain from 1 to 6% organic matter. The organic matter content of subsoils is even smaller. However, the influence of organic matter on soil properties, and consequently on plant growth, is far greater than the low percentage would indicate.

Organic matter binds mineral particles into a granular soil structure that is largely responsible for the loose, easily managed condition of productive soils. Part of the soil organic matter that is especially effective in stabilizing these granules consists of certain gluelike substances produced by various soil organisms, including plant roots (Figure 1.20).

Organic matter also increases the amount of water a soil can hold and the proportion of water available for plant growth (Figure 1.21). In addition, it is a major soil source of the plant nutrients phosphorus and sulfur, and the primary source of nitrogen for most plants. As soil organic matter decays, these nutrient elements, which are present in organic combinations, are released as soluble ions that can be taken up by plant roots. Finally, organic matter, including plant and animal residues, is the main food that supplies carbon and energy to soil organisms. Without it, biochemical activity so essential for ecosystem functioning would come to a near standstill.

Humus, usually black or brown in color, is a collection of very complex organic compounds which accumulate in soil because they are relatively resistant to decay. Just as clay is the colloidal fraction of soil mineral matter, so humus is the colloidal fraction of soil organic matter. Because of their charged surfaces, both humus and clay act as contact bridges between larger soil particles; thus, both play an important role in the formation of soil structure. The surface charges of humus, like those of clay, attract and hold both nutrient ions and water molecules. However, gram for gram, the capacity of humus to hold nutrients and water is far greater than that of clay. Unlike clay, humus contains certain components that can have a hormone-like stimulatory effect on plants. All in all, small amounts of humus may remarkably increase the soil's capacity to promote plant growth.

FIGURE 1.20 Abundant organic matter, including plant roots, helps create physical conditions favorable for the growth of higher plants as well as microbes (left). In contrast, soils low in organic matter, especially if they are high in silt and clay, are often cloddy (right) and not suitable for optimum plant growth.

FIGURE 1.21 Soils higher in organic matter are darker in color and have greater water-holding capacities than soils low in organic matter. The soil in each container has the same texture, but the one on the right has been depleted of much of its organic matter. The same amount of water was applied to each container. As the photo shows, the depth of water penetration was less in the high organic matter soil (left) because of its greater water-holding capacity. It required a greater volume of the low organic matter soil (right) to hold the same amount of water.

1.13 SOIL WATER: A DYNAMIC SOLUTION

Water is of vital importance in the ecological functioning of soils. The presence of water in soils is essential for the survival and growth of plants and other soil organisms. The soil moisture regime, often reflective of climatic factors, is a major determinant of the productivity of terrestrial ecosystems, including agricultural systems. Movement of water, and substances dissolved in it, through the soil profile is of great consequence to the quality and quantity of local and regional water resources. Water moving through the regolith is also a major driving force in soil formation.

Two main factors help explain why **soil water** is different from our everyday concept of, say, drinking water in a glass.

1. Water is held within soil pores with varying degrees of tenacity depending on the amount of water present and the size of the pores. The attraction between water and the surfaces of soil particles greatly restricts the ability of water to flow as it would flow in a drinking glass.

2. Because soil water is never pure water, but contains hundreds of dissolved organic and inorganic substances, it may be more accurately called the *soil solution.* An important function of the soil solution is to serve as a constantly replenished, dilute nutrient solution bringing dissolved nutrient elements (e.g., calcium, potassium, nitrogen, and phosphorus) to plant roots.

When the soil moisture content is optimum for plant growth (Figure 1.17), the water in the large and intermediate-sized pores can move in the soil and can be used by plants. As some of the moisture is removed by the growing plants, however, that which remains is in the tiny pores and in thin films around soil particles. The soil solids strongly attract this soil water and consequently compete with plant roots for it. Thus, not all soil water is *available* to plants. Depending on the soil, one-fourth to two-thirds of the moisture remains in the soil after the plants have wilted or died for lack of water.

Soil Solution

The soil solution contains small but significant quantities of soluble inorganic compounds, some of which supply elements that are essential for plant growth. Refer to Table 1.1 for a listing of the 18 **essential elements**, along with their sources. The soil solids, particularly the fine organic and inorganic colloidal particles, release these ele-

Acid	Neutral	Alkaline
pH = 6.0	pH = 7.0	pH = 8.0

FIGURE 1.22 Diagrammatic representation of acidity, neutrality, and alkalinity. At neutrality the H^+ and OH^- ions of a solution are balanced, their respective numbers being the same (pH 7). At pH 6, the H^+ ions are dominant, being 10 times greater, whereas the OH^- ions have decreased proportionately, being only one-tenth as numerous. The solution therefore is acid at pH 6, there being 100 times more H^+ ions than OH^- ions present. At pH 8, the exact reverse is true; the OH^- ions are 100 times more numerous than the H^+ ions. Hence, the pH 8 solution is alkaline. This mutually inverse relationship must always be kept in mind when pH data are used.

ments to the soil solution, from which they are taken up by growing plants. Such exchanges, which are critical for higher plants, are dependent on both soil water and the fine soil solids.

One other critical property of the soil solution is its *acidity* or *alkalinity*. Many chemical and biological reactions are dependent on the levels of the H^+ and OH^- ions in the soil. The levels of these ions also influence the solubility, and in turn the availability, of several essential nutrient elements (including iron and manganese) to plants.

The concentrations of H^+ and OH^- ions in the soil solution are commonly ascertained by measuring the *pH* of the soil solution. Technically, the pH is the negative logarithm of the activity of H^+ ions in the soil solution. Figure 1.22 shows very simply the relationship between pH and the concentration of H^+ and OH^- ions. It should be studied carefully along with Figure 1.23, which shows ranges in pH commonly encountered

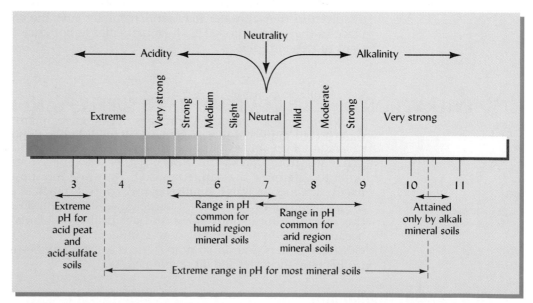

FIGURE 1.23 Extreme range in pH for most mineral soils and the ranges commonly found in humid region and arid region soils. Also indicated are the maximum alkalinity for alkali soils and the minimum pH likely to be encountered in very acid peat soils.

in soils from different climatic regions. Sometimes referred to as a *master variable*, the pH controls the nature of many chemical and microbial reactions in the soil. It is of great significance in essentially all aspects of soil science.

1.14 SOIL AIR: A CHANGING MIXTURE OF GASES

Approximately half of the volume of the soil consists of pore spaces of varying sizes (refer to Figure 1.17), which are filled with either water or air. When water enters the soil, it displaces air from some of the pores; the air content of a soil is therefore inversely related to its water content. If we think of the network of soil pores as the ventilation system of the soil connecting airspaces to the atmosphere, we can understand that when the smaller pores are filled with water the ventilation system becomes clogged. Think how stuffy the air would become if the ventilation ducts of a classroom became clogged. Because oxygen could not enter the room, nor carbon dioxide leave it, the air in the room would soon become depleted of oxygen and enriched in carbon dioxide and water vapor by the respiration (breathing) of the people in it. In an air-filled soil pore surrounded by water-filled smaller pores, the metabolic activities of plant roots and microorganisms have a similar effect.

Therefore, soil air differs from atmospheric air in several respects. First, the composition of soil air varies greatly from place to place in the soil. In local pockets, some gases are consumed by plant roots and by microbial reactions, and others are released, thereby greatly modifying the composition of the soil air. Second, soil air generally has a higher moisture content than the atmosphere; the relative humidity of soil air approaches 100% unless the soil is very dry. Third, the content of carbon dioxide (CO_2) is usually much higher, and that of oxygen (O_2) lower, than contents of these gases found in the atmosphere. Carbon dioxide in soil air is often several hundred times more concentrated than the 0.035% commonly found in the atmosphere. Oxygen decreases accordingly and, in extreme cases, may be 5 to 10%, or even less, compared to about 20% for atmospheric air.

The amount and composition of air in a soil are determined to a large degree by the water content of the soil. The air occupies those soil pores not filled with water. As the soil drains from a heavy rain or irrigation, large pores are the first to be filled with air, followed by medium-sized pores, and finally the small pores, as water is removed by evaporation and plant use. This explains the tendency for soils with a high proportion of tiny pores to be poorly aerated. In such soils, water dominates, and the soil air content is low, as is the rate of diffusion of the air into and out of the soil from the atmosphere. The result is high levels of CO_2 and low levels of O_2, unsatisfactory conditions for the growth of most plants. In extreme cases, lack of oxygen both in the soil air and dissolved in the soil water may fundamentally alter the chemical reactions that take place in the soil solution. This is of particular importance to understanding the functions of wetland soils.

1.15 INTERACTION OF FOUR COMPONENTS TO SUPPLY PLANT NUTRIENTS

As you read our discussion of each of the four major soil components, you may have noticed that the impact of one component on soil properties is seldom expressed independently from that of the others. Rather, the four components interact with each other to determine the nature of a soil. Thus, soil moisture, which directly meets the needs of plants for water, simultaneously controls much of the air and nutrient supply to the plant roots. The mineral particles, especially the finest ones, attract soil water, thus determining its movement and availability to plants. Likewise, organic matter, because of its physical binding power, influences the arrangement of the mineral particles into clusters and, in so doing, increases the number of large soil pores, thereby influencing the water and air relationships.

Essential Element Availability

Perhaps the most important interactive process involving the four soil components is the provision of essential nutrient elements to plants. Plants absorb essential nutrients, along with water, directly from one of these components: the soil solution. However, the amount of essential nutrients in the soil solution at any one time is far less than is needed to produce a mature plant. Consequently, the soil solution nutrient levels must be constantly replenished from the inorganic or organic parts of the soil and from fertilizers or manures added to agricultural soils.

Fortunately, relatively large quantities of these nutrients are associated with both inorganic and organic soil solids. By a series of chemical and biochemical processes, nutrients are released from these solid forms to replenish those in the soil solution. For example, through *ion exchange,* essential elements such as Ca^{2+} and K^+ are released from the colloidal surfaces of clay and humus to the soil solution. The intimate contact between soil solution ions and adsorbed[4] ions makes this exchange possible. In the example below, an H^+ ion in the soil solution is shown to exchange readily with an adsorbed K^+ ion on the colloidal surface. The K^+ ion is then available in the soil solution for uptake by crop plants.

$$\boxed{\text{colloid}}\, K^+ + H^+ \text{ ion} \longrightarrow \boxed{\text{colloid}}\, H^+ + K^+ \text{ ion}$$

Adsorbed Soil solution Adsorbed Soil solution

The K^+ ion thus released can be readily taken up (absorbed) by plants. Some scientists consider that this ion exchange process is among the most important of chemical reactions in nature.

Nutrient ions are also released to the soil solution as soil microorganisms decompose organic tissues. Plant roots can readily absorb all of these nutrients from the soil solution, provided there is enough O_2 in the soil air to support root metabolism.

Most soils contain large amounts of plant nutrients relative to the annual needs of growing vegetation. However, the bulk of most nutrient elements is held in the structural framework of primary and secondary minerals and organic matter. Only a small fraction of the nutrient content of a soil is present in forms which are readily available to plants. Table 1.2 will give you some idea of the quantities of various essential elements present in different forms in typical soils of humid and arid regions.

Figure 1.24 illustrates how the two solid soil components interact with the liquid component (soil solution) to provide essential elements to plants. Plant roots do not ingest soil particles, no matter how fine, but are able to absorb only nutrients that are dissolved in the soil solution. Because elements in the coarser soil framework of the soil are only slowly released into the soil solution over long periods of time, the bulk of most nutrients in a soil is not readily available for plant use. Nutrient elements in the framework of colloid particles are somewhat more readily available to plants, as these particles break down much faster because of their greater surface area. Thus, the struc-

---

[4] *Adsorption* refers to the attraction of ions to the surface of particles, in contrast to *absorption,* the process by which ions are taken *into* plant roots. The adsorbed ions are exchangeable with ions in the soil solution.

TABLE 1.2 Quantities of Six Essential Elements Found in Upper 15 cm of Representative Soils in Temperate Regions

Essential element	Humid region soil			Arid region soil		
	In solid framework, kg/ha	Exchangeable, kg/ha	In soil solution, kg/ha	In solid framework, kg/ha	Exchangeable, kg/ha	In soil solution, kg/ha
Ca	8,000	2,250	60–120	20,000	5,625	140–280
Mg	6,000	450	10–20	14,000	900	25–40
K	38,000	190	10–30	45,000	250	15–40
P	900	—	0.05–0.15	1,600	—	0.1–0.2
S	700	—	2–10	1,800	—	6–30
N	3,500	—	7–25	2,500	—	5–20

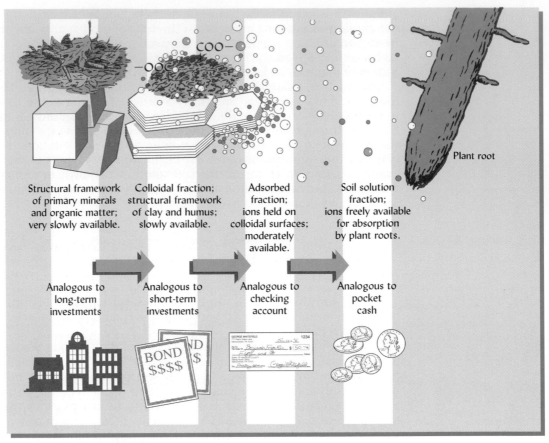

FIGURE 1.24 Nutrient elements exist in soils in various forms characterized by different accessibility to plant roots. The bulk of the nutrients is locked up in the structural framework of primary minerals, organic matter, clay, and humus. A smaller proportion of each nutrient is adsorbed in a swarm of ions near the surfaces of soil colloids (clay and organic matter). From the swarm of adsorbed ions, a still smaller amount is released into the bulk soil solution, where uptake by plant roots can take place.

tural framework is the major storehouse and, to some extent, a significant source of essential elements in many soils.

The distribution of nutrients among the various components of a fertile soil, as illustrated in Figure 1.24, may be likened to the distribution of financial assets in the portfolio of a wealthy individual. In such an analogy, nutrients readily available for plant use would be analogous to cash in the individual's pocket. A millionaire would likely keep most of his or her assets in long-term investments such as real estate or bonds (the coarse fraction solid framework), while investing a smaller amount in short-term stocks and bonds (colloidal framework). For more immediate use, an even smaller amount might be kept in a checking account (exchangeable nutrients), while a tiny fraction of the overall wealth might be carried to spend as currency and coins (nutrients in the soil solution). As the cash is used up, the supply is replenished by making a withdrawal from the checking account. The checking account, in turn, is replenished occasionally by the sale of long-term investments. It is possible for a wealthy person to run short of cash even though he or she may own a great deal of valuable real estate. In an analogous way, plants may use up the readily available supply of a nutrient even though the total supply of that nutrient in the soil is very large. Luckily, in a fertile soil, the process described in Figure 1.24 can help replenish the soil solution as quickly as plant roots remove essential elements.

1.16 NUTRIENT UPTAKE BY PLANT ROOTS

To be taken up by a plant, a nutrient element must be in a soluble form and must be located *at the root surface.* Often, parts of a root are in such intimate contact with soil particles (see Figure 1.25) that a direct exchange may take place between nutrient ions adsorbed on the surface of soil colloids and H^+ ions from the surface of root cell walls. In any case, the supply of nutrients in contact with the root will soon be depleted. This fact raises the question of how a root can obtain additional supplies once the nutrient ions at the root surface have all been taken up into the root. There are three basic mechanisms by which the concentration of nutrient ions at the root surface is maintained (Figure 1.26).

First, **root interception** comes into play as roots continually grow into new, undepleted soil. For the most part, however, nutrient ions must travel some distance in the soil solution to reach the root surface. *Second,* movement of ions can take place by **mass flow,** as when dissolved nutrients are carried along with the flow soil water toward a root that is actively drawing water from the soil. In this type of movement, the nutrient ions are somewhat analogous to leaves floating down a stream. *Third,* nutrient ions continually move by **diffusion** from areas of greater concentration toward the nutrient-depleted areas of lower concentration around the root surface.

In the diffusion process, the random movements of ions in all directions causes a *net* movement from areas of high concentration to areas of lower concentrations, independent of any mass flow of the water in which the ions are dissolved. Factors such as soil compaction, cold temperatures, and low soil moisture content, which reduce root interception, mass flow, or diffusion, can result in poor nutrient uptake by plants even in soils with adequate supplies of soluble nutrients. Furthermore, the availability of nutrients for uptake can also be negatively or positively influenced by the activities of microorganisms that thrive in the immediate vicinity of roots. Maintaining the supply of available nutrients at the plant root surface is thus a process that involves complex interactions among different soil components.

It should be noted that the plant membrane separating the inside of the root cell from the soil solution is permeable to dissolved ions only under special circumstances. Plants do not merely take up, by mass flow, those nutrients that happen to be in the water that roots are removing from the soil. Nor do dissolved nutrient ions brought to the root's outer surface by mass flow or diffusion cross the root cell membrane and enter the root passively by diffusion. On the contrary, a nutrient is taken up when a chemical carrier molecule in the root cell membrane forms an activated complex with the nutrient and then travels across the membrane to the interior of the root cell before releasing the nutrient. The carrier mechanism, activated by root metabolic energy, allows the plant to accumulate concentrations of nutrients inside the root cell that far exceed the nutrient concentrations in the soil solution. Because different nutrients are taken up by specific types of carrier molecules, the plant is able

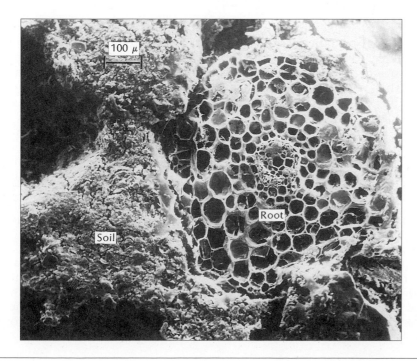

FIGURE 1.25 Scanning electron micrograph of a cross section of a peanut root surrounded by soil. Note the intimacy of contact. [Courtesy Tan and Nopamornbodi (1981)]

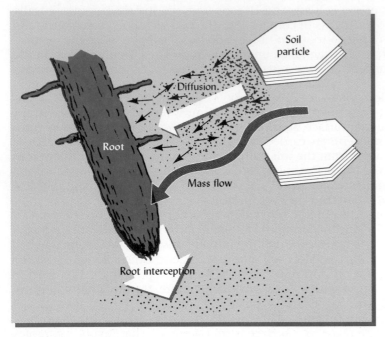

FIGURE 1.26 Three principal mechanisms by which nutrient ions dissolved in the soil solution come into contact with plant roots. All three mechanisms may operate simultaneously, but one mechanism or another may be most important for a particular nutrient. For example, in the case of calcium, which is generally plentiful in the soil solution, mass flow alone can usually bring sufficient amounts to the root surface. However, in the case of phosphorus, diffusion is needed to supplement mass flow because the soil solution is very low in this element in comparison to the amounts needed by plants.

to exert some control over how much and in what relative proportions essential elements are taken up.

Since nutrient uptake is an active metabolic process, conditions that inhibit root metabolism may also inhibit nutrient uptake. Examples of such conditions include excessive soil water content or soil compaction resulting in poor soil aeration, excessively hot or cold soil temperatures, and aboveground conditions which result in low translocation of sugars to plant roots. We can see that plant nutrition involves biological, physical, and chemical processes and interactions among many different components of soils and the environment.

1.17 SOIL QUALITY, DEGRADATION, AND RESILIENCE

Soil is a basic resource underpinning all terrestrial ecosystems. Managed carefully, soils are a reusable resource, but in the scale of human lifetimes they cannot be considered a renewable resource. As we shall see in the next chapter, most soil profiles are thousands of years in the making. In all regions of the world, human activities are destroying some soils far faster than nature can rebuild them. Soils completely washed away by erosion or excavated and paved over by urban sprawl are lost, for all practical purposes. More often, soils are degraded in quality rather than totally destroyed.

Soil quality is a measure of the ability of a soil to carry out particular ecological functions, such as those described in Sections 1.2 to 1.6. Soil quality reflects a combination of *chemical, physical,* and *biological* properties. Some of these properties are relatively unchangeable, inherent properties that help define a particular type of soil. Soil texture and mineral makeup (Section 1.11) are examples. Other soil properties, such as structure (Section 1.11) and organic matter content (Section 1.12) can be significantly changed by management. These more changeable soil properties can indicate the status of a soil's quality relative to its potential, much the way water turbidity or oxygen content indicate the water quality status of a river.

Mismanagement of forests, farms, and rangeland causes widespread degradation of soil quality by erosion that removes the topsoil, little by little (see Chapter 15). Another widespread cause of soil degradation is the accumulation of salts in improperly irrigated soils in arid regions. When people cultivate soils and harvest the crops without returning organic residues and mineral nutrients, the soil supply of organic matter and nutrients becomes depleted (see Chapter 11). Such depletion is particularly widespread in sub-Saharan Africa, where degrading soil quality is reflected in diminished capacity to produce food. Contamination of a soil with toxic substances from industrial processes or chemical spills can degrade its capacity to provide habitat for soil organisms, to grow plants that are safe to eat, or to safely recharge ground and surface waters. Degradation of soil quality by pollution is usually localized, but the environmental impacts and costs involved are very large.

While protecting soil quality must be the first priority, it is often necessary to attempt to restore the quality of soils that have already been degraded. Some soils have sufficient **resilience** to recover from minor degradation if left to revegetate on their own. In other cases, more effort is required to restore degraded soils. Organic and inorganic amendments may have to be applied, vegetation may have to be planted, physical alterations by tillage or grading may have to be made, or contaminants may have to be removed. As societies around the world assess the damage already done to their soil resources, the job of **soil restoration** is becoming serious business.

1.18 CONCLUSION

The earth's soil is comprised of numerous soil individuals, each of which is a three-dimensional natural body in the landscape. Each individual soil is characterized by a unique set of properties and soil horizons as expressed in its profile. The nature of the soil layers seen in a particular profile is closely related to the nature of the environmental conditions at a site.

Soils perform five broad ecological functions: they act as the principal medium for plant growth, regulate water supplies, recycle raw materials and waste products, and serve as a major engineering medium for human-built structures. They are also home to many kinds of living organisms. Soil is thus a major ecosystem in its own right. The soils of the world are extremely diverse, each type of soil being characterized by a unique set of soil horizons. A typical surface soil in good condition for plant growth consists of about half solid material (mostly mineral, but with a crucial organic component, too) and half pore spaces filled with varying proportions of water and air. These components interact to influence a myriad of complex soil functions, a good understanding of which is essential for wise management of our terrestrial resources.

STUDY QUESTIONS

1. As a society, is our reliance on soils likely to increase or decrease in the decades ahead? Explain.

2. Discuss how *a soil,* a natural body, differs from *soil,* a material that is used in building a roadbed.

3. What are the five main roles of soil in an ecosystem? For each of these ecological roles, suggest one way in which interactions occur with another of the five roles.

4. Think back over your activities during the past week. List as many incidents as you can in which you came into direct or indirect contact with soil.

5. Figure 1.17 shows the volume composition of a loam surface soil in ideal condition for plant growth. To help you understand the relationships among the four components redraw this pie chart to represent what the situation might be after the soil has been compacted by heavy traffic. Redraw the original pie chart again, but show how the four components would be related on a mass (weight) basis rather than on a volume basis.

6. Explain in your own words how the soil's nutrient supply is held in different forms, much the way that a person's financial assets might be held in different forms.

7. List the essential nutrient elements that plants derive mainly from the soil.

8. Are all elements contained in plants essential nutrients? Explain.
9. Define these terms: *soil texture, soil structure, soil pH, humus, soil profile, B horizon, soil quality, solum,* and *saprolite.*
10. Describe four processes that commonly lead to degradation of soil quality.

REFERENCE

Tan, K. H., and O. Nopamornbodi. 1981. "Electron microbeam analysis and scanning electron microscopy of soil–root interfaces," *Soil Sci.,* **131**:100–106.

It is a poem of existence . . . not a lyric but a slow-moving epic whose beat has been set by eons of the world's experience. . . .
—JAMES MICHENER, CENTENNIAL

The first astronauts to explore the moon labored in their clumsy pressurized suits to collect samples of rocks and dust from the lunar surface. These they carried back to Earth for analysis. It turned out that moon rocks are similar in composition to those found deep in the Earth—so similar that scientists concluded that the moon itself began as a large chunk of molten Earth that broke away eons ago, when the young planet nearly melted in a stupendous collision with a Mars-sized object, leaving the Pacific Ocean as a scar. On the moon, this rock remained unchanged or crumbled into dust with the impact of meteors. On Earth, the rock at the surface, coming in contact with water, air, and living things, was transformed into something new, into many different kinds of living soils. This chapter reveals the story of how rock and dust become "the ecstatic skin of the Earth."[1]

We will study the processes of soil formation that develop, from relatively uniform regolith, the variegated layers of the soil profile. We will also learn about the environmental factors that influence these processes to produce soils in Belgium so different from those in Brazil, soils on limestone so different from those on sandstone, and soils in the valley bottoms so different from those on the hills.

Every landscape is comprised of a suite of different soils, each influencing ecological processes in its own way. Whether we intend to modify, exploit, preserve, or simply understand the landscape, our success will depend on our knowing how soil profile properties relate to the environment on each site and to the landscape as a whole.

2.1   WEATHERING OF ROCKS AND MINERALS

The influence of **weathering**, the physical and chemical breakdown of particles, is evident everywhere. Nothing escapes it. It breaks up rocks and minerals, modifies or destroys their physical and chemical characteristics, and carries away the soluble products. Likewise, it synthesizes new minerals of great significance in soils. The nature of the rocks and minerals being weathered determines the rates and results of the breakdown and synthesis (Figure 2.1).

---

[1] From Logan (1995).

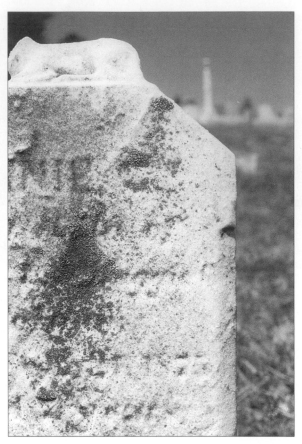

FIGURE 2.1 Two stone markers, photographed on the same day in the same cemetery, illustrate the effect of rock type on weathering rates. The date and initials carved in the slate marker in 1798 are still sharp and clear, while the date and figure of a lamb carved in the marble marker in 1875 have weathered almost beyond recognition. The slate rock consists largely of resistant silicate clay minerals, while the marble consists mainly of calcite, which is much more easily attacked by acids in rainwater. (Photo courtesy of R. Weil)

Characteristics of Rocks and Minerals

The rocks in the earth's outer surface are commonly classified as **igneous, sedimentary,** and **metamorphic.** Those of igneous origin are formed from molten magma and include such common rocks as granite and diorite (Figure 2.2).

Igneous rock is composed of primary minerals[2] such as light-colored quartz, muscovite, and feldspars and dark-colored biotite, augite, and hornblende. In general, dark-

---

[2] *Primary minerals* have not been altered chemically since they formed as molten lava solidified. *Secondary minerals* are recrystallized products of the chemical breakdown and/or alteration of primary minerals.

Rock texture	Quartz	Light-colored minerals (e.g., feldspars, muscovite)		Dark-colored minerals (e.g., hornblende, augite, biotite)
Coarse	Granite	Diorite	Gabbro	Peridotite / Hornblendite
Intermediate	Rhyolite	Andesite	Basalt	
Fine	Felsite / Obsidian		Basalt glass	

FIGURE 2.2 Classification of some igneous rocks in relation to mineralogical composition and the size of mineral grains in the rock (rock texture). Worldwide, light-colored minerals and quartz are generally more prominent than are the dark-colored minerals.

colored minerals contain iron and magnesium and are more easily weathered. Therefore, dark-colored igneous rocks such as gabbro and basalt are more easily broken down than are granites and other lighter-colored igneous rocks.

When weathering products from old rocks are compacted or cemented, they form new sedimentary rocks. For example, quartz sand weathered from a granite and deposited near the shore of a prehistoric sea may become cemented into a solid mass called *sandstone*. Similarly, clays may be compacted into *shale* (Table 2.1).

Metamorphic rocks are those that have formed by the metamorphism or change in form of other rocks. As the Earth's continental plates shift, and sometimes collide, forces are generated that can uplift great mountain ranges. As a result, igneous and sedimentary masses are subjected to tremendous heat and pressure that compress, distort, and/or partially remelt the original rocks. Igneous rocks are commonly modified to form schist or gneiss in which light and dark minerals have been reoriented into bands. Sedimentary rocks, such as limestone and shale, may be metamorphosed to marble and slate, respectively (Table 2.1). As is the case for igneous and sedimentary rock, the particular minerals that dominate a given metamorphic rock influence its resistance to chemical weathering.

Weathering: A General Case

Weathering combines the processes of destruction and synthesis (Figure 2.3). Moving from left to right in the weathering diagram, the original rocks and minerals are destroyed by both *physical disintegration* and *chemical decomposition*. Three groups of minerals that remain in well-weathered soils are shown on the right side of Figure 2.3: (1) silicate clays, (2) very resistant end products, including iron and aluminum oxide clays, and (3) very resistant primary minerals, such as quartz. In highly weathered soils of humid tropical and subtropical regions, the oxides of iron and aluminum and certain silicate clays with low Si/Al ratios predominate because most other constituents have been broken down and removed (see Table 2.2).

2.2   PHYSICAL WEATHERING (DISINTEGRATION)

TEMPERATURE.   Rocks heat up during the day and cool down at night, causing alternate expansion and contraction of their constituent minerals. As some minerals expand more than others, temperature changes set up differential stresses that eventually cause the rock to crack apart.

Because the outer surface of a rock is often warmer or colder than the inner, more protected portions, some rocks may weather by *exfoliation*—the peeling away of outer layers (Figure 2.4). This process may be sharply accelerated if ice forms in the surface cracks.

ABRASION BY WATER, ICE, AND WIND.   When loaded with sediment, water has tremendous cutting power, as is amply demonstrated by the gorges, ravines, and valleys around the

TABLE 2.1   Some of the More Important Sedimentary and Metamorphic Rocks and the Minerals Commonly Dominant in Them

	Type of rock	
Dominant mineral	Sedimentary	Metamorphic
Calcite ($CaCO_3$)	Limestone	Marble
Dolomite ($CaCO_3 \cdot MgCO_3$)	Dolomite	Marble
Quartz (SiO_2)	Sandstone	Quartzite
Clays	Shale	Slate
Variable	Conglomerate[a]	Gneiss[b]
Variable		Schist[b]

[a] Small stones of various mineralogical makeup are cemented into conglomerate.
[b] The minerals present are determined by the original rock, which has been changed by metamorphism. Primary minerals present in the igneous rocks commonly dominate these rocks, although some secondary minerals are also present.

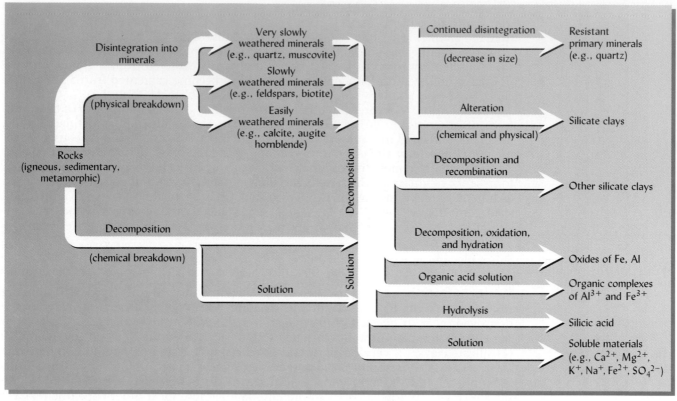

FIGURE 2.3 Pathways of weathering that occur under moderately acid conditions common in humid temperate regions. The disintegration of rocks into small individual mineral grains is a physical process, whereas decomposition, recombination, and solution are chemical processes. Alteration of minerals involves both physical and chemical processes. Note that resistant primary minerals, newly synthesized secondary minerals, and soluble materials are products of weathering. In arid regions the physical processes predominate, but in humid topical areas decomposition and recombination are most prominent.

world. The rounding of riverbed rocks and beach sand grains is further evidence of the abrasion that accompanies water movement.

Windblown dust and sand also can wear down rocks by abrasion, as can be seen in the many picturesque rounded rock formations in certain arid regions. In glacial areas, huge moving ice masses embedded with soil and rock fragments grind down rocks in their path and carry away large volumes of material (see Section 2.10).

TABLE 2.2   Partial Elemental Analysis of an Igneous Rock (a Granite) and the Mature Soil Developed from It in a Warm Humid Region (Percent of Total Composition)

Note that during soil formation there was a relative loss of phosphorus, calcium, potassium and silicon, but a relative increase in iron and aluminum. A declining ratio of silicon to aluminum is considered to be an indicator of more complete weathering.

Element[a]	In rock	In soil	Percent change
P	0.057	0.035	−38
K	3.11	0.44	−86
Ca	1.07	0.23	−79
Si	32.9	21.8	−34
Al	7.68	14.36	+87
Fe	2.50	8.54	+242
Si/Al	4.28	1.52	−64

[a] Elemental analysis is given here, though many geologists report these values in terms of the oxides of the element: for example, Al_2O_3, rather than simply Al. [Data adapted from Jackson (1964).]

FIGURE 2.4 Two illustrations of rock weathering. (Left) An illustration of concentric weathering called *exfoliation*. A combination of physical and chemical processes stimulate the mechanical breakdown, which produces layers that appear much like the leaves of a cabbage. (Right) Concentric bands of light and dark colors indicate that chemical weathering (oxidation and hydration) has occurred from the outside inward, producing iron compounds that differ in color. (Right photo courtesy of R. Weil)

PLANTS AND ANIMALS.   Plant roots sometimes enter cracks in rocks and pry them apart, resulting in some disintegration. Burrowing animals may also help disintegrate rock somewhat. However, such influences are of little importance in producing parent material when compared to the drastic physical effects of water, ice, wind, and temperature change.

2.3   CHEMICAL WEATHERING (DECOMPOSITION)

While physical weathering is accentuated in dry, cool regions, chemical weathering is most intense where the climate is wet and hot. However, both types of weathering occur together, and each tends to accelerate the other.

Chemical weathering is enhanced by water (with its omnipresent solutes), oxygen, and the organic and inorganic acids that result from biochemical activity. These agents act in concert to convert primary minerals (e.g., feldspars and micas) to secondary minerals (e.g., silicate clays) and release plant nutrient elements in soluble forms. Note the importance of water in each of the six main types of chemical weathering reactions:

HYDRATION.   Intact water molecules may bind to a mineral by the process called **hydration.**

$$\underset{\text{Hematite}}{5Fe_2O_3} + \underset{\text{Water}}{9H_2O} \xrightarrow{\text{Hydration}} \underset{\text{Ferrihydrite}}{Fe_{10}O_{15} \cdot 9H_2O}$$

HYDROLYSIS.   In hydrolysis reactions, water molecules split into their hydrogen and hydroxyl components and the hydrogen often replaces a cation from the mineral structure. A simple example is the action of water on microcline, a potassium-containing feldspar.

$$\underset{\text{(solid)}}{KAlSi_3O_8} + \underset{\text{Water}}{H_2O} \xrightleftharpoons{\text{Hydrolysis}} \underset{\text{(solid)}}{HAlSi_3O_8} + \underset{\text{(solution)}}{K^+ + OH^-}$$

DISSOLUTION.   Water is capable of dissolving many minerals by hydrating the cations and anions until they become dissociated from each other and surrounded by water molecules. Gypsum dissolving in water provides an example.

$$CaSO_4 \cdot 2H_2O + 2H_2O \xrightleftharpoons{\text{Dissolution}} Ca^{2+} + SO_4^{2-} + 4H_2O$$

(solid)　　　　　Water　　　　　　　　　(solution)　　　　Water

CARBONATION AND OTHER ACID REACTIONS.　Weathering is accelerated by the presence of acids, which increase the activity of hydrogen ions in water. For example, when carbon dioxide dissolves in water (a process enhanced by microbial and root respiration) the carbonic acid (H_2CO_3) produced hastens the chemical dissolution of calcite in limestone:

$$H_2CO_3 + CaCO_3 \xrightleftharpoons{\text{Carbonation}} Ca^{2+} + 2HCO_3^-$$

Carbonic acid　　Calcite　　　　　　(solution)　　(solution)
　　　　　　　　　(solid)

OXIDATION-REDUCTION.　Minerals that contain iron, manganese, or sulfur are especially susceptible to oxidation-reduction reactions. Iron is usually laid down in primary minerals in the divalent Fe(II) (ferrous) form. When exposed to air and water during soil formation, the iron is easily oxidized (loses an electron) and becomes trivalent Fe(III) (ferric), changing ionic radius and destabilizing the crystal structure of the mineral.

$$4FeO + O_2 + 2H_2O \xrightleftharpoons{\text{Oxidation}} 4FeOOH$$

Fe(II)　　　　　　　　　　　　　　Goethite
oxide　　　　　　　　　　　　　[Fe(III) oxide]

The oxidation and/or removal of iron during weathering is often made visible by changes in the colors of the resulting altered minerals (see Figure 2.4, right).

COMPLEXATION.　Soil biological processes produce organic acids such as oxalic, citric, and tartaric acids, as well as the much larger fulvic and humic acid molecules (see Section 11.4). These can form organic complexes (chelates) with the Al^{3+} ions held within the structure of silicate minerals. By so doing they remove the Al^{3+} from the mineral, which then is subject to further disintegration:

$$K_2[Si_6Al_2]Al_4O_{20}(OH)_4 + 6C_2O_4H_2 + 8H_2O \xrightleftharpoons{\text{Complexation}} 2K^+ + 8OH^- + 6C_2O_4Al^+ + 6Si(OH)_4^0$$

Muscovite　　　　　　Oxalic acid　　Water　　　　Potassium hydroxide　Complex　　(solution)
(solid)　　　　　　　　　　　　　　　　　　　　(solution)　　(solution)

INTEGRATED WEATHERING PROCESSES.　The various chemical weathering processes occur simultaneously and are interdependent. The released substances can also be recombined to form silicate clays and other secondary silicate minerals. In this manner, the biochemical processes of weathering transform primary geologic materials into the compounds of which soils are made.

2.4 FACTORS INFLUENCING SOIL FORMATION

We learned in Chapter 1 that *the soil* is a collection of *individual soils*, each with distinctive profile characteristics. This concept of soils as organized natural bodies derived initially from late-19th-century field studies by a brilliant Russian team of soil scientists led by V. V. Dukochaev. They noted similar profile layering in soils hundreds of kilometers apart, provided that the climate and vegetation were similar at the two locations. Many of our modern concepts concerning the factors of soil formation are derived from the work of the late Hans Jenny, an American soil scientist who recognized the five major factors that control the formation of soils:

1. *Parent materials* (geological or organic precursors to the soil)
2. *Climate* (primarily precipitation and temperature)
3. *Biota* (living organisms, especially native vegetation, microbes, soil animals, and human beings)

4. *Topography* (slope, aspect, and landscape position)
5. *Time* (the period of time since the parent materials became exposed to soil formation)

Soils are often defined in terms of these factors as "dynamic natural bodies having properties derived from the combined effects of climate and biotic activities, as modified by topography, acting on parent materials over periods of time."

We will now examine how each of these five factors affects the outcome of soil formation. However, as we do, we must keep in mind that these factors do not exert their influences independently. Indeed, interdependence is the rule. For example, contrasting climatic regimes are likely to be associated with contrasting types of vegetation, and perhaps differing topography and parent material, as well. Nonetheless, in certain situations one of the factors has had the dominant influence in determining differences among a set of soils. Soil scientists refer to such a set of soils as a *lithosequence, climosequence, biosequence, toposequence,* or *chronosequence.*

2.5 PARENT MATERIALS

Geological processes have brought to the earth's surface numerous parent materials in which soils form (Figure 2.5). The nature of the parent material profoundly influences soil characteristics. For example, a soil might inherit a sandy texture (see Section 4.2) from a coarse-grained, quartz-rich parent material such as granite or sandstone. Soil texture, in turn, helps control the percolation of water through the soil profile, thereby affecting the translocation of fine soil particles and plant nutrients.

The chemical and mineralogical composition of parent material also influences both chemical weathering and the natural vegetation. For example, the presence of limestone in parent material will delay the development of acidity that typically occurs in humid climates. In addition, trees growing in limestone materials produce leaf litter that is relatively high in calcium. Incorporation of the calcium-rich litter into the soil further delays the process of acidification and, in humid temperate areas, the progress of soil development.

Parent material also influences the quantity and type of clay minerals present in the soil profile. First, the parent material itself may contain varying amounts and types of clay minerals, perhaps from a previous weathering cycle. Second, the nature of the parent material greatly influences the kinds of clays that can develop as the soil evolves (see Section 8.5). In turn, the nature of the clay minerals present markedly affects the kind of soil that develops.

Classification of Parent Materials

Inorganic parent materials can either be formed in place as residual material weathered from rock, or they can be transported from one location and deposited at another (Figure 2.6). In wet environments (such as swamps and marshes), incomplete decomposition may allow organic parent materials to accumulate from the residues of many generations of vegetation. Although it is their chemical and physical properties that most influence soil development, parent materials are often classified with regard to the mode of placement in their current location, as seen on the right side of Figure 2.6.

2.6 RESIDUAL PARENT MATERIAL

Residual parent material develops in place from weathering of the underlying rock. In stable landscapes it may have experienced long and possibly intense weathering. Where the climate is warm and humid, residual parent materials are typically thoroughly leached and oxidized, and show the red and yellow colors of various oxidized iron compounds (see Plates 9, 11, and 15). In cooler and especially drier climates, the color and chemical composition of residual parent material tends to resemble more closely the rock from which it formed.

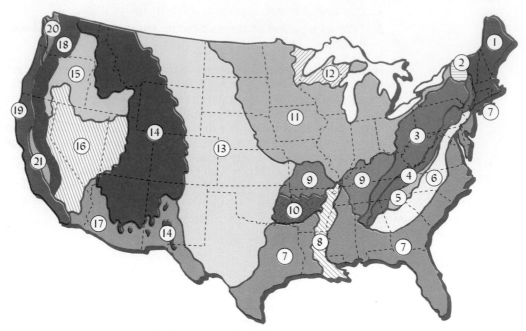

FIGURE 2.5 Generalized physiographic and regolith map of the United States. The regions are as follows (major areas of residual parent materials are italicized).

1. New England: mostly glaciated metamorphic rocks.
2. Adirondacks: glaciated metamorphic and sedimentary rocks.
3. *Appalachian Mountains and plateaus:* shales and sandstones.
4. *Limestone valleys and ridges:* mostly limestone.
5. Blue Ridge mountains: sandstones and shales.
6. *Piedmont Plateau:* metamorphic rocks.
7. Atlantic and Gulf coastal plain: sedimentary rocks with sands and clays.
8. Mississippi floodplain and delta: alluvium.
9. *Limestone uplands:* mostly limestone and shale.
10. *Sandstone uplands:* mostly sandstone and shale.
11. Central lowlands: mostly glaciated sedimentary rocks with till and loess.
12. Superior uplands: glaciated metamorphic and sedimentary rocks.
13. *Great Plains region:* sedimentary rocks.
14. *Rocky Mountain region:* sedimentary, metamorphic, and igneous rocks.
15. Northwest intermountain: mostly igneous rocks; loess in river basins.
16. Great Basin: gravels, sands, alluvial fans; igneous and sedimentary rocks.
17. Southwest arid region: gravel, sand, and other debris of desert and mountain.
18. *Sierra Nevada and Cascade mountains:* igneous and volcanic rocks.
19. *Pacific Coast province:* mostly sedimentary rocks.
20. Puget Sound lowlands: glaciated sedimentary.
21. California central valley: alluvium and outwash.

2.7 COLLUVIAL DEBRIS

Colluvial debris, or colluvium, is made up of poorly sorted rock fragments detached from the heights above and carried downslope, mostly by gravity, assisted in some cases by frost action. Rock fragment (talus) slopes, cliff rock debris (detritus), and similar heterogeneous materials are good examples. Avalanches are made up largely of such accumulations.

Colluvial parent materials are frequently coarse and stony because physical rather than chemical weathering has been dominant. Stones, gravel, and fine materials are interspersed (not layered), and the coarse fragments are rather angular (Figure 2.7). Packing voids, spaces created when tumbling rocks come to rest against each other (sometimes at precarious angles), help account for the easy drainage of many colluvial deposits and also for their tendency to be unstable and prone to slumping and landslides, especially if disturbed by excavations.

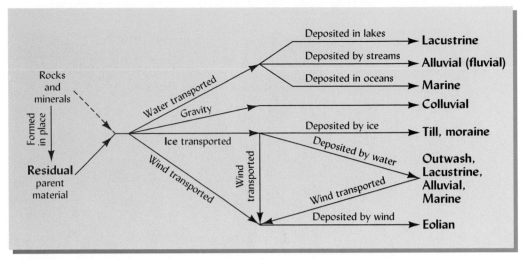

FIGURE 2.6   How various kinds of parent material are formed, transported, and deposited.

2.8   ALLUVIAL STREAM DEPOSITS

There are three general classes of alluvial deposits: *floodplains, alluvial fans,* and *deltas.* They will be considered in order.

Floodplains

That part of a river valley that is inundated during floods is a floodplain. Sediment carried by the swollen stream is deposited during the flood, with the coarser materials being laid down near the river channel where the water is deeper and flowing with more turbulence and energy. Finer materials settle out in the calmer flood waters farther from the channel. Each major flooding episode lays down a distinctive layer of sediment, creating the stratification that characterizes alluvial soils (Figure 2.8).

FIGURE 2.7   A productive soil formed in colluvial parent material in the Appalachian mountains of the eastern United States. The ridge in the background was at one time much taller, but material from the crest tumbled downslope and came to rest in the configuration seen in the soil profile. Note the unstratified mix of particle sizes and the rather angular nature of the coarse fragments. (Photo courtesy of R. Weil)

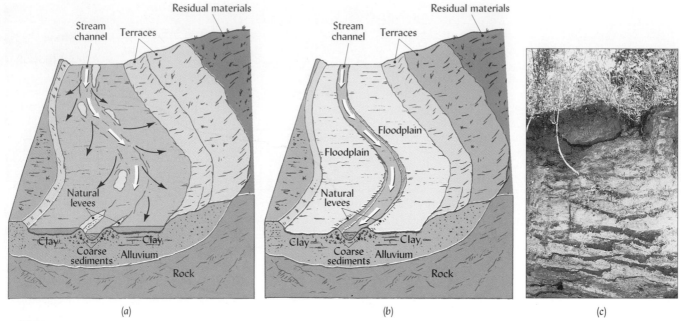

FIGURE 2.8 Illustration of floodplain development. (*a*) A stream is at flood stage, has overflowed its banks, and is depositing sediment in the floodplain. The coarser particles are being deposited near the stream channel where water is moving most rapidly, while the finer clay particles are being deposited where water movement is slower. (*b*) After the flood the sediments are in place and vegetation is growing. (*c*) Contrasting layers of sand, silt, and clay characterize the alluvial floodplain. Each layer resulted from separate flooding episodes. (Redrawn from *Physical Geology*, 2d ed., by F. R. Flint and B. J. Skinner, copyright © 1977 John Wiley & Sons, Inc. Reprinted by permission of John Wiley & Sons, Inc.) (Photo courtesy of R. Weil)

If, over a period of time, there is a change in grade, a stream may cut down through its already well-formed alluvial deposits. This cutting action leaves **terraces** above the floodplain on one or both sides. Some river valleys feature two or more terraces at different elevations, each reflecting a past period of alluvial deposition and stream cutting.

Alluvial Fans

Streams that leave a narrow valley in an upland area and suddenly descend to a much broader valley below deposit sediment in the shape of a fan, as the water spreads out and slows down. The rushing water tends to sort the sediment particles by size, first dropping the gravel and coarse sand, then depositing the finer materials toward the bottom of the alluvial fan.

Delta Deposits

Much of the finer sediment carried by streams is not deposited in the floodplain but is discharged into the lake, reservoir, or ocean into which the streams flow. Some of the suspended material settles near the mouth of the river, forming a delta. Such delta deposits are by no means universal, being found at the mouths of only a few rivers of the world. A delta often is a continuation of a floodplain (its front, so to speak). It is clayey in nature and is likely to be poorly drained as well.

2.9 MARINE SEDIMENTS

Streams eventually deposit much of their sediment loads in oceans, estuaries, and gulfs. The coarser fragments settle out near the shore and the finer particles at a distance (Figure 2.9). Over long periods of time, these underwater sediments build up, in some cases becoming hundreds of meters thick. Changes in the relative elevations of sea and land may later raise these marine deposits above sea level, creating a coastal plain. The deposits are then subject to a new cycle of weathering and soil formation.

A coastal plain usually has only moderate slopes, being more level in the low-lying parts nearer the coastline and more hilly farther inland, where streams and rivers flowing down the steeper grades have more deeply dissected the landscape. The land surface in the lower coastal portion may be only slightly above the water table during at least part of the year, so wetland forests and marshes often characterize such parent materials.

Marine deposits are quite variable in texture. Some are sandy, as is the case in much of the Atlantic seaboard coastal plain. Others are high in clay, as are deposits found in the Atlantic and Gulf coastal flatwoods and in the interior pinelands of Alabama and Mississippi. Where streams have cut down through layers of marine sediments (as in the detailed block diagram in Figure 2.9), clays, silts, and sand may be encountered side by side.

FIGURE 2.9 Diagrams showing sediments laid down by marine waters adjacent to coastal residual igneous and metamorphic rocks. Note that the marine sediments are alternate layers of fine clay and coarse-textured sands and gravels. The photo in the lower right shows such layering on coastal marine sediment. The soils developing from the marine and residual parent materials are also shown (names in capital letters). The river had cut through the marine sediments and has deposited some alluvial material in its basin. This diagram and photo illustrate the relationship between marine sediments and residual materials in the Coastal Plain of southeastern United States. [Photo courtesy of R. Weil; soils diagram from Miller (1976)]

2.10 PARENT MATERIALS TRANSPORTED BY GLACIAL ICE AND MELTWATERS

During the Pleistocene epoch (about 10^4 to 10^7 years ago), up to 20% of the world's land surface—northern North America, northern and central Europe, and parts of northern Asia—was invaded by a succession of great ice sheets, some more than 1 km thick.

In North America, Pleistocene-epoch glaciers covered most of what is now Canada, southern Alaska, and the northern part of the contiguous United States. The southernmost extension went down the Mississippi Valley, where the least resistance was met because of the lower and smoother topography (Figure 2.10).

Europe and central North America apparently sustained several distinct ice invasions over a period of 1 to 1.5 million years. As the glacial ice pushed forward, the existing regolith with much of its mantle of soil was swept away, hills were rounded, valleys were filled, and, in some cases, the underlying rocks were severely ground and gouged. Thus, the glacier became filled with rock and all kinds of unconsolidated materials, carrying great masses of these materials as it pushed ahead. Finally, as the ice melted and the glacier retreated, a mantle of glacial debris or **drift** remained. This provided a new regolith and fresh parent material for soil formation.

Glacial Till and Associated Deposits

The materials deposited directly by the ice, called **glacial till**, are heterogeneous (unstratified) mixtures of debris, which vary in size from boulders to clay. Glacial till may therefore be somewhat similar in appearance to colluvial materials, except that the coarse fragments are more rounded from their grinding journey in the ice, and the deposits are often much more densely compacted because of the great weight of the overlying ice sheets. Much glacial till is deposited in irregular ridges called **moraines.** Figure 2.11 shows how glacial sheets deposited several types of soil parent materials.

Glacial Outwash and Lacustrine Sediments

The torrents of water gushing forth from melting glaciers carried vast loads of sediment. In valleys and on plains, where the glacial waters were able to flow away freely, the sediment formed an **outwash plain** (Figure 2.11). Such sediments, with sands and gravels

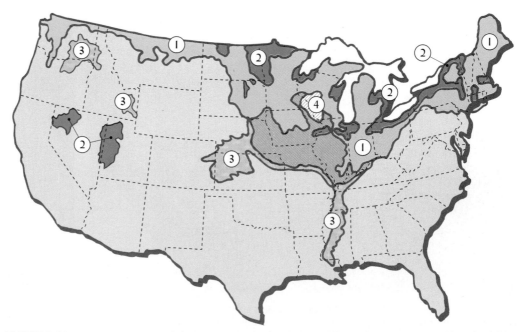

FIGURE 2.10 Areas in the United States covered by the continental ice sheet and the deposits either directly from, or associated with, the glacial ice. (1) Till deposits of various kinds; (2) glacial-lacustrine deposits; (3) the loessial blanket (note that the loess overlies great areas of till in the Midwest); (4) an area, mostly in Wisconsin, that escaped glaciation and is partially loess covered.

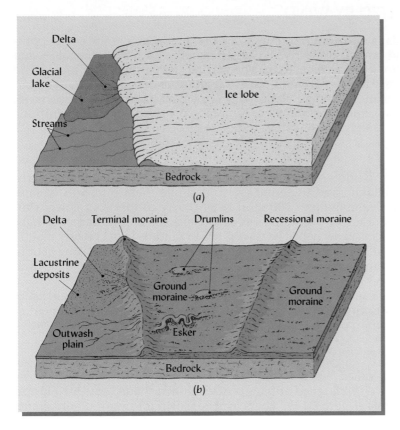

FIGURE 2.11 Illustration of how several glacial materials were deposited. (*a*) A glacier ice lobe moving to the left, feeding water and sediments into a glacial lake and streams, and building up glacial till near its front. (*b*) After the ice retreats, terminal, ground, and recessional moraines are uncovered along with cigar-shaped hills (drumlins), the beds of rivers that flowed under the glacier (eskers), and lacustrine, delta, outwash deposits.

sorted by flowing water, are common **valley fills** in parts of the United States. Figure 2.12 shows the sorted layering of coarse and fine materials in **glacial outwash** overlaid by mixed materials of glacial till.

When the ice front came to a standstill where there was no ready escape for the water, ponding began; ultimately, very large lakes were formed (Figure 2.11). Particularly prominent in North America were those south of the Great Lakes and in the Red River Valley of Minnesota and Manitoba (see Figure 2.10). The **lacustrine deposits** formed in these glacial lakes range from coarse delta materials and beach deposits near the shore to larger areas of fine silts and clay deposited from the deeper, more still waters at the center of the lake. Areas of inherently fertile (though not always well-drained) soils developed from these materials as the lakes dried.

FIGURE 2.12 The stratification of different-size particles can be seen in the glacial outwash in the lower part of this soil profile in North Dakota. The outwash is overlain by a layer of glacial till containing a random assortment of particles, ranging in size from small boulders to clays. Note the rounded edges of the rocks, evidence of the churning action within the glacier. Scale has 10-cm marks. (Photo courtesy of R. Weil)

2.11 PARENT MATERIALS TRANSPORTED BY WIND

Wind is capable of picking up an enormous quantity of material at one site and depositing it at another. Wind can most effectively pick up material from soil or regolith that is loose, dry, and unprotected by vegetation. The smaller the particles, the higher and farther the wind will carry them. Wind-transported (**eolian**) materials important as parent material for soil formation include, from largest to smallest particle size: **dune sand, loess** (pronounced "luss"), and **aerosolic dust.** Windblown **volcanic ash** from erupting volcanoes is a special case that is also worthy of mention.

Dune Sand

Along the beaches of the world's oceans and large lakes, and over vast barren deserts, strong winds pick up medium and fine sand grains and pile them into hills of sand called *dunes.* The dunes, ranging up to 100 m in height, may continue to slowly shift their locations in response to the prevailing winds. Because most other minerals have been broken down and carried away by the waves, beach sand usually consists mainly of quartz, which is devoid of plant nutrients and highly resistant to weathering action. Nonetheless, over time dune grasses and other pioneering vegetation may take root and soil formation may begin.

Desert sands, too, are usually dominated by quartz, but they may also include substantial amounts of other minerals that could contribute more to the establishment of vegetation and the formation of soils, should sufficient rainfall occur.

Loess

The windblown materials called *loess* are composed primarily of silt with some very fine sand and clay. They cover wide areas in the central United States, eastern Europe, Argentina, and central Asia (Figure 2.13*b*). Loess may be blown for hundreds of kilometers. The deposits farthest from the source are thinnest and consist of the finest particles.

In the United States (Figure 2.13*a*), the main sources of loess were the great barren expanses of till and outwash left in the Missouri and Mississippi River valleys by the

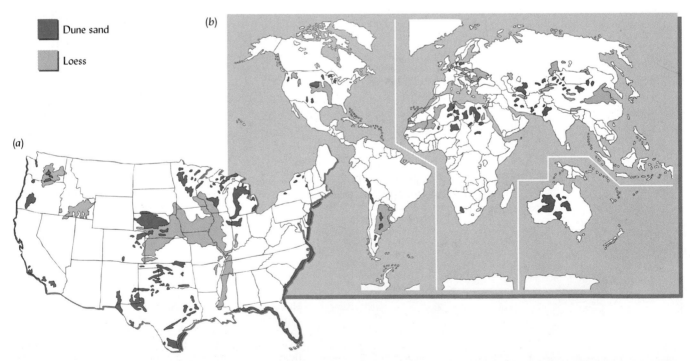

FIGURE 2.13 (*a*) Approximate distribution of loess and dune sand in the United States. The soils that have developed from loess are generally silt loams, often quite high in fine sands. Note especially the extension of the central loess deposit down the eastern side of the Mississippi River and the smaller areas of loess in Washington, Oregon, and Idaho. The most prominent areas of dune sands are the Sand Hills of Nebraska and the dunes along the eastern shore of Lake Michigan. (*b*) Other major eolian deposits of the world include the loess deposits in Argentina, eastern Europe, and northern China, and the large areas of dune sands in north Africa and Australia.

retreating glaciers of the last Ice Age. During the winter months, winds picked up fine materials and moved them southward, covering the existing soils and parent materials with a blanket of loess that accumulated to as much as 8 m thick.

Aerosolic Dust

Very fine particles (about 1 to 10 μm) carried high into the air may travel for thousands of kilometers before being deposited, usually with rainfall. These fine particles are called *aerosolic dust* because they can remain suspended in air, due to their very small size. While this dust has not blanketed the receiving landscapes as thickly as is typical for loess, it does accumulate at rates that make significant contributions to soil formation. Much of the calcium carbonate in soils of the western United States probably originated as windblown dust. Recent studies have shown that dust, originating in the Sahara Desert of northern Africa and transported over the Atlantic Ocean in the upper atmosphere, is the source of much of the calcium and other nutrients found in the highly leached soils of the Amazon basin in South America.

Volcanic Ash

During volcanic eruptions cinders fall in the immediate vicinity of the volcano, while fine, often glassy, ash particles may blanket extensive areas downwind. Soils developed from volcanic ash are most prominent within a few hundred kilometers of the volcanoes that ring the Pacific Ocean. Important areas of volcanic ash parent materials occur in Japan, Indonesia, New Zealand, western United States (in Hawaii, Montana, Oregon, Washington, and Idaho), Mexico, and Chile. The soils formed are uniquely light and porous, and tend to accumulate organic matter more rapidly than other soils in the area.

2.12 ORGANIC DEPOSITS

Organic material accumulates in wet places where plant growth exceeds the rate of residue decomposition. These residues sink into the water, where their decomposition is limited by lack of oxygen. As a result, organic deposits often accumulate up to several meters in depth (Figure 2.14). Collectively, these organic deposits are called **peat.**

Distribution and Accumulation of Peats

Peat deposits are found all over the world, but most extensively in the cool climates and in areas that have been glaciated. About 75% of the 340 million hectares of peat lands in the world are found in Canada and northern Russia. The United States is a distant third, with about 20 million hectares.

The rate of peat accumulation varies from one area to another. In the warm climate of Florida, peat deposits of the everglades accumulated for 3000 to 4000 years at the rate of about 0.8 mm/yr. In Wisconsin, accumulation rates have been estimated to be 0.2 to 0.4 mm/yr.

Types of Peat Materials

Based on the nature of the parent materials, four kinds of peat are recognized:

1. Moss peat, the remains of mosses such as sphagnum
2. Herbaceous peat, residues of herbaceous plants such as sedges, reeds, and cattails
3. Woody peat, from the remains of woody plants, including trees and shrubs
4. Sedimentary peat, remains of aquatic plants (e.g., algae) and of fecal material of aquatic animals

Organic deposits generally contain two or more of these kinds of peats. Alternating layers of different peats are common, as are mixtures of the peats. Because the succession of plants, as the residues accumulate, tends to favor trees (see Figure 2.14), woody

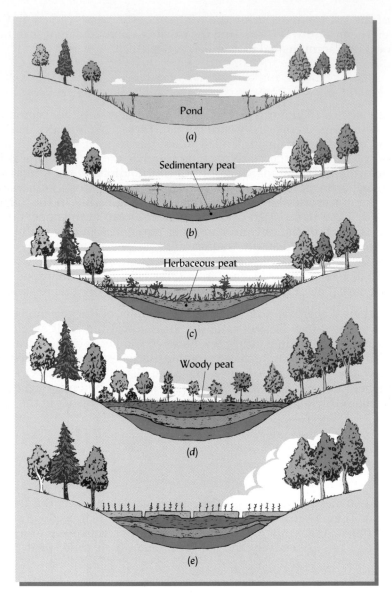

FIGURE 2.14 Stages in the development of a typical woody peat bog and the area after clearing and draining. (*a*) Nutrient runoff from the surrounding uplands encourages aquatic plant growth, especially around the pond edges. (*b, c*) Organic debris fills in the bottom of the pond. (*d*) Eventually trees cover the entire area. (*e*) If the land is cleared and a drainage system installed, the area becomes a most productive muck soil.

peats often dominate the surface layers of organic materials. In cases where a wetland area has been drained, woody peats tend to make very productive agricultural soils that are especially well suited for vegetable production. While moss peats have high water-holding capacities, they tend to be quite acid. Sedimentary peat is generally undesirable as an agricultural soil.

The organic material is called **peat** or **fibric** if the residues are sufficiently intact to permit the plant fibers to be identified. If most of the material has decomposed sufficiently so that little fiber remains, the term **muck** or **sapric** is used. In mucky peats (**hemic** materials) some of the plant fibers can be recognized and some cannot.

WETLAND PRESERVATION. Wetland areas are important natural habitats for wildlife, and their drainage reduces the area of such habitats. While organic material is the foundation for some very productive agricultural soils, environmentalists argue that such use is unsustainable because, once drained, the organic deposits will decompose and disappear after a century or so; therefore, these areas might be better left in (or returned to) their natural state (see Section 7.7).

Recognizing that the effects of **parent materials** on soil properties are modified by the combined influences of **climate, biotic activities, topography,** and **time,** we will now turn to these other four factors of soil formation, starting with climate.

2.13 CLIMATE

Climate is perhaps the most influential of the four factors acting on parent material because it determines the nature and intensity of the weathering that occurs over large geographic areas. The principal climatic variables influencing soil formation are ***effective precipitation*** (see Box 2.1) and *temperature,* both of which affect the rates of chemical, physical, and biological processes.

Effective Precipitation

We have already seen (Section 2.3) that water is essential for all the major chemical weathering reactions. To be effective in soil formation, water must penetrate into the regolith. The greater the depth of water penetration, the greater the depth of weathering soil and development. Surplus water percolating through the soil profile transports soluble and suspended materials from the upper to the lower layers. It may also carry away soluble materials in the drainage waters. Thus, percolating water stimulates weathering reactions and helps differentiate soil horizons.

Likewise, a deficiency of water is a major factor in determining the characteristics of soils of dry regions. Soluble salts are not leached from these soils, and in some cases they build up to levels that curtail plant growth. Soil profiles in arid and semiarid regions are also apt to accumulate carbonates and certain types of cracking clays.

Temperature

For every 10°C rise in temperature, the rates of biochemical reactions more than double. Temperature and moisture both influence the organic matter content of soil through their effects on the balance between plant growth and microbial decomposition (see Figure 11.17). If warm temperatures and abundant water are present in the profile at the same time, the processes of weathering, leaching, and plant growth will be maximized. The very modest profile development characteristic of cold areas contrasts sharply with the deeply weathered profiles of the humid tropics (Figure 2.15).

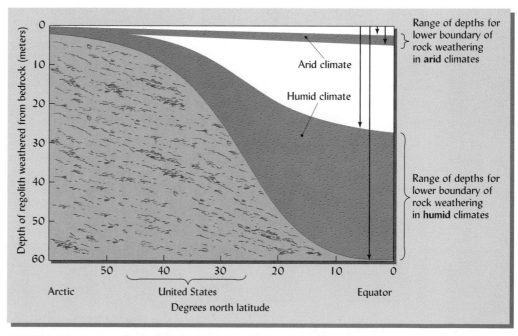

FIGURE 2.15 A generalized illustration of the effects of two climatic variables, temperature and precipitation, on the depths of regolith weathered in bedrock. The stippled areas represent the range of depths to which the regolith typically extends. In cold climates (arctic regions), the regolith is shallow under both humid and arid conditions. In the warmer climates of the lower latitudes, the depth of the residual regolith increases sharply in humid regions, but is little affected in arid regions. Under humid tropical conditions, the regolith may be 50 or more meters in depth. The vertical arrows represent depths of weathering near the Equator. Remember that soil depth may not be as great as regolith depth.

BOX 2.1   EFFECTIVE PRECIPITATION FOR SOIL FORMATION

Water from rain and melting snow is a primary requisite for parent material weathering and soil development. To fully promote soil development, water must not only enter the profile and participate in weathering reactions; it must also percolate through the profile and translocate soluble weathering products.

Let's consider a site that receives an average of 600 mm of rainfall per year. The amount of water leaching through a soil is determined not only by the total annual precipitation but by at least four other factors, as well.

a. **Seasonal distribution** of precipitation. The 600 mm of rainfall distributed evenly throughout the year, with about 50 mm each month, is likely to cause less soil leaching or erosion than the same annual amount of rain falling at the rate of 100 mm per month during a six-month rainy season.

b. **Temperature and evaporation.** In a hot climate, evaporation from soils and vegetation is much higher than in a cool climate. Therefore, in the hot climate, much less of the 600 mm will be available for percolation and leaching. Most or all will evaporate soon after it falls on the land. Thus, 600 mm of rain may cause more leaching and profile development in a cool climate than in a warmer one. Similar reasoning would suggest that rainfall concentrated during a mild winter (as in California) may be more effective in leaching the soil than the same amount of rain concentrated in a hot summer (as in the Great Plains).

c. **Topography.** Water falling on a steep slope will run downhill so rapidly that only a small portion will enter the soil where it falls. Therefore, even though they receive the same rainfall, a level or concave site will experience more percolation and leaching than a steeply sloping site. The effective rainfall can be said to be greater on the level site than on the sloping one. The concave site will receive the greatest effective rainfall because, in addition to direct rainfall, it will collect the runoff from the adjacent sloping site, as well.

d. **Permeability.** Even if all three of the above conditions are the same, more rain water will infiltrate and leach through a coarse, sandy profile than a tight, clayey one. Therefore the sandy profile can be said to experience a greater effective precipitation, and more rapid soil development may be expected.

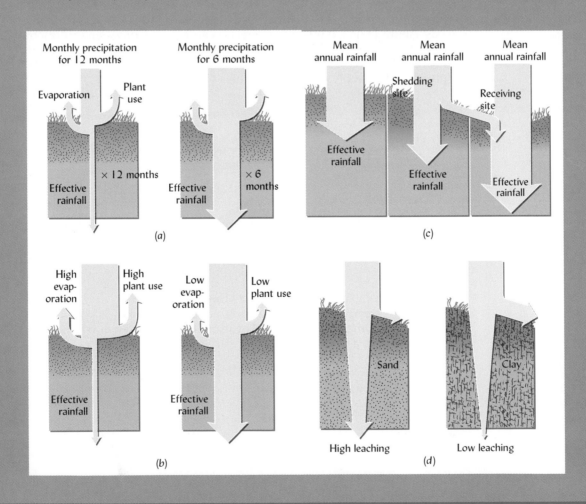

Climate also influences the natural vegetation. Humid climates favor the growth of trees. In contrast, grasses are the dominant native vegetation in subhumid and semiarid regions, while shrubs and brush of various kinds dominate in arid areas. Thus, climate also exerts its influence through a second soil-forming factor, the living organisms.

Considering soils with similar temperature regime, parent material, topography, and age, increasing effective annual precipitation generally leads to increasing clay and organic matter contents, higher acidity, and lower ratio of Si/Al (an indication of more highly weathered minerals; see Table 2.2). However, many places have experienced climates in past geologic epochs that were not at all similar to the climate evident today. This fact is illustrated in certain old landscapes in arid regions, where highly leached and weathered soils stand as relics of the humid tropical climate that prevailed there many thousands of years ago.

2.14 BIOTA: LIVING ORGANISMS

Organic matter accumulation, biochemical weathering, profile mixing, nutrient cycling, and aggregate stability are all enhanced by the activities of organisms in the soil. Vegetative cover reduces natural soil erosion rates, thereby slowing down the rate of mineral surface soil removal. Organic acids produced from certain types of plant leaf litter bring iron and aluminum into solution by complexation and accelerate the downward movement of these metals and their accumulation in the B horizon.

Role of Natural Vegetation

A-HORIZON DEVELOPMENT. The effect of vegetation on soil formation can be seen by comparing properties of soils formed under grassland and forest vegetation near the boundary between these two ecosystems (Figure 2.16). In the grassland, much of the organic matter added to the soil is from the deep fibrous grass root systems. By contrast, tree leaves falling on the forest floor are the principal source of soil organic matter in the forest. As a result, the soils under grassland generally have a thicker A horizon with a deeper distribution of organic matter than in comparable soils under forests, which characteristically store most of their organic matter in the forest floor and a thin A horizon. The light-colored, leached E horizon typically found under the O or A horizon of a forested soil is generally not found in a grassland soil.

CATION CYCLING BY TREES. The ability of natural vegetation to take up mineral elements strongly influences the characteristics of the soils that develop, especially their acidity. Litter falling from coniferous trees (e.g., pines and firs) will recycle only small quantities of calcium, magnesium, and potassium compared to those recycled by litter from some deciduous trees (e.g., beech, oaks, and maples) that take up and store much larger amounts of such metallic cations. Because conifer tree roots allow more of these base-forming cations to be lost by leaching, soil acidity is more strongly developed under most coniferous vegetation than under most deciduous trees. The acidic, resinous needles from coniferous trees tend to accumulate in a thick O horizon with distinct layers of fibric (undecomposed) and sapric (highly decomposed) material. The more readily broken down leaves of deciduous trees generally form a thinner forest floor with less distinct layers.

HETEROGENEITY IN RANGELANDS. In arid and semiarid rangelands, competition for limited soil water does not permit vegetation dense enough to completely cover the soil surface. Scattered shrubs or bunch grasses are interspersed with openings in the plant canopy where the soil is bare or partially covered with plant litter. The widely scattered vegetation alters soil properties in several ways. Plant canopies trap windblown dust that is often relatively rich in silt and clay. Roots scavenge nutrients such as nitrogen, phosphorus, potassium and sulfur from the interplant areas. These nutrients are then deposited with the leaf litter under the plant canopies. The decaying litter adds organic acids, which lower the soil pH and stimulate mineral weathering. As time goes on, the relatively bare soil areas between plants decline in fertility and may increase in size as they become impoverished and even less inviting for the establishment of plants. Simultaneously, the vegetation creates "islands" of enhanced fertility, thicker A horizons, and often more deeply leached calcium carbonate (see Figure 2.17).

Forests—natural vegetation

Organic matter accumulation
Zone of maximum leaching

Clay, hydrous oxide accumulation

Disintegrated weathered soil material

Relatively unweathered bedrock

| Fresh unweathered rock | 100 | 1,000 | 10,000 | 100,000 | Well-developed Ultisol |

Time (years) ⟶

(a)

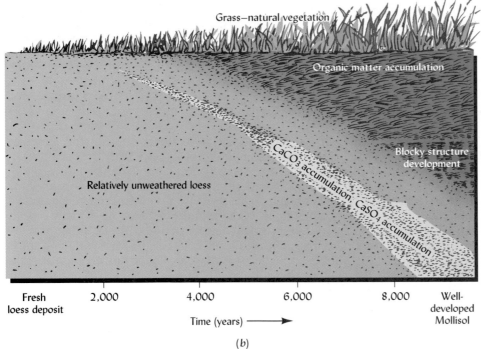

Grass—natural vegetation

Organic matter accumulation

Blocky structure development

CaCO₃ accumulation, CaSO₄ accumulation

Relatively unweathered loess

| Fresh loess deposit | 2,000 | 4,000 | 6,000 | 8,000 | Well-developed Mollisol |

Time (years) ⟶

(b)

FIGURE 2.16 How two soil profiles may have developed in climatic areas that encouraged as natural vegetation either (*a*) forests or (*b*) grasslands. Organic matter accumulation in the upper horizons occurs in time, the amount and distribution depending on the type of natural vegetation present. Clay and iron oxide accumulate and characteristic structures develop in the lower horizons. The end products differ markedly from the soil materials from which they formed. Note that the time scales for soil development differ markedly for the two parent materials.

Role of Animals

The role of animals in soil-formation processes must not be overlooked. For example, large animals such as gophers, moles, and prairie dogs bore into the lower soil horizons, bringing materials to the surface. Their tunnels are often open to the surface, encouraging movement of water and air into the subsurface layers. Old animal burrows in the lower horizons often become filled with soil material from the overlying A horizon, creating profile features known as *crotovinas* (Figure 2.18).

FIGURE 2.17   The scattered bunch grasses of this semiarid rangeland in the Patagonia region of Argentina have created "islands" of soil with enhanced fertility and thicker A horizons. A lens cap placed at the edge of one of these islands provides scale and highlights the increased soil thickness under the plant canopies. Such small-scale, plant-associated soil heterogeneity is common where soil water limitations prevent complete plant ground cover. (Photo courtesy of Ingrid C. Burke, Short-Grass Steppe Long-Term Ecological Research Program, Colorado State University)

Earthworms and other small animals also bring about considerable soil mixing as they burrow through the soil, significantly affecting soil formation. In general, the mixing activities of animals, sometimes called **pedoturbation**, tends to undo or counteract the tendency of other soil-forming processes to accentuate the differences among soil horizons.

HUMAN INFLUENCES.   Human activities also influence soil formation. For example, it is believed that Native Americans regularly set fires to maintain several large areas of

FIGURE 2.18   Abandoned animal burrows in one horizon filled with soil material from another horizon are called *crotovinas*. In this Illinois prairie soil, dark, organic-matter-rich material from the A horizon has filled in old prairie dog burrows that extend into the B horizon. The dark circular shapes in the subsoil mark where the pit excavation cut through these burrows. Scale marked every 10 cm. (Photo courtesy of R. Weil)

prairie grasslands in Indiana and Michigan. In more recent times, human destruction of natural vegetation (trees and grass) and subsequent tillage of the soil for crop production has abruptly modified soil formation. Likewise, irrigating an arid region or draining a wetland soil drastically influences the soil-forming factors, as does adding fertilizer and lime to soils of low fertility. In surface mining and urbanizing areas today, bulldozers may have an effect on soils almost akin to that of the ancient glaciers; they level and mix soil horizons and set the clock of soil formation back to zero.

2.15 TOPOGRAPHY

Topography relates to the configuration of the land surface and is described in terms of differences in elevation, slope, and landscape position—in other words, the lay of the land. Topography can hasten or delay the work of climatic forces (see Box 2.1). Steep slopes generally encourage erosion of the surface layers and allow less rainfall to enter the soil before running off, thus preventing soil formation from getting very far ahead of soil destruction. In semiarid regions, less effective moisture on the steeper slopes also results in a more sparse, less diverse plant cover. Therefore, soils on steep terrain have relatively shallow, poorly differentiated soil profiles (Figure 2.19) compared to nearby soils on more level terrain. The three soils shown in Figure 2.19 could be called a topo sequence or **catena** of soils.

INTERACTION WITH VEGETATION.   Topography often interacts with vegetation to influence soil formation. In grassland–forest transition zones, trees are commonly confined to the slight depressions where soil is generally wetter than in upland positions. As would be expected, the nature of the soil in the depressions is quite different from that in the uplands. If water stands for part or all of the year in a depression in the landscape, climate is less influential in regulating soil development. Low-lying areas may give rise to peat bogs and, in turn, to organic soils (see Figure 2.14).

SLOPE ASPECT.   Topography affects the absorbance of solar energy in a given landscape. In the northern hemisphere, south-facing slopes are more perpendicular to the sun's rays and are generally warmer and thereby commonly lower in moisture than their north-

FIGURE 2.19 Topography influences soil properties, including soil depth. The diagram on the left shows the effect of slope on the profile characteristics and the depth of a soil on which forest trees are the natural vegetation. The photo on the right illustrates the same principle under grassland vegetation. Often a relatively small change in slope can have a great effect on soil development. See Section 2.18 for explanation of horizon symbols. (Photo courtesy of R. Weil)

facing counterparts (see also Figure 7.17). Consequently, soils on the south slopes tend to be lower in organic matter and are not so deeply weathered.

SALT BUILDUP. In arid and semiarid areas, topography influences the buildup of soluble salts. Dissolved salts from surrounding upland soils move on the surface and through the underground water table to the lower-lying areas. There they rise to the surface as the water evaporates, often accumulating to plant-toxic levels.

PARENT MATERIAL INTERACTIONS. Topography can also interact with parent material. For example, in areas of tilted beds of sedimentary rock, the ridges often consist of resistant sandstone, while the valleys are underlaid by more weatherable limestone. In many landscapes, topography reflects the distribution of residual, colluvial, and alluvial parent materials, with residual materials on the upper slopes, colluvium covering the lower slopes, and alluvium filling the valley bottom (Figure 2.20).

2.16 TIME

Soil-forming processes take time to show their effects. The clock of soil formation starts ticking when a landslide exposes new rock to the weathering environment at the surface, when a flooding river deposits a new layer of sediment on its floodplain, when a glacier melts and dumps its load of mineral debris, or when a bulldozer cuts and fills a landscape to level a construction or mine-reclamation site.

RATES OF WEATHERING. Where the other factors of soil formation are favorable, organic matter may accumulate to form a darkened A horizon in a mere decade or two. In some cases, an incipient B horizon has become discernible on humid-region mine spoils in as few as 40 years. However, the formation of a simple B horizon with altered colors and structure is generally likely to require centuries. The accumulation of silicate clays usually becomes noticeable only after thousands of years. A mature, deeply weathered soil may be hundreds of thousands of years in the making.

When we speak of a "young" or a "mature" soil, we are not so much referring to the age of the soil in years, as to the degree of weathering and profile development. Time interacts with the other factors of soil formation. On a site in a warm climate, with much rain falling on a permeable parent material that is rich in weatherable minerals,

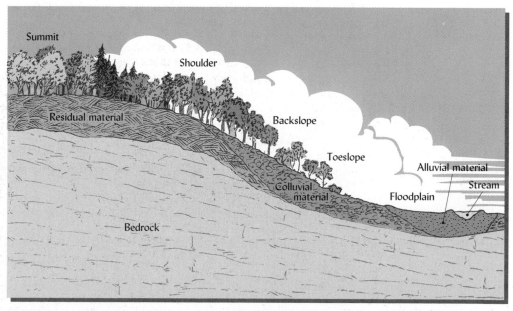

FIGURE 2.20 An interaction of topography and parent material as factors of soil formation. The soils on the summit, toe-slope, and floodplain in this idealized landscape have formed from residual, colluvial, and alluvial parent materials, respectively.

weathering and soil profile differentiation will proceed far more rapidly than on a site with steep slopes and resistant parent material in a cold, dry climate.

There are a few instances in which soil formation is so rapid that the process can be observed, and the effect of time measured, in a human life span. Examples include the rapid formation of an A horizon on freshly deposited, fertile alluvium and the mineralogical, color, and structural changes that occur within a few years when certain sulfide-containing materials are first exposed to air by mining, land drainage, or dredging operations (see Sections 9.4 and 12.21).

CHRONOSEQUENCE. Most soil-profile features develop so slowly that it is not possible to directly measure time-related changes in their formation. Indirect methods, such as carbon dating or the presence of fossils and human artifacts, must be turned to for evidence about the time required for different aspects of soil development to occur. In a different approach to studying the effects of time on soil development, soil scientists look for a **chronosequence**—a set of soils that share a common community of organisms, climate, parent material, and slope, but differ with regard to the length of time that the materials have been subjected to weathering and soil formation. A chronosequence can sometimes be found among the soils forming on alluvial terraces of differing age. The highest terraces have been exposed for the longest time, those in lower positions have been more recently exposed by the cutting action of the stream, and those on the current flood plain are the youngest, being still subject to periodic additions of new material.

We will now focus on the processes that change parent material into a soil, processes that operate under the influence of the five major factors of soil formation we have just considered.

2.17 SOIL FORMATION IN ACTION[3]

Accumulation of regolith by the breakdown of bedrock or the deposition (by wind, water, ice, etc.) of unconsolidated geologic materials may precede or, more commonly, occur simultaneously with the development of the distinctive horizons of a soil profile. During the formation (**genesis**) of a soil from parent material, the regolith undergoes many profound changes.

Four Broad Processes of Soil Genesis

Soil genesis is brought about by a series of specific changes that can be grouped into four broad processes (Figure 2.21): (1) **transformations**, such as mineral weathering and organic matter breakdown, by which some soil constituents are modified or destroyed and others are synthesized; (2) **translocations** or movements of inorganic and organic materials from one horizon up or down to another, the materials being moved mostly by water but also by soil organisms; (3) **additions** of materials to the developing soil profile from outside sources, such as organic matter from leaves, dust from the atmosphere, or soluble salts from the groundwater; and (4) **losses** of materials from the soil profile by leaching to groundwater, erosion of surface materials, or other forms of removal. Often, transformation and translocation result in the accumulation of materials in a particular horizon.

These processes of soil genesis, operating under the influence of the environmental factors discussed previously, give us a logical framework for understanding the relationships between particular soils and the landscapes and ecosystems in which they function. In analyzing these relationships for a given site, ask yourself: What are the materials being added to this soil? What transformations and translocations are taking place in this profile? What materials are being removed? And how have the climate, organisms, topography, and parent material at this site affected these processes over time?

---

[3] For the original discussion of many of the concepts in this section, see Simonson (1958). In addition, detailed discussions of soil genesis can be found in Boul, Hole, McCracken, and Southard (1998).

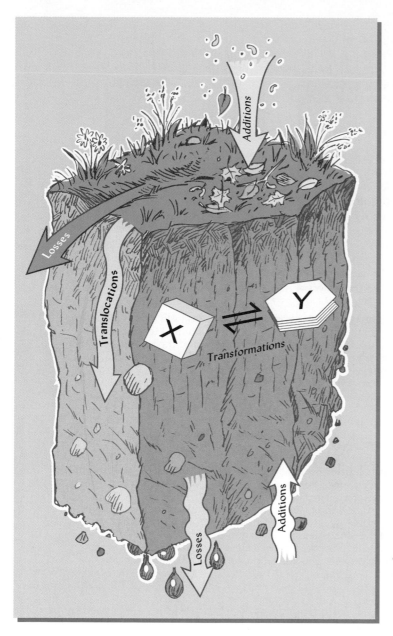

FIGURE 2.21 A schematic illustration of additions, losses, translocations, and transformations as the fundamental processes driving soil-profile development.

A Simplified Example

A-HORIZON DEVELOPMENT. Consider the changes that might take place as a soil develops from a thick layer of relatively uniform parent material. Although some physical weathering and leaching may be necessary to allow plants to grow in certain parent materials, soil formation truly begins when plants become established and begin to deposit their litter and root residues on and in the surface layers of the parent material. The plant residues are disintegrated and partly decomposed by soil organisms that also synthesize humus and other new organic substances. The accumulation of humus enhances the availability of water and nutrients, providing a positive feedback for accelerated plant growth and further humus buildup. The resulting organic-mineral mixture near the soil surface, which comes into being rather quickly, is commonly the first soil horizon developed, the A horizon. It is darker in color and its chemical and physical properties differ from those of the original parent material. Individual soil particles in this horizon commonly clump together under the influence of organic substances to form granules, differentiating this layer from the deeper layers and from the original parent material (see Figure 2.22).

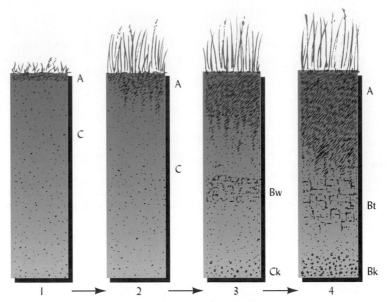

FIGURE 2.22 Stages of development of a hypothetical soil forming from relatively uniform parent material such as loess in a humid temperate climate (see Section 2.18 and Table 2.3 for explanation of the horizon symbols). At stage 1, no layering exists. Organic matter has just begun to accumulate from plant residues near the surface, producing an A horizon. By stage 2, the activity of earthworms and the decay of plant roots have incorporated organic matter into the upper few centimeters of soil, and soluble ions (e.g. Ca^{2+}) have moved downward. Clays are beginning to form from minerals in the loess and to disperse in the percolating waters. By stage 3, more organic matter has accumulated and moved deeply into the soil. Further leaching of soluble ions has occurred, and additional clay is forming and beginning to translocate into lower horizons. A Bw horizon is now distinguished by a change in color and the development of some blocklike structure. Deeper in the regolith, ions leached from the upper layers have precipitated, forming a Ck horizon enriched in calcium carbonates. By stage 4, even higher levels of organic matter are found deeper in the profile. Clays have accumulated sufficiently to form a Bt horizon, and have stimulated the formation of distinct blocklike structures. The pedogenic processes have extended the B horizon down to the upper part of the carbonate accumulation, forming a Bk horizon.

LEACHING.   Organic acids that form from decaying plant residues are carried by percolating waters into the soil, where they stimulate weathering reactions. The acid-charged percolating water dissolves various chemicals and translocates (leaches) them from upper to lower horizons where they may precipitate, creating respective zones of depletion and accumulation. The dissolved substances include both positively charged ions (cations; e.g., Ca^{2+}) and negatively charged ions (anions; e.g., CO_3^- and SO_4^{2-}) released from the breakdown of minerals and organic matter. In semiarid and arid regions, precipitation of these ions at the maximum depth of water penetration commonly produces horizons enriched in calcite ($CaCO_3$) or gypsum ($CaSO_4 \cdot 2H_2O$).

MINERAL WEATHERING.   Biochemical processes accelerate the weathering of primary minerals, disintegrating and altering some to form different kinds of silicate clays. As other primary minerals decompose, the decomposition products recombine into new minerals that include additional types of silicate clays and hydrous oxides of iron and aluminum (see Figure 2.3).

These newly formed clay minerals may accumulate where they are formed, or they may move downward and accumulate deeper in the profile. As materials are removed from one layer and accumulate in another, adjacent layers become more distinct from each other and soil horizons are formed. In the zones of clay accumulation, blocklike or prismatic units of soil structure commonly develop when the clayey mass periodically dries out and cracks (see Section 4.7). As the soil matures, the various horizons within the profile generally become more numerous and more distinctly different from each other.

2.18   THE SOIL PROFILE

At each location on the land, the earth's surface has experienced a particular combination of influences from the five soil-forming factors, causing a different set of layers (horizons) to form in each part of the landscape, thus slowly giving rise to the natural bodies we call **soils.** Each soil is characterized by a given sequence of these horizons. A vertical exposure of this sequence is termed a **soil profile.** We will now consider the major horizons making up soil profiles and the terminology used to describe them.

The Master Horizons and Layers

Five **master** soil horizons are recognized and are designated using the capital letters O, A, E, B, and C (Figure 2.23). Subordinate horizons may occur within some master horizons and these are designated by lowercase letters following the capital master horizon letter (e.g., Bt, Ap, or Oi).

O HORIZONS.   The O group is comprised of organic horizons that generally form above the mineral soil or occur in an organic soil profile. They derive from dead plant and animal residues. Generally absent in grassland regions, O horizons usually occur in forested areas and are commonly referred to as the **forest floor.**

A HORIZONS.   The topmost mineral horizons, designated A horizons, generally contain enough partially decomposed (humified) organic matter to give the soil a color darker than that of the lower horizons. The A horizons are often coarser in texture, having lost some of the finer materials by translocation to lower horizons and by erosion.

E HORIZONS.   These are zones of maximum leaching or *eluviation* (from Latin *ex* or *e*, out, and *lavere*, to wash) of clay, iron, and aluminum oxides, which leaves a concentration of resistant minerals, such as quartz, in the sand and silt sizes. An E horizon is usually found underneath the A horizon and is generally lighter in color than either the A horizon above it or the horizon below. Such E horizons are quite common in soils developed under forests, but they rarely occur in soils developed under grassland. Distinct E horizons can be seen in the soils in Plates 10 and 11 (after page 178).

FIGURE 2.23 Hypothetical mineral soil profile showing the major horizons that may be present in a well-drained soil in the temperate humid region. Any particular profile may exhibit only some of these horizons, and the relative depths vary. In addition, however, a soil profile may exhibit more detailed subhorizons than indicated here. The solum includes the A, E, and B horizons plus some cemented layers of the C horizon.

B Horizons. B horizons form below an O, A, or E horizon and have undergone sufficient changes during soil genesis so that the original parent material structure is no longer discernable. In many B horizons materials have accumulated, typically by illuviation (from the Latin *il,* in, and *lavere,* to wash) from the horizons above. In humid regions, B horizons are the layers of maximum accumulation of materials such as iron and aluminum oxides (Bo or Bs horizons) and silicate clays (Bt horizons), some of which may have illuviated from upper horizons and some of which may have formed in place. Such B horizons can be clearly seen in the middle depths of the profiles shown in Plates 1, 10, and 11 (after page 178). In arid and semiarid regions, calcium carbonate or calcium sulfate may accumulate in the B horizon (giving Bk and By horizons, respectively).

C Horizon. The C horizon is the unconsolidated material underlying the solum (A and B horizons). It may or may not be the same as the parent material from which the solum formed. The C horizon is below the zones of greatest biological activity and has not been sufficiently altered by soil genesis to qualify as a B horizon. While loose enough to be dug with a shovel, C horizon material often retains some of the structural features of the parent rock or geologic deposits from which it formed (see, for example, the lower third of the profiles shown in Plates 7 and 11 after page 178). Its upper layers may in time become a part of the solum as weathering and erosion continue.

R Layers. These are consolidated rock, with little evidence of weathering.

Subdivisions within Master Horizons

Often distinctive layers exist *within* a given master horizon, and these are indicated by a numeral *following* the letter designation. For example, if three different combinations of structure and colors can be seen in the B horizon, then the profile may include a B1–B2–B3 sequence.

If two different geologic parent materials (e.g., loess over glacial till) are present within the soil profile, the numeral 2 is placed in front of the master horizon symbols for horizons developed in the second layer of parent material. For example, a soil would have a sequence of horizons designated O–A–B–2C if the C horizon developed in glacial till while the upper horizons developed in loess.

Transition Horizons

Transitional layers between the master horizons (O, A, E, B, and C) may be dominated by properties of one horizon but also have prominent characteristics of another. The two applicable capital letters are used to designate the transition horizons (e.g., AE, EB, BE, and BC), the dominant horizon being listed before the subordinate one. Letter combinations such as E/B are used to designate transition horizons where distinct parts of the horizon have properties of E while other parts have properties of B.

Subordinate Distinctions

Since the capital letter designates the nature of a master horizon in only a very general way, specific horizon characteristics may be indicated by a lowercase letter following the master horizon designation. For example, three types of O horizons (Oi, Oe, and Oa) are indicated in the profile shown in Figure 2.23, which presents a commonly encountered sequence of horizons. Other subordinate distinctions include special physical properties and the accumulation of particular materials, such as clays and salts. A list of the recognized subordinate letter designations and their meanings is given in Table 2.3. We suggest that you mark this table for future reference, and study it now to get an idea of the distinctive soil properties that can be indicated by horizon designations. By way of illustration, a Bt horizon is a B horizon characterized by clay accumulation (t from the German *ton,* meaning clay); likewise, in a Bk horizon, carbonates (k) have accumulated. We have already used the suffixes i, a, and e to distinguish different types of O horizons. The significance of several other subordinate horizon designations will be discussed in the next chapter.

TABLE 2.3   Lowercase Letter Symbols to Designate Subordinate Distinctions Within Master Horizons

Letter	Distinction	Letter	Distinction
a	Organic matter, highly decomposed	n	Accumulation of sodium
b	Buried soil horizon	o	Accumulation of Fe and Al oxides
c	Concretions or nodules	p	Plowing or other disturbance
d	Dense unconsolidated materials	q	Accumulation of silica
e	Organic matter, intermediate decomposition	r	Weathered or soft bedrock
f	Frozen soil	s	Illuvial accumulation of O.M.[a] and Fe and Al oxides
g	Strong gleying (mottling)	ss	Slickensides
h	Illuvial accumulation of organic matter	t	Accumulation of silicate clays
i	Organic matter, slightly decomposed	v	Plinthite (high iron, red material)
j	Jarosite	w	Distinctive color or structure
jj	Cryoturbation (frost churning)	x	Fragipan (high bulk density, brittle)
k	Accumulation of carbonates	y	Accumulation of gypsum
m	Cementation or induration	z	Accumulation of soluble salts

[a] O.M. = organic matter.

Horizons in a Given Profile

It is not likely that the profile of any one soil will show all of the horizons that collectively are shown in Figure 2.23. The ones most commonly found in well-drained soils are Oi and Oe (or Oa) if the land is forested; A or E (or both, depending on circumstances); Bt or Bw; and C. Conditions of soil genesis will determine which others are present and their clarity of definition.

When a virgin (never-cultivated) soil is plowed and cultivated for the first time, the upper 15 to 20 cm becomes the plow layer or Ap horizon. Cultivation, of course, obliterates the original layered condition of the upper portion of the profile, and the Ap horizon becomes more or less homogeneous. An example of an Ap horizon can be seen in the upper 20 cm of the soil shown in Plate 1 (following page 178). In some soils, the A and E horizons are deeper than the plow layer. In other cases, where the upper horizons are quite thin, the plow line is just at the top of, or even down in, the B horizon.

A sometimes perplexing profile feature is the presence of a buried soil resulting from natural or human action. In profile study and description, such a situation requires careful analysis.

2.19   CONCLUSION

The parent materials from which soils develop vary widely around the world and from one location to another only a few meters apart. Knowledge of these materials, their sources or origins, mechanisms for their weathering, and means of transport and deposition are essential to understanding soil genesis.

Soil formation is stimulated by *climate* and living *organisms* acting on *parent materials* over periods of *time* and under the modifying influence of *topography*. These five major factors of soil formation determine the kinds of soil that will develop at a given site. When all of these factors are the same at two locations, the kind of soil at these locations should be the same.

Soil genesis starts when layers or horizons not present in the parent material begin to appear in the soil profile. Organic matter accumulation in the upper horizons, the downward movement of soluble ions, the synthesis and downward movement of clays, and the development of specific soil particle groupings (structure) in both the upper and lower horizons are signs that the process of soil formation is under way.

The four general processes of soil formation (gains, losses, transformations, and translocations) and the five major factors influencing these processes provide us with an invaluable logical framework in site selection and in predicting the nature of soil bodies likely to be found on a particular site. Conversely, analysis of the horizon properties of a soil profile can tell us much about the nature of the climatic, biological, and geological conditions (past and present) at the site.

Characterization of the horizons in the profile leads to the identity of a soil individual, which is then subject to classification—the topic of our next chapter.

STUDY QUESTIONS

1. What is meant by the statement, *weathering combines the processes of destruction and synthesis?* Give an example of these two processes in the weathering of a primary mineral.
2. How is water involved in the main types of chemical weathering reactions?
3. Explain the significance of the ratio of silicon to aluminum in soil minerals.
4. Give an example of how parent material may vary across large geographic regions on one hand, but may also vary within a small parcel of land, on the other.
5. Name the five factors affecting soil formation. With regard to each of these factors of soil formation, compare a forested Rocky Mountain slope to the semiarid grassland plains far below.
6. How do *colluvium, glacial till,* and *alluvium* differ in appearance and agency of transport?
7. What is *loess,* and what are some of its properties as a parent material?
8. Give two specific examples for each of the four broad processes of soil formation.
9. Assuming a level area of granite rock was the parent material in both cases, describe in general terms how you would expect two soil profiles to differ, one in a warm, semiarid grassland and the other in a cool, humid pine forest.
10. For the two soils described in question 5, make a profile sketch using master horizon symbols and subordinate suffixes to show the approximate depths, sequence, and nature of the horizons you would expect to find in each soil.

REFERENCES

Boul, S. W., F. D. Hole, R. J. McCracken, and R. J. Southard. 1998. *Soil Genesis and Classification,* 4th ed. (Ames, Iowa: Iowa State University Press).

Jackson, M. L. 1964. "Chemical composition of soils," in F. E. Bear (ed.), *Chemistry of the Soil,* 2nd ed. (New York: Reinhold).

Jenny, Hans. 1941. *Factors of Soil Formation: A System of Quantitative Pedology.* (originally published by McGraw-Hill; Mineola, N.Y.: Dover).

Jenny, Hans. 1980. *The Soil Resource—Origins and Behavior.* Ecological Studies, vol. 37. (New York: Springer-Verlag).

Logan, William Bryant. 1995. *Dirt: The Ecstatic Skin of the Earth.* (New York: Riverhead Books).

Miller, F. P. 1976. *Maryland Soils Bulletin 212.* (College Park, Md.: University of Maryland Cooperative Extension Service).

Simonson, R. W. 1958. "Outline of a generalized theory of soil genesis," *Soil Sci. Soc. Amer. Proc.,* **22**:152–156.

Soil Survey Division Staff. 1993. *Soil Survey Manual.* Agricultural Handbook 18. (Washington, D.C.: U.S. Government Printing Office).

Soil Survey Staff. 1998. *Keys to Soil Taxonomy,* 8th ed. (Washington, D.C.: USDA Natural Resources Conservation Service).

3

SOIL CLASSIFICATION

It is embarrassing not to be able to agree on what soil is. In this the pedologists are not alone. Biologists cannot agree on a definition of life and philosophers on philosophy.
—HANS JENNY, THE SOIL RESOURCE: ORIGIN AND BEHAVIOR

People classify things in order to make sense of their world. We do it all the time, whenever we call things by group names, giving them labels that inform us on important properties. Imagine a world without classifications. Imagine surviving in the woods knowing only that each plant was a plant, not which are edible by people, which attract wildlife, or which are poisonous to touch. Imagine surviving in a city knowing only that each person was a person, not a child or an adult, a male or a female, a police officer, a hoodlum, a friend, a teacher, a potential date, or any of the other categories into which we classify people. So, too, our understanding and management of soils and terrestrial systems would be hobbled if we knew only that a soil was a soil. How could we organize our information about soils? How could we learn from others' experience or communicate our knowledge to clients, colleagues, or workers?

Throughout history people have used various systems to name and classify soils. From the time crops were first cultivated, humans noticed differences in soils and classified them, grouping them according to their suitability for different uses by giving them descriptive names such as *black cotton soils, rice soils,* or *olive soils.* Other soil names still in common use today have geological connotations, suggesting the parent materials from which the soils formed: *limestone soils, piedmont soils,* and *alluvial soils.* Such terms, while they may convey some meaning to local users, do little to help us organize our knowledge of soils or to see the relationships among soils in the world.

In this chapter we will learn how soils are classified as natural bodies, on the basis of their profile characteristics, not merely on the basis of their suitability for a particular use. Such a soil classification system is essential to foster global communications about soils among soil scientists and all people concerned with the management of land and the conservation of the soil resource. Through such systems we can take advantage of research and experience at one location to predict the behavior of similarly classified soils at another location. Soil names such as *Mollisols* or *Oxisols* conjure up similar mental images in the minds of soil scientists everywhere, whether they live in the United States, Europe, Japan, developing countries, or elsewhere. Through the classification system, we can create a universal language of soils that enhances communication among users of soils around the world.

3.1 CONCEPT OF INDIVIDUAL SOILS

The natural body concept of soils recognizes the existence of individual entities, each of which we call *a soil*. Just as human individuals differ from one another, soil individuals have characteristics distinguishing each from the others. However, unlike for humans or trees, there are no sharp demarcations between one soil individual and another. Rather, there is a gradation in properties as one moves from one soil individual to an adjacent one. The gradation in soil properties can be compared to the gradation in the wavelengths of light as you move from one color to another in a rainbow. The change is gradual, and yet we identify a boundary differentiating between what we call green and what we call blue.

Pedon

Soils in the field are heterogenous; that is, the profile characteristics are not exactly the same in any two points within the soil individual you may choose to examine. Consequently, it is necessary to characterize a soil individual in terms of an imaginary three-dimensional unit called a **pedon** (rhymes with "head on," from the Greek *pedon*, ground; see Figure 3.1). It is the smallest sampling unit that displays the full range of properties characteristic of a particular soil.

Polypedon

Pedons occupy from about 1 to 10 m² of land area. Because it is what is actually examined during field investigation of soils, the pedon serves as the fundamental unit of soil classification. However, a soil unit in a landscape usually consists of a group of very similar pedons, closely associated together in the field. Such a group of similar pedons, or **polypedon,** is of sufficient size to be recognized as a landscape component termed a **soil individual.**

Soil Series

All the soil individuals in the world that have in common a suite of soil profile properties and horizons that fall within a particular range are said to belong to the same **soil series.** A soil series, then, is class of soils, not a soil individual, in the same way that *Pinus sylvestrus* is a species of tree, not a particular individual tree.

Groupings of Soils

We have now identified the most specific and most general extremes in the concept of soils. One extreme is that of a natural body called *a soil,* characterized by a three-dimensional sampling unit (pedon), related groups of which (polypedons) are included in a soil individual. At the other extreme is *the soil,* a collection of all these natural bodies that is distinct from water, solid rock, and other natural parts of the earth's crust. To understand all these soils we will study a hierarchical soil classification scheme developed by the U.S. Department of Agriculture[1] that groups soils into classes at increasing levels of generality between these two extremes.

3.2 COMPREHENSIVE CLASSIFICATION SYSTEM: *SOIL TAXONOMY*[2]

A comprehensive soil classification system, called *Soil Taxonomy,*[3] provides a hierarchical grouping of natural soil bodies. The system is based on *soil properties* that can

---

[1] See Appendix B for summaries of the Canadian and United Nations Food and Agriculture Organization systems.

[2] For a complete description of *Soil Taxonomy* see Soil Survey Staff (1999). The first edition of *Soil Taxonomy* was published as Soil Survey Staff (1975).

[3] Taxonomy is the science of the principles of classification. For a review of the achievements and challenges of *Soil Taxonomy,* see SSSA (1984).

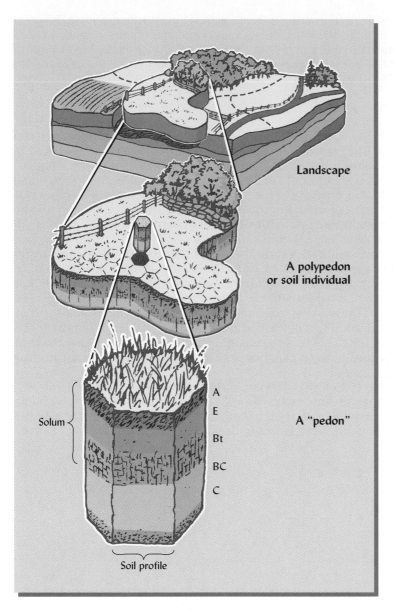

FIGURE 3.1 A schematic diagram to illustrate the concept of pedon and of the soil profile that characterizes it. Note that several contiguous pedons with similar characteristics are grouped together in a larger area (outlined by broken lines) called a *polypedon* or soil *individual*. Several soil individuals are present in this landscape.

be objectively observed or measured rather than on presumed mechanisms of soil formation.

Soil Taxonomy employs a unique *nomenclature* which gives a definite connotation of the major characteristics of the soils in question. It is truly international since it is not based on any one national language. Consideration will be given to the nomenclature used after brief reference is made to the major criteria for the system—soil properties.

Bases of Soil Classification

Most of the chemical, physical, and biological properties presented in this text are used as criteria for classifying soils in *Soil Taxonomy*. A few examples are moisture and temperature status throughout the year, as well as color, texture, and structure of the soil. Chemical and mineralogical properties, such as the contents of organic matter, clay, iron and aluminum oxides, silicate clays, salts, the pH, the percentage base saturation,[4]

---

[4] The percentage base saturation is the percentage of the soil's negatively charged sites (cation exchange capacity) that are satisfied by attracting metal cations (such as Ca^{2+}, Mg^{2+} and K^+) that form strong bases (e.g., $Ca(OH)_2$) upon hydrolysis (see Section 8.9).

and soil depth are other important criteria for classification. While many of the properties used may be observed in the field, others require precise measurements on samples taken to a sophisticated laboratory. Precise measurements are also used to define certain *diagnostic soil horizons,* the presence or absence of which help determine the place of a soil in the classification system.

Diagnostic Surface Horizons

The diagnostic horizons that occur at the soil surface are called *epipedons* (from the Greek *epi,* over, and *pedon,* soil). The epipedon includes the upper part of the soil darkened by organic matter, the upper eluvial horizons, or both. It may include part of the B horizon (see Figure 3.2) if the latter is significantly darkened by organic matter. Seven epipedons are recognized (Table 3.1), but only five occur naturally over wide areas. The other two, anthropic and plaggen, are the result of intensive human use. They are common in parts of Europe and Asia where soils have been utilized for many centuries.

The **mollic epipedon** (Latin *mollis,* soft) is a mineral surface horizon noted for its dark color associated with its accumulated organic matter (>0.6% organic C throughout the thickness), for its thickness (generally >25 cm), and for its softness even when dry. It has a high base saturation (greater than 50% base-forming cations). These epipedons are characteristic of soils developed under native prairies (Figure 3.2 and Plate 8, after page 178).

The **umbric epipedon** (Latin *umbra,* shade; hence, dark) has the same general characteristics as the mollic epipedon except the percentage base saturation is less than 50%.

The **ochric epipedon** (Greek *ochros,* pale) is a mineral horizon that is either too thin, too light in color, or too low in organic matter to be either a mollic or umbric horizon (see Plates 1, 7, and 11).

The **melanic epipedon** (Greek *melas,* melan, black) is a mineral horizon that is very black in color due to its high organic matter content (organic carbon >6%). It is characteristic of soils high in such minerals as allophane, developed from volcanic ash. It is more than 30 cm thick, and is extremely low in density (see Plate 2).

FIGURE 3.2 The mollic epipedon (a diagnostic horizon) in this soil includes genetic horizons designated Ap, A2, and Bt1, all darkened by the accumulation of organic matter. Scale marked every 10 cm. (Photo courtesy of R. Weil).

TABLE 3.1 Major Features of Diagnostic Surface Horizons (Epipedons) in Mineral Soils Used to Differentiate at the Higher Levels of *Soil Taxonomy*

Diagnostic horizon	Major features
Surface horizons = epipedons	
Mollic	Thick, dark-colored, high base saturation, strong structure
Umbric	Same as mollic except low base saturation
Ochric	Light-colored, low organic content, may be hard and massive when dry
Melanic	Thick, black, high in organic matter (>6% organic C), common in volcanic ash soils
Histic	Very high in organic content, wet during some part of year
Anthropic	Human-modified molliclike horizon, high in available P
Plaggen	Human-made sodlike horizon created by years of manuring

The **histic epipedon** (Greek *histos,* tissue) is a relatively thin layer of **organic soil materials** overlying a mineral soil. Formed in wet areas, the histic epipedon is commonly a layer of peat or muck about 20 to 30 cm thick, with a black to dark brown color, and a very low density.

Diagnostic Subsurface Horizons

Many subsurface horizons are used to characterize different soils in *Soil Taxonomy*. The 18 that are considered diagnostic horizons are listed in Figure 3.3, along with major processes causing their formation. We will briefly discuss a few of the more commonly encountered subsurface diagnostic horizons.

The **argillic horizon** is a subsurface accumulation of high-activity silicate clays that have moved downward from the upper horizons or have formed in place. An example is shown in Plate 1 between 25 and 70 cm on the scale in the photograph. The clays often are found as coatings on pore walls (as shown in Figure 4.3) and surfaces of the structural groupings. The coatings usually appear as shiny surfaces or as clay bridges between sand grains. Termed **argillans** or **clay skins,** they are concentrations of clay translocated from upper horizons (see Plate 23, before page 179).

The **natric horizon** likewise has silicate clay accumulation (with clay skins), but the clays are accompanied by more than 15% exchangeable sodium on the colloidal complex and by columnar or prismatic soil structural units. The natric horizon is found mostly in arid and semiarid areas. An example is shown in Figure 4.10*e*.

The **oxic horizon** is a highly weathered subsurface horizon that is very high in Fe and Al oxides, and in low-activity silicate clays (e.g., kaolinite). The cation-holding capacity is <16 cmol$_c$/kg clay. The horizon is at least 30 cm deep and has <10% weatherable minerals in the fine fraction. It is generally physically stable, crumbly, and not very sticky, despite its high clay content. It is found mostly in humid tropical and subtropical regions (see Plate 9, between about 1 and 3 feet on the scale).

The **spodic horizon** is an illuvial horizon that is characterized by the accumulation of colloidal organic matter and aluminum oxide (with or without iron oxide). It is commonly found in highly leached forest soils of cool humid climates, typically on sandy-textured parent materials (see Plate 10, reddish brown and black layers below the whitish layer).

The **albic horizon** is a light-colored eluvial horizon that is low in clay and oxides of Fe and Al. These materials have largely been moved downward from this horizon (see Plate 10, at about 10 cm depth).

A number of horizons have accumulations of saltlike chemicals that have leached from upper horizons in the profile. **Calcic horizons** contain an accumulation of carbonates (mostly $CaCO_3$) that often appear as white chalklike nodules (see the Bk horizon in the lower part of the profile shown in Figure 3.2). **Gypsic horizons** have an accumulation of gypsum ($CaSO_4 \cdot 2H_2O$), and **salic horizons** an accumulation of soluble salts. These are found mostly in soils of arid and semiarid regions.

In some subsurface diagnostic horizons, the materials are cemented or densely packed, resulting in relatively impermeable layers called *pans* (**duripan, fragipan,** and **placic horizons**).

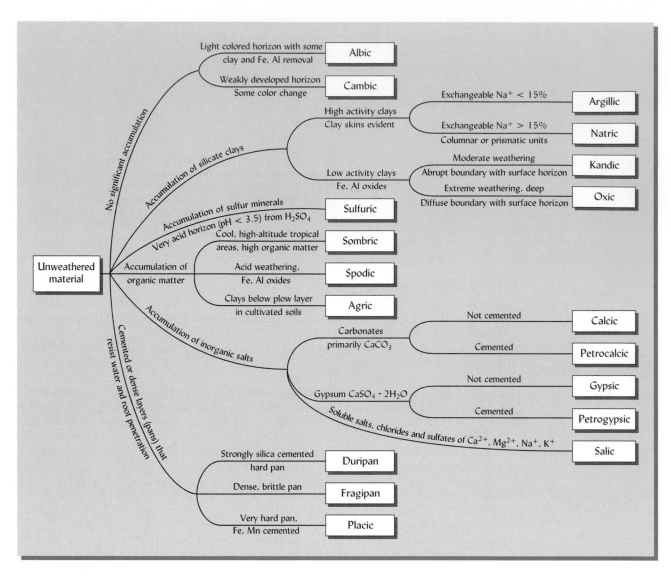

FIGURE 3.3 Names and major distinguishing characteristics of subsurface diagnostic horizons. Note that the focus is primarily on soil properties, not on how the properties have presumably evolved. Among the characteristics emphasized is the accumulation of silicate clays, organic matter, Fe and Al oxides, calcium compounds, and soluble salts, as well as materials that become cemented or highly acidified, thereby constraining root growth. The presence or absence of these horizons plays a major role in determining in which class a soil falls in *Soil Taxonomy*. See Chapter 8 for discussion of low- and high-activity clays.

Soil Moisture Regimes (SMR)

A soil moisture regime refers to the presence or absence of either water-saturated conditions (usually groundwater) or plant-available soil water during specified periods in the year.

Aquic. Soil is saturated with water and virtually free of gaseous oxygen for sufficient periods of time for evidence of poor aeration (gleying and mottling) to occur.

Udic. Soil moisture is sufficiently high year-round in most years to meet plant needs. This regime is common for soils in humid climatic regions. An extremely wet moisture regime with excess moisture for leaching throughout the year is termed **perudic**.

Ustic. Soil moisture is intermediate between Udic and Aridic regimes—generally there is some plant-available moisture during the growing season, although significant periods of drought may occur.

Aridic. The soil is dry for at least half of the growing season and moist for less than 90 consecutive days. This regime is characteristic of arid regions. The term *torric* is used to indicate the same moisture condition in certain soils that are both hot and dry in summer, though they may not be hot in winter.

Xeric. This soil moisture regime is found in typical Mediterranean-type climates, with cool, moist winters and warm, dry summers. Like the Ustic regime, it is characterized by having long periods of drought in the summer.

Soil Temperature Regimes

Soil temperature regimes, such as frigid, mesic, and thermic, are used to classify soils at some of the lower levels in *Soil Taxonomy*. The cryic (Greek *kryos,* very cold) temperature regime distinguishes some higher-level groups. These regimes are based on mean annual soil temperature, mean summer temperature, and the difference between mean summer and winter temperatures, all at 50 cm depth. The specific temperature regimes will be described in the discussion of soil families (Section 3.17).

3.3 CATEGORIES AND NOMENCLATURE OF *SOIL TAXONOMY*

There are six categories of classification in *Soil Taxonomy:* (1) *order,* the highest (broadest) category, (2) *suborder,* (3) *great group,* (4) *subgroup,* (5) *family,* and (6) *series* (the most specific category). These categories are hierarchical because the lower categories fit within the higher categories. Thus, each order has several suborders, each suborder has several subgroups, and so forth. This system may be compared with those used for the classification of plants or animals, as shown in Table 3.2. Just as *Trifolium repens* identifies a specific kind of plant, the Miami series identifies a specific kind of soil.

Nomenclature of Soil Taxonomy

Although unfamiliar at first sight, the nomenclature system of *Soil Taxonomy* has a very logical construction and conveys a great deal of information about the nature of the soils named. The system is easy to learn after a bit of study. The nomenclature is used throughout this book, especially to identify the kinds of soils shown in illustrations. When reading, if you make a conscious effort to identify the parts of each soil class mentioned in the text and figure captions, and recognize the level of category indicated, the system will become second nature.

The names of orders are combinations of (1) formative elements, which generally define the characteristics of the soils, and (2) the ending *sols* (from the Latin *solum,* soil). For example, soils of the order **Aridisols** (from the Latin *aridus,* dry) are characteristically dry soils in arid regions. Those of the order **Inceptisols** (from the Latin *inceptum,* beginning) are soils with only the beginnings or inception of profile development.

The names of suborders automatically identify the order of which they are a part. For example, soils of the suborder **Aquolls** are the wetter soils (from the Latin *aqua,* water) of the Mollisols order. Likewise, the name of the great group identifies the sub-

TABLE 3.2 Comparison of the Classification of a Common Cultivated Plant, White Clover (*Trifolium repens*), and a Soil, Miami Series

Plant classification			Soil classification	
Phylum	Pterophyta		Order	Alfisols
Class	Angiospermae		Suborder	Udalfs
Subclass	Dicotyledoneae	Increase specificity	Great Group	Hapludalfs
Order	Rosales		Subgroup	Oxyaquic Hapludalfs
Family	Leguminosae		Family	Fine loamy, mixed, mesic, active
Genus	*Trifolium*		Series	Miami
Species	*repens*		Phase[a]	Miami silt loam

[a] Technically not a category in *Soil Taxonomy* but used in field surveying. *Silt loam* refers to the texture of the A horizon.

order and order of which it is a part. **Argiaquolls** are Aquolls with clay or argillic (Latin *argilla,* white clay) horizons. In the following illustration, note that the three letters *oll* identify each of the lower categories as being in the M**oll**isols order.

M**oll**isols	Order
Aqu**oll**s	Suborder
Argiaqu**oll**s	Great group
Typic Argiaqu**oll**s	Subgroup

If one is given only the subgroup name, the great group, suborder, and order to which the soil belongs are automatically known.

Family names in general identify subsets of the subgroup that are similar in texture, mineral composition, and mean soil temperature at a depth of 50 cm. Thus the name **Typic Argiaquolls, fine, mixed, mesic, active** identifies a family in the Typic Argiaquolls subgroup with a fine texture, mixed clay mineral content, mesic (8 to 15°C) soil temperature, and clays active in cation exchange.

Soil series are named after a geographic feature (town, river, etc.) near where they were first recognized. Thus, names such as *Fort Collins, Cecil, Miami, Norfolk,* and *Ontario* identify soil series first described near the town or geographic feature named. Approximately 19,000 soil series have been classified in the United States alone.

With this brief explanation of the nomenclature of the new system, we will now consider the general nature of soils in each of the soil orders.

3.4 SOIL ORDERS

Each of the world's soils is assigned to one of 12 **orders,** largely on the basis of soil properties that reflect a major course of development, with considerable emphasis placed on the presence or absence of major diagnostic horizons (Table 3.3). The names and major characteristics of each soil order are shown in Table 3.3. Note that all order names have a common ending, *sols* (from the Latin *solum,* soil)

The general conditions that enhance the formation of soils in the different orders are shown in Figure 3.4. From soil profile characteristics, soil scientists can ascertain the relative degree of soil development in the different orders, as shown in this figure. Note that soils with essentially no profile layering (Entisols) have the least development, while the deeply weathered soils of the humid tropics (Oxisols and Ultisols) show the greatest soil development. The effect of climate (temperature and moisture) and of vegetation (forests or grasslands) on the kinds of soils that develop is also indicated in Figure 3.4.

TABLE 3.3   **Names of Soil Orders in *Soil Taxonomy* with Their Derivation and Major Characteristics**

The bold letters in the order names indicate the formative element used as the ending for suborders and lower taxa within that order.

Name	Formative element	Derivation	Pronunciation	Major characteristics
Alfisols	alf	Nonsense symbol	Ped*alf*er	Argillic, natric, or kandic horizon; high to medium base saturation
Andisols	and	Jap. *ando,* blacksoil	*And*esite	From volcanic ejecta, dominated by allophane or Al-humic complexes
Ar**id**isols	id	L. *aridus,* dry	A*rid*	Dry soil, ochric epipedon, sometimes argillic or natric horizon
Entisols	ent	Nonsense symbol	Rec*ent*	Little profile development, ochric epipedon common
Gelisols	el	Gk. *gelid,* very cold	J*el*ly	Permafrost, often with cryoturbation (frost churning)
Histosols	ist	Gk. *histos,* tissue	*Hist*ology	Peat or bog; >20% organic matter
Inc**ept**isols	ept	L. *inceptum,* beginning	Inc*ept*ion	Embryonic soils with few diagnostic features, ochric or umbric epipedon, cambic horizon
M**oll**isols	oll	L. *mollis,* soft	M*oll*ify	Mollic epipedon, high base saturation, dark soils, some with argillic or natric horizons
Oxisols	ox	Fr. *oxide,* oxide	O*x*ide	Oxic horizon, no argillic horizon, highly weathered
Sp**od**osols	od	Gk. *spodos,* wood ash	P*od*zol; odd	Spodic horizon commonly with Fe, Al oxides and humus accumulation
Ultisols	ult	L. *ultimus,* last	*Ult*imate	Argillic or kandic horizon, low base saturation
V**ert**isols	ert	L. *verto,* turn	In*vert*	High in swelling clays; deep cracks when soil dry

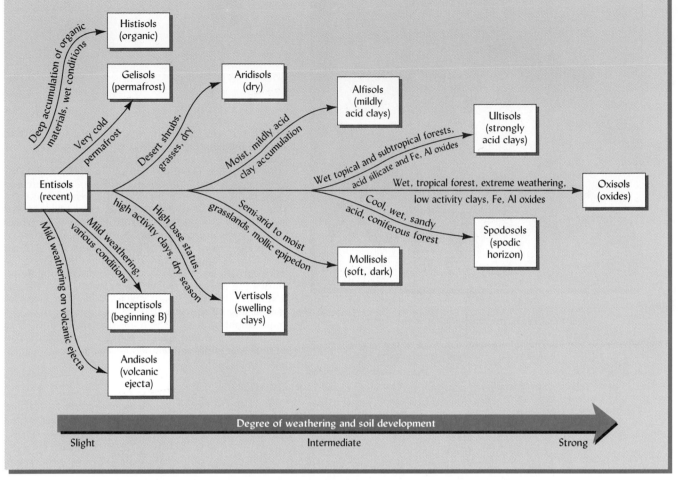

FIGURE 3.4 Diagram showing general degree of weathering and soil development in the different soil orders classified in *Soil Taxonomy*. Also shown are the general climatic and vegetative conditions under which soils in each order are formed.

To some degree, most of the soil orders occur in climatic regions that can be described by moisture and temperature regimes. Figure 3.5 illustrates some of the relationships among the soil orders with regard to these climatic factors. While only the Gelisols and Aridisols orders are defined directly in relation to climate, Figure 3.5 indicates that the most highly weathered soils tend to be associated with the warmer and wetter climates. Figure 3.6 is a simplified general soil map showing the major areas of the United States dominated by each order. Profiles for each soil order are shown in color on Plates 1 to 12 (after page 178). A more detailed soil map for the United States can be found in Appendix A and on the web at www.statlab.iastate.edu/soils/photogal/ orders/ordmap.htm. A general world map of the 12 soil orders is printed in color on the front end papers.

Although a detailed description of all the lower levels of soil categories is far beyond the scope of this book, a general knowledge of the 12 soil orders is essential for understanding the nature and function of soils in different environments. The simplified key given in Figure 3.7 helps illustrate how *Soil Taxonomy* can be used to key out the order of any soil based on properties of the soil profile. Because certain diagnostic properties take precedence over others, the key must always be used starting at the top, and working down. It will be useful to review this key after reading about the general characteristics, nature, and occurrence of each soil order.

We will now consider each of the soil orders, beginning with those characterized by little profile development and progressing to those with the most highly weathered profiles (as represented from left to right in Figure 3.4).

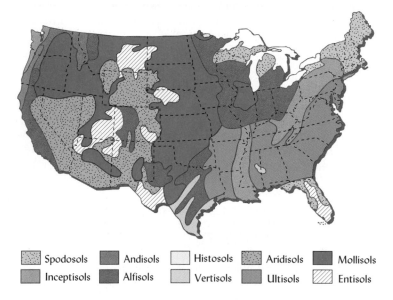

FIGURE 3.5 Diagram showing the general soil moisture and soil temperature regimes that characterize the most extensive soils in each of eight soil orders. Soils of the other four orders (Andisols, Entisols, Inceptisols, and Histosols) may be found under any of the soil moisture and temperature conditions (including the area marked EIH). Major areas of Vertisols are found only where clayey materials are in abundance and are most extensive where the soil moisture and temperature conditions approximate those shown inside the box with broken lines. Note that these relationships are only approximate and that less extensive areas of soils in each order may be found outside the indicated ranges. For example, some Ultisols (Ustults) and Oxisols (Ustox) have soil moisture levels for at least part of the year that are much lower than this graph would indicate. (The terms used at the bottom to describe the soil temperature regimes are those used in helping to identify soil families.)

FIGURE 3.6 Simplified general soil map of the United States showing patterns of soil orders based on *Soil Taxonomy*. For more detailed soil maps showing the distribution of both soil orders and suborders for the United States, see Appendix A.

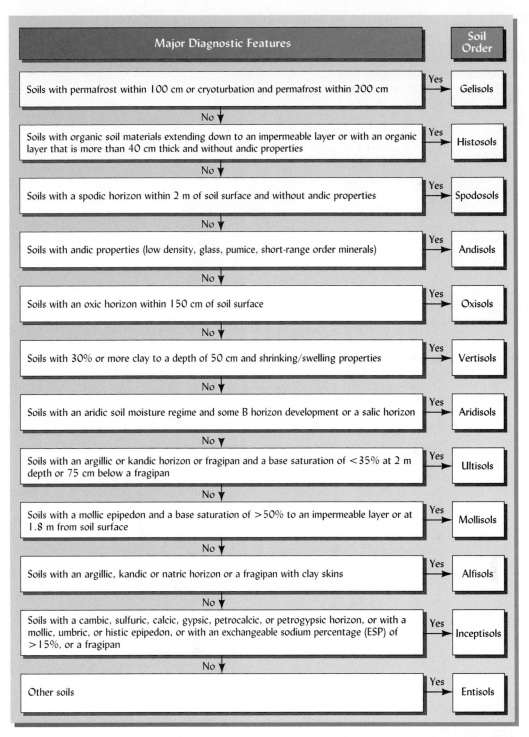

FIGURE 3.7 A simplified key to the 12 soil orders in *Soil Taxonomy*. In using the key, always begin at the top. Note how diagnostic horizons and other profile features are used to distinguish each soil order from the remaining orders. Entisols, having no such special diagnostic features, key out last. Also note that the sequence of soil orders in this key bears no relationship to the degree of profile development and adjacent soil orders may not be more similar than nonadjacent ones. See Section 3.2 for explanations of the diagnostic horizons.

3.5 ENTISOLS (RECENT: LITTLE IF ANY PROFILE DEVELOPMENT)

Global Distribution of Entisols

Suborders
Aquents (wet)
Arents (mixed horizons)
Fluvents (alluvial deposits)
Orthents (typical)
Psamments (sandy)

Weakly developed mineral soils without natural genetic (subsurface) horizons, or with only the beginnings of such horizons (see Plate 4, after page 178) belong to the Entisols order. Most have an ochric epipedon and a few have human-made anthropic or agric epipedons. Some have albic subsurface horizons. Soil productivity ranges from very high for certain Entisols formed in recent alluvium to very low for those forming in shifting sand or on steep rocky slopes.

This is an extremely diverse group of soils with little in common, other than the lack of evidence for all but the earliest stages of soil formation (see Figure 3.8). Entisols are either young in years or their parent materials have not reacted to soil-forming factors. On parent materials such as fresh lava flows or recent alluvium (Fluvents), there has been too little time for much soil formation. In extremely dry areas, scarcity of water and vegetation may inhibit soil formation. Likewise, frequent saturation with water (Aquents) may delay soil formation. Some Entisols occur on steep slopes, where the rates of erosion may exceed the rates of soil formation, preventing horizon development. Others occur on construction sites where bulldozers destroy or mix together the soil horizons, causing the existing soils to become Entisols (some have suggested that these be called *Urbents*, or urban Entisols).

FIGURE 3.8 Profile of a Psamment formed on sandy alluvium in Virginia. Note the accumulation of organic matter in the A horizon but no other evidence of profile development. The knife is 24 cm long. (Photo courtesy of R. Weil)

3.6 INCEPTISOLS (FEW DIAGNOSTIC FEATURES: INCEPTION OF B HORIZON)

Global Distribution of Inceptisols

Suborders

Anthrepts (human-made, high phosphorus, dark surface)

Aquepts (wet)

Cryepts (very cold)

Udepts (humid climate)

Ustepts (semiarid)

Xerepts (dry summers, wet winters)

In Inceptisols the beginning or *inception* of profile development is evident, and some diagnostic features are present. However, the well-defined profile characteristics of soils thought to be more mature have not yet developed. For example, a cambic horizon showing some color or structural change is common in Inceptisols, but a more mature illuvial B horizon such as an argillic cannot be present. Other subsurface diagnostic horizons that may be present in Inceptisols include duripans, fragipans, and calcic, gypsic, and sulfidic horizons. The epipedon in most Inceptisols is an ochric, although a plaggen or weakly expressed mollic or umbric epipedon may be present. Inceptisols show more significant profile development than Entisols, but are defined to exclude soils with diagnostic horizons or properties that characterize certain other soil orders. Thus, soils with only slight profile development occurring in arid regions, or containing permafrost or andic properties, are excluded from the Inceptisols. They fall, instead, in the soil orders Aridisols, Gelisols, or Andisols, as discussed in later sections. Inceptisols are found in almost every climatic region, and their natural productivity varies widely.

3.7 ANDISOLS (VOLCANIC ASH SOILS)

Global Distribution of Andisols

Suborders

Aquands (wet)

Cryands (cold)

Torrands (hot, dry)

Udands (humid)

Ustands (moist/dry)

Vitrands (volcanic glass)

Xerands (dry summers, moist winters)

Andisols are usually formed on volcanic ash and cinders and are commonly found near the volcano source or in areas downwind from the volcano, where a sufficiently thick layer of ash has been deposited during eruptions (Figure 3.9). They are generally found in volcanic materials deposited in recent geological time. Andisols have not had time to become highly weathered. The principal soil-forming process has been the rapid weathering (transformation) of volcanic ash to produce amorphous or poorly crystallized silicate minerals such as **allophane** and **imogolite** and the iron oxy-hydroxide, **ferrihydrite**. The accumulation of organic matter is also quite pronounced (but not enough to form a histic epipedon), due largely to its protection in aluminum–humus complexes. Little downward translocation of these colloids, or other profile development, has yet taken place. Like the Entisols and Inceptisols, Andisols are young soils, usually having devel-

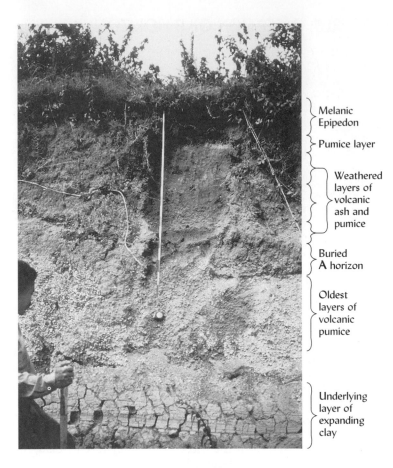

Melanic
Epipedon

Pumice layer

Weathered
layers of
volcanic
ash and
pumice

Buried
A horizon

Oldest
layers of
volcanic
pumice

Underlying
layer of
expanding
clay

FIGURE 3.9 An Andisol developed on layers of volcanic ash and pumice in central Africa. (Photo courtesy of R. Weil)

oped for only 5000 to 10,000 years. Unlike the previous two orders of immature soils, Andisols have a unique set of ***andic properties*** due to a common type of parent materials.

Andisols have andic soil properties in at least 35 cm of the upper 60 cm of soil. Materials with andic properties are characterized by a high content of volcanic glass and/or a high content of amorphous or poorly crystalline iron and aluminum minerals. The combination of these minerals and the high organic matter results in light, fluffy soils that are easily tilled, but have a high water-holding capacity and resist erosion by water. They are mostly found in regions where rainfall keeps them from being susceptible to erosion by wind. Andisols are usually of high natural fertility, except that phosphorus availability is severely limited by the extremely high phosphorus retention capacity of the andic materials (see Section 13.8). Fortunately, proper management of plant residues and fertilizers can usually overcome this difficulty. Some Andisols have a melanic epipedon, a surface diagnostic horizon that has a high organic matter content and dark color (see Plate 2, after page 178).

3.8 GELISOLS (PERMAFROST AND FROST CHURNING)

Global Distribution of Gelisols

Suborders
Histels (organic)
Orthels (no special features)
Turbels (cryoturbation)

Gelisols are young soils with little profile development. Cold temperatures and frozen conditions for much of the year slow the process of soil formation. The principal defining feature of these soils is the presence of a **permafrost** layer (see Plates 5 and 14 after page 178). Permafrost is a layer of material that remains at temperatures below 0°C for more than two consecutive years. It may be a hard, ice-cemented layer of soil material (e.g., designated Cfm in profile descriptions), or, if dry, it may be uncemented (e.g., designated Cf). In Gelisols, the permafrost layer lies within 100 cm of the soil surface, unless **cryoturbation** is evident within the upper 100 cm, in which case the permafrost may begin as deep as 200 cm from the soil surface.

Cryoturbation is the physical disturbance of soil materials caused by the formation of ice wedges and by the expansion and contraction of water as it freezes and thaws. This *frost churning* action moves the soil material so as to orient rock fragments along the lines of force and to form broken, convoluted horizons (e.g., designated Cjj), and/or accumulations of organic matter at the top of the permafrost. The frost churning also may form patterns on the ground surface, such as hummocks and ice-rich polygons that may be several meters across. In some cases rocks forced to the surface form rings or net-like patterns.

Gelisols showing evidence of cryoturbation are called *Turbels*. Other Gelisols, often found in wet environments, have developed in accumulations of mainly organic materials, making them **Histels** (Greek *histos,* tissue; Figure 3.10). Most of the soil-forming processes that occur take place above the permafrost in the **active layer** that thaws every year or two. Various types of diagnostic horizons may have developed in different Gelisols, including mollic, histic, umbric, calcic, and, occasionally, argillic horizons.

Blanketed under snow and ice for much of the year, most Gelisols support tundra vegetation of lichens, grasses, and low shrubs that grow during the brief summers. Large areas of Gelisols consist of bogs, some literally floating on layers of frozen or unfrozen water. Gelisols present unique problems for construction projects. Many of the Histels

Active layer

Permafrost

FIGURE 3.10 A Gelisol with a histic epipedon and permafrost. This Histel was photographed in Alaska in July. Scale in centimeters. (Photo courtesy of Bockheim, University of Wisconsin)

are very wet and have little bearing strength to hold up roads or foundations. The permafrost in some Gelisols is unstable once disturbed.

3.9 HISTOSOLS (ORGANIC SOILS WITHOUT PERMAFROST)

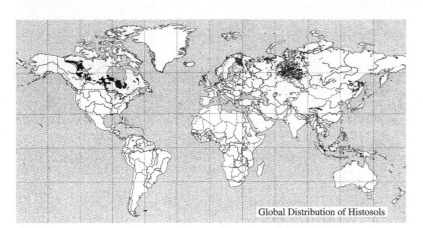
Global Distribution of Histosols

Suborders

Fibrists (fibers of plants obvious)

Folists (leaf mat accumulations)

Hemists (fibers partly decomposed)

Saprists (fibers not recognizable)

Histosols are soils that have undergone little profile development because of the anaerobic environment in which they form. The main process of soil formation evident in Histosols is the accumulation of partially decomposed organic parent material without permafrost (which would cause the soil to be classified in the Histels suborder of Gelisols). Histosols consist of one or more thick layers of *organic soil material*. Generally, Histosols have organic soil materials in more than half of the upper 80 cm of soil (Plate 6 after page 178), or in two-thirds of the soil overlying shallow rock.

While not all wetlands contain Histosols, all Histosols (except Folists) occur in wetland environments. They can form in almost any moist climate in which plants can grow, from equatorial to arctic regions, but they are most prevalent in cold climates, up to the limit of permafrost. Horizons are differentiated by the type of vegetation contributing the residues, rather than by translocations and accumulations within the profile.

Organic deposits accumulate in marshes, bogs, and swamps, which are habitats for water-loving plants such as pond-weeds, cattails, sedges, reeds, mosses, shrubs, and even some trees. Generation after generation, the residues of these plants sink into the water, which inhibits their oxidation by reducing oxygen availability and, consequently, acts as a partial preservative.

Because of their high content of organic matter, Histosols are generally black to dark brown in color. They are extremely lightweight when dry, being only about 10 to 20% as dense as most mineral soils. Histosols also have high water-holding capacities on a mass basis. While a mineral soil will absorb and hold from 20 to 40% of its weight of water, a cultivated Histosol may hold a mass of water equal to 200 to 400% of its dry weight. However, because of the low density of the organic material, most Histosols will hold only about the same amount of plant-available water as will a good mineral soil on a per-hectare or per-unit-volume basis.

Because the ecological roles of natural wetland environments have not always been appreciated (or protected by law), more than 50% of the original wetland area in the United States has been drained for agricultural or other uses, especially for vegetable and flower production (Figure 3.11). Some Histosols make very productive farmlands, but the organic nature of the materials requires liming, fertilization, tillage, and drainage practices quite different from those applied to soils in the other 11 orders.

In some places, Histosols are also mined for their peat, which is sold for use in potting media, as a mulch, and to make peat-fiber pots. Peat deposits are also used for fuel in some countries, especially in Russia, where several power stations are fueled by this material.

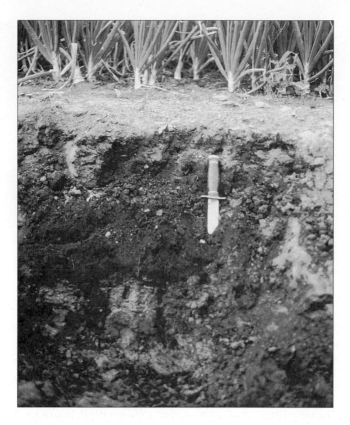

FIGURE 3.11 Profile of a drained Histosol on which onions are being produced in New York State. The organic soil rests on lake-laid (lacustrine) mineral material. (Photo courtesy of R. Weil)

3.10 ARIDISOLS (DRY SOILS)

Global Distribution of Aridisols

Suborders

Argids (clay)

Calcids (carbonate)

Cambids (typical)

Cryids (cold)

Durids (duripan)

Gypsids (gypsum)

Salids (salty)

Aridisols occupy a larger area globally than any other soil order (more than 12%) except Entisols. Water deficiency is a major characteristic of these soils. The soil moisture level is sufficiently high to support plant growth for no longer than 90 consecutive days. The natural vegetation consists mainly of scattered desert shrubs and short bunchgrasses. Soil properties, especially in the surface horizons, may differ substantially between interspersed bare and vegetated areas (see Section 2.14).

Aridisols are characterized by an ochric epipedon that is generally light in color and low in organic matter (see Plate 3 after page 178). The processes of soil formation have brought about a redistribution of soluble materials, but there is generally not enough water to leach these materials completely out of the profile. Therefore, they often accumulate at a lower level in the profile. These soils may have a horizon of accumulation of calcium carbonate (calcic), gypsum (gypsic), soluble salts (salic), or exchangeable sodium (natric). Under certain circumstances, carbonates may cement together the soil particles and coarse fragments in the layer of accumulation, producing hard layers known as ***petrocalcic*** horizons (Figure 3.12). These hard layers act as impediments to plant root growth and also greatly increase the cost of excavations for buildings.

FIGURE 3.12 Two features characteristic of some Aridisols. (Left) Wind-rounded pebbles have given rise to a desert pavement. (Right) A petrocalcic horizon of cemented calcium carbonate. (Photo courtesy of R. Weil)

Some Aridisols (the *Argids*) have an argillic horizon, most probably formed under a wetter climate that long ago prevailed in many areas that are deserts today. With time, and the addition of carbonates from calcareous dust and other sources, many argillic horizons become engulfed by carbonates (*Calcids*). On steeper land surfaces subject to erosion, argillic horizons do not get a chance to form, and the dominant soils are often *Cambids* (Aridisols with only weakly differentiated cambic subsurface B horizons).

In rocky soils, erosion may remove all the fine particles from the surface layers, leaving behind a layer of wind-rounded pebbles that is called **desert pavement** (see Figure 3.12). The surfaces of the pebbles in desert pavement often have a shiny coating called **desert varnish.** This coating is thought to be produced by algae that extract iron and manganese from the minerals and leave an oxide coating on the pebbles.

Except where there is groundwater or irrigation, the soil layers are only moist for short periods during the year. These short moist periods may be sufficient for drought-adapted desert shrubs and annual plants, but not for conventional crop production. If groundwater is present near the soil surface, soluble salts may accumulate in the upper horizons to levels that most crop plants cannot tolerate.

Without irrigation, Aridisols are not suitable for growing cultivated crops. Some areas are used for low-intensity grazing, especially with sheep or goats, but the production per unit area is low. The overgrazing of Aridisols leads to increased heterogeneity of both soils and vegetation. The animals graze the relatively even cover of palatable grasses, giving a competitive advantage to various shrubs not eaten by the grazing animals. The scattered shrubs compete against the struggling grasses for water and nutrients. The once-grassy areas become increasingly bare, and the soils between the scattered shrubs succumb to erosion by the desert winds and occasional thunderstorms. (Figure 3.13).

FIGURE 3.13 It is easy to overgraze the vegetation on an Aridisol, and an overgrazed Aridisol is subject to ready wind and water erosion. This gully was formed during infrequent but heavy rainstorms in an area in Arizona that at one time had been well covered with desert plants such as alkali sacaton and vine mesquite, plants that were destroyed by the overgrazing. (Photo courtesy U.S. Natural Resource Conservation Service)

3.11   VERTISOLS (DARK, SWELLING AND CRACKING CLAYS)

Global Distribution of Vertisols

Suborders

Aquerts (wet)

Cryerts (cold)

Torrerts (hot summer, very dry)

Uderts (humid)

Usterts (moist/dry)

Xererts (dry summers, moist winters)

The main soil-forming process affecting Vertisols is the shrinking and swelling of clay as these soils go through periods of drying and wetting. Vertisols have a high content (>30%) of sticky, swelling and shrinking-type clays to a depth of 1 m or more. Most Vertisols are dark, even blackish in color, to a depth of 1 m or more (Plate 12 after page 178). However, unlike for most other soils, the dark color of Vertisols is not necessarily indicative of a high organic matter content. The organic matter content of dark Vertisols typically ranges from as much as 5 or 6% to as little as 1%.

Vertisols typically develop from limestone, basalt or other calcium and magnesium rich parent materials. In east Africa, they typically form in landscape depressions which collect the calcium and magnesium leached out of the surrounding upland soils. The presence of these cations encourages the formation of swelling-type clays (see Section 8.5).

Vertisols are found mostly in subhumid to semiarid environments in warm regions, but a few (Cryerts) occur where the average soil temperatures are as low as 0°C. The native vegetation is usually grassland. Vertisols generally occur where the climate features dry periods of several months. In dry seasons the clay shrinks, causing the soils to develop deep, wide cracks that are diagnostic for this order (Figure 3.14).

FIGURE 3.14   Wide cracks formed during the dry season in the surface layers of this Vertisol in India. Surface debris can slough off into these cracks and move to subsoil. When the rains come, water can move quickly to the lower horizons, but the cracks are soon sealed, making the soils relatively impervious to the water.

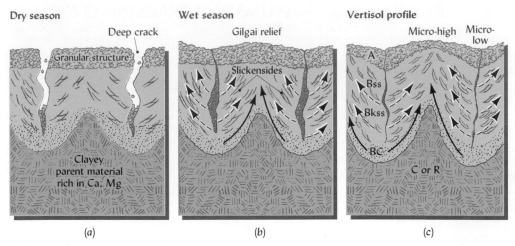

Dry season | Wet season | Vertisol profile

(a) | (b) | (c)

FIGURE 3.15 Vertisols typically are high in swelling-type clay and have developed wedgelike structures in the subsoil horizons. (*a*) During the dry season, large cracks appear as the clay shrinks upon drying. Some of the surface soil granules fall into the cracks, causing a partial mixing, or *inversion,* of the horizons. (*b*) During the wet season, rainwater pours down the cracks, wetting the soil near the bottom of the cracks first, and then the entire profile. As the clay absorbs water, it swells the cracks shut, entrapping the collected granular soil. The increased soil volume causes lateral and upward movement of the soil mass. The soil is pushed up between the cracked areas. As the subsoil mass shears from the strain, smooth surfaces or *slickensides* form. (*c*) These processes result in a Vertisol profile that typically exhibits gilgai, cracks more than 1 m deep, and slickensides in a Bss horizon. Calcium carbonate concretions often accumulate in a Bkss horizon.

The surface soil generally forms granules, of which a significant number may slough off into the cracks, giving rise to a partial inversion of the soil (Figure 3.15). This accounts for the association with the term *invert,* from which this order derives its name.

When the rains come, water entering the large cracks moistens the clay in the subsoils, causing it to swell. The repeated shrinking and swelling of the subsoil clay results in a kind of imperceptively slow "rocking" movement of great masses of soil. As the subsoil swells, blocks of soil shear off from the mass and rub past each other under pressure, giving rise in the subsoil to shiny, grooved, tilted surfaces called *slickensides* (Figure 3.16). Eventually, this back-and-forth motion may form bowl-shaped

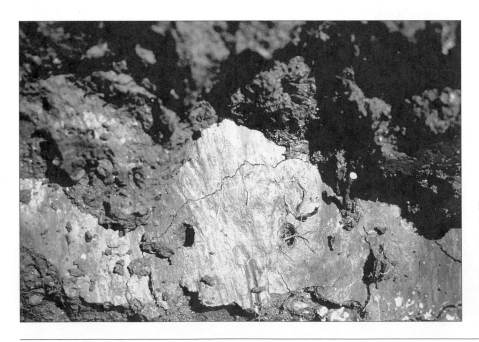

FIGURE 3.16 An example of a slickenside in a Vertisol. Massive rubbing of soil blocks against each other in response to the swelling clays have brought about this shiny appearance. The white concretions of carbonate minerals are evidence of the high base status of these soils. (Photo courtesy of R. Weil)

depressions with relatively deep profiles surrounded by slightly raised areas in which little soil development has occurred and in which the parent material remains close to the surface (see Figure 3.15). The resulting pattern of microhighs and microlows on the land surface, called *Gilgai,* is usually discernable only where the soil is untilled.

The high shrink-swell potential of Vertisols makes them extremely problematic for any kind of highway or building construction (Figure 8.12). This property also makes agricultural management very difficult. Because they are very sticky and plastic when wet and become very hard when dry, the timing of tillage operations is critical. Some farmers refer to Vertisols as *24-hour soils,* because they are said to be too wet to plow one day and too dry the next.

3.12 MOLLISOLS (DARK, SOFT SOILS OF GRASSLANDS)

Global Distribution of Mollisols

Suborders
Albolls (albic horizon)
Aquolls (wet)
Cryolls (very cold)
Rendolls (calcareous)
Udolls (humid)
Ustolls (moist/dry)
Xerolls (dry summers, moist winters)

The principal process in the formation of Mollisols is the accumulation of calcium-rich organic matter, largely from the dense root systems of prairie grasses, to form the thick, soft mollic epipedon that characterizes soils in this order (Plate 8 after page 178). This humus-rich surface horizon is often 60 to 80 cm in depth and high in calcium and magnesium. Its cation exchange capacity is more than 50% saturated with base-forming cations (Ca^{2+}, Mg^{2+}, etc.). Mollisols in humid regions generally have higher organic matter and darker, thicker mollic epipedons than their lower-moisture-regime counterparts (see Section 12.10).

The surface horizon generally has granular or crumb structures, largely resulting from an abundance of organic matter and swelling-type clays. The aggregates are not hard when the soils are dry, hence the name *Mollisol,* implying softness (Table 3.3). In addition to the mollic epipedon, Mollisols may have an argillic (clay), natric, albic, or cambic subsurface horizon, but not an oxic or spodic horizon.

Most Mollisols have developed under grass vegetation (Figure 3.17). Grassland soils of the central part of the United States, lying between Aridisols on the west and the Alfisols on the east, make up the central concept of this order. However, a few soils developed under forest vegetation (primarily in depressions) have a mollic epipedon and are included among the Mollisols. Two Mollisols profiles are included in Figure 3.18.

Mollisols cover a larger land area in the United States than any other soil order (about 22%). They are dominant in the Great Plains of the United States (Figure 3.19), as well as in Illinois. Because the high native fertility of Mollisols makes them among the world's most productive soils, few Mollisols have been left uncultivated in regions with sufficient rainfall for crop production. In the United States, efforts are underway to preserve the few remnants of the once vast and diverse prairie ecosystem.

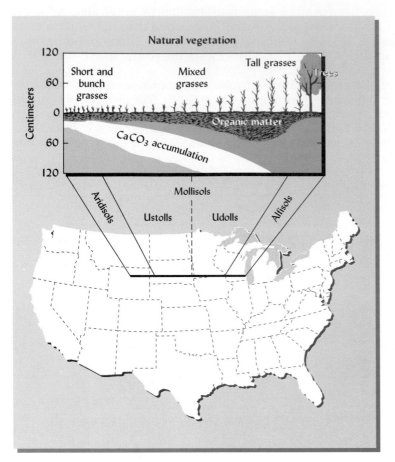

FIGURE 3.17 Correlation between natural grassland vegetation and certain soil orders is graphically shown for a transect across north central United States. The control, of course, is climate. Note the deeper organic matter and deeper zone of calcium accumulation, sometimes underlain by gypsum, as one proceeds from the drier areas in the west toward the more humid region where prairie soils are found. Alfisols may develop under grassland vegetation, but more commonly occur under forests and have lighter-colored surface horizons.

FIGURE 3.18 Monoliths of profiles representing three soil orders. The suborder names are in parentheses. Genetic (not diagnostic) horizon designations are also shown. Note the spodic horizons in the Spodosol characterized by humus (Bh) and iron (Bs) accumulation. In the Alfisol is found the illuvial clay horizon (Bt), and the structural B horizon (Bw) is indicated in the Mollisols. The thick dark surface horizon (mollic epipedon) characterizes both Mollisols. Note that the zone of calcium carbonate accumulation (Bk) is near the surface in the Ustoll, which has developed in a dry climate. The E/B horizon in the Alfisol has characteristics of both E and B horizons.

FIGURE 3.19 Typical landscape dominated by Ustolls. These productive soils produce much of the wheat and feed grains in the United States and other subhumid areas of the world. The dark, bare soil strips are in summer fallow to conserve water. (Photo courtesy of R. Weil)

3.13 ALFISOLS (ARGILLIC OR NATRIC HORIZON, MEDIUM TO HIGH BASES)

Global Distribution of Alfisols

Suborders

Aqualfs (wet)

Cryalf (cold)

Udalfs (humid)

Ustalfs (moist/dry)

Xeralfs (dry summers, moist winters)

The Alfisols appear to be more strongly weathered than soils in the orders just discussed, but less so than Spodosols (see following). They are formed in cool to hot humid areas (see Figure 3.5) but also are found in the semiarid tropics and Mediterranean climates. Most often, Alfisols develop under native deciduous forests, although in some cases, as in California and parts of Africa, grass savanna is the native vegetation.

Alfisols are characterized by a subsurface diagnostic horizon in which silicate clay has accumulated by illuviation (see Plate 1, after page 178). Clay skins or other signs of clay movement are present in such a B horizon (see Plate 23). In Alfisols, this clay-rich horizon is only moderately leached, and its cation exchange capacity is more than 35% saturated with base-forming cations (Ca^{2+}, Mg^{2+}, etc.). In most Alfisols this horizon is termed *argillic* because of its accumulation of silicate clays. The horizon is termed *natric* if, in addition to having an accumulation of clay, it is more than 15% saturated with sodium and has prismatic or columnar structure (see Figure 4.10). In some Alfisols in subhumid tropical regions, the accumulation is termed a *kandic* (from the mineral kandite) horizon because the clays have a low cation exchange capacity.

Alfisols very rarely have a mollic epipedon, for such soils would be classified in the Argiudolls or other suborder of Mollisols with an argillic horizon. Instead, Alfisols typically have a relatively thin, gray to brown ochric epipedon (Plate 1, after page 178, shows an example) or an umbric epipedon. Those formed under deciduous temperate forests commonly have a light-colored, leached *albic* E horizon immediately under the A horizon (see Figures 1.3 and 3.18).

In general, Alfisols are productive soils. Good hardwood forest growth and crop yields are favored by their medium- to high-base status, generally favorable texture, and location (except for some Xeralfs) in regions with sufficient rainfall for plants for at least part of the year. In the United States these soils rank favorably with the Mollisols and Ultisols in their productive capacity. Many, especially the sandier ones, are quite susceptible to erosion by heavy rains if deprived of their natural surface litter. Alfisols in udic moisture regimes are sufficiently acidic to require amendment with limestone for many kind of plants (see Chapter 9).

3.14 ULTISOLS (ARGILLIC HORIZON, LOW BASES)

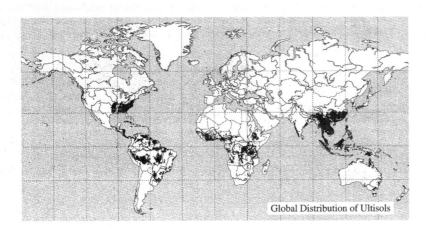

Global Distribution of Ultisols

Suborders

Aquults (wet)

Humults (high humus)

Udults (humid)

Ustults (moist/dry)

Xerults (dry summers, moist winters)

The principal processes involved in forming Ultisols are clay mineral weathering, translocation of clays to accumulate in an argillic or kandic horizon, and leaching of base-forming cations from the profile. Most Ultisols have developed under moist conditions in warm to tropical climates. Ultisols are formed on old land surfaces, usually under forest vegetation, although savanna or even swamp vegetation is also common. They often have an ochric or umbric epipedon, but are characterized by a relatively acidic B horizon that has less than 35% of the exchange capacity satisfied with base-forming cations. The clay accumulation may be either an argillic horizon or, if the clay is of low activity, a kandic horizon. Ultisols commonly have both an epipedon and a subsoil that is quite acid and low in plant nutrients.

Ultisols are more highly weathered and acidic than Alfisols, but less acid than Spodosols and less highly weathered than the Oxisols. Except for the wetter members of the order, their subsurface horizons are commonly red or yellow in color, evidence of accumulations of oxides of iron (see Plate 11). Certain Ultisols that formed under fluctuating wetness conditions have horizons of iron-rich mottled material called **plinthite** (see Plate 15). This material is soft and can be easily dug from the profile so long as it remains moist. When dried in the air, however, plinthite hardens irreversibly into a kind of ironstone that is virtually useless for cultivation, but can be used to make durable bricks for building.

Although Ultisols are not naturally as fertile as Alfisols or Mollisols, they respond well to good management. They are located mostly in regions of long growing seasons and of ample moisture for good forest or crop production (Figure 3.20). The silicate clays of Ultisols are usually of the nonsticky type, which, along with the presence of iron oxides and aluminum, assures ready workability.

FIGURE 3.20 The soils in this high-elevation, tropical area of south Asia are Ultisols in the suborder Humults. These soils are being intensively used both for house construction and for market gardens. The combination of a favorable climate and soils that are high in organic matter (Humults have at least 9% down to the upper part of the B horizon) and respond well to fertilizer has encouraged local residents to use every bit of the land in producing vegetables to supplement their incomes. (Photo courtesy of R. Weil)

3.15 SPODOSOLS (ACID, SANDY, FOREST SOILS, LOW BASES)

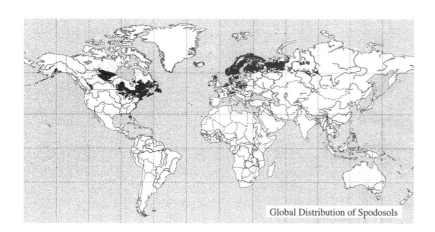

Global Distribution of Spodosols

Suborders
Aquods (wet)
Cryods (icy cold)
Humods (humus)
Orthods (typical)

Spodosols occur mostly on coarse-textured, acid parent materials subject to ready leaching. They occur only in moist to wet areas, commonly where it is cold or temperate (see Figure 3.5), but also in some tropical and subtropical areas. Intensive acid leaching is the principal soil-forming process. They are mineral soils with a *spodic* horizon, a subsurface accumulation of illuviated organic matter, and an accumulation of aluminum oxides with or without iron oxides (see Plate 10 and Figure 3.18). This usually thin, dark, illuvial horizon typically underlies a light, ash-colored, eluvial *albic* horizon.

Spodosols form under forest vegetation, especially under coniferous species whose needles are low in base-forming cations like calcium and high in acid resins. As this acid litter decomposes, strongly acid organic compounds are released and carried down into the permeable profile by percolating waters. These acids bind with iron and aluminum and carry them downward until they precipitate in the spodic horizon. Similarly, soluble organic compounds move downward and are likewise precipitated above or intermixed with the iron and aluminum compounds. Most minerals except quartz are removed by this acid leaching, generally removing the coloring agents from the E horizon. Deeper in the profile, precipitated oxides of aluminum and iron form a Bs spodic horizon that is often reddish brown in color. The leaching organic compounds may precipitate and form a black-colored Bh spodic horizon. The precipitation often occurs along wavy wetting fronts, thus yielding striking Spodosol profiles (Figure 3.21). The depth of these weathering and leaching processes, hence the spodic horizons, can vary from less than 20 cm to several meters.

Spodosols are not naturally fertile. When properly fertilized, however, these soils can become quite productive. For example, most potato-producing soils of northern Maine are Spodosols, as are some of the vegetable- and fruit-producing soils of Florida, Michigan, and Wisconsin. Because of their sandy nature and occurrence in regions of high rainfall, groundwater contamination by leaching of soluble fertilizers and pesticides has proved to be a serious problem where these soils are used in crop production. The best use for most Spodosols is forestry and wildlife habitat. Because they are already quite acid and poorly buffered, watersheds dominated by soils of this order are susceptible to damage from acid rain (see Section 9.4).

3.16   OXISOLS (OXIC HORIZON, HIGHLY WEATHERED)

Suborders

Aquox (wet)

Perox (very humid)

Torrox (hot, dry)

Udox (humid)

Ustox (moist/dry)

Global Distribution of Oxisols

The Oxisols are the most highly weathered soils in the classification system (see Figure 3.4). They form in hot climates with nearly year-round moist conditions; hence, the native vegetation is generally thought to be tropical rain forest. However, some Oxisols (Ustox) are found in areas which are today much drier than was the case when the soils were forming their oxic characteristics. Their most important diagnostic feature is a deep oxic subsurface horizon. This horizon is generally very high in clay-size particles dominated by hydrous oxides of iron and aluminum. Weathering and intense leaching have removed a large part of the silica from the silicate materials in this horizon. Some quartz and 1:1-type silicate clay minerals remain, but the hydrous oxides are dominant (see Chapter 8 for information on the various clay minerals). The epipedon in most Oxisols is either ochric or umbric. Usually the boundaries between subsurface horizons are indistinct, giving the subsoil a relatively uniform appearance with depth.

The clay content of Oxisols is generally high, but the clays are of the low-activity, nonsticky type. Consequently, when the clay dries out it is not hard and cloddy, but is easily worked. Also, Oxisols are resistant to compaction, so water moves freely through the profile. The depth of weathering in Oxisols is typically much greater than for most of the other soils, 20 m or more having been observed. The low-activity clays have a very limited capacity to hold nutrient cations such as Ca^{2+}, Mg^{2+}, and K^+, so they are typically of low natural fertility and moderately acid. The high concentration of iron and aluminum oxides also gives these soils a capacity to bind so tightly with what little phosphorus is present, that phosphorus deficiency often limits plant growth once the natural vegetation is disturbed.

FIGURE 3.21 Two Spodosol profiles. The Spodosol in the Upper Penninsula of Michigan (left) shows a relatively deep, very wavy spodic horizon under a discontinuous albic horizon. It has a uniform ochric Ap horizon, indicating that it has been cleared and plowed. The Spodosol in Quebec (right) has a very shallow spodic horizon and an undisturbed O horizon consisting of conifer needless in various states of decay. Typical of Spodosols, both soils are sandy, very acidic, and best suited to supporting coniferous forests such as influenced their formation. The bar in each photo is 10 cm long. (Photos courtesy of R. Weil)

Road and building construction is relatively easily accomplished on most Oxisols because these soils are easily excavated, do not shrink and swell, and are physically very stable on slopes. The very stable aggregation of the clays, stimulated largely by iron compounds, makes these soils quite resistant to erosion. Millions of people in the tropics depend on them for food and fiber production. However, because of their low natural fertility, most Oxisols have been left under forest vegetation or are farmed by shifting cultivation methods. Nutrient cycling by deep-rooted trees is especially important to the productivity of these soils. Probably the best use of Oxisols, other than supporting rain forests, is the culture of mixed-canopy perennial crops, especially tree crops. Such cultures can restore the nutrient cycling system that characterized the soil–plant relationships before the rain forest was removed.

3.17 LOWER-LEVEL CATEGORIES IN *SOIL TAXONOMY*

Suborders

Within each soil order just described, soils are grouped into suborders on the basis of soil properties that reflect major environmental controls on current soil-forming processes. The suborders for each order are listed next to the global soil order maps in the previous sections. Many suborders are indicative of the moisture regime or, less frequently, the temperature regime under which the soils are found. Thus, soils formed under wet conditions generally are identified under separate suborders (e.g., Aquents, Aquerts, and Aquepts), as being wet soils.

To determine the relationship between suborder names and soil characteristics, refer to Table 3.4. Here the formative elements for suborder names are identified and their connotations given. Thus, the Ustolls are dry Mollisols. Likewise, soils in the Udults suborder (from the Latin *udus,* humid) are moist Ultisols.

Great Groups

The great groups are subdivisions of suborders. More than 240 great groups are recognized. They are defined largely by the presence or absence of diagnostic horizons and

TABLE 3.4   Formative Elements in Names of Suborders in *Soil Taxonomy*

Formative element	Derivation	Connotation of formative element
alb	L. *albus,* white	Presence of albic horizon (a bleached eluvial horizon)
anthr	Gk. *anthropos,* human	Presence of anthropic or plaggen epipedon
aqu	L. *aqua,* water	Characteristics associated with wetness
ar	L. *arare,* to plow	Mixed horizons
arg	L. *argilla,* white clay	Presence of argillic horizon (a horizon with illuvial clay)
calc	L. *calcis,* lime	Presence of calcic horizon
camb	L. *cambriare,* to change	Presence of cambric horizon
cry	Gk. *kryos,* icy cold	Cold
dur	L. *durus,* hard	Presence of a duripan
fibr	L. *fibra,* fiber	Least decomposed stage
fluv	L. *fluvius,* river	Floodplains
fol	L. *folia,* leaf	Mass of leaves
gyps	L. *gypsum,* gypsum	Presence of gypsic horizon
hem	Gk. *hemi,* half	Intermediate stage of decomposition
hist	Gk. *histos,* tissue	Presence of histic epipedon
hum	L. *humus,* earth	Presence of organic matter
orth	Gk. *orthos,* true	The common ones
per	L. *per,* throughout time	Of year-round humid climates, perudic moisture regime
psamm	Gk. *psammos,* sand	Sand textures
rend	Modified from Rendzina	Rendzinalike—high in carbonates
sal	L. *sal,* salt	Presence of salic (saline) horizon
sapr	Gk. *sapros,* rotten	Most decomposed stage
torr	L. *torridus,* hot and dry	Usually dry
turb	L. *turbidus,* disturbed	Cryoturbation
ud	L. *udus,* humid	Of humid climates
ust	L. *ustus,* burnt	Of dry climates, usually hot in summer
vitr	L. *vitreus,* glass	Resembling glass
xer	Gk. *xeros,* dry	Dry summers, moist winters

the arrangements of those horizons. These horizon designations are included in the list of formative elements for the names of great groups shown in Table 3.5. Note that these formative elements refer to epipedons such as umbric and ochric (see Table 3.1), to subsurface horizons such as argillic and natric, and to certain diagnostic impervious layers such as duripans and fragipans (see Figure 3.22).

Remember that the great group names are made up of these formative elements attached as prefixes to the names of suborders in which the great groups occur. Thus, Ustolls with a natric horizon (high in sodium) belong to the Natrustolls great group. As can be seen in the example discussed in Box 3.1, soil descriptions at the great group level can provide important information not indicated at the higher, more general levels of classification.

Subgroups

Subgroups are subdivisions of the great groups. More than 1300 subgroups are recognized. The central concept of a great group makes up one subgroup, termed *Typic.* Thus, the Typic Hapludolls subgroup typifies the Hapludolls great group. Other subgroups may have characteristics that intergrade between those of the central concept and soils of other orders, suborders, or great groups. A Hapludoll with restricted drainage would be classified as an Aquic Hapludoll. One with evidence of intense earthworm activity would fall in the Vermic Hapludolls subgroup. Some intergrades may have properties in common with other orders or with other great groups. Thus, soils in the Entic Hapludolls subgroup are very weakly developed Mollisols, close to being in the Entisols order. The subgroup concept illustrates very well the flexibility of this classification system.

Families

Within a subgroup, soils fall into a particular family if, at a specified depth, they have similar physical and chemical properties affecting the growth of plant roots. About 8000 families have been identified. The criteria used include broad classes of particle

TABLE 3.5 Formative Elements for Names of Great Groups and Their Connotation

These formative elements combined with the appropriate suborder names give the great group names.

Formative element	Connotation	Formative element	Connotation	Formative element	Connotation
acr	Extreme weathering	fol	Mass of leaves	petr	Cemented horizon
agr	Agric horizon	fragi	Fragipan	plac	Thin pan
al	High aluminum, low iron	fragloss	See *frag* and *gloss*	plagg	Plaggen horizon
alb	Albic horizon	fulv	light-colored melanic horizon	plinth	Plinthite
and	Ando-like	gyps	Gypsic horizon	psamm	Sand texture
anhy	Anhydrous	gloss	Tongued	quartz	High quartz
aqu	Water saturated	hal	Salty	rhod	Dark red colors
argi	Argillic horizon	hapl	Minimum horizon	sal	Salic horizon
calc, calci	Calcic horizon	hem	Intermediate decomposition	sapr	Most decomposed
camb	Cambic horizon	hist	Presence of organic materials	somb	Dark horizon
chrom	High chroma	hum	Humus	sphagn	Sphagnum moss
cry	Cold	hydr	Water	sulf	Sulfur
dur	Duripan	kand	Low-activity 1:1 silicate clay	torr	Usually dry and hot
dystr, dys	Low base saturation	lithic	Near stone	ud	Humid climates
endo	Fully water saturated	luv, lu	Illuvial	umbr	Umbric epipedon
epi	Perched water table	melan	Melanic epipedon	ust	Dry climate, usually hot in summer
eutr	High base saturation	molli	With a mollic epipedon	verm	Wormy, or mixed by animals
ferr	Iron	natr	Presence of a natric horizon	vitr	Glass
fibr	Least decomposed	pale	Old development	xer	Dry summers, moist winters
fluv	Floodplain				

size, mineralogy, cation exchange activity of the clay, temperature, and depth of the soil penetrable by roots. Table 3.6 gives examples of the classes used. Terms such as *loamy, sandy,* and *clayey* are used to identify the broad particle size classes. Terms used to describe the mineralogical classes include *smectitic, kaolinitic, siliceous, carbonatic,* and *mixed.* The clays are described as *superactive, active, semiactive,* or *subactive* with regard to their capacity to hold cations. For temperature classes, terms such as *cryic, mesic,* and *thermic* are used. The terms *shallow* and *micro* are sometimes used at the family level to indicate unusual soil depths.

Thus, a Typic Argiudoll from Iowa, loamy in texture, having a mixture of moderately active clay minerals and with annual soil temperatures (at 50 cm depth) between 8 and 15°C, is classed in the *Typic Argiudolls loamy, mixed, active, mesic* family. In contrast, a

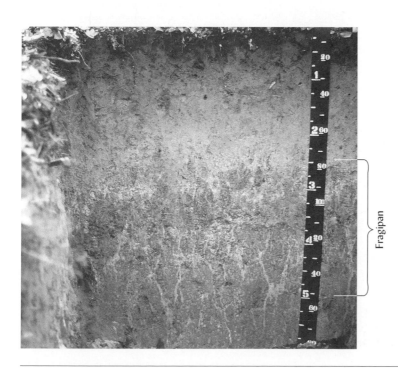

FIGURE 3.22 A forested Fragiudalf in Missouri containing a typical well-developed fragipan with coarse prismatic structure (outlined by gray, iron-depleted coatings). Fragipans (usually Bx or Cx horizons) are extremely dense and brittle. They consist mainly of silt, often with considerable sand, but not very much clay. One sign of encountering a fragipan in the field is the ringing noise that your shovel will make when you attempt to excavate it. Digging through a fragipan is almost like digging concrete. Plant roots cannot penetrate this layer. Yet, once a piece of a fragipan is broken loose, it fairly easily crushes with hand pressure. But it does not squash or act in a plastic manner as a claypan would; instead, it bursts in a *brittle* manner. (Photo courtesy of Fred Rhoton, Agricultural Research Service, U.S. Department of Agriculture)

BOX 3.1 GREAT GROUPS, FRAGIPANS, AND ARCHAEOLOGIC DIGS

Soil Taxonomy is a communications tool that helps scientists and land managers share information. In this box we will see how misclassification, even at a lower level in *Soil Taxonomy* such as the great group, can have costly ramifications.

In order to preserve our national heritage, both federal and state laws require that an archaeological impact statement be prepared prior to starting major construction works on the land. The archaeological impact is usually assessed in three phases. Selected sites are then studied by archaeologists, with the hope that at least some of the artifacts can be preserved and interpreted before construction activities obliterate them forever. Only a few relatively small sites can be subjected to actual archaeological digs because of the expensive skilled hand labor involved (Figure 3.23).

Such an archaeological impact study was ordered as a precursor to construction of a new highway in a mid-Atlantic state. In the first phase, a consulting company gathered soils and other information from maps, aerial photographs, and field investigations to determine where neolithic people may have occupied sites. Then the consultants identified about 12 ha of land where artifacts indicated significant neolithic activities. The soils in one area were mapped mainly as Typic Dystrudepts. These soils formed in old colluvial and alluvial materials that, many thousands of years ago, had been along a river bank. Several representative soil profiles were examined by digging pits with a backhoe. The different horizons were described, and it was determined in which horizons artifacts were most likely to be found. What was not noted was the presence in these soils of a **fragipan**, a dense, brittle layer that is extremely difficult to excavate using hand tools.

A fragipan is a subsurface diagnostic horizon used to classify soils, usually at the great group or subgroup level (see Figure 3.22). Its presence would distinguish Fragiudepts from Dystrudepts.

When it came time for the actual hand excavation of sites to recover artifacts, a second consulting company was awarded the contract. Unfortunately, their bid on the contract was based on soil descriptions that did not specifically classify the soils as Fragiudepts—soils with very dense, brittle, hard fragipans in the layer that would need to be excavated by hand. So difficult was this layer to excavate and sift through by hand, that it nearly doubled the cost of the excavation—an additional expense of about $1 million. Needless to say, there ensued a controversy as to whether this cost would be borne by the consulting firm that bid with faulty soils data, the original consulting firm that failed to adequately describe the presence of the fragipan, or the highway construction company that was paying for the survey.

This episode gives us an example of the practical importance of soil classification. The formative element *Fragi* in a soil great group name warns of the presence of a dense, impermeable layer that will be very difficult to excavate, will restrict root growth (often causing trees to topple in the wind or become severely stunted), may cause a perched water table (epiaquic conditions), and will interfere with proper percolation in a septic drain field.

FIGURE 3.23 *An archaeological dig. (Photo courtesy of Antonio Segovia, University of Maryland)*

TABLE 3.6 Some Commonly Used Particle-Size, Mineralogy, Cation Exchange Activity, and Temperature Classes Used to Differentiate Soil Families

The characteristics generally apply to the subsoil or 50 cm depth. Other criteria used to differentiate soil families (but not shown here) include the presence of calcareous or highly aluminum toxic (allic) properties, extremely shallow depth (shallow or micro), degree of cementation, coatings on sand grains, and the presence of permanent cracks.

Particle-size class	Mineralogy class	Cation exchange activity class[b] Term	Cation exchange activity class[b] CEC / % clay	Soil temperature regime class Mean annual temperature, °C	Soil temperature regime class >6°C difference between summer and winter	Soil temperature regime class <6°C difference between summer and winter
Ashy	Mixed	Superactive	0.60	<−10	Hypergelic[c]	—
Fragmental	Micaceous	Active	0.4 to 0.6	−4 to −10	Pergelic[c]	—
Sandy-skeletal[a]	Siliceous	Semiactive	0.24 to 0.4	+1 to −4	Subgelic[c]	—
Sandy	Kaolinitic	Subactive	<0.24	<+8	Cryic	—
Loamy	Smectitic			<+8	Frigid[d]	Isofrigid
Clayey	Gibbsitic			+8 to +15	Mesic	Isomesic
Fine-silty	Gypsic			+15 to +22	Thermic	Isothermic
Fine-loamy	Carbonic			>+22	Hyperthermic	Isohyperthermic
Etc.	Etc.					

[a] Skeletal refers to presence of up to 35% rock fragments by volume.
[b] Cation exchange activity class is not used for taxa already defined by low CEC (e.g., kandic or oxic groups).
[c] Permafrost present.
[d] Frigid is warmer in summer than Cryic.

sandy-textured Typic Haplorthod, high in quartz and located in a cold area in eastern Canada, is classed in the *Typic Haplorthods sandy, siliceous, frigid* family (note that clay activity classes are not used for soils in sandy textural classes).

Series

The series category is the most specific unit of the classification system. It is a subdivision of the family, and each series is defined by a specific range of soil properties involving primarily the kind, thickness, and arrangement of horizons. Features such as a hardpan within a certain distance below the surface, a distinct zone of calcium carbonate accumulation at a certain depth, or striking color characteristics greatly aid in series identification.

In the United States, each series is given a name, usually from some town, river, or county, such as Fargo, Muscatine, Cecil, Mohave, or Ontario. There are about 19,000 soil series in the United States.

The complete classification of a Mollisol, the Kokomo series, is given in Figure 3.24. This figure illustrates how *Soil Taxonomy* can be used to show the relationship between *the soil,* a comprehensive term covering all soils, and a specific soil series. The figure deserves study because it reveals much about the structure and use of *Soil Taxonomy.* If a soil series name is known, the complete *Soil Taxonomy* classification of the soil may be found on the World Wide Web at http://www.statlab.iastate.edu/soils-info/osd/.

Phase

Although technically not included as a class in *Soil Taxonomy,* a **phase** is a subdivision based on some important deviation that influences the use of the soil, such as surface texture, erosion, slope, stoniness, or soluble salt content. Thus, a Cecil sandy loam, 3 to 5% slope, and a Hagerstown silt loam, stony phase, are examples of phases of soil series.

3.18 SOIL SURVEYS AND MAPS

Even a small tract of land such as a suburban home lot or farm field is likely to contain several soil series that may be quite different from one another. They may even be members of different soil orders. A map showing where these different soils occur would provide valuable information for making the best use of the land. In fact, **soil maps** are in great demand as tools that communicate geographic information about

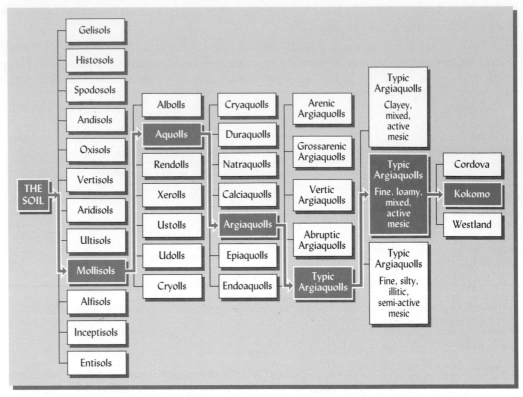

FIGURE 3.24 Diagram illustrating how one soil (Kokomo) keys out in the overall classification scheme. The shaded boxes show that this soil is in the Mollisols order, Aquolls suborder, Argiaquolls great group, and so on. In each category, other classification units are shown in the order in which they key out in *Soil Taxonomy*.

soils for practical land planning and management. Many soil scientists therefore specialize in mapping soils. Their basic task is threefold: (1) to define each soil unit to be mapped, (2) to compile information about the nature of each soil, and (3) to delineate the boundaries where each soil unit occurs in the landscape.

While soil mappers may use such high-technology tools as computers and geopositioning satellites, they mainly rely on simple field tools, hands, and eyes to study soils in the field. The thorough study of a soil requires a soil pit to be dug by hand or with a backhoe. One or more soil scientists enter the pit and study a typical pedon (see Section 3.1) as exposed on the pit face. They write a soil description in a standard format that includes the texture, colors, structure, and other observable properties for each horizon.

Of course, in making a soil map, soil scientists cannot dig pits all over the landscape. Instead, they bring up soil material from small boreholes made with a hand-turned **soil auger.** The nature of the soils and the locations of the boundary lines surrounding them are inferred from information obtained by auger borings at numerous locations across a landscape. The soil mappers usually use an aerial photo as a base on which to draw the soil boundaries, coordinating their field observations with the landscape features visible on the aerial photo (see Figure 3.25).

Mapping Units

Soil Taxonomy (or some other classification system) is usually the basis for preparing a soil survey. Local features, planned land use, and map scale[5] (level of detail) will dictate the specific soil units that are mapped. The field **mapping units** may be somewhat different from the classification units of *Soil Taxonomy*. The mapping units may represent some further differentiation below the soil series level—namely, phases of

---

[5] The scale tells us the ratio of distance on a map to distance on the ground. On a relatively small-scale map, 1 cm may represent 100 kilometers (scale = 1:10,000,000). On a relatively large-scale map, 1 cm may represent only 100 m (scale = 1:10,000). The map scale determines the level of detail that can be represented. Detailed (large-scale) soil maps in the United States are commonly made at scales of 1:12,000 (1 in. = 1000 ft) and 1:15,840 (4 in. = 1 mile).

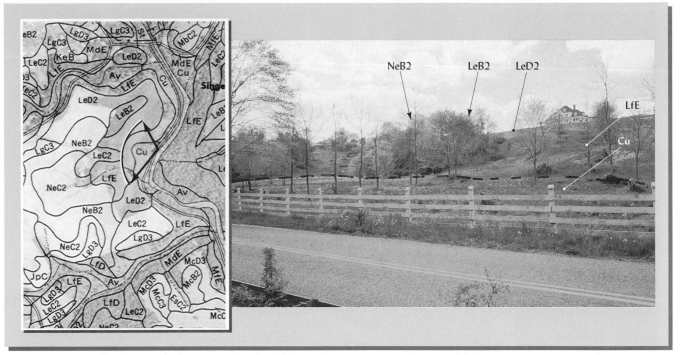

FIGURE 3.25 A small section of detailed soil survey map of Harford County, Maryland (left), and a ground-level view of the part of the landscape represented on the map by the view arrows (right). The map is reproduced here at the original scale (1:15,840) and represents an area of about 1.1 km². The map symbols represent soil consociations such as LeD2, which is named for the soil phase "Legore silt loam, 12 to 15% moderately eroded," a soil in the Alfisols order. The level land in the foreground is an alluvial soil in the Inceptisols order, the Codorus silt loam. [Map from Smith and Matthews (1975); photo courtesy of R. Weil]

soil series; or the soil mappers may choose to group together similar or associated soils into conglomerate mapping units. Examples of such soil mapping units follow.

Soil Consociation.   A **soil consociation** is the smallest practical mapping unit for most detailed soil surveys. It is an area that contains primarily one phase of a soil series. For example, a mapping unit may be labeled as the consociation "Legore silt loam, 12 to 15% slopes, moderately eroded." Although small areas of other soils may exist in this mapping unit, these so-called impurities should be so similar to the named soil that the differences do not affect land management.

Soil Complex.   Sometimes contrasting rather than similar soils occur adjacent to each other in a pattern so intricate that it cannot be drawn on a soil map. In such cases, a **soil complex** is delineated on the soil map to indicate an area containing two or three distinctly different soil series.

SOIL ASSOCIATION.   A **soil association** is a more general grouping of individual soils that occur together in a landscape. It is used mainly in small-scale, regional soil maps. Soil associations are named after the two or three dominant soils in the group, but may contain several additional, less extensive soils. The soils may have formed in the same or in different parent materials and may be from the same or different soil orders (see Figure 3.26). The only requirement is that the soils occur together in the same area. A given soil association represents a defined range of soil properties and landscape relationships, even though the range of conditions included may be quite large.

Soil Survey Interpretations

In addition to the soil map, a modern **soil survey report** contains descriptive information on the soil mapping units and their suitability for various land uses. In the United States the USDA Natural Resource Conservation Service produces most soil survey reports on a county by county basis in collaboration with state and local governments.

To be most useful for land use or site planning, the geographic soils information on the map should be integrated with the descriptive information in the rest of the report.

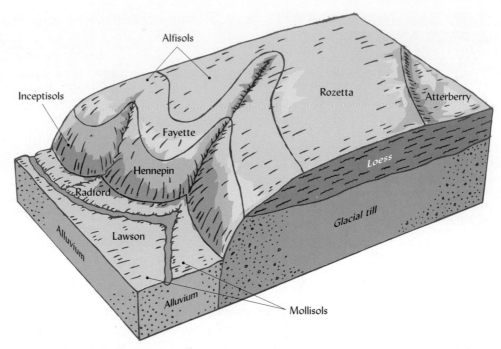

FIGURE 3.26 Diagram of the Rozetta–Fayette–Hennepin soil association in Bureau County, Illinois. This soil association includes three types of parent material (loess, glacial till, and alluvium) and a range of topography and vegetation types which give rise to five main soil series belonging to three different soil orders. [Based on Zwicker (1992)]

The report usually contains detailed soil profile descriptions for all of the mapping units, as well as tables that provide soil characterization data and interpretive rankings for the mapping units. For each mapping unit, the soil survey offers interpretive information on agricultural (e.g. yield potentials, suitability for irrigation, drainage requirements, and land capability classification) and nonagricultural (e.g., wildlife habitat, forestry, landscaping, waste disposal, building construction, and source of roadbed materials) land uses. Table 3.7 lists a few of the many types of interpretive ratings given in the soil survey report for the soils mapped in Figure 3.25. If you were planning to develop a park on this site, the information in Table 3.7 would suggest that the Cu mapping unit would be suitable for a nature trail, but not for a visitor center with rest rooms.

For complex land-planning analyses the enormous quantity of information stored in a county soil survey report can be used to best advantage with the help of a computerized **geographic information system** (**GIS**). In many land-planning projects, soil properties comprise just one of several types of geographic information that must be integrated by a GIS in order to take best advantage of a given site or to find the site best suited for the proposed land use. Planning the best use of different farm fields, finding a site suitable for a sanitary landfill, and determining where ponds can be constructed for a wildlife refuge are some examples of projects that could make good use of both soils and nonsoils geographic information. Nonsoils information to be considered might include topography, streams, vegetation, present land use, and legal zoning categories. Efforts are underway to digitize soil survey maps so they can be entered into a GIS to help in planning rational use of the land.

3.19 CONCLUSION

It is fortunate that scientists have recognized that *the soil* which covers the earth is in fact comprised of a large number of individual *soils,* each with distinctive properties. Among the most important of these properties are those associated with the horizontal layers, or *horizons,* found in a soil profile. These horizons reflect the physical, chemical, and biological processes that have stimulated the genesis of individual soils, and are among the most significant properties that influence how soils can and should be used. Knowledge of the kinds and properties of soils around the world is critical to humanity's struggle for survival and well-being.

TABLE 3.7   A Sampling of Some of the Interpretive Information Provided by the Soil Survey of Harford County Area, Maryland

The four mapping units described are those in the site shown in Figure 3.25. The interpretations included are some of those that would be of interest for developing the site as a county park.

Mapping unit code	Name of mapping unit	Land use Capability Classification[a]	Limitations for use of soil			Suitability for open land wildlife
			Camp areas	Septic filter fields	Paths and trails	
Cu	Codorus silt loam	Ilw-7	Severe—flood hazard	Severe—high water table, flood hazard	Moderate—flood hazard	Good
LeB2	Legore silt loam, 3 to 8% slopes, moderately eroded	Ile-10	Slight	Slight	Slight	Good
LeD2	Legore silt loam, 15 to 25% slopes, moderately eroded	IVe-10	Severe—slope	Severe—slope	Moderate—slope	Fair
LfE	Legore very stony silt loam, 25 to 45% slopes	VIIs-3	Severe—slope	Severe—slope	Severe—slope	Poor
NeB2	Neshaminy silt loam, 3 to 8% slopes, moderately eroded	Ile-4	Slight	Moderate—mod. permeability	Slight	Good

[a] See Section 15.9 for an explanation of the USDA Land Capability Classification System.
Data abstracted from Smith, et al. (1975).

A soil classification system based on these properties is equally critical if we expect to use knowledge gained at one location to solve problems at other locations where similarly classed soils are found. *Soil Taxonomy,* a classification system based on measurable soil properties, helps fill this need in some 55 countries. It is a system that is being constantly updated to better serve these needs as scientists learn more about the nature and properties of the world's soils and the relationships among them.

Making soil surveys is both a science and an art by which many soil scientists apply their understanding of soil classification to landscapes in the real world. Mapping soils is not only a profession; many would say that it is a way of life. Soil survey reports are used in countless practical ways by soil scientists and nonscientists alike. The soil survey, combined with powerful geographic information systems, enables planners to make rational decisions about *what* should go *where* on the land. The challenge for soil scientists and concerned citizens is to develop the foresight and fortitude to use the planning process to preserve our most valuable soils, rather than hasten their destruction under shopping malls and landfills.

STUDY QUESTIONS

1. Diagnostic horizons are used to classify soils in *Soil Taxonomy*. Explain the difference between a diagnostic horizon (such as an argillic horizon) and a genetic horizon designation (such as a Bt1 horizon). Give an field example of a diagnostic horizon that contains several genetic horizon designations.

2. Explain the relationships among a *soil individual,* a *polypedon,* a *pedon,* and a *landscape.*

3. Rearrange the following soil orders from the *least* to the *most* highly weathered: Oxisols, Alfisols, Mollisolls, Entisols, and Inceptisols.

4. What is the principal soil property by which Ultisols differ from Alfisols? Inceptisols from Entisols?

5. Use the key given in Figure 3.7 to determine the soil order of a soil with the following characteristics: a spodic horizon at 30 cm depth, permafrost at 80 cm depth. Explain your choice of soil order.

6. Of the five soil forming factors discussed in Chapter 2 (parent material, climate, organisms, topography, and time), choose *two* that have had the dominant influence on developing soil properties characterizing each of the following soil orders: Vertisols, Mollisols, Spodosols, and Oxisols.

7. To which soil order does each of the following belong: Psamments, Udolls, Argids, Udepts, Fragiudalfs, Haplustox, and Calciusterts.

8. What's in a name? Write a hypothetical soil profile description and land-use suitability interpretation for a hypothetical soil that is classified in the Aquic Argixerolls subgroup.

9. Explain why *Soil Taxonomy* is said to be a hierarchical classification system.

10. Name the soil taxonomy category and discuss the engineering implications of these soil taxonomy classes: Aquic Paleudults, Fragiudults, Haplusterts, Saprists, and Turbels.

11. A soil mapper drew a boundary around an area in which he made six randomly located auger borings, two in soil *A*, with an argillic horizon more than 60 cm thick and strong brown in color, and the other four in soil *B*, with a somewhat lighter brown argillic horizon between 50 and 70 cm thick. Other soil properties, as well as management considerations, were similar for the two types of soils. Would the map unit delineated likely be a *soil association*, a *soil consociation*, or a *soil complex?* Explain.

12. Assume you are planning to buy a 4-ha site on which to start a small orchard. Explain how you could use the county soil survey to help determine if the prospective sight was suitable for your intended use.

13. If you were hired by a state government to produce a GIS-based map showing where investments should be made to protect farmland from suburban development, what "layers" of information would you want to include in the GIS?

REFERENCES

Eswaran, H. 1993. "Assessment of global resources: Current status and future needs," *Pedologie,* **43**(1):19–39.

McCracken, R. J., et al. 1985. "An appraisal of soil resources in the USA," in R. F. Follet and B. A. Stewart (eds.), *Soil Erosion and Crop Productivity.* (Madison, Wis.: Amer. Soc. Agron.).

Riecken, F. F., and G. D. Smith. 1949. "Principal upland soils of Iowa, their occurrence and important properties," *Agron.,* **49** (revised). Iowa Agr. Exp. Sta.

Smith, H., and E. Matthews, 1975. *Soil Survey of Harford County Area, Maryland.* (Wahington, D.C.: U.S. Soil Conservation Service).

Soil Survey Staff. 1993a. *National Ssoil Survey Handbook.* Title 430 VI. (Washington D.C.: U.S. Department of Agriculture).

Soil Survey Staff. 1993b. *Soil Survey Manual.* Handbook no. 18 (Washinton, D.C.; U.S. Department of Agriculture).

Soil Survey Staff. 1975. *Soil Taxonomy: A Basic System of Soil Classification for Making and Interpreting Soil Surveys.* (Washington, D.C.: USDA Natural Resources Conservation Service).

Soil Survey Staff. 1998. *Keys to Soil Taxonomy,* 8th ed. (Washington, D.C.: USDA Natural Resources Conservation Service).

Soil Survey Staff. 1999. *Soil Taxonomy: A Basic System of Soil Classification for Making and Interpreting Soil Surveys,* 2nd ed. (Washington, D.C.: USDA Natural Resources Conservation Service).

SSSA. 1984. *Soil Taxonomy, Achievements and Challenges.* SSSA Special Publication no. 14. (Madison, Wis.: Soil Sci. Soc. Amer.).

Star, J., and J. Estes. 1990. *Geographic Information Systems: An Introduction.* (Englewood Cliffs, N.J.: Prentice Hall).

Zwicker, S. E. 1992. *Soil Survey of Bureau County, Illinois.* (Washington, D.C.: USDA Natural Resources Conservation Service).

4

SOIL ARCHITECTURE AND PHYSICAL PROPERTIES

And when that crop grew, and was harvested, no man had crumbled a hot clod in his fingers and let the earth sift past his fingertips.
—JOHN STEINBECK, THE GRAPES OF WRATH

Soil physical properties profoundly influence how soils function in an ecosystem and how they can best be managed. Success or failure of both agricultural and engineering projects often hinges on the physical properties of the soil used. The occurrence and growth of many plant species and the movement of water and solutes over and through the soil are closely related to soil physical properties.

Soil scientists use the color, texture, and other physical properties of soil horizons in classifying soil profiles and in making field determinations about soil suitability for agricultural and environmental projects. Knowledge of the basic soil physical properties is not only of great practical value in itself, but will also help in understanding many aspects of soils considered in later chapters.

The physical properties discussed in this chapter relate to the solid particles of the soil and the manner in which they are aggregated together. If we think of the soil as a house, the solid particles in soil are the building blocks from which the house is constructed. **Soil texture** describes the size of the soil particles. The larger mineral fragments usually are embedded in, and coated with, clay and other colloidal materials. Where the larger mineral particles predominate, the soil is gravelly, or sandy; where the mineral colloids are dominant, the soil is claylike. All gradations between these extremes are found in nature.

In building a house, the manner in which the building blocks are put together determines the nature of the walls, rooms, and passageways. Organic matter and other substances act as cement between individual particles, encouraging the formation of clumps or aggregates of soil. **Soil structure** describes the manner in which soil particles are aggregated together. This property, therefore, defines the nature of the system of pores and channels in a soil.

The physical properties considered in this chapter directly describe the nature of soil solids and their impacts on the soil water and air which reside in the pore spaces between the solid particles. Together, soil texture and structure help determine the nutrient-supplying ability of soil solids, as well as the ability of the soil to hold and conduct the water and air necessary for plant root activity. These factors also determine how soils behave when used for highways, building construction and foundations, or when manipulated by tillage. In fact, through their influence on water movement through and off soils, soil physical properties also exert considerable control over the destruction of the soil itself by erosion.

4.1 SOIL COLOR

The main reason for studying soil colors is that they provide valuable clues to the nature of other soil properties and conditions. Because of the importance of color in soil classification and interpretation, a standard system for accurate color description has been developed using Munsell color charts (Figure 4.1). In this system, a small piece of soil is compared to standard color chips in a soil color book. Each color chip is described by the three components of color: the **hue** (in soils, usually redness or yellowness), the **chroma** (intensity or brightness, a chroma of 0 being neutral gray), and the **value** (lightness or darkness, a value of 0 being black).

Soils display a wide range of reds, browns, yellows, and even greens (see Plates 16 and 19, following page 178). Some soils are nearly black, others nearly white. Some soil colors are very bright, others are dull grays. Soil colors may vary from place to place in the landscape, as when adjacent soils have differing surface horizon colors (e.g., Plate 17). Colors also typically change with depth through the various layers (horizons) within a soil profile. In many soils, the horizons in a given profile have colors that are similar in hue, but vary with respect to chroma and value. Even within a single horizon or clod of soil, colors may vary from spot to spot (see Plate 15).

Causes of Soil Colors

Most soil colors are derived from the colors of iron oxides and organic matter that coat the surfaces of soil particles. Organic coatings tend to darken and mask the colors derived from iron oxides (see Plate 8, after page 178).

Because of the color changes that take place when various iron-containing minerals undergo oxidation and reduction, soil colors can provide extremely valuable insights into the hydrologic regime or drainage status of a soil. Bright (high-chroma) colors throughout the profile are typical of well-drained soils through which water easily passes and in which oxygen is generally plentiful. Prolonged anaerobic conditions can cause iron oxide coatings to become chemically reduced, changing high-chroma (red or brown) colors to low-chroma (gray, bluish or gray-green) colors, a condition referred to as **gley** (as in Plate 20, after page 178).

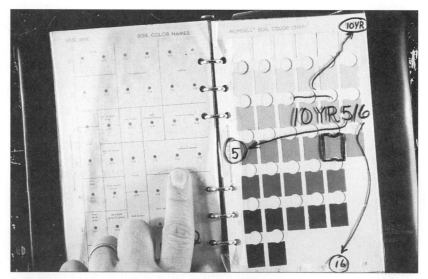

FIGURE 4.1 Determining soil color by comparison to color chips in a Munsell Color Book. Each page shows a different hue, ranging from 5R (red) to 5Y (yellow). On a single page, higher-value (lighter) colors are nearer the top and lower-value (darker) colors are near the bottom. Higher-chroma (brighter) colors are nearer the right-hand side and duller, grayish (low-chroma) colors are nearer the left-hand side. The complete Munsell color description of the soil pictured is 10YR 5/6 (yellowish brown), moist. (Photos by R. Weil)

4.2 SOIL TEXTURE (SIZE DISTRIBUTION OF SOIL PARTICLES)

The size of mineral particles in soil may seem too mundane a subject to warrant much attention, yet knowledge of the proportions of different-sized particles in a soil (i.e., the **soil texture**) is critical for understanding soil behavior and management. When investigating soils on a site, the texture of various soil horizons is often the first and most important property to determine, for a soil scientist can draw many conclusions from this information. Furthermore, the texture of a soil in the field is not readily subject to change, so it is considered a basic property of a soil.

Nature of Soil Separates

Diameters of individual soil particles range over six orders of magnitude, from boulders (1 m) to submicroscopic clays ($<10^{-6}$ m). Scientists group these particles into soil **separates** according to several classification systems, as shown in Figure 4.2. The classification established by the U.S. Department of Agriculture is used in this text. The size ranges for these separates are not purely arbitrary, but reflect major changes in how the particles behave and in the physical properties they impart to soils.

Gravels, cobbles, boulders and other **coarse fragments** >2 mm in diameter may affect the behavior of a soil, but they are not considered to be part of the **fine earth fraction** to which the term *soil texture* properly applies.

SAND. Sand particles are those smaller than 2 mm but larger than 0.05 mm. They may be rounded or angular (Figure 4.3), depending on the extent to which they have been worn down by abrasive processes during soil formation. The coarsest sand particles may be rock fragments containing several minerals, but most sand grains consist of a single mineral, usually quartz (SiO_2) or other primary silicate (Figure 4.4).

Sand feels gritty between the fingers and the particles are generally visible to the naked eye. As sand particles are relatively large, so, too, the voids between them are relatively large and promote free drainage of water and entry of air into the soil. The relationship between particle size and **specific surface area** (the surface area for a given volume or mass of particles) is illustrated in Figure 4.5. Because of their large size, particles of sand have relatively low specific surface area. Therefore, sand particles can hold little water, and soils dominated by sand are prone to drought. Sand particles are considered noncohesive; that is, they do not tend stick together in a mass (see Section 4.9).

SILT. Particles smaller than 0.05 mm but larger than 0.002 mm in diameter are classified as silt. Individual silt particles are not visible to the unaided eye (see Figure 4.3), nor do they feel gritty when rubbed between the fingers. These are essentially microsand

FIGURE 4.2 Classification of soil particles according to their size. The shaded scale in the center and the names on the drawings of particles follow the United States Department of Agriculture system, which is widely used throughout the world. The USDA system is also used in this book. The other two systems shown are also widely used by soil scientists and by highway construction engineers. The drawing illustrates the size of soil separates (note scale).

FIGURE 4.3 (top) A thin section of a loamy soil as seen through a microscope using polarized light (empty pores appear black). Both the sand and silt particles shown are irregular in size and shape, the silt being only smaller. Although quartz (*q*) dominates the sand and silt fractions in this soil, several other silicate minerals can be seen (*p* = plagioclase, *k* = feldspar). Clay films can be seen to coat the walls of the large pores. Photo courtesy of Martin Rabenhorst, University of Maryland

particles, with quartz generally the dominant mineral. Where silt is composed of weatherable minerals, the smaller size of the particles allows weathering to proceed rapidly enough to release significant amounts of plant nutrients.

Although silt is composed of particles similar in shape to sand, it feels smooth or silky, like flour. The pores between silt particles are much smaller (and much more numerous) than those in sand, so silt retains more water and lets less drain through. However, even when wet, silt itself does not exhibit much **stickiness** or **plasticity** (malleability). What little plasticity, cohesion (stickiness), and adsorptive capacity some silt fractions exhibit is largely due to a film of adhering clay.

CLAY. Particles smaller than 0.002 mm are classified as clay and have a very large specific surface area, giving them a tremendous capacity to adsorb water and other sub-

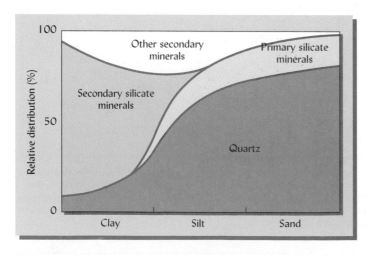

FIGURE 4.4 General relationship between particle size and kinds of minerals present. Quartz dominates the sand and coarse silt fractions. Primary silicates such as the feldspars, hornblende, and micas are present in the sands and, in decreasing amounts, in the silt fraction. Secondary silicates dominate the fine clay. Other secondary minerals, such as the oxides of iron and aluminum, are prominent in the fine silt and coarse clay fractions.

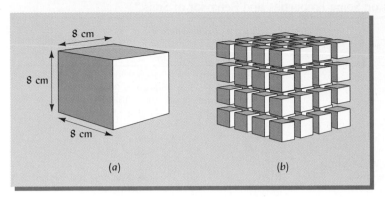

FIGURE 4.5 The relationship between the surface area of a given mass of material and the size of its particles. In the single large cube (*a*) each face has 64 cm² of surface area. The cube has six faces, so the cube has a total of 384 cm² surface area (6 faces × 64 cm² per face). If the same cube of material was cut into smaller cubes (*b*) so that each cube was only 2 cm on each side, then the same mass of material would now be present as 64 smaller cubes (4 × 4 × 4). Each face of each small cube would have 4 cm² (2 × 2 cm) of surface area, giving 24 cm² of surface area for each cube (6 faces × 4 cm² per face). The total mass would therefore have 1536 cm² (24 cm² per cube × 64 cubes) of surface area. This is four times as much surface area as the single large cube. Since clay particles are very very small and usually platelike in shape, their surface area is thousands of times greater than that of the same mass of sand particles.

stances. A spoonful of clay may have a surface area the size of a football field (see Section 8.1). This large adsorptive surface causes clay particles to cohere together in a hard mass after drying. When wet, clay is sticky and can be easily molded.

Clay size particles are so small that they behave as *colloids*—if suspended in water they do not readily settle out. Unlike most sand and silt particles, clay particles tend to be shaped like tiny flakes or flat platelets. The pores between clay particles are very small and convoluted, so movement of both water and air is very slow. Each unique clay mineral (see Chapter 8) imparts very different properties to the soils in which they are prominent. Therefore, soil properties such as shrink-swell behavior, plasticity, water-holding capacity, soil strength, and chemical adsorption depend on the *kind* of clay present as well as the *amount*.

Influence of Surface Area on Other Soil Properties

When particle size decreases, specific surface area and related properties increase greatly, as shown graphically in Figure 4.6. Fine colloidal clay has about 10,000 times as much surface area as the same weight of medium-sized sand. Soil texture influences many other soil properties in far-reaching ways (see Table 4.1) as a result of five fundamental phenomena that increase with surface area:

1. Water is retained in soils as thin films on the surfaces of soil particles.
2. Both gases and dissolved chemicals are attracted to and adsorbed by mineral particle surfaces.

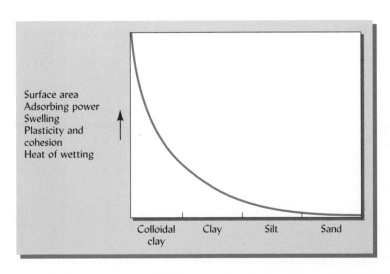

Surface area
Adsorbing power
Swelling
Plasticity and cohesion
Heat of wetting

Colloidal clay   Clay   Silt   Sand

FIGURE 4.6 The finer the texture of a soil, the greater is the effective surface exposed by its particles. Note that adsorption, swelling, and the other physical properties cited follow the same general trend and that their intensities go up rapidly as the colloidal size is approached.

TABLE 4.1   Generalized Influence of Soil Separates on Some Properties and Behavior of Soils.[a]

Property/behavior	Rating associated with soil separates		
	Sand	*Silt*	*Clay*
Water-holding capacity	Low	Medium to High	High
Aeration	Good	Medium	Poor
Drainage rate	High	Slow to Medium	Very slow
Soil organic matter level	Low	Medium to High	High to Medium
Decomposition of organic matter	Rapid	Medium	Slow
Warm-up in spring	Rapid	Moderate	Slow
Compactability	Low	Medium	High
Susceptibility to wind erosion	Moderate (high if fine sand)	High	Low
Susceptibility to water erosion	Low (unless fine sand)	High	Low if aggregated, high if not
Shrink-swell potential	Very Low	Low	Moderate to very high
Sealing of ponds, dams, and landfills	Poor	Poor	Good
Suitability for tillage after rain	Good	Medium	Poor
Pollutant leaching potential	High	Medium	Low (unless cracked)
Ability to store plant nutrients	Poor	Medium to high	High
Resistance to pH change	Low	Medium	High

[a] Exceptions to these generalizations do occur, especially as a result of soil structure and clay minerology.

3. Weathering takes place at the surface of mineral particles, releasing constituent elements including plant nutrients into the soil solution.

4. The surfaces of mineral particles often carry both negative and some positive electromagnetic charges so that particle surfaces and the water films between them tend to attract each other (see Section 4.7) and stick together in a coherent mass.

5. Microorganisms tend to grow on and carry out reactions on particle surfaces.

4.3   SOIL TEXTURAL CLASSES

Textural class names convey an idea of the size distribution of particles (Figure 4.7) and indicate the general nature of soil physical properties.

Sands and loamy sands are dominated by the properties of sand, for the sand separate comprises at least 70% of the material by weight (less than 15% of the material is clay). Characteristics of the clay separate are distinctly dominant in clays, sandy clays, and silty clays.

LOAMS.   An ideal **loam** may be defined as a mixture of sand, silt, and clay particles that exhibits the *properties* of those separates in about equal proportions. This definition does not mean that the three separates are present in equal *amounts* (as will be revealed by careful study of Figure 4.7). This anomaly exists because a relatively small percentage of clay is required to engender clayey properties in a soil, whereas small amounts of sand and silt have a lesser influence on how a soil behaves.

Most soils are some type of loam. They may possess the ideal makeup of equal proportions previously described and be classed simply as loam. However, a loam in which sand is dominant is classified as a *sandy loam*. In the same way, there may occur silt loams, silty clay loams, sandy clay loams, and clay loams.

COARSE FRAGMENT MODIFIERS.   For some soils, qualifying factors such as stone, gravel, and the various grades of sand become part of the textural class name. Fragments that range from 2 to 75 mm along their greatest diameter are termed *gravel* or *pebbles;* those ranging from 75 to 250 mm are called *cobbles* (if round) or *flags* (if flat); and those more than 250 mm across are called *stones* or *boulders*. A *cobbly, fine sandy loam* is an example of such a modified textural class.

Alteration of Soil Textural Class

Over very long periods of time, pedologic processes (see Chapter 2) such as erosion, deposition, illuviation, and weathering can alter the textures of various soil horizons. However,

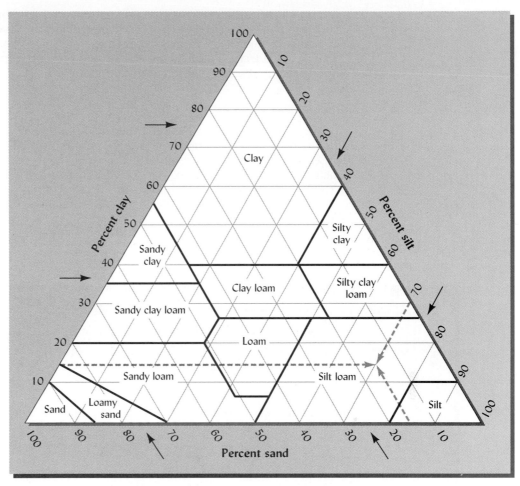

FIGURE 4.7 The major soil textural classes are defined by the percentages of sand, silt, and clay according to the heavy boundary lines shown on the textural triangle. If these percentages have been determined for a soil sample by particle size analysis, then the triangle can be used to determine the soil textural class name that applies to that soil sample. To use the graph, first find the appropriate clay percentage along the left side of the triangle, then draw an imaginary line from that location across the graph going parallel to the base of the triangle. Next find the sand percentage along the base of the triangle, then draw a line inward going parallel to the triangle side labeled "Percent silt." The small arrows indicate the proper direction in which to draw the lines. The name of the compartment in which these two lines intersect indicates the textural class of the soil sample. Percentages for any two of the three soil separates is all that is required. Consider, as an example, a soil which has been determined to contain 15% sand, 15% clay, and 70% silt. This example is indicated by the light dashed lines that intersect in the compartment labeled "Silt loam."

management practices generally do not alter the textural class of a soil on a field scale. The texture of a given soil can be changed only by mixing it with another soil of a different textural class. For example, the incorporation of large quantities of sand to change the physical properties of a clayey soil for use in greenhouse pots or for turf would bring about such a change. However, where specifications (as for a landscape design) call for soil materials of a certain textural class, it is advisable to find a naturally occurring soil that meets the specification, rather than attempt to alter the textural class by mixing in sand or clay.

Also, it should be noted that adding peat or compost to a soil while mixing a potting medium does not constitute a change in texture, since this property refers only to the mineral particles. The term *soil texture* is not relevant to artificial media that contain mainly perlite, peat, styrofoam, or other nonsoil materials.

Determination of Textural Class by the "Feel" Method

Textural class determination is one of the first field skills a soil scientist should develop. Determining the textural class of a soil by its feel is of great practical value in soil survey, land classification, and any investigation in which soil texture may play a role. Accuracy depends largely on experience, so practice whenever you can, beginning with soils of known texture to calibrate your fingers.

The textural triangle (see Figure 4.7) should be kept in mind when determining the textural class by the feel method as explained in Box 4.1.

Laboratory Particle-Size Analyses

While a set of sieves can be used to separate out the sand, a sedimentation procedure is usually used to determine the amounts of silt and clay. The principle involved is simple. Because soil particles are more dense than water, they tend to sink, settling at a velocity that is proportional to their size. In other words: "The bigger they are, the faster they fall." The equation that describes this relationship is referred to as *Stokes' law*. In its simplest form it tells us that the velocity of settling V is proportional to the square of a particle's diameter d

$$V = kd^2$$

BOX 4.1   A METHOD FOR DETERMINING TEXTURE BY FEEL

The first, and most critical, step in the texture-by-feel method is to knead a walnut-sized sample of moist soil into a uniform puttylike consistency, slowly adding water if necessary. This step may take a few minutes, but a premature determination is likely to be in error as hard clumps of clay and silt may feel like sand grains. The soil should be moist, but not quite glistening. Try to do this with only one hand so as to keep your other hand clean for writing in a field notebook (and shaking hands with your client).

While squeezing and kneading the sample, note its maleability, stickiness, and stiffness, all properties associated with the clay content. A high silt content makes a sample feel smooth and silky, with little stickiness or resistance to deformation. A soil with a significant content of sand feels rough and gritty, and makes a grinding noise when rubbed near one's ear.

Get a feel for the amount of clay by attempting to squeeze a ball of properly moistened soil between your thumb and the side of your forefinger, making a ribbon of soil. Make the ribbon as long as possible until it breaks from its own weight (see Figure 4.8).

Interpret your observations as follows:

1. Soil will not cohere into a ball, falls apart: **sand**
2. Soil forms a ball, but will not form a ribbon: **loamy sand**
3. Soil ribbon is dull and breaks off when less than 2.5 cm long and
   a. Grinding noise is audible; grittiness is prominent feel: **sandy loam**
   b. Smooth, floury feel prominent; no grinding audible: **silt loam**
   c. Only slight grittiness and smoothness; grinding not clearly audible: **loam**
4. Soil exhibits moderate stickiness and firmness, forms ribbons 2.5 to 5 cm long, and
   a. Grinding noise is audible; grittiness is prominent feel: **sandy clay loam**
   b. Smooth, floury feel prominent; no grinding audible: **silty clay loam**
   c. Only slight grittiness and smoothness; grinding not clearly audible: **clay loam**
5. Soil exhibits dominant stickiness and firmness, forms shiny ribbons longer than 5 cm, and
   a. Grinding noise is audible; grittiness is dominant feel: **sandy clay**
   b. Smooth, floury feel prominent; no grinding audible: **silty clay**
   c. Only slight grittiness and smoothness; grinding not clearly audible: **clay**

A more precise estimate of sand content (and hence more accurate placement in the horizontal dimension of the textural class triangle) can be made by wetting a pea-sized clump of soil in the palm of your hand and smearing it around with your finger until it your palm becomes coated with a souplike suspension of soil. The sand grains will stand out visibly and their volume as compared to the original "pea" can be estimated, as can their relative size (fine, medium, coarse etc.).

It is best to learn the method using samples of known textural class. With practice, accurate textural class determinations can be made on the spot.

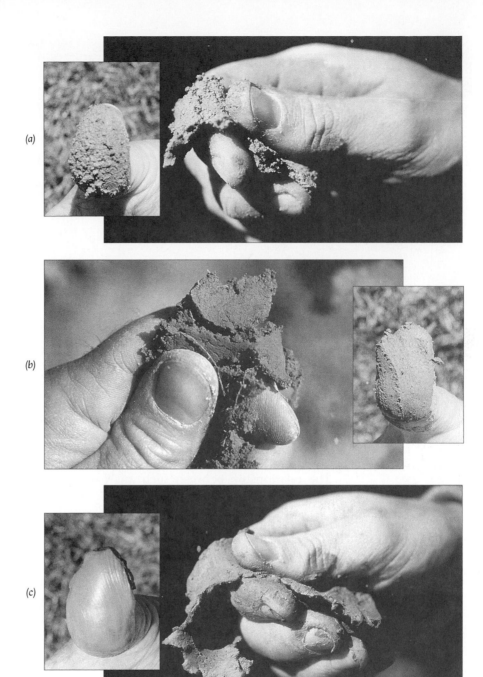

FIGURE 4.8 The "feel" method for determining soil textural class. A moist soil sample is rubbed between the thumb and forefingers (insets) and squeezed out to make a "ribbon." (*a*) The gritty, noncohesive appearance and short ribbon of a sandy loam with about 15% clay. (*b*) The smooth, dull appearance and crumbly ribbon characteristic of a silt loam. (*c*) The smooth, shiny appearance and long, flexible ribbon of a clay. See Box 4.1 for details. (Large photos courtesy of R. Weil)

where *k* is a constant related to the acceleration due to gravity and the density and viscosity of water. By measuring the amount of soil still in suspension after various amounts of settling time, the percentages of each size fraction can be determined so as to identify the soil textural class and generate particle-size distribution curves such as those shown in Figure 4.9.

Figure 4.9 presents particle-size distribution curves for soils representative of three textural classes. The fact that these curves are smooth emphasizes that there is no sharp line of demarcation in the distribution of sand, silt, and clay fractions, and suggests a gradual change of properties with change in particle size.

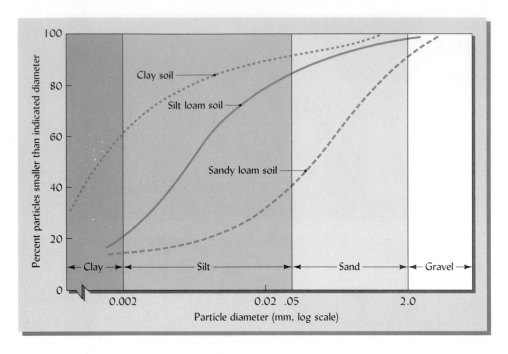

FIGURE 4.9 Particle-size distribution in three soils varying widely in their textures. Note that there is a gradual transition in the particle-size distribution in each of these soils.

It is important to note that soils are assigned to textural classes *solely* on the basis of the mineral particles of sand size and smaller; therefore, the percentages of sand, silt, and clay always add up to 100%. The amounts of stone and gravel are rated separately. Organic matter is usually removed from a soil sample by oxidation before the mechanical separation.

The relationship between such analyses and textural class names is commonly shown diagrammatically as a triangular graph (see Figure 4.7). This **textural triangle** also enables us to use laboratory particle-size analysis data to check the accuracy of field textural determinations by feel.

4.4   STRUCTURE OF MINERAL SOILS

The term **structure** relates to the arrangement of primary soil particles into groupings called **aggregates** or **peds.** The pattern of pores and peds defined by soil structure greatly influences water movement, heat transfer, aeration, and porosity in soils. Activities such as timber harvesting, grazing, tillage, trafficking, drainage, liming, and manuring impact soils largely through their effect on soil structure, especially in the surface horizons.

The processes involved in the formation, stabilization, and management of soil structure will be discussed in Sections 4.7 and 4.8. Here we will examine the nature of the various types of structural peds.

Many types of structural peds occur in soils, often within different horizons of a particular soil profile. Soil structure is characterized in terms of the shape (or *type*), size, and distinctness (or *grade*) of the peds. The four principal shapes of soil structure are *spheroidal, platy, prismlike,* and *blocklike.* These structural types (and some subtypes) are illustrated in Figure 4.10.

In describing soil structure, soil scientists note not only the *type* (shape) of the structural peds present, but also the relative *size* (fine, medium, coarse) and degree of development or distinctness of the peds (*grades* such as strong, moderate, or weak). For example, the soil horizon shown in Figure 4.10*d* might be described as having "weak, fine, subangular blocky structure." Generally, the structure of a soil is easier to observe when the soil is relatively dry. When wet, structural peds may swell and press closer together, making the individual peds less well defined. In any case, the structural arrangement of soil particles and the pore spaces between structural peds greatly influence soil density, an aspect of soil architecture that we will now examine in detail.

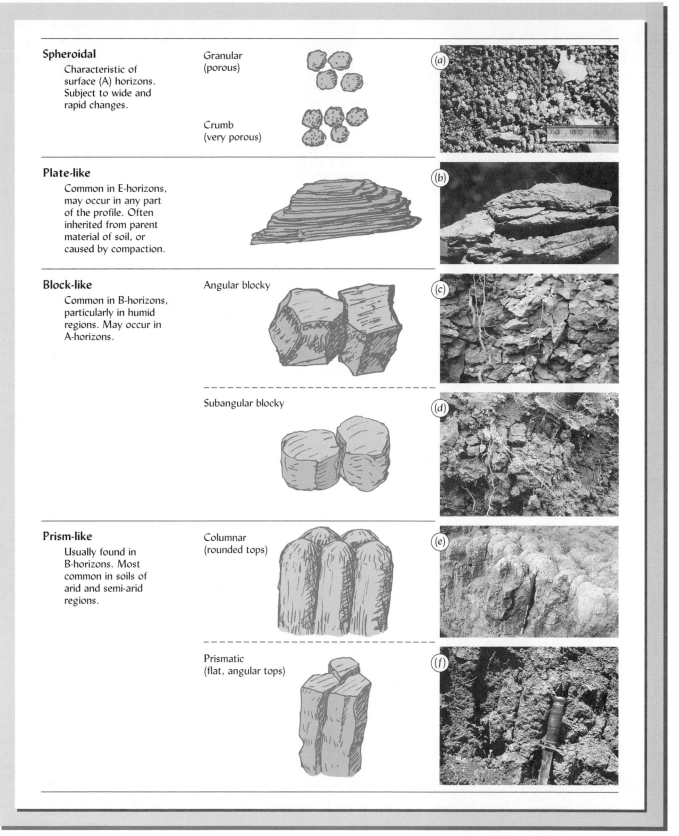

Spheroidal

Characteristic of surface (A) horizons. Subject to wide and rapid changes.

Granular (porous)

Crumb (very porous)

(a)

Plate-like

Common in E-horizons, may occur in any part of the profile. Often inherited from parent material of soil, or caused by compaction.

(b)

Block-like

Common in B-horizons, particularly in humid regions. May occur in A-horizons.

Angular blocky

(c)

Subangular blocky

(d)

Prism-like

Usually found in B-horizons. Most common in soils of arid and semi-arid regions.

Columnar (rounded tops)

(e)

Prismatic (flat, angular tops)

(f)

FIGURE 4.10 The various structure types (shapes) found in mineral soils. Their typical location is suggested. The drawings illustrate their essential features and the photos indicate how they look in situ. For scale, note the 15-cm-long pencil in (*e*) and the 3-cm-wide knife blade in (*d*) and (*f*). (Photo (*e*) courtesy of J. L. Arndt, now with Petersen Environmental Consulting; North Dakota State University. Others courtesy of R. Weil)

4.5 SOIL DENSITY

Particle Density

Soil **particle density** D_p is defined as the mass per unit volume of soil *solids* (in contrast to the volume of the *soil*, which would also include spaces between particles). Thus, if 1 cubic meter (m^3) of soil solids weighs 2.6 megagrams (Mg), the particle density is 2.6 Mg/m^3 (which can also be expressed as 2.6 grams per cubic centimeter).

Particle density is essentially the same as the **specific gravity** of a solid substance. The chemical composition and crystal structure of a mineral determines its particle density. Particle density is *not* affected by pore space, and therefore is not related to particle size or to the arrangement of particles (soil structure).

Particle densities for most mineral soils vary between the narrow limits of 2.60 to 2.75 Mg/m^3 because quartz, feldspar, micas, and the colloidal silicates that usually make up the major portion of mineral soils all have densities within this range. For general calculations concerning arable mineral surface soils (1 to 5% organic matter), a particle density of about 2.65 Mg/m^3 may be assumed if the actual particle density is not known.

Bulk Density

A second important mass measurement of soils is **bulk density** D_b, which is defined as the mass of a unit volume of dry soil. This volume includes both solids and pores. A careful study of Figure 4.11 should make clear the distinction between particle and bulk density. Both expressions of density use only the mass of the solids in a soil; therefore, any water present is excluded from consideration.

There are several methods of determining soil bulk density by obtaining a known volume of soil, drying it to remove the water, and weighing the dry mass. A special coring instrument (Figure 4.12) can obtain a sample of known volume without disturbing the natural soil structure. For surface soils, perhaps the simplest method is to dig a small hole, collect all the excavated soil, and then line the hole with plastic film and fill it completely with a measured volume of water.

Factors Affecting Bulk Density

Soils with a high proportion of pore space to solids have lower bulk densities than those that are more compact and have less pore space. Consequently, any factor that influences soil pore space will affect bulk density. Typical ranges of bulk density for various

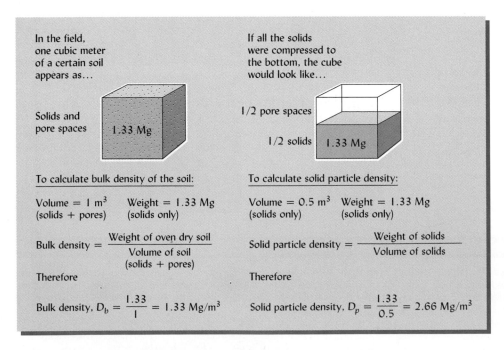

FIGURE 4.11 Bulk density D_b and particle density D_p of soil. Bulk density is the weight of the solid particles in a standard volume of field soil (solids plus pore space occupied by air and water). Particle density is the weight of solid particles in a standard volume of those solid particles. Follow the calculations through carefully and the terminology should be clear. In this particular case the bulk density is one-half the particle density, and the percent pore space is 50.

FIGURE 4.12 A special sampler designed to remove a cylindrical core of soil without causing disturbance or compaction. The sampler head contains an inner cylinder and is driven into the soil with blows from a drop hammer. The inner cylinder containing an undisturbed soil core is then removed and trimmed on the end with a knife to yield a core whose volume can easily be calculated from its length and diameter. The weight of this soil core is then determined after drying in an oven. (Photo courtesy of R. Weil)

soil materials and conditions are illustrated in Figure 4.13. It would be worthwhile to study this figure until you have a good feel for these ranges of bulk density.

EFFECT OF SOIL TEXTURE.   As illustrated in Figure 4.13, fine-textured soils such as silt loams, clays, and clay loams generally have lower bulk densities than do sandy soils.[1] This is true because the solid particles of the fine-textured soils tend to be organized in porous granules, especially if adequate organic matter is present. In these aggregated soils, pores exist both between *and* within the granules. This condition assures high total pore space and a low bulk density. In sandy soil, however, organic matter contents generally are low, the solid particles are less likely to be aggregated together, and the bulk densities are commonly higher than in the finer-textured soils. Figure 4.14 illustrates the concept that similar amounts of large pores are present in both sandy and well-aggregated fine-textured soils, but that sandy soils have few of the fine, within-ped pores, and so have less total porosity.

While sandy soils generally have high bulk densities, the packing arrangement of the sand grains also affects their bulk density (see Figure 4.15). Loosely packed grains may fill as little as 52% of the bulk volume, while tightly packed grains may fill as much as 75% of the volume. The bulk density is generally lower if the sand particles are mostly of one size class (i.e., *well-sorted* sand), while a mixture of different sizes (i.e., *well-graded* sand) is likely to have an especially high bulk density. In the latter case, the smaller particles partially fill in the spaces between the larger particles. The most dense materials are those characterized by both a mixture of sand sizes and a tight packing arrangement.

DEPTH IN SOIL PROFILE.   Deeper in the soil profile, bulk densities are generally higher, probably as a result of lower organic matter contents, less aggregation, fewer roots, and compaction caused by the weight of the overlying layers. Very compact subsoils may have bulk densities of 2.0 Mg/m^3 or even greater. Many soils formed from glacial till (see Section 2.10) have extremely dense subsoils as a result of past compaction by the enormous mass of ice.

Useful Density Figures

For engineers involved with moving soil during construction, or for landscapers bringing in topsoil by the truckload, a knowledge of the bulk density of various soils is use-

---

[1] This fact may seem counterintuitive at first because sandy soils are commonly referred to as "light" soils, while clays and clay loams are referred to as "heavy" soils. The terms *heavy* and *light* in this context refer not to the mass per unit volume of the soils, but to the amount of effort that must be exerted to manipulate these soils with tillage implements—the sticky clays being much more difficult to till.

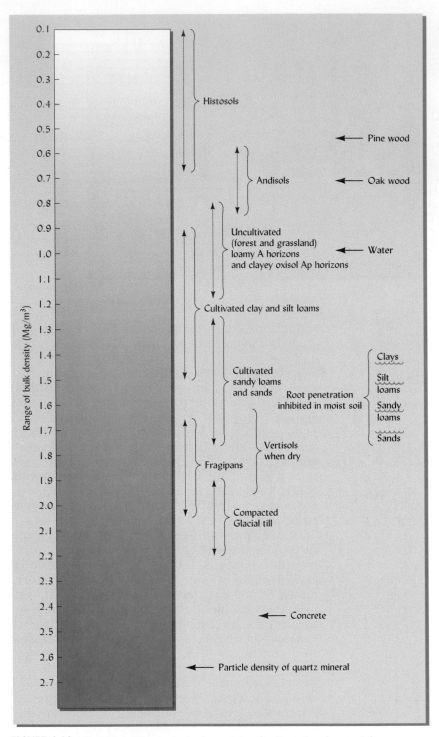

FIGURE 4.13 Bulk densities typical of a variety of soils and soil materials.

ful in estimating the weight of soil to be moved. A typical medium-textured mineral soil might have a bulk density of 1.25 Mg/m³, or 1250 kilograms in a cubic meter.[2] People are often surprised by how heavy soil is. Imagine driving your pickup truck to a nursery where natural topsoil is sold in bulk and filling your truck bed with a nice, rounded

---

[2] Most commercial landscapers and engineers in the United States still use English units. To convert values of bulk density given in units of Mg/m³ into values of lb/yd³, multiply by 1686. Therefore, 1 yd³ of a typical medium-textured mineral soil with a bulk density of 1.25 Mg/m³ would weigh over a ton (1686 * 1.25 = 2108 lb/yd³).

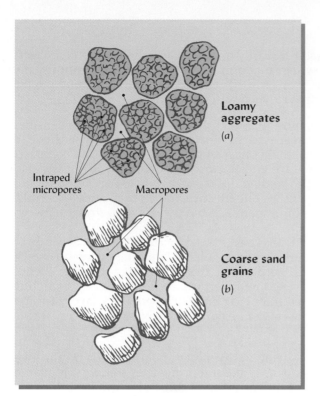

FIGURE 4.14 A schematic comparison of sandy and clayey soils showing the relative amounts of large (macro-) pores and small (micro-) pores in each. There is less total pore space in the sandy soils than in the clayey one because the clayey soil contains a large number of fine pores within each aggregate (*a*), but the sand particles (*b*), while similar in size to the clayey aggregates, are solid and contain no pore spaces within them. This is the reason why, among surface soils, those with coarse texture are usually more dense than those with finer textures.

load. Of course, you would not really want to do this as the load might break your rear axle; you certainly would not be able to drive away with the load. A typical "half-ton" (1000 lb or 454 kg) load capacity pickup truck could carry less than 0.4 m³ of this soil even though the truck bed has room for about six times this volume of material.

The mass of soil in 1 ha to a depth of normal plowing (15 cm) can be calculated from soil bulk density. If we assume a bulk density of 1.3 Mg/m³ for a typical arable surface soil, such a hectare–furrow slice 15 cm deep weighs about 2 million kg.[3] This estimate of the mass of surface soil in a hectare of land is very useful in calculating lime and fertilizer application rates and organic matter mineralization rates (see Boxes 8.1, 9.2, 9.6, and 12.1 for detailed examples). However, this estimated mass must be adjusted if the bulk density is other than 1.3 Mg/cm³ or the depth of the layer under consideration is more or less than 15 cm.

---

[3] 10,000 m²/ha × 1.3 Mg/m³ × 0.15 m = 1950 Mg/ha ≈ 2 million kg per ha 15 cm deep. A comparable figure in the English system is 2 million lb per acre–furrow slice 6 to 7 in. deep.

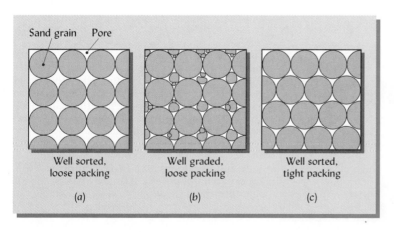

FIGURE 4.15 The uniformity of grain size and the type of packing arrangement significantly affect the bulk density of sandy materials. Materials consisting of all similar-sized grains are termed *well sorted* (or poorly graded). Those with a variety of grain sizes are *well graded* (or poorly sorted). In either case, compaction of the particles into a tight packing arrangement markedly increases the bulk density of the material and decreases its porosity. Note that the size distribution of sand and gravel particles can be described as either graded or sorted, the two terms having essentially opposite meanings. Geologists usually speak of rivers having sorted the sand grains by size as deposits were laid down. Engineers usually are concerned as to whether or not sand consists of a gradation of sizes (i.e., is well graded or not).

Management Practices Affecting Bulk Density

Changes in bulk density for a given soil are easily measured and can alert soil managers to changes in soil quality and ecosystem function. Increases in bulk density usually indicate a poorer environment for root growth, reduced aeration, and undesirable changes in hydrologic function, such as reduced water infiltration.

FOREST LANDS. The surface horizons of most forested soils have rather low bulk densities (see Figure 4.13). Tree growth and forest ecosystem function are particularly sensitive to increases in bulk density. Conventional timber harvest generally disturbs and compacts 20 to 40% of the forest floor (Figure 4.16) and is especially damaging along the skid trails where logs are dragged and at the landing decks—areas where logs are piled and loaded onto trucks (Table 4.2). An expensive, but effective, means of moving logs while minimizing compactive degradation of forest lands is the use of cables strung between towers or hung from large balloons.

Intensive recreational use of forested soils can also lead to increased bulk densities in areas with heavy foot traffic, such as campsites and trails. An important consequence of increased bulk density is a diminished capacity of the soil to take in water, hence increased losses by surface runoff. Damage from hikers can be minimized by restricting foot traffic to well-designed, established trails that may include a thick layer of wood chips, or even a raised boardwalk in the case of heavily traveled paths over very fragile soils, such as in wetlands.

AGRICULTURAL LAND. Although tillage may temporarily loosen the surface soil, in the long term intense tillage increases soil bulk density because it depletes soil organic matter and weakens soil structure (see Section 4.8). The effect of cultivation can be minimized by adding crop residues or farm manure in large amounts and rotating cultivated crops with a grass sod.

In modern agriculture, heavy machines used to pull implements, apply amendments, or harvest crops can create yield-limiting soil compaction. Traffic is particularly damaging on wet soil. Generally, with heavier loads and on wetter soils, compactive effects are more pronounced and penetrate more deeply into the profile. Certain tillage implements, such as the moldboard plow and the disk harrow, compact the soil below their working depth even as they lift and loosen the soil above. Use of these implements or repeated trips over the field by heavy machinery can form **plow pans** or **traffic pans**, dense zones immediately below the plowed layer (Figure 4.17). Other tillage implements, such as the chisel plow and the spring-tooth harrow, do not press down upon the soil beneath them, and so are useful in breaking up plow pans and stirring the soil with a minimum of compaction. Large chisel-type plows can be used in **subsoiling** to break up dense subsoil layers (Figure 4.18). However, in many soils the effects of subsoiling are quite temporary. Any tillage tends to reduce soil strength, thus making the soil less resistant to subsequent compaction.

FIGURE 4.16 Timber harvest with a conventional rubber-tired skidder in a boreal forest in western Alberta, Canada results in disturbance to the surface soil horizons (forest floor). Such practices cause significant soil compaction that can impair soil ecosystem functions for many years. Timber harvest practices that can reduce such damage to forest soils include selective cutting, use of flexible-track vehicles and overhead cable transport of logs, and abstaining from harvest during wet conditions. (Photo courtesy of Andrei Startsev, Alberta Environmental Center).

TABLE 4.2   Effects of Timber Harvest on Bulk Density at Different Depths in Two Forested Ultisols in Georgia

Rubber-wheeled skidders were used to harvest the logs.
Note the generally higher bulk densities of the sandy loam soil compared to the clay loam, and the greater effect of timber harvest on the skidder trails.

Soil Depth, cm	Bulk density, Mg/m³		
	Preharvest	Postharvest, off trails	Postharvest, skidder trails
Upper coastal plain, sandy loam			
0–8	1.25	1.50	1.47
8–15	1.40	1.55	1.71
15–23	1.54	1.61	1.81
23–30	1.58	1.62	1.77
Piedmont, clay loam			
0–8	1.16	1.36	1.52
8–15	1.39	1.49	1.67
15–23	1.51	1.51	1.66
23–30	1.49	1.46	1.61

Data from Gent, et al. (1984, 1986).

Another approach for minimizing compaction is to carefully restrict all wheel traffic to specific lanes, leaving the rest of the field (usually 90% or more of the area) free from compaction. Such **controlled traffic** systems are widely used in Europe, especially on clayey soils. Gardeners can practice controlled traffic by establishing permanent foot paths between planting beds. The paths may be enhanced by covering with a thick mulch, planting to sod grass, or paving with flat stones.

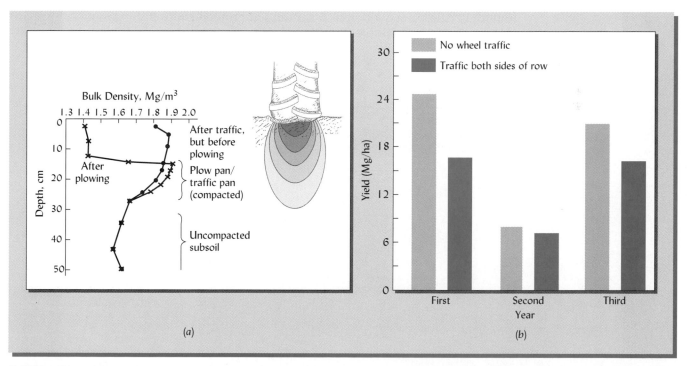

FIGURE 4.17 Tractors and other heavy equipment compact the soil to considerable depths, increasing bulk density and reducing plant growth and crop yields. The effects are especially damaging if the soil is wet when trafficked. (*a*) The tires of a heavy vehicle compact a sandy loam soil to about 30 cm, creating a traffic pan. Plowing temporarily loosens the compacted surface soil (plow layer), but increases compaction just below the plowed layer, creating a combined traffic pan and plow pan. Bulk densities in excess of 1.8 Mg/m³ prevented the penetration of cotton roots in this case. (*b*) The yield of potatoes was reduced in two out of three years in this test on a clay loam in Minnesota. Yield reductions are often most pronounced in relatively dry years when plants have the greatest need for subsoil moisture. [Based on data from Camp and Lund (1964) and Voorhees (1984)]

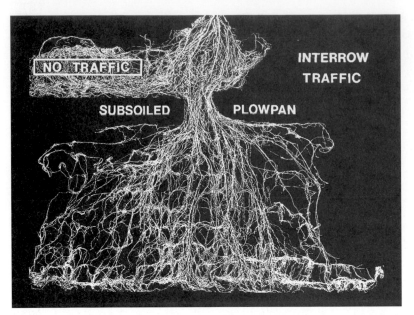

FIGURE 4.18 Root distribution of a cotton plant. On the right, interrow tractor traffic and plowing have caused a plowpan that restricts root growth. Roots are more prolific on the left where there had been no recent tractor traffic. The roots are seen to enter the subsoil through a loosened zone created by a subsoiling chisel-type implement. (Courtesy USDA National Tillage Machinery Laboratory)

Some managers attempt to reduce compaction using an opposite strategy in which special wide tires are fitted to heavy equipment so as to spread the weight over more soil surface, thus reducing the force applied per unit area (Figure 4.19a). Home gardeners can avoid concentrating their body weight on just the few square centimeters of their footprints by standing on wooden boards when preparing seedbeds in relatively wet soil (Figure 4.19b).

Influence of Bulk Density on Soil Strength and Root Growth

High bulk density may occur as a natural soil profile feature (for example, a fragipan), or it may be an indication of human-induced soil compaction. In any case, root growth

(a)

(b)

FIGURE 4.19 One approach to reducing soil compaction is to spread the applied weight over a larger area of the soil surface. Examples are extra-wide wheels on heavy vehicles used to apply soil amendments (left) and standing on a wooden board while preparing a garden seedbed in early spring (right). (Photos courtesy of R. Weil) is inhibited by excessively dense soils for a number of reasons, including the soil's resistance to penetration, poor aeration, slow movement of nutrients and water, and the buildup of toxic gases and root exudates.

Roots penetrate the soil by pushing their way into pores. If a pore is too small to accommodate the root cap, the root must push the soil particles aside and enlarge the pore. To some degree, the density per se restricts root growth, as the roots encounter fewer and smaller pores. However, root penetration is also limited by **soil strength**, the property of the soil that causes it to resist deformation. One way to quantify soil strength is to measure the force needed to push a standard cone-tipped rod (a **penetrometer**) into the soil (see also Section 4.9). Compaction generally increases both bulk density and soil strength.

Soil water content affects *soil strength* (see Section 4.9 and Figure 5.26). Soil strength is increased when a soil is compacted to a higher bulk density, and also when a soil dries out and hardens. Therefore, the effect of bulk density on root growth is most pronounced in a relatively dry soil, a higher bulk density being necessary to prevent root penetration when the soil is moist. For example, a traffic pan having a bulk density of 1.6 Mg/m^3 may completely prevent the penetration of roots when the soil is rather dry, yet roots may readily penetrate this same layer when it is in a moist condition.

The more clay present in a soil, the smaller the average pore size, and the greater the resistance to penetration at at a given bulk density. Therefore, if the bulk density is the same, roots more easily penetrate a moist sandy soil than a moist clayey one. The growth of roots into moist soil is generally limited by bulk densities ranging from 1.45 Mg/m^3 in clays to 1.85 Mg/m^3 in loamy sands (see Figure 4.13).

4.6 PORE SPACE OF MINERAL SOILS

One of the main reasons for measuring soil bulk density is that this value can be used to calculate pore space. For soils with the same particle density, the lower the bulk density, the higher the percent pore space (**total porosity**). See Box 4.2 for the formula expressing this relationship.

Factors Influencing Total Pore Space

In Chapter 1 (Figure 1.17) we noted that for an "ideal" medium-textured, well-granulated surface soil in good condition for plant growth, approximately 50% of the soil volume would consist of pore space, and that the pore space would be about half-filled with air and half-filled with water. Actually, total porosity varies widely among soils for the same reasons that bulk density varies. Values range from as low as 25% in compacted subsoils to more than 60% in well-aggregated, high-organic-matter surface soils. As is the case for bulk density, management can exert a decided influence on the pore space of soils. Cultivation tends to lower the total pore space compared to that of uncultivated soils. This reduction usually is associated with a decrease in organic matter content and a consequent lowering of **granulation**.

Size of Pores

Bulk density values help us predict only *total* porosity. However, soil pores occur in a wide variety of sizes and shapes. The size of a pore largely determines what role the pore can play in the soil. We will simplify our discussion at this point by referring only to **macropores** (larger than about 0.05 mm) and **micropores** (smaller than about 0.05 mm).

MACROPORES.   The macropores characteristically allow the ready movement of air and the drainage of water. They also are large enough to accommodate plant roots and the wide range of tiny animals that inhabit the soil (see Chapter 10). Several types of macropores are illustrated in Figure 4.20.

In well-structured soils, the macropores are generally found between peds. These **interped pores** may occur as spaces between loosely packed granules or as the planar cracks between tight-fitting blocky and prismatic peds. Figure 4.21 shows that the decrease in organic matter and increase in clay that occur with depth in many profiles are associated with a shift from macropores to micropores. Macropores created by roots, earthworms and other organisms constitute a very important type of pores termed **biopores**. In some clayey soils, biopores are the principal form of macropores, greatly facilitating the growth of plant roots (Table 4.3).

BOX 4.2   CALCULATION OF PERCENT PORE SPACE IN SOILS

The bulk density of a soil can be easily measured and particle density can usually be assumed to be 2.65 Mg/m³ for most silicate-dominated mineral soils. Direct measurement of the pore space in soil requires the use of much more tedious and expensive techniques. Therefore, when information on the percent pore space is needed, it is often desirable to calculate the pore space from data on bulk and particle densities using the following:

$$\% \text{ pore space} = 100 - \left(\frac{D_b}{D_p} \times 100\right)$$

EXAMPLE

Consider a cultivated clay soil for which the bulk density is determined to be 1.28 Mg/m³. If we have no information on the particle density, we assume that the particle density is approximately that of the common silicate minerals (i.e., 2.65 Mg/m³). We calculate the percent pore space using the formula derived above:

$$\% \text{ pore space} = 100 - \left(\frac{1.28 \text{ Mg/m}^3}{2.65 \text{ Mg/m}^3} \times 100\right) = 100 - 48.3 = 51.7$$

This value of pore space, 51.7%, is quite close to the typical percentage of air and water space described in Figure 1.17 for a well-granulated, medium- to fine-textured soil in good condition for plant growth. This simple calculation tells us nothing about the relative amounts of large and small pores, however, and so must be interpreted with caution.

For certain soils it is inaccurate to assume that the soil particle density is 2.65 Mg/m³. For instance, a soil with a high organic matter content can be expected to have a particle density somewhat lower than 2.65. Similarly, a soil rich in iron oxide minerals will have a particle density greater than 2.65 because these minerals have particle densities as high as 3.5.

MICROPORES.   In contrast to macropores, micropores are usually filled with water in field soils. Even when not water-filled, they are too small to permit much air movement. Water movement in micropores is slow, and much of the water retained in these pores is not available to plants (see Chapter 5). Fine-textured soils, especially those without a stable granular structure, may have a preponderance of micropores, thus allowing relatively slow gas and water movement, despite the relatively large volume of total pore

(a)　　　　*(b)*　　　　*(c)*

FIGURE 4.20   Various types of soil pores. (*a*) Many soil pores occur as *packing pores,* spaces left between primary soil particles. The size and shape of these spaces is largely dependent on the size and shape of the primary sand, silt, and clay particles and their packing arrangement. (*b*) In soils with structural peds, the spaces between the peds form *interped pores.* These may be rather planar in shape, as with the cracks between prismatic peds, or they may be more irregular, like those between loosely packed granular aggregates. (*c*) *Biopores* are formed by organisms such as earthworms, insects, and plant roots. Most of these are long, sometimes branched channels, but some are round cavities left by insect nests and the like.

TABLE 4.3   Distribution of Different-Sized Loblolly Pine Roots in the Soil Matrix and in Old Root Channels in the Uppermost Meter of an Ultisol in South Carolina

The root channels were generally from 1 to 5 cm in diameter and filled with loose surface soil and decaying organic matter. They are easy for roots to penetrate and have better fertility and aeration than the surrounding soil matrix.

	Numbers of roots counted per 1 m² of the soil profile		
Root size, diameter	Soil matrix	Old root channels	Comparative increase in root density in the old channels, %
Fine roots, <4 mm	211	3617	94
Medium roots, 4–20 mm	20	361	95
Coarse roots, >20 mm	3	155	98

Calculated from Parker and Van Lear (1996).

space. Aeration, especially in the subsoil, may be inadequate for satisfactory root development and desirable microbial activity. While the larger micropores accommodate plant root hairs and microorganisms, the smaller micropores (sometimes termed *ultramicropores* and *cryptopores*) are too small to allow the entry of even the smallest bacteria. They therefore can serve as protected sites in which organic compounds may remain untouched for centuries (see Chapter 11).

In recent years, conservation tillage practices, which minimize plowing and associated soil manipulations, have been widely adopted in the United States (see Sections 4.8 and 15.5). Because of increased accumulation of organic matter near the soil surface and the development of a long-lived network of macropores (especially biopores), some conservation tillage systems lead to greater macroporosity of the surface layers. These benefits are particularly likely to accrue in soils with extensive production of earthworm burrows, which may remain undisturbed in the absence of tillage. Unfortunately, such improvements in porosity do not always occur. Under some circumstances, less pore space has been found with conservation tillage than with conventional tillage, a matter of some concern in soils with poor internal drainage.

4.7   FORMATION AND STABILIZATION OF SOIL AGGREGATES

The formation and maintenance of a high degree of aggregation is one of the most difficult tasks of soil management, yet it is also one of the most important, since it is a potent means of influencing ecosystem function. The organization of surface soils into relatively large structural aggregates provides for the low bulk density and a high proportion of macropores so desirable for most soil uses.

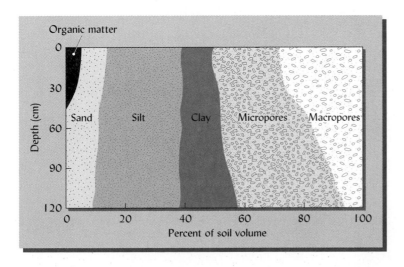

FIGURE 4.21 Volume distribution of organic matter, sand, silt, clay, and pores of macro- and microsizes in a representative silt loam with good soil structure. Note that at the lowest depths, about one-third of the mineral matter is clay, giving the lower horizons enough clay to be classified as a silty clay loam.

Some aggregates readily succumb to the beating of rain and the rough and tumble of plowing and tilling the land. Others resist disintegration, thus making the maintenance of a suitable soil structure comparatively easy (see Figures 4.26 and 4.27, pages 119 and 121). Generally, the smaller aggregates are more stable than the larger ones, so maintaining the much-prized larger aggregates requires much care.

We will discuss practical means of managing soil structure after we consider those factors responsible for aggregate formation and those that give the aggregates stability once they are formed. Since both sets of factors are operating simultaneously, it is sometimes difficult to distinguish their relative effects on the development of stable aggregates in soils.

Hierarchical Organization of Soil Aggregates[4]

The large aggregates (>1 mm) so desirable for most soil uses are typically composed of smaller aggregates, which in turn are composed of still smaller units, down to clusters of clay and humus less than 0.001 mm in size. You may easily demonstrate the existence of this *hierarchy of aggregation* by selecting a few of the largest aggregates in a soil and gently crushing or picking them apart to separate them into many smaller-sized aggregates. Then try rubbing the tiniest of these granules between your thumb and forefinger. You will find that most of these break down into a smear of still smaller aggregates composed of silt, clay, and humus. The hierarchical organization of aggregates (Figure 4.22) seems to be characteristic of most soils, with the exception of certain Oxisols and some very young Entisols. At each level in the hierarchy of aggregates, different factors are responsible for binding together the subunits.

Factors Influencing Aggregate Formation and Stability in Soils

Both biological and physical-chemical (abiotic) processes are involved in the formation of soil aggregates. The physical-chemical processes tend to be most important at the smaller end of the scale, biological processes at the larger end. Also, the physical-chemical processes of aggregate formation are mainly associated with clays and, hence, tend to be of greater importance in finer-textured soils. In sandy soils that have little clay, aggregation is almost entirely dependent on biological processes.

Physical-Chemical Processes

Most important among the physical-chemical processes are (1) the mutual attraction among clay particles and (2) the swelling and shrinking of clay masses.

FLOCCULATION OF CLAYS AND THE ROLE OF ADSORBED CATIONS.   Except in very sandy soils that are almost devoid of clay, aggregation begins with the **flocculation** of clay particles into microscopic clumps or *floccules* (Figure 4.23). If two clay platelets come close enough to each other, the cations compressed in a layer between them will attract the negative charges on both platelets, thus serving as bridges to hold the platelets together. This process is repeated until a small "stack" of parallel clay platelets, termed a *clay domain,* is formed (Figure 4.22d). Other types of clay domains are more random in orientation, resembling a house of cards. These form when the positive charges on the edges of the clay platelets attract the negative charges on the planar surfaces (Figure 4.23). Clay floccules or domains, along with charged organic colloids (humus), provide much of the long-term stability for the smaller (< 0.03 mm) **microaggregates.** The cementing action of inorganic compounds, such as iron oxides, produces very stable small aggregates sometimes called **pseudosand** in certain clayey soils (Ultisols and Oxisols) of hot, humid regions.

When Na^+ (rather than polyvalent cations such as Ca^{2+} or Al^{3+}) is a prominent adsorbed ion, as in some soils of arid and semiarid areas, the attractive forces are not able to overcome the natural repulsion of one negatively charged clay by another. The clay platelets cannot approach closely enough to flocculate, so remain dispersed as far apart from one another as possible. Clay in this dispersed, gel-like condition causes the

---

[4] The role of organic matter and biological processes in the hierarchical organization of soil aggregates was originally put forward by Tisdall and Oades (1982) and elaborated on by Oades (1993) and Tisdall (1994).

Microaggregate
Root — Fungal hyphae

3.0 mm

Macroaggregate
• Roots
• Hyphae

(a)

Fungal hyphae — Spore
Root

0.3 mm

Microaggregate
• Root hairs
• Hyphae
• Polysaccharides

(b)

Silt encrusted with microbial debris | Plant and microbial debris encrusted with clay

0.03 mm

Submicroaggregate
• Mineral grains encrusted with plant and microbial debris
• Plant debris coated with clay

(c)

Clay-humus domains — Microbial debris
Silt
Iron oxides — Water retention pores
Silt

0.003 mm

Primary particles
of silt, clay and humus
• Clay and clay-humus domains

(d)

FIGURE 4.22 Larger aggregates are often composed of an agglomeration of smaller aggregates. This illustration shows four levels in this hierarchy of soil aggregates. The different factors important for aggregation at each level are indicated. (*a*) A *macroaggregate* composed of many microaggregates bound together mainly by a kind of sticky network formed from fungal hyphae and fine roots. (*b*) A *microaggregate* consisting mainly of fine sand grains and smaller clumps of silt grains, clay, and organic debris bound together by root hairs, fungal hyphae, and microbial gums. (*c*) A very small submicroaggregate consisting of fine silt particles encrusted with organic debris and tiny bits of plant and microbial debris (called *particulate organic matter*) encrusted with even smaller packets of clay, humus, and Fe or Al oxides. (*d*) Clusters of parallel and random clay platelets interacting with Fe or Al oxides and organic polymers at the smallest scale. These organoclay clusters or *domains* bind to the surfaces of humus particles and the smallest of mineral grains.

soil to become almost structureless, impervious to water and air, and very undesirable from the standpoint of plant growth (See Section 9.16).

VOLUME CHANGES IN CLAYEY MATERIALS. As a soil dries out and water is withdrawn, the platelets in clay domains move closer together, causing the domains and, hence, the soil mass to shrink in volume. As a soil mass shrinks, cracks will open up along planes of

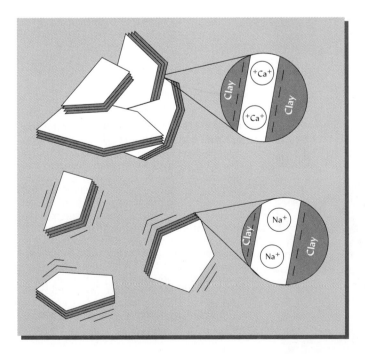

FIGURE 4.23 The role of cations in the flocculation of soil clays. The di- and trivalent cations, such as Ca^{2+} and Al^{3-}, are tightly adsorbed and can effectively neutralize the negative surface charge on clay particles. These cations can also form bridges that bring clay particles close together. Monovalent ions, especially Na^+, with relatively large hydrated radii, can cause clay particles to repel each other and create a dispersed condition. Two things contribute to the dispersion: (1) the large hydrated sodium ion does not get close enough to the clay to effectively neutralize the negative charges, and (2) the single charge on sodium is not effective in forming a bridge between clay particles.

weakness. Over the course of many cycles (as occur between rain or irrigation events in the field) the network of cracks becomes more extensive and the aggregates between the cracks better defined. Water uptake, especially by fibrous-rooted perennial grasses, accentuates the physical aggregation processes associated with wetting and drying. This effect is but one of many examples of ways in which physical and biological soil processes interact.

Freezing and thawing cycles have a similar effect, since the formation of ice crystals is a drying process that also draws water out of clay domains.

Biological Processes

ACTIVITIES OF SOIL ORGANISMS.   Among the biological processes of aggregation, the most prominent are: (1) the burrowing and molding activities of earthworms, (2) the enmeshment of particles by sticky networks of roots and fungal hyphae, and (3) the production of organic glues by microorganisms, especially bacteria and fungi.

Plant roots (particularly root hairs) and fungal hyphae exude sugarlike polysaccharrides and other organic compounds, forming sticky networks that bind together individual soil particles and tiny microaggregates into larger agglomerations called **macroaggregates** (see Figure 4.22*a*). The threadlike fungi that associate with plant roots (called *mycorrhizae;* see Section 10.8) are especially effective in providing this type of relatively short-term stabilization of large aggregates, because they secrete a gooey protein called **glomalin**, which is very effective as a cementing agent (Figure 4.24).

(a)

(b)

Hypha

Spore

Fine root

(c)

FIGURE 4.24 Fungal hyphae binding soil particles into aggregates. (*a*) Close-up of a hyphae growing over the surface of a mineral grain encrusted with microbial cells and debris. Bar = 10 μm. (*b*) An advanced stage of aggregation during the formation of soil from dune sands. Note the net of fungal hyphae and the encrustation of the mineral grains with organic debris. Bar = 50 μm. (*c*) Hyphae of the root-associated fungus from the genus *Gigaspora* interconnecting particles in a sandy loam from Oregon. Note also the fungal spore and the plant root. Bar = 320 μm. [Photos (*a*) and (*b*) courtesy of Sharon L. Rose, Willamette University; photo (*c*) courtesy of R. P. Schreiner, USDA-ARS, Corvallis, Ore.]

Bacteria also produce polysaccharides and other organic glues as they decompose plant residues. Bacterial polysaccharides are shown intermixed with clay domains at a very small scale in Figure 4.25.

INFLUENCE OF ORGANIC MATTER. In most soils, organic matter is *the* major agent stimulating the formation and stabilization of granular and crumb-type aggregates (see Figure 4.26). In the first place, organic matter provides the energy substrate that makes possible the previously mentioned activities of the fungi, bacteria, and soil animals. Second, as organic residues decompose, gels and other viscous microbial products, along with associated bacteria and fungi, encourage crumb formation. Organic exudates from plant roots also participate in this aggregating action.

INFLUENCE OF TILLAGE. Tillage can have both favorable and unfavorable effects on aggregation. If the soil is not too wet or too dry when the tillage is performed, the short-term effect of tillage is generally favorable. Tillage implements break up large clods, incorporate organic matter into the soil, kill weeds, and generally create a more favorable seedbed (See Box 4.3). Immediately after plowing, the surface soil is loosened (its cohesive strength is decreased) and total porosity is increased.

FIGURE 4.25 An ultrathin section illustrating the interaction among organic materials and silicate clays in a water-stable aggregate. The dark-colored materials (C) are groups of clay particles that are interacting with organic polysaccharides (P). A bacterial cell (B) is also surrounded by polysaccharides. Note the generally horizontal orientation of the clay particles, an orientation encouraged by the organic materials. [From Emerson, et al. (1986); photograph provided by R. C. Foster, CSIRO, Glen Osmond, Australia])

FIGURE 4.26 The aggregates of soils high in organic matter are much more stable than are those low in this constituent. The low-organic-matter soil aggregates fall apart when they are wetted; those high in organic matter maintain their stability.

BOX 4.3   PREPARING A GOOD SEEDBED

Early in the growing season, one of the main activities of a farmer or gardener is the preparation of a good seedbed to assure that the sowing operation goes smoothly, and that the plants come up quickly, evenly, and well spaced.

A good seedbed consists of soil loose enough to allow easy root elongation and seedling emergence. [see photo (a)]. At the same time, a seedbed should be packed firmly enough to ensure good contact between the seed and moist soil so that the seed can easily imbibe water to begin the germination process. The seedbed should also be relatively free of large clods. Seeds could fall between such clods and become lodged too deeply for proper emergence and without sufficient soil contact.

Tillage may be needed to loosen compacted soil, help control weeds, and, in cool climates, help the soil dry out so it will warm more rapidly. On the other hand, the objectives of seedbed preparation may be achieved with little or no tillage if soil and climatic conditions are favorable and a mulch or herbicide is used to control early weeds.

Mechanical planters can assist in maintaining a good seedbed. Most planters are equipped with coulters (sharp steel disks) designed to cut a path through plant residues on the soil surface. No-till planters usually follow the coulter with a pair of sharp cutting wheels called a **double disk opener** that opens a groove in the soil into which the seeds can be dropped [see photos (b) and (c)].

(a) A bean seed emerges from a seedbed. (Photo courtesy R. Weil)

Most planters also have a press wheel that follows behind the seed dropper and packs the loosened soil just enough to ensure that the groove is closed and the seed is pressed into contact with moist soil.

Ideally, only a narrow strip in the seed row is packed down to create a seed *germination* zone, while the soil between the crop rows is left as loose as possible to provide a *good rooting* zone. The surface of the interrow rooting zone may be left in a rough condition to encourage water infiltration and discourage erosion. The above principles also apply to the home gardener who may be sowing seeds by hand.

(b) A no-till planter in action and (c) a diagram showing how it works. [Photo (b) and diagram (c) courtesy of Deere & Company, Moline, Ill.]

FIGURE 4.27 Overly tilled, puddled soil (left) and well granulated untilled, grassland soil (right). Excess tillage depleted the soil organic matter and weakened the structure. Plant roots and especially humus play the major role in soil granulation. Thus a sod tends to encourage development of a granular structure in the surface horizon of cultivated land. (Courtesy USDA Natural Resources Conservation Service)

Over longer periods, tillage greatly hastens the oxidation of soil organic matter, thus reducing the aggregating effects of this soil component. Tillage operations, especially if carried out when the soil is wet, also tend to crush or smear stable soil aggregates, resulting in loss of macroporosity and the creation of a *puddled* condition (Figure 4.27).

4.8 TILLAGE AND STRUCTURAL MANAGEMENT OF SOILS

When protected under dense vegetation and undisturbed by tillage, most soils (except perhaps some sparsely vegetated soils in arid regions) possess a surface structure sufficiently stable to allow rapid infiltration of water and to prevent crusting. However, for the manager of cultivated soils, the development and maintenance of stable surface soil structure is a major challenge. Many studies have shown that aggregation and associated desirable soil properties such as water infiltration rate decline under long periods of tilled row-crop cultivation (Table 4.4).

Tillage and Soil Tilth

Simply defined, **tilth** refers to the physical condition of the soil in relation to plant growth. Tilth depends not only on aggregate formation and stability, but also on such factors as bulk density, soil moisture content, degree of aeration, rate of water infiltration, drainage, and capillary water capacity. As might be expected, tilth often changes

TABLE 4.4 Effect of Period of Corn Cultivation on Soil Organic Matter, Aggregate Stability, and Water Infiltration[a] in Five Silt Loam Inceptisols from Southwest France

In soils with depleted organic-matter levels, aggregates easily broke down under the influence of water, forming smaller aggregates and dispersed materials that sealed the soil surface and inhibited infiltration. A level of 3% soil organic matter seems to be sufficient for good structural stability in these temperate region silt loam soils.

			Infiltration	
Period of cultivation, years	Organic matter, %	Aggregate stability, mm MWD[b]	Of total rain, %	Prior to ponding, mm
100	0.7	.35	25	6
47	1.6	.61	34	9
32	2.6	.76	38	15
27	3.1	1.38	47	25
15	4.2	1.52	44	23

[a] Infiltration is the amount of water that entered the soil when 64 mm of "rain" was applied during a 2-hour period.
[b] MWD = mean weighted diameter or average size of the aggregates that remained intact after sieving under water. [Data from Le Bissonnais and Arrouays (1997)]

rapidly and markedly. For instance, the workability of fine-textured soils may be altered abruptly by a slight change in moisture.

Clayey soils are especially prone to puddling and compaction because of their high plasticity and cohesion. When puddled clayey soils dry, they usually become dense and hard. Proper timing of trafficking is more difficult for clayey than for sandy soils, because the former take much longer to dry to a suitable moisture content and may also become too dry to work easily.

Some clayey soils of humid tropical regions are much more easily managed than those just described. The clay fraction of these soils is dominated by hydrous oxides of iron and aluminum, which are not as sticky, plastic, and difficult to work. These soils may have very favorable physical properties, since they hold large amounts of water but have such stable aggregates that they respond to tillage after rainfall much like sandy soils.

Farmers in temperate regions typically find their soils too wet for tillage just prior to planting time (early spring), while farmers in tropical regions may face the opposite problem of soils too dry for easy tillage just prior to planting (end of dry season). In tropical and subtropical regions with a long dry season, soil often must be tilled in a very dry state in order to prepare the land for planting with the onset of the first rains. Tillage under such dry conditions can be very difficult and can result in hard clods if the soils contain much sticky-type silicate clay.

Conventional Tillage and Crop Production

Since the Middle Ages, the moldboard plow has been the primary tillage implement most used in the Western world.[5] Its purpose is to lift, twist, and invert the soil while incorporating crop residues and animal wastes into the plow layer (Figure 4.28). The moldboard plow is often supplemented by the disk plow, which is used to cut up residues and partially incorporate them into the soil. In conventional practice, such primary tillage has been followed by a number of secondary tillage operations, such as harrowing to kill weeds and to break up clods, thereby preparing a suitable seedbed. After the crop is planted, the soil may receive further secondary tillage to control weeds and to break up crusting of the immediate soil surface.

Conservation Tillage and Soil Tilth

In recent years, land-management systems have been developed that minimize the need for soil tillage. No-till is a system by which special equipment plants seed into

---

[5] For an early but still valuable critique of the moldboard plow, see Faulkner (1943).

FIGURE 4.28 While the action of the moldboard plow lifts, turns, and loosens the upper 15 to 20 cm of soil (the furrow slice), the counterbalancing downward force compacts the next lower layer of soil. This compacted zone can develop into a *plowpan*. Compactive action can be understood by imagining that you are lifting a heavy weight—as you lift the weight your feet press down on the floor below. (Photo courtesy of R. Weil)

residue-covered soil with virtually no soil disturbance (see Box 4.3). Other minimum-tillage systems permit some stirring of the soil, but still leave a high proportion of the crop residues on the surface. These organic residues protect the soil from the beating action of raindrops and the abrasive action of the wind, thereby reducing water and wind erosion and maintaining soil structure (see Section 15.5). For this reason, the tillage practices followed in these systems are called *conservation tillage*.[6]

Soil Crusting

Falling drops of water during heavy rains or sprinkler irrigation beat apart the aggregates exposed at the soil surface. In some soils the dilution of salts by this water stimulates the dispersion of clays. Once the aggregates become dispersed, small particles and dispersed clay tend to wash into and clog the soil pores. Soon the soil surface is covered with a thin layer of fine, structureless material called a **surface seal.** The surface seal inhibits water infiltration and increases erosion losses.

As the surface seal dries, it forms a hard **crust.** Seedlings, if they emerge at all, can do so only through cracks in the crust. A crust-forming soil is compared to one with stable aggregates in Figure 4.29. Formation of a crust soon after a crop is sown may allow so few seeds to emerge that the crop has to be replanted. In arid and semiarid regions, soil sealing and crusting can have disastrous consequences because high runoff losses leave little water available to support plant growth.

Crusting can be minimized by keeping some vegetative or mulch cover on the land to reduce the impact of raindrops. Improved management of soil organic matter and use of certain soil amendments can "condition" the soil and help prevent clay dispersion and crust formation (see also Section 9.20).

Soil Conditioners

GYPSUM.   Gypsum (calcium sulfate) can improve the physical condition of many types of soils, especially high-sodium soils of semiarid regions (see Chapter 9). The more sol-

---

[6] The U.S. Department of Agriculture defines *conservation tillage* as any system that leaves at least 30% of the soil surface covered by residues.

(c)

FIGURE 4.29 Scanning electron micrographs of the upper 1 mm of a soil with stable aggregation (a) compared to one with unstable aggregates (b). Note that the aggregates in the immediate surface have been destroyed and a surface crust has formed. The bean seedling (c) must break the soil crust as it emerges from the seedbed. [Photos (a) and (b) from O'Nofiok and Singer (1984), used with permission of Soil Science Society of America; photo (c) courtesy of R. Weil]

(a)   (b)

uble gypsum products provide enough electroytes (cations and anions) to promote flocculation and inhibit the dispersion of aggregates, thus preventing surface crusting. Field trials have shown that gypsum-treated soils permit greater water infiltration and are less subject to erosion than untreated soils. Similarly, gypsum can reduce the strength of hard subsurface layers, thereby allowing greater root penetration and subsequent plant uptake of water from the subsoil.

ORGANIC POLYMERS.   Certain synthetic organic polymers can stabilize soil structure in much the same way as do natural organic polymers such as polysaccharides. Figure 4.30 shows the dramatic stabilizing effect of synthetic polyacrylamides (PAM) used in irrigation water. A number of research reports indicate that the best results can be obtained by combining the use of PAM and gypsum products.

Several species of algae that live near the soil surface are known to produce quite effective aggregate-stabilizing compounds. Application of small quantities of commercial preparations containing such algae may bring about a significant improvement in surface soil structure.

Various humic materials are marketed for their soil conditioning effects when incorporated at low rates (< 500 kg/ha). However, carefully conducted research at many universities has failed to show that these materials have significantly affected aggregate stability or crop yield, as claimed.

General Guidelines for Managing Soil Tilth

Although each soil presents unique problems and opportunities, the following principles are generally relevant to managing soil tilth:

1. Minimizing tillage, especially moldboard plowing, disk harrowing or rototilling, reduces the loss of aggregate-stabilizing organic matter.

2. Timing traffic activities to occur when the soil is as dry as possible and restricting tillage to periods of optimum soil moisture conditions will minimize destruction of soil structure.

3. Mulching the soil surface with crop residues or plant litter adds organic matter, encourages earthworm activity, and protects aggregates from beating rain.

4. Adding crop residues, composts, and animal manures to the soil is effective in stimulating microbial supply of the decomposition products that help stabilize soil aggregates.

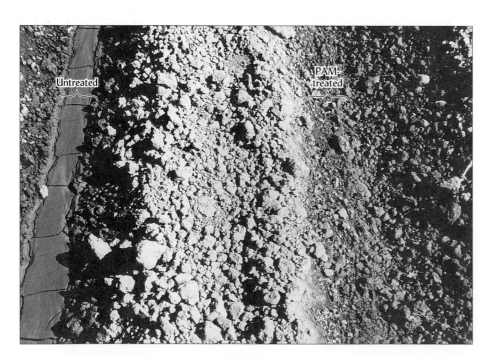

FIGURE 4.30 The remarkable stabilizing effect of a synthetic polyacrylamide is seen in the furrow on the right compared to the left, untreated, furrow. Irrigation water broke down much of the structure of the untreated soil but had no effect on the treated row. [From Mitchell (1986)]

5. Including sod crops in the rotation favors stable aggregation by helping to maintain soil organic matter, providing maximal aggregating influence of fine plant roots, and assuring a period without tillage.

6. Using cover crops and green manure crops, where practical, provides another good source of root action and organic matter for structural management.

7. Applying gypsum (or calcareous limestone if the soil is acidic) by itself or in combination with synthetic polymers can be very useful in stabilizing surface aggregates, especially in irrigated soils.

4.9 SOIL PROPERTIES RELEVANT TO ENGINEERING USES

Field Rating of Soil Consistence and Consistency

CONSISTENCE. Soil **consistence** is a term used by soil scientists to describe the resistance of a soil to mechanical stresses or manipulations at various moisture contents. Soils are rated for consistence as part of describing a soil profile and for estimating suitability for traffic and tillage. This property is a composite expression of those forces of mutual attraction among soil particles, and between particles and pore water, that determine the ease with which a soil can be reshaped or ruptured.

Moisture content greatly influences how a soil responds to stress; hence, moist and dry soils are given separate consistence ratings (Table 4.5). A dry clayey soil that cannot be crushed between the thumb and forefinger but can be crushed easily underfoot would be designated as *hard*. If a clod of moist soil crumbles into aggregates when crushed with only light pressure, it is said to be **friable.** Friable soils are easily tilled or excavated.

The degrees of *stickiness* and *plasticity* (malleability) of soil in the wet condition are often included in describing soil consistence (although not shown in Table 4.5).

CONSISTENCY. The term **consistency** is used in a similar manner by soil engineers to describe the degree to which a soil resists deformation when a force is applied. However, consistency is determined by the soil's resistance to *penetration* by an object, while the soil scientist's consistence describes resistence to *rupture*. Instead of crushing a clod of

TABLE 4.5 Some Field Tests and Terms Used to Describe the Consistence and Consistency of Soils

The consistency of cohesive materials is closely related to, but not exactly the same as, their consistence. Conditions of least coherence are represented by terms at the top of each column, those of greater coherence near the bottom.

Soil consistence[a]				Soil consistency[b]	
Dry soil	Moist to wet soil	Soil dried then submerged in water	Field rupture (crushing) test	Soil at in situ moisture	Field penetration test
Loose	Loose	Not applicable	Specimen not obtainable	Soft	Blunt end of pencil penetrates deeply with ease
Soft	Very friable	Noncemented	Crumbles under very slight force between thumb and forefinger	Medium firm	Blunt end of pencil can penetrate about 1.25 cm with moderate effort
Slightly hard	Friable	Extremely weakly cemented	Crumbles under slight force between thumb and forefinger	Firm	Blunt end of pencil can penetrate about 0.5 cm
Hard	Firm	Weakly cemented	Crushes with difficulty between thumb and forefinger	Very firm	Blunt end of pencil makes slight indentation; thumbnail easily penetrates
Very hard	Extremely firm	Moderately cemented	Cannot be crushed between thumb and forefinger, but can be crushed slowly underfoot	Hard	Blunt end of pencil makes no indentation; thumbnail barely penetrates
Extremely hard	Slightly rigid	Strongly cemented	Cannot be crushed by full body weight underfoot		

[a] Abstracted from Soil Survey Division Staff (1993).
[b] Modified from McCarthy (1993).

soil, the engineer attempts to penetrate it with either the blunt end of a pencil (some use their thumb) or a thumbnail. For example, if the blunt end of a pencil makes only a slight indentation, but the thumbnail penetrates easily, the soil is rated as *very firm* (Table 4.5).

Field observations of both consistence and consistency provide valuable information to guide decisions about loading and manipulating soils. For construction purposes, however, soil engineers usually must make more precise measurements of a number of related soil properties that help predict how a soil will respond to applied stress.

Soil Strength and Sudden Failure

Perhaps the most important property of a soil for engineering uses is its **strength.** This is a measure of the capacity of a soil mass to withstand stresses without giving way to those stresses by rupturing or becoming deformed. The failure of a soil to withstand stress can be seen where a structure topples as its weight exceeds the soil's bearing strength, where earthen dams give way under the pressure of impounded water, or where pavements and structures slide down unstable hillsides (Figure 1.9).

The strength of **cohesive soils** (essentially soils with a clay content of more than about 15%) declines dramatically if the material is very wet and the pores are nearly filled with water. The water forces the clay particles apart, making the soil prone to failure, often with catastrophic results such as the mudslide shown in Figure 4.31.

NONCOHESIVE SOILS.   The strength of dry, noncohesive soil materials such as sand depends entirely on frictional forces, including the interlocking of rough particle surfaces. One reflection of the strength of a noncohesive material is its **angle of repose,** the steepest angle to which it can be piled without slumping. Smooth rounded sands grains cannot be piled as steeply as can rough, interlocking sands. If a small amount of water bridges the gaps between particles, electrostatic attraction of the water for the mineral surfaces will increase the soil strength. Interparticle water bridges explain why cars can drive along the edge of the beach where the sand is moist, but their tires sink in and lose traction on loose dry sand or in saturated quicksand (see Figure 5.20 for an example of the latter).

COLLAPSIBLE SOILS.   Certain soils that exhibit considerable strength at low in situ water contents lose their strength suddenly if they become wet. Such soils may collapse without warning under a roadway or building foundation. A special case of soil collapse is **thixotropy,** the sudden liquification of a wet soil mass subjected to vibrations, such as those accompanying earthquakes and blasting.

FIGURE 4.31 Houses damaged by a mudslide that occurred when the soils of a steep hillside in Oregon became saturated with water after a period of heavy rains. The weight of the wet soil exceeded its shear strength, causing the slope to fail. Excavations for roads and houses near the foot of a slope can contribute to the lack of slope stability, as can removal of tree roots by large-scale clear-cutting on the slope itself. (Photo courtesy of John Griffith, Coas Bay, Ore.)

Settlement—Gradual Compression

While embankments and hill slopes commonly fail due to stresses that exceed the soil's strength, most buildings and roads are unlikely to provide loads that cause the soil to rupture. Instead, most foundation problems result from slow, often uneven, vertical subsidence or **settlement** of the soil.

COMPACTION CONTROL. For the purposes of growing plants, soil compaction is to be avoided; however, soils are purposely compacted prior to being used as a roadbed or for a building foundation. Compaction occurring later under the heavy load would result in uneven settlement and a cracked pavement or foundation. Compaction to an optimum density is usually achieved on clayey soils by heavy sheepsfoot rollers that knead the soil like a baker kneads bread dough (Figure 4.32). For sandy soils, vibrating rollers or impact hammers do a better job because they can jar the particles into a tight packing arrangement (see Figure 4.15). The **Proctor test** is the most common method used to determine how wet a soil should be to allow optimal compaction.

COMPRESSIBILITY. A **consolidation test** may be conducted on a soil specimen to determine its **compressibility**—how much its volume will be reduced by a given applied force. Because of the relatively low porosity and equidimensional shape of the individual mineral grains, very sandy soils resist compression once the particles have settled into a tight packing arrangement. They make excellent soils to bear foundations. The high porosity of clay floccules and the flakelike shape of clay particles gives clayey soils much greater compressibility. Soils consisting mainly of organic matter (peats) have the highest compressibilities and generally are unsuitable for foundations.

Expansive Soils

Damage caused by expansive soils in the United States rarely makes the evening news programs, although the total cost annually exceeds that caused by tornados, floods, earthquakes, or any other type of natural disaster. Expansive clays occur on about 20% of the land area in the United States and cause upwards of $4 billion in damages annually to pavements, foundations, and utility lines. The damages can be severe in certain sites in all parts of the country, but are most extensive in regions that have long dry periods alternating with periods of rain (e.g., California, Texas, Wyoming, and Colorado).

Some clays, particularly the smectites, swell when wet and shrink when dry (see Section 8.4). Expansive soils are rich in these types of clay (Figure 3.14, p. 77).

FIGURE 4.32 Compaction of soils used as foundations and roadbeds is accomplished by heavy equipment such as this sheepsfoot roller. The knobs ("sheepsfeet") concentrate the mass on a small impact area, punching and kneading the loose, freshly graded soil to optimum density. (Photo courtesy of R. Weil)

As a dry, clayey soil takes on increasing amounts of water, it undergoes dramatic and distinct changes in behavior and consistency. A hard, rigid solid in the dry state, it becomes a crumbly (friable) semisolid when a certain moisture content (termed the **shrinkage limit**) is reached. If it contains expansive clays, the soil also begins to swell in volume as this moisture content is exceeded. Increasing the water content beyond the **plastic limit** will transform the soil into a malleable, plastic mass and cause additional swelling. The soil will remain in this plastic state until its **liquid limit** is exceeded, causing it to transform into a viscous liquid that will flow when jarred. These critical water contents (measured in units of percent) are termed the **Atterberg limits**, after the German engineer who developed the system. Soil engineers determine the Atterberg limits as part of their investigation of soil consistency, to help predict the behavior of particular soils and the suitability of these materials for different construction purposes.

PLASTICITY INDEX.   The plasticity index (PI) is the difference between the plastic limit (PL) and liquid limit (LL) and indicates the water-content range over which the soil has plastic properties:

$$PI = LL - PL$$

Soils with a high plasticity index (greater than about 25) are usually expansive clays that make poor roadbeds or foundations. Figure 4.33 shows the relationship among the Atterberg limits and the changes in soil volume associated with increasing water contents for a hypothetical soil. Smectite clays (see Section 8.4) generally have high liquid limits and plasticity indices, especially if saturated with sodium. Kaolinite and other nonexpansive clays have low liquid limit values.

COEFFICIENT OF LINEAR EXTENSIBILITY.   The expansiveness of a soil (and therefore the hazard of its destroying foundations and pavements) can be quantified as the *coefficient of linear extensibility* (COLE). Suppose a sample of soil is moistened to its plastic limit and molded into the shape of a bar with length LM. If the bar of soil is allowed to air dry, it will shrink to length LD. The COLE is the percent reduction in length of the soil bar upon shrinking:

$$COLE = \frac{LM - LD}{LM} \times 100$$

Figure 4.33 indicates how the volume change used to calculate the COLE relates to the Atterberg limits.

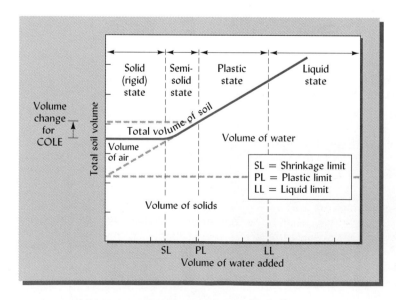

FIGURE 4.33   A common depiction of The Atterberg limits, which mark major shifts in the behavior of a cohesive soil as its water content changes (from left to right). As water is added to a certain volume of dry solids, first air is displaced; then, as more water is added, the total volume of the soil increases (if the soil has some expansive properties). When the shrinkage limit (SL) is reached, the once rigid, hard solid becomes a crumbly, semisolid. With more water, the plastic limit (PL) is reached, after which the soil becomes plastic and can be molded. It remains in a plastic stage over a range of water contents until the liquid limit (LL) is exceeded, at which point the soil begins to behave as a viscous liquid that will flow when jarred. The volume change for calculating the coefficient of linear extensibility (COLE) is shown at the left.

Unified Classification System for Soil Materials

The U.S. Army Corps of Engineers and the U.S. Bureau of Reclamation established the Unified System of Classification in order to aid in predicting the engineering behavior of different soil materials. Each type of soil material is given a two-letter designation based primarily on its particle-size distribution (texture), Atterberg limits, and organic-matter content.

The system first groups soils into coarse-grained soils and fine-grained soils. The coarse materials are further divided on the basis of grain size (gravels and sands), amount of fines present, and uniformity of grain size (well or poorly graded). The fine-grained soils are divided into silts, clays, and organic materials. These classes are further subdivided on the basis of their liquid limit (above or below 50) and their plasticity index.

This classification of soil materials helps engineers predict the soil strength, expansiveness, compressibility, and other properties so that appropriate engineering designs can be made for the soil at hand.

4.10 CONCLUSION

Physical properties exert a marked influence on the behavior of soils with regard to plant growth, hydrology, environmental management, and engineering uses. The nature and properties of the individual particles, their size distribution, and their arrangement in soils determine the total volume of nonsolid pore space, as well as the pore sizes, thereby impacting on water and air relationships.

The properties of individual particles and their proportionate distribution (soil texture) are subject to little human control in field soils. However, it is possible to exert some control over the arrangement of these particles into aggregates (soil structure) and on the stability of these aggregates. Tillage and traffic must be carefully controlled to avoid undue damage to soil tilth, especially when soils are rather wet. Generally, nature takes good care of soil structure, and humans can learn much about soil management by studying natural systems. Vigorous and diverse plant growth, generous return of organic residues, and minimal physical disturbance are attributes of natural systems worthy of emulation. Proper plant species selection, crop rotation, and management of chemical, physical, and biological factors can help assure maintenance of soil physical quality. In recent years, these management goals have been made more practical by the advent of conservation tillage systems that minimize soil manipulations while decreasing soil erosion and water runoff.

Particle size, moisture content, and plasticity of the colloidal fraction all help determine the stability of soil in response to loading forces from traffic, tillage, or building foundations. The physical properties presented in this chapter greatly influence nearly all other soil properties and uses, as discussed throughout this book.

STUDY QUESTIONS

1. If you were investigating a site for a proposed housing development, how could you use soil colors to help predict where problems might be encountered?

2. You are considering the purchase of some farmland in a region with variable soil textures. The soils on one farm are mostly of sandy loams and loamy sands, while those on a second farm are mostly clay loams and clays. List the potential advantages and disadvantages of each farm as suggested by the texture of its soils.

3. Revisit your answer to question 2. Explain how soil structure in both the surface and subsurface horizons might modify your opinion of the merits of each farm.

4. Two different timber-harvest methods are being tested on adjacent forest plots with clay loam surface soils. Initially, the bulk density of the surface soil in both plots was 1.1 Mg/m^3. One year after the harvest operations, plot *A* soil had a bulk density of 1.48 Mg/m^3, while that in plot *B* was 1.29 Mg/m^3. Interpret these values with regard to the relative merits of systems *A* and *B*, and the likely effects on the soil's function in the forest ecosystem.

5. What are the textural classes of two soils, the first with 15% clay and 45% silt, and the second with 80% sand and 10% clay? (Hint: Use Figure 4.7)

6. For the forest plot *B* in question 4, what was the change in percent pore space of the surface soil caused by timber harvest? Would you expect that most of this change was in the micropores or in the macropores? Explain.

7. Discuss the positive and negative impacts of tillage on soil structure. What is another physical consideration would you have to take into account in deciding whether or not to change from a conventional to a conservation tillage system?

8. What would you, as a home gardener, consider to be the three best and three worst things that you could do with regard to managing the soil structure in your home garden?

9. What does the plasticity index tell an engineer about a soil, and why would this information be important?

10. In a humid region characterized by expansive soils, a homeowner experienced burst water pipes, doors that no longer closed properly, and large vertical cracks in the brick walls. The house had had no problems for over 20 years, and a consulting soil scientist blamed the problems on a large tree that was planted near the house some 10 years before the problems began to occur. Explain.

REFERENCES

Camp, C. R., and J. F. Lund. 1964. "Effects of soil compaction on cotton roots," *Crops and Soils,* **17**:13–14.

Emerson, W. W., R. C. Foster, and J. M. Oades. 1986. "Organomineral complexes in relation to soil aggregation and structure," in P. M. Huang and M. Schnitzer (eds.), *Interaction of Soil Minerals with Natural Organics and Microbes.* SSSA Special Publication no. 17. (Madison, Wis.: Soil Sci. Soc. Amer.).

Faulkner, E. H. 1943. *Plowman's Folly* (Norman, Okla.: University of Oklahoma Press).

Gent, J. A., Jr., and L.A. Morris. 1986. "Soil compaction from harvesting and site preparation in the Upper Gulf Coastal Plain," *Soil Sci. Soc. Amer. J.,* **50**:443–446.

Gent, J. A., Jr., R. Ballard, A. E. Hassan, and D. K. Cassel. 1984. "Impact of harvesting and site preparation on physical properties of Piedmont forest soils," *Soil Sci. Soc. Amer. J.,* **48**:173–177.

Le Bissonnais, Y., and D. Arrouays. 1997. "Aggregate stability and assessment of soil crustability and erodibility: II. Application to humic loamy soils with various organic carbon contents," *European J. Soil Sci.,* **48**:39–48.

McCarthy, D. F. 1993. *Essentials of Soil Mechanics and Foundations,* 4th ed. (Englewood Cliffs, NJ: Prentice Hall).

Mitchell, A. R. 1986. "Polyacrylamide application in irrigation water to increase infiltration," *Soil Sci.,* **141**:353–358.

Oades, J. M. 1993. "The role of biology in the formation, stabilization, and degradation of soil structure," *Geoderma,* **56**:377–400.

O'Nofiok, O., and M. J. Singer. 1984. "Scanning electron microscope studies of surface crusts formed by simulated rainfall," *Soil Sci. Soc. Amer. J.,* **48**:1137–1143.

Parker, M. M., and D. H. Van Lear. 1996. "Soil heterogeneity and root distribution of mature loblolly pine stands in Piedmont soils," *Soil Sci. Soc. Amer. J.,* **60**:1920–1925.

Soil Survey Division Staff. 1993. *Soil Survey Manual.* USDA-NRSC Agricultural Handbook 18. (Washington, D.C.: U.S. Government Printing Office), pp. 174–175.

Tisdall, J. M. 1994. "Possible role of soil microorganisms in aggregation in soils," *Plant and Soil,* **159**:115–121.

Tisdall, J. M., and J. M. Oades. 1982. "Organic matter and water-stable aggregates in soils," *J. Soil Sci.,* **33**:141–163.

Unger, P. W., and T. C. Kaspar. 1994. "Soil compaction and root growth: A review," *Agron. J.,* **86**:759–766.

Voorhees, W. B. 1984. "Soil compaction, a curse or a cure?" *Solutions,* **28**:42–47 (Peoria, Ill.: Solutions Magazine Inc.).

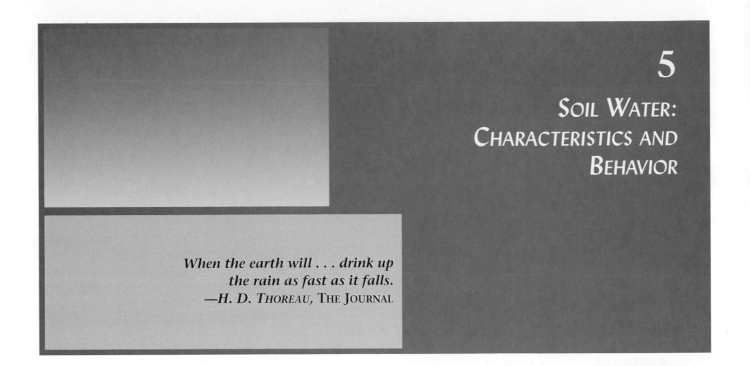

5

SOIL WATER: CHARACTERISTICS AND BEHAVIOR

*When the earth will . . . drink up
the rain as fast as it falls.*
—*H. D. THOREAU, THE JOURNAL*

Water is a vital component of every living thing. Although it is one of nature's simplest chemicals, water has unique properties that promote a wide variety of physical, chemical, and biological processes. These processes greatly influence almost every aspect of soil development and behavior, from the weathering of minerals to the decomposition of organic matter, from the growth of plants to the pollution of groundwater.

We are all familiar with water. We drink it, wash with it, and swim in it. But water in the soil is something quite different from water in a drinking glass. In the soil, water is intimately associated with solid particles, particularly those that are colloidal in size. The interaction between water and soil solids changes the behavior of both.

Water causes soil particles to swell and shrink, to adhere to each other, and to form structural aggregates. Water participates in innumerable chemical reactions that release or tie up nutrients, create acidity, and wear down minerals so that their constituent elements eventually contribute to the saltiness of the oceans.

Attraction to solid surfaces restricts some of the free movement of water molecules, making it less liquid and more solidlike in its behavior. In the soil, water can flow up as well as down. Plants may wilt and die in a soil whose profile contains a million kilograms of water in a hectare. A layer of sand and gravel in a soil profile may actually inhibit drainage, causing the upper horizons to become muddy and saturated with water during much of the year. These and other soil water phenomena seem to contradict our intuition about how water ought to behave.

Soil–water interactions influence how much rainwater runs into and through the soil and how much runs off the surface. Control of these processes in turn determines the movement of chemicals to the groundwater, and of both chemicals and eroded soil particles to streams and lakes. The interactions affect the rate of water loss through leaching and evapotranspiration, the balance between air and water in soil pores, the rate of change in soil temperature, the rate and kind of metabolism of soil organisms, and the capacity of soil to store and provide water for plant growth.

The characteristics and behavior of water in the soil is a common thread that interrelates nearly every chapter in this book. The principles contained in this chapter will help us understand why mudslides occur in water-saturated soils (Chapter 4), why earthworms improve soil quality (Chapter 10), and why rice paddies contribute to global ozone depletion (Chapter 12). Mastery of the principles presented in this chapter are fundamental to a working knowledge of the soil system.

5.1 STRUCTURE AND RELATED PROPERTIES OF WATER

The ability of water to influence so many soil processes is determined primarily by the structure of the water molecule. This structure also is responsible for the fact that water is a liquid, not a gas, at temperatures found on earth. Water is, with the exception of mercury, the *only* inorganic (not carbon-based) liquid found on Earth. Water is a simple compound, its individual molecules containing one oxygen atom and two much smaller hydrogen atoms. The elements are bonded together covalently, each hydrogen atom sharing its single electron with the oxygen.

Polarity

The arrangement of the three atoms in a water molecule is not symmetrical, as one might expect. Instead of the atoms being arranged linearly (H-O-H), the hydrogen atoms are attached to the oxygen in a V-shaped arrangement at an angle of only 105°. As shown in Figure 5.1, this results in an asymmetrical molecule with the shared electrons spending most of the time nearer to the oxygen than to the hydrogen. Consequently, the water molecule exhibits *polarity*; that is, the charges are not evenly distributed. Rather, the side on which the hydrogen atoms are located tends to be electropositive and the opposite side electronegative. The fact that water consists of polar molecules accounts for many of the properties that allow water to play its unique roles in the soil environment.

Hydrogen Bonding

Through a phenomenon called **hydrogen bonding**, a hydrogen atom may be shared between two electronegative atoms forming a relatively low-energy link. Because of its high electronegativity, an O atom in one water molecule exerts some attraction for the H atom in a neighboring water molecule. This type of bonding accounts for the polymerization of water.

Cohesion versus Adhesion

Hydrogen bonding accounts for two basic forces responsible for water retention and movement in soils: the attraction of water molecules for each other (**cohesion**) and the attraction of water molecules for solid surfaces (**adhesion**). By adhesion (also called *adsorption*), some water molecules are held rigidly at the surfaces of soil-solids. In turn, these tightly bound water molecules hold by cohesion other water molecules farther removed from the solid surfaces (Figure 5.2). Together, the forces of adhesion and cohesion make it possible for the soil solids to retain water and control its movement and use.

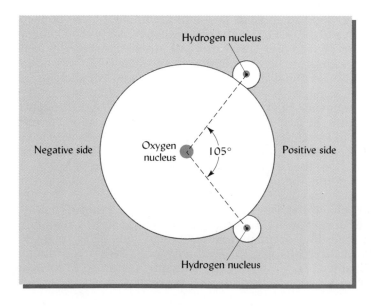

FIGURE 5.1 Two-dimensional representation of a water molecule showing a large oxygen atom and two much smaller hydrogen atoms. The HOH angle of 105° results in an asymmetrical arrangement. One side of the water molecule (that with the two hydrogens) is electropositive; the other is electronegative. This accounts for the polarity of water.

FIGURE 5.2 The forces of adhesion (between water molecules) and cohesion (between water and solid surface) in a soil–water system. The forces are largely a result of H-bonding shown as broken lines. The adhesive or adsorptive force diminishes rapidly with distance from the solid surface. The cohesion of one water molecule to another results in water molecules forming temporary clusters that are constantly changing in size and shape as individual water molecules break free or join up with others. The cohesion between water molecules also allows the solid to indirectly restrict the freedom of water for some distance beyond the solid–liquid interface.

Surface Tension

Another important property of water that markedly influences its behavior in soils is that of **surface tension.** This property is commonly evidenced at liquid–air interfaces and results from the greater attraction of water molecules for each other (cohesion) than for the air above. The net effect is an inward force at the surface that causes water to behave as if its surface were covered with a stretched elastic membrane, an observation familiar to those who have seen insects walking on water in a pond (Figure 5.3). As we shall see, surface tension is an important factor in the phenomenon of capillarity, which determines how water moves and is retained in soil.

FIGURE 5.3 Everyday evidences of water's surface tension (left) as insects land on water and do not sink, and of forces of cohesion and adhesion (right) as a drop of water is held between the fingers. (Photos courtesy of R. Weil)

The movement of water up a wick typifies the phenomenon of capillarity. Two forces cause capillarity: (1) the attraction of water for the solid (adhesion or adsorption), and (2) the surface tension of water, which is due largely to the attraction of water molecules for each other (cohesion).

Capillarity can be demonstrated by placing one end of a fine, clean glass tube in water. The water rises in the tube; the smaller the tube bore, the higher the water rises. The water molecules are attracted to the sides of the tube (adhesion) and start to spread out along the glass in response to this attraction. At the same time, the cohesive forces hold the water molecules together and create surface tension, causing a curved surface (called a *meniscus*) to form at the interface between water and air in the tube (Figure 5.4). Lower pressure under the meniscus in the glass tube (P2) allows the higher pressure (P1) on the free water to push water up the tube. The process continues until the water in the tube has risen high enough that its weight just balances the pressure differential across the meniscus.

The height of rise in a capillary tube is inversely proportional to the tube radius r. Capillary rise is also inversely proportional to the density of the liquid, and is directly proportional to the liquid's surface tension and the degree of its adhesive attraction to the soil surface. If we limit our consideration to water at a given temperature (e.g., 20°C), then these factors can be combined into a single constant, and we can use a simple capillary equation to calculate the height of rise h:

$$h = \frac{0.15}{r}$$

where both h and r are expressed in centimeters. This equation tells us that the smaller the tube bore, the greater the capillary force and the higher the water rise in the tube (Figure 5.5a).

The upward movement due to capillarity in soils is illustrated in Figure 5.5. Usually the eventual height of rise resulting by capillarity is greater with fine-textured soils, but the rate of flow may be very slow because of frictional forces in the tiny pores. The large pores in sandy soils present little frictional resistance to rapid capillary water move-

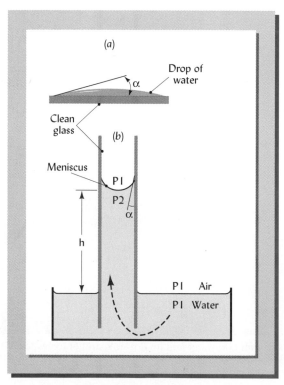

FIGURE 5.4 The interaction of water with a hydrophillic surface (*a*) results in a characteristic *contact angle* (α). If the solid surface surrounds the water as in a tube, a curved water–air interface termed the *meniscus* forms because of adhesive and cohesive forces. When air and water meet in a curved meniscus, pressure on the convex side of the curve is lower than on the concave side. (*b*) Capillary rise occurs in a fine hydrophillic (e.g., glass) tube because pressure under the meniscus (P2) is less than pressure on the free water.

Unsaturated flow by capillarity

$$h \ (cm) = \frac{0.15}{r \ (cm)}$$

Radius of tube

(a)

Coarse sand | Fine sand

(b)

(c)

FIGURE 5.5 Upward capillary movement of water through tubes of different bore and soils with different pore sizes. (a) The capillary equation can be graphed to show that the height of rise h doubles when the tube inside radius is halved. The same relationship can be demonstrated using glass tubes of different bore size. (b) The same principle also relates pores sizes in a soil and height of capillary rise, but the rise of water in a soil is rather jerky and irregular because of the tortuous shape and variability in size of the soil pores (as well as because of pockets of trapped air). (c) The finer the soil texture, the smaller the average pore diameter and, hence, the higher the ultimate rise of water above a free water table. However, because of the much greater frictional forces in the smaller pores, the capillary rise is much slower in the finer-textured soil than in the sand.

ment. However, as expected from our discussion of the capillary equation, the large radii of the pores between sand grains result in relatively little height of capillary rise.[1]

Although traditionally illustrated as an upward adjustment, capillary movement takes place in any direction (Figure 5.6). The significance of capillarity in controlling water movement in small pores will become evident as we turn to soil water energy concepts.

5.3 SOIL WATER ENERGY CONCEPTS[2]

As we consider energy, we should keep in mind that all substances, including water, tend to move or change from a higher to a lower energy state. Therefore, if we know the pertinent energy levels at various points in a soil, we can predict the direction of water movement. It is the *differences* in energy levels from one contiguous site to another that influence this water movement.

---

[1] For example, if water rises by capillarity to a height of 25 cm above a free-water surface in a sand (as shown in the example in Figure 5.5c), then it can be estimated (by rearranging the capillary equation to $r = 0.15/h$) that the smallest continuous pores must have a radius of about 0.006 cm (0.15/25 = 0.006). This calculation gives an approximation of the minimum effective pore radius in a soil.

[2] For a comprehensive treatise on these and other physical soil concepts see Hillel (1998).

FIGURE 5.6 As this field irrigation scene in Arizona shows (left), water has moved up by capillarity from the irrigation furrow toward the top of the ridge. The photo on the right illustrates some horizontal movement to both sides and away from the irrigation water.

Forces Affecting Potential Energy

The discussion of the structure and properties of water in the previous section suggests three important forces affecting the energy level of soil water. First, adhesion, or the attraction of water to the soil solids (matrix), provides a **matric** force (responsible for adsorption and capillarity) that markedly reduces the energy state of water near particle surfaces. Second, the attraction of water to ions and other solutes, resulting in **osmotic** forces, tends to reduce the energy state of water in the soil solution. Osmotic movement of pure water across a semipermeable membrane into a solution (osmosis) is evidence of the lower energy state of water in the solution. The third major force acting on soil water is **gravity**, which always pulls the water downward. The energy level of soil water at a given elevation in the profile is thus higher than that of water at some lower elevation. This difference in energy level causes water to flow downward.

Soil Water Potential

The *difference* in energy level of water from one site or one condition (e.g., in wet soil) to another (e.g., in dry soil) determines the direction and rate of water movement in soils and in plants. In a wet soil, most of the water is retained in large pores or thick water films around particles. Therefore, most of the water molecules in a wet soil are not very close to a particle surface and so are not held very tightly by the soil solids (the matrix). In this condition, the water molecules have considerable freedom of movement, so their energy level is near that of water molecules in a pool of pure water outside the soil. In a drier soil, however, the water that remains is located in small pores and thin water films, and is therefore held tightly by the soil solids. Thus the water molecules in a drier soil have little freedom of movement, and the energy level of the water is much lower than that of the water in wet soil. If wet and dry soil samples are brought in touch with each other, water will move from the wet soil (higher energy state) to the drier soil (lower energy).

To predict how water will move in soils the energy status of soil water in a particular location in the profile is compared to that of pure water at standard pressure and temperature, unaffected by the soil and located at some reference elevation. The *difference* in energy levels between this pure water in the reference state and that of the soil water is termed **soil water potential** (Figure 5.7), the term *potential*, like the term *pressure*, implying a difference in energy status. Water will move from a soil zone having a high soil water potential to one having a lower soil water potential (Figure 5.8). This fact should be always kept in mind in considering the behavior of water in soils.

The soil water potential is due to several forces, each of which is component of the **total soil water potential** ψ_t. These components are due to differences in energy levels

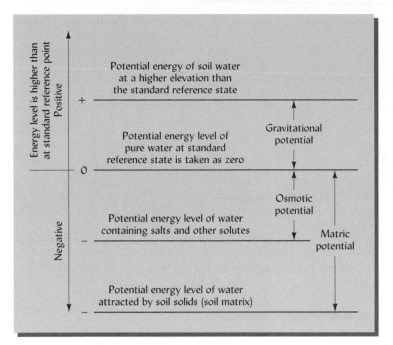

Potential energy of soil water at a higher elevation than the standard reference state

Potential energy level of pure water at standard reference state is taken as zero

Gravitational potential

Potential energy level of water containing salts and other solutes

Osmotic potential

Matric potential

Potential energy level of water attracted by soil solids (soil matrix)

FIGURE 5.7 Relationship between the potential energy of pure water at a standard reference state (pressure, temperature, and elevation) and that of soil water. If the soil water contains salts and other solutes, the mutual attraction between water molecules and these chemicals reduces the potential energy of the water, the degree of the reduction being termed *osmotic potential*. Similarly, the mutual attraction between soil solids (soil matrix) and soil water molecules also reduces the water's potential energy. In this case the reduction is called *matric potential*. Since both of these interactions reduce the water's potential energy level compared to that of pure water, the changes in energy level (osmotic potential and matric potential) are both considered to be negative. In contrast, differences in energy due to gravity (gravitational potential) are always positive. This is because the reference elevation of the pure water is purposely designated at a site in the soil profile below that of the soil water. A plant root attempting to remove water from a moist soil would have to overcome all three forces simultaneously.

resulting from gravitational, matric, submerged hydrostatic and osmotic forces and are termed **gravitational potential** ψ_g, **matric potential** ψ_m, **submergence potential** ψ_s, and **osmotic potential** ψ_o, respectively. All of these components act simultaneously to influence water behavior in soils. The general relationship of soil water potential to potential energy levels is shown in Figure 5.7 and can be expressed as:

$$\psi_t = \psi_g + \psi_m + \psi_o + \ldots$$

FIGURE 5.8 Whether concerning matric potential, osmotic potential, or gravitational potential (as shown here), water always moves to where its energy state will be lower. In this case the energy lost by the water is used to turn the historic Mabry Mills waterwheel and grind flour. (Photo courtesy of R. Weil)

where the ellipses (. . .) indicate the possible contribution of additional potentials such as ψ_s.

Gravitational Potential

The force of gravity acts on soil water the same as it does on any other body (Figure 5.8), the attraction being toward the earth's center. The gravitational potential ψ_g of soil water may be expressed mathematically as:

$$\psi_g = gh$$

where g is the acceleration due to gravity and h is the height of the soil water above a reference elevation. The reference elevation is usually chosen within the soil profile or at its lower boundary to ensure that the gravitational potential of soil water above the reference point will always be positive.

Following heavy precipitation or irrigation, gravity plays an important role in removing excess water from the upper horizons and in recharging groundwater below the soil profile. It will be given further attention when the movement of soil water is discussed (see Section 5.5).

Pressure Potential (Including Submergence and Matric Potentials)

This component accounts for the effects on soil water potential of all factors other than gravity and solute levels. It most commonly includes (1) the positive hydrostatic pressure due to the weight of water in saturated soils and aquifers, and (2) the negative pressure due to the attractive forces between the water and the soil solids or the soil matrix.

The hydrostatic pressures give rise to what is often termed the **submergence potential** ψ_s, a component that is operational only for water in saturated zones. Anyone who has dived to the bottom of a swimming pool has felt hydrostatic pressure on the eardrums.

The attraction of water to solid surfaces gives rise to the **matric potential** ψ_m, which is always negative because the water attracted by the soil matrix has an energy state lower than that of pure water. (These negative pressures are sometimes referred to as *suction* or *tension*.) The matric potential is operational in unsaturated soil above the water table, while the submergence potential applies to water in saturated soil or below the water table (Figure 5.9).

While each of these pressures is significant in specific field situations, the matric potential is important in all unsaturated soils because there are omnipresent interactions between soil solids and water. The movement of soil water, the availability of water to plants, and the solutions to many civil engineering problems are determined to a considerable extent by matric potential. Consequently, matric potential will receive primary attention in this text, along with gravitational and osmotic potential.

FIGURE 5.9 The matric potential and submergence potential are both pressure potentials that may contribute to total water potential. The matric potential is always negative and the submergence potential is positive. When water is in unsaturated soil above the water table (top of the saturated zone) it is subject to the influence of matric potentials. Water below the water table in saturated soil is subject to submergence potentials. In the example shown here, the matric potential decreases linearly with elevation above the water table, signifying that water rising by capillary attraction up from the water table is the only source of water in this profile. Rainfall or irrigation (see dotted line) would alter or curve the straight line, but would not change the fundamental relationship described.

Matric potential ψ_m, which results from the phenomena of adhesion (or adsorption) and of capillarity, influences soil moisture retention as well as soil water movement. Differences between the ψ_m of two adjoining zones of a soil encourage the movement of water from moist zones (high energy state) to dry zones (low energy state) or from large pores to small pores. Although this movement may be slow, it is extremely important, especially in supplying water to plant roots.

Osmotic Potential

The osmotic potential ψ_o is attributable to the presence of solutes in the soil solution. The solutes may be inorganic salts or organic compounds. Their presence reduces the potential energy of water, primarily because of the reduced freedom of movement of the water molecules that cluster around each solute ion or molecule. The greater the concentration of solutes, the more the osmotic potential is lowered. As always, water will tend to move to where its energy level will be lower, in this case to the zone of higher solute concentration. However, liquid water will move in response to differences in osmotic potential (the process termed **osmosis**) only if a *semipermeable membrane* exists between the zones of high and low osmotic potential, allowing water through but *preventing the movement of the solute.* If no membrane is present, the solute, rather than the water, generally moves to equalize concentrations.

Because soil zones are *not* generally separated by membranes, the osmotic potential ψ_o has little effect on the mass movement of water in soils. Its major effect is on the uptake of water by plant root cells that *are* isolated from the soil solution by their semipermeable cell membranes. In soils high in soluble salts, ψ_o may be lower (have a greater negative value) in the soil solution than in the plant root cells. This leads to constraints in the uptake of water by the plants. In very salty soil, the soil water osmotic potential may be low enough to cause cells in young seedlings to collapse (plasmolyze) as water moves from the cells to the lower osmotic potential zone in the soil.

Methods of Expressing Energy Levels

Several units can be used to express differences in energy levels of soil water. One is the *height of a water column* (usually in centimeters) whose weight just equals the potential under consideration. We have already encountered this means of expression since the h in the capillary equation (page 134) tells us the matric potential of the water in a capillary pore. A second unit is the standard *atmosphere* pressure at sea level, which is 760 mm Hg or 1020 cm of water. The unit termed *bar* approximates the pressure of a standard atmosphere. Energy may be expressed per unit of mass (**joules/kg**) or per unit of volume (**newtons/m²**). In the International System of Units (SI), 1 pascal (Pa) equals 1 newton (N) acting over an area of 1 m². In this text we will use Pa or kilopascals (kPa) to express soil water potential. Since other publications may use other units, Table 5.1 is provided to show the equivalency among common means of expressing soil water potential.

TABLE 5.1 Approximate Equivalents Among Expressions of Soil Water Potential

Height of unit column of water, cm	Soil water potential, bars	Soil water potential, kPa[a]
0	0	0
10.2	−0.01	−1
102	−0.1	−10
306	−0.3	−30
1,020	−1.0	−100
15,300	−15	−1,500
31,700	−31	−3,100
102,000	−100	−10,000

[a] The SI unit kilopascal (kPa) is equivalent to 0.01 bars.

5.4 SOIL MOISTURE CONTENT AND SOIL WATER POTENTIAL

The previous discussions suggest an inverse relationship between the water content of soils and the tenacity with which the water is held in soils. Water is more likely to flow out of a wet soil than from one low in moisture. Many factors affect the relationship between soil water potential ψ and moisture content θ. A few examples will illustrate this point.

Soil Moisture versus Energy Curves

The relationship between soil water potential ψ and moisture content θ of three soils of different textures is shown in Figure 5.10. Such curves are sometimes termed *water release characteristic curves,* or simply *water characteristic curves.* The clay soil holds much more water at a given potential than does the loam or sand. Likewise, at a given moisture content, the water is held much more tenaciously in the clay than in the other two soils (note that soil water potential is plotted on a log scale).

Soil structure also influences soil water content–energy relationships. A well-granulated soil has more total pore space and greater overall water-holding capacity than one with poor granulation or one that has been compacted. The compacted soil will hold less total water but is likely to have a higher proportion of small- and medium-sized pores, which hold water with greater tenacity than do larger pores.

Measurement of Soil Water Status

The soil water characteristic curves just discussed highlight the importance of making two general kinds of soil water measurements: the *amount* of water present (water content) and the *energy status* of the water (soil water potential). In order to understand or manage water supply and movement in soils it is essential to have information (directly measured or inferred) on *both* types of measurements. For example, a soil water potential measurement might tell us whether water will move toward the groundwater, but without a corresponding measurement of the soil water content, we would not know the potential significance of the contribution to groundwater.

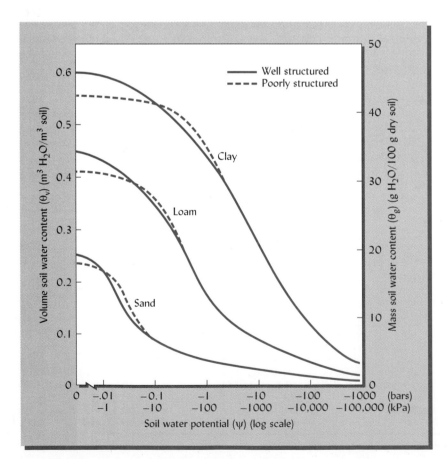

FIGURE 5.10 Soil water characteristic curves for three representative mineral soils. The curves show the relationship obtained by slowly drying completely saturated soils. The dashed lines show the effect of compaction or poor structure. The soil water potential ψ (which is negative) is expressed in terms of bars (upper scale) and kilopascals (kPa) (lower scale). Note that the soil water potential is plotted on a log scale.

Generally, the behavior of soil water is most closely related to the energy status of the water, not to the amount of water in a soil. Thus, a clay soil and a loamy sand will both feel moist and will easily supply water to plants when the ψ_m is, say, −10 kPa. However, the amount of water held by the clay loam, and thus the length of time it could supply water to plants, would be far greater at this potential than would be the case for the loamy sand.

Measuring Water Content or Potential

The **volumetric water content** θ is defined as the volume of water associated with a given volume (usually 1 m³) of dry soil (see Figure 5.10). A comparable expression is the **mass water content** θ_m, or the mass of water associated with a given mass (usually 1 kg) of dry soil. Both of these expressions have advantages for different uses. In most cases we shall use the volumetric water content θ in this text.

Because in the field we think of plant root systems as exploring a certain depth of soil, and because we express precipitation (and sometimes irrigation) as a depth of water (e.g., mm of rain), it is often convenient to express the volumetric water content as a *depth ratio* (depth of water per unit depth of soil). Conveniently, the numerical values for these two expression are the same. For example, for a soil containing 0.1 m³ of water per m³ of soil (10% by volume) the depth ratio of water is 0.1 m of water per m of soil depth.[3]

Of the three methods of measuring soil water *content* listed in Table 5.2, the **gravimetric method** is the most straightforward and is often used to calibrate the others. As illustrated in Box 5.1, a sample of moist soil is weighed before and after complete drying, and the mass of water lost is divided by the dry soil mass to calculate grams of water per g of dry soil.[4] Although inexpensive and simple, this method does involve remov-

---

[3] When measuring amounts of water added to soil by irrigation, it is customary to use units of volume such as m³ and hectare-meter (the volume of water that would cover a hectare of land to a depth of 1 m). Generally, farmers and ranchers in the irrigated regions of the United States use the English units ft³ or acre-foot (the volume of water needed to cover an acre of land to a depth of 1 ft).

[4] Enough drying time must be allowed (usually 24 to 48 hours at 105°C) so that the soil has stopped losing water and has reached a constant weight. To save time, a microwave oven may be used. About a dozen small samples of soil (about 20 g each) in glass beakers may be dried on a turntable in a 1000-W microwave oven using three or more consecutive 3-minute periods with the oven power set at high and the oven door opened between periods to allow moisture to escape. For less soil, the exposure time should be reduced correspondingly.

TABLE 5.2 Some Methods of Measuring Soil Water

Note that more than one method may be needed to cover the entire range of soil moisture conditions.

Method	Measures soil water		Useful range, kPa	Used mainly in		Comments
	Content	Potential		Field	Lab	
1. Gravimetric	x		0 to <−10,000		x	Destructive sampling; slow (1 to 2 days) unless microwave used. The standard for calibration.
2. Neutron scattering	x		0 to <−1,500	x		Radiation permit needed; expensive equipment; not good in high-organic-matter soils; requires access tube.
3. Time domain reflectometry (TDR)	x		0 to <−10,000	x	x	May be automated; accurate to 1 kPa; requires wave guides; expensive instrument.
4. Tensiometer		x	0 to −85	x		Accurate to 0.1 to 1 kPa; limited range; inexpensive; can be automated; needs periodic servicing.
5. Resistance blocks		x	−100 to <−1,500	x		Can be automated; not sensitive near optimum plant water contents, must be individually calibrated.
6. Thermocouple psychrometer		x	−50 to <−10,000	x	x	Moderately expensive; wide range; accurate only to ±50 kPa.
7. Thermal dissipation blocks		x	−50 to <−1,500	x	x	Moderately expensive; must be individually calibrated.
8. Pressure membrane apparatus		x	−50 to <−10,000		x	Used in conjunction with gravimetric method to construct water characteristic curve.

ing a sample of soil and bringing it back to the lab. Instrumental methods, such as **neutron scattering** or **time-domain reflectometry** (TDR) (Figure 5.11), involve certain limitations and greater initial expense, but are capable of making repeated measurements on the same soil once the instrument probe has been installed.

Several instruments for measuring soil water *potential* are also described in Table 5.2, each with limitations and advantages. The most straightforward of these methods is the **tensiometer,** which is essentially a water-filled tube with one end sealed and the other end capped by a porous ceramic tip. It is inserted into the soil, and water flows out through the tip into the adjacent soil until the water potential is the same in the tensiometer tube and in the soil. As the water is drawn out, a vacuum develops under the sealed end of the tube and is measured by an attached vacuum gauge (Figure 5.12). The **electrical resistance**

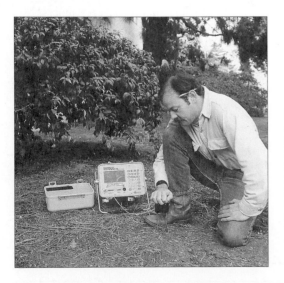

FIGURE 5.11 Instrumental measurement of soil water content using time domain reflectometry (TDR). The electronic instrument sends a pulse of electromagnetic energy down the two parallel metal rods of a waveguide that the hydrologist is pushing into the soil. The TDR instrument makes precise picosecond measurements of the speed at which the pulse travels down the rods, a speed influenced by the nature of the surrounding soil. Microprocessors in the instrument analyze the wave patterns generated and calculate the apparent dielectric constant of the soil. Since the dielectric constant of a soil is mainly influenced by its water content, the instrument can accurately convert its measurements into volumetric water content of the soil. (Photo courtesy of Soilmoisture Equipment Corporation, Goleta, Calif., and Dr. Stephen J. Cullen, hydrologist)

FIGURE 5.12 Tensiometer method of determining water potential in the field. Cross section showing the essential components of a tensiometer. Water moves through the porous end of the instrument in response to the pull (matric potential) of the soil. The vacuum so created is measured by a gauge that reads in kPa of tension (–kPa water potential).

block method uses small porous blocks with embedded electrodes. When the blocks are placed in a moist soil they absorb water, decreasing the resistance to the flow of electricity between the embedded electrodes. Both of these instruments can be automated to turn on irrigation systems when the soil reaches a predetermined water potential.

5.5 THE FLOW OF LIQUID WATER IN SOIL

Three types of water movement within the soil are recognized: (1) saturated flow, (2) unsaturated flow, and (3) vapor movement. In all cases water flows in response to energy gradients, with water moving from a zone of higher to one of lower water potential. *Saturated flow* takes place when the soil pores are completely filled (or saturated) with water. *Unsaturated flow* occurs when the larger pores in the soil are filled with air, leaving only the smaller pores to hold and transmit water. *Vapor movement* occurs as vapor pressure differences develop in relatively dry soils.

Saturated Flow Through Soils

Under some conditions, at least part of a soil profile may be completely saturated; that is, all pores, large and small, are filled with water. The lower horizons of poorly drained soils are often saturated, as are portions of well-drained soils above stratified layers of clay. During and immediately following a heavy rain or irrigation, pores in the upper soil zones are often filled entirely with water.

The quantity of water per unit of time Q/t that flows through a column of saturated soil can be expressed by Darcy's law, as follows:

$$\frac{Q}{t} = A \cdot K_{sat} \cdot \frac{\Delta \psi}{L}$$

where A is the cross-sectional area of the column through which the water flows, K_{sat} is the **saturated hydraulic conductivity**, $\Delta \psi$ is the change in water potential between the ends of the column (for example, $\psi_1 - \psi_2$) and L is the length of the column. For a given column, the rate of flow is determined by the ease with which the soil transmits water

(K_{sat}) and the amount of force driving the water, namely the **water potential gradient** $\Delta\psi/L$. For saturated flow this force may also be called the **hydraulic gradient.** By analogy, think of pumping water through a garden hose, with K_{sat} representing the size of the hose (water flows more readily through a larger hose) and $\Delta\psi/L$ representing the size of the pump that drives the water through the hose.

The units in which K_{sat} is measured are length/time, typically cm/s or cm/h. The K_{sat} is an important property that helps determine how well a soil or soil material will perform in such uses as irrigated cropland, sanitary landfill cover material, wastewater storage lagoon lining, and septic tank drain field (Table 5.3) Figure 5.13 illustrates how K_{sat} can be determined by measuring the rate of downward vertical saturated flow, but keep in mind that saturated flow in other situations may be horizontal or even upwards (as when groundwater under pressure wells up under a stream).

Factors Influencing Saturated Hydraulic Conductivity

Any factor affecting the size and configuration of soil pores will influence hydraulic conductivity. The total flow rate in soil pores is proportional to the fourth power of the radius. Thus, flow through a pore 1 mm in radius (say, a small earthworm channel) is equivalent to that in 10,000 pores with a radius of 0.1 mm even though it takes only 100 pores of radius 0.1 mm to give the same cross-sectional area as a 1-mm pore. As a result, macropores (radius > 0.05 mm) account for most water movement in saturated soils. The presence of biopores, such as root channels and earthworm burrows (typically >1 mm in radius), may have a marked influence on the saturated hydraulic conductivity of different soil horizons (Table 5.4). Because they usually have more macropore space, sandy soils generally have higher saturated conductivities than finer-textured soils. Likewise, soils with stable granular structure conduct water much more rapidly than do those with unstable structural units, which break down upon being wetted.

PREFERENTIAL FLOW. In natural field soil, water may flow rapidly through certain pathways in preference to slower, more uniform movement through the bulk of the soil. These **preferential flow** pathways may allow for rapid movement of chemical-laden water deep into the soil profile, increasing the likelihood of groundwater contamination. Deep clay shrinkage cracks, deep earthworm burrows, and old root channels may serve as preferential flow paths.

Unsaturated Flow in Soils

In unsaturated soils, the macropores are filled with air, leaving only the finer pores to accommodate water movement. The driving force for unsaturated water flow is mainly the **matric potential gradient**, the difference in the matric potential of the moist soil areas and nearby drier areas into which the water is moving. Movement will be from a

TABLE 5.3   Some Approximate Values of Saturated Hydraulic Conductivity (in Various Units) and Interpretations for Soil Uses

K_{sat}, cm/s	K_{sat}, cm/h	K_{sat}, in./h	Comments
1×10^{-2}	36	14	Typical of beach sand.
5×10^{-3}	18	7	Typical of very sandy soil, too rapid to effectively filter pollutants in wastewater.
5×10^{-4}	1.8	0.7	Typical of moderately permeable soils, K_{sat} between 1.0 and 15 cm/h considered suitable for most agricultural, recreational, and urban uses calling for good drainage.
5×10^{-5}	0.18	0.07	Typical of fine-textured, compacted or poorly structured soils. Too slow for proper operation of septic tank drain fields, most types of irrigation, and many recreational uses such as playgrounds.
$<1 \times 10^{-8}$	$<3.6 \times 10^{-5}$	$<1.4 \times 10^{-5}$	Extremely slow; typical of compacted clay. K_{sat} of 10^{-5} to 10^{-8} cm/h may be required where nearly impermeable material is needed, as for wastewater lagoon lining or landfill cover material.

FIGURE 5.13 Saturated flow (percolation) in a column of soil with cross-sectional area A, cm². All soil pores are filled with water. Water is shown running off into a container to indicate that water is actually moving down the column. The force driving the water through the soil is the water potential gradient, $\psi_1 - \psi_2/L$, where both water potentials and length are expressed in cm (see Table 5.1). If we measure the quantity of water flowing out Q/t as cm³/s we can rearrange Darcy's law (from page 143) to calculate the saturated hydraulic conductivity of the soil K_{sat} in cm/s as:

$$K_{sat} = \frac{Q}{A} \cdot \frac{L}{\psi_1 - \psi_2}$$

Remember that the same principles apply where the water potential gradient moves the water in a horizontal direction.

zone of thick moisture films (high matric potential, e.g., –1 kPa) to one of thin films (lower matric potential, e.g., –100 kPa).

INFLUENCE OF TEXTURE.   Figure 5.14 shows the general relationship between matric potential ψ_m (and, in turn, water content) and hydraulic conductivity of a sandy loam and clay soil. Note that at or near zero potential (which characterizes the saturated flow region), the hydraulic conductivity is thousands of times greater than at potentials that characterize typical unsaturated flow (–10 kPa and below).

TABLE 5.4   The Saturated Hydraulic Conductivity K_{sat} and Related Properties of Various Horizons in a Typic Hapludult Profile

The upper horizons had many biopores (mainly earthworm burrows), which resulted in high values of K_{sat} as well as extreme variability from sample to sample. The presence of a clay-enriched argillic horizon resulted in reduced K_{sat} values. Apparently most large biopores in this soil did not extend below 30 cm.

Horizon	Depth, cm	Clay, %	Bulk density, Mg/m³	Mean K_{sat}, cm/h	Range of K_{sat} values[a], cm/h
Ap	0–15	12.6	1.42	22.4	0.80–70
E	15–30	11.1	1.44	7.9	0.50–24
E/B	30–45	14.5	1.47	0.93	0.53–1.33
Bt	45–60	22.2	1.40	0.49	0.19–0.79
Bt	60–75	27.2	1.38	0.17	0.07–0.27
Bt	75–90	24.1	1.28	0.04	0.01–0.07

[a] For each soil layer K_{sat} was determined on 5 soil cores of 7.5 cm diameter.
Data from Waddell and Weil (1996).

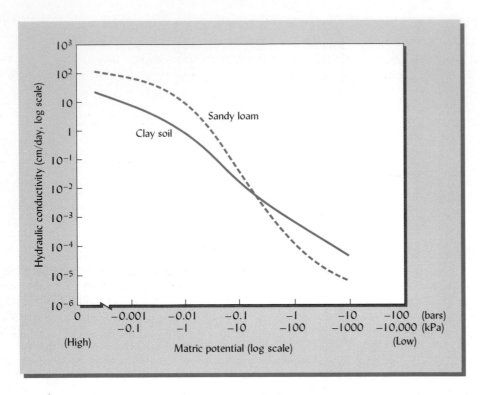

FIGURE 5.14 Generalized relationship between matric potential and hydraulic conductivity for a sandy soil and a clay soil (note log scales). Saturation flow takes place at or near zero potential, while much of the unsaturated flow occurs at a potential of –0.1 bar (–10 kPa) or below.

At high potential levels (high moisture contents), hydraulic conductivity is higher in the sand than in the clay. The opposite is true at low potential values (low moisture contents). This relationship is to be expected because the sandy soil contains many large pores which are water-filled when the soil water potential is high (and the soil is quite wet), but most of these have been emptied by the time the soil water potential becomes lower than about –10 kPa. The clay soil has many more micropores which are still water-filled at lower soil water potentials (drier soil conditions) and can participate in unsaturated flow.

5.6 INFILTRATION AND PERCOLATION

The process by which water enters the soil pore spaces and becomes soil water is termed **infiltration**, and the rate at which water can enter the soil is termed the **infiltration capacity**. This is a pivotal process in landscape hydrology that greatly influences the moisture regime for plants and the potential for soil degradation, chemical runoff, and down-valley flooding. The units of infiltration are m/s or, more commonly, cm/h. The infiltration capacity is not constant over time, but generally decreases during an irrigation or rainfall episode (Figure 5.15).

Like hydraulic conductivity, infiltration capacity is a characteristic of each particular soil and depends mainly upon the texture and structure at the soil surface, but also upon the presence of any layers in the soil profile that might restrict the downward movement of water. The infiltration capacity of a soil may be easily measured using a simple device known as a **double ring infiltrometer** (see Figure 5.15).

Once the water has infiltrated the soil, the water moves downward into the profile by the process termed **percolation**. Both saturated and unsaturated flow are involved in percolation of water down the profile, and rate of percolation is related to the soil's hydraulic conductivity. In the case of water that has infiltrated a relatively dry soil, the progress of water movement can be observed by the darkened color of the soil as it becomes wet (Figure 5.16). There usually appears to be a sharp boundary, termed a **wetting front**, between the dry underlying soil and the soil already wetted (see Figure 5.17). During an intense rain or heavy irrigation, water movement near the soil surface occurs mainly by saturated flow in response to gravity. At the wetting front, however, water is moving into the underlying drier soil in response to matric potential gradients as well

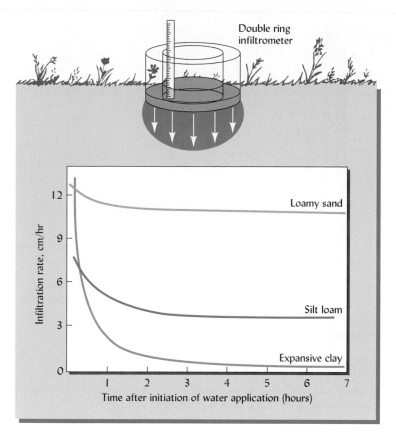

FIGURE 5.15 The potential rate of water entry into the soil, or infiltration capacity, can be measured by recording the drop in water level in a double ring infiltrometer (top). Changes in the infiltration rate of several soils during a period of water application by rainfall or irrigation are shown (bottom). Generally, water enters a dry soil rapidly at first, but its infiltration rate slows as the soil becomes saturated. The decline is least for very sandy soils with macropores that do not depend on stable structure or clay shrinkage. In contrast, a soil high in expansive clays may have a very high initial infiltration rate when large cracks are open, but a very low infiltration rate once the clays swell with water and close the cracks. Most soils fall between these extremes, exhibiting a pattern similar to that shown for the silt loam soil.

FIGURE 5.16 The wetting front 24 hours after a 5 cm rainfall. Water removal by plant roots had dried the upper 70 to 80 cm of this humid-region (Alabama) profile during a previous three-week dry spell. The clearly visible boundary results from the rather abrupt change in soil water content at the wetting front between the dry, lighter-colored soil and the soil darkened by the percolating water. The wavy nature of the wetting front in this natural field soil is evidence of the heterogeneity of pore sizes. Scale in 10-cm intervals. (Photo courtesy of R. Weil)

FIGURE 5.17 Water infiltration into a relatively dry soil after one and three hours of a steady rain. The wetting fronts indicate the depth of water penetration. After three hours, the upper 30 to 40 cm of soil is saturated with water. There is a transition zone of near saturation with water above the wetting zone which, in turn, is above the dry zone.

as gravity. The relative influence of these forces on wetting fronts in a sandy loam and a clay loam is illustrated in Figure 5.18.

Water Movement in Stratified Soils

The fact that, at the wetting front, water is moving by unsaturated flow has important ramifications for how percolating water behaves when it encounters an abrupt change in pore sizes due to such layers as fragipans or claypans, or sand and gravel lenses. In some cases, such pore-size stratification may be created by soil managers, as when

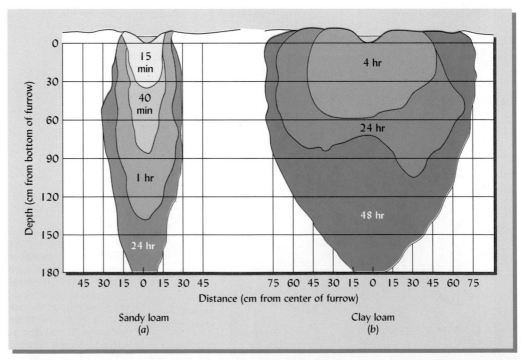

FIGURE 5.18 Comparative rates of irrigation water movement into a sandy loam and a clay loam. Note the much more rapid rate of movement in the sandy loam, especially in a downward direction. [Redrawn from Cooney and Peterson (1955)]

coarse plant residues are plowed under in a layer, or a layer of gravel is placed under finer soil in a planting container. In all cases, the effect on water percolation is similar—that is, the downward movement is impeded—even though the causal mechanism may vary. It is not surprising that percolating water should slow down markedly when it reaches a layer with finer pores, which therefore has a lower hydraulic conductivity. However, the fact that a layer of *coarser* pores will temporarily stop the movement of water may not be obvious. It has this effect because the macropores of the sand offer less attraction for the water than do the finer pores of the overlying material (Figure 5.19). Because it retards downward movement, stratification markedly increases the amount of water the upper part of the soil holds in the field (Figure 5.20). Interestingly, a coarse sand layer in an otherwise fine-textured soil profile would also inhibit the *rise* of water from moist subsoil layers up to the surface soil, a situation that could be illustrated by turning Figure 5.19*b* upside down. This principle also allows a layer of gravel to act as a capillary barrier under a concrete slab foundation to prevent water from soaking up from the soil and through the concrete floor of a home basement.

(a)   (b)   (c)

FIGURE 5.19 Downward water movement in soils with a stratified layer of coarse material. (*a*) Water is applied to the surface of a medium-textured topsoil. Note that after 40 min, downward movement is no greater than movement to the sides, indicating that in this case the gravitational force is insignificant compared to the matric potential gradient between dry and wet soil. (*b*) The *downward* movement stops when a coarse-textured layer is encountered. After 110 min, no movement into the sandy layer has occurred. The macropores of the sand provide less attraction for water than the finer-textured soil above. Only when the water content (and in turn the matric potential gradient) is raised sufficiently will the water move into the sand. (*c*) After 400 min, the water content of the overlying layer becomes sufficiently high to give a water potential of about –1 kPa or more, and downward movement into the coarse material takes place. (Courtesy W. H. Gardner, Washington State University)

FIGURE 5.20 One result of soil layers with contrasting texture. This North Carolina soil has about 50 cm of loamy sand coastal plain material atop deeper layers of silty clay-loam-textured material derived from the piedmont. Rainwater rapidly infiltrates the sandy surface horizons, but its downward movement is arrested at the finer-textured layer, resulting in saturated conditions near the surface and a quick-sandlike behavior. (Photo courtesy of R. Weil)

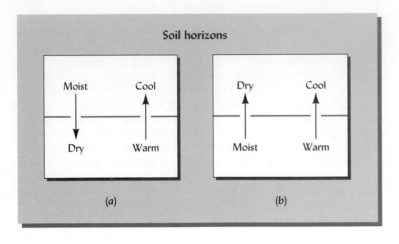

FIGURE 5.21 Vapor movement tendencies that may be expected between soil horizons differing in temperature and moisture. In (*a*) the tendencies more or less negate each other, but in (*b*) they are coordinated and considerable vapor transfer might be possible if the liquid water in the soil capillaries does not interfere.

5.7 WATER VAPOR MOVEMENT IN SOILS

Water vapor moves from one point to another within the soil in response to differences in vapor pressure. Thus, water vapor will move from a moist soil where the soil air is nearly 100% saturated with water vapor (high vapor pressure) to a drier soil where the vapor pressure is somewhat lower. Also, water vapor will move from a zone of low salt content to one with a higher salt content (e.g., around a fertilizer granule). The salt lowers the vapor pressure of the water and encourages water movement from the surrounding soil.

If the temperature of one part of a uniformly moist soil is lowered, the vapor pressure will decrease and water vapor will tend to move toward this cooler part. Heating will have the opposite effect in that heating will increase the vapor pressure and the water vapor will move away from the heated area. Figure 5.21 illustrates these relationships.

Because the amount of water vapor is small, its movement in soils is of limited practical importance if the soil moisture is kept near optimum for plant growth. In dry soils, however, water vapor movement may be of considerable significance to drought-resistant desert plants (*xerophytes*).

5.8 QUALITATIVE DESCRIPTION OF SOIL WETNESS

The measurement of soil water potential and the observable behavior of soil water always depend on that portion of the soil water that is farthest from a particle surface and therefore has the highest potential. As an initially water-saturated soil dries down, both the soil as a whole and the soil water it contains undergo a series of gradual changes in physical behavior and in their relationships with plants. These changes are due mainly to the fact that the water remaining in the drying soil is found in smaller pores and thinner films where the water potential is lowered principally by the action of matric forces. Matric potential therefore accounts for an increasing proportion of the total soil water potential, while the proportion attributable to gravitational potential decreases.

Maximum Retentive Capacity

When all soil pores are filled with water from rainfall or irrigation, the soil is said to be **saturated** with respect to water (Figure 5.22) and at its **maximum retentive capacity.** The matric potential is close to zero, nearly the same as that of pure water. The volumetric water content is essentially the same as the total porosity. The soil will remain at maximum retentive capacity only so long as water continues to infiltrate, for the water in the largest pores (sometimes termed **gravitational water**) will percolate downward, mainly under the influence of gravitational forces (hydrostatic and gravitational potentials).

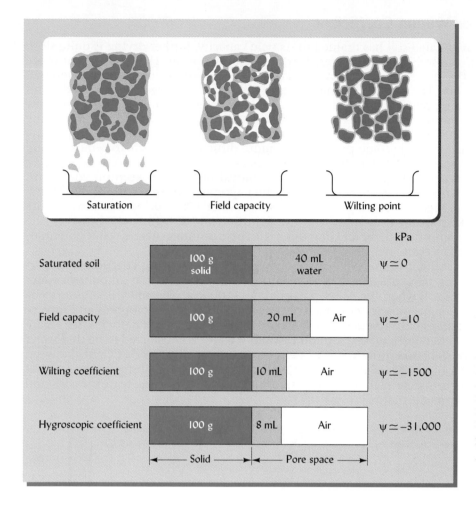

FIGURE 5.22 Volumes of water and air associated with 100 g of a well-granulated silt loam at different moisture levels. The top bar shows the situation when a representative soil is completely saturated with water. This situation will usually occur for short periods of time during a rain or when the soil is being irrigated. Water will soon drain out of the larger pores (*macropores*). The soil is then said to be at the *field capacity*. Plants will remove water from the soil quite rapidly until they begin to wilt. When permanent wilting of the plants occurs, the soil water content is said to be at the *wilting coefficient*. There is still considerable water in the soil, but it is held too tightly to permit its absorption by plant roots. A further reduction in water content to the *hygroscopic coefficient* is illustrated in the bottom bar. At this point the water is held very tightly, mostly by the soil colloids.

Field Capacity

Once the rain or irrigation has ceased, water in the largest soil pores will drain downward quite rapidly in response to the hydraulic gradient (mostly gravity). After one to three days, this rapid downward movement will gradually become negligible as matric forces play a greater role in the movement of the remaining water. The soil then is said to be at its **field capacity.** In this condition, water has moved out of the macropores and air has moved in to take its place. The micropores or capillary pores are still filled with water and can supply plants with needed water. The matric potential will vary slightly from soil to soil but is generally in the range of −10 to −30 kPa, assuming drainage into a less-moist zone of similar porosity (soil in a flower pot will cease drainage while much wetter than field capacity). Water movement will continue to take place by unsaturated flow, but the rate of movement is very slow since it now is due primarily to capillary forces, which are effective only in micropores. The water found in pores small enough to retain it against rapid gravitational drainage, but large enough to allow capillary flow in response to matric potential gradients, is sometimes termed **capillary water.**

Field capacity is a very useful term because it refers to an approximate degree of soil wetness at which several important soil properties are in transition. At field capacity, a soil is holding the maximal amount of water useful to plants. Additional water would remain in the soil for only a short time before draining, and would reduce soil aeration. At field capacity, the soil is near its lower plastic limit—that is, the soil behaves as a crumbly semisolid at water contents below field capacity, and as a plastic puttylike material that easily turns to mud at water contents above field capacity (see Section 4.9). Finally, at field capacity, sufficient pore space is filled with air to allow optimal aeration for most aerobic microbial activity and for the growth of most plants (see Section 7.6).

Permanent Wilting Percentage or Wilting Coefficient

Once an unvegetated soil has drained to its field capacity, further drying is quite slow, especially if the soil surface is covered to reduce evaporation. However, if plants are growing in the soil they will remove water from their rooting zone, and the soil will continue to dry. The roots will remove water first from the largest water-filled pores where the water potential is relatively high. As these pores are emptied, roots will draw their water from the progressively smaller pores and thinner water films in which the matric water potential is lower and the forces attracting water to the solid surfaces are greater. Hence, it will become progressively more difficult for plants to remove water from the soil at a rate sufficient to meet their needs.

As the soil dries, the rate of plant water removal may fail to keep up with plant needs, and plants may begin to wilt during the daytime to conserve moisture. At first the plants will regain their turgor at night when water is not being lost through the leaves and the roots can catch up with the plant's demand. Ultimately, however, the plant will remain wilted night and day when the roots cannot generate water potentials low enough to coax the remaining water from the soil.

The water content of the soil at this stage is called the **wilting coefficient** or **permanent wilting percentage** and by convention is taken to be that amount of water retained by the soil when the water potential is –1500 kPa. The soil will appear to be dusty dry, although some water remains in the smallest of the micropores and in very thin films (perhaps only 10 molecules thick) around individual soil particles (see Figure 5.22). As illustrated in Figure 5.23, **plant available water** is considered to be that water retained in soils between the states of field capacity and wilting coefficient (between –10 to –30 kPa and –1500 kPa). The amount of capillary water remaining in the soil that is unavailable to higher plants can be substantial, especially in fine-textured soils and those high in organic matter.

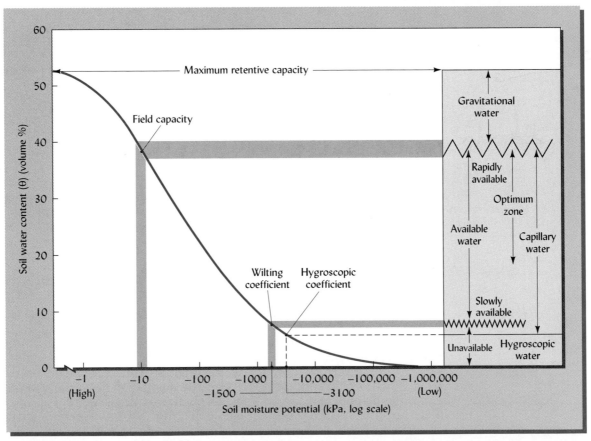

FIGURE 5.23 Water content–matric potential curve of a loam soil as related to different terms used to describe water in soils. The wavy lines in the diagram to the right suggest that measurements such as field capacity are only approximations. The gradual change in potential with soil moisture change discourages the concept of different "forms" of water in soils. At the same time, such terms as *gravitational* and *available* assist in the qualitative description of moisture utilization in soils.

Hygroscopic Coefficient

Although plant roots do not generally dry the soil beyond the permanent wilting percentage, if the soil is exposed to the air, water will continue to be lost by evaporation. When soil moisture is lowered below the wilting point, the water molecules that remain are very tightly held, mostly being adsorbed by colloidal soil surfaces. This state is approximated when the atmosphere above a soil sample is essentially saturated with water vapor (98% relative humidity) and equilibrium is established at a water potential of –3100 kPa. The water is thought to be in films only 4 or 5 molecules thick and is held so rigidly that much of it is considered nonliquid and can move only in the vapor phase. The moisture content of the soil at this point is termed the **hygroscopic coefficient.** Soils high in colloidal materials (clay and humus) will hold more water under these conditions than will sandy soils that are low in clay and humus. Soil water considered to be **unavailable** to plants includes the hygroscopic water as well as that portion of capillary water retained at potentials below –1500 kPa (see Figure 5.23).

Availability of Soil Water to Plants

EFFECT OF TEXTURE AND ORGANIC MATTER.   For soils with greater contents of silt and clay (finer textures), the amount of water held at field capacity tends to be greater. The water held at potentials below the wilting coefficient (water unavailable to plants) also increases with finer soil texture, but not as sharply. Therefore, the amount of water held at potentials between those of field capacity and wilting point (i.e., plant-available water) tends to be greatest in medium-textured soils such as silt loams. For most temperate-region soils, higher clay contents increase mainly the water held too tightly for plants to use (Figure 5.24). However, with increasing contents of low-activity-type clays (typical of humid tropical regions) and/or organic matter, soils tend to hold more plant-available water because the greater amount of water held at field capacity more than compensates for that held at potentials below wilting point (Figure 5.25). Increased soil organic matter can also enhance water availability to plants by improving soil structure and infiltration capacity.

EFFECT OF COMPACTION.   Soil compaction (increased bulk density) and weakening of soil structure can reduce the amount of water available to plants for several reasons. The compression of macropores into micropores results in less water being held at field capacity, more water being held at the wilting coefficient, fewer air-filled pores when

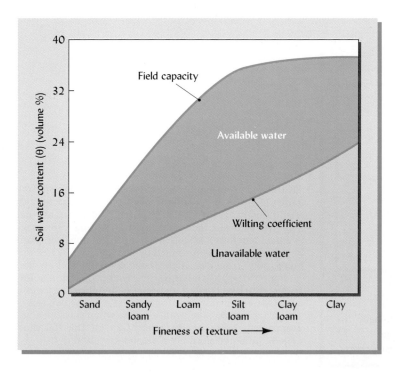

FIGURE 5.24   General relationship between soil water characteristics and soil texture. Note that the wilting coefficient increases as the texture becomes finer. The field capacity increases until we reach the silt loams, then levels off. Remember these are representative curves; individual soils would probably have values different from those shown.

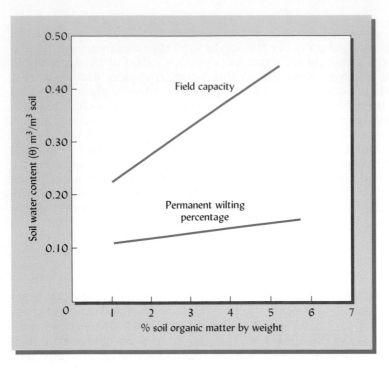

FIGURE 5.25 The effects of organic matter content on the field capacity and permanent wilting percentage of a number of silt loam soils. The differences between the two lines shown is the available soil moisture content, which was obviously greater in the soils with higher organic matter levels. [Redrawn from Hudson (1994); used with permission of the Soil & Water Conservation Society]

the soil is wet, and more cohesion of clay particles when the soil is relatively dry. The latter effect may increase soil strength enough to prevent roots from being able to penetrate soil layers which contain otherwise "available" water (see Figure 5.26).

EFFECT OF SOIL DEPTH AND LAYERING. We have already seen (Section 5.6) how soil profiles stratified with layers of abruptly differing texture can retard downward water movement and increase the amount of water held. In any field soil, each layer or horizon will have different characteristics that affect the amount of water held and the availability of that water to growing plants. The total depth of soil into which roots can explore also helps determine how much water will be available for plant use. A quantitative example involving some of these considerations is given in Box 5.2.

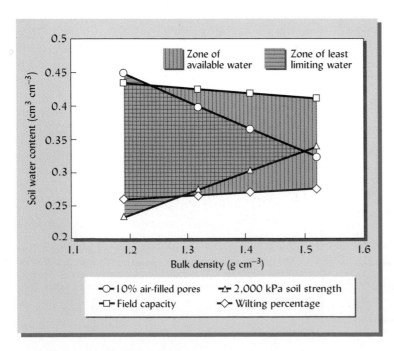

FIGURE 5.26 Influence of increased bulk density on the range of soil water contents available for plant uptake. Traditionally, plant-available water is defined as that retained between field capacity and wilting percentage (vertical hatching). If soils are compacted, however, plant use of water may be restricted by poor aeration (<10% air-filled pore space) at high water contents and by soil strength (>2000 kPa) that restricts root penetration at low water contents. The latter criteria define the *least limiting water range* shown by horizontal hatching. The two sets of boundary criteria give similar results when the soil is not compacted (bulk density about 1.25 for the soil illustrated). [Adapted from Da Silva and Kay (1997)]

BOX 5.2 CALCULATION OF THE TOTAL AVAILABLE WATER HOLDING CAPACITY OF A SOIL PROFILE

The total amount of water available to a plant growing in a field soil is a function of the rooting depth of the plant and the sum of the water held between field capacity and wilting percentage in each of the soil horizons explored by the roots. For each soil horizon, then, the mass available water holding capacity is estimated as the difference between the mass water contents θ_m (Mg water per 100 Mg soil) at field capacity and permanent wilting percentage. This value can be converted into a volume water content θ by multiplying by the ratio of the bulk density of the soil to the density of water. Finally, this volume ratio is multiplied by the thickness of the horizon to give the total centimeters of available water capacity (AWC) in that horizon. For the first horizon described in Table 5.5 the calculation (with units) is:

$$\left(\frac{22 \text{ g}}{100 \text{ g}} - \frac{8 \text{ g}}{100 \text{ g}} \right) * \frac{1.2 \text{ Mg}}{m^3} * \frac{1 \text{ m}^3}{1 \text{ Mg}} * 20 \text{ cm} = 3.36 \text{ cm AWC}$$

Note that all units cancel out except cm, resulting in the depth of available water (cm) held by the horizon. In Table 5.5, the available water holding capacity of all horizons within the rooting zone are summed to give a total AWC for the soil–plant system. Since no roots penetrated to the last horizon (1.0 to 1.25 m), this horizon was not included in the calculation. We can conclude that for the soil–plant system illustrated, 14.13 cm of water could be stored for plant use. At a typical summertime water-use rate of 0.5 cm of water per day, this soil could hold about a four-week supply.

TABLE 5.5 Calculation of Estimated Soil Profile Available Water Holding Capacity

Soil depth, cm	Relative root length	Soil bulk density, Mg/m³	Field capacity (FC), g/100g	Wilting percentage (WP), g/100g	Available water holding capacity (AWC), cm
0–20	xxxxxxxxx	1.2	22	8	$20 * 1.2 \left(\frac{22}{100} - \frac{8}{100} \right) = 3.36$ cm
20–40	xxxx	1.4	16	7	$20 * 1.4 \left(\frac{16}{100} - \frac{7}{100} \right) = 2.52$ cm
40–75	xx	1.5	20	10	$35 * 1.5 \left(\frac{20}{100} - \frac{10}{100} \right) = 5.25$ cm
75–100	xx	1.5	18	10	$25 * 1.5 \left(\frac{18}{100} - \frac{10}{100} \right) = 3.00$ cm
100–125	—	1.6	15	11	No roots
Total					$3.36 + 2.52 + 5.25 + 3.00 = 14.13$ cm

How Plant Roots Are Supplied with Water.   At any one time, only a small proportion of the soil water is adjacent to the absorptive root surfaces. How then do roots get access to the immense amount of water needed by growing plants? One answer is that water slowly moves to the roots by capillary flow in response to the matric potential gradient set up as roots dry out the soil in their immediate vicinity. At the same time, the root tips of growing plants are continually extending into wetter soil areas. The roots of most plants proliferate profusely in the upper 20 to 30 cm of soil (Table 5.6), but roots exploring pores and crevices 1 or 2 or more meters deep in the profile play a critical role in obtaining water, especially during droughty periods when the soil near the surface may become quite dry. Roots that spread widely near the soil surface and roots that penetrate many meters deep are two of the many types of root distribution patterns that plants have evolved in order to solve their water supply problems under less-than-ideal soil water conditions.

5.9 CONCLUSION

Water impacts all life. The interactions and movement of this simple compound in soils help determine whether these impacts are positive or negative. An understanding of the

TABLE 5.6 Percentage of Root Mass of Three Crops and Two Trees Found in the Upper 30 cm Compared with Deeper Depths (30–180 cm)

Plant species	Percentage of roots	
	Upper 30 cm	30–180 cm
Soybeans	71	29
Corn	64	36
Sorghum	86	14
Pinus radiata	82	18
Eucalyptus marginata	86	14

Data for crops from Mayaki, et al. (1976); for trees, estimated from Bowen (1985).

principles that govern the attraction of water for soil solids and for dissolved ions can help maximize the positive impacts while minimizing the less-desirable ones.

The water molecule has a polar structure that results in electrostatic attraction of water to both soluble cations and soil solids. These attractive forces tend to reduce the potential energy level of soil water below that of pure water. The extent of this reduction, called soil water potential ψ, has a profound influence on a number of soil properties, but especially on the movement of soil water and its uptake by plants.

The water potential due to the attraction between soil solids and water (the matric potential ψ_m) combines with the force of gravity ψ_g to largely control water movement. This movement is relatively rapid in soils high in moisture and with an abundance of macropores. In drier soils, however, the adsorption of water on the soil solids is so strong that its movement in the soil and its uptake by plants are greatly reduced. As a consequence, plants die for lack of water—even though there are still significant quantities of water in the soil—because that water is unavailable to plants.

Water is supplied to plants by capillary movement toward the root surfaces and by growth of the roots into moist soil areas. In addition, vapor movement may be of significance in supplying water for drought-resistant desert species (xerophytes). The osmotic potential ψ_o becomes important in soils with high soluble salt levels that can impede plant uptake of water from the soil. Such conditions occur most often in soils with restricted drainage in areas of low rainfall and in potted indoor plants.

The characteristics and behavior of soil water are very complex. As we have gained more knowledge, however, it has become apparent that soil water is governed by relatively simple, basic physical principles. Furthermore, researchers are discovering the similarity between these principles and those governing the movement of groundwater and the uptake and use of soil moisture by plants—the subject of the next chapter.

STUDY QUESTIONS

1. What is the role of the *reference state of water* in defining soil water potential? Describe the properties of this reference state of water.

2. Imagine a root of a cotton plant growing in the upper horizon of an irrigated soil in California's Imperial Valley. As the root attempts to draw water molecules from this soil, what forces (potentials) must it overcome? If this soil were compacted by a heavy vehicle, which of these forces would be most affected? Explain.

3. Using the terms *adhesion, cohesion, meniscus, surface tension, atmospheric pressure,* and *hydrophilic surface,* write a brief essay to explain why water rises up from the water table in a mineral soil.

4. Suppose you were hired to design an automatic irrigating system for a wealthy homeowner's garden. You determine that the flower beds should be kept at a water potential above –60 kPa, but not wetter than –10 kPa as the annual flowers here are sensitive to both drought and lack of good aeration. The rough turf areas, however, can do well if the soil dries to as low as –300 kPa. Your budget allows either tensiometers or electrical resistance blocks to be hooked up to electronic switching valves. Which instruments would you use and where? Explain.

5. Suppose the homeowner referred to in question 4 increased your budget and asked to use the TDR method to measure soil water contents. What additional information about the soils, not necessary for using the tensiometer, would you have to obtain to use either the resistance blocks or the TDR instrument? Explain.

6. A greenhouse operator was growing ornamental woody plants in 15-cm-tall plastic containers filled with a loamy sand. He watered the containers daily with a sprinkler system. His first batch of 1000 plants yellowed and died from too much water and not enough air. As an employee of the greenhouse, you suggest that he use 30-cm-tall pots for the next batch of plants. Explain your reasoning.

7. Suppose you measured the following data for a soil. A horizon from 0 to 30 cm with bulk density = 1.2 Mg/m^3: θ_m = 28% at –10 kPa, 20% at –100 kPa, and 8% at –1500 kPa. Bt horizon 30 to 70 cm with bulk density = 1.4 Mg/m^3: θ_m = 30% at –10 kPa, 25% at –100 kPa, and 15% at –1500 kPa. Bx horizon from 70 to 120 cm with bulk density = 1.95 Mg/m^3: θ_m = 20% at –10 kPa, 15% at –100 kPa, and 5% at –1500 kPa. Estimate the total available water holding capacity (AWC) of the soil in question.

8. A forester obtained a cylindrical core (L = 15 cm, r = 3.25 cm) of soil from a field site. She placed all the soil in a metal can with a tight-fitting lid. The empty metal can weighed 300 g and when filled with the field-moist soil weighed 972 g. Back in the lab, she placed the can of soil, with lid removed, in an oven for several days until it ceased to lose weight. The weight of the dried can with soil (including the lid) was 870 g. Calculate both θ_m and θ.

9. Give four reasons why compacting a soil is likely to reduce the amount of water available to growing plants.

10. Since even rapidly growing, finely branched root systems rarely contact more than 1 or 2% of the soil particle surfaces, how is it that the roots can utilize much more than 1 or 2% of the water held on these surfaces?

REFERENCES

Bowen, G. D. 1985. "Roots as components of tree productivity," in M. G. R. Cannell and J. E. Jackson (eds.), *Attributes of Trees as Crop Plants*. (Midlothian, Scotland: Institute of Terrestrial Ecology).

Cooney, J. J., and J. E. Peterson. 1955. *Avocado Irrigation*. Leaflet 50. California Agricultural Extension Service.

Da Silva, A. P., and B. D. Kay. 1997. "Estimating the least limiting water range of soil from properties and management," *Soil Sci. Soc. Amer. J.,* **61**:877–883.

Hillel, D. 1998. *Environmental Soil Physics*. Orlando, FL: Academic Press. 500 p.

Huck, M. G., B. Klepper, and H. M. Taylor. 1970. "Diurnal variations in root diameter," *Plant Physiol.,* **45**:529.

Hudson, B. D. 1994. "Soil organic matter and available water capacity," *J. Soil and Water Cons.,* **49**:189–194.

MacFall, J. S., and G. A. Johnson. 1994. "Use of magnetic resonance imaging of plants and soils," in S. H. Anderson and J. W. Hopmans (eds.), *Tomography of Soil–Water–Root Processes*. SSSA Special Publication no. 36. (Madison, Wis.: Soil Sci. Soc. Amer.), pp. 99–113.

Mayaki, W. C., L. R. Stone, and I. D. Teare. 1976. "Irrigated and nonirrigated soybean, corn, and grain sorghum root systems," *Agron. J.,* **68**:532–534.

Waddell, J. T. and Weil, R. R. 1996. "Water distribution in soil under ridge-till and no-till corn," *Soil Sci. Soc. Amer. J.,* **60**:230–237.

6

SOIL AND THE HYDROLOGIC CYCLE

Both soil and water belong to the biosphere, to the order of nature, and—as one species among many, as one generation among many yet to come—we have no right to destroy them.
—DANIEL HILLEL, OUT OF EARTH

One of the most striking—and troubling—features of human society is the yawning gap in wealth between the world's rich and poor. The life experience of the one group is quite incomprehensible to the other. So it is with the distribution of the world's water resources. The rain forests of the Amazon and Congo basins are drenched by more than 2000 mm of rain each year, while the deserts of North Africa and central Asia get by with less than 200. Nor is the supply of water distributed evenly throughout the year. Rather, periods of high rainfall and flooding alternate with dry spells or periods of drought.

Yet, one could say that everywhere the supply of water is adequate to meet the needs of the plants and animals native to the natural communities of the area. Of course, this is so only because the plants and animals have adapted to the local availability of water. Early human populations, too, adapted to the local water supplies by settling where water was plentiful from rain or rivers, by learning which underground gourds could quench one's thirst, by developing techniques to harvest water for agriculture and store it in underground cisterns, and by adopting nomadic lifestyles that allowed them and their herds to follow the rains and the grass supply.

But "civilized" humans have seemingly lost the ability and desire to adapt their needs and cultures to their environment. Rather, it is a mark of civilization that humans join in organized efforts bent on adapting their environment to their desires. Hence, the ancients tamed the flows of the Tigris and Euphrates. We moderns dig wells in the Sahel, bottle up the mighty Nile at Aswan, channel the waters of the Colorado to arid southern California, pump out the aquifers under farms and suburbs, and create sprawling cities (with swimming pools and bluegrass lawns!) in the deserts of the American Southwest or on the sands of Arabia. Truly, cities like Las Vegas are gambling in more ways than one.

There is plenty of room for improvement in managing water resources, and many of the improvements are likely to come as a result of better management of soils. The soil plays a central role in the cycling and use of water. By serving as a massive reservoir, soil helps moderate the adverse effects of excesses and deficiencies of water. The soil can help us treat and reuse waste waters from animal, domestic, and industrial sources. The flow of water through the soil in these and other circumstances connects the chemical pollution of soils to the possible contamination of groundwater.

In Chapter 5 we considered the nature and movement of water in soils. In this chapter we will see how those principles apply to the cycling of water between the soil, the atmosphere, and vegetation. We will then examine the unique role of the soil in water-resource management.

6.1 THE GLOBAL HYDROLOGIC CYCLE

There are nearly 1400 million km³ of water on the earth—enough (if it were all above-ground at a uniform depth) to cover the earth's surface to a depth of some 3 km. Most of this water, however, is relatively inaccessible and is not active in the annual cycling of water that supplies rivers, lakes, and living things.

The more actively cycling water is in the surface layer of the oceans, in shallow groundwater, in lakes and rivers, in the atmosphere, and in the soil (see Figure 6.1). Although the combined volume is a tiny fraction of the water on Earth, these pools of water are accessible for movement in and out of the atmosphere and from one place on the Earth's surface area to another.

The Hydrologic Cycle

Solar energy drives the cycling of water from the earth's surface to the atmosphere and back again in what is termed the ***hydrologic cycle*** (Figure 6.2). Solar energy stimulates *evaporation*—the conversion of liquid water into water vapor. The water vapor moves up

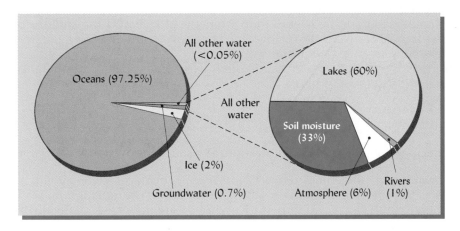

FIGURE 6.1 The sources of the earth's water. The preponderance of water is found in the oceans, glaciers and ice caps, and deep groundwater (left) but most of these waters are inaccessible for rapid exchange with the atmosphere and the land. The sources on the right, though much smaller in quantity, are actively involved in water movement through the hydrologic cycle. (Data from several sources)

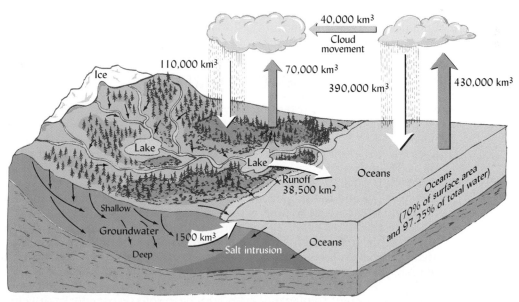

FIGURE 6.2 The hydrologic cycle upon which all life depends is very simple in principle. Water evaporates from the earth's surface, both the oceans and continents, and returns in the form of rain or snow-fall. The net movement of clouds brings some 40,000 km³ of water to the continents and an equal amount of water is returned through runoff and groundwater seepage that is channeled through rivers to the ocean. About 86% of the evaporation and 78% of the precipitation occurs in the ocean areas. However, the processes occurring on land areas where the soils are influential have impacts not only on humans but on all other forms of life, including those residing in the sea.

into the atmosphere, eventually forming clouds that can move from one region of the globe to another. Within an average of about 10 days, pressure and temperature differences in the atmosphere cause the water vapor to condense into liquid droplets or solid particles, which return to the earth as rain or snow.

About 500,000 km³ of water are evaporated from the earth surfaces and vegetation each year, some 110,000 km³ of which falls as rain or snowfall on the continents. Some of the water falling on land moves over the surface of the soil (surface runoff) and some infiltrates into the soil and drains into the groundwater. Both the surface runoff and groundwater seepage enter streams and rivers that, in turn, flow downstream into the oceans. The volume of water returned in this way is about 40,000 km³, which balances the same quantity of water that is transferred annually in clouds from the oceans to the continents.

Water Balance Equation

It is often useful to consider the components of the hydrologic cycle as they apply to a given **watershed,** an area of land drained by a single system of streams and bounded by ridges that separate it from adjacent watersheds. All the precipitation falling on a watershed is either stored in the soil, returned to the atmosphere (see Section 6.2), or discharged from the watershed as surface or subsurface flow (runoff). Water is returned to the atmosphere either by **evaporation** from the land surface (vaporization of soil water) or, after plant uptake and use, by vaporization from the stomata on the surfaces of leaves (a process termed **transpiration**). Together, these two pathways of evaporative loss to the atmosphere are called **evapotranspiration.**

The disposition of water in a watershed is often expressed by the **water-balance equation**, which in its simplest form is:

$$P = ET + SS + D$$

where P = precipitation, ET = evapotranspiration, SS = soil storage, and D = discharge.

For a forested watershed (sometimes termed a *catchment*), management may aim to maximize D so as to provide more water to downstream users. The equation makes it clear that discharge can be increased only if ET and/or SS are decreased, changes that may or may not be desirable. In the case of an irrigated field, water applied in irrigation would be included on the left side of the equation. Irrigation managers may want to save water applied by minimizing unnecessary losses in D and allowing negative values for soil storage (withdrawals of soil water) during parts of the year.

6.2 FATE OF PRECIPITATION AND IRRIGATION WATER

Some precipitation is intercepted by plant foliage and returned to the atmosphere by evaporation without ever reaching the soil. In forested areas, such **interception** may prevent 30 to 50% of the precipitation from reaching the soil. Most of the water that does reach the soil penetrates downward by the process of **infiltration**, especially if the soil surface structure is loose and open. If the rate of rainfall or snowmelt exceeds the infiltration capacity of the soil, considerable **surface runoff** and erosion may take place, thereby reducing the proportion of the water that moves into the soil and causing the pollution of lakes and streams with dissolved chemicals and detached soil particles.

Once the water penetrates the soil, some of it is subject to downward percolation and eventual loss from the root zone by **drainage.** In humid areas and in some irrigated areas of arid and semiarid regions, up to 50% of the precipitation may be lost as drainage below the root zone. However, during subsequent periods of low rainfall, some of this water may move back up into the plant-root zone by **capillary rise.** Such movement is important to plants in areas with deep soils, especially in dry climates.

The water retained by the soil is referred to as **soil storage** water, some of which eventually moves upward by capillarity and is lost by **evaporation** from the soil surface. Much of the remainder is absorbed by plants and then moves through the roots and stems to the leaves, where it is lost by **transpiration.** The water thus lost to the atmosphere by **evapotranspiration** may later return to the soil as precipitation or irrigation water, and the cycle starts again.

Factors Affecting Water Partitioning

TIMING OF PRECIPITATION.   The amount of water moving through each of the channels just discussed is influenced greatly by the timing, rate, and form of the precipitation. Heavy rainfall, even if of short duration, can supply water faster than most soils can absorb it. This accounts for the fact that in some arid regions with only 200 mm of annual rainfall, a cloudburst that brings 20 to 50 mm of water in a few minutes can result in flash flooding and gully erosion. A larger amount of precipitation spread over several days of gentle rain could move slowly into the soil, thereby increasing the stored water available for plant absorption, as well as replenishing the underlying groundwater. As illustrated in Figure 6.3, the timing of snowfall can affect the partitioning between surface runoff and infiltration.

TYPE OF VEGETATION.   Plants help determine the proportion of water that runs off and that which penetrates the soil. The vegetation and surface residues of perennial grasslands and dense forests protect the porous soil structure from the beating action of raindrops. Therefore, they encourage water infiltration and reduce the likelihood that soil will be carried off by any runoff that does occur. In general, very little runoff occurs from land under undisturbed forests or well-managed turf grass.

STEM FLOW.   Many plant canopies direct rainfall toward the plant stem, thus altering the spatial distribution of rain reaching the soil. Under a forest canopy, more than half of the rainfall may trickle down the leaves, twigs, and branches to the tree trunk, there to progress downward as **stem flow.** Likewise, certain crop canopies, such as that of corn, funnel a large proportion of the rainfall to the soil in the crop row (Figure 6.4).

SOIL MANAGEMENT.   A major objective of soil- and water-management systems is to encourage water infiltration rather than runoff. This may be achieved by enhancing soil surface storage to allow more time for infiltration to take place (Figure 6.5a). Another

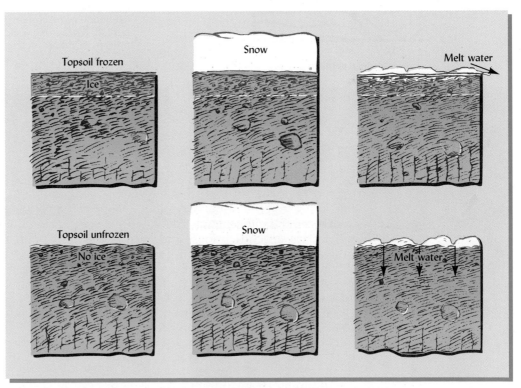

FIGURE 6.3   The relative timing of freezing temperatures and snowfall in the fall in some temperate regions drastically influences water runoff and infiltration into soils in the spring. The upper three diagrams illustrate what happens when the surface soil freezes before the first heavy snowfall. The snow insulates the soil so that it is still frozen and impermeable as the snow melts in the spring. The lower sequence of diagrams illustrates the situation when the soil is unfrozen in the fall when it is covered by the first deep snowfall.

FIGURE 6.4 Vertical and horizontal distribution of soil water resulting from stem flow. The contours indicate the soil water potential in kPa between two corn rows in a sandy loam soil. During the previous two days, 26 mm of rain fell on this field. Many plant canopies, including that of the corn crop shown, direct a large proportion of rainfall toward the plant stem. Stem flow results in uneven spatial distribution of water. In cropland this may have ramifications for the leaching of chemicals such as fertilizers, depending on whether they are applied in or away from the zone of highest wetting near the plant stems. The concentration of water by stem flow may also increase the likelihood of macropore flow in soils after only moderate rainfall. [Data from Waddell and Weil (1996); used by permission of the Soil Science Society of America]

(a)

(b)

FIGURE 6.5 Managing soils to increase infiltration of rainwater. (a) Small furrow dikes on the right side of this field in Texas retain rainwater long enough for it to infiltrate rather than run off. (b) This saturated soil under a winter cover crop of hairy vetch is riddled with earthworm burrows that greatly increased the infiltration of water from a recent heavy rain. Scale in centimeters. [Photo (a) courtesy of O. R. Jones, USDA Agricultural Research Service, Bushland, Texas; photo (b) courtesy of R. Weil] approach is to establish dense vegetation during periods of high rainfall. This may include the use of **cover crops**, plants established between the principle crop-growing seasons (Figure 6.5*b*). Such cover crops encourage the activities of earthworms and protect the soil structure, leading to greatly enhanced water infiltration. However, remember that cover crops also transpire water. If the following crop will be dependent on soil storage water, care may be needed to kill the cover crop before it can dry out the soil profile.

Compaction by heavy equipment used in farming and forestry can sharply reduce water infiltration rates, thus increasing runoff losses.

SOIL PROPERTIES. If the soil is loose and open (e.g., sands and well-granulated soils), a high proportion of the incoming water will infiltrate the soil, and relatively little will run off. In contrast, heavy clay soils with unstable soil structures resist infiltration and encourage runoff. These differences, attributable to soil properties as well as vegetation, are illustrated in Figure 6.6.

Having considered the overall hydrologic cycle, we will now focus our attention on the component of the cycle for which soils and plants play the most prominent roles.

6.3 THE SOIL–PLANT–ATMOSPHERE CONTINUUM

The flow of water through the soil–plant–atmosphere continuum (SPAC) is a major component of the overall hydrologic cycle. Figure 6.7 ties together many of the processes we have just discussed: *interception, surface runoff, percolation, drainage, evaporation, plant water uptake, water movement to plant leaves,* and *transpiration* of water from the leaves back into the atmosphere.

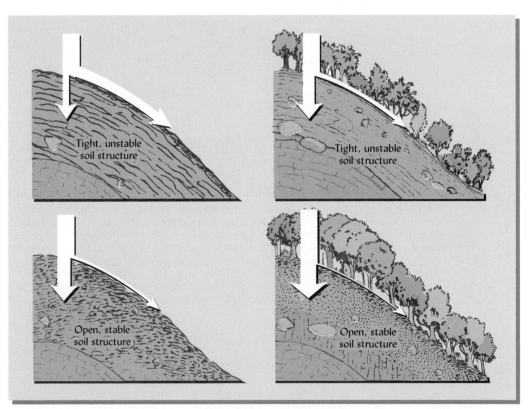

FIGURE 6.6   Influence of soil structure and vegetation on the partitioning of rainfall into infiltration and runoff. The upper two diagrams show soils with tight, unstable structure that resists infiltration and percolation. The bare soil is especially prone to surface sealing and resulting high losses by runoff. Even with forest cover, the low-permeability soils cannot accept all the rain in an intense storm. The two lower diagrams show much greater infiltration into soils that have open, stable structures with significant macropore space. The more open soil structure combined with the protective effects of the forest floor and canopy nearly eliminate surface runoff.

FIGURE 6.7 Soil–plant–atmosphere continuum (SPAC) showing water movement from soil to plants to the atmosphere and back to the soil in a humid to subhumid region. Water behavior through the continuum is subject to the same energy relations covering soil water that were discussed in Chapter 5. Note that the moisture potential in the soil is −50 kPa, dropping to −70 kPa in the root, declining still further as it moves upward in the stem and into the leaf, and is very low (−500 kPa) at the leaf–atmosphere interface, from whence it moves into the atmosphere where the moisture potential is −20,000 kPa. Moisture moves from a higher to lower moisture potential. Note the suggested ranges for partitioning of the precipitation and irrigation water as it moves through the continuum.

If a plant is to absorb water from the soil, the water potential must be lower in the plant root than in the soil adjacent to the root. Likewise, movement up the stem to the leaf cells is in response to differences in water potential, as is the movement from leaf surfaces to the atmosphere.

Two primary factors determine whether plants are well supplied with water: (1) the rate at which water is supplied by the soil to the absorbing roots, and (2) the rate at which water is evaporated from the plant leaves. Since factors affecting the soil's ability to supply water were discussed in Section 5.8, we will now address the loss of water by evaporation from soil–plant systems.

Evapotranspiration

It is quite difficult to determine just how much of the loss of soil water occurs directly from the soil (by **evaporation**, E) and how much occurs from the leaf surfaces after plant uptake (by **transpiration**, T). **Evapotranspiration** (ET), the combined loss resulting from these two processes, is more readily determined. The evaporation component of ET may be viewed as a "waste" of water from the standpoint of plant productivity. However, at least some of the transpiration component is essential for plant growth, providing the water that plants need for cooling, nutrient transport, photosynthesis and turgor maintenance.

POTENTIAL ET. The **potential evapotranspiration** rate (PET) tells us how fast water vapor *would* be lost from a densely vegetated plant–soil system *if* soil water content were continuously maintained at an optimal level. The PET is largely determined by climatic variables, such as temperature, relative humidity, cloud cover, and wind speed, that influence the *vapor pressure gradient* between a wet soil, leaf, or body of water and the atmosphere.

A number of mathematical models have been devised to estimate PET from climatological data, but in practice, PET can be most easily estimated by applying a correction factor to the amount of water evaporated from an open pan of water of standard design (a class-A evaporation pan—see Figure 6.8). Dense, well-watered vegetation typically transpires water about 65% as rapidly as water evaporates from an open pan; hence, the correction factor for dense vegetation such as a lawn is typically 0.65 (it is lower for less-dense vegetation):

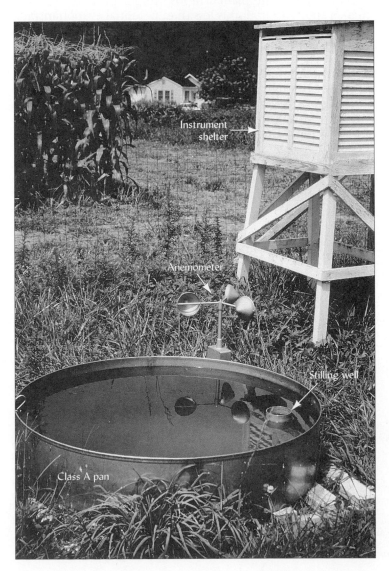

FIGURE 6.8 A class-A evaporation pan used to help estimate potential evapotranspiration (PET). Once a day, the water level is determined in the stilling well (small cylinder) and a measured amount of water is added to bring the level back up to the original mark. Evaporation from the pan integrates the effects of relative humidity, temperature, wind speed, and other climatic variables related to the vapor pressure gradient. Also shown are an anemometer to measure wind speed and a shelter containing instruments to measure temperature and relative humidity. (Photo courtesy of R. Weil)

$$PET \approx 0.65 * \text{pan evaporation}$$

Values of PET range from more than 1500 mm per year in hot, arid areas, to less than 40 mm in very cold regions. During the winter in temperate regions, PET may be less than 1 mm per day. By contrast, hot, dry wind will continually sweep away water vapor from a wet surface, creating a particularly steep vapor pressure gradient and PET levels as high as 10 to 12 mm per day.

EFFECT OF SOIL MOISTURE.   Evaporation from the soil surface E at a given temperature is determined to a large extent by soil surface wetness and by the ability of the soil to replenish this surface water as it evaporates. In most cases, the upper 15 to 25 cm of soil provides most of the water for surface evaporation E. Unless a shallow water table exists, the upward capillary movement of water is very limited and the surface soil soon dries out, greatly reducing further evaporation loss.

However, because plant roots penetrate deep into the profile, a significant portion of the water lost by evapotranspiration ET comes from the subsoil layers (see Figure 6.9). As was illustrated in Figure 1.5, water stored deep in the profile is especially important to vegetation in regions having alternating moist and dry seasons (such as Ustic or Xeric moisture regimes).

PLANT WATER STRESS.   For dense vegetation growing in soil well supplied with water, ET will nearly equal PET. When soil water content is less than optimal, the plant will not be able to withdraw water from the soil fast enough to satisfy the PET. If water evaporates from the leaves faster than it enters the roots, the plant will lose turgor pressure and wilt. Under these conditions, actual evapotranspiration is less than potential evapotranspiration and the plant experiences **water stress**. The difference between PET and actual ET is termed the **water deficit**. A large deficit is indicative of high water stress and aridity.

INFLUENCE OF PLANT CHARACTERISTICS.   In vegetated areas, incoming solar radiation is either absorbed by plant leaves on its way through the plant canopy, or it reaches the ground and is absorbed by the soil. Therefore, for annual plants, as the leaf area per unit land area (a ratio termed the **leaf area index**) increases during the growing season, more radiation will be absorbed by the foliage to stimulate transpiration (T), and less will reach the soil to promote evaporation (E).

In contrast, perennial vegetation, such as pastures and forests, have very high leaf area indices both early and late in the growing season. Where the forest floor (leaf litter) is undisturbed, even less direct sunlight strikes the soil and E is very low throughout the year.

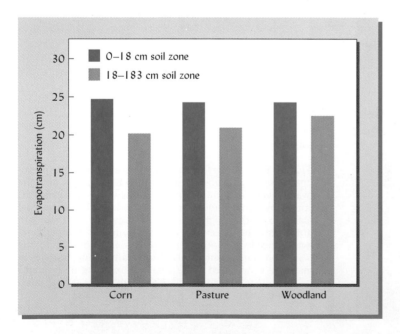

FIGURE 6.9 Evapotranspiration loss from the surface layer (0 to 18 cm) compared with loss from the subsoil (18 to 183 cm). Note that more than half the water loss came from the upper 18 cm and only half from the *next 165 cm* of depth. Periods of measurement: corn, May 23 to September 25; pasture, April 15 to August 23; woodland, May 25 to September 28. [Calculated from Dreibelbis and Amerman (1965)]

In summary, water losses from the soil surface and from transpiration are determined by: (1) climatic conditions, (2) plant cover in relation to soil surface (leaf area index), (3) efficiency of water use by different plants, and (4) length and season of the plant growing period.

6.4 EFFICIENCY OF WATER USE

The vegetative dry matter produced while using a given amount of water is an important measure of efficiency, especially in areas where moisture is scarce. This efficiency may be expressed in terms of dry matter yield per unit of water transpired (*T efficiency*) or the dry matter yield per unit of water lost by evapotranspiration (*ET efficiency*).

TRANSPIRATION RATIO. Another expression of water use efficiency is the **transpiration ratio,** which is the inverse of T efficiency and is expressed as kilograms of water transpired to produce 1 kg of dry matter. The transpiration ratio for a given plant is markedly affected by climatic conditions, ranging from 200 to 500 in humid regions and almost twice as much in arid regions (Table 6.1).

At a given location there are considerable differences in transpiration ratio among different plant species. For example, the data in Table 6.1 show that in semiarid Akron, Colorado, millet and maize (corn) have relatively low transpiration ratios compared to clover.

We can see the significance of these figures by considering a wheat crop growing in a semiarid region and having a transpiration ratio of 500. For every kilogram of above-ground dry matter produced, about 500 kg (or liters) of water were transpired. If we assume that only about 40% of the dry matter produced by the time of harvest is in the grain (the rest is in the straw), we see that it takes about 1250 L of water to produce 1 kg of wheat grain. When we take into account the water used to grow fruits, vegetables, and feed for cattle, as well as grains, no wonder it is estimated that almost 7000 L (1700 gal) of water are required to grow a single day's food supply for one adult in the United States.

Indeed, the amount of water necessary to bring a crop to maturity is very large. For example, a representative crop of wheat containing 5000 kg/ha of dry matter and having a transpiration ratio of 500 will withdraw water from the soil during the growing season equivalent to about 250 mm of rain. This amount of water, *in addition* to that evaporated from the surface, must be supplied during the growing season. It is not surprising that water—and the ability of the soil to store it—is often the most critical factor in plant growth.

ET EFFICIENCY. Since evapotranspiration (ET) includes both transpiration (T) from plants and evaporation (E) from the soil surface, ET efficiency is more subject to management than T efficiency. Highest ET efficiency is attained where plant density and other growth factors are optimum for plant growth. Increases in leaf area index (LAI) lead to increases in plant production, up to the point at which virtually all available solar radiation is absorbed by leaf surfaces. Furthermore, as shown in Figure 6.10, when the LAI is increased over the range of 0 to about 4, plant production increases by a greater amount than does ET. The plant production per unit of water vaporized (ET efficiency) is therefore increased.

TABLE 6.1 Transpiration Ratios of Various Crops as Determined at Different Locations

Kilograms of water used in production of 1 kg of dry matter.

Crop	Harpenden, England	Dahme, Germany	Madison, Wisconsin	Pusa, India	Akron, Colorado
Beans	209	282	—	—	736
Clover	269	310	576	—	797
Maize (corn)	—	—	271	337	368
Millet	—	—	—	—	310
Oats	—	376	503	469	597
Peas	259	273	477	563	788
Potatoes	—	—	385	—	636
Wheat	247	338	—	544	513

Data compiled by Lyon, et al. (1952).

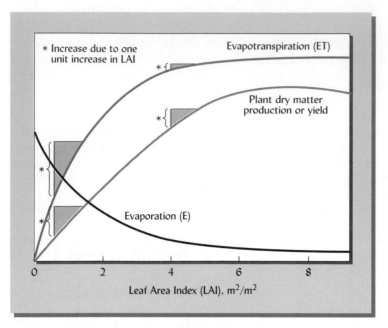

FIGURE 6.10 Generalized effect of vegetative leaf area index (LAI) on plant production, evapotranspiration (ET), and evaporation (E). The stippled areas show the effects of increasing LAI by one unit. Note that ET increases with increasing LAI up to a point and then levels off when the foliage intercepts nearly all the incoming solar radiation. Simultaneously, E decreases and levels off at a very low value. Except at either very low or very high LAI levels, practices (such as fertilization or closer plant spacing) that increase LAI are likely to increase plant productivity by a greater percentage than the corresponding increase in ET, resulting in improved water-use efficiency.

EFFECT OF CROP MANAGEMENT.   Figure 6.11 illustrates the effect on ET efficiency of fertilizing a nutrient-deficient soil. The increased production per unit of water vaporized is the result of the fertilized, more vigorous plants (1) sending their roots deeper into the soil profile, and (2) producing a greater leaf area, which intercepts a larger portion of the solar radiation, leaving less radiation to evaporate water from the soil (see also Figure 6.10). Therefore, as plant production is increased by improved nutrition, T is increased substantially, but E is decreased, so that ET is increased only a little. We can conclude that so long as the supply of water is not too limited, maintaining optimum conditions for plant growth (by closer plant spacing, fertilization, or selection of more vigorous varieties) increases the efficiency of water use by plants.

In harsh semiarid environments, one should apply this principle with caution, however, if irrigation water is not available and the period of rainfall is very short. The modest increase in total water use (ET) by the more vigorously growing plants may deplete the stored soil water before the life cycle of the plant is complete, resulting in serious water stress, or even plant death before any harvestable yield has been produced.

FIGURE 6.11 Water-use efficiency of wheat as affected by application of phosphorus fertilizer with and without nitrogen fertilizer (12-year average). ET efficiency increased with increased fertilizer rates. [Data from Black (1982)]

6.5   CONTROL OF EVAPOTRANSPIRATION (ET)

As we have just seen, water loss by ET is closely related to the total leaf area exposed to solar radiation. To some degree, adjusting the total leaf area exposed can bring ET into balance with PET (and thus lessen water stress). Five approaches to accomplish this are the following:

1. If water can be added by irrigation (an option that will be discussed in Section 6.11) both ET and plant production are likely to increase, often dramatically. Of course, for reasons of cost or lack of a water source, this practice is often not feasible.

2. Where rapid growth might prematurely deplete the available water, it may be wise to limit plant-growth factors such as nutrient supply to only moderate levels, thus keeping LAI in check.

3. The LAI of desired plants can be limited by sowing fewer seeds per unit land area or by spacing plants farther apart. It should be noted, however, that plants growing farther apart tend to individually produce a greater leaf area, compensating somewhat for the lower density. Also, wider spacing allows greater evaporation losses from the soil.

4. Elimination of undesired plants (i.e., weeds) can remove a major cause of T losses that could greatly reduce the soil water supplies available for the desired species.

5. In semiarid regions, elimination of all vegetation for a period of time may allow rainwater to accumulate in the soil profile for use by subsequent plantings.

Figure 3.19 (page 81) shows typical alternating strips of summer **fallow** (unvegetated period) and wheat in a region of Ustolls.

6.6   CONTROL OF SURFACE EVAPORATION (E)

More than half the precipitation in semiarid and subhumid areas is usually returned to the atmosphere by evaporation (E) directly from the soil surface. In natural rangeland systems, E is a large part of ET because plant communities tend to self-regulate toward low plant densities and leaf area configurations that minimize the deficit between PET and ET. In addition, plant residues on the soil surface are sparse. Evaporation losses are also high in arid-region irrigated soil, especially if inefficient practices are used (see Section 6.11). Even in humid-region rain-fed areas, E losses are significant during hot rainless periods. Such moisture losses rob the plant community of much of its growth potential and reduce the water available for discharge to streams. Careful study of Figure 6.12 will clarify these relationships and the principles discussed in Sections 6.4 and 6.5.

For arable soils, the most effective practices aimed at controlling E are those that provide some cover for the soil. This cover can best be provided by mulches and by selected conservation tillage practices that leave plant residues on the soil surface, mimicking the soil cover of natural ecosystems.

VEGETATIVE MULCHES.   A *mulch* is any material placed on the soil surface primarily for the purpose of reducing evaporation or controlling weeds. Examples include sawdust, manure, straw, leaves, and crop residues. Mulches can be highly effective in checking evaporation, but they may be expensive and labor-intensive to produce or purchase, transport to the field, and apply to the soil. Mulching is therefore most practical for small areas (gardens and landscaping beds) and for high-value crops such as cut flowers, berries, fruit trees, and certain vegetables. In addition to reducing evaporation, vegetative mulches may provide these benefits: (1) reducing the spread of soilborne diseases; (2) providing a clean path for foot traffic; (3) reducing weed growth (if applied thickly); (4) moderating soil temperatures, especially preventing overheating in summer months (see Section 7.11); (5) increasing water infiltration; (6) providing organic matter and, possibly, plant nutrients to the soil; (7) encouraging earthworm populations; and (8) reducing soil erosion. Most of these ancillary benefits do not accrue from the use of plastic or paper mulches, discussed next.

PAPER AND PLASTIC MULCHES.   Specially prepared paper and plastics are also used as mulches to control evaporative water losses. In the case of black plastic and dark paper, they also effectively control weeds. The mulch is often applied by machine and plants grow

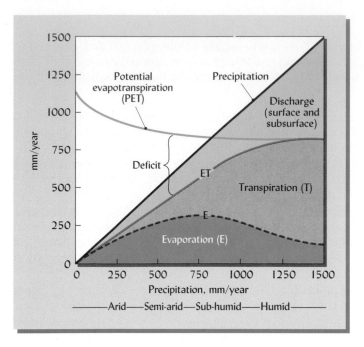

FIGURE 6.12 Partitioning of liquid water losses (discharge) and vapor losses (evaporation and transpiration) in regions varying from low (arid) to high (humid) levels of annual precipitation. The example shown assumes that temperatures are constant across the regions of differing rainfall. Potential evapotranspiration (PET) is somewhat higher in the low-rainfall zones because the lower relative humidity there increases the vapor pressure gradient at a given temperature. Evaporation (E) represents a much greater proportion of total vapor losses ET in the drier regions due to sparse plant cover caused by interplant competition for water. The greater the gap between PET and ET, the greater the deficit and the more serious the water stress to which plants are subject.

through slits or holes in the sheeting (Figure 6.13). These mulches are widely used for vegetable and small fruit crops and in landscaping beds, where they are often covered with a layer of tree-bark mulch or gravel for a more pleasing appearance. As long as the ground is covered, evaporation is checked, and in some cases remarkable increases in plant growth have been reported. Unless rainfall is very heavy, the mulch does not seriously interfere with the infiltration of rainwater into the soil.

The temperature of the soil under plastic mulch is commonly 8 to 10°C higher than under a straw mulch. The higher temperature may be helpful for crops established early in spring, but it can reduce the yields of heat-sensitive plants in summer (see Section 7.11). A common problem with plastic mulches is that they can rarely be completely removed at the end of the growing season, so after a number of years scraps of plastic accumulate in the soil, causing an unsightly mess and interfering with water movement and cultivation.

CROP RESIDUE AND CONSERVATION TILLAGE.   Plant residues left on the soil surface are effective in reducing E and, in turn, in conserving soil moisture. *Conservation tillage* practices leave a high percentage of the residues from the previous crop on or near the surface (Figure 6.14). A conservation tillage practice widely used in subhumid and semiarid

FIGURE 6.13   For crops with high cash value, plastic mulches are commonly used. The plastic is installed by machine (left) and at the same time the plants are transplanted (right). Plastic mulches help control weeds, conserve moisture, encourage rapid early growth, and eliminate need for cultivation. The high cost of plastic makes it practical only with the highest-value crops. (Courtesy K. Q. Stephenson, Pennsylvania State University)

FIGURE 6.14 Conservation tillage leaves plant residues on the soil surface, reducing both evaporation losses and erosion. (Left) In a semiarid region (South Dakota) the straw from the previous year's wheat crop was only partially buried to anchor it against the wind while still allowing it to cover much of the soil surface. In the next year the left half of the field, now shown growing wheat, will be thusly *stubble mulched* and the right half sown to wheat. (Right) *No-till* planted corn in a more humid region grows up through the straw left on the surface by a previous wheat crop. Note that with no-till, almost no soil is directly exposed to solar radiation, rain or wind. (Photos courtesy of R. Weil, left, and USDA Natural Resources Conservation Service, right)

regions is **stubble mulch** tillage. With this method, residues such as wheat stubble or cornstalks from the previous crop are uniformly spread on the soil surface. The land is then tilled with special implements that permit much of the plant residue to remain on or near the surface. Conservation tillage planters are capable of planting through the stubble and allow much of it to remain on the surface during the establishment of the next crop (see Figure 6.14). Combining summer fallow with stubble mulching saves soil moisture in dry areas, as shown in Table 6.2. Unfortunately, plant growth in dry regions is usually insufficient to produce residue mulches at the levels needed to reduce E.

Other conservation tillage systems that leave residues on the soil surface include no-tillage (see Figure 6.14 and Box 4.3, p. 120), where the new crop is planted directly into the sod or residues of the previous crop, with no plowing or disking. This and other systems will receive further attention in Section 15.5.

6.7 PERCOLATION-EVAPORATION BALANCE

In our discussion of the hydrologic cycle we noted two types of liquid losses of water from soils: (1) percolation or drainage water, and (2) runoff water (see Figure 6.2). *Percolation water* recharges the groundwater and moves chemicals out of the soil. *Runoff water* often carries appreciable amounts of soil (erosion) as well as dissoved chemicals.

TABLE 6.2 Gains in Soil Water from Different Rates of Straw Mulch During Fallow Periods at Four Great Plains Locations

Location	Av. annual precipitation, mm	Soil water gain, mm, at each mulch rate			
		0 Mg/ha	2.2 Mg/ha	4.4 Mg/ha	6.6 Mg/ha
Bushland, Tex.	508	71	99	99	107
Akron, Colo.	476	134	150	165	185
North Platte, Nebr.	462	165	193	216	234
Sidney, Mont.	379	53	69	94	102
Average		107	127	145	157
Average gain by mulching			20	38	50

From Greb (1983).

When the amount of rainfall entering a soil exceeds the water-holding capacity of the soil, losses by percolation will occur. Percolation losses are influenced by (1) the amount of rainfall and its distribution, (2) runoff from the soil, (3) evaporation, (4) the character of the soil, and (5) the nature of the vegetation.

Figure 6.15 illustrates the relationships among precipitation, runoff, soil storage, soil water depletion, and percolation for representative humid and semiarid regions and for an irrigated arid region.

HUMID REGION.   In the *humid temperate* region, the rate of water infiltration into the soil (precipitation minus runoff) is greater, at least during certain seasons, than the rate of evapotranspiration. As soon as the soil field capacity is reached, percolation into the substrata occurs. Maximum percolation occurs during the winter and early spring, when evaporation is lowest. This period is when most leaching and chemical pollution of groundwater takes place. During the summer, evapotranspiration exceeds precipita-

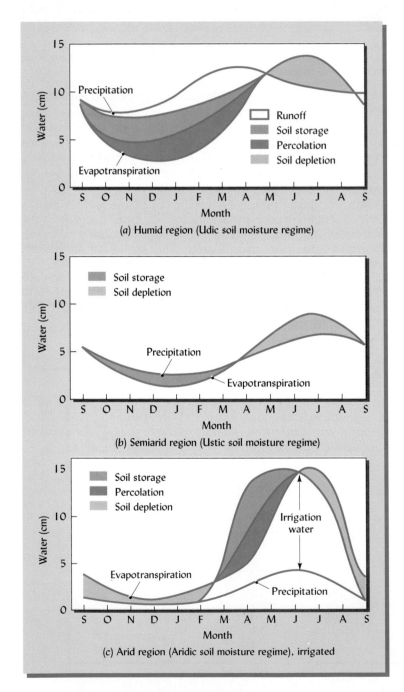

FIGURE 6.15 Generalized curves for precipitation and evapotranspiration for three temperate zone regions: (*a*) a humid region, (*b*) a semiarid region, and (*c*) an irrigated arid region. Note the absence of percolation through the soil in the semiarid region. In each case water is stored in the soil. This water is released later when evapotranspiration demands exceed the precipitation. In the semiarid region evapotranspiration would likely be much higher if ample soil moisture were available. In the irrigated arid region soil, the very high evapotranspiration needs are supplied by irrigation. Soil moisture stored in the spring is utilized by later summer growth and lost through evaporation during the late fall and winter.

tion, resulting in a depletion of soil water. Normal plant growth is possible only because of water stored in the soil from the previous winter and early spring.

SEMIARID REGION. In the *semiarid region* water is stored in the soil during the winter months and is used to meet the moisture deficit in the summer. But because of the low rainfall, little runoff and percolation out of the profile occurs. Water may move to the lower horizons, but it is absorbed by plant roots and ultimately is lost by transpiration.

IRRIGATED SOIL. The irrigated soil in the arid region shows a unique pattern. Irrigation in the early spring, along with a little rainfall, provides more water than is being lost by evapotranspiration. The soil is charged with water, and some percolation may occur. As we shall see in Section 9.19, irrigation systems must provide enough water for some percolation, in order to remove excess soluble salts. During the summer, fall, and winter months, the stored water is depleted because the amount being added is less than the very high evapotranspiration that takes place in response to large vapor pressure gradients.

The comparative losses of water by evapotranspiration and percolation through soils found in different climatic regions are shown in Figure 6.16. These differences should be kept in mind while reading the following section on percolation and groundwaters.

6.8 PERCOLATION AND GROUNDWATERS

When drainage water moves downward and out of the soil, it eventually encounters a zone in which the pores are all saturated with water. Often this saturated zone lies above a layer of impermeable rock or clay. The upper surface of this zone of saturation is known as the **water table**, and the water within the saturated zone is termed **groundwater.** The water table (Figure 6.17) is commonly only 1 to 10 m below the soil surface in humid regions but may be several hundred or even thousands of meters deep in arid regions. In swamps it is essentially at the land surface.

The unsaturated zone above the water table is termed the **vadose zone** (see Figures 6.17 and 6.18). The vadose zone may include unsaturated materials underlying the soil profile, and so may be considerably deeper than the soil itself. In some cases, however, the saturated zone may be sufficiently high to include the lower soil horizons, with the vadose zone confined to the upper soil horizons.

Shallow groundwater receives downward-percolating drainage water. Most of the groundwater, in turn, seeps laterally through porous geological materials (termed **aquifers**) until it is discharged into springs and streams. Groundwater may also be

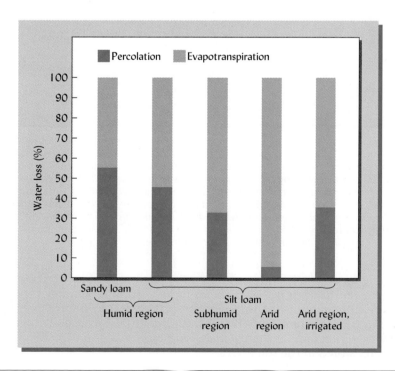

FIGURE 6.16 Percentage of the water entering the soil that is lost by downward percolation and by evapotranspiration. Representative figures are shown for different climatic regions.

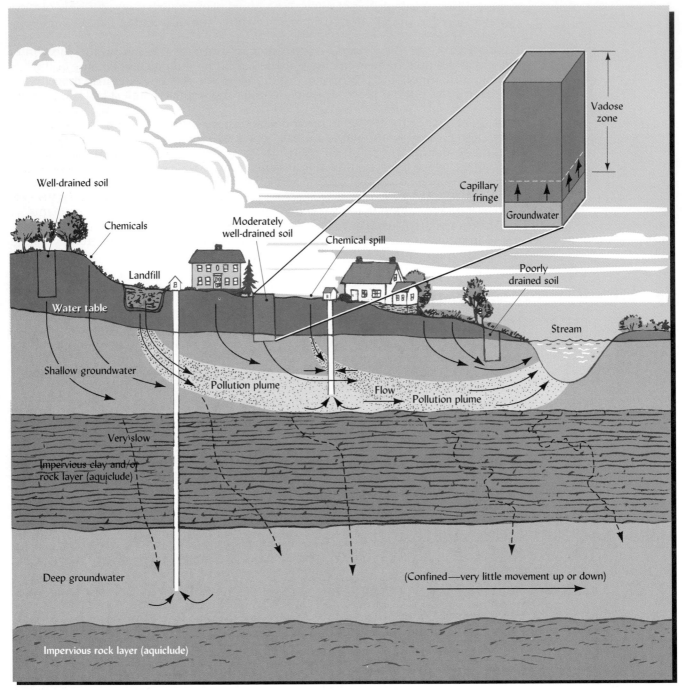

FIGURE 6.17 Relationship of the water table and groundwater to water movement into and out of the soil. Precipitation and irriga- tion water move down the soil profile under the influence of gravity (gravitational water), ultimately reaching the water table and underlying shallow groundwater. The unsaturated zone above the water table is known as the *vadose zone*. As water is removed from the soil by evapotranspiration, groundwater moves up from the water table by capillarity in what is termed the *capillary fringe*. Groundwa- ter also moves horizontally down the slope toward a nearby stream, carrying with it chemicals that have leached through the soil, including essential plant nutrients (N, P, Ca, etc.) as well as pesticides and other pollutants from domestic, industrial, and agricultural wastes. Groundwaters are major sources of water for wells, the shallow ones removing water from the groundwater near the surface, while deep wells exploit deep and usually large groundwater reserves. Two plumes of pollution are shown, one originating from land- fill leachate, the other from a chemical spill. The former appears to be contaminating the shallow well.

removed by pumping for domestic and irrigation uses. The water table will move up or down in response to the balance between the amount of drainage water coming in through the soil and the amount lost through pumped wells and natural seepage to springs and streams. In humid temperate regions, the water table is usually highest in early spring following winter rains and snow melt, but before evapotranspiration begins to withdraw the water stored in the soil above.

Groundwater that is near the surface can serve as a reciprocal water reservoir for the soil. As plants remove water from the soil, it may be replaced by upward capillary movement from a shallow water table. The zone of wetting by capillary movement is known as the **capillary fringe** (see Figure 6.18). Such movement can provide a steady and significant supply of water that enables plants to survive during periods of low rainfall.[1]

LEACHING.[2]   Percolation of water through the soil to the water table not only replenishes the groundwater, it also dissolves and carries downward a variety of inorganic and organic chemicals found in the soil or on the land surface. Chemicals *leached* from the soil to the groundwater (and eventually to streams and rivers) in this manner include elements weathered from minerals, natural organic compounds resulting from the decay of plant residues, plant nutrients derived from natural and human sources, and various synthetic chemicals applied intentionally or inadvertently to soils.

In the case of plant nutrients, especially nitrogen, downward movement through the soil and into underlying groundwaters has two serious implications. First, the leaching of these chemicals represents a depletion of plant nutrients from the root zone (see Section 14.2). Second, accumulation of these chemical nutrients in ponds, lakes, reservoirs, and groundwater downstream may stimulate a process called *eutrophication,* which ultimately depletes the oxygen content of the water, with disastrous effects on fish and other aquatic life (see also Section 13.2). Also, in some areas underground sources of drinking water may become contaminated with excess nitrates to levels unsafe for human consumption (see Section 12.8).

Of even more concern is the contamination of groundwater with various highly toxic synthetic compounds, such as pesticides and their breakdown products or chemicals leached out of waste disposal sites. Figure 6.17 illustrates how the groundwater is charged with these chemicals, and how a plume of contamination spreads to downstream wells and bodies of water.

Chemical Movement Through Macropores

Macropores play a critical role in determining field hydraulic conductivity (see Section 5.5). Old root channels, earthworm burrows, and clay shrinkage cracks commonly con-

---

[1] Capillary rise from shallow groundwater may also bring a steady supply of salts to the surface if the groundwater is brackish (see Section 9.14 for details on this soil-degrading process).

[2] For a review of chemical transport through field soils see Jury and Fluhler (1992).

FIGURE 6.18   The water table, capillary fringe, zone of unsaturated material above the water table (vadose zone), and groundwater are illustrated in this photograph. The groundwater can provide significant quantities of water for plant uptake. (Photo courtesy of R. Weil)

tribute large macropores that may provide channels for rapid water and chemical flow from the soil surface to depths of 1 m or more.

Once chemicals are carried below the zone of greatest root and microbial activity, they are less likely to be removed or degraded before being carried further down to the groundwater. This means that chemicals that are normally broken down in the soil by microorganisms within a few weeks may move down through large macropores to the groundwater before their degradation can occur.

BYPASS FLOW. In some cases, leaching of chemicals is most serious if the chemicals are merely applied on the soil surface. As shown in Figure 6.19, chemicals may be washed from the soil surface into large pores, through which they can quickly move downward. Research suggests that most of the water flowing through large macropores does not come into contact with the bulk of the soil. Such flow is sometimes termed **bypass flow** as it tends to move rapidly around, rather than through, the soil matrix. As a result, if chemicals have been incorporated into the upper few centimeters of soil, their movement into the larger pores is reduced, and downward leaching is greatly curtailed.

INTENSITY OF RAIN OR IRRIGATION. Flow of water and chemicals through macropores, such as the vertical cracks and channels just mentioned, is much more pronounced when water is applied by high-intensity rainfall or irrigation. Apparently, when a large volume of water is rapidly applied, some localized saturation occurs that encourages flow through macropores. Table 6.3 illustrates the effects of rainfall *intensity* on the leaching of pesticides applied to turf grass.

Manipulations of soil macropores and irrigation intensity provide us with two examples of how environmental quality can be impacted by human activities that influence the hydrologic cycle. Likewise, we have considered steps that can be taken to influence the infiltration of water into soils and to maximize plant biomass production from the water stored in soils. The picture would not be complete if we did not consider briefly three other anthropogenic modifications to the hydrologic cycle: (1) the use of artificial drainage to enhance the downward percolation of excess water from some soils; conversely, (2) the application of additional water to soils as a means of wastewater disposal; and (3) the use of irrigation to supplement water available for plant growth. We shall consider artificial drainage first.

6.9 ENHANCING SOIL DRAINAGE

Some soils tend to be water-saturated in the upper part of their profile for extended periods during part or all of the year. The prolonged saturation may be due to the low-lying landscape position in which the soil is found, such that the regional water table is at or near the soil surface for extended periods (**endoaquic**). In other soils, water may accumulate above an impermeable layer in the soil profile (**epiaquic**), creating a **perched water table** (see Figure 6.20). In some cases, soils with either type of saturation may be components of wetlands (see Section 7.7), transitional ecosystems between land and water.

TABLE 6.3 Influence of Water Application Intensity on the Leaching of Pesticides Through the Upper 50 cm of a Grass Sod Covered Mollisol

The tests were conducted on 20-cm-diameter soil columns with the natural structure, including worm channels, undisturbed. All columns received 2.54 cm of simulated rain per week for four weeks, but the heavy rain was applied in four doses of 2.54 cm each, while the light rain was applied as 16 evenly spaced doses of 0.64 cm each. Metalaxyl is much more highly water soluble than Isazofos, but in each case the heavy rain stimulated much more pesticide leaching through the macropores.

| | Pesticide leached as percentage of that applied to surface | |
Pesticide	Heavy rains	Light rains
Isazofos	8.8	3.4
Metalaxyl	23.8	13.9

Data from Starrett, et al. (1996).

Before rain

Chemical on soil surface — Chemical mixed with soil

After rain

▢ Chemical infiltration

FIGURE 6.19 Preferential or bypass flow in macropores transports soluble chemicals downward through a soil profile. Where the chemical is on the soil surface (left), and can dissolve in surface-ponded water when it rains, it may rapidly be transported down cracks, earthworm channels, and other macropores. Where the chemical is dispersed within the soil matrix in the upper horizon (right), most of the water moving down through the macropores will bypass the chemical, and thus little of the chemical will be carried downward.

Reasons for Enhancing Soil Drainage

Water-saturated, poorly aerated soil conditions are essential to the normal functioning of wetland ecosystems and to the survival of many wetland plant species. However, for most other land uses, these conditions are a distinct detriment.

ENGINEERING PROBLEMS. During construction, the muddy, low-bearing-strength conditions of saturated soils make it very difficult to operate machinery (see Section 4.9). Houses built on poorly drained soils may suffer from uneven settlement and flooded basements during wet periods. Similarly, a high water table will result in capillary rise of water into roadbeds and around foundations, lowering the soil strength and leading to damage from frost-heaving (see Section 7.8) if the water freezes in winter. Heavy trucks traveling

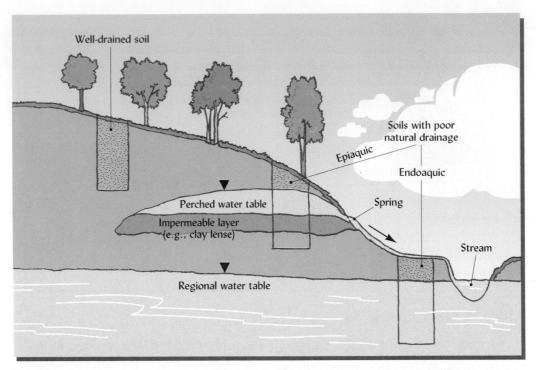

FIGURE 6.20 Cross-section of a landscape showing the regional and perched water tables in relation to three soils, one well-drained and two with poor internal drainage. By convention, a triangle (▼) is used to identify the level of the water table. The soil containing the perched water table is wet in the upper part, but unsaturated below the impermeable layer, and therefore is said to be *epiaquic* (Greek *epi,* upper), while the soil saturated by the regional water table is said to be *endoaquic* (*Greek* endo, under). Artificial drainage can help to lower both types of water tables.

over a paved road underlaid by a saturated soil create potholes and eventually destroy the pavement.

PLANT PRODUCTION.   Water-saturated soils make the production of most upland crops and forest species difficult, if not impossible. Prolonged saturation makes it difficult to carry out operations in a timely fashion as tractors and other equipment used for planting, tillage, and harvest operations tend to form deep ruts and bog down frequently. Even if a crop or forest stand can be established, the poorly aerated conditions may stunt its growth because the roots of all but a few specially adapted plants (bald cypress trees, rice, cattails, etc.) require aerated soil for respiration (see Section 7.1). Furthermore, a high water table early in the growing season will confine the plant roots to a shallow layer of partially aerated soil; the resulting restricted root system can lead to water stress later in the year, when the weather turns dry and the water table drops rapidly.

For these and other reasons, artificial drainage systems have been widely used to remove excess (gravitational) water and lower the water table in poorly drained soils. Land drainage is practiced in select areas in almost every climatic region, but is most widely used to enhance the agricultural productivity of clayey alluvial and lacustrine soils. Drainage systems are also a vital, if sometimes neglected, component of arid region irrigation systems, where they are needed to remove excess salts and prevent waterlogging.

Artificial drainage is a major alteration of the soil system, and the following list of potential beneficial and detrimental effects should be carefully considered. Also take note that in many instances laws designed to protect wetlands require that a special permit be obtained for the installation of a new artificial drainage system.

Benefits of Artificial Drainage

1. Increased bearing strength and improved soil workability, which allow more timely field operations and greater access to vehicular or foot traffic (e.g., in recreational facilities).

2. Reduced frost heaving of foundations, pavements, and crop plants (e.g., see Figure 7.14).

PLATE 1 Alfisols—an Aeric Epiaqualf from western New York. Argillic horizon between 20–80 cm. Scale in centimeters.

PLATE 2 Andisols—a Typic Melanudand from western Tanzania. Scale in 10 cm.

PLATE 3 Aridisols—a Typic Haplocambid from western Nevada. Scale in feet.

PLATE 4 Entisols—a Typic Quartzipsamment from eastern Texas. Scale in feet.

PLATE 5 Gelisols—a Typic Aquaturbel from Alaska. Permafrost below 32 cm on scale.

PLATE 6 Histosols—a Limnic Haplosaprist from southern Michigan. Buried mineral soil at bottom of scale. Scale in feet.

PLATE 7 Inceptisols—a Typic Dystrudept from West Virginia. Knife is in cambic horizon.

PLATE 8 Mollisols—a Typic Hapludoll from central Iowa. Mollic epipedon to 1.8 ft. Scale in feet.

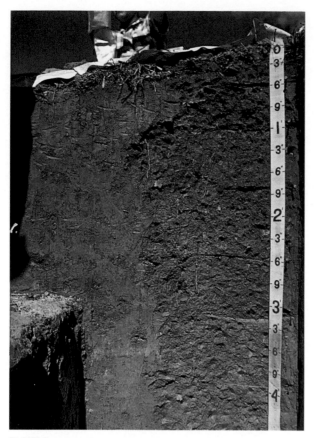

PLATE 9 Oxisols—a Udeptic Hapludox from central Puerto Rico. Scale in feet.

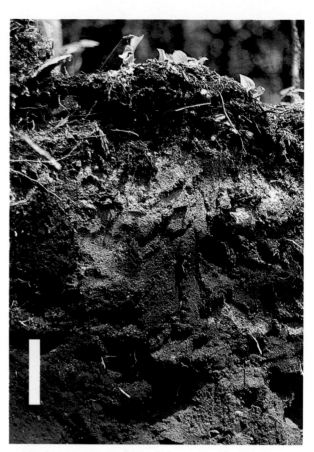

PLATE 10 Spodosols—a Humic Cryorthod from southern Quebec. Albic horizon at about 10 cm. Bar = 10 cm.

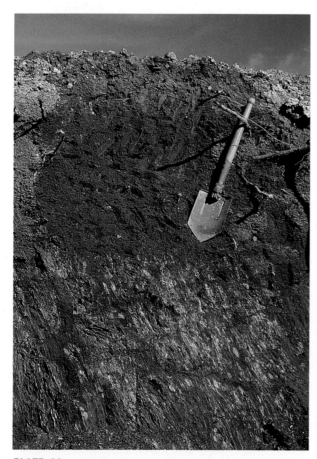

PLATE 11 Ultisols—a Typic Hapludult from central Virginia showing metamorphic rock structure in the saprolite below the 60-cm-long shovel.

PLATE 12 Vertisols—a Typic Haplustert from Queensland, Australia, during wet season. Scale in meters.

PLATE 13 Typic Argiustolls in eastern Montana with a chalky white calcic horizon (Bk and Ck) overlain by a Mollic epipedon (Ap, A2, and Bt).

PLATE 14 Ice wedge and permafrost underlying Gelisols in the Seward Peninsula of Alaska. Shovel is 1 m long.

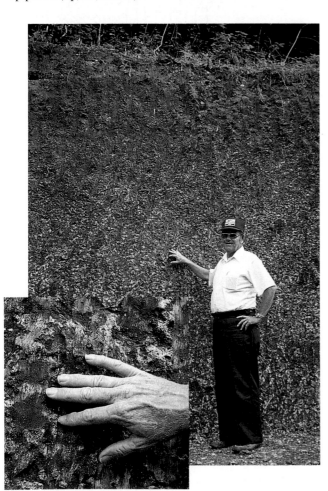

PLATE 15 A Typic Plinthudult in central Sri Lanka. Mottled zone is plinthite, in which ferric iron concentrations will harden irreversibly if allowed to dry.

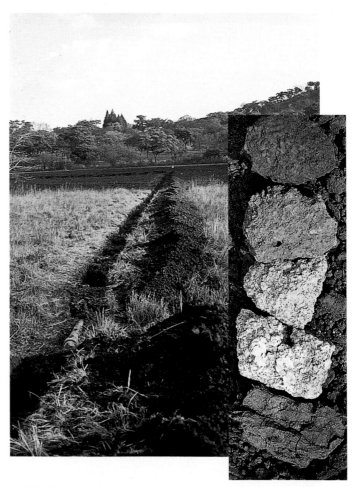

PLATE 16 A soil catena or toposequence in central Zimbabwe. Redder colors indicate better internal drainage. Inset: B-horizon clods from each soil in the catena.

PLATE 17 Connecticut River valley in western Massachusetts. Note variable alluvial soils and presence of riparian forest buffer along the river bank.

PLATE 18 Redox concentrations (red) and depletions (gray) in a Btg horizon from an Aquic Paleudalf.

PLATE 19 Green-gray colors from reduced structural iron around roots in an A horizon from a forested Aquic Hapludalf are an indicator of hydric conditions.

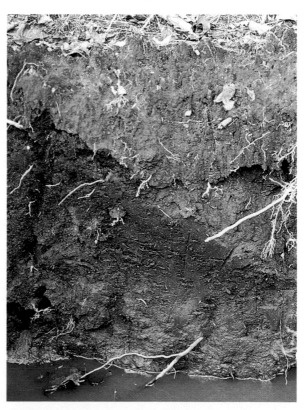

PLATE 20 Oxidized (red) root zones in the A and E horizons indicate a hydric soil. They result from oxygen diffusion out from roots of wetland plants having aerenchyma tissues (air passages).

PLATE 21 Dark (black) humic accumulation and gray humus depletion spots in the A horizon are indicators of a hydric soil. Water table is 30 cm below the soil surface.

PLATE 22 Nitrogen-deficient corn on Udolls in central Illinois. Ponded water after heavy rains resulted in nitrogen loss by denitrification and leaching.

PLATE 23 The waxlike ped surfaces in this Bt horizon from an Ultisol are clay skins (argillans). Bar = 1 cm.

PLATE 24 Acid sulfate weathering in a Sulfudept forming from pyritic mine spoil. A dark sulfidic horizon underlies a lighter-colored sulfuric horizon. Oxidation of sulfides may produce acid drainage waters. Scale marks = 10 cm.

PLATE 25 Early stages of soil formation in material dredged from Baltimore Harbor. Sulfidic materials (black), acid drainage (orange liquid), salt accumulations (whitish crust), and initiation of prismatic structure (cracks) are all evident.

PLATE 26 Stream polluted by acid coal-mine drainage caused by sulfurization in mine-spoil soils. Oxidation of $FeSO_4$ and precipitation of iron oxides on the rocks cause the orange color.

PLATE 27 Excessive inputs of nitrogen and phosphorus from upstream farmland resulted in the algal bloom that caused this slow-moving coastal plain stream to become choked with a green scum.

PLATE 28 Normal (left) and phosphorus-deficient (right) corn plants. Note stunting and purple color.

PLATE 29 Eroded calcareous soil (Ustolls) with iron-deficient sorghum.

PLATE 30 Zinc deficiency on peach tree. Note whorl of small, misshaped leaves.

PLATE 31 Zinc deficiency on sweet corn. Note broad whitish bands.

PLATE 32 Boron deficiency on alfalfa. Note reddish foliage.

PLATE 33 Nitrogen-deficient pine with normal trees in background.

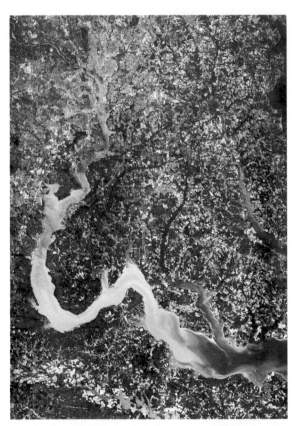

PLATE 34 Landsat Thematic Mapper image of Washington, D.C. (upper left), and sediment-laden Potomac River (center). Composite image with natural colors.

PLATE 35 Landsat Thematic Mapper image of Palo Verde Valley, California, irrigation scheme. Composite image using bands 2, 3, and 4. Lush vegetation appears bright red.

Photo 1 courtesy Bill Waltman and D. J. Lathwell, Cornell University; photos 2, 10, 11, 13, and 15–31 courtesy of R. Weil; photos 3, 4, 8, and 12 courtesy of R. W. Simonson; photos 5 and 14 courtesy of Chien-Lu Ping, Agriculture and Forest Experimental Station, University of Alaska—Fairbanks; photos 6, 7 and 9 courtesy of Soil Science Society of America; photo 34 copyright Space Imaging, Inc..; photo 35 courtesy of Earth Satellite Corp, Rockville, Md.

3. Enhanced rooting depth, growth, and productivity of most upland plants due to improved oxygen supply and, in acid soils, lessened toxicity of manganese and iron (see Sections 7.3 and 7.5).

4. Reduced levels of fungal disease infestation in seeds and on young plants.

5. More rapid soil warming in spring, resulting in earlier maturing crops (see Section 7.11).

6. Less production of methane and nitrogen gases that cause global environmental damages (see Sections 11.11 and 12.9).

7. Removal of excess salts from irrigated soils, and prevention of salt accumulation by capillary rise in areas of salty groundwater (see Section 9.14).

Detrimental Effects of Artificial Drainage

1. Loss of wildlife habitat, especially waterfowl breeding and overwintering sites.

2. Reduction in nutrient assimilation and other biochemical functions of wetlands (see Section 7.7).

3. Increased leaching of nitrates and other contaminants to groundwater.

4. Accelerated loss of soil organic matter, leading to subsidence of certain soils (see Sections 11.10 and 11.12).

5. Increased frequency and severity of flooding due to loss of runoff water retention capacity.

6. Greater cost of damages when flooding occurs on alluvial lands developed after drainage.

Drainage Systems

Artificial drainage systems are designed to promote two general types of drainage: (1) *surface drainage,* and (2) internal or *subsurface drainage.* Each will be discussed briefly.

Poor surface drainage is a problem where the landscape is nearly level and soils are fine-textured.

SURFACE DRAINAGE DITCHES. The removal of surface water may be hastened by constructing shallow ditches with gentle side slopes that do not interfere with equipment traffic. If there is some slope on the land, the shallow ditches are usually oriented across the slope and across the direction of planting and cultivating, thereby permitting the interception of water as it runs off down the slope. For removing surface water from landscaped lawns, this system of drainage can be modified by constructing gently sloping swales rather than ditches.

LAND FORMING. Often, surface drainage ditches are combined with **land forming** or smoothing to eliminate the ponding of water and facilitate its removal from the land (Figure 6.21). Small ridges are cut down and depressions are filled in using precision,

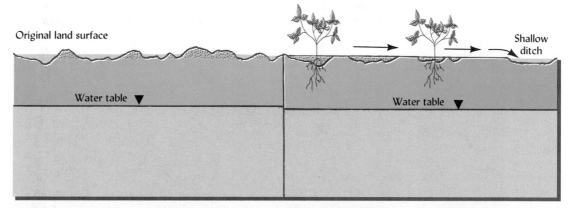

FIGURE 6.21 Land surface before (left) and after (right) land forming or smoothing. The original soil surface (left) is uneven and water ponds in depressions. After land smoothing (right), the land has a uniform slope and water runs off the surface at a controlled rate, to be carried away by a shallow ditch. Note that the soil from the ridges (stippled) is used to fill in the depressions.

BOX 6.1 SUCCESS OR FAILURE IN LANDSCAPE DRAINAGE DESIGN

A hedge of hemlock trees, all carefully pruned into ornamental shapes as part of an intricate landscape design, were dying again because of poor drainage. A very expensive effort to improve the drainage under the hemlocks had proved to be a failure and the replanted hemlocks were again causing an unsightly blemish in an otherwise picture-perfect garden.

Finally, the landscape architect for the world-famous ornamental garden called in a soil scientist to assist her in finding a solution to the dying hemlock problem. Records showed that in the previous, failed attempt to correct the drainage problem, contractors had removed all the hemlock trees in the hedge and had dug a trench under the hedge some 3 m deep (a). They had then backfilled the trench with gravel up to about 1 m from the soil surface, completing the backfill with a high-organic-matter, silt loam topsoil. It was into this silt loam that the new hemlock trees had been planted. Finally a bark mulch had been applied to the surface.

The landscaper who had designed and installed the drainage system apparently had little understanding of the various soil horizons and their relation to the local hydrology. When the soil scientist examined the problem site, he found an impermeable claypan that was causing water from upslope areas to move laterally into the hemlock root zone.

Basic principles of soil water movement also told him that water would not drain from the fine pores in the silt loam topsoil into the large pores of the gravel, and therefore the gravel in the trench would do no good in draining the silt loam topsoil (compare the situation to that in Figure 5.19). In fact, the water moving laterally over the impermeable layer created a perched water table that poured water into the gravel-filled trench, soon saturating both the gravel and the silt loam topsoil.

To cure the problem, the previous "solution" had to be undone (b). The dead hemlocks were removed, the ditches were reexcavated, and the gravel was removed from the trenches. Then the trench, except for the upper ½ m, was filled with a sandy loam subsoil to provide a suitable rooting medium for the replacement evergreen trees (a different species was chosen for reasons unrelated to drainage). The upper ½ m of the trench was filled with a sandy loam topsoil, which was also acid but contained a higher level of organic matter. The interface between the subsoil and surface soil was mixed so that there would be no abrupt change in pore configuration. This would allow an unsaturated wetting front to move down from the upper to the lower layers, drawing down any excess water.

About 1 m uphill from the trench an *interceptor drain* was installed (b). This involved digging a small trench through the impermeable clay layer that was guiding water to the area. A perforated drainage pipe surrounded by a layer of gravel was laid in the bottom of this trench with about a 1% slope to allow water to flow away from the area to a suitable outlet. The interceptor drain prevented the water moving laterally over the impermeable soil layer from reaching the evergreen hedge root zone.

Even though the replanting of the hedge was followed by an exceptionally rainy year, the new drainage system kept the soil well aerated, and the trees thrived. The principles of water movement explained in Sections 5.6 and 6.9 of this text were applied successfully in the field.

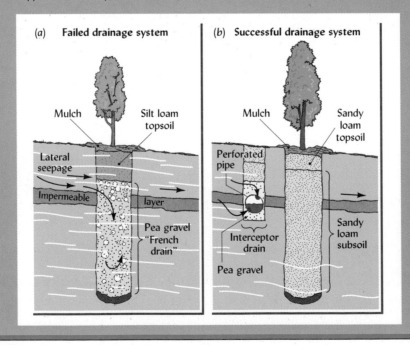

(a) **Failed drainage system** Mulch — Silt loam topsoil — Lateral seepage — Impermeable layer — Pea gravel "French drain"

(b) **Successful drainage system** Mulch — Sandy loam topsoil — Perforated pipe — Interceptor drain — Pea gravel — Sandy loam subsoil laser-guided field-leveling equipment. The resulting land configuration permits excess water to move at a controlled rate over the soil surface to the outlet ditch and then on to a natural drainage channel. Land smoothing is also commonly used to prepare a field for flood irrigation (see Section 6.11).

Subsurface (Internal) Drainage

The water table rises when the rate of water input (usually rain or irrigation water percolating down from the soil surface) exceeds the rate of groundwater movement away from the area. Therefore, the purpose of a subsurface drainage system is to provide a series of "giant pores" or pathways that vastly increase the rate of groundwater movement and subsequently lower the level of the water table.

The design of a drainage system must be compatible with the basic principle that water moves to where its energy level will be lower. That is, water will enter a pipe or ditch where its potential will be essentially zero only if it is under a positive pressure ($\psi \geq 0$) (due to the sum of the positive gravitational and submergence potentials). Therefore, internal drainage will occur only when the pathway for drainage is located *below* the level of the water table (Figure 6.22). No water will be removed if the pipe or ditch bottom is in wet but not saturated soil, because the matric component in such soil results in a negative water potential ($\psi < 0$). Box 6.1 provides an example of how knowledge of basic soil properties and water-movement principles can be appplied in designing a system to alleviate drainage problems. In practice, subsurface drainage is enhanced by open ditches or buried perforated pipes (drain tiles) as shown in Figures 6.22 and 6.23.

BUILDING FOUNDATION DRAINS. Surplus water around foundations and underneath basement floors of houses and other buildings can cause serious damage. The removal of this excess water is commonly accomplished using buried perforated pipe, placed alongside and slightly below the foundation or underneath the floor (Figure 6.22c). The perforated pipe must be sloped to allow water to move rapidly to an outlet ditch or sewer. If the drain successfully prevents the water table from rising above the floor level, water will not seep into the basement for the same reason that it will not seep into a drainpipe placed above the water table.

6.10 SEPTIC TANK DRAIN FIELDS

Our discussion of practical water management in soil systems would be incomplete without consideration of the role of soil and, in particular, water movement through soil in treating wastewater in areas not served by a centralized sewage treatment system.

Thousands of ordinary suburbanites get their first exposure to the importance of water movement through soils when they choose a beautiful home site "in the country," make plans to build a dream home, and apply for the required local government building permit. The local authorities will usually not allow a home to be built until arrangements are made for wastewater treatment. Typically, a soil scientist will come out to inspect the soils at the homesite and judge their suitability for use as a septic tank **drain field.** If the soils fall short in some of the attributes described later in this section, the landowner may be denied the permit to build.

Operation of a Septic System

The most common type of on-site wastewater treatment for homes not connected to municipal sewage systems is the *septic tank* and associated *drain field* (sometimes called *filter field* or *absorption field*). Over 10 million homes in the United States use septic tank drain fields to treat sewage and wastewater. This method of sewage and wastewater treatment depends on several soil processes, of which the most fundamental is the movement of water through the soil.

In essence, a septic drain field operates like artificial soil drainage in reverse. A network of perforated underground pipes is laid in trenches, very much like the network of drainage pipes used to lower the water table in a poorly drained soil. But instead of draining water away from the soil, the pipes in a septic drain field carry wastewater *to* the soil, the water entering the soil via slits or perforations in the pipes. In a properly

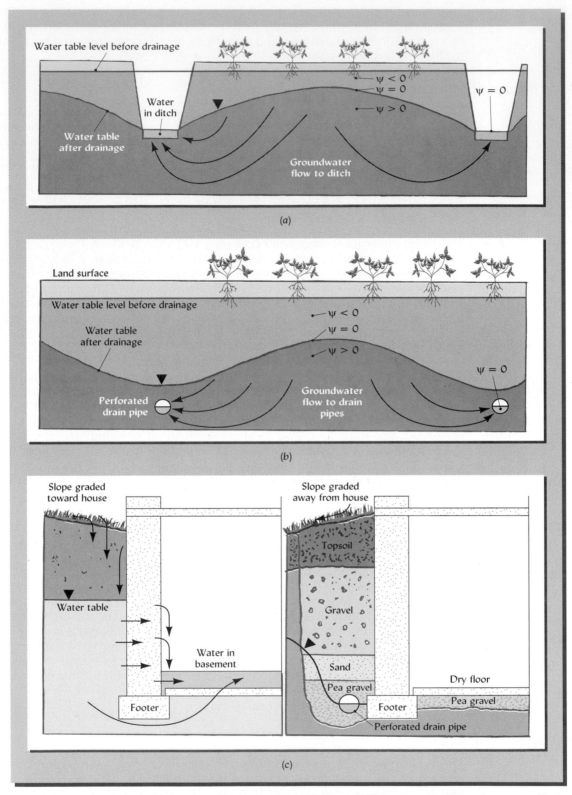

FIGURE 6.22 Three types of subsurface drainage systems. (*a*) Open ditches are used to lower the water table in a poorly drained soil. The wet season levels of the water table before and after ditch installation are shown. The water table is deepest next to the ditch, and the drainage effect diminishes with distance from the ditch. (*b*) Buried "tile lines" made of perforated plastic pipe act very much as the ditches in *a*, but have two advantages: They are not visible after installation and they do not present any obstacle for surface equipment. Note the flow lines indicating the paths taken by water moving to the drainage ditches or pipes in response to the water potential gradients between the submerged water ($\psi > 0$) and free water in the drainage ditch and pipes ($\psi = 0$). (*c*) The water table around a building foundation before (left) and after (right) installation of a *footer drain* and correction of surface grading. The principles of water movement in soils are applied to keep the basement dry.

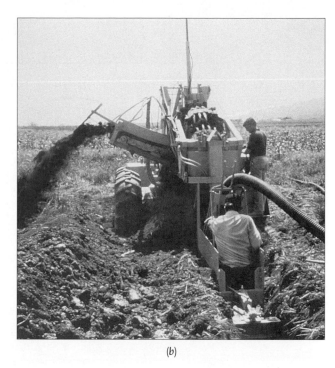

(a)   (b)

FIGURE 6.23   (*a*) An open-ditch drainage system designed to lower the water table during the wet season. The water flowing in the ditch has seeped in from the saturated soil. Such ditches also speed the removal of surface runoff water. The capillary fringe above the water table can be seen as a dark band of soil. In order to protect water quality from chemicals and sediments that might reach the ditch via surface runoff, buffer strips of grass or forest vegetation should be established on both sides. Note in the background the piles of *spoil*, subsoil excavated to make the ditches. This material is usually spread thinly and mixed by tillage with the surface soil in the field, but it should first be tested for extreme acidity or other detrimental properties that could impair the quality of the surface soil in which it is to be mixed. (*b*) Specialized, laser-guided equipment laying a drain line made of corrugated plastic pipe. The pipe has perforations on the underside that allow water to seep in from a saturated soil. The trench will be backfilled with the soil shown piled on the left. The drain line must be laid deeper than the seasonal high water table if it is to remove any water. [Photo (*a*) courtesy of R. Weil; photo (*b*) courtesy of USDA Natural Resources Conservation Service]

functioning septic drain field the wastewater will enter the soil and percolate downward, undergoing several purifying processes before it reaches the groundwater. One of the advantages of this method of sewage treatment is that it has the potential to replenish groundwater supplies for other uses.

THE SEPTIC TANK.   The basic design of a standard septic tank and drain field is shown in Figure 6.24*a*. Water carrying wastes from toilets, sinks, and bathtubs flows by gravity through sealed pipes to a large underground concrete box , called the *septic tank*. Baffles in the tank cause the inflowing wastewater to slow down and drop most (70%) of its load of suspended solid materials. As these organic solids partially decompose by microbial action in the septic tank, their volume is reduced so that many years usually pass before the septic tank becomes too full and the accumulated **sludge** (also called *septage*) has to be pumped out.

THE DRAIN FIELD.   The water exiting the septic tank via a pipe near the top is termed the septic tank **effluent.** Although its load of suspended solids has been much reduced, it still carries organic particles, dissolved chemicals (including nitrogen), and microorganisms (including pathogens). The flow is directed to one or more buried pipes that comprise the **drain field.** Blanketed in gravel and buried in trenches about 0.6 to 1 m under the soil surface, these pipes are perforated on the bottom to allow the wastewater to seep out and enter the soil.

It is at this point that soil properties play a crucial role. Septic systems depend on the soil in the drain field to (1) keep the effluent out of sight and out of contact with people, (2) treat or purify the effluent, and (3) conduct the purified effluent to the groundwater.

FIGURE 6.24 (a) A septic tank and drain field comprising a standard system for on-site wastewater treatment. Most of the solids suspended in the household wastewater settle out in the concrete septic tank. Effluent from the tank flows to the drain field, where it seeps out of the perforated pipes and into the soil. In the soil, the effluent is purified by microbial, chemical, and physical processes as it percolates toward the groundwater. (b) Photo shows the tell-tale dark strips of lawn where poorly functioning septic tank drain lines are stimulating the grass with wastewater and nitrogen. (Photo courtesy of R. Weil)

Soil Properties Influencing Suitability for a Septic Drain Field

The soil should have a *saturated hydraulic conductivity* (see Section 5.5) that will allow the wastewater to enter and pass through the soil profile rapidly enough to avoid backups that might saturate the surface soil with effluent, but slowly enough to allow the soil to purify the effluent before it reaches the groundwater. The soil should be sufficiently *well aerated* to encourage *microbial breakdown* of the wastes and *destruction of pathogens*. The soil should have some fine pores and clay or organic matter to adsorb and filter contaminants from the wastewater.

Properties that can be observed in the soil profile that indicate the presence of a high water table at some time of the year, such as mottled and gleyed colors, may disqualify a site for use as a septic drain field. Likewise, impermeable layers such as a fragipan or a heavy claypan could eliminate a building site from consideration.

Septic tank drain fields installed where soil properties are not appropriate may result in extensive pollution of groundwater and in health hazards caused by seepage of untreated wastewater (Figure 6.24b).

SUITABILITY RATING. The suitability of a site for septic drain-field installation largely depends on soil properties that affect water movement and the ease of installation (see Table 6.4). For example, a septic drain field laid out on a slope greater than 15% may allow considerable lateral movement of the percolating water such that at some point downslope, the wastewater will seep to the surface and present a potential health hazard.

PERC TEST. One of the most important tests conducted to determine the suitability of a soil for the installation of a septic tank is called the **perc test**. This is a test that determines the *percolation rate* (which is related to the saturated hydraulic conductivity described in Section 5.5). The percolation rate is expressed in millimeters (or other unit of depth) of water entering the soil per hour and indicates whether or not the soil can accept wastewater rapidly enough to provide a practical disposal medium (Table 6.4).

TABLE 6.4 **Soil Properties Influencing Suitability for a Septic Tank Drain Field**

Note that most of these soil properties pertain to the movement of water through the soil profile.

	Limitations		
Soil property[a]	Slight	Moderate	Severe
Flooding	—	—	Floods frequent to occasional
Depth to bedrock or impermeable pan, cm	>183	102–183	<102
Ponding of water	No	No	Yes
Depth to seasonal high water table, cm	>183	122–183	<122
Permeability (perc test) at 60 to 152 cm soil depth, mm/h	50–150	15–50	<15 or >150[b]
Slope of land, %	<8	8–15	>15
Stones >7.6 cm, % of dry soil by weight	<25	25–50	>50

[a] Assumes soil does not contain permafrost and has not subsided more than 60 cm.
[b] Soil permeability (as determined by a perc test) greater than 150 mm/h is considered too fast to allow for sufficient filtering and treatment of wastes.
Adapted from Soil Survey Staff (1993), Table 620-17.

The test is simple to conduct (Figure 6.25) and should be carried out during the wettest part of the year.

6.11 IRRIGATION PRINCIPLES AND PRACTICES[3]

In most regions of the world, insufficient water is the prime limitation to agricultural productivity. In semiarid and arid regions intensive crop production is all but impossible without supplementing the meager rainfall provided by nature. However, if given supplemental water through irrigation, the sunny skies and fertile soils of some arid regions stimulate extremely high crop yields. It is no wonder, then, that many of the earliest civilizations and city-states depended on irrigated agriculture (and vice versa). During the 20th century, the area of irrigated cropland expanded greatly in many parts of the world, including the 17 semiarid states of the western United States.

---

[3] For a fascinating account of water resources and irrigation management in the Middle East, see Hillel (1995). For a practical manual on small-scale irrigation with simple but efficient microtechnology, see Hillel (1997). CAST (1988 and 1996) provides overviews of methods and prospects for commercial irrigated agriculture in the Western United States.

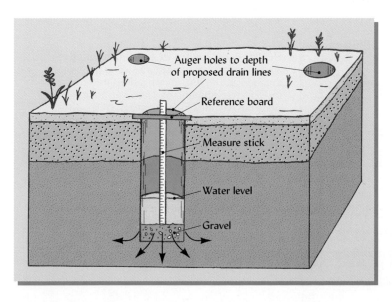

FIGURE 6.25 The perc test used to help determine soil suitability for a septic tank drain field. On the site proposed for the drain field, a number of holes are drilled to the depth where the perforated pipes are to be laid. The bottom of each hole is lined with a few centimeters of gravel, and the holes are filled 75 cm deep with water as a pretreatment to assure that the soil is wet when the test is conducted. After the water has drained, the hole is refilled with water and a measuring rod is used to determine how rapidly the water level drops during a period of several hours or a day.

The total area under irrigation seems to be leveling off at about 250 million hectares worldwide, about 16% of the world's agricultural land. The high productivity induced by irrigation is evident in the fact that this irrigated land is responsible for some 40% of global crop production.

Expanded and improved irrigation, especially in Asia, has been a major factor in helping global food supplies keep up with, and even surpass, the global growth in population. As a result, irrigated agriculture remains the largest **consumptive use** of water resources, accounting for about 80% of all water consumed, both worldwide and in the United States (Figure 6.26).

Water resources for irrigation are most rapidly becoming scarce and/or expensive in arid regions where they are most needed. Reducing waste and achieving greater efficiency of water use in irrigation are increasingly important aspects that will be emphasized in this section. One of the other major problems associated with irrigation, the salinization of soils and drainage waters, is considered in Chapter 9.

Water-Use Efficiency

Various measures of water-use efficiency are used to compare the relative benefits of different irrigation practices and systems. The most meaningful overall measure of efficiency would compare the output of a system (crop biomass or value of marketable product) to the amount of water allocated as an input into the system. There are many factors to consider (such as type of plants grown, reuse of "wasted" water by others downstream, etc.), so such comparisons must be made with caution.

APPLICATION EFFICIENCY. A simpler measure of water-use efficiency, sometimes termed the **water application efficiency**, compares the amount of water allocated to irrigate a field to the amount of water actually used by the irrigated plants. In this regard, most irrigation systems are very inefficient, with as little as 30 to 50% of the water that is taken from the source ever reaching the plant roots. Much of the water loss occurs in the canal and ditch system used to deliver the water to the irrigated field. To reduce these losses during water distribution, ditches can be lined with concrete or plastic (see Figure 6.27).

FIELD WATER EFFICIENCY. Water-use efficiency *in the field* may be expressed as:

$$\text{Field water efficiency} = \frac{\text{Water transpired by the crop}}{\text{Water applied to the field}} \times 100$$

Values are usually quite low, especially if traditional systems are used. Commonly, more than half of the water actually delivered to the field is not transpired by the crop, but is lost as surface runoff, deep percolation below the root zone, and/or evaporation from the soil surface. Since some deep percolation is necessary to remove excess salts (see Section 9.19), the best **field water efficiency** possible is probably not much greater than 80 to 90%. Achieving a high level of field water efficiency (or a low level of water

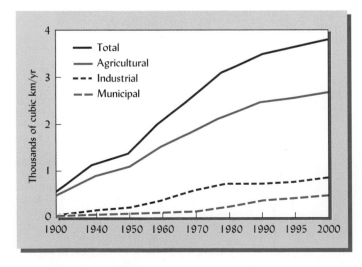

FIGURE 6.26 Trends in global consumptive water use by major categories of users. Note that agriculture (mainly irrigation) uses by far the greatest amount of water, although its share has declined in recent years. Water withdrawn for irrigation and lost to the atmosphere by evapotranspiration is a consumptive use, but water withdrawn to cool an electric generation turbine and then returned to the river from which it came is not a consumptive use. On average, about 30 to 35% of the water withdrawals for irrigation are returned to water sources for reuse and are not part of the consumptive use, while 90 to 95% of the withdrawals for industrial use are returned and so are not considered consumptive use. [Redrawn from Kuylenstierna, et al. (1997)]

FIGURE 6.27 Concrete-lined irrigation ditches (center) and standard-sized siphon pipes (right) can increase the efficiency of water delivery to the field. Unlined ditches (left) lose much of the water to adjacent soil areas or to the groundwater. Note the evidence of capillary movement above the water level in the unlined ditch.

wastage) is very dependent on the skill of the irrigation manager and on the methods of irrigation used (Table 6.5).

Surface Irrigation

In these systems the entire land surface is moistened to at or near saturation, so evaporation losses from the soil are quite large. Water is applied to the upper end of a field and allowed to distribute itself by gravity flow. Usually the land must be leveled and shaped so that the water will flow uniformly across the field. The water may be distributed in **furrows** graded to a slight slope so that water applied to the upper end of the field will flow down the furrows at a controlled rate (Figure 6.28). In **border irrigation** systems, the land is shaped into broad strips 10 to 30 m wide, bordered by low dikes.

WATER CONTROL. Water is usually brought to surface-irrigated fields in supply ditches or gated pipes (such as are shown in Figures 6.27 and 6.28). The amount of water that enters the soil is determined by the permeability of the soil and by the length of time a given spot in the field is inundated with water. Achieving a uniform infiltration of exactly the required amount of water is very difficult, and depends on controlling the slope and length of the irrigation runs across the field. If the soil is highly permeable (e.g., some loamy sands), too much water may infiltrate near the upper end of the field and too little may reach the lower end (Figure 6.29). By contrast, in clayey soils, erosion, runoff, and waterlogging may be problems at the lower end.

The **level basin** technique of surface irrigation, as is used for paddy rice and certain tree crops, alleviates these problems because each basin has no slope and is completely surrounded by dikes that allow water to stand on the area until infiltration is complete. This method is not practical for highly permeable soils into which water would infiltrate so fast that the basin would never fill. On sloping land, terraces can be built in a modification of the level-basin method (Figure 6.30).

Sprinkler Systems

In sprinkler irrigation, water is sprayed through the air onto a field, simulating rainfall. Thus the entire soil surface, as well as plant foliage (if present), is wetted. This leads to

TABLE 6.5 Some Characteristics of the Three Principal Methods of Irrigation

Methods and specific examples	Direct costs of installation, 1996 dollars[a]	Labor requirements	Field water efficiency, %[b]	Suitable soils
Surface: basin, flood, furrow	400–700	High to low, depending on system	40–50	Nearly level land; not too sandy or rocky
Sprinkler: center pivot, movable pipe, solid set	600–1200	Medium to low	60–70	Level to moderately sloping; not too clayey
Microirrigation: drip, porous pipe, spitter, bubbler	700–1500	Low	80–90	Steep to level slopes; any texture, including rocky or gravelly soils

[a] Average ranges from many sources. Costs of required drainage systems not included.
[b] Values from Hillel (1997). Field water efficiency = $100 \times$ (water transpired by crop/water applied to field).

FIGURE 6.28 Typical irrigation scene. The use of easily installed siphons or gated pipes reduces the labor of irrigation and makes it easier to control the rate of application of water. Note the upward capillary movement of water along the sides of the rows. (Courtesy of USDA Natural Resources Conservation Service)

evaporative losses similar to those described for surface systems. Furthermore, an additional 5 to 20% of the applied water may be lost by evaporation or windblown mist as the drops fly through the air. One advantage is that plants often respond positively to the cooler, better-aerated sprinkler water. A disadvantage is that wet leaves may increase the incidence of fungal diseases in some plants, such as grapes, fruit trees, and roses, so sprinkler systems are not often used for these plants.

WATER CONTROL. A sprinkler system should be designed to deliver water at a rate that is less than the infiltration capacity of the soil, so that runoff or excessive percolation will not occur. In practice, runoff and erosion may be problems if the soil infiltration capacity is low, the land is relatively steep, or too much water is applied in one place. Because of better control over application rates, the field water-use efficiency is generally higher for sprinkler systems than for surface systems, especially on coarse-textured soils.

FIGURE 6.29 Penetration of water into a coarse-textured soil under surface irrigation. The high infiltration rate causes most of the water to soak in near the gated pipe at the upper end of the field. The uneven penetration of water results in the potential for leaching losses of water and dissolved chemicals at the upper end of the field, while plants at the lower end may receive insufficient water to moisten the entire potential root zone. On a less-permeable soil, or on a field with a steep slope, the tail water runoff losses would likely be greater at the lower end and leaching potential less at the upper end.

FIGURE 6.30   The level-basin type of surface irrigation modified for sloping land in South Asia by construction of terraces. Paddy rice is growing in the flooded terrace basins. This type of irrigation is practical only on soils of low permeability. (Photo courtesy of R. Weil)

SUITABLE SOILS.   Sprinkler irrigation is practical on a wider range of soil conditions than is the case for the surface systems. Various types of sprinkler systems are adapted to moderately sloping as well as level land (Figure 6.31). They can be used on soils with a wide range of textures, even those too sandy for surface irrigation systems.

Microirrigation

The "micro" in microirrigation suggests that only a small portion of the soil is wetted by these systems (in contrast to the complete wetting accomplished by most surface and sprinkler systems). The name may also refer to the tiny amounts of water applied at any one time and the miniature size of the equipment involved.

Perhaps the best-established microirrigation system is *drip* (or *trickle*) *irrigation,* in which tiny emitters attached to plastic tubing apply water to the soil surface alongside individual plants. In some cases the tubing and emitters are buried 20 to 50 cm deep so the water soaks directly into the root zone. In either case, water is applied at a low rate (sometimes drop by drop) but at a high frequency, with the objective of maintaining optimal soil water availability in the immediate root zone while leaving most of the soil volume dry (Figure 6.32*a*).

Other forms of microirrigation that are especially well adapted for irrigating individual trees include *spitters* (microsprayers) and *bubblers* (small vertical standpipes); (Fig-

FIGURE 6.31   A center-pivot irrigation system in eastern Colorado. The system is rotating toward the left (note the dry soil at the extreme left), and the center pivot of the system is located approximately 0.5 km in the distance. Each self-propelled tower moves in coordination with all the others. The shed at the right houses a diesel-powered water pump that taps a deep aquifer. The tank next to the shed holds liquid fertilizer that can be metered into the irrigation water. (Photo courtesy of R. Weil)

(a)   (b)

FIGURE 6.32 Two examples of microirrigation. (*a*) *Drip* or *trickle* irrigation with a single emitter for each seedling in a cabbage field in Africa. (*b*) A *microsprayer* or *spitter* irrigating an individual tree in a home garden in Arizona. In both cases, irrigation wets only the small portion of the soil in the immediate root zone. Small quantities of water applied at high frequency (such as once or twice a day) assures that the root zone is kept almost continuously at an optimal moisture content. (Left photo courtesy of R. Weil)

ure 6.32*b*). The bubblers (and usually the spitters) require that a small level basin be formed in the soil under each tree.

WATER CONTROL.   Water is normally carried to the field in pipes, run through special filters to remove any grit or chemicals that might clog the tiny holes in the emitters (filtering is not necessary for bubblers), and then distributed throughout the field by means of a network of plastic pipes. Soluble fertilizers may be added to the water as needed.

If properly maintained and managed, microirrigation allows much more control over water application rates and spatial distribution than do either surface or sprinkler systems. Losses by supply-ditch seepage, sprinkler-drop evaporation, runoff, drainage (in excess of that needed to remove salts), soil evaporation, and weed transpiration can be greatly reduced or eliminated. Once in place, the labor required for operation is modest.

Microirrigation often produces healthier plants and higher crop yields because the plant is never stressed by low water potentials or low aeration conditions that are associated with the feast or famine regime of infrequent, heavy water applications made by all surface irrigation systems and most sprinkler systems. A disadvantage or risk is that there is very little water stored in the soil at any time, so even a brief breakdown of the system could be disastrous in hot, dry weather.

EQUIPMENT.   The capital costs for microirrigation tend to be higher than for other systems (see Table 6.5), but the differences are not so great if the cost of drainage systems for control of salinity and waterlogging in surface systems, the costs of high-pressure pumping in sprinkler systems, and the real value of wasted water is taken into account. Because of its high water-use efficiency, microirrigation is most profitable where water supplies are scarce and expensive, and where high-valued plants such as fruit trees are being grown.

Irrigation Water Management

The two most serious irrigation management problems relate to the quality of the water being applied and to the efficiency of the irrigation water in stimulating plant production.

SALINITY BUILDUP.   Most irrigation systems are located in semiarid and arid regions where the levels of soluble salts in the drainage water and, in turn, in the streams and rivers are relatively high. When this water is added to the soil and percolation takes place, still more salts are dissolved from the soil itself, making the drainage water even more saline than the originally added water. As the drainage water is repeatedly reused downstream, the salt buildup in the water can become very damaging to both the physical and chemical properties of the soils to which the water is applied (see Box 9.4).

EFFICIENT WATER USE. Microirrigation systems are the most efficient in water use, but whether or not they are most economical to use for a given situation depends on many factors. The use of crop residues or mulches to reduce evaporation from the soil surface while simultaneously reducing the soil temperature (see Section 7.11) can enhance overall productivity and water-use efficiency. It is also wise to concentrate irrigation on those crops that produce high-value products with relatively low levels of water use. Much water is currently squandered in producing low-value, high-transpiration-ratio crops (such as forages, cereals, and cotton) that would be more economically imported from rain-fed areas if all the costs of obtaining irrigation water were actually paid for by the irrigators (as opposed to being subsidized by governments).

Likewise, arid-region homeowners' attempts to sustain humid-region landscaping (e.g., green lawns and high-water-use shrubs) around their homes rather than using xerophytic (desert) plants, mulches and stones are equally wasteful.

Finally, to maintain or increase plant production in the face of dwindling water supplies, the application of water must be scheduled according to plant needs, which change with the weather, the LAI, and other factors. The basic principles of water retention and movement in soils and water use by plants, as outlined in this and the previous chapter, must be applied in developing efficient irrigation schedules.

6.12 CONCLUSION

The hydrologic cycle encompasses all movements of water on or near the earth's surface. It is driven by solar energy, which evaporates water from the ocean, the soil and vegetation. The water cycles into the atmosphere, returning elsewhere to the soil and the oceans in rain and snow.

The soil is an essential component of the hydrologic cycle. It receives precipitation from the atmosphere, rejecting some of it, which is then forced to run off into streams and rivers, and absorbing the remainder, which then moves downward, to be either transmitted to the groundwater, taken up and later transpired by plants, or evaporated directly from soil surfaces and returned to the atmosphere.

The behavior and movement of water in soils and plants are governed by the same set of principles: Water moves in response to differences in energy levels, moving from higher to lower water potential. These principles can be used to manage water more effectively and to increase the efficiency of its use.

Management practices should encourage movement of water into well-drained soils while minimizing evaporative (E) losses from the soil surface. These two objectives will provide as much water as possible for plant uptake and groundwater recharge. Water from the soil must satisfy the transpiration (T) requirements of healthy leaf surfaces; otherwise, plant growth will be limited by water stress. Practices that leave plant residues on the soil surface and that maximize plant shading of this surface will help achieve high efficiency of water use.

Extreme soil wetness, characterized by surface ponding and saturated conditions, is a natural and necessary condition for wetland ecosystems. However, for most other land uses extreme wetness is detrimental. Drainage systems have therefore been developed to hasten the removal of excess water from soil and lower the water table so that upland plants can grow without aeration stress, and so the soil can better bear the weight of vehicular and foot traffic.

A septic tank drain field operates as a drainage system in reverse. Septic waste waters can be disposed of and treated by soils if the soils are freely draining. Soils with low permeability or high water tables may indicate good conditions for wetland creation or appropriate sites for installation of artificial drainage for agricultural use, but they are not generally suited for septic tank drain fields.

Irrigation waters from streams or wells greatly enhance plant growth, especially in regions with scarce precipitation. With increasing competition for limited water resources, it is essential that irrigators manage water with maximal efficiency so that the greatest production can be achieved with the least waste of water resources. Such efficiency is encouraged by practices that favor transpiration over evaporation, such as mulching and the use of microirrigation.

As the operation of the hydrologic cycle causes constant changes in soil water content, other soil properties are also affected, most notably soil aeration and temperature, the subjects of the next chapter.

STUDY QUESTIONS

1. You know that the forest vegetation that covers a 120 km² wildland watershed uses an average of 4 mm of water per day during the summer. You also know that the soil averages 150 cm in depth and at field capacity can store 0.2 mm of water per mm of soil depth. However, at the beginning of the season the soil was quite dry, holding a average of only 0.1 mm/mm. As the watershed manager, you are asked to predict how much water will be carried by the streams draining the watershed during the 90-day summer period when 450 mm of precipitation falls on the area. Use the water balance equation to make a rough prediction of the stream discharge as a percentage of the precipitation and in cubic meters of water.

2. Draw a simple diagram of the hydrologic cycle using a separate arrow to represent these processes: *evaporation, transpiration, infiltration, interception, percolation, surface runoff,* and *soil storage.*

3. Describe and give an example of the *indirect* effects of plants on the hydrologic balance though their effects on the soil.

4. State the basic principle that governs how water moves through the SPAC. Give two examples, one at the soil–root interface and one at the leaf–atmosphere interface.

5. Define *potential evapotranspiration* and explain its significance to water management.

6. What is the role of evaporation from the soil (E) in determining water-use efficiency, and how does it affect ET? List three practices that can be used to control losses by E.

7. Weed control should reduce water losses by what process?

8. Comment on the relative advantages and disadvantages of organic versus plastic mulches.

9. What does conservation tillage conserve? How does it do it?

10. Explain under what circumstances earthworm channels might increase downward saturated water flow, but not have much effect on the leaching of soluble chemicals applied to the soil.

11. What will be the effect of placing a perforated drainage pipe in the capillary fringe zone just above the water table in a wet soil? Explain in terms of water potentials.

12. What soil features may limit the use of a site for a septic tank drain field?

13. Which irrigation systems are likely to be used where: (1) Water is expensive and the market value of crops produced per hectare is high, and (2) The cost of irrigation water is subsidized and the value of crop products that can be produced per hectare is low. Explain.

REFERENCES

Black, A. L. 1982. "Long-term N-P fertilizer and climate influences on morphology and yield components of spring wheat," *Agron. J.,* **74**:651–57.

CAST. 1988. *Effective Use of Water in Irrigated Agriculture.* Task Force Report no. 113. (Ames, Iowa: Council for Agricultural Science and Technology).

CAST. 1996. *Future of Irrigated Agriculture.* Task Force Report no. 127. (Ames, Iowa: Council for Agricultural Science and Technology).

Dreibelbis, F. R., and C. R. Amerman. 1965. "How much topsoil moisture is available to your crops?," *Crops and Soils,* **17**:8–9.

Fouss, J. L. 1974. "Drain tube materials and installation," in J. Van Schilfgaarde (ed.), *Drainage for Agriculture*. Agronomy Series no. 17. (Madison, Wis.: Amer. Soc. Agron.), pp. 147–177.

Greb, B. W. 1983. "Water conservation: Central Great Plains," in H. E. Dregue and W. O. Willis (eds.), *Dryland Agriculture*. Agronomy Series no. 23. (Madison, Wis.: Amer. Soc. of Agron.).

Hillel, D. 1995. *The Rivers of Eden* (New York: Oxford University Press).

Hillel, D. 1997. *Small-Scale Irrigation for Arid Zones*. FAO Development Series 2. (Rome: U.N. Food and Agriculture Organization).

Jury, W. A., and H. Fluhler. 1992. "Transport of chemicals through soil: Mechanisms, models, and field applications," *Advances in Agronomy,* **47**:141–201.

Kuylenstierna, J. L., G. Björklund, and P. Najlis. 1997. "Future sustainable water use: Challenges and constraints," *J. Soil Water Conserv.,* **52**:151–156.

Lyon, T. L., H. O. Buckman, and N. C. Brady. 1952. *The Nature and Properties of Soils,* 5th ed. (New York: Macmillan).

Soil Survey Staff. 1993. *National Soil Survey Handbook*. Title 430-VI. (Washington, D.C.: USDA Natural Resources Conservation Service).

Starrett, S. K., N. E. Christians, and T. Al Austin. 1996. "Movement of pesticides under two irrigation regimes applied to turfgrass," *J. Environ. Qual.,* **25**:566–571.

Waddell, J., and R. Weil. 1996. "Water distribution in soil under ridge-till and no-till corn," *Soil Sci. Soc. Amer. J.,* **60**:230–237.

7

SOIL AERATION AND
TEMPERATURE

The naked earth is warm with Spring. . . .
—JULIAN GRENFELL, INTO BATTLE

It is a central maxim of ecology that "everything is connected to everything else." This interconnectedness is one of the reasons that soils are such fascinating (and challenging) objects of study. In this chapter we shall explore two aspects of the soil environment, aeration and temperature, that are not only closely connected to each other, but are both also intimately influenced by many of the soil properties discussed in other chapters.

Since air and water share the pore space of soils, it is not surprising that much of what we learned about the texture, structure, and porosity of soils (Chapter 4) and the retention and movement of water in soils (Chapters 5 and 6) will have direct bearing on soil aeration. These are some of the physical parameters affecting aeration status, but chemical and biological processes also affect, and are affected by, soil aeration.

For the growth of plants and the activity of microorganisms, soil aeration status can be just as important as soil moisture status, and can sometimes be even more difficult to manage. In most forest, range, agricultural, and ornamental applications, a major management objective is to maintain a high level of oxygen in the soil for root respiration. Yet it is also vital that we understand the chemical and biological changes that take place when the oxygen supply in the soil is depleted.

Soil temperatures affect plant and microorganism growth and also influence soil drying by evaporation. The movement and retention of heat energy in soils are often ignored but hold the key to understanding many important soil phenomena, from frost-damaged pipelines and pavements to the spring awakening of biological activity in soils. The unusually high soil temperatures that result from fires on forest-, range-, or croplands can markedly change critical physical and chemical soil properties.

We will see that increasing soil temperatures influence soil aeration largely through their stimulating effects on the growth of plants and soil organisms and on the rates of biochemical reactions. Nowhere are these interrelationships more critical than in the water-saturated soils of wetlands, ecosystems that will, therefore, receive special attention in this chapter.

7.1 SOIL AERATION

Aeration involves the ventilation of the soil, with gases moving both into and out of the soil. Aeration determines the rate of gas exchange with the atmosphere, the proportion of pore spaces filled with air, the composition of that soil air, and the resulting chemical oxidation or reduction potential in the soil environment.

For respiration to be carried out by soil organisms, including plant roots, oxygen (O_2) must be supplied and carbon dioxide (CO_2) removed. In a well-aerated soil, the exchange of these two gases between the soil and the atmosphere is sufficiently rapid to prevent the deficiency of oxygen or the toxicity of excess carbon dioxide. For most upland plants, the supply of oxygen in the soil air must be kept above 0.1 L/L (as compared to 0.2 L/L in the atmosphere). In turn, the concentrations of CO_2 and other potentially toxic gases, such as methane or ethylene, must not be allowed to build up excessively.

Oxygen availability in field soils is regulated by three principal factors: (1) *soil macroporosity* (as affected by texture and structure), (2) *soil water content* (as it affects the proportion of porosity that is filled with air), and (3) *O_2 consumption* by respiring organisms (including plant roots and microorganisms). The term *poor soil aeration* refers to a condition in which the availability of O_2 in the root zone is insufficient to support optimal growth of most plants and aerobic microorganisms.

Excess Moisture

Typically, poor aeration becomes a serious impediment to plant growth when more than 80 to 90% of the soil pore space is filled with water (leaving less than 10 to 20% of the pore space filled with air). The high soil water content not only leaves little pore space for air storage, but, more important, the water blocks the pathways by which gases could exchange with the atmosphere. Compaction can also cut off gas exchange, even if the soil is not very wet and has a large percentage of air-filled pores.

The extreme case of excess water occurs when all or nearly all of the soil pores are filled with water. The soil is then said to be **water saturated** or **waterlogged.** Waterlogged soil conditions are typical of wetlands, and may also occur for short periods of time in depressions and flat areas on upland sites. In well-drained soils, saturated conditions may occur temporarily when water is being added.

Certain plants adapted to life in waterlogged soils are termed **hydrophytes.** Examples include a number of grass species, such as rice, eastern gama grass, and spartina marsh grasses that can transport oxygen for respiration down to their roots via hollow structures in their stems and roots known as **aerenchyma** tissues. Mangroves and other hydrophytic trees produce aerial roots and other structures that allow their roots to obtain O_2 while growing in water-saturated soils.

Most plants, however, are dependent on a supply of oxygen from the soil, and so suffer dramatically if good soil aeration is not maintained by drainage or other means (Figure 7.1). Some plants succumb to O_2 deficiency or toxicity of other gases within hours after the soil is saturated.

Gaseous Interchange

The more rapidly roots and microbes use up oxygen and release carbon dioxide, the greater is the need for the exchange of gases between the soil and the atmosphere. This exchange is facilitated by two mechanisms, **mass flow** and **diffusion.** Mass flow of air is much less important than diffusion in determining the total exchange that occurs. It is enhanced, however, by fluctuations in soil moisture content that force air in or out of the soil or by wind and changes in barometric pressure.

The great bulk of the gaseous interchange in soils occurs by *diffusion*. Through this process, each gas moves in a direction determined by its own partial pressure. The *partial pressure* of a gas in a mixture is simply the pressure this gas would exert if it alone were present in the volume occupied by the mixture. Thus, if the pressure of air is 1 atmosphere (~100 kPa), the partial pressure of oxygen, which makes up about 21% (.21 L/L) of the air by volume, is approximately 21 kPa.

Diffusion allows extensive gas movement from one area to another even though there is no overall pressure gradient for the total mixture of gases. There is, however, a concentration gradient for each individual gas, which may be expressed as a *partial pressure gradient*. As a consequence, the higher concentration of oxygen in the atmosphere will result in a net movement of this particular gas into the soil. Carbon dioxide and water vapor normally move in the opposite direction, since the partial pressures of these two gases are generally higher in the soil air than in the atmosphere. A representation of the principles involved in diffusion is given in Figure 7.2.

FIGURE 7.1 Most plants depend on the soil to supply oxygen for root respiration, and therefore are disastrously affected by even relatively brief periods of soil saturation during which oxygen becomes depleted. (Left) Sugar beets on a clay loam soil dying where the soil has become water-saturated in a compacted area. (Right) Pine trees dying in a sandy soil area that has become saturated as a result of flooding by beavers. A new community of plants better adapted to poorly aerated soil conditions is taking over the site. (Photos courtesy of R. Weil)

7.2 MEANS OF CHARACTERIZING SOIL AERATION

The aeration status of a soil can be characterized in several ways, including (1) the content of oxygen and other gases in the soil atmosphere, (2) the air-filled soil porosity, and (3) the chemical oxidation-reduction (redox) potential.

Gaseous Composition of the Soil Air

OXYGEN. The atmosphere above the soil contains nearly 21% O_2, 0.035% CO_2, and more than 78% N_2. In comparison, soil air has about the same level of N_2 but is consistently lower in O_2 and higher in CO_2. The O_2 content may be only slightly below 20%

FIGURE 7.2 The process of diffusion between gases in a soil pore and in the atmosphere. The total gas pressure is the same on both sides of the boundary. The partial pressure of oxygen is greater, however, in the atmosphere. Therefore, oxygen tends to diffuse into the soil pore where fewer oxygen molecules per unit volume are found. The carbon dioxide molecules, on the other hand, move in the opposite direction owing to the higher partial pressure of this gas in the soil pore. This diffusion of O_2 into the soil pore and of CO_2 into the atmosphere will continue as long as the respiration of root cells and microorganisms consumes O_2 and releases CO_2.

in the upper layers of a soil with a stable structure and an abundance of macropores. It may drop to less than 5% or even to near zero in the lower horizons of a poorly drained soil if oxygen is being rapidly consumed by actively growing plant roots or by microbes. When all the soil pores are filled with water, soil microorganisms extract most of the O_2 dissolved in the water for metabolic purposes. Once the supply of O_2 is virtually exhausted, the soil environment is said to be **anaerobic**.

CARBON DIOXIDE. Since the N_2 content of soil air is relatively constant, there is a general inverse relationship between the contents of the other two major components of soil air—O_2 and CO_2—with O_2 decreasing as CO_2 increases. Although the actual differences in CO_2 amounts may not be impressive, they are significant, comparatively speaking. Thus, when the soil air contains only 0.35% CO_2, this gas is about 10 times as concentrated as it is in the atmosphere. In cases where the CO_2 content becomes as high as 10%, it may be toxic to some plant processes.

OTHER GASES. Soil air usually is much higher in water vapor than is the atmosphere, being essentially saturated except at or very near the surface of the soil (see Section 5.7). Also, under waterlogged conditions, the concentrations of gases such as methane (CH_4) and hydrogen sulfide (H_2S), which are formed as organic matter decomposes, are notably higher in soil air. Another gas produced by anaerobic microbial metabolism is ethylene (C_2H_4). This gas is particularly toxic to plant roots, even in concentrations lower than 1 $\mu L/L$ (0.0001%). Root growth of a number of plants has been shown to be inhibited by ethylene that accumulates when gas exchange rates between the atmosphere and the soil are too slow.

Air-Filled Porosity

In Chapter 1 (Figure 1.17) we noted that the ideal soil composition for plant growth would include close to a 50:50 mix of air and water in the soil pore space, or about 25% air in the soil, by volume (assuming a total porosity of 50%). Many researchers believe that, in most soils, microbiological activity and plant growth become severely inhibited when air-filled porosity falls below 20% of the pore space or 10% of the total soil volume (with correspondingly high water contents).

One of the principal reasons that high water contents cause oxygen deficiencies for roots is that water-filled pores block the diffusion of oxygen into the soil to replace that used by respiration. In fact, oxygen diffuses 10,000 times faster through a pore filled with air than through a similar pore filled with water.

7.3 OXIDATION-REDUCTION (REDOX) POTENTIAL[1]

Redox Potential (E_h)

The reaction that takes place as the reduced state of an element is changed to the oxidized state may be illustrated by the oxidation of two-valent iron [Fe^{2+} or Fe(II)] in FeO to the trivalent form [Fe^{3+} or Fe(III)] in FeOOH.

$$\overset{(2+)}{\underset{Fe(II)}{2FeO}} + 2H_2O \rightleftharpoons \overset{(3+)}{\underset{Fe(III)}{2FeOOH}} + 2H^+ + 2e^-$$

Note that Fe(II) loses an electron e^- as it changes to Fe(III), and that H^+ ions are formed in the process. The loss of an electron suggests that there are potentials for the transfer of electrons from one substance to another. This **redox potential** can be measured using a platinum electrode.

The redox potential E_h provides a measure of the tendency of a substance to accept or donate electrons. It is usually measured in volts or millivolts. As is the case for water

---

[1] For a review of redox reactions in soils, see Bartlett and James (1993).

potential (Section 5.3), redox potential is related to a reference state, in this case the hydrogen couple $\frac{1}{2} H_2 \rightleftharpoons H^+ + e^-$, whose redox potential is arbitrarily taken as zero. If a substance will accept electrons easily, it is known as an *oxidizing agent*; if a substance supplies electrons easily, it is a *reducing agent*.

Role of Oxygen (O_2)

As its name suggests, oxygen gas (O_2) is a strong oxidizing agent. It rapidly accepts electrons from many other elements. All aerobic respiration requires O_2 to serve as the electron acceptor as living organisms oxidize organic carbon to release energy for life.

Oxygen can oxidize both organic and inorganic substances. Keep in mind, however, that as it oxidizes another substance, O_2 is in turn reduced. This reduction process can be seen in the following reaction.

$$\overset{(0)}{\frac{1}{2} O_2} + 2H^+ + 2e^- \rightleftharpoons \overset{(2-)}{H_2O}$$

Note that the oxygen atom having zero charge in O_2 accepts two electrons, taking on a charge of –2 when it becomes part of the water molecule. These electrons could have been donated by two molecules of FeO undergoing oxidation, as shown in the previous reaction. If we combine the two equations we can see the overall effect of oxidation and reduction.

$$
\begin{aligned}
2FeO + 2H_2O &\rightleftharpoons 2FeOOH + 2H^+ + 2e^- \\
\tfrac{1}{2}O_2 + 2H^+ + 2e^- &\rightleftharpoons H_2O \\
\hline
2FeO + \tfrac{1}{2}O_2 + H_2O &\rightleftharpoons 2FeOOH
\end{aligned}
$$

The donation and acceptance of electrons (e^-) and H^+ ions on each side of the equation have balanced each other and therefore do not appear in the combined reaction, but for the specific reduction and oxidation reactions they are both very important.

The redox potential E_h of a soil is dependent on both the presence of electron acceptors (oxygen or other oxidizing agents) and pH. The O_2 content of soil air and E_h (redox potential) are generally inversely related. In a well-drained soil, the E_h is in the 0.4 to 0.7 volt (V) range. As aeration is reduced, the E_h declines to a level of about 0.3 to 0.35 V when gaseous oxygen is depleted. Under flooded conditions, in warm, organic-matter-rich soils, E_h values as low as –0.3 V can be found.

Other Electron Acceptors

Other elements in addition to oxygen can act as terminal electron acceptors (oxidizers). For example, N(V) in nitrate accepts two electrons when it is reduced to N(III) in nitrite:

$$\underset{N(V)}{\overset{(5+)}{NO_3}} + 2e^- + 2H^+ \rightleftharpoons \underset{N(III)}{\overset{(3+)}{NO_2^-}} + H_2O$$

Similar reactions involve the reduction or oxidation of Fe, Mn, and S (see reactions in Figure 7.3).

The effect of pH on redox potentials relating to several important reactions taking place in soils is shown in Figure 7.3. Note that in all cases the E_h decreases as the pH rises from 2 to 8. Since both pH and E_h are easily measured, it is not too difficult to ascertain in a given soil whether a specified reaction would be likely to occur. For example, at pH 6 the E_h would need to be somewhat lower than +0.5 V to encourage the reduction of nitrates to nitrites, and about +0.2 volts to stimulate the reduction of FeOOH to the Fe^{2+} ion. For methane to form in a waterlogged soil at the same pH (6), an E_h of about –0.2 V would be required.

The E_h value at which oxidation-reduction reactions occur varies with the specific chemical to be oxidized or reduced. In Table 7.1 are listed oxidized and reduced forms of several elements important in soils, along with the approximate redox potentials at which the oxidation-reduction reactions occur. The E_h values explain the sequence of reactions that are known to occur when a well-aerated soil becomes saturated with water.

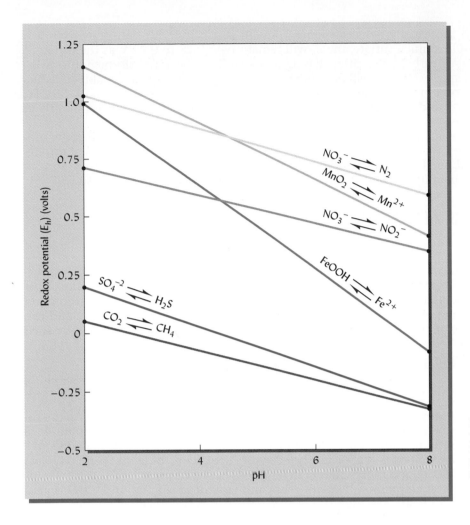

FIGURE 7.3 The effect of pH on the redox potential E_h at which several important reduction-oxidation reactions take place in soil. [From McBride (1994); used with permission of Oxford University Press]

At first, respiration will reduce the concentrations of O_2 in the soil air and dissolved in the soil water. As the concentration of this electron acceptor is lowered, so, too, the redox potential of the soil is lowered. Since O_2 is reduced to water at E_h levels of 0.38 to 0.32 V, once the soil redox potential falls below this level the soil is essentially devoid of O_2. At lower E_h values, the only microorganisms that can function are those able to use elements other than oxygen as their metabolic electron acceptors.

TABLE 7.1 Oxidized and Reduced Forms of Certain Elements in Soils and the Redox Potentials E_h at Which Changes in Form Occur in a Soil at pH 6.5

Note that gaseous oxygen is depleted at E_h levels of 0.38 to 0.32 V. At lower E_h levels microorganisms utilize elements other than oxygen as the electron acceptor in their metabolism. By donating electrons they transform these elements into their reduced valence state.

Element	Oxidized forms common in well-oxidized soils	Reduced forms common in waterlogged soils	E_h at which change of form occurs, V
Oxygen	O_2	H_2O	0.38 to 0.32
Nitrogen	NO_3^-	N_2, NH_4^+	0.28 to 0.22
Manganese	Mn^{4+} [Mn(IV) oxides]	Mn^{2+} [Mn(II) oxides]	0.22 to 0.18
Iron	Fe^{3+} [Fe(III) oxides]	Fe^{2+} [Fe(II) oxides]	0.11 to 0.08
Sulfur	SO_4^{2-}	S^{2-}, H_2S	−0.14 to −0.17
Carbon	CO_2	CH_4, C_2H_2, CH_3CH_2OH	−0.20 to −0.28

E_h data from Patrick and Jugsujinda (1992).

The next most easily reduced element present is usually N(V) (in nitrate, NO_3^-). If the soil contains much nitrate, the E_h will remain near 0.28 to 0.22 V as the nitrate is reduced. Once nearly all the nitrate has disappeared [N(V) been transformed into N(III) and other N species], the E_h will drop further. At this point, organisms capable of reducing Mn will become active, and so on. Thus as E_h values fall, the elements N, Mn, Fe, and S (in SO_4^{-2}) and C (in CO_2) accept electrons and become reduced, predominantly in the order listed.

In other words, the soil E_h must be lowered to zero or less before methane is produced, but an E_h value of 0.28 to 0.22 is low enough to result in the reduction of nitrate-N. Thus, soil aeration helps determine the specific chemical species present and, in turn, the availability, mobility, and possible toxicity of various elements in soils.

7.4 FACTORS AFFECTING SOIL AERATION

Drainage of Excess Water

Drainage of gravitational water out of the soil is necessary for good aeration. Two factors commonly impede such drainage: (1) the presence of the water table in the upper part of the soil (see Section 6.9) and/or (2) a poorly developed network of macropores (see Section 4.6). A shallow water table is usually a function site topography and can often be ameliorated by artificial drainage. Macropore content is determined by such factors as soil texture, bulk density, aggregate stability, organic matter content and formation of biopores (see Section 4.6).

Rates of Respiration in the Soil

The concentrations of both oxygen and carbon dioxide are largely dependent on microbial activity, which in turn depends on the availability of organic carbon compounds as food. Incorporation of large quantities of manure, crop residues, or sewage sludge may alter the soil air composition appreciably. Likewise, the cycling of plant residues by leaf fall, root mass decay, and root excretion in natural ecosystems provides the substrate for microbial activity. Respiration by plant roots and enhanced respiration by soil organisms near the roots are also significant processes. All these processes are very much enhanced as soil temperature increases (see Section 7.8).

Soil Heterogeneity

Subsoils are usually more deficient in oxygen than are topsoils. Not only is the water content usually higher (in humid climates), but the total pore space, as well as the macropore space, is generally much lower in the deeper horizons. In addition, the pathway for diffusion of gases into and out of the soil is longer for deeper horizons. However, if organic substrates are in low supply in the subsoil, it may still be aerobic. For this reason, certain recently flooded soils are anaerobic in the upper 50 to 100 cm and are aerobic below.

PROFILE. As seen in Figure 7.4, the aeration status varies greatly in different locations in a soil profile. In well-drained, relatively uniform soils the trend is a general reduction in O_2 and increase in CO_2 as one moves down the profile. However, poorly aerated zones or pockets may be found in any horizon of an otherwise well-drained and well-aerated soil.

TILLAGE. One cause of soil heterogeneity is tillage, which has both short-term and long-term effects on soil aeration. In the short term, stirring the soil often allows it to dry out faster and mixes in large quantities of air. These effects are especially evident on somewhat compacted, fine-textured soils, on which plant growth often responds immediately to such soil disturbances as a cultivation to control weeds or "knife in" fertilizer. In the long term, however, tillage may reduce macroporosity (see Section 4.6).

LARGE MACROPORES. Poorly aerated zones may result from a heavy-textured or compacted soil layer, or it may be merely the inside of a soil ped (structural unit) where the smallness of pores may limit ready air exchange (Figure 7.5). In well-drained soils, the large pores (cracks) between peds, and old root channels in the subsoil, may periodically fill with water, causing localized zones of poor aeration. In saturated soil, such large pores may cause the opposite effect, as they facilitate O_2 diffusion into the soil during periods of drying.

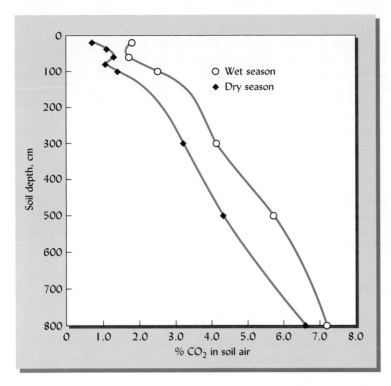

FIGURE 7.4 Changes in the concentration of soil air with depth into the profile of a Haplustox soil under a tropical rainforest in the Brazilian Amazon region. The source of the CO_2 was likely a combination of root and microbial respiration. Although by far the highest rates of CO_2 production were in the upper soil layers, gas produced there had little distance to travel to reach the atmosphere and many large pores to travel through. The concentration of CO_2 increased with depth because the increasing travel distance to the surface and the much lower macroporosity at depth greatly slowed the movement of gases, causing CO_2 to accumulate. During the wet season, more macropores were filled with water, slowing the diffusion of CO_2 out of the soil (and of O_2 into the soil). [Data from Davidson and Trumbore (1995)]

PLANT ROOTS. The roots of growing plants may either reduce or increase the O_2 concentration in their immediate vicinity. In somewhat poorly drained soils, respiration by roots of upland plants may deplete the O_2 in nearby soil (as can be seen in Plate 19, after page 178). In contrast, hydrophytic plants with aerenchyma tissues may transport surplus O_2 into their roots, allowing some to diffuse into the soil and produce an oxidized zone in an otherwise anaerobic soil (see for example, Plate 20).

For these reasons, oxidation reactions may be occurring within a few centimeters or millimeters from another location where reducing conditions exist. This heterogeneity of soil aeration should be kept in mind when attempting to understand the role that soil plays in elemental cycling and ecosystem function.

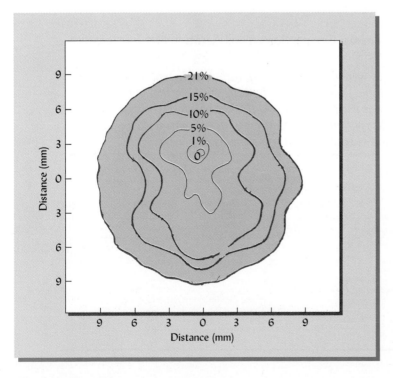

FIGURE 7.5 A map showing the oxygen content of soil air in a wet aggregate from an Aquic Hapludoll (Muscatine silty clay loam) from Iowa. The measurements were made with a unique microelectrode. Note that the oxygen content near the aggregate center was zero, while that near the edge of the aggregate was 21%. Thus pockets of oxygen deficiency can be found in a soil whose overall oxygen content may not be low. [From Sexstone, et al. (1985)]

Seasonal Differences

There is marked seasonal variation in the composition of soil air. In the springtime in temperate humid regions, the soils are often wet and opportunities for ready gas exchange are poor. But due to low soil temperatures, the respiration of plant roots and soil microorganisms is restricted, so the utilization of oxygen and release of CO_2 are also restrained. In the summer months, the soils are commonly lower in moisture content, and the opportunity for gaseous exchange is increased. However, more favorable temperatures also stimulate vigorous respiration by plant roots and microorganisms, releasing copious quantities of carbon dioxide.

Effects of Vegetation

In addition to the root respiration effects mentioned previously, vegetation may affect soil aeration by removing large quantities of water via transpiration, enough to lower the water table in some poorly drained soils. The data in Table 7.2 illustrate this effect for a pine forest. The effects of soil depth and season are also evident.

7.5 ECOLOGICAL EFFECTS OF SOIL AERATION

Effects on Organic Residue Degradation

Soil aeration influences many soil reactions and, in turn, many soil properties. The most obvious of these reactions are associated with microbial activity, especially the breakdown of organic residues and other microbial reactions. Poor aeration slows down the rate of decay, as evidenced by the relatively high levels of organic matter that accumulate in poorly drained soils.

The presence or absence of oxygen gas completely modifies the nature of the decay process and its effect on plant growth. Where O_2 is present, aerobic organisms are active, and oxidation reactions such as shown on page 326 occur. In the absence of gaseous oxygen, anaerobic organisms take over, which are much slower at organic matter breakdown.

Poorly aerated soils tend to contain a wide variety of only partially oxidized products such as ethylene gas (C_2H_4), alcohols, and organic acids, many of which can be toxic to higher plants and to many decomposer organisms. The latter effect helps account for the formation of Histosols in wet areas where inhibition of decomposition allows thick layers of organic matter to accumulate.

TABLE 7.2   **Effect of Timber Harvest with Minimal Compaction on Soil Aeration and Temperature Regimes in a Subtropical Pine Forest**

Once the 55-year-old loblolly pine trees were cut, evapotranspiration and shading were reduced, resulting in a higher water table, lower redox potentials, and warmer spring temperatures. The latter stimulated microbial use of oxygen and further lowered redox potentials in this Vertic Ochraqualf. Note that the redox potentials were lower with the warmer temperatures in spring, even though the soil was not as wet in spring as in winter.

Site treatment	Time soil is saturated, %	Soil temperature, °C		Soil redox potentials E_h, V	
		Winter	Spring	Winter	Spring
Measured at 50-cm depth					
Undisturbed pine stand	31	11.8	18.3	0.83	0.65
Trees cut, not compacted	64	11.7	20.5	0.51	0.11
Measured at 100-cm depth					
Undisturbed pine stand	54	13.3	17.3	0.83	0.49
Trees cut, not compacted	46	13.2	18.7	0.54	0.22

Data from Tiarks, et al. (1996).

NUTRIENTS. Through its effects on the redox potential, the level of soil oxygen largely determines the forms of several inorganic elements, as shown in Table 7.1. The oxidized states of nitrogen and sulfur are readily utilizable by higher plants. In alkaline soils reduction of iron and manganese may be desirable to alleviate plant deficiencies of these elements, but in acid soils oxidized conditions are desirable for iron and manganese nutrition of most plants because, in these soils, the reduced forms are so soluble that toxicities may occur. However, some reduction of iron may be beneficial as it will release phosphorus from insoluble iron-phosphate compounds. Such phosphorus release has implications for eutrophication (see Section 13.2) when it occurs in saturated soils or in underwater sediments.

Such differences illustrate the interaction of aeration and soil pH in supplying available nutrients to plants (see Chapters 9 and 14).

OTHER ELEMENTS. Redox potential determines the species of such toxic elements as chromium, arsenic, and selenium. In contrast to iron and managanese, these elements tend to become less toxic and environmentally damaging when in their reduced states. Therefore, encouraging reducing conditions by waterlogging soils or adding easily decomposed organic materials can help reduce environmental problems associated with soils contaminated by these elements (see, for example, Box 9.4).

Hexavalent Cr^{6+} compounds are mobile under a wide range of pH conditions and are highly toxic to humans. Trivalent (reduced) Cr^{3+} forms oxides and hydroxides that are quite immobile except in very acid soils. Therefore, it would be desirable to reduce Cr^{6+} to Cr^{3+} in soil contaminated with chromium from industrial processes or mining wastes. Easily decomposed organic matter in soils is quite effective in promoting this reduction (Figure 7.6), and fortunately the Cr^{3+}, once reduced, does not tend to reoxidize.

SOIL COLORS. As was discussed in Section 4.1, soil color is influenced markedly by the oxidation status of iron and manganese. Colors such as red, yellow, and reddish brown are characteristic of well-oxidized conditions. More subdued shades such as grays and blues predominate if insufficient oxygen is present. Soil color can be used in field methods for determining the status of soil drainage. Imperfectly drained soils are characterized by contrasting streaks of oxidized and reduced materials (see Plates 15 and 18, after page 178). Such a mottled condition indicates a zone of alternating good and poor aeration, a condition not conducive to the optimum growth of most plants.

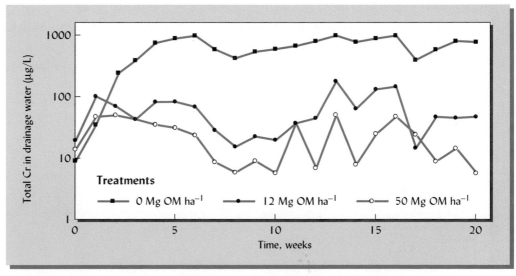

FIGURE 7.6 Effect of adding decomposable organic matter (OM) on the concentration of chromium in water draining from a chromium-contaminated soil. Here dried cattle manure was added as the decomposable OM. As the manure oxidized, it caused the reduction of the toxic, mobile Cr^{6+} to the relatively immobile, nontoxic Cr^{3+}. Note the log scale for the Cr in the water, indicating that the high level of manure addition caused the Cr level to be lowered approximately 100-fold. The coarse textured soil was a Typic Torripsamment in California. [Data from Losi, et al. (1994)]

METHANE PRODUCTION. Methane gas (CH_4) is produced by the reduction of CO_2 (Table 7.1). Its formation occurs when the E_h is reduced to about -0.2 V, a condition common in natural wetlands and in rice paddies. It is estimated that wetlands (see Section 7.7) in the United States emit about 100 million metric tons of methane annually. The production of the organic compound methane in submerged soils is of universal significance, since this gas is one of those contributing to the greenhouse effect and global warming (see Section 11.11).

Effects on Activities of Higher Plants

Plants are adversely affected in at least three ways by conditions of poor aeration: (1) the growth of the plant, particularly the roots, is curtailed; (2) the absorption of nutrients and water is decreased; and (3), as discussed in the previous section, the formation of certain inorganic compounds toxic to plant growth is favored.

Different plant species vary in their ability to tolerate poor aeration (Table 7.3). Sugar beets and barley are examples of crop species that require high air porosities for best growth (see Figure 7.1 *Left*). In contrast, ladino clover and reed canary grass can grow with very low air porosity. Rice and cranberries are crops that can grow with their roots submerged in water.

Furthermore, the tolerance of a given plant to low porosity may be different for seedlings than for rapidly growing plants. A case in point is the tolerance of red pine to restricted drainage during its early development and its poor growth or even death on the same site at later stages (see Figure 7.1 *Right*).

Knowledge of plant tolerance to poor aeration is useful in choosing appropriate species to revegetate wet sites. The occurrence of plants specially adapted to anaerobic conditions is useful in identifying wetland sites (see Section 7.7).

While soil compaction does decrease the exchange of gases, the negative effects of soil compaction are not all owing to poor aeration. Soil layers can become so dense as to impede the growth of roots even if an adequate oxygen supply is available (see Section 4.5).

7.6 AERATION IN RELATION TO CONTAINERIZED AND ORNAMENTAL PLANTS

Container-Grown Plants

Potted plants frequently suffer from waterlogging because it is difficult to supply the exact amount of water the plant needs. To prevent waterlogging, most containers have holes at the bottom to drain excess water. As was the case with stratified soils in the field (see Section 5.6), water drains out of the holes at the bottom of the container *only* when the soil at the bottom is saturated with water. If the growing medium is mostly mineral

TABLE 7.3 **Examples of Plants with Varying Degrees of Tolerance to a High Water Table and Accompanying Restricted Aeration**

The plants in the leftmost column commonly thrive in wetlands. Those in the rightmost column are very sensitive to poor aeration.

		Plants adapted to grow well with a water table at the stated depth		
<10 cm	*15 to 30 cm*	*40 to 60 cm*	*75 to 90 cm*	*>100 cm*
Bald cypress	Alsike clover	Birdsfoot trefoil	Beech	Arborvitae
Black spruce	Black Willow	Black locust	Birch	Barley
Common cattail	Cottonwood	Bluegrass	Cabbage	Beans
Cranberries	Deer tongue	Linden	Corn	Cherry
Duckgrass	Eastern gama grass	Mulberry	Hairy vetch	Hemlock
Fragmites grass	Ladino clover	Mustard	Millet	Oats
Maiden cane	Loblolly pine	Red maple	Peas	Peach
Mangrove	Orchard grass	Sorghum	Red oak	Sand lovegrass
Pitcher plant	Redtop grass	Sycamore		Sugar beets
Reed canarygrass	Tall fescue	Weeping lovegrass		Walnut
Rice		Willow oak		Wheat
Skunk cabbage				White pine
Spartina grass				
Swamp white oak				
Swamp rosemalow				
Water tupelo				

soil, the fine soil pores remain filled with water, leaving no room for air, and anaerobic conditions soon prevail. Use of taller containers will allow for better aeration in the upper part of the medium.

To manage these problems, potting mixes are engineered to meet the requirements of the containerized plants. Mineral soil generally makes up no more than one-third of the volume of most potting mixes, the remainder being composed of inert lightweight but coarse-grained materials such as perlite (expanded volcanic glass), vermiculite (expanded mica—not soil vermiculite), or pumice (porous volcanic rock). Most modern mixes also contain some stable organic material, such as peat, shredded bark, wood chips, or compost, that holds water as well as adding macroporosity.

Tree and Lawn Management

In transplanting a young tree seedling or any woody species, special caution must be taken to prevent poor aeration or waterlogging immediately around the young roots. Figure 7.7 illustrates the right and the wrong way to manage the transplanting of trees in compacted soils.

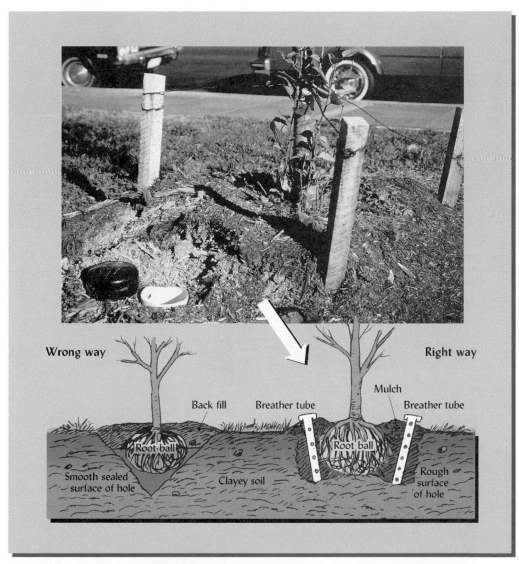

FIGURE 7.7 Providing a good supply of air to tree roots can be a problem, especially when trees are planted in fine-textured, compacted soils of urban areas. A machine-dug hole with smooth sides will act as a "tea cup" and fill with water, suffocating tree roots. Breather tubes, a larger rough-surfaced hole, and a layer of surface mulch in which some fine tree roots can grow are all measures that can improve the aeration status of the root zone. (Photo courtesy of R. Weil)

The aeration of well-established, mature trees must also be safeguarded. If operators push surplus excavated soil around the base of a tree during landscape grading, serious consequences are soon noticed (Figure 7.8). The tree's feeder roots near the original soil surface are soon deficient in oxygen even if the overburden is no more than 5 to 10 cm in depth. Building a protective wall (a *dry well*) around the base of a valuable tree before grading operations begin will preserve enough of the original surface to allow the tree roots access to the O_2 they need, thereby saving the tree.

Management systems for heavily trafficked lawns commonly have components relating to soil aeration. For example, one means of increasing the aeration in compacted lawn areas is to use *core cultivation* that actually removes small cores of soil from the surface horizon, thereby permitting gas exchange to take place more easily (Figure 7.9).

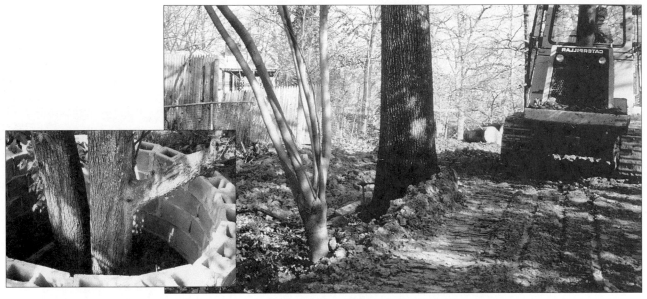

FIGURE 7.8 Protection of valuable trees during landscape grading operations. Even a thin layer of soil spread over a large tree's root system can suffocate the roots and kill the tree. (Inset) In order to preserve the original ground surface so that tree feeder roots can obtain sufficient oxygen, a dry well may be constructed of brick or any decorative material. The dry well may be incorporated into the final landscape design, or filled in at a rate of a few centimeters per year. (Photos courtesy of R. Weil)

FIGURE 7.9 One means of increasing the aeration of compacted soil is by core cultivation. The machine removes small cores of soil, leaving holes about 2 cm in diameter and 5 to 8 cm deep. This method is commonly used on high-traffic turf areas. Note that the machine *removes* the cores and does not simply punch holes in the soil, a process that would increase compaction around the hole and impede air diffusion into the soil. (Photos courtesy of R. Weil)

7.7 WETLANDS AND THEIR POORLY AERATED SOILS[2]

Poorly aerated areas called **wetlands** cover approximately 14% of the world's ice-free land, with the greatest areas occurring in the cold regions of Canada, Alaska, and Russia. In the continental United States, about 400,000 km² exist today, less than half of the area that is estimated to have existed when European settlement of the nation began. Most wetland losses occurred as farmers used artificial drainage (see Section 6.9) to convert them into cropland. This conversion process was assisted by some of the same government agencies (e.g., U.S. Army Corps of Engineers and USDA) that are now working to protect wetlands from further damage. In recent decades, filling and drainage for urban development has also taken its toll of wetland areas. Since environmental consciousness has become a force in modern societies, wetland preservation has become a major issue, and the loss of wetlands has been slowed.

Defining a Wetland[3]

Wetland is a scientific term for ecosystems that are transitional between land and water. These systems are neither strictly terrestrial (land-based) nor aquatic (water-based). While there are many different types of wetlands, they all share a key feature, namely *soils that are water-saturated near the surface for prolonged periods when the soil temperature is high enough to result in anaerobic conditions*. It is largely the prevalence of anaerobic conditions that determines the kinds of plants, animals, and soils found in these areas. There is widespread agreement that the wetter end of a wetland occurs where the water is too deep for rooted, emergent vegetation to take hold. The difficulty is in precisely defining the so-called *drier end* of the wetland, the boundary beyond which exist nonwetland, upland systems in which the plant–soil–animal community is no longer predominantly influenced by the presence of water-saturated soils.

What are the characteristics these scientists look for to indicate the existence of a wetland system? Most authorities agree that three characteristics can be found in any wetland:

1. A wetland hydrology or water regime
2. Hydric soils
3. Hydrophytic plants

We shall briefly examine each.

Wetland Hydrology

WATER BALANCE. Water flows into wetlands from surface runoff (e.g., bogs and marshes), from groundwater seepage, and from direct precipitation. It flows out by surface and subsurface flows, as well by evaporation and transpiration (see Section 6.3 and Table 7.2). The balance between inflows and outflows, as well as the water storage capacity of the wetland itself, determines how wet it will be and for how long.

Tidal marshes in coastal areas may be flooded for part of each day (Figure 7.10). Other wetlands may be flooded for only a month or so each year, while some may never be flooded, although they are saturated within the upper soil horizons. If the period of saturation occurs when the soil is too cold for microbial or plant-root activity to take place, anaerobic conditions may not develop, even in flooded soils. This temperature requirement is sometimes referred to as the *growing season* in reference to the time of year during which plants are actively growing. As we shall see in Section 7.8, microbial activity continues down to about 5°C, so O_2 may be depleted when the soil is warmer than this (even if plants are dormant). Remember, it is the anaerobic condition, not just saturation, that makes a wetland a wetland.

---

[2] Two well-illustrated, nontechnical, yet informative publications on wetlands are Welsh, et al. (1995), and CAST (1994). For a compilation of technical papers on hydric soils and wetlands, see Rabenhorst, et al. (1999).

[3] In 1987, the United States Army Corps of Engineers and the Environmental Protection Agency agreed on the following definition to be used in enforcing the Clean Water Act: "The term wetlands means those areas that are inundated or saturated by surface or ground water at a frequency and duration sufficient to support, and that under normal circumstances do support, a prevalence of vegetation typically adapted for life in saturated soil conditions."

FIGURE 7.10 An example of a wetland with a daily hydroperiod that follows the rise and fall of the slightly brackish estuary tides. This tidal marsh with a beaver house is seen at low tide when some saturated soils and emergent plants are exposed. The drier-end boundary of the wetland is probably just beyond the treeline in the background. (Photo courtesy of R. Weil).

In the field, even during dry periods, there are many signs one can look for to indicate where saturated conditions frequently occur. Past periods of flooding will leave water stains on trees and rocks, and a coating of sediment on the plant leaves and litter. Drift lines of once-floating branches, twigs, and other debris also suggest previous flooding. Trees with extensive root masses above ground indicate an adaptation to saturated conditions. But perhaps the best indicator of saturated conditions is the presence of **hydric soils**.

Hydric Soils[4]

In order to assist in delineating wetlands, soil scientists developed the concept of hydric soils. In *Soil Taxonomy* (see Chapter 3) these soils are mostly (but not exclusively) classified either in the order Histosols, in Aquic suborders such as Aquents, Aquepts, and Aqualfs, or in Aquic subgroups. These soils generally have an aquic or peraquic moisture regime (see Section 3.2).

Three properties help define hydric soils. First, they are subject to *periods of saturation* that inhibit the diffusion of O_2 into the soil. Second, for substantial periods of time they undergo *reduced conditions* (see Section 7.3); that is, electron acceptors other than O_2 are reduced. Third, they exhibit certain features termed *hydric soil indicators* (Box 7.1).

Hydrophytic Vegetation

Although the vast majority of plant species cannot survive the conditions characteristic of wetlands, there do exist varied and diverse communities of plants that have evolved special mechanisms to adapt to life in saturated, anaerobic soils. These plants comprise the **hydrophytic vegetation** that distinguishes wetlands from other systems.

Typical adaptive features include hollow aerenchyma tissues that allow plants like spartina grass to transport O_2 down to their roots. Certain trees (such as bald cypress) produce adventitious roots, buttress roots, or knees. Other species spread their roots in a shallow mass on or just under the soil surface, where some O_2 can diffuse even under a layer of ponded water. The leftmost column in Table 7.3 lists a few common hydrophytes. Not all the plants in a wetland are likely to be hydrophytes, but the majority usually are.

---

[4] The U.S. Department of Agriculture Natural Resources Conservation Service defines a hydric soil as one "that formed under conditions of saturation, flooding or ponding long enough during the growing season to develop anaerobic conditions in the upper part." For an illustrated field guide to features that indicate hydric soils, see Hurt, et al. (1996). For a current list of soil series considered to be Hydric soils, see the Website http://www.statlab.iastate.edu/soils/hydric/.

BOX 7.1 HYDRIC SOIL INDICATORS

Hydric soil indicators are features associated (sometimes only in specific geographic regions) with the occurrence of saturation and reduction. Most of the indicators can be observed in the field by digging a small pit to a depth of about 50 cm. They principally involve the loss or accumulation of various forms of Fe, Mn, S, or C. The carbon (organic matter) accumulations are most evident in Histosols, but thick, *dark surface layers* in other soils can also be indicators of hydric conditions in which organic matter decomposition has been inhibited (see, for example, Plates 6 and 21, after page 178).

Iron, when reduced to Fe(II), becomes sufficiently soluble that it migrates away from reduced zones and may precipitate as Fe(III) compounds in more aerobic zones. Zones where reduction has removed or depleted the iron coatings from mineral grains are termed **redox depletions**. They commonly exhibit the gray, low-chroma colors of the bare, underlying minerals (see Section 4.1 for an explanation of chroma). Also, iron itself turns gray to blue-green when reduced. The contrasting colors of redox depletions or reduced iron and zones of reddish oxidized iron result in unique mottled **redoximorphic features** (see, for example, Plates 15 and 18). Other redoximorphic features involve reduced Mn. These include the presence of hard black *nodules* that sometimes resemble shotgun pellets. Under severely reduced conditions the entire soil matrix may exhibit *low-chroma colors*, termed *gley*. Colors with a chroma of 1 or less quite reliably indicate reduced conditions (Figure 7.11).

Always keep in mind that redoximorphic features are indicative of hydric soils only when they occur in the upper horizons. Many soils of upland areas exhibit redoximorphic features only in their deeper horizons, due to the presence of a fluctuating water table at depth. Upland soils that are saturated or even flooded for short periods, especially if during cold weather, are *not* wetland (hydric) soils.

A unique redoximorphic feature associated with certain wetland plants is the presence, in an otherwise gray matrix, of reddish oxidized iron around root channels where O_2 diffused out from the aerenchyma-fed roots of a hydrophyte (see Plate 20). These *oxidized root zones* exemplify the close relationship between hydric soils and *hydrophytic vegetation*.

FIGURE 7.11 The relationship between the occurrence of some soil features and the annual duration of water-saturated conditions. The *absence* of iron concentrations (mottles) with colors of chroma >4, and the *presence* of strong expressions of the other features are indications that a soil may be hydric. [Adapted from Veneman (1999)]

Wetland Chemistry

Even in a flooded wetland, O_2 will be able to diffuse from the atmosphere or from oxygenated water into the upper 1 or 2 cm of soil, creating a thin *oxidizing zone* (see Figure 7.12). The diffusion of O_2 within the saturated soil is extremely limited, so that a few centimeters deeper into the profile, O_2 is eliminated and the redox potential becomes low enough for reactions such as nitrate reduction to take place. The close proximity of the oxidized and anaerobic zones allows water passing through wetlands to be stripped of N by the sequential oxidation of ammonium N to nitrate N, and then the reduction of the nitrate to various nitrogen gases that escape into the atmosphere (see Section 12.9).

Redox potentials may become low enough for iron reduction to produce redoximorphic features, and for sulfate reduction to produce rotten-egg-smelling hydrogen sulfide (H_2S) gas. The anaerobic zone may extend downward, or in some cases may be

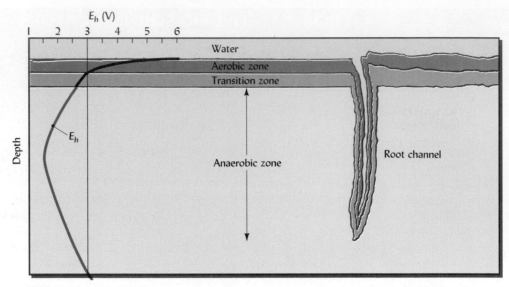

FIGURE 7.12 Representative redox potentials within the profile of an inundated hydric soil. Many of the biological and chemical functions of wetlands depend on the close proximity of reduced and oxidized zones in the soil. The changes in redox potential at the lower depths depend largely on the vertical distribution of organic matter. In some cases, low subsoil organic matter results in a second oxidized zone beneath the reduced zone.

limited to the upper horizons where microbial activity is high. The anaerobic carbon reactions discussed in Section 7.5 are characteristic of this zone, including methane (swamp gas) production. These and other chemical reactions involving the cycling of C, N, and S are explained in Sections 11.2, 12.9, and 12.21. Toxic elements such as chromium and selenium undergo redox reactions that may help remove them from the water before it leaves the wetland. Acids from industry or mine drainage may also be neutralized by reactions in hydric soils.

Constructed Wetlands

Realizing all the beneficial functions of wetlands, scientists and engineers have begun not only to find ways to preserve natural wetlands, but to construct artificial ones for specific purposes, such as wastewater treatment (see, for example, Box 13.2).

Another reason for attempting to construct wetlands is the provision in several regulations that allows for the destruction of certain natural wetland areas, provided that an equal or larger area of new wetlands is constructed or that previously degraded wetlands are restored. As would be expected, this process, termed **wetland mitigation,** has been only partially successful, as scientists cannot be expected to create what they do not fully understand.

We have seen the influence of soil water on soil aeration. We now turn to another soil physical property, soil temperature, that is also greatly influenced by the content of water in a soil.

7.8 PROCESSES AFFECTED BY SOIL TEMPERATURE

The temperature of a soil greatly affects the physical, biological, and chemical processes occurring in that soil, and in plants growing on it (Figure 7.13).

Plant Processes

The growth rates of most plants are actually more sensitive to *soil* temperature than to aboveground *air* temperature, but this is not often appreciated since air temperature is more commonly measured. Most plants have a rather narrow range of soil temperatures for optimal growth (Figure 7.13).

In cool temperate regions, soil temperature often limits the productivity of crops and natural vegetation. The yields of some vegetables and small fruits can be markedly

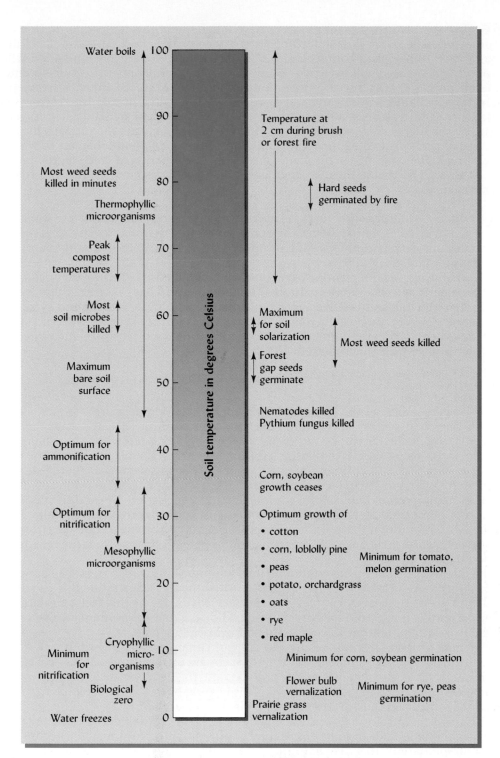

FIGURE 7.13 Soil temperature ranges associated with a variety of soil processes.

increased by warming the soil. The life cycles of plants are also influenced greatly by soil temperature. For example, tulip bulbs require chilling in early winter, to develop flower buds although flower development is suppressed until the soil warms up the following spring.

In warm regions, and in the summer in temperate regions, soil temperatures may be too high for optimal plant growth, especially in the upper few centimeters of soil. Even plants of tropical origin, such as corn, are adversely affected by soil temperatures higher than 35°C. Root growth near the surface may be encouraged by shading the soil with either live vegetation, plant residues, or the use of a mulch.

SEED GERMINATION. Different plant processes have different optimal temperatures. One of the processes most sensitive to soil temperature is seed germination. For example, farmers know that if they plant corn seed into soils cooler than 7 to 10°C, germination will not occur and the seed will likely rot. Many herbaceous annual plants require specific soil temperatures to trigger seed germination, accounting for much of the difference in species composition between early and late season weeds in cultivated land. Likewise, the seeds of certain plants adapted to open gaps in a forest stand are stimulated to germinate by the higher daily maximum soil temperatures and greater fluctuations in soil temperature that occur when the forest canopy is disturbed by timber harvest or wind-thrown trees. The seeds of certain prairie grasses require a period of cold soil temperatures (2 to 4°C) to enable them to germinate the following spring, a process termed *vernalization*.

ROOT FUNCTIONS. Root functions such as nutrient uptake and water uptake are sluggish in cool soils with temperatures below the optimum for the particular species. One result is that nutrient deficiencies, especially of phosphorus, often occur in young plants in early spring, only to disappear when the soil warms later in the season. The phenomenon of *winter burn* of plant foliage is another consequence of low soil temperature that particularly affects evergreen shrubs. On bright sunny days in winter and early spring when the soil is still cold, evergreen plants may become desiccated and even die because the slow water uptake by roots in the cold soils cannot keep up with the high evaporative demand of bright sun on the foliage. The problem can be prevented by covering the shrubs with a shade cloth.

Microbial Processes

Microbiological processes are influenced markedly by soil temperature changes. General soil microbial activity and organic matter decomposition virtually ceases below about 5°C, a benchmark temperature sometimes referred to as *biological zero*. The rates of microbial processes such as respiration typically more than double for every 10°C rise in temperature up to an optimum of about 35 to 40°C (considerably higher than the optimum for plant growth). The dependence of microbial respiration on warm soil temperatures has important implications for soil aeration (see Section 7.7) and the decomposition of plant residues and, hence, the cycling of the nutrients they contain. The productivity of northern (boreal) forests is probably limited by the inhibiting effect of low soil temperatures on microbial recycling and release of nitrogen from the tree litter and soil organic matter.

The microbial oxidation of ammonium ions to nitrate ions, which occurs most readily at temperatures near 30°C, is also negligible when the soil temperature is low, below about 8 to 10°C. Farmers in cool regions can take advantage of this fact by injecting ammonia fertilizers into cold soils in the spring, knowing that ammonium ions will not be readily oxidized to nitrate ions until the soil temperature rises.

High soil temperatures can be used to control certain plant diseases. In environments with hot, sunny summers—maximum daily air temperatures >35°C—covering the ground with transparent plastic sheeting can raise the temperature of the upper few centimeters of soil to as high as 50 to 60°C. This heating process, called *soil solarization*, is used to control pests and diseases in some high-value crops.

Warm soil temperatures are critical for the microbial destruction of toxic organic pesticides and pollutants in soil. Temperature control is critical for some new technologies (**bioremediation**) that take advantage of the ability of certain microorganisms to degrade petroleum products, pesticides, and other compounds.

Freezing and Thawing

When soil temperatures fluctuate above and below 0°C, the water in the soil undergoes cycles of freezing and thawing. Alternate freezing and thawing subject the soil aggregates to pressures as zones of pure ice, called *ice lenses,* form within the soil and as ice crystals form and expand. These pressures alter the physical structure of the soil. In a saturated soil with a puddled structure, the frost action breaks up the large masses and greatly improves granulation. In contrast, for soils with good aggregation to begin with, freeze-thaw action when the soil is very wet can lead to structural deterioration.

Alternate freezing and thawing can force objects upward in the soil, a process termed *frost heaving* (Figure 7.14). Objects subject to heaving include stones, fence posts, and perennial tap-rooted plants. This action, which is most severe where the soil is silty in texture, wet, and lacking a covering of snow or dense vegetation, can drastically reduce the stand of alfalfa (Figure 7.14*f*), some clovers, and trefoil.

Freezing can also heave shallow foundations, roads, and runways which have fine material as a base. Gravels and pure sands are normally resistant to frost damage, but silts and sandy soils with modest amounts of finer particles are particularly susceptible. Very clay-rich soils do not usually exhibit much frost heave, but ice-lens segregation can still occur and can lead to severe loss of strength during thawing when the soil turns to mud. To avoid damage by freezing soil temperatures, foundation footings (as well as water pipelines) should be set into the soil below the maximum depth to which the soil freezes—a depth that ranges from less than 10 cm in subtropical zones, such as South Texas and Florida, to more than 200 cm in very cold climates.

FIGURE 7.14 How frost heaving moves objects upward. (*a*) Position of the object (stone, plant, or fence post) before the soil freezes. (*b*) As lenses of pure ice form in the freezing soil by attraction of water from the unfrozen soil below, the frozen soil tightens around the upper part of the object, lifting it somewhat—enough to break the root in the case of the plant. (*c*) the objects are lifted upwards as ice-lens formation continues with deeper penetration of the freezing front. (*d*) As for freezing, thawing commences from the surface downward. Water from thawing ice lenses escapes to the surface because it cannot drain downward through the frozen soil. The soil surface subsides while the heaved objects are held in the "jacked-up" position by the still-frozen soil around their lower parts. (*e*) After complete thaw, the stone is closer to the surface than previously (although rarely at the surface unless erosion of the thawed soil has occurred), and the upper part of the broken plant's root is exposed, so that is likely to die. (*f*) Alfalfa plants lifted out of the ground by frost action. (*g*) Fence posts encased in concrete that have been progressively "jacked out" of the ground by frost action over several years. [Photo (*f*) courtesy of R. Weil; photo (*g*) courtesy of R. L. Berg, Corps of Engineers, Cold Regions Research and Engineering Laboratory, Hanover, N.H.]

FIGURE 7.15 (Left) Wildfires of a lodgepole pine stand heat up the surface layers of this sandy soil (an Inceptisol) in Oregon. (Center) Note that the soil temperature is increased sufficiently near the surface to volatilize organic compounds, some of which then move down into the soil and condense (solidify) on the surface of cooler soil particles. These condensed compounds are waxlike hydrocarbons that are water repellent. As a consequence (right) the infiltration of water into the soil is drastically reduced and remains so for a period of at least six years. [From Dryness (1976)]

Soil Heating by Fire

Fire is one of the most far-reaching ecosystem disturbances in nature. In addition to the obvious aboveground effects of forest, range, or crop-residue fires, the brief, but sometimes dramatic, changes in soil temperature also may have lasting impacts. Unless the fire is artificially stoked with added fuel, the temperature rise itself is usually very brief and is limited to the upper few centimeters of soil.

The heat may affect the breakdown and movement of organic compounds. Figure 7.15 shows the results of wildfire on a lodgepole pine stand in Oregon. The high temperatures (>125°C being common) essentially distill various fractions of the organic matter, with some of the volatilized hydrocarbon compounds moving quickly through the soil pores to deeper, cooler areas. As these compounds reach cooler soil particles deeper in the soil, they condense (solidify) on the surface of the soil particles and fill some of the surrounding pore spaces. Some of the condensed compounds are water-repellent (hydrophobic) hydrocarbons. Consequently, when rain comes, water infiltration in even a sandy soil is greatly reduced over that of unburned areas. This effect of soil temperature is quite common on chaparral lands in semiarid regions and may be responsible for the disastrous mudslides that occur when the layer of soil above the hydrophobic zone becomes saturated with rainwater.

Fires also promote the germination of certain seeds, which have hard coatings that prevent them from germinating until they are heated above 70 to 80°C. On the other hand, burning of straw in wheat fields generates similar soil temperatures, but with the effect of killing most of the weed seeds near the surface and thus greatly reducing subsequent weed infestation. The heat and ash may also hasten the cycling of nutrient plant nutrients (see Chapter 14). Fires set to clear land of timber slash may burn long and hot enough to seriously deplete soil organic matter and kill so many soil organisms that forest regrowth is inhibited.

7.9 ABSORPTION AND LOSS OF SOLAR ENERGY[5]

The temperature of soils in the field is directly or indirectly dependent on at least three factors: (1) the net amount of heat energy the soil absorbs; (2) the heat energy required to bring about a given change in the temperature of a soil; and (3) the energy required for processes such as evaporation, which are constantly occurring at or near the surface of soils.

---

[5] For a review of models that describe these processes on a global scale, see Sellers, et al. (1997).

Solar radiation is the primary source of energy to heat soils. But clouds and dust particles intercept the sun's rays and absorb, scatter, or reflect most of the energy (Figure 7.16). Only about 35 to 40% of the solar radiation actually reaches the earth in cloudy humid regions, and 75% in cloud-free arid areas. The global average is about 50%.

Little of the solar energy reaching the earth actually results in soil warming. The energy is used primarily to evaporate water from the soil or leaf surfaces, or is radiated or reflected back to the sky. Only about 10% is absorbed by the soil and can be used to warm it. Even so, this energy is of critical importance to soil processes and to plants growing on the soils.

ALBEDO. The fraction of incident radiation that is reflected by the land surface is termed the **albedo**, and ranges from as low as 0.1 to 0.2 for dark-colored, rough soil surfaces to as high as 0.5 or more for smooth, light-colored surfaces. Vegetation may affect the surface albedo either way, depending on whether it is dark green and growing or yellow and dormant.

The fact that dark-colored soils absorb more energy than lighter-colored ones does not necessarily imply, however, that dark soils are always warmer. In fact, in most landscapes, the darkest soils are also usually the wettest and therefore the slowest to warm.

ASPECT. The angle at which the sun's rays strike the soil also influences soil temperature. If the sun is directly overhead, the incoming path of the rays is perpendicular to the soil surface and energy absorption (and soil temperature increase) is greatest (Figure 7.17). The effect of the direction of slope, or **aspect**, on forest species is illustrated in the photo in Figure 7.17. Forming the soil surface into ridges is one method of controlling the soil aspect on a microscale.

SOIL COVER. Whether the soil is bare or is covered with vegetation, mulch or snow is another factor markedly influencing the amount of solar radiation reaching the soil.

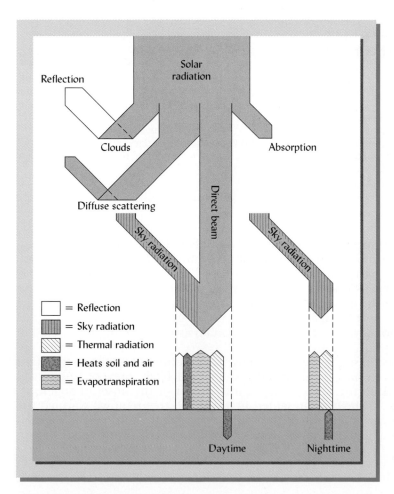

FIGURE 7.16 Schematic representation of the radiation balance in daytime and nighttime in the spring or early summer in a temperate region. About half the solar radiation reaches the earth, either directly or indirectly, from sky radiation. Most radiation that strikes the earth in the daytime is used as energy for evapotranspiration or is radiated back to the atmosphere. Only a small portion, perhaps 10%, actually heats the soil. At night the soil loses some heat, and some evaporation and thermal radiation occur.

FIGURE 7.17 (Upper) Effect of the angle at which the sun's rays strike the soil on the area of soil that is warmed. (*a*) If a given amount of radiation from the sun strikes the soil at right angles, the radiation is concentrated in a relatively small area, and the soil warms quite rapidly. (*b*) If the same amount of radiation strikes the soil at a 45° angle, the area affected is larger by about 40%, the radiation is not so concentrated, and the soil warms up more slowly. This is one of the reasons why north slopes tend to have cooler soils than south slopes. It also accounts for the colder soils in winter than in summer. (Lower) A view looking eastward of a forested area in Virginia illustrates the temperature effect. The main ridge (left to right) is running north and south and the smaller side ridges east and west (up and down). The dark patches are pine trees in this predominantly hardwood deciduous forest. The pines dominate the southern (warmer and drier) slopes on each east-west ridge. (Photo courtesy of R. Weil)

Bare soils warm up more quickly and cool off more rapidly than those covered with vegetation, with snow, or with plastic mulches. Frost penetration during the winter is considerably greater in bare, noninsulated land.

Even low-growing vegetation such as turf grass has a very noticeable influence on soil temperature and on the temperature of the surroundings (Table 7.4). Much of the cooling effect is due to heat dissipated by transpiration of water. To experience this effect, on a blistering hot day, try having a picnic on an asphalt parking lot instead of on a growing green lawn!

The effect of a dense forest is universally recognized. Timber-harvest practices that leave sufficient canopy to provide about 50% shade will likely prevent undue soil warming that could hasten the loss of soil organic matter or the onset of anaerobic conditions in wet soils. In the case presented in Table 7.2, tree removal warmed the soil in spring, even though it also raised the soil water content. However, as we shall see in the next section, a higher water content normally slows the warming of soils in spring.

TABLE 7.4 Maximum Surface Temperatures for Four Types of Surfaces on a Sunny August Day in College Station, Texas

	Maximum temperature, °C	
Type of surface	Day	Night
Green, growing turf grass	31	24
Dry, bare soil	39	26
Brown, summer-dormant grass	52	27
Dry synthetic sports turf	70	29

Data from Beard and Green (1994).

7.10 THERMAL PROPERTIES OF SOILS

Specific Heat of Soils

A dry soil is more easily heated than a wet one. This is because the amount of energy required to raise the temperature of water by 1°C (its heat capacity) is much higher than that required to warm soil solids by 1°C. When heat capacity is expressed per unit mass—for example, in calories per gram (cal/g)—it is called **specific heat** or specific heat capacity c. The specific heat of pure water is about 1.00 cal/g (or 4.18 joules per gram, J/g); that of dry soil is about 0.2 cal/g (0.8 J/g). The specific heat largely controls the degree to which soils warm up in the spring, wetter soils warming more slowly than drier ones (see Box 7.2).

Temperature control systems referred to as *heat pumps,* which are designed to both warm and cool buildings, take advantage of the high specific heat of soils. A network of pipes is laid underground near the building to be heated and cooled to maximize heat-exchange contact with the soil. Advantage is taken of the fact that subsoils are generally warmer than the atmosphere in the winter and cooler than the atmosphere in the summer. Water circulating through the network of pipes absorbs heat from the soil during the winter and releases it to the soil in the summer. The high specific heat of soils permits a large exchange of energy to take place without greatly modifying the soil temperature.

Heat of Vaporization

The evaporation of water from soil surfaces requires a large amount of energy, 540 kilocalories (kcal) or 2.257 J for every kilogram of water vaporized. This energy must be provided by solar radiation or it must come from the surrounding soil. In either case, evaporation has the potential of cooling the soil, much the way it chills a person who comes out from swimming on a windy day. The low temperature of wet soils in spring is due partially to evaporation and partially to high specific heat.

Thermal Conductivity of Soils

As shown in Section 7.9, some of the solar radiation that reaches the earth slowly penetrates the profile largely by conduction; this is the same process by which heat moves along an iron rod when one end is placed in a fire. The movement of heat in soil is analogous to the movement of water (see Section 5.5), the rate of flow being determined by a driving force and by the ease with which heat flows through the soil. This can be expressed as Fourier's Law:

$$Q_h = K * \frac{\Delta T}{x}$$

where Q_h is the *thermal flux,* the quantity of heat transferred across a unit cross-sectional area in a unit time; K is the **thermal conductivity** of the soil; and $\Delta T/x$ is the temperature gradient over distance x that serves as the driving force for the conduction of heat.

BOX 7.2 CALCULATING THE SPECIFIC HEAT OR HEAT CAPACITY OF MOIST SOILS

Soil water content markedly impacts soil temperature changes through its effect on the specific heat or heat capacity c of a soil. For example, consider two soils with comparable characteristics, *soil A*, a relatively dry soil with only 10 g water/100 g soil solids, and *soil B*, a wetter soil with 30 g water/100 g soil solids.

We can assume the following values for specific heat:

$$\text{Water} = 1.0 \text{ cal/g}$$
$$\text{Dry mineral soil} = 0.2 \text{ cal/g}$$

For soil A with 10 g water/100 g dry soil, or 0.1 g water/g dry soil, the number of calories required to raise the temperature of 0.1 g of water by 1°C is

$$0.1 \text{ g} \times 1.0 \text{ cal/g} = 0.1 \text{ cal}$$

The corresponding figure for the 1.0 g of soil solids is

$$1 \text{ g} \times 0.2 \text{ cal/g} = 0.2 \text{ cal}$$

Thus, a total of 0.3 cal (0.1 + 0.2) is required to raise the temperature of 1.1 g (1.0 + 0.1) of the moist soil by 1°C. Since the specific heat is the number of calories required to raise the temperature of 1 g of moist soil by 1°C, we can calculate the specific heat of soil A as follows:

$$c_{\text{soil A}} = \frac{0.3}{1.1} = 0.273 \text{ cal/g}$$

These calculations can be expressed as a simple equation to calculate the weighted average specific heat of a mixture of substances:

$$c_{\text{moist soil}} = \frac{c_1 m_1 + c_2 m_2}{m_1 + m_2}$$

where c_1 and m_1 are the specific heat and mass of substance 1 (the dry mineral soil, in this case), and c_2 and m_2 are the specific heat of substance 2 (the water, in this case).

Applying this equation to soil A, we again calculate that $c_{\text{soil A}}$ is 0.273 cal/g, as follows:

$$c_{\text{soil A}} = \frac{0.2 \text{ cal/g} * 1.0 \text{ g} + 1.0 \text{ cal/g} * 0.10 \text{ g}}{1.0 \text{ g} + 0.10 \text{ g}}$$

$$= \frac{0.30 \text{ cal/g}}{1.1 \text{ g}}$$

$$= 0.273 \text{ cal/g}$$

In the same manner, we calculate the specific heat of the wetter soil B:

$$c_{\text{soil B}} = \frac{0.2 \text{ cal/g} * 1.0 \text{ g} + 1.0 \text{ cal/g} * 0.30 \text{ g}}{1.0 \text{ g} + 0.30 \text{ g}}$$

$$= \frac{0.50 \text{ cal/g}}{1.3 \text{ g}}$$

$$= 0.385 \text{ cal/g}$$

The wet soil B has a specific heat c_B of 0.385 cal/g, whereas the drier soil A has a specific heat c_A of 0.273 cal/g. Because it must absorb an additional 0.112 cal (0.385 − 0.273) of solar radiation for every degree of temperature rise, the wet soil will warm up much more slowly than the drier soil.

The thermal conductivity K of soil is influenced by a number of factors, the most important being the moisture content of the soil and the degree of compaction. Heat passes through water many times faster than through air. As the water content increases in a soil, the air content decreases, and the transfer resistance is decidedly lowered. When sufficient water is present to form a bridge between most of the soil particles, further additions will have little effect on heat conduction. Heat moves through mineral particles even faster than through water, so when particle-to-particle contact is increased by soil compaction, heat-transfer rates are also increased. Therefore, a wet, compacted soil would be the poorest insulator or the best conductor of heat. Here again the interconnectedness of soil properties is demonstrated.

Relatively dry, loose soil makes a good insulating material. Buildings built mostly underground can take advantage of both the low thermal conductivity and relatively high heat capacity of large volumes of soil. Soil thermal conductivity can also affect air temperature above the soil, as shown in Figure 7.18.

Vertical and Seasonal Temperature Fluctuations

The temperature of the soil at any time depends on the ratio of the energy absorbed to that being lost. The constant change in this relationship is reflected in the seasonal, monthly, and daily temperatures. The accompanying data (Figures 7.19 and 7.20) from College Station, Texas, and Lincoln, Nebraska, are representative of average seasonal and monthly temperatures in relation to soil depth in subhumid temperate regions.

It is apparent from Figures 7.19 and 7.20 that considerable seasonal and monthly variations of soil temperature occur, even at the lower depths. The surface layer temperatures of bare soil vary more or less according to the temperature of the air, although these layers are generally warmer than the air throughout the year.

In the subsoil, the seasonal temperature increases and decreases are smaller and lag behind changes registered in the surface soil and in the air. Accordingly, the temperature data for March at College Station suggest that the surface soil temperatures have

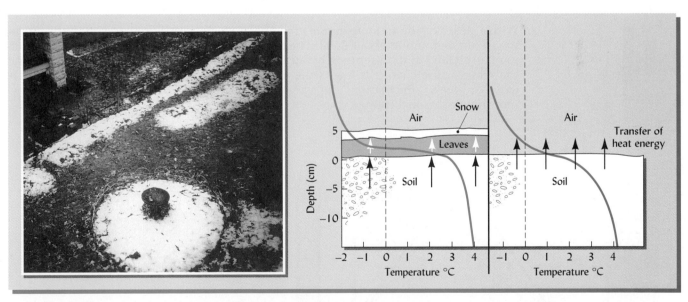

FIGURE 7.18 Transfer of heat energy from soil to air. The scene, looking down on a garden after an early fall snow storm, shows snow on the leaf-mulched flower beds, but not on areas where the soil is bare or covered with thin turf. The reason for this uneven accumulation of snow can be seen in the temperature profiles. Having stored heat from the sun, the soil layers are often warmer than the air as temperatures drop in fall (this is also true at night during other seasons). On bare soil, heat energy is transferred rapidly from the deeper layers to the surface, the rate of transfer being enhanced by high moisture content or compaction, which increase the *thermal conductivity* of the soil. As a result, the soil surface and the air above it are warmed to above freezing, so snow melts and does not accumulate. The leaf mulch, which has a low thermal conductivity, acts as an insulating blanket that slows the transfer of stored heat energy from the soil to the air. The upper surface of the mulch is therefore hardly warmed by the soil, and the snow remains frozen and accumulates. A heavy covering of snow can itself act as an insulating blanket. (Photo courtesy of R. Weil)

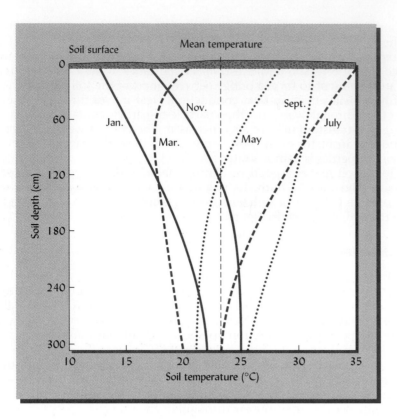

FIGURE 7.19 Average monthly soil temperatures for various months at different soil depths at College Station, Tex. (5-year average). Note the lag in soil temperature change at the lower depths. [From Fluker (1958)]

already begun to respond to the warming of the spring, while temperatures of the deep subsoil seem to still be responding to the cold winter weather.

Daily Variations

With a clear sky, the air temperature in temperate regions rises from lowest in the morning to a maximum at about 2 P.M. The surface soil, however, does not reach its maximum until later in the afternoon because of the usual lag. This retardation is greater and the temperature change is less as the depth increases. The lower subsoil shows little daily or weekly fluctuation; the variation there, as already emphasized, is a slow monthly or seasonal change.

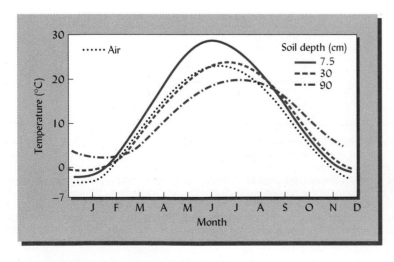

FIGURE 7.20 Average monthly air and soil temperatures at Lincoln, Nebr. (12 years). Note that the 7.5-cm soil layer is consistently warmer than the air above and that the 90-cm soil horizon is cooler in spring and summer, but warmer in the fall and winter, than surface soil.

7.11 SOIL TEMPERATURE CONTROL

The temperature of field soils is not subject to radical human regulation. However, two kinds of management practice have significant effects on soil temperature: those that affect the cover or mulch on the soil, and those that reduce excess soil moisture. These effects have meaningful biological implications.

Organic Mulches and Plant-Residue Management

Soil temperatures are influenced by soil cover and especially by organic residues or other types of mulch on the soil surface. The mulch may be spread on the surface or it may be the result of natural leaf fall or conservation tillage that leaves plant residues on the soil surface. Figure 7.21 shows that mulches effectively buffer extremes in soil temperatures. In periods of hot weather, they keep the surface soil cooler than where no cover is used; in contrast, during cold weather they keep the soil warmer than it would be if bare.

The forest floor is a prime example of a natural temperature-modifying mulch. It is not surprising, therefore, that timber harvest practices can markedly affect forest soil temperature regimes (Figure 7.22). Disturbance of the leaf mulch, changes in water content due to reduced evapotranspiration, and compaction by machinery are all factors that influence soil temperatures through thermal conductivity. Reduced shading after tree removal also lets in more solar radiation.

CONCERNS IN COOL CLIMATES.   While mulch provides great control over erosion (see Section 15.5), the soil-temperature-depressing effects of some mulch practices have a serious negative impact on the production of crops like corn in cold climates. The lower soil temperatures in May and early June resulting from these practices inhibit seed germination, seedling performance, and, often, the yields of corn. The effect of the residue mulch is most pronounced in lowering the midday maximum temperature, and has much less effect on the minimum temperature reached at night. This effect is well illustrated by the data in Figure 7.23, which also features an innovative way to alleviate this problem by pushing aside the residues in just a narrow band over the seed row in the no-tillage system. Another approach to solving this problem is to ridge the soil, permit water to drain out of the ridge, and then plant on the drier, warmer ridgetop (or on the south side of the ridge—see Section 7.9).

ADVANTAGES IN WARM CLIMATES.   In warm regions, delayed planting is not a problem. In fact, the cooler near-surface soil temperatures under a mulch may reduce heat stress on roots

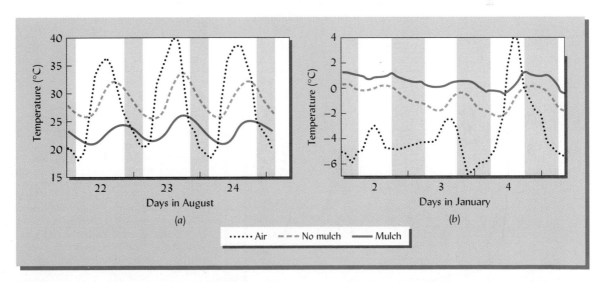

FIGURE 7.21   (a) Influence of straw mulch (8 tons/ha) on air temperature at a depth of 10 cm during an August hot spell in Bushland, Tex. Note that the soil temperatures in the mulched area are consistently lower than where no mulch was applied. (b) During a cold period in January the soil temperature was higher in the mulched than in the unmulched area. Note that in all cases fluctuations in air temperature caused changes in soil temperature, but the soil temperature changes were subdued and delayed. The shaded bars represent nighttime. [Redrawn from Unger (1978); used with permission of American Society of Agronomy]

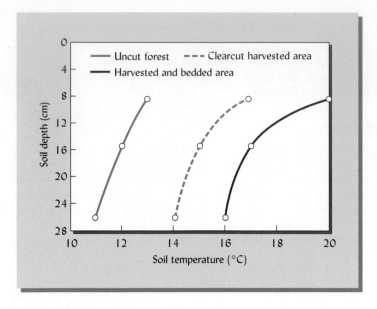

FIGURE 7.22 Influence of timber harvest and site preparation on soil temperature profiles in June on a forested wetland in Michigan. Clear-cut timber harvest greatly disturbed the site and raised soil temperatures by 4 to 5°C. Tillage to form beds for tree replanting eliminated the remaining forest floor mulch and dried the soil, thus causing temperatures to rise another 2 to 3°C. The higher temperatures increased the rate of soil organic matter decomposition. [Redrawn from Trettin, et al. (1996)]

during summer. Plant-residue mulches also conserve soil moisture by decreasing evaporation. The resulting cooler, moist surface layer of soil is an important part of no-tillage systems because it allows roots to proliferate in this zone, where nutrient and aeration conditions are optimal.

Plastic Mulches

One of the reasons for the popularity of plastic mulches for gardens and high-value specialty crops is their effect on soil temperature. In contrast to the organic mulches, plastic mulches generally increase soil temperature, clear plastic having a greater heating effect than black plastic. In temperate regions, this effect can be used to extend the growing season or to hasten production to take advantage of the higher prices offered by early-season markets (Figure 7.24).

In warmer climates, and during the summer months, the soil-heating effect of plastic mulches may be quite detrimental, inhibiting root growth in the upper soil layers and sometimes seriously decreasing crop yields (see Table 7.5, for example). Obviously, the weed-control and moisture-conservation effects of plastic mulches may be outweighed in some circumstances by their excessive heating effects.

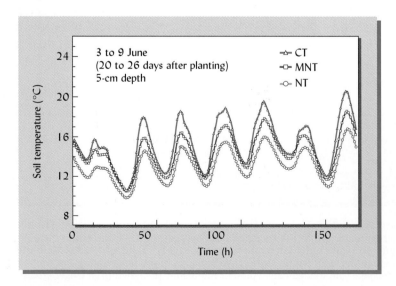

FIGURE 7.23 Tillage effects on hourly temperature changes near the surface of a cold Alfisol in northern British Columbia. The soil had been managed to grow barley under no-tillage (NT) and conventional (clean surface without residues) tillage (CT) systems for the previous 14 years. In the clean-tilled soil, midafternoon temperatures peaked at 4°C higher than those in the residue mulch covered no-tillage soil. A modification of the no-tillage system (MNT) that pushed aside the residues in a narrow (7.5-cm-wide) band over the seeding row eliminated much of the temperature depression while keeping most of surface covered by the soil- and water-conserving mulch. Note the daily temperature changes and the general warming trend during the seven days shown. [From Arshad and Azooz (1996)]

FIGURE 7.24 These irrigated southern California strawberries will come to market in winter when prices are still high because of the effect of the clear plastic mulch on soil temperature. (Photo courtesy of R. Weil)

Moisture Control

Another means of exercising some control over soil temperature is by controlling soil moisture. Poorly drained soils in temperate regions that are wet in the spring have temperatures 3 to 6°C lower than comparable well-drained soils. Removing the excess water by installing drainage systems (see Section 6.9) or using ridging systems of tillage can cause the soil to warm more quickly in spring.

7.12 CONCLUSION

Soil aeration and soil temperature critically affect the quality of soils as habitats for plants and other organisms. Most plants have definite requirements for soil oxygen along with limited tolerance for carbon dioxide, methane, and other such gases found in poorly aerated soils. Some microbes, such as the nitrifiers and general-purpose decay organisms, are also constrained by low levels of soil oxygen.

TABLE 7.5 Soil Temperature and Tomato Yield with Straw or Black Plastic Mulch

The data are averages for two years of tomato production on a sandy loam Ultisol near Griffin, Georgia. The straw kept the surface soil from rising to detrimentally high temperatures, while it also increased infiltration of rainwater and reduced soil compaction. Daily drip irrigation supplied plenty of water, but could not overcome the temperature effects of the black plastic mulch.

	Not irrigated		Irrigated daily	
	Straw mulch	Plastic mulch	Straw mulch	Plastic mulch
Average soil temperature, °C[a]	24	37	24	35
Tomato yield, Mg/ha	68	30	70	24

[a] Soil temperature measured at 5 cm below the soil surface, average of weeks 2–10 of the growing season.
Data calculated from Tindall, et al. (1991).

Soils with extremely wet moisture regimes are unique with respect to their morphology and chemistry and to the plant communities they support. Such hydric soils are characteristic of wetlands and help these ecosystems perform a myriad of valuable functions.

Plants as well as microbes are also quite sensitive to differences in soil temperature, particularly in temperate climates where low temperatures can limit essential biological processes. Soil temperature also impacts on the use of soils for engineering purposes, again primarily in the cooler climates. Frost action, which can move perennial plants such as alfalfa out of the ground, can do likewise to building foundations, fence posts, sidewalks, and highways.

Soil water competes with soil air for the occupancy of soil pores and interferes with the diffusion of gases into and out of the soil. Soil water also resists changes in soil temperature by virtue of its high specific heat and its high energy requirement for evaporation. Regulation of soil water is an important means of regulating soil aeration and temperature.

STUDY QUESTIONS

1. What are the two principle gases involved with soil aeration, and how do their relative amounts change as one samples deeper into a soil profile?

2. It is sometimes said that organisms in anaerobic environments will use the combined oxygen in nitrate or sulfate instead of the oxygen in O_2. Why is this statement incorrect? What actually happens when organisms reduce sulfate or nitrate?

3. If an alluvial forest soil were flooded for 10 days and you sampled the gases evolving from the wet soil, what gases would you expect to find (other than oxygen and carbon dioxide)? In what order of appearance? Explain.

4. Explain why warm weather during periods of saturation is required in order to form a hydric soil.

5. If you were in the field trying to delineate the so-called drier end of a wetland area, what are three soil properties and three other indicators that you might look for?

6. What are the three major components that define a wetland?

7. Discuss four plant processes that are influenced by soil temperature.

8. Explain how a brush fire might lead to subsequent mudslides, as often occurs in California.

9. If you were to build a house below ground in order to save heating and cooling costs, would you firmly compact the soil around the house? Explain your answer.

10. If you measured a daily maximum air temperature of 28°C at 1 P.M., what might you expect the daily maximum temperature to be at a 15-cm depth in the soil? At about what time of day would the maximum temperature occur at this depth? Explain.

11. In relation to soil temperature, explain why conservation tillage has been more popular in Missouri than in Minnesota.

REFERENCES

Arshad, A., and R. H. Azooz. 1996. "Tillage effects on soil thermal properties in a semi-arid cold region," *Soil Sci. Soc. Am. J.*, **60**:561–567.

Bartlett, R. J., and B. R. James. 1993. "Redox chemistry of soils," *Advances in Agronomy*, **50**:151–208.

Beard, J. B. and R. L. Green. 1994. "The role of turfgrasses in environmental protection and their benefits to humans," *J. Environ. Qual.*, **23**:452–460.

CAST. 1994. *Wetland Policy Issues*. Publication no. CC1994-1. (Ames, Iowa: Council for Agricultural Science and Technology).

Davidson, E. A., and S. E. Trumbore. 1995. "Gas diffusivity and production of CO_2 in deep soils of the eastern Amazon," *Tellus,* **47B:**550–565.

Dryness, C. T. 1976. "Effects of wildfire on soil wetability in the high cascades of Oregon," USDA Forest Service Research Paper PNW-202. (Washington, D.C.: USDA).

Fluker, B. J. 1958. "Soil temperature," *Soil Sci.,* **86:**35–46.

Hurt, G. W., P. M. Whited, and R. F. Pringle (eds.). 1996. *Field Indicators of Hydric Soils in the United States.* (Fort Worth, Tex.: USDA Natural Resources Conservation Service).

Losi, M. E., C. Amrheim, and W. T. Frankenberger, Jr. 1994. "Bioremediation of chromate-contaminated groundwater by reduction and precipitation in surface soils," *J. Environ. Qual.,* **23:**1141–1150.

McBride, M. B. 1994. *Environmental Chemistry of Soils.* (New York: Oxford University Press).

Patrick, W. H., Jr., and A. Jugsujinda. 1992. "Sequential reduction and oxidation of inorganic nitrogen, manganese, and iron in flooded soil," *Soil Sci. Soc. Amer. J.,* **56:**1071–1073.

Rabenhorst, M. C., J. Bell, and P. McDaniel (eds.). 1999. *Quantifying Soil Hydromorphology.* Special Publication No. 54. (Madison, Wis.: Soil Sci. Soc. Amer.).

Sellers, P., R. Dickinson, D. Randall, K. Betts, F. Hall, J. Berry, G. Collatz, A. Denning, H. Mooney, C. Nobre, N. Sato, C. Field, and A. Henderson-Sellers. 1997. "Modeling the exchange of energy, water, and carbon between continents and the atmosphere," *Science,* **275:**502–509.

Sexstone, A. J., et al. 1985. "Direct measurement of oxygen profiles and denitrification rates in soil aggregates," *Soil Sci. Soc. Amer. J.,* **49:**645–651.

Tiarks, A. E., W. H. Hudnall, J. F. Ragus, and W. B. Patterson. 1996. "Effect of pine plantation harvesting and soil compaction on soil water and temperature regimes in a semi-tropical environment," in A. Schulte and D. Ruhiyat (eds.), *Proceedings of International Congress on Soils of Tropical Forest Ecosystems 3rd Conference on Forest Soils: Vol. 3, Soil and Water Relationships.* (Samarinda, Indonesia: Mulawarmon University Press).

Tindall, J. A., R. B. Beverly, and D. E. Radcliff. 1991. "Mulch effect on soil properties and tomato growth using micro-irrigation," *Agron. J.,* **83:**1028–1034.

Trettin, C. C., M. Davidian, M. F. Jurgensen, and R. Lea. 1996. "Organic matter decomposition following harvesting and site preparation of a forested wetland," *Soil Sci. Soc. Am. J.,* **60:**1994–2003.

Unger, P. W. 1978. "Straw mulch effects on soil temperatures and sorghum germination and growth," *Agron. J.,* **70:**858–864.

Veneman, P. L. M., D. L. Lindbo, and L. A. Spokas. 1999. "Soil moisture and redoximorphic features: A historical perspective," in M. J. Rabenhorst, J. C. Bell, and P. A. McDaniel (eds.), *Quantifying Soil Hydromorphology.* Special Publication No. 54.

Welsh, D., D. Smart, J. Boyer, P. Minkin, H. Smith, and T. McCandless (eds.). 1995. *Forested Wetlands: Functions, Benefits, and Use of Best Management Practices.* (Radnor, Pa.: USDA Forest Service).

8

SOIL COLLOIDS: THEIR NATURE AND PRACTICAL SIGNIFICANCE

The landscape of the clays is like—the intricate folds of the womb—whose activity is to receive, contain, enfold, and give birth.
—WILLIAM BRYANT LOGAN

Next to photosynthesis and respiration, no process in nature is more vital to plant and animal life than the exchange of ions between soil particles and plant roots. These **cation** and **anion exchanges** occur mostly on the surfaces of the finer or **colloidal**[1] **fractions** of both the inorganic and organic matter—clays and humus.

Such colloidal particles serve very much as does a modern bank. They are the sites within the soil where ions of essential mineral elements such as calcium, potassium, and sulfur are held and protected from excessive loss by percolating rain or irrigation water. Subsequently, the essential ions can be withdrawn from the colloidal bank sites and taken up by plant roots. In turn, these elements can be deposited or returned to the colloids through plant residues and through the addition of commercial fertilizers, lime, and manures.

Cation and anion exchange mechanisms are also significant in controlling the movement of some organic chemicals through the soil and into nearby streams or well water. Any chemical that is attracted by positive or negative charges will be held to some extent by the soil colloids, later to be released slowly into the surrounding environment.

But cation and anion exchange capabilities are not the only soil properties dependent upon soil colloids. These fine soil particles markedly influence soil structure formation and stability, and the retention and movement of water, with their implications for the control of soil air and soil temperature. Likewise, the use of soils for building sites and for highway construction is dependent to a considerable degree on the amount and nature of the soil colloids.

No understanding of soils is complete without at least a general knowledge of the nature and constitution of colloidal particles (clays and humus). Their common properties will be covered first and then their individual characteristics will receive attention.

---

[1] Some of the clay and organic particles are somewhat larger then the upper limit of the colloidal state (1 μm). However, they do have some colloidlike properties, such as positive and negative surface charges, the chemical attraction of ions to their surfaces, and the ability to disperse. Accordingly, the fine fractions of soils (clays and humus) will be referred to as *soil colloids*, recognizing that technically some of these particles are not small enough to be classed as colloids. For a review of the composition and properties of clays and humus, see Dixon and Weed (1989).

8.1   PROPERTIES OF SOIL COLLOIDS

General Properties

SIZE.   The most important common property of colloids is their extremely small size. They are too small to be seen with an ordinary light microscope. Only with an electron microscope can they be photographed. Most are smaller than 2 millionths of a meter [micrometers (μm)] in diameter.

SURFACE AREA.   Because of their small size, all soil colloids expose a large **external** surface per unit mass. The external surface area of 1 g of colloidal clay is at least 1000 times that of 1 g of coarse sand. Some colloids, especially certain silicate clays, have extensive **internal** surfaces which commonly greatly exceed the external surface area. The total surface area of soil colloids ranges from 10 m^2/g for clays with only external surfaces to more than 800 m^2/g for clays with extensive internal surfaces. The colloidal surface area in the upper 15 cm of 1 ha of a clay soil could be as high as 700,000 km^2 (270,000 mi^2).

SURFACE CHARGES.   Soil colloidal surfaces, both external and internal, carry negative and/or positive charges. For most soil colloids, electronegative charges predominate, although some mineral colloids in very acid soils have a net electropositive charge. The presence and intensity of the particle charges influence the attraction and repulsion of the particles toward each other, thereby influencing both physical and chemical properties. The source of these charges will be considered later.

ADSORPTION OF IONS AND WATER.   The charges on the surfaces of soil colloids attract ions of the opposite charge to the colloidal surfaces. A negatively charged colloidal particle, referred to as a *micelle* (microcell), attracts hundreds of thousands of positively charged ions or *cations*,[2] such as H^+, Al^{3+}, Ca^{2+} and Mg^{2+}. This gives rise to an **ionic double layer** (Figure 8.1). The colloidal particle constitutes the *inner* ionic layer, being essentially a huge anion, the external and internal surfaces of which are negative in charge. The *outer* ionic layer is made up of a swarm of rather loosely held (adsorbed) cations that are attracted to the negatively charged surfaces. Thus, a colloidal particle is accompanied by a swarm of cations that are **adsorbed** or held on the particle surfaces.

In addition to the adsorbed cations, a large number of water molecules are associated with soil colloidal particles. Some are attracted to the adsorbed cations, each of which is hydrated. Others are held in the internal surfaces of the colloidal particles. These water molecules play a critical role in determining both the physical and chemical properties of soil.

Anions such as NO_3^-, Cl^-, and SO_4^{2-} may also be adsorbed on some soil colloids, this time the attraction being positive charges on the colloidal surfaces. While the adsorption of anions is not as extensive as that of cations, it is an important mechanism for holding negatively charged constituents.

Properties of Specific Types of Colloids

There are four major types of colloids present in soils: (1) layer silicate clays, (2) iron and aluminum oxide clays, (3) allophane and associated clays, and (4) humus. Each group possesses the general colloidal characteristics described in the preceding section, but each has specific characteristics that make it distinctive and useful.

LAYER SILICATE CLAYS.   Silicate clay minerals are the dominant inorganic colloids in most all soils, especially those of temperate areas. Their most important properties are their layerlike, crystalline structures and their negative charges. Each particle is comprised of a series of layers, much like the pages of a book (Figure 8.2). The layers are comprised of planes of closely packed oxygen atoms held together by silicon, aluminum, magne-

---

[2] In water solutions the cations are always hydrated, giving rise, in the case of hydrogen, to the **hydronium** ion H_3O^+ [sometimes written H^+ (aq)]. For simplicity, however, the H^+ ion designation will be used in this text.

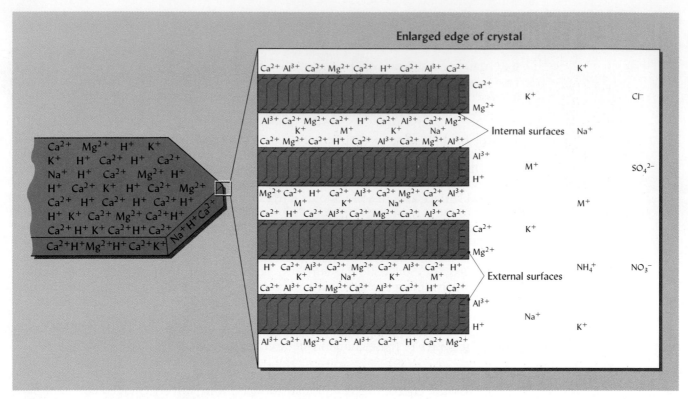

FIGURE 8.1 Diagrammatic representation of a silicate clay crystal with its sheetlike structure, negative surface charges, and associated adsorbed cations. The enlarged schematic view of the edge of the crystal illustrates the negatively charged internal as well as external surfaces to which cations and water molecules are attracted. Note that the orderly structured particle provides an inner negatively charged ionic layer that attracts a swarm of positively charged cations. Together they constitute what is known as an *ionic double layer*. Some clays exhibit positive charges (not shown) that can attract negative ions, such as NO_3^- and SO_4^{2-}.

sium, hydrogen, and/or iron atoms that occupy spaces between the oxygens. The formula of one of these clays, kaolinite $[Si_2Al_2O_5(OH)_4]$ illustrates their general chemical makeup.

ALLOPHANE AND IMOGOLITE. In many soils there are significant quantities of colloidal silicates whose crystalline structure is not sufficiently ordered to be detected easily by X-ray diffraction. The most significant of such poorly crystallized silicate colloids are **allophane** and its more weathered companion **imogolite**. These minerals are somewhat poorly defined aluminum silicates with a general composition approximating $Al_2O_3 \cdot 2SiO_2H_2O$. They are most prevalent in soils developed from volcanic ash (Andisols), such as those found in the Northwestern part of the United States.

IRON AND ALUMINUM OXIDE CLAYS. These clays are most prominent in the highly weathered Ultisols and Oxisols of the tropics and semitropics. Examples of iron and aluminum oxides common in soils are goethite (FeOOH), hematite (Fe_2O_3), and gibbsite $[Al(OH)_3]$. For simplicity, they will be referred to as the **Fe, Al oxide** clays. Some have definite crystalline structures, but others are amorphous. They are not as sticky and plastic as the layer silicate clays on which they are commonly found as coatings.

ORGANIC SOIL COLLOIDS: HUMUS. The colloidal organization of humus has some similarities to that of clay. A highly charged micelle is surrounded by a swarm of cations. The humus colloids are not crystalline, however. They are convoluted chains of carbon bonded to hydrogen, oxygen, and nitrogen. The organic colloidal particles vary in size, but they may be at least as small as the silicate clay particles.

The negative charges of humus are associated with partially dissociated enolic (—OH), carboxyl (—COOH), and phenolic (◯—OH) groups; these groups, in turn, are associated with central units of varying size and complexity.

FIGURE 8.2 Crystals of three silicate clay minerals and a photomicrograph of humic acid found in soils. (*a*) Kaolinite from Illinois magnified about 1900 times (note hexagonal crystal at upper right). (*b*) A fine-grained mica from Wisconsin magnified about 17,600 times. (*c*) Montmorillonite (a smectite group mineral) from Wyoming magnified about 21,000 times. (*d*) Fulvic acid (a humic acid) from Georgia magnified about 23,000 times. [(*a*)–(*c*) Courtesy Dr. Bruce F. Bohor, Illinois State Geological Survey; (*d*) from Dr. Kim H. Tan, University of Georgia; used with permission of Soil Science Society of America]

8.2 ADSORBED CATIONS

In humid regions, the cations of calcium, aluminum, and, to a lesser extent, hydrogen are the most numerous; whereas in an arid-region soil, calcium, magnesium, potassium, and sodium predominate. A colloidal complex may be represented in the following simple and convenient way for each region:

$$
\begin{array}{ll}
a\ Ca^{2+} & e\ Ca^{2+} \\
b\ Al^{3+} & f\ Mg^{2+} \\
c\ H^{+}\quad \boxed{\text{Micelle}} & g\ K^{+}\quad \boxed{\text{Micelle}} \\
d\ M^{+} & h\ M^{+}
\end{array}
$$

$$\text{Humid region} \qquad\qquad \text{Arid region}$$

The M^{+} stands for the small amounts of other base-forming cations (e.g., Na^{+} and NH_4^{+}) adsorbed by the colloids. The *a* through *h* indicate that the numbers of cations are variable.

These examples illustrate that soil colloids and their associated exchangeable ions can be considered, although perhaps in an oversimplified way, as *complex salts* with the large anion (micelle) surrounded by numerous cations.

Cation Prominence

Two major factors will determine the relative proportion of the different cations adsorbed by clays. First, these ions are not all held with equal tightness by the soil colloids. The order of strength of adsorption[3] when they are present in equivalent quantities is $Al^{3+} > Ca^{2+} > Mg^{2+} > K^+ = NH_4^+ > Na^+$. Second, the relative concentration of the cations in the soil solution will help determine the degree to which adsorption occurs, Al^{3+} and H^+ being dominant in very acid soils, and metallic cations, particularly Ca^{2+} and Mg^{2+}, in neutral and alkaline soils.

Cation Exchange

As pointed out in Chapter 1, the surface-held adsorbed cations are subject to exchange with other cations found in the soil solution. For example, a calcium ion held on the surface of a colloidal particle (micelle) is subject to exchange with two H^+ ions in the soil solution.

$$\boxed{Micelle}\ Ca^{2+} + 2H^+ \rightleftharpoons \boxed{Micelle}\ {}^{H^+}_{H^+} + Ca^{2+}$$

(colloid)   (soil solution)   (colloid)   (soil solution)

Thus, soil colloids are focal points for cation exchange reactions, which have profound effects on soil–plant relations. These will be discussed in greater detail (see Section 8.7) after we give more detailed consideration to silicate clays.

8.3 FUNDAMENTALS OF LAYER SILICATE CLAY STRUCTURE[4]

The use of X-ray diffraction, electron microscopy, and other techniques has demonstrated that silicate clay particles are crystalline. The mineralogical organization of these particles varies from one type of clay to another, but there are some common features that we will first examine.

Silica Tetrahedral and Aluminium-Magnesium Octahedral Sheets

The most important silicate clays are known as **phyllosilicates** (Greek *phyllon,* leaf), because of their leaflike or planar structure. As shown in Figure 8.3, they are comprised of two kinds of **sheets**, one consisting of two planes of oxygen with mainly silicon in the spaces between the oxygens, and the other having two planes of oxygen and hydroxyls with aluminum or magnesium in the spaces between the oxygens and hydroxyls.

The basic building block for the silica-dominated sheet is a unit composed of one silicon atom surrounded by four oxygen atoms. It is called the silica **tetrahedron** because the centers of the four oxygens define the apices of a four-sided geometric solid (see Figure 8.3). An array of tetrahedra that share oxygens with one another give a **tetrahedral sheet** as shown in Figure 8.3.

Aluminum and/or magnesium ions are the key cations in the second type of sheet. In the basic units, an aluminum or magnesium ion is surrounded by six oxygen atoms or hydroxy groups, the centers of which define the apices of an eight-sided solid or **octahedron** (see Figure 8.3). Numerous octahedra linked together comprise the **octahedral sheet.** An aluminum-dominated sheet is known as a *dioctahedral* sheet, whereas one dominated by magnesium is called a *trioctahedral* sheet. The distinction is due to the fact that *two* aluminum ions in a *di*octahedral sheet satisfy the same negative charges from surrounding oxygens and hydroxys as *three* magnesium ions in a *tri*octahedral sheet.

---

[3] The strength of adsorption of the H^+ ion is difficult to determine since hydrogen-dominated mineral colloids break down to form aluminum-saturated colloids.

[4] The authors are indebted to Dr. Darrell G. Schultze of Purdue University for kindly providing the structural models for the silicate clay minerals.

FIGURE 8.3 The basic molecular and structural components of silicate clays. (*a*) A single *tetrahedron,* a four-sided building block comprised of a silicon ion surrounded by four oxygen atoms; and a single eight-sided *octahedron,* in which an aluminum (or magnesium) ion is surrounded by six hydroxy groups or oxygen atoms. (*b*) In clay crystals thousands of these tetrahedral and octahedral building blocks are connected to give planes of silicon and aluminum (or magnesium) ions. These planes alternate with planes of oxygen atoms and hydroxy groups. Note that apical oxygen atoms are common to adjoining tetrahedral and octahedral sheets. The silicon plane and associated oxygen-hydroxy planes make up a *tetrahedral sheet.* Similarly, the aluminum-magnesium plane and associated oxygen-hydroxy planes comprise the *octahedral sheet.* Different combinations of tetrahedral and octahedral sheets are termed *layers.* In some silicate clays these layers are separated by *interlayers* in which water and adsorbed cations are found. Many layers are found in each *crystal* or *micelle* (microcell).

The tetrahedral and octahedral sheets are the fundamental structural units of silicate clays. They, in turn, are bound together within the crystals by shared oxygen atoms into different **layers.** The specific nature and combination of sheets in these layers vary from one type of clay to another and largely control the physical and chemical properties of each clay. The relationship between sheets and layers shown in Figure 8.3 is important and should be well understood.

Isomorphous Substitution

The structural arrangements just described suggest a very simple relationship among the elements making up silicate clays. In nature, however, the weathering of a wide variety of rocks and minerals permits cations of comparable size to substitute for silicon, aluminum, and magnesium ions in the respective tetrahedral and octahedral sheets.

TABLE 8.1 Ionic Radii of Elements Found in Silicate Clays and an Indication of Which Are Found in the Tetrahedral and Octahedral Sheets

Note that Al, Fe, O, and OH can fit in either.

Ion	Radius, nm[a]	Found in
Si^{4+}	0.042	Tetrahedral sheet
Al^{3+}	0.051	Tetrahedral sheet
Fe^{3+}	0.064	
Mg^{2+}	0.066	Octahedral sheet
Zn^{2+}	0.074	Exchange sites
Fe^{2+}	0.070	
Na^+	0.097	
Ca^{2+}	0.099	
K^+	0.133	
O^{2-}	0.140	Both sheets
OH^-	0.155	Both sheets

[a] 1 nm = 10^{-9}m.

The ionic radii of a number of ions common in clays are listed in Table 8.1 to illustrate this point. Note that aluminum is only slightly larger than silicon. Consequently, aluminum can fit into the center of the tetrahedron in the place of the silicon without changing the basic structure of the crystal. The process, called ***isomorphous substitution,*** is common and accounts for the wide variability in the nature of silicate clays.

Isomorphous substitution also occurs in the octahedral sheet. Note from Table 8.1 that ions such as iron and zinc are not too different in size from aluminum and magnesium ions. As a result, these ions can fit into the position of either the aluminum or magnesium as the central ion in the octahedral sheet. It should be emphasized that some layer silicates are characterized by isomorphous substitution in either or both of the tetrahedral or octahedral sheets.

Source of Charges

Isomorphous substitution is of vital importance because it is the primary source of both negative and positive charges of silicate clays. For example, the substitution of one Al^{3+} for a Si^{4+} in the tetrahedral sheet leaves one unsatisfied negative charge. Likewise, the substitution of one Al^{3+} for one Mg^{2+} in a trioctahedral sheet results in one excess positive charge. The **net charge** associated with a clay micelle is the balance between the positive and negative charges. In layer silicate clays, the net charge is almost always negative. This subject will receive more attention later (see Section 8.6).

8.4 MINERALOGICAL ORGANIZATION OF SILICATE CLAYS

On the basis of the number and arrangement of tetrahedral (silica) and octahedral (aluminium-magnesium) sheets contained in the crystal units or layers, silicate clays are classified into two different groups, 1:1-type minerals and 2:1-type minerals. For illustrative purposes, each of these is discussed briefly in terms of one member of each group.

1:1-Type Minerals

The layers of the 1:1-type minerals are made up of one tetrahedral (silica) sheet and one octahedral (alumina) sheet—hence, the terminology *1:1-type crystal* (Figure 8.4). In soils, **kaolinite** is the most prominent member of this group, which includes *halloysite, nacrite,* and *dickite.*

The tetrahedral and octahedral sheets in a layer of a kaolinite crystal are held together tightly by oxygen atoms, which are mutually shared by the silicon and aluminum cations in their respective sheets. These layers, in turn, are bound to other adjacent layers by **hydrogen bonding** (see Section 5.1). Consequently, the structure is *fixed*. The effective surface of kaolinite is thus restricted to its outer faces or to its **external surface**

FIGURE 8.4 Models of ions that constitute a layer of the 1:1-type clay kaolinite. The primary elements of the octahedral (upper left) and tetrahedral (lower left) sheets are depicted as they might appear separately. In the crystal structure, however, these sheets are held together by common oxygen ions that are known as *apical* oxygen ions. Note that each layer consists of alternating octahedral (alumina) and tetrahedral (silica) sheets—hence, the designation 1:1. Aluminum ions surrounded by six hydroxy groups and/or oxygen atoms (counting the apical oxygen) make up the octahedral sheet (upper left). Smaller silicon ions associated with four oxygen atoms (counting the apical oxygen) constitute the tetrahedral sheet (lower left). The octahedral and tetrahedral sheets are bound together (center) by mutually shared (apical) oxygen atoms. The result is a layer with hydroxys on one surface and oxygens on the other. A schematic drawing of the ionic arrangement (right) shows a cross-sectional view of a crystal layer. Note the common apical oxygen ions that hold the sheets together. The kaolinite mineral is comprised of a series of these flat layers tightly held together with no interlayer spaces. [Note: To permit us to view the front silicon atoms in the silicon plane (center drawing), we have not shown the bottom oxygen atoms that are normally present].

area. Also, there is little isomorphous substitution in this 1:1-type mineral. Along with the relatively low surface area of kaolinite, this accounts for its low capacity to adsorb cations.[5]

Some major physical and chemical properties of kaolinite are shown in Table 8.2. Other properties of kaolinite include low plasticity (capacity of being molded), low stickiness, low cohesion, and little shrinkage or swelling. Because of these properties, some kaolinite-containing soils make good bases for roadbeds and building foundations, and are commonly used in making bricks. They are easy to cultivate for agriculture and, with some nutrient supplementation from manure and fertilizer, can be very productive.

2:1-Type Minerals

The crystal units (layers) of these minerals are characterized by an octahedral sheet sandwiched between two tetrahedral sheets. Four general groups have this basic crystal structure. Two of them, **smectite** and **vermiculite**, include expanding-type minerals; the other two, **fine-grained micas** (**illite**) and **chlorite**, are relatively nonexpanding.

---

[5] Since the adsorbed cations may be freely exchanged with other cations, the capacity to adsorb cations is usually referred to as **cation exchange capacity** (see Section 8.8).

TABLE 8.2   Major Properties of Selected Silicate Clay Minerals[a] and of Humus

Property	2:1-type clays				1:1-type clay	Humus
	Smectite	Vermiculite	Fine mica	Chlorite	Kaolinite	
Size, μm	0.01–1.0	0.1–5.0	0.2–2.0	0.1–2.0	0.1–5.0	0.1–1.0
Shape	Flakes	Plates; flakes	Flakes	Variable	Hexagonal crystals	Variable
External surface, m²/g	80–140	70–120	70–100	70–100	10–30	Variable[d]
Internal surface, m²/g	570–660	600–700	—	—	—	
Intralayer spacing[b], nm	1.0–2.0	1.0–1.5	1.0	1.4	0.7	—
Net negative charge, cmol$_c$/kg[c]	80–120	100–180	15–40	15–40	2–5	100–550

[a] Both dioctahedral and trioctahedral forms of the 2:1-type clays are found in soils, but trioctahedral sheets are most prominent in chlorites while dioctahedral sheets are generally more common in the others.
[b] From the top of one layer to the top of the next similar layer; 1 nm = 10^{-9} m.
[c] Centimoles charge per kilogram (1 cmol$_c$ = 0.01 mol$_c$), a measure of the cation exchange capacity (see Section 8.8).
[d] It is very difficult to accurately determine the surface area of organic matter. Different procedures give values ranging from 20 to 800 m²/g.

EXPANDING MINERALS. The **smectite** group is noted for **interlayer expansion**, which occurs by swelling when the minerals are wetted, the water entering the interlayer space and forcing the layers apart. **Montmorillonite** is the most prominent member of this group in soils, although *beidellite, nontronite,* and *saponite* also are found.

The flakelike crystals of smectites (see Figure 8.2) are composed of 2:1-type layers, as shown in Figure 8.5. In turn, these layers are loosely held together by very weak oxygen-to-oxygen and cation-to-oxygen linkages. Exchangeable cations and associated water molecules are attracted between layers (the interlayer space), causing **expansion** of the crystal lattice. The **internal surface** thus exposed by far exceeds the external surface area of these minerals. For example, the **specific surface** or total surface area per unit mass (external and internal) of one smectite mineral (montmorillonite) is 650 to 800 m^2/g. A comparable figure for kaolinite is only 10 to 30 m^2/g (see Table 8.2).

Isomorphous substitution of Mg^{2+} for some of the Al^{3+} ions in the dioctahedral sheet accounts for most of the negative charge for smectites, although some substitution of Al^{3+} for Si^{4+} has occurred in the tetrahedral sheet. The smectites commonly show a high cation exchange capacity, perhaps 20 to 40 times that of kaolinite (see Table 8.2).

Smectites also are noted for their high plasticity and cohesion and their marked swelling when wet and shrinkage on drying. Wide cracks commonly form as smectite-dominated soils (e.g., Vertisols) are dried. The dry aggregates or clods are very hard, making such soils difficult to till. Also, soils high in smectite make very poor bases for building foundations or roads (see Section 8.11).

Vermiculites are also 2:1-type minerals, an octahedral sheet being found between two tetrahedral sheets. In most soil vermiculites, the octahedral sheet is aluminum-dominated (**dioctahedral**), although magnesium-dominated (**trioctahedral**) vermiculites are also found. In the tetrahedral sheet of most vermiculites, considerable substitution of aluminum for silicon has taken place. This accounts for most of the very high net negative charge associated with these minerals (see Table 8.2).

FIGURE 8.5 Model of two crystal layers and an interlayer characteristic of montmorillonite, a smectite expanding-lattice 2:1-type clay mineral. Each layer is made up of an octahedral (alumina) sheet sandwiched between two tetrahedral (silica) sheets with shared apical oxygen ions that hold the sheets together. There is little attraction between oxygen atoms in the bottom tetrahedral sheet of one unit and those in the top tetrahedral sheet of another. This permits a variable space between layers, which is occupied by water and exchangeable cations. The internal surface area thus exposed far exceeds the surface around the outside of the crystal. Note that magnesium has replaced aluminum in some sites of the octahedral sheet. Likewise, some silicon atoms in the tetrahedral sheet may be replaced by aluminum (not shown). These substitutions give rise to a negative charge, which accounts for the high cation exchange capacity of this clay mineral. A schematic drawing of the ionic arrangement is shown at the right.

Water molecules, along with magnesium and other ions, including Al-hydroxy ions, are strongly adsorbed in the interlayer space of vermiculites (Figure 8.6). However, these interlayer constituents act primarily as bridges holding the units together rather than as wedges driving them apart. The degree of swelling is, therefore, considerably less for vermiculites than for smectites. For this reason, vermiculites are considered to be **limited-expansion** clay minerals, expanding more than kaolinite but much less than the smectites.

NONEXPANDING MINERALS. **Micas** and **chlorites** are the types of minerals in this group. Muscovite and biotite are examples of unweathered micas often found in the sand and silt separates. Weathered minerals similar in structure to these micas are found in the clay fraction of soils. They are called *fine-grained micas*.[6]

Like smectites, fine-grained micas have a 2:1-type crystal. However, the particles are much larger than those of the smectites. Also, the major source of charge is in the tetrahedral sheet, where about 20% of the silicon sites are occupied by aluminum atoms. This results in a high net negative charge in the tetrahedral sheet, even higher than that found in vermiculites. To satisfy this charge, potassium ions are strongly attracted in the interlayer space. They are just the right size to fit snugly into certain spaces in the adjoining tetrahedral sheets (Figure 8.6). The potassium, being strongly attracted to the adjacent sheets, acts as a binding agent, preventing expansion of the crystal. Hence, fine-grained micas are quite **nonexpansive.** The hydration, cation adsorption, swelling, shrinkage, plasticity, specific surface area, and size of fine-grained mica crystals are all intermediate between similar characteristics of kaolinite and smectite.

Soil **chlorites** are basically iron-magnesium silicates with some aluminum present. In a typical chlorite clay crystal, 2:1 layers, such as are found in vermiculites, alternate with a hydroxide sheet (see Figure 8.6) that can be either dioctahedral or trioctahedral.

---

[6] In the past, the term **illite** has been used to identify these clay minerals. Current nomenclature identifies illite, along with others such as **glauconite**, as more specific clay minerals in the fine-grained mica group.

FIGURE 8.6 Schematic drawing illustrating the organization of tetrahedral and octahedral sheets in one 1:1-type mineral (kaolinite) and four 2:1-type minerals. The octahedral sheets in each of the 2:1-type clays can be either aluminum dominated (dioctahedral) or magnesium dominated (trioctahedral). However, in most chlorites the trioctahedral sheets are dominant while the dioctahedral sheets are generally most prominent in the other three 2:1-types. Note that kaolinite is nonexpanding, the layers being held together by hydrogen bonds. Maximum interlayer expansion is found in smectite, with somewhat less expansion in vermiculite because of the moderate binding power of numerous Mg^{2+} ions. Fine-grained mica and chlorite do not expand because K^+ ions (fine-grained mica) or an octahedral-like sheet of hydroxides of Al, Mg, Fe, and so forth (chlorite) tightly bind the 2:1 layers together. The interlayer spacings are shown in nanometers (1 nm = 10^{-9} m).

However, magnesium commonly dominates both the interlayer hydroxide (trioctahedral) sheet as well as the trioctahedral sheet of the 2:1 layer of chlorites.

The negative charge of chlorites is about the same as that of fine-grained micas and considerably less than that of the smectites or vermiculites. Like fine micas, chlorites may be interstratified with vermiculites or smectites in a single crystal. Particle size and surface area for chlorites are also about the same as for fine-grained micas. There is no water adsorption between the chlorite crystal units, which is in keeping with the non-expansive nature of this mineral.

Mixed and Interstratified Layers

Specific groups of clay minerals do not occur independently of one another. In a given soil, it is common to find several clay minerals in an intimate mixture. Furthermore, some mineral colloids have properties and compositions intermediate between those of any two of the well-defined minerals just described. Such minerals are termed **mixed layer** or **interstratified** because the individual layers within a given crystal may be of more than one type. Terms such as *chlorite-vermiculite* and *fine-grained mica-smectite* are used to describe mixed-layer minerals. In some soils, they are more common than single-structured minerals such as montmorillonite.

8.5 GENESIS OF SOIL COLLOIDS

Silicate Clays

The silicate clays are developed from the weathering of a wide variety of minerals by at least two distinct processes. *First* is a slight physical and chemical *alteration* of primary minerals. For example, as weathering occurs, muscovite mica breaks down in size to the colloidal range and is altered chemically by losing some of its potassium and gaining other base-forming cations (M^+). The sites formerly occupied by potassium provide negative charges. The fine mica colloid that emerges still has the 2:1-type structure, only having been altered in the process.

Intense weathering that stimulates the *decomposition* of primary minerals with the subsequent *recrystallization* of certain of their products into silicate clays offers the *second* means of clay formation. An example of recrystallization is the formation of kaolinite, a 1:1-type clay mineral, from solutions containing soluble aluminum and silicon that came from the breakdown of primary minerals having a 2:1-type structure. The specific clay mineral that forms depends on weathering conditions and the specific ions present in the weathering solution as crystallization occurs.

The more specific conditions conducive to the formation of important clay types are shown in Figure 8.7. Note that fine-grained micas and magnesium-rich chlorites represent earlier weathering stages of the silicates, and kaolinite and (ultimately) iron and aluminum oxides the most advanced stages. The smectites (e.g., montmorillonite) represent intermediate stages. Figure 8.7 indicates a variety of processes by which individual clays are formed.

Iron and Aluminum Oxides

Goethite (FeOOH) and hematite (Fe_2O_3) are the two most common iron oxides in soils. They are both produced by the weathering of iron-containing primary silicate minerals. Goethite tends to be dominant in the temperate and more moist zones, while hematite with its deep red color is more prominent (even dominant) in tropical or drier conditions.

Gibbsite [$(AlOH)_3$], the most common oxide of aluminum found in soils, is a product of weathering of a variety of primary aluminosilicates. It represents the most advanced stage of weathering in soils.

Allophane and Imogolite

Relatively little is known of factors influencing the formation of allophane and imogolite. While they are commonly associated with soils of volcanic origin (Andisols), they are also formed from igneous rocks and are found in some Spodosols. Apparently, volcanic ashes release significant quantities of $Si(OH)_x$ and $Al(OH)_x$ materials that precipi-

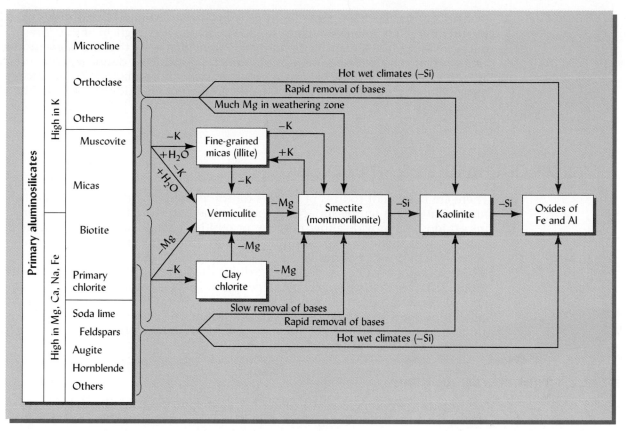

FIGURE 8.7 General conditions for the formation of the various layer silicate clays and oxides of iron and aluminum. Fine-grained micas, chlorite, and vermiculite are formed through rather mild weathering of primary aluminosilicate minerals, whereas kaolinite and oxides of iron and aluminum are products of much more intense weathering. Conditions of intermediate weathering intensity encourage the formation of smectite. In each case silicate clay genesis is accompanied by the removal in solution of such elements as K, Na, Ca, and Mg.

tate as gels in a relatively short period of time. These minerals are generally poorly crystalline in nature, imogolite being the product of a more advanced state of weathering than that which produces allophane.

Humus

The breakdown and alteration of plant residues by microorganisms and the concurrent synthesis of new, more stable, organic compounds results in the formation of the dark-colored colloidal organic material called *humus* (see Section 11.4 for details). The various organic structural units associated with the decay and synthesis provide charged sites for the attraction of both cations and anions.

Geographic Distribution of Clays

The clay of any particular soil is generally made up of a mixture of different colloidal minerals. In a given soil, the mixture may vary from horizon to horizon. This occurs because the kind of clay that develops depends not only on climatic influences and profile conditions but also on the nature of the parent material. The situation may be further complicated by the presence in the parent material itself of clays that were formed under a preceding and perhaps an entirely different type of climatic regime. Nevertheless, some general deductions seem possible, based on the relationships shown in Figure 8.7.

The well-drained and well-weathered Oxisols of humid and subhumid tropics tend to be dominated by kaolinite, along with oxides of iron and aluminum. These clays also dominate Ultisols, such as those in the southeastern part of the United States.

The 2:1-type clays are more prominent in Alfisols, Mollisols, and Vertisols, where weathering is less intense. Where the parent material is high in micas, fine-grained micas such as illite are apt to be formed. Parent materials that are high in metallic cations (particularly magnesium) or are subject to restricted drainage, which discourages the leaching of these cations, encourage smectite formation. Thus, fine-grained micas and smectites are more likely to be prominent in Aridisols than in the more humid areas.

8.6 SOURCES OF CHARGES ON SOIL COLLOIDS

There are two major sources of charges on soil colloids: (1) hydroxyls and other such functional groups on the surfaces of the colloidal particles that by releasing or accepting H⁺ ions can provide either negative or positive charges, and (2) the charge imbalance brought about by the isomorphous substitution in some clay crystal structures of one cation by another of similar size but differing in charge.

All colloids, organic or inorganic, exhibit the surface charges associated with OH groups, charges that are largely **pH dependent**. In the case of the 2:1-type clays, however, these surface charges are complemented by a much larger number of charges emanating from the isomorphous substitution of one cation for another. Since they are not dependent on the pH, they are termed **permanent** or **constant charges**. We will consider these constant charges first.

Constant Charges on Silicate Clays

We noted in Section 8.3 that isomorphous substitution could be the source of both negative and positive charges. Examples of specific substitutions will now be considered.

NEGATIVE CHARGES.   A net negative charge is found in minerals where there has been an isomorphous substitution of a lower-charged ion (e.g., Mg^{2+}) for a higher-charged ion (e.g., Al^{3+}). Such substitution commonly occurs in some aluminum-dominated dioctahedral sheets and leaves an unsatisfied negative charge.

The substitution of Mg^{2+} for Al^{3+} is an important source of the negative charge on the smectite, vermiculite, and chlorite clay micelles.

A second example is the substitution of an Al^{3+} for an Si^{4+} in the tetrahedral sheet, which also leaves one negative charge unsatisfied. This can be illustrated as follows, assuming that on an average two silicon atoms are associated with four oxygens in a charge-neutral unit.

Such a substitution is common in several of the important soil silicate clay minerals, such as the fine-grained micas, vermiculites, and even some smectites.

POSITIVE CHARGES.   Isomorphous substitution can also be a source of positive charges if the substituting cation has a higher charge than the ion for which it substitutes. In a trioctahedral sheet, there are three magnesium ions surrounded by oxygen and hydroxy groups, and the sheet has no charge. However, if an Al^{3+} ion substitutes for one of the Mg^{2+} ions, a positive charge results.

Trioctahedral sheet (no substitution)	Trioctahedral sheet (Al^{3+} substituted for Mg^{2+})
$Mg_3O_2(OH)_2$	$Mg_2AlO_2(OH)_2{}^+$
No charge	One excess positive charge

Such positive charges are characteristic of the trioctahedral hydroxide sheet in the interlayer of clay minerals such as chlorites, a charge that is overbalanced by negative charges in the tetrahedral sheet. Indeed, in several 2:1-type silicate clays, including chlorites and smectites, substitutions in both the tetrahedral and octahedral sheets can occur. The net charge in these clays is the balance between the negative and positive charges. In all 2:1-type silicate clays, however, the **net charge** is negative since those substitutions leading to negative charges far outweigh those producing positive charges.

pH-Dependent Charges

The second source of charges noted on some layer silicate clays (e.g., kaolinite) and on humus, allophane, and Fe, Al oxides, is dependent on the soil pH and consequently is termed **variable** or **pH-dependent.** Both negative and positive charges come from this source.

NEGATIVE CHARGES. The pH-dependent charges are associated primarily with hydroxyl (OH) groups on the edges and surfaces of the inorganic and organic colloids. The OH groups are attached to iron and/or aluminum in the inorganic colloids (e.g., $>$Al—OH) and to the carbon in CO groups in humus (e.g., —CO—OH). Under moderately acid conditions, there is little or no charge on these particles, but as the pH increases, the hydrogen dissociates from the colloid OH group, and negative charges result.

$$>Al-OH + OH^- \rightleftharpoons >Al-O^- + H_2O$$

$$-CO-OH + OH^- \rightleftharpoons -CO-O^- + H_2O$$

| No charge (soil solids) | (soil solution) | Negative charge (soil solids) | (soil solution) |

As indicated by the $\rightleftharpoons$ arrows, the reactions are reversible. If the pH increases, more OH^- ions are available to force the reactions to the right, and the negative charge on the particle surfaces increases. If the pH is lowered, OH^- ion concentrations are reduced, the reaction goes back to the left, and the negativity is reduced.

Another source of increased negative charges as the pH is increased is the removal of positively charged complex aluminum hydroxy ions [e.g., $Al(OH)_2{}^+$]. At low pH levels, these ions block negative sites on the silicate clays (e.g., vermiculite) and make them unavailable for cation exchange. As the pH is raised, the $Al(OH)_2{}^+$ ions react with the OH^- ion in the soil solution to form insoluble $Al(OH)_3$, thereby freeing the negatively charged sites.

$$>Al-(OH)_2{}^- Al(OH)_2{}^+ + OH^- \longrightarrow >Al-(OH)_2{}^- + Al(OH)_3$$

| Negative charged site is blocked | | Negative charge site is freed | No charge |

This mechanism of increasing negative charges is of special importance in soils high in Fe, Al oxides, such as are found in the southeastern United States and in the tropics.

POSITIVE CHARGES. Under moderate to extreme acid soil conditions, some silicate clays and Fe, Al oxides may exhibit net positive charges. Once again, exposed OH groups are involved. In this case, however, as the soils become more acid, **protonation**—the attachment of H^+ ions to the surface OH groups—takes place. The reaction for silicate clays may be shown simply as:

$$\diagdown\hspace{-4pt}\diagup Al-OH \ + \ H^+ \ \rightleftharpoons \ \diagdown\hspace{-4pt}\diagup Al-OH_2{}^+$$

No charge   (soil solution)   Positive charge
(soil solids)   (soil solids)

Thus, in some cases, the same site on the inorganic soil colloid may be responsible for negative charge (high pH), no charge (intermediate pH), or positive charge (very low pH). The reaction as the pH of a highly alkaline soil is reduced may be illustrated as follows:

$$\diagdown\hspace{-4pt}\diagup Al-O^- + H^+ \ \rightleftharpoons \ \diagdown\hspace{-4pt}\diagup AlOH \ + \ H^+ \ \rightleftharpoons \ \diagdown\hspace{-4pt}\diagup AlOH_2{}^+$$

Negative charge   No charge   Positive charge
(high pH)   (intermediate pH)   (very low pH)

Since a mixture of humus and several inorganic colloids is usually found in soil, it is not surprising that positive and negative charges may be exhibited at the same time. In most soils of temperate regions, the negative charges far exceed the positive ones (Table 8.3). However, in some acid soils high in Fe, Al oxides or allophane, the overall net charge may be positive. The effect of soil pH on positive and negative charges on such soils is illustrated in Figure 8.8.

The charge characteristics of selected soil colloids are shown in Table 8.3. Note the high percentage of constant negative charges in some 2:1-type clays (e.g., smectites and vermiculites). Humus, kaolinite, allophane and Fe, Al oxides have mostly variable (pH-dependent) negative charges and exhibit modest positive charges at low pH values.

Cation and Anion Adsorption[7]

The charges associated with soil particles attract simple and complex ions of opposite charge. Thus, a given colloidal mixture may exhibit not only a maze of positive and negative surface charges, but an equally complex group of simple cations and anions such as Ca^{2+} and $SO_4{}^{2-}$ that are attracted by the particle charges.

---

[7] In addition to the simple inorganic cations and anions, more complex organic compounds as well as charged organomineral complexes are adsorbed by the charged soil colloids (see Section 8.11).

TABLE 8.3 Charge Characteristics of Representative Colloids Showing Comparative Levels of Permanent (Constant) and pH-Dependent Negative Charges as Well as pH-Dependent Positive Charges

	Negative charge			
Colloid type	Total at pH 7, $cmol_c/kg$	Constant, %	pH dependent, %	Positive charge, $cmol_c/kg$
Organic	200	10	90	0
Smectite	100	95	5	0
Vermiculite	150	95	5	0
Fine-grained micas	30	80	20	0
Chlorite	30	80	20	0
Kaolinite	8	5	95	2
Gibbsite (Al)	4	0	100	5
Goethite (Fe)	4	0	100	5
Allophane	30	10	90	15

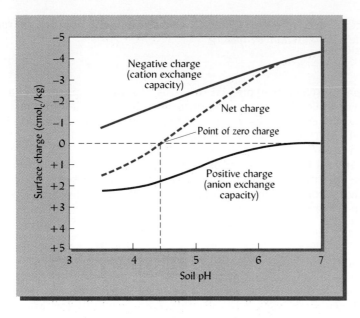

FIGURE 8.8 Relationship between soil pH and positive and negative charges on an Oxisol surface horizon in Malaysia. The negative charges (cation exchange capacity) increase and the positive charges (anion exchange capacity) decrease with increasing soil pH. The point of zero charge is about pH 4.4. [Redrawn from Shamshuddin and Ismail (1995)]

8.7 CATION EXCHANGE

In Section 8.2, the various cations adsorbed by the **exchange complex**[8] were shown to be subject to replacement by other cations through a process called *cation exchange.* The reaction of carbonic acid with the exchange complex of a temperature-region surface soil illustrates cation exchange in nature. For the sake of simplicity, we can assume that the numbers of Ca^{2+}, Al^{3+}, H^+, and other metallic cations such as Mg^{2+} and K^+ (represented as M^+) are in the ratio of 40, 20, 5, and 20 per micelle. The exchange with the hydrogen in carbonic acid can be represented as follows:

$$\boxed{\text{Micelle}} \begin{array}{l} Ca_{40} \\ Al_{20} \\ H_5 \\ M_{20} \end{array} + 5H_2CO_3 \rightleftharpoons \boxed{\text{Micelle}} \begin{array}{l} Ca_{38} \\ Al_{20} \\ H_{10} \\ M_{19} \end{array} + 2Ca(HCO_3)_2 + M(HCO_3)$$

(soil solids) (soil solution) (soil solids) (soil solution)

Where sufficient rainfall is available to leach calcium and other metallic cations, the reaction tends to go toward the right and the soils tend to become more acid. However, the metallic cations drive the reaction to the left in acid soils to which calcium-containing lime has been added, and in low rainfall areas where metallic cations tend to build up because there is insufficient water to leach the metallic cations from the soil.

Cations such as K^+, Ca^{2+}, Mg^{2+}, Na^+, and NH_4^+ that are contained in manure or chemical fertilizer also take part in cation exchange reactions. By cation exchange they are adsorbed by the soil colloids, a form in which they are protected from rapid leaching, and yet one from which they can be absorbed by plants.

8.8 CATION EXCHANGE CAPACITY

Previous sections have dealt qualitatively with exchange. We now turn to a consideration of the quantitative cation exchange or the **cation exchange capacity** (CEC). This property, which is defined simply as "the sum total of the exchangeable cations that a soil can adsorb," is easily determined by standard methods. A sample of soil is thoroughly leached with a solution containing a common ion such as NH_4^+ which replaces all other exchangeable cations from the soil colloids. The adsorbed NH_4^+ is then replaced by leaching with a solution containing a second cation such as K^+, and the

---

[8] This term includes all soil colloids, inorganic and organic, capable of holding and exchanging cations.

NH_4^+ in the leachate water is determined. This gives a measure of the positive charges the soil can adsorb, or the *cation exchange capacity.*

Means of Expression[9]

The cation exchange capacity (CEC) is expressed in terms of moles of positive charge adsorbed per unit mass. For the convenience of expressing CEC in whole numbers, we shall use **centimoles of positive charge per kilogram of soil** ($cmol_c/kg$). Thus, if a soil has a cation exchange capacity of 10 $cmol_c/kg$, 1 kg of this soil can adsorb 10 $cmol_c$ of H^+ ion, for example, and can exchange it with 10 $cmol_c$ of any other cation. This emphasizes that in cation exchange it is the **number of charges**, not the number of ions or molecules with which we are most concerned. The mole concept of charge is reviewed briefly in Box 8.1 along with examples relating moles to mass or weight of ions taking part in cation exchange reactions.

Cation Exchange Capacities of Soils

The cation exchange capacity (CEC) of a given soil horizon is determined by the relative amounts of different colloids in that soil and by the CEC of each of these colloids. Figure 8.9 illustrates the common range in CEC among different soils and other organic and inorganic exchange materials. Note that sandy soils, which are generally low in all colloidal material, have low CECs compared to those exhibited by silt loams and clay loams. Also note the very high CECs associated with humus compared to those exhibited by the

---

[9] Until recent years, cation exchange capacity was expressed as milliequivalents per 100 g of soil. In this text, the International System of Units (SI) is being used. Fortunately, it is easy to compare soil data using either of these methods of expression since 1 milliequivalent per 100 g of soil is equal to 1 centimole ($cmol_c$) of positive or negative charge per kilogram of soil.

BOX 8.1   CHEMICAL EXPRESSION OF CATION EXCHANGE

One mole of any atom, molecule, or charge is defined as 6.02×10^{23} (Avogadro's number) of atoms, molecules, or charges. Thus, 6.02×10^{23} negative charges associated with the soil colloidal complex would attract 1 mole of positive charge from adsorbed cations such as Ca^{2+}, Mg^{2+}, and H^+. The number of moles of the positive charge provided by the adsorbed cations in any soil gives us a measure of the *cation exchange capacity* (CEC) of that soil. The CEC of soils commonly ranges from 3 to 50 *centimoles per kilogram soil* (3 to 50 $cmol_c/kg$).

Using the mole concept, it is easy to relate the mole charges to the mass of ions or compounds involved in cation or anion exchange. Consider, for example, the exchange that takes place when adsorbed sodium ions in an alkaline arid-region soil are replaced by hydrogen ions:

$$\boxed{\text{Micelle}}\ Na^+ + H^+ \rightleftharpoons \boxed{\text{Micelle}}\ H^+ + Na^+$$

If 2 $cmol_c$ of adsorbed Na^+ ions per kilogram of soil were replaced by H^+ ions in this reaction, how many grams of Na^+ ions would be replaced?

Since the Na^+ ion is singly charged, the mass of Na^+ needed to provide 1 mole of charge (1 mol_c) is the gram atomic weight of sodium, or 23 g (see periodic table in Appendix C). The mass providing 1 *centimole* of charge ($cmol_c$) is $1/100$ of this amount, or 0.23 g. Thus, the mass of the 2 $cmol_c$ Na^+ replaced from 1 kg of soil is:

$$2\ cmol_c\ Na^+/kg \times 0.23\ g\ Na^+/cmol_c = 0.46g\ Na^+/kg\ soil$$

The 0.46 g Na^+ would be replaced by only 0.02 g H, which is the mass of 2 $cmol_c$ of this much lighter element.

In this example, note that the number of charges provided by the replacing ion is *equivalent* to the number associated with the ion being replaced. Thus, 1 mole of negative charges attracts 1 mole of positive charges whether the charges come from H^+, K^+, Na^+, NH_4^+, Ca^{2+}, Mg^{2+}, Al^{3+}, or any other cation. Keep in mind, however, that only one-half the atomic weights of divalent cations, such as Ca^{2+} or Mg^{2+}, and only one-third the atomic weight of trivalent Al^{3+} are needed to provide 1 mole of charge. This *chemical equivalency* principle applies to both cation and ion exchange.

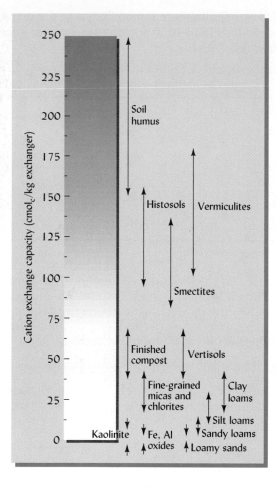

FIGURE 8.9 Ranges in the cation exchange capacities (at pH 7) that are typical of a variety of soils and soil materials. The high CEC of humus shows why this colloid plays such a prominent role in most soils, and especially those high in kaolinite and Fe, Al oxides, clays that have low CECs.

inorganic clays, especially kaolinite and Fe, Al oxides. The CEC coming from humus generally plays a very prominent role, and sometimes a dominant one, in cation exchange reactions in surface soils.

Using the CEC range from Figure 8.9 it is possible to estimate the CEC of a soil if the quantities of the different soil colloids in the soil are known. Box 8.2 illustrates how this can be done.

Data in Table 8.4 show the average CEC values for nine different soil orders compiled by the U.S. Soil Conservation Service. Note the very high CEC for the Histosols, verifying the high CEC of the organic colloids. The Vertisols, which are very high in swelling-type clays (mostly smectite), have the highest average CEC of the mineral soils. Next come the Aridisols and Mollisols, which are also commonly high in 2:1-type clays.

BOX 8.2 ESTIMATING CATION EXCHANGE CAPACITIES

Assume you want to estimate the cation exchange capacities (CEC) of the surface horizon of a cultivated Mollisol from Iowa (pH = 7.0, 20% clay, and 4% organic matter)

The most dominant clays in the Mollisol are likely the 2:1 types, and their average CEC would be about 80 $cmol_c$/kg. At pH 7.0, the CEC of the organic matter is about 200 $cmol_c$/kg (see Table 8.3, page 240). Since 1 kg of the soil has 0.2 kg (20%) of clay and 0.04 kg (4%) of organic matter, we can calculate the CEC associated with each of these sources.

From the clays:	0.2 kg × 80 $cmol_c$/kg = 16 $cmol_c$
From the O.M.:	0.04 kg × 200 $cmol_c$/kg = 8 $cmol_c$
The total CEC of the Mollisol is:	16 + 8 = 24 $cmol_c$/kg soil

In a very general way, the type of clay mineralogy can be estimated by reversing the preceding steps when CEC is known but the type of clay is uncertain.

TABLE 8.4 The Average Cation Exchange Capacities (CEC) and pH Values of More Than 3000 Surface Soil Samples Representing Nine Different Soil Orders

Note the very high CEC of the Histosols and among the mineral soils of the Vertisols. Ultisols have the lowest CEC.

Soil order	CEC, cmol$_c$/kg	pH
Ultisols	3.5	5.60
Alfisols	9.0	6.00
Spodosols	9.3	4.93
Entisols	11.6	7.32
Inceptisols	14.6	6.08
Aridisols	15.2	7.26
Mollisols	18.7	6.51
Vertisols	35.6	6.72
Histosols	128.0	5.50

From Holmgren, et al. (1993).

The Ultisols, whose clays are dominantly kaolinite and hydrous oxides of iron and aluminum, have relatively low CEC values. The Entisols and Inceptisols include soils developed on recent alluvium and lacustrine materials that are quite high in clay. Despite large variations in soil organic matter and texture, these data appear to reflect the quantities and kinds of soil colloids found in the soils.

pH and Cation Exchange Capacity

In previous sections it was pointed out that the cation exchange capacity of most soils increases with pH. At very low pH values, the cation exchange capacity is also generally low (Figure 8.10). Under these conditions, only the permanent charges of the 2:1-type

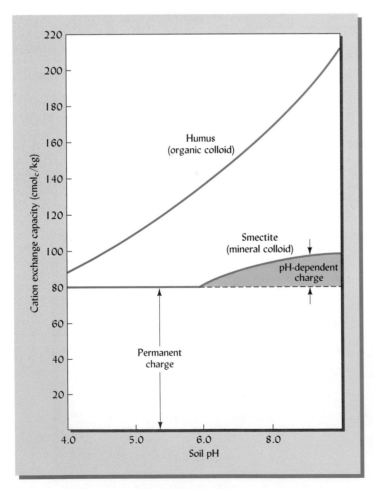

FIGURE 8.10   Influence of pH on the cation exchange capacity of smectite and humus. Below pH 6.0 the charge for the clay mineral is relatively constant. This charge is considered permanent and is due to ionic substitution in the crystal unit. Above pH 6.0 the charge on the mineral colloid increases slightly because of ionization of hydrogen from exposed hydroxyl groups at crystal edges. In contrast to the clay, essentially all of the charges on the organic colloid are considered pH dependent. [Smectite data from Coleman and Mehlich (1957); organic colloid data from Helling, et al. (1964)]

clays (see Section 8.6) and a small portion of the pH-dependent charges of organic colloids, allophane, and some 1:1-type clays hold exchangeable ions. As the pH is raised, the negative charges on some 1:1-type silicate clays, allophane, humus, and even Fe, Al oxides increase, thereby increasing the cation exchange capacity. To obtain a measure of this maximum retentive capacity, the CEC is commonly determined at a pH of 7.0 or above. At neutral or slightly alkaline pH, the CEC reflects most of those pH-dependent charges as well as the permanent ones.

8.9 AVAILABILITY OF EXCHANGEABLE CATIONS AS PLANT NUTRIENTS

Cation Saturation

The specific exchangeable cations associated with soil colloids differ from one climatic region to another—Ca^{2+}, Al^{3+}, complex aluminum hydroxy ions, and H^+ being most prominent in humid regions, and Ca^{2+}, Mg^{2+}, and Na^+ dominating in low-rainfall areas (Table 8.5). The cations that dominate the exchange complex have a marked influence on soil properties.

In a given soil, the proportion of the cation exchange capacity satisfied by a particular cation is termed the **percentage saturation** for that cation. Thus, if 50% of the CEC is satisfied by Ca^{2+} ions, the exchange complex is said to have a **percentage calcium saturation** of 50.

This terminology is especially useful in identifying the relative proportions of sources of acidity and alkalinity in the soil solution. Thus, the percentage saturation with Al^{3+} and H^+ ions gives an indication of the acid conditions, while increases in the nonacid cation percentage (commonly referred to as the **percentage base saturation**[10]) indicate the tendency toward neutrality and alkalinity. The percentage base saturation is an important soil property, especially because it is inversely related to soil acidity. These relationships will be discussed further in Chapter 9.

Cation Exchange and Nutrient Availability

Exchangeable cations generally are available to both higher plants and microorganisms. By cation exchange, hydrogen ions from the root hairs and microorganisms replace nutrient cations from the exchange complex. The nutrient cations are forced into the soil solution, where they can be assimilated by the adsorptive surfaces of roots and soil organisms, or they may be removed by drainage water.

CATION SATURATION.   Several factors operate to expedite or retard the release of nutrients to plants. First, there is the percentage saturation of the exchange complex by the nutrient cation in question. For example, if the percentage calcium saturation of a soil is high, the displacement of this cation is comparatively easy and rapid. Thus, 6 cmol/kg of exchangeable calcium in a soil whose exchange capacity is 8 cmol/kg (75%

---

[10] Technically speaking, nonacid cations such as Ca^{2+}, Mg^{2+}, K^+, and Na^+ are not bases. When adsorbed by soil colloids in the place of H^+ ions, however, they reduce acidity and increase the soil pH. For that reason, they are referred to as *bases* and the portion of the CEC that they satisfy is usually termed *percentage base saturation*.

TABLE 8.5   **Cation Exchange Data for Representative Mineral Surface Soils in Different Areas**

Characteristics	Warm humid-region soil (Ultisol)	Cool humid-region soil (Alfisol)	Semiarid-region soil (Ustoll)	Arid-region soil (Natrargids)[a]
Exchangeable calcium, $cmol_c$/kg	2–5	6–9	14–17	12–14
Other exchangeable bases, $cmol_c$/kg	1–2	2–3	5–7	8–12
Exchangeable aluminum and/or hydrogen, $cmol_c$/kg	3–7	4–8	0–2	0
Cation exchange capacity, $cmol_c$/kg	3–12	12–18	20–26	20–26
Base saturation, %	25	50–75	90–100	100
Probable pH	5.0–5.4	5.6–6.0	7	8–10

[a] Significant sodium saturation.

calcium saturation) probably would mean ready availability, but 6 cmol/kg when the total exchange capacity of a soil is 30 cmol/kg (20% calcium saturation) would produce lower availability. This is one reason that for calcium-loving plants, such as alfalfa, the calcium saturation of at least part of the soil should approach or even exceed 80%.

INFLUENCE OF COMPLEMENTARY ADSORBED CATIONS. A second factor influencing the plant uptake of a given cation is the complementary ions held on the colloids. As was discussed in Section 8.2, the strength of adsorption of different cations is in the following order:

$$Al^{3+} > Ca^{2+} > Mg^{2+} > K^+ = NH_4^+ > Na^+$$

Consequently, a nutrient cation such as K^+ is less tightly held by the colloids if the complementary ions are Al^{3+} and H^+ (acid soils) than if they are Mg^{2+} and Na^+ (neutral to alkaline soils). The loosely held K^+ ions are more readily available for absorption by plants or for leaching in acid soils (see also Section 13.13).

There are also some nutrient antagonisms, which in certain soils cause inhibition of uptake of some cations by plants. Thus, potassium uptake by plants is limited by high levels of calcium in some soils. Likewise, high potassium levels are known to limit the uptake of magnesium even when significant quantities of magnesium are present in the soil.

EFFECT OF TYPE OF COLLOID. Differences exist in the tenacity with which several types of colloidal micelles hold specific cations and in the ease with which they exchange cations. At a given percentage base saturation, smectites—which have a high charge density per unit of colloid surface—hold calcium much more strongly than does kaolinite (low charge density). As a result, smectite clays must be raised to about 70% base saturation before calcium will exchange easily and rapidly enough to satisfy most plants. In contrast, a kaolinite clay exchanges calcium much more readily, serving as a satisfactory source of this constituent at a much lower percentage base saturation. The need to add limestone to the two soils will be somewhat different, partly because of this factor.

8.10   ANION EXCHANGE

Anions are held by soil colloids in two major ways. First, they are held by anion adsorption mechanisms similar to those responsible for cation adsorption. Second, they may actually react with surface oxides or hydroxides, forming more definitive **inner-sphere complexes.** We shall consider anion adsorption first.

The basic principles of **anion exchange** are similar to those of cation exchange, except that the charges on the colloids are positive and the exchange is among negatively charged anions. The positive charges associated with the surfaces of kaolinite, iron and aluminum oxides, and allophane attract anions such as SO^{4+} and NO^{3-}. A simple example of an anion exchange reaction is as follows:

$$\boxed{\text{Micelle}}\ NO_3^- + Cl^- \rightleftharpoons \boxed{\text{Micelle}}\ Cl^- + NO_3^-$$

(positively charged   (soil solution)   (positively charged   (soil solution)
soil solid)   soil solid)

Just as in cation exchange, *equivalent* quantities of NO_3^- and Cl^- are exchanged; the reaction can be reversed; and plant nutrients so released can be absorbed by plants.

In contrast to cation exchange capacities, anion exchange capacities of soils generally *decrease* with increasing pH. In some very acid tropical soils that are high in kaolinite and in iron and aluminum oxides, the anion exchange capacity may actually exceed the cation exchange capacity.

Anion exchange is very important in making anions available for plant growth and at the same time in retarding the leaching of such anions from the soil. For example, anion exchange restricts the loss of sulfates from subsoils in the southern United States (see Section 12.22). Even the leaching of nitrate may be retarded by anion exchange in the subsoil of certain highly weathered soils of the humid tropics. Similarly, the downward movement into groundwater of some charged organic pollutants found in organic wastes can be retarded by such anion and/or cation exchange reactions.

Inner-Sphere Complexes

Some anions, such as phosphates, arsenates, molybdates, and sulfates, can react with particle surfaces, forming **inner-sphere complexes.** For example, the $H_2PO_4^-$ ion may react with the protonated hydroxyl group rather than remain as an easily exchanged anion.

$$\text{>Al—OH}_2^+ \ + \ H_2PO_4^- \longrightarrow \text{>Al—H}_2PO_4 \ + \ H_2O$$

| (soil solid) | (soil solution) | (soil solid) | (soil solution) |

This reaction actually reduces the net positive charge on the soil colloid. Also, the $H_2PO_4^-$ is held very tightly by the soil solids and is not readily available for plant uptake (see Section 13.8).

Anion adsorption and exchange provide important though complex mechanisms for interactions in the soil and between the soil and plants. Together with cation exchange they largely determine the ability of soils to hold nutrients in a form that is accessible to plants, and to retard movement of pollutants in the environment.

Weathering and CEC/AEC Levels

The range of CEC levels of different clay minerals shown in Figure 8.11 shows that clays developed under mild weathering conditions (e.g., smectites and vermiculites) have much higher CEC levels than those developed under more extreme weathering pressures. It also illustrates the general effect of increasing weathering intensities on the magnitude of negative charges that underpin the CEC levels. Also shown are the AEC levels that, in turn, tend to be much higher in clays developed under strong weathering conditions (e.g., kaolinite) than in those found under more mild weathering. This generalized figure is helpful in obtaining a first approximation of CEC and AEC levels in soils of different climatic regions. It must be used with caution, however, since some soils high in 2:1-type clays are found in areas currently undergoing intensive weathering. The nature of the parent material and the time allowed for the weathering to occur also influence the clay types present and the CEC/AEC relations.

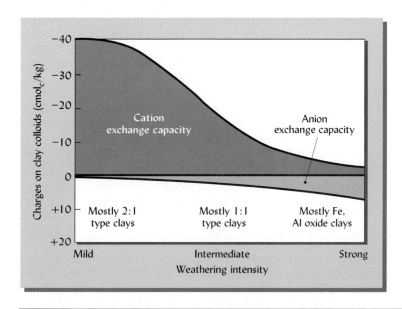

FIGURE 8.11 The effect of weathering intensity on the charges on clay minerals and, in turn, on their cation and anion exchange capacities (CECs and AECs). Note the high CEC and very low AEC associated with mild weathering, which has encouraged the formation of 2:1-type clays such as fine-grained micas, vermiculites, and smectites. More intense weathering destroys the 2:1-type clays and leads to the formation of first kaolinite and then oxides of Fe and Al. These have much lower CECs and considerably higher AECs. Such changes in clay dominance account for the curves shown.

8.11 SOIL COLLOIDS AND ENVIRONMENTAL QUALITY

The physical and chemical properties of soil colloids have an impact on soil quality in several ways. Three examples of such effects will be discussed briefly.

Groundwater Contamination

Soil colloids help control the movement of pesticides and other organic compounds into groundwater. The retention of these chemicals by soil colloids can prevent their downward movement through the soil or can delay that movement until the compounds are broken down by soil microbes.

By accepting or releasing protons (H$^+$ ions), groups such as —OH, —NH$_2$, and —COOH in the chemical structures of some organic compounds provide positive or negative charges that stimulate anion or cation exchange reactions. Other organic compounds participate in inner-sphere complexation and adsorption reactions just as do the nutrients we have discussed. However, most organic compounds are more commonly **absorbed** within the soil organic colloids by a process termed **partitioning**. The soil organic colloids tend to act as a solvent for the applied chemicals, thereby partitioning their concentrations between those held on the soil colloids and those left in the soil solution. These nonionic or organic compounds are **hydrophobic**, being repelled by water. As a result, moist clays contribute little to partitioning since their adsorbed water molecules prevent the movement of the nonionic organic chemicals into or around the clay particles.[11] Since we seldom know for certain the exact involvement of the adsorption, complexation, or partitioning processes, we use the general term **sorption** to describe the retention by soils of these organic compounds. The tendency of a pesticide or other organic compound to leach into the groundwater is determined by the solubility of the compound and by the ratio of the amount of chemical sorbed by the soil to that remaining in solution.

Building and Road Construction

Figure 8.12 reminds us of the adverse effects of the shrinking and swelling properties of soils high in smectites. The trememdous pressures these clays can engender when wetted result in broken and cracked building foundations and pavements. If steps are not taken to provide proper footings prior to construction, disaster often occurs. This is just one example of the critical role soil colloids play in determining the usefulness of our soils.

Clays as Sealants

The physical and chemical properties of swelling-type clays, and particularly the smectites, make them useful tools in preventing unwanted movement of water and water contaminants. Such clays are widely used in the construction of liners to seal ponds, sewage lagoons, industrial waste lagoons, and landfills. A layer of the smectite placed on the bottom and sides of the pond or lagoon expands when wetted and forms a relatively impenetrable barrier to the movement not only of water but of organic and inorganic contaminants contained in the water. These contaminants are held in the containment area and are prevented from moving downward into the groundwater.

Smectites are also used to prevent the undesirable upward movement of organic chemicals through boreholes drilled to monitor the presence of organic contaminants in groundwater. As shown in Figure 8.13, the wetted smectite (bentonite) forms an impenetrable plug between the sampling tube and the larger borehole, thereby preventing the upward movement of chemicals from the groundwater. Increasingly, environmental scientists are using swelling-type clays for such environmental monitoring, as well as for the removal of organic chemicals from water by partitioning.

---

[11] The hydrated metal cations (e.g., Ca^{2+}) that are adsorbed on the surface of smectites can be replaced with large organic cations, giving rise to what are termed **organoclays**. Such clay surfaces are more friendly toward applied organic compounds, making it possible for the clay to participate in partitioning. Advantage is taken of this phenomenon when smectite organoclays are used to remove organic contaminants from wastewaters and from contaminated groundwaters.

FIGURE 8.12 The different swelling tendencies of two types of clay are illustrated in the lower left. All four cylinders initially contained dry, sieved clay soil, the two on the left from the B horizon of a soil high in kaolinite, the two on the right from one of a soil high in montmorillonite. An equal amount of water was added to the two center cylinders. The kaolinitic soil settled a bit and was not able to absorb all the water. The montmorillonitic soil swelled about 25% in volume and absorbed nearly all the added water. The scenes to the right and above show a practical application of knowledge about these clay properties. (Upper) Soils containing large quantities of smectite undergo pronounced volume changes as the clay swells and shrinks with wetting and drying. Such soils (e.g., the California Vertisol shown here) make very poor building sites. The normal-appearing homes (upper) are actually built on deep, reinforced-concrete pilings (lower right) that rest on nonexpansive substrata. Construction of the 15 to 25 such pilings needed for each home more than doubles the cost of construction. (Photo courtesy of R. Weil)

8.12 CONCLUSION

Colloidal materials control most of the chemical and physical properties of soils. These materials include five major classes of crystalline silicate clays, other aluminosilicates such as allophane, the oxides of iron and aluminum, and the organic colloids (humus). Due to their extremely small particle sizes and platelike structures, these colloids possess enormous surface areas, much of which are internal surfaces that are much like the pages of a book.

The mineralogical and chemical constitutions of these colloids result in electrostatic charges on or near the surfaces of the colloidal particles. Negative charges predominate on most silicate clays. Positive charges characterize some colloids, such as the oxides of Fe and Al, especially if the pH is low. The charges on colloids attract ions and other sub-

Cap removable for sampling

Waterproof box

Concrete for stability

Smectite (Bentonite) plug

Sampling tube (10 cm diameter)

Bore hole (25 cm diameter)

Sand through which chemical can move

Perforations to permit movement of gas or liquid into monitoring well

FIGURE 8.13 Illustration of the use of a smectite (bentonite) as a plug or sealant to prevent upward leakage through a borehole that was drilled to permit the installation of a well for monitoring the presence of organic chemicals in soils. Air-dry bentonite is tapped in place and then wetted. When the wetted clay swells, it fits tightly around the sampling tube and the bore hole walls. The wet smectite resists the upward movement of organic materials, forcing them to enter the perforated sampling tube from which they can be removed for analysis. The plug also protects the well from the possibility of chemicals washing down from the soil surface and therefore being mistaken for groundwater contaminants. [Redrawn from Reid and Ulery (1998)]

stances with opposite charges: cations being attracted to negatively charged sites and anions to the positively charged ones.

At the charge sites on soil colloids, exchanges between one ion or substance and others can take place. This enables plants to exchange H^+ ions for nutrient ions, such as Ca^{2+}, K^+, and OH^- ions for SO_4^{2-} and NO_3^-. A colloid's capacity to hold exchangeable cations (*cation exchange capacity*) and exchangeable anions (*anion exchange capacity*) measures its ability to provide nutrients to plant roots and to retard the downward movement of pollutants to drainage water.

Cation exchange joins photosynthesis as a fundamental life-supporting process. Without this soil property, terrestrial ecosystems would not be able to retain sufficient nutrients to support natural or introduced vegetation, especially in the event of such disturbances as timber harvest, fire, or cultivation.

STUDY QUESTIONS

1. Describe the *soil colloidal complex,* indicate its various components, and explain how it tends to serve as a "bank" for plant nutrients.

2. How do you account for the difference in surface area associated with a grain of kaolinite clay compared to that of montmorillonite, a smectite?

3. Contrast the difference in crystalline structure among *kaolinite, smectites, fine-grained micas, vermiculites,* and *chlorites.*

4. There are two basic processes by which silicate clays are formed by weathering of primary minerals. Which of these would likely be responsible for the formation of (1) fine-grained mica, and (2) kaolinite from muscovite mica? Explain.

5. If you wanted to find a soil high in kaolinite, where would you go? The same for (1) smecitite and (2) vermiculite?

6. Which of the silicate clay minerals would be *most* and *least* desired if one were interested in (1) a good foundation for a building, (2) a high cation exchange capacity, (3) an adequate source of potassium, and (4) a soil on which hard clods form after plowing?

7. Which of the following would you expect to be *most* and *least* sticky and plastic when wet: (1) a soil with significant sodium saturation in a semiarid area, (2) a soil high in exchangeable calcium in a subhumid temperate area, or (3) a well-weathered acid soil in the tropics? Explain your answer.

8. A soil contains 4% humus, 10% montmorillonite, 10% vermiculite, and 10% Fe, Al oxides. What is its approximate cation exchange capacity?

9. Calculate the number of grams of Al^{3+} ions needed to replace 10 $cmol_c$ of Ca^{2+} ion from the exchange complex of 1 kg of soil.

10. Explain how the properties of colloids in natural soils affect environmental quality.

11. Explain how expanding 2:1 clays are used in environmental monitoring and pollution control.

REFERENCES

Buseck, P. R. 1983. "Electron microscopy of minerals," *Amer. Scientist,* **71**:175–185.

Coleman, N. T., and A. Mehlich. 1957. "The chemistry of soil pH," in *The Yearbook of Agriculture (Soil)*. (Washington, D.C.: U.S. Department of Agriculture).

Dixon, J. B., and S. B. Weed. 1989. *Minerals in Soil Environments,* 2d ed. (Madison, Wis.: Soil Sci. Soc. Amer.).

Helling, C. S., et al. 1964. "Contribution of organic matter and clay to soil cation exchange capacity as affected by the pH of the saturated solution," *Soil Sci. Soc. Amer. Proc.,* **28**:517–520.

Holmgren, G. G. S., M. W. Meyer, R. L. Chaney, and R. B. Daniels. 1993. "Cadmium, lead, zinc, copper, and nickel in agricultural soils in the United States of America," *J. Environ. Qual.,* **22**:335–348.

Parker, J. C., D. F. Amos, and L. W. Zelanzny. 1982. "Water adsorption and swelling of clay minerals in soil systems," *Soil Sci. Soc. Amer. J.,* **46**:450–456.

Reid, D. A. and A. L. Ulery. 1998. "Environmental applications of smectites," in J. Dixon, D. Schultze, W. Bleam, and J. Amonette (eds.), *Environmental Soil Mineralogy.* (Madison, Wis.: Soil Sci. Soc. Amer.).

Shamshuddin, J., and H. Ismail. 1995. "Reactions of ground magnesium limestone and gypsum in soils with variable-charged minerals," *Soil Sci. Soc. Amer. J.,* **59**:106–112.

9

SOIL ACIDITY, ALKALINITY, AND SALINITY

What have they done to the rain?
—PETER, PAUL, AND MARY

The degree of acidity or alkalinity is a *master variable* that affects nearly all soil properties—chemical, biological, and physical. Commonly expressed as **soil pH**, this variable largely controls plant nutrient availability and microbial reaction in soils. It affects which trees, shrubs, or grasses will dominate the landscape under natural conditions and determines which cultivated crops will grow well or even grow at all in a given field site. The soil pH also determines the fate of many soil pollutants, affecting their breakdown and possible movement from the soil into groundwater and streams.

Climate and associated vegetation influences soil pH level, with acidity dominating humid areas and alkalinity prevalent in low-rainfall areas. But human activities, often unintended, also affect the pH of the soil. Chemical fertilizers and organic wastes react in the soil to form strong inorganic acids such as HNO_3 and H_2SO_4. These lead to increased soil acidity. These same two acids are found in **acid rain**, which originates from gases emitted into the atmosphere primarily by the combustion of fossil fuel (power plants, automobiles, etc.) and the burning of trees and other biomass. Concerns are increasing about the possible damage of acid rain to forests, crops, and lakes.

In some low-lying and imperfectly drained areas of arid and semiarid regions, neutral salts tend to build up to levels that adversely affect plant growth. The natural tendency of these soils to become saline is accelerated when salt-bearing irrigation waters are applied and the salts are not allowed to drain out of the soil profile. Thriving civilizations of the past disintegrated when their otherwise productive soils were allowed to become saline. The global area of salt-affected soils is growing daily as a result of faulty irrigation practices, especially in low-income countries of arid regions. These soils and their management are of vital concern to all of humanity.

Before we consider sources of H^+ and OH^- ions in nature, briefly review in Box 9.1 the concept of pH and how it helps us understand the reciprocal relationship between the concentrations of these two ions in water systems. Also see the chart in Figure 9.1 listing the pH values of some substances we use daily. These are compared with the pH of different acid and alkaline soils to which we will be referring in this chapter.

BOX 9.1   SOIL PH, SOIL ACIDITY, AND ALKALINITY

Whether a soil is acid, neutral, or alkaline is determined by the comparative concentrations of H^+ and OH^- ions.[a] Pure water provides these ions in equal concentrations:

$$H_2O \rightleftharpoons H^+ + OH^-$$

The equilibrium for this reaction is far to the left, only an infinitesimal amount of H^+ or OH^- ions being formed from 1 mole of water. The *ion product* of the concentrations of the H^+ and OH^- ions is a constant (K_w), which at 25°C is known to be 1×10^{-14}:

$$[H^+] \times [OH^-] = K_w = 10^{-14}$$

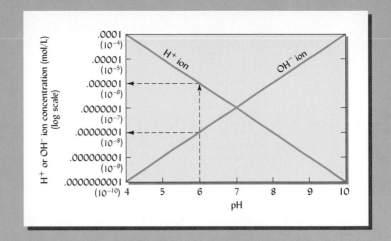

Since in pure water the concentration of H ions [H^+] must be equal to that of OH^- ions [OH^-], this equation shows that the concentration of each is 10^{-7} ($10^{-7} \times 10^{-7} = 10^{-14}$). It also shows the inverse relationship between the concentrations of these two ions. As one increases, the other must decrease proportionately. Thus, if we were to increase the H^+ ion concentration [H^+] by 10 times (from 10^{-7} to 10^{-6}), the [OH^-] would be decreased by 10 times (from 10^{-7} to 10^{-8}) since the product of these two concentrations must equal 10^{-14}:

$$10^{-6} \times 10^{-8} = 10^{-14}$$

Scientists have simplified the means of expressing the very small concentrations of H^+ and OH^- ions by using the *negative logarithm of the H^+ ion concentration*, termed the *pH*. Thus, if the H^+ concentration in an acid medium is 10^{-5}, the pH is 5; if it is 10^{-9} in an alkaline medium, the pH is 9.

Note that the pH also gives us an indirect measure of the OH^- ion concentration since the product of [H^+] × [OH^-] must always equal 10^{-14}. Thus at pH 5 the [OH^-] is 10^{-9} ($10^{-5} \times 10^{-9} = 10^{-14}$); at pH 8 it is 10^{-6} ($10^{-8} \times 10^{-6} = 10^{-14}$).

The figure above shows the relationship between pH and the concentrations of H^+ and OH^- ions. Note that as one goes down, the other goes up, and vice versa. The dotted line illustrates the concentrations of these two ions at pH = 6.0: [H^+] = 10^{-6}; [OH^-] = 10^{-8}. This reciprocal relationship between H^+ and OH^- ions should always be kept in mind in studying soil acidity and alkalinity.

---

[a] Technically speaking, chemical reactions are influenced by the *activity* of an ion rather than by its *concentration*. This is because of the electrostatic effect of one ion on the activity of its nearby neighbors. Ionic activities are thus essentially effective concentrations. Since the difference between ionic activities and concentrations in soils are not so great, we will use the term *concentration* in this text.

9.1   SOURCES OF ACIDITY

Aluminum and Hydrogen[1]

Two adsorbed cations—aluminum[2] and hydrogen—are largely responsible for soil acidity. The bulk of these two elements are tightly bound by soil organic matter and clay minerals. However, under very acid conditions (Figure 9.2) some of the aluminum becomes soluble, and is held by soil colloids as exchangeable Al^{3+} ions.[3] This adsorbed aluminum contributes to soil acidity by providing to the soil solution Al^{3+} ions that are, in turn, hydrolyzed to produce H^+ ions as follows:

---

[1] In water hydrogen ions are always hydrated, existing as hydronium ions (H_3O^+) rather than as simple H^+ ions. However, for simplicity this text will use the unhydrated H^+ to represent this ion.

[2] Iron (Fe^{3+}) also is solubilized under acid conditions and forms hydroxy cations just as does aluminum. However, since iron is solubilized only in extremely acid soils, we shall focus primarily on the involvement of aluminum.

[3] As is the case for most metallic cations in soils, the Al^{3+} ion is highly hydrated, being present in forms such as $Al(H_2O)_6^{3+}$. For simplicity, however, it will be shown as the simple Al^{3+} ion.

FIGURE 9.1 Chart showing a range of pH from 1 to 12 and the approximate pH of products commonly used in our society every day (left). Comparable pH ranges are shown (right) for soils we will be studying in this text.

$$\boxed{\text{Micelle}}\ \text{Al}^{3+} \rightleftharpoons \text{Al}^{3+}$$

Adsorbed   Soil solution

$$\text{Al}^{3+} + \text{H}_2\text{O} \rightleftharpoons \text{AlOH}^{2+} + \text{H}^+$$

If the pH is raised, the Al^{3+} is converted to aluminum hydroxy ions by reactions such as the following:

$$\text{Al}^{3+} \underset{\text{H}^-}{\overset{\text{OH}^-}{\rightleftharpoons}} \text{AlOH}^{2+} \underset{\text{H}^+}{\overset{\text{OH}^-}{\rightleftharpoons}} \text{Al(OH)}_2^+ \underset{\text{H}^+}{\overset{\text{OH}^-}{\rightleftharpoons}} \text{Al(OH)}_3$$

Note that the reactions are reversible, being shifted to the right with added OH^- ions and to the left with added H^+ ions.

In some 2:1-type clays, particularly vermiculite, the aluminum hydroxy ions (as well as iron hydroxy ions) play another role. They move into the interlayer space of the crystal units and become very tightly bound, preventing intracrystal expansion and blocking some of the exchange sites. Raising the soil pH results in the removal of these ions and the freeing up of some of the exchange sites. Thus aluminum (and iron) hydroxy ions are partly responsible for the pH-dependent or **variable charge** of soil colloids (see Section 8.6).

In moderately acid soils, some hydrogen ions that at low pH would be bound tenaciously by the organic matter, Fe, Al oxides, and 1:1-type clays now are present in exchangeable forms. The exchangeable H^+ ions are in equilibrium with the soil solution H^+ ions.

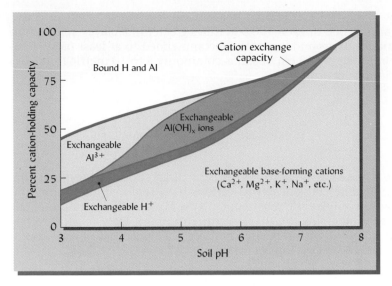

FIGURE 9.2 General relationship between soil pH and the cations held as exchangeable ions or bound by organic matter and clay minerals. The cation exchange capacity line is estimated using average data from 60 Wisconsin soils from Helling, et al. (1964). Note evidence of the variable charge on the colloids since the CEC increased with increasing pH. Probable levels of exchangeable and bound cations are shown. Under very acid conditions, exchangeable aluminum and hydrogen ions and bound hydrogen and aluminum dominate. At higher pH values, the exchangeable base-forming cations predominate, while at intermediate values, aluminum hydroxy ions such as $Al(OH)^{2+}$ are prominent. This diagram shows typical conditions; any particular soil would likely give a modified distribution.

$$\boxed{\text{Micelle}} \begin{array}{c} H^+ \\ H^+ \end{array} \rightleftharpoons 2H^+$$

Exchangeable H^+ Soil solution H^+

Again, the colloidal control of soil acidity is demonstrated, as is the dominant role of the aluminum and hydrogen ions.

Soil pH and Cation Associations

Figure 9.2 summarizes the effect of pH on the distribution of ions in soils containing some variable charge colloids. Note that the cation exchange capacity of the soil increases with soil pH. Also note the change in cation dominance on the exchange complex as one moves from very acid soils, to moderate acidity, and to neutrality and above. Study this figure carefully, keeping in mind that for any particular soil the distribution of ions might be somewhat different.

Note that in Figure 9.2 two forms of hydrogen and aluminum are shown: (1) that tightly held by the pH-dependent sites (*bound*), and (2) that associated with permanent negative charges on the colloids (*exchangeable*). Only the exchangeable ions have an immediate effect on soil pH, but, as we shall see later, both forms are very much involved in determining how much lime or sulfur is needed to change soil pH.

While the factors responsible for soil acidity are far from simple, two dominant groups of elements are in control. The different aluminum-containing ions and H^+ ions generate acidity, and most of the other cations combat it. This simple statement is worth remembering.

Soil Acidity Classes

Research suggests three kinds of acidity: (1) **active acidity** due to the H^+ and Al^{3+} ions in the soil solution; (2) **salt-replaceable acidity**, involving the aluminum and hydrogen that are *easily exchangeable* by other cations in a simple unbuffered salt solution such as KCl; and (3) **residual acidity**, which is associated with the large quantities of Al^{3+}, H^+, and $Al(OH)_x$ ions that are bound in nonexchangeable forms by organic matter and silicate clays.

The quantity of H^+ ion to be neutralized varies among the acidity classes. Only 2 to 3 kg of calcium carbonate is required to neutralize the active or soil solution acidity in the upper 15 cm of 1 ha of a typical mineral soil. The quantity of salt-replaceable (exchangeable) acidity is much higher, commonly more than 100 times that needed to neutralize the soil solution (active) acidity.

The residual acidity that is associated with nonexchangeable bound forms of aluminum and hydrogen (see Figure 9.2) is commonly far greater than either the active or salt-replaceable acidity. It may be 1000 times greater than the soil solution or active acidity in a sandy soil and 50,000 or even 100,000 times greater in a clayey soil high in organic matter. The amount of ground limestone recommended to at least partly neutralize residual acidity in the upper 15 cm of soil is commonly 5 to 10 metric tons (Mg) per hectare (2.25 to 4.5 tons per acre).

9.2 SOURCES OF ALKALINITY

Base-Forming Cations

In arid and semiarid areas, base-forming cations (e.g., Ca^{2+}, Mg^{2+}, and Na^+) dominate the exchangeable complex of soils. These adsorbed cations have replaced aluminum and hydrogen on the exchange complex, thereby reducing the H^+ ion concentration of the soil solution (and inversely increasing the OH^- ion concentration). They also have a second and more direct effect on the OH^- ion concentration. The hydrolysis of colloids saturated with these cations releases OH^- ions as follows:

$$\boxed{\text{Micelle}}\ Ca^{2+} + 2H_2O \rightleftharpoons \boxed{\text{Micelle}}\ \begin{matrix} H^+ \\ H^+ \end{matrix} + Ca^{2+} + 2OH^-$$

(soil solid)   (soil solution)   (soil solid)   (soil solution)

Soils whose exchange complexes are saturated with cations such as calcium, magnesium, and sodium and are said to have a high **percentage base saturation** and have high pH values (see Figure 9.3).

Role of Carbonates and Bicarbonates

In some soils of dry areas, carbonates and bicarbonates tend to accumulate, even in the surface horizons. Often a distinct whitish *calcic* horizon forms in the profile (see Plate 13). These materials definitely reinforce the tendency toward alkalinity and resist rapid changes in soil pH.

The primary source of carbonates and bicarbonates in soils is carbonic acid (H_2CO_3) that forms when CO_2 from microbial and root respiration reacts with water:

$$CO_2 + H_2O \rightleftharpoons H_2CO_3 \underset{H^+}{\overset{OH^-}{\rightleftharpoons}} HCO_3^- + H_2O \underset{H^+}{\overset{OH^-}{\rightleftharpoons}} CO_3^{2-} + H_2O$$

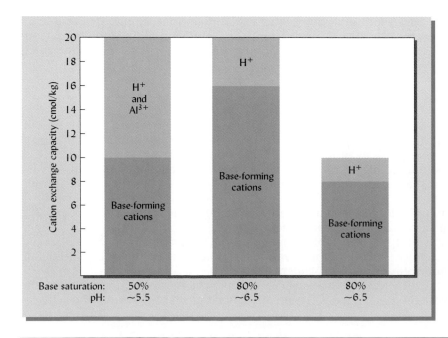

FIGURE 9.3 Three soils with similar colloid types but with different percentage base saturations—50, 80, and 80, respectively. The first (left) is a clay loam (CEC of 20 cmol$_c$/kg); the second, (center) the same soil satisfactorily limed; and the third (right), a sandy loam with a CEC of only 10 cmol$_c$/kg. Note especially that soil pH is correlated more or less closely with percentage base saturation. Also note that the sandy loam (right), because of its low cation exchange capacity, has a higher pH than the acid clay loam (left), even though the clay contains more exchangeable bases.

The pH of a solution that is in equilibrium with CO_2 at a concentration found in the soil atmosphere is about 4.6. In soils with higher pH levels, the abundant OH^- ions react with the H_2CO_3 to form first the bicarbonate (HCO_3^-) and then the carbonate (CO_3^{2-}) ion. The relationship between the concentrations of HCO_3^- and CO_3^{2-} ions and those of H^+ and OH^- ions is shown in Figure 9.4. Note that an increase of either HCO_3^- or CO_3^{2-} ions in the soil solution results in an increase of OH^- ions and a decrease of H^+ ions.

Role of the Cations (Na$^+$ versus Ca^{2+})

The particular cation associated with carbonate anions also influences the pH level. For example, if adsorbed Na^+ is prominent on the colloidal complex and in the soil solution, and if ample HCO_3^- and CO_3^{2-} are present, $NaHCO_3$ and Na_2CO_3 will form. Both these salts are water soluble and highly ionized, thus assuring continued high levels of HCO_3^- and CO_3^{2-} ions and, in turn, high pH values. Soils containing sodium carbonate commonly have pH values as high as 10 or above.

Fortunately, Ca^{2+} is the dominant cation in most alkaline soils. This cation reacts with carbonic acid to form $Ca(HCO_3)_2$ and $CaCO_3$. The $Ca(HCO_3)_2$ is quite soluble and provides sufficient HCO_3^- ions to give pH values up to about 8.2. In contrast, $CaCO_3$ is not very soluble and provides a low concentration of CO_3^{2-} ions. As a result, $CaCO_3$-laden (*calcareous*) horizons commonly have pH values in the range of 7.5 to 8.0, a level that can be tolerated by most plants.

Influences of Salts

Neutral salts such as NaCl, KCl, $CaCl_2$, and $MgCl_2$ can moderate the effects of adsorbed cations and carbonates on the pH of some alkaline soils. For example, high levels of NaCl constrain the ionization of $NaHCO_3$.

$$NaHCO_3 \rightleftharpoons Na^+ + HCO_3^-$$

The high Na^+ ion concentration coming from the NaCl drives the reaction to the left. The result is a lower HCO_3^- concentration and associated lowering of OH^- ion concentration and soil pH values.

Soluble salts also influence the OH^- ion concentration arising from high Na^+ adsorption on the colloidal complex.

$$\boxed{Micelle}\ Na^+ + H_2O \rightleftharpoons \boxed{Micelle}\ H^+ + Na^+ + OH^-$$

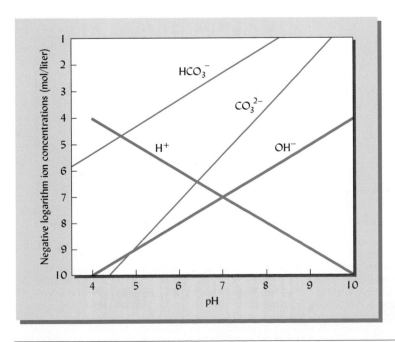

FIGURE 9.4 Effect of carbonates (HCO_3^- and CO_3^{2-}) on the concentrations of H^+ and OH^- ions expressed in negative logarithms. (The H^+ line represents increase in pH ($-\log H^+$ ions) from about 4 to nearly 10). This diagram shows that the levels of HCO_3^- and CO_3^2 ions in the soil solution plays a major role in determining the pH of soils. [Modified from Rowell (1988)]

This reaction likewise is driven to the left in the presence of soluble salts such as NaCl or Na$_2$SO$_4$. The consequence is a reduction in the concentration of OH$^-$ ions and a lower pH. This section reminds us of the major role of the exchange complex in controlling soil pH.

9.3 BUFFERING OF SOILS

As previously indicated, soils tend to resist changes in the pH of the soil solution. This resistance, called **buffering,** is important since it prevents drastic fluctuations in soil pH that could be detrimental to plants and microorganisms. It also largely determines the amount of amendments such as lime or sulfur required to effect a desired change in soil pH.

Buffering is stimulated by soil reactions that either replenish hydrogen ions as they are removed from the soil solution, or remove excess hydrogen ions that are added to the soil solution. Figure 9.5 illustrates in a simple manner how buffering may be effective.

Three primary mechanisms are thought to be responsible for this soil buffering: (1) interactions with aluminum and iron compounds at low pH, (2) cation exchange reactions at intermediate pH levels, and (3) reactions with carbonates and bicarbonates at high pH. The pH ranges and percentage base saturations at which these mechanisms are effective in one group of soils are shown in Figure 9.6, which can be referred to as the mechanisms are considered.

Very Acid Soils

In very acid soils, iron and aluminum hydroxy ions provide a buffer mechanism through reactions such as the following:

$$Al(OH)^{2+} + H_2O \rightleftharpoons Al(OH)_2^+ + H^+$$

$$Al(OH)_2^+ + H_2O \rightleftharpoons Al(OH)_3 + H^+$$

If OH$^-$ ions were added, initially the H$^+$ ions would be neutralized, but the reactions would shift to the right, more H$^+$ ions would be generated, and the pH change would be small. Likewise, if more H$^+$ ions were to be added, the reaction would shift to the left and only modest change would take place in the soil solution pH. In other words, buffering would occur.

Intermediate pH Soils

For soils with intermediate pH levels, buffering can be explained in terms of the equilibrium that exists among the active, salt-replaceable, and residual acidities (see Figure 9.5). This general relationship can be illustrated as follows:

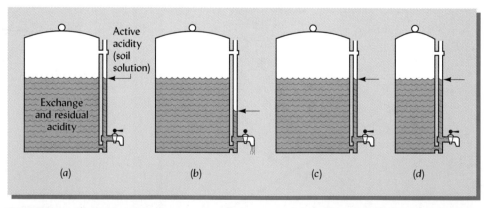

FIGURE 9.5   The buffering action of a soil can be likened to that of a coffee dispenser. (*a*) The active acidity, which is represented by the coffee in the indicator tube on the outside of the urn, is small in quantity. (*b*) When hydrogen ions are removed, this active acidity falls rapidly. (*c*) The active acidity is quickly restored to near the original level by movement from the exchange and residual acidity. By this process, there is considerable resistance to the change of active acidity. A second soil with the same active acidity (pH) level but a much smaller exchange and residual acidity (*d*) would have a lower buffering capacity.

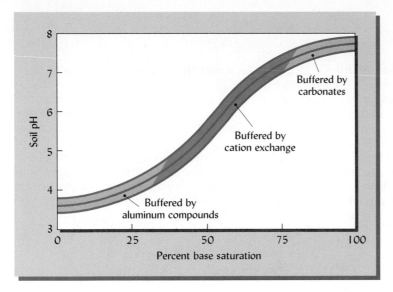

FIGURE 9.6 The relationship between percent base saturation and soil pH of Vermont soils, the appropriate pH range where three major buffering mechanisms are most effective: aluminum compounds at low pH, cation exchange at intermediate pH, and carbonates at high pH values. Although there is much variation from one soil to another in the pH–base saturation relationship, the three major mechanisms shown for buffering soils usually pertain. [Magdoff and Bartlett (1985)]

$$\begin{array}{c}\boxed{\begin{array}{l} \text{H} \\ \text{Micelle} \quad \text{Al} \\ \text{AlOH} \end{array}} \quad \rightleftharpoons \quad \boxed{\text{Micelle}} \begin{array}{l} \text{H}^+ \\ \text{Al}^{3+} \\ \text{AlOH}^{2+} \end{array} \quad \rightleftharpoons \quad \text{H}^+ + \text{Al}^{3+} + \text{Al(OH)}^{2+}\end{array}$$

Bound forms   Exchangeable forms   Soil solution
(residual acidity)   (salt-replaceable   (active acidity)
                or exchange acidity)

Once again the stage is set for buffering. If the released H^+ ions are neutralized by adding lime, the reaction shifts to the right to replenish the neutralized H^+ ions, and the pH changes only modestly. Similar resistance to drastic pH changes is noted if more H^+ ions are added to the system. The reaction merely shifts to the left, adsorbing or binding part of the H^+ ions and resisting the pH change.

Alkaline Soils

Similar buffer reactions involving carbonates, bicarbonates, carbonic acid, and water take place in alkaline soils. For example:

$$CaCO_3 + 2H_2O \rightleftharpoons Ca^{2+} + 2OH^- + H_2CO_3$$

$$Ca(HCO_3)_2 + 2H_2O \rightleftharpoons Ca^{2+} + 2OH^- + 2H_2CO_3$$

The addition of H^+ ions to neutralize the OH^- ions would shift these reactions to the right while increased OH^- ions would shift them to the left. In both cases, however, there would be resistance to pH change (buffering). Thus, throughout the entire pH range there are mechanisms that buffer the soil solution and prevent rapid changes in soil pH.

Buffering Capacity of Soils

Soils vary greatly in their buffering capacity. Other things being equal, the higher the cation exchange capacity (CEC) of a soil, the greater is its buffering capacity. In a soil with a high CEC, more reserve and exchangeable acidity must be neutralized or increased to affect a given change in soil pH. Thus, a clay loam soil containing 6% organic matter and 30% of a 2:1-type clay would be more highly buffered than a sandy loam with 2% organic matter and 10% kaolinite.

The buffer curve shown in Figure 9.6 suggests that some soils are most highly buffered at pH values below 5 and above 7. This would be in those ranges where aluminum compounds (low pH) and carbonates (high pH) tend to control buffer reactions. The soils are least well buffered at intermediate pH levels where cation exchange is the primary buffer mechanism. Even at these intermediate pH levels, however, the soil is able to resist rapid changes in pH.

9.4   VARIABILITY IN SOIL pH

Natural Changes

Natural weathering and organic decay processes result in the formation of both acid- and base-forming chemicals. The base-forming cations (Ca^{2+}, Mg^{2+}, etc.) are released as rocks and minerals break down. Hydrogen ions are generated from acids such as carbonic, nitric, sulfuric, and strong organic acids that are formed as plant residues decompose. The H^+ ions so generated help solubilize aluminum from mineral surfaces, a key factor in the more acid soils.

In low-rainfall areas there is little leaching of the base-forming cations, and pH values of near 7 and above prevail. In more humid areas, leaching depletes the upper horizons of base-forming cations; hydrogen and aluminum ions accumulate, the percentage of base saturation declines, and the pH is lowered. The effect of increasing precipitation on soil acidity is shown in Figure 9.7.

Human-Induced Changes

In recent times, these long-term natural changes have been increasingly augmented by changes stemming from human activities, a few examples of which we will discuss.

CHEMICAL FERTILIZER USAGE.   The worldwide expansion in recent years of the use of chemical fertilizers has brought about significant changes in soil pH. Widely used ammonium-based fertilizers, such as $(NH_4)_2SO_4$ and $(NH_4)_2HPO_4$, are oxidized in the soil by microbes to produce strong inorganic acids by reactions such as the following:

$$(NH_4)_2SO_4 + 4O_2 \rightarrow 2HNO_3 + H_2SO_4 + 2H_2O$$

These strong acids provide H^+ ions that result in lower pH values.

TILLAGE PRACTICES.   The pH of the upper soil horizons is influenced by tillage practices. In no-tillage systems, the decay of organic residues at or near the soil surface produces organic and inorganic acids that can acidify the upper soil layers. Likewise, nitrogen fertilizer that is commonly applied to the surface layers in the no-tillage systems enhances acidification of these layers. The combined effects on pH of tillage practices and fertilizer applications are shown in Figure 9.8.

ACID DEPOSITION FROM THE ATMOSPHERE.   Acid rain (used here to include snow, fog, and dry deposition) is a second significant worldwide source of nitric and sulfuric acids. Nitrogen- and sulfur-containing gases are emitted into the atmosphere from the combustion of coal, gasoline, and other fossil fuels, as well as from the burning of forests and crop residues (Figure 9.9). These gases react with water and other substances in the atmo-

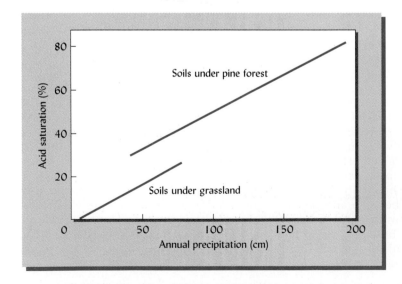

FIGURE 9.7   Effect of annual precipitation on the percent acid saturation of untilled California soils under grassland and pine forests. Note that the degree of acidity goes up as the precipitation increases. Also note that the forest produced a higher degree of soil acidity than did the grassland. [From Jenny, et al. (1968)]

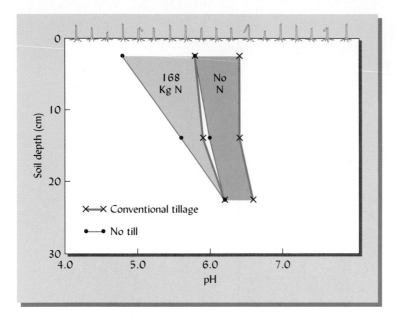

FIGURE 9.8 The effects of 10 years of two tillage practices (conventional and no-till) with and without nitrogen fertilizer on the pH of an Alfisol in a Kentucky corn field. The no-till plots were consistently lower in pH than those that were conventionally tilled. This difference was accentuated on the nitrogen-treated plots, especially in the upper few centimeters of soil. The differences were likely due primarily to the formation of organic acids from residue decay in the no-till plots, and to inorganic acids (e.g., HNO_3) formed from the surface-applied nitrogen fertilizers. [Drawn from data in Blevins, et al. (1983); used with permission of Elsevier Science Publishing Company]

sphere to form HNO_3 and H_2SO_4 that are then returned to the earth in rainwater. This precipitation is called *acid rain* since its pH is commonly between 4.0 and 4.5 and may be a low as 2.0. Normal rainfall that is in equilibrium with atmospheric carbon dioxide has a pH of 5.0 to 6.0. A map showing the pH of acid rain in North America is presented in Figure 9.10.

The quantity of H_2SO_4 and HNO_3 brought to the earth in acid rain globally is enormous, but the amount falling on a given hectare in a year is not enough to immediately change soil pH. In time, however, the cumulative effects of annual deposition can influence both the soils and the plants growing in them.

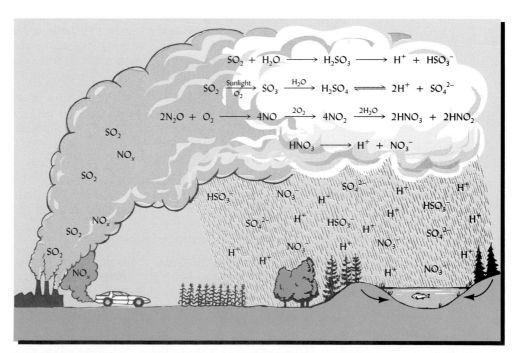

FIGURE 9.9 Illustration of the formation of nitrogen and sulfur oxides from the combustion of fuel in sulfide ore processing and from motor vehicles. The further oxidation of these gases and their reaction with water to form sulfuric acid and nitric acid are shown. These help acidify rainwater, which falls on the soil as acid rain. NO_x indicates a mixture of nitrogen oxides, primarily N_2O and NO. [Modified from National Research Council (1983)]

FIGURE 9.10 Annual mean value of pH in precipitation weighted by the amount of precipitation in the United States and Canada for 1980. [From U.S./Canada Work Group No. 2 (1982)]

Acid rain is particularly worrisome in humid forested areas near urbanized and industrialized sites that provide the S- and N-containing pollutants.[4] Forests in the eastern part of the United States and much of Europe are affected (Figure 9.11). Studies have shown that the leaching of base-forming cations and the mobilization of toxic levels of aluminum are enhanced by the presence in the soil solution of the strong acid anions SO_4^{2+} and NO_3^- along with the H^+ ion provided by the acid rain.

DISPOSAL OF ACID-FORMING ORGANIC WASTES. Deposition of wastes such as sewage sludge on agricultural and forested lands can decrease soil pH. Organic and inorganic acids are formed as the large quantities of sludge are decomposed and soil pH is lowered. This is not true, however, for some *lime-stabilized* sludges, which may have a final pH of 7.5 to 8.5, making them useful in combating soil acidity.[5]

IRRIGATION PRACTICES. In arid and semiarid regions, irrigation waters bring salts from upstream watersheds into the downstream irrigated areas. When the water evaporates, the salts are left behind and can accumulate in the surface horizon. If the salts contain the Na^+ ion, and especially if the bicarbonate ion is present, undesirable pH levels as well as a destabilized soil structure can result.

OXIDATION OF SULFIDIC SOIL MATERIALS. Soils derived from parent materials originally associated with marine environments sometimes contain large quantities of reduced sulfur in the form of pyrite (FeS_2), iron sulfide (FeS), and elemental sulfur (S). Some soils in coastal areas of the southeastern United States, Southeast Asia, and West Africa are formed on such materials. When these soils are exposed to oxygen by drainage, dredging, or exca-

---

[4] See Joslin et al. (1992), Robarge and Johnson (1992), and Likens, et al. (1996) for discussions of chemical changes associated with forest soil acidification.

[5] Certain lime-stabilized sludges have received large quantities of lime to control pathogens and odors. If the high lime content of these sludges is not taken into account, their application to soil with low buffering capacities could bring about **overliming**.

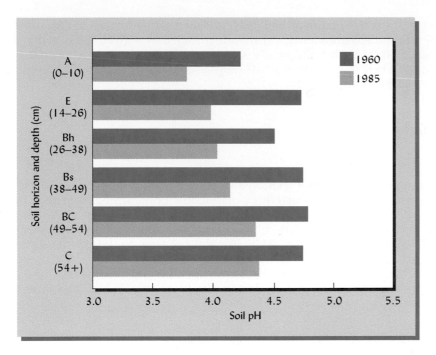

FIGURE 9.11 Reduction in soil pH of 67 forested Spodosol sites in northern Belgium thought to be due to 25 years of atmospheric deposition as well as natural acidification. [From Ronse, et al. (1988); used with permission of Williams and Wilkins, Baltimore, Md.]

vation for mining, the S and FeS are oxidized (see Plates 24 and 25), ultimately resulting in the formation of sulfuric acid, which accounts for their being called *acid sulfate* soils.

$$4FeS + 9O_2 + 4H_2O \rightarrow 2Fe_2O_3 + 4H_2SO_4$$

$$2S + 3O_2 + 2H_2O \rightarrow 2H_2SO_4$$

The oxidation processes produce extreme acidity, with pH values as low as 1.5. Such soils would best be left undisturbed or returned to anaerobic conditions, not only for agricultural and economic reasons, but to avoid contamination of water with acid drainage (see Plate 26) and to maintain these areas as natural habitats for wildlife and natural flora.

Hydrogen Ion Variability

There is considerable variation in the pH of the soil solution at sites only a few centimeters apart. Such variation may result from microbial decay of unevenly distributed organic residues in the soil, or from the effects of plant roots (see Section 10.6). For example, the pH of the rhizosphere immediately around the roots is lower than in the surrounding soil when NH_4^+ ions are adsorbed (Figure 9.12). Apparently, the NH_4^+ ions replace H^+ ions on the root surface, thereby bringing about the pH reduction.

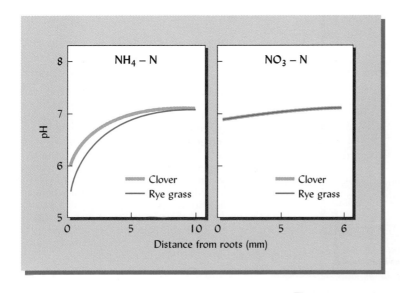

FIGURE 9.12 The pH of rhizosphere at different distances from the roots of ryegrass and clover receiving ammonium and nitrate sources of nitrogen. The exchange of NH_4^+ ions for H^+ ions on the root surfaces released H^+ ions to the rhizosphere, thereby reducing the pH. No such exchange took place with NO_3^- and no pH reduction resulted. [From Hinsinger and Gilkes (1996), with permission of Blackwell Science Ltd.]

The variability of the soil pH is not all bad. It affords microorganisms and plant roots a great variety of solution environments. Organisms curtailed by a given hydrogen ion concentration may find, at an infinitesimally short distance away, a pH level that is more satisfactory. Such variety of environments helps account for the great diversity in microbial species present in normal soils.

9.5   SOIL REACTION: CORRELATIONS

Nutrient Availability And Microbial Activity

Figure 9.13 shows in general terms the relationship between soil pH and the availability of plant nutrients, as well as the activities of soil microorganisms. Note that in strongly acid soils the availability of the macronutrients (Ca, Mg, K, P, N, and S) as well as molybdenum and boron is curtailed. In contrast, availability of micronutrient cations (Fe, Mn, Zn, Cu, and Co) is increased by low soil pH, even to the extent of toxicity to higher plants and microorganisms.

Moving into the alkaline range, we find that molybdenum and all the macronutrients (except phosphorus) are amply available, but levels of available Fe, Mn, Zn, Cu, and

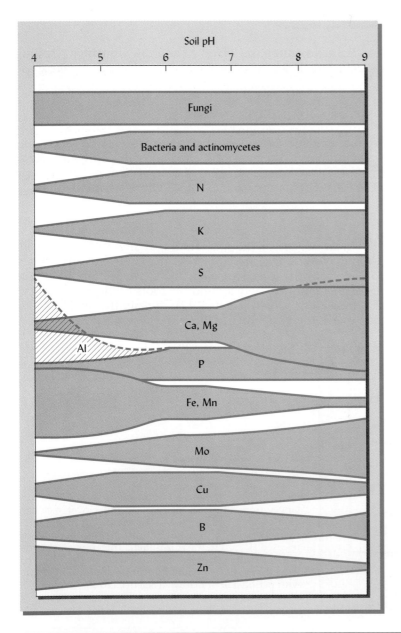

FIGURE 9.13   Relationships existing in mineral soils between pH on the one hand and the activity of microorganisms and the availability of plant nutrients on the other. The wide portions of the bands indicate the zones of greatest microbial activity and the most ready availability of nutrients. Note the hatched area showing that low pH greatly increases the solubility of aluminum, a nonnutrient element which is often toxic to many plants in very acid soils. When the correlations as a whole are taken into consideration, a pH range of about 5.5 to 7.0 seems to best promote the availability of plant nutrients. In short, if soil pH is suitably adjusted for phosphorus, the other plant nutrients, if present in adequate amounts, will be satisfactorily available in most cases.

Co may be so low that plant growth is constrained. Phosphorus and boron availability is likewise reduced in alkaline soil—commonly to a deficiency level.

It is difficult to generalize about nutrient–pH relationships. However, it appears from Figure 9.13 that the pH range of 5.5 to 6.5 or perhaps 7.0 may provide the most satisfactory plant nutrient levels overall.

Figure 9.13 suggests that most general-purpose bacteria and actinomycetes function well at intermediate and high pH values. Fungi seem to be particularly versatile, flourishing satisfactorily over a wide pH range. Therefore, fungal activity tends to predominate in acid soils, whereas at intermediate and higher pH fungi meet stiff competition from actinomycetes and bacteria.

Higher Plants and pH

Plants vary considerably in their tolerance to acid and/or alkaline conditions (Figure 9.14). For example, certain legume crops such as alfalfa and sweet clover grow best in near-neutral or alkaline soils, and most humid-region mineral soils must be limed to grow these crops satisfactorily. Most economic plants suffer from aluminum toxicity if the soil pH drops below 5.0.

Rhododendrons and azaleas are at the other end of the scale. They require a considerable amount of iron, which is abundantly available only at low pH values. If the pH and the percentage base saturation are not low enough, these plants will show chlorosis (yellowing of the leaves) and other symptoms indicative of iron deficiency.

Many forest trees grow well over a wide range of soil pH levels, which indicates they have some tolerance to soil acidity. This is to be expected since some forest trees (e.g., conifers) generally tend to enhance soil acidity, and forests exist in humid regions where acid soils are common. There are some differences, however, among tree species in their tolerance of high soil acidity and aluminum. For example, the pines are generally very tolerant to highly acid soils and to high soluble aluminum levels. The spruces are somewhat less tolerant but more so than most hardwoods (e.g., maples, oaks, and beeches). Elm, poplar, honey locust and lucaena are known to be less tolerant to acid soils than are most other forest species.

The most productive arable soils in use today have intermediate pH values, being not too acidic and not too alkaline. In terms of pH, a range of perhaps 5.5 to 7.0 is most suitable for most crop plants.

Soil pH and Environmental Quality

There are many examples of the influence of soil pH on environmental quality, but only one will be cited—the influence of pH on groundwater contamination by an herbicide, atrazine, that is widely used on corn.

In a soil with pH < 5.7, the atrazine molecule attracts a proton (H^+ ion), giving the herbicide molecule a positive charge—a process called **protonation.** The positively charged molecule is then adsorbed on the negatively charged soil colloids, where it is held until it can be decomposed by soil organisms. The adsorbed pesticide is less likely to move downward and into the groundwater. At pH values above 5.7, however, the adsorption is greatly reduced and the tendency for the herbicide to move downward in the soil is increased. Of course, the adsorption in acidic soils also reduces the availability of atrazine to weed roots, thus reducing its effectiveness as a weed killer.

9.6 DETERMINATION OF SOIL pH

Soil pH is tested routinely, and the determination is easy and rapid to make. Soil samples are collected in the field and the pH is measured directly, or the samples are brought to the laboratory for more accurate pH determination.

Potentiometric Method

The most accurate method of determining soil pH is with a pH meter (Figure 9.15). In this method a sensing glass electrode (referenced with a standard calomel electrode) is inserted into a soil–water mixture that simulates the soil solution. The difference between the H^+ ion activities in the wet soil and inside the glass electrode gives rise to a electrometric potential difference that is related to the soil solution pH. The instrument gives very consistent results and its operation is simple.

| | | Soil pH | | |
| | | 4   5   6   7+ | | |
Herbaceous plants	Trees and shrubs	Strongly acid and very strongly acid soils	Range of moderately acid soils	Slightly acid and slightly alkaline soils
Alfalfa Sweet clover Asparagus Buffalo grass Wheatgrass (tall)	Walnut Alder Eucalyptus Arborvitae			▓▓▓▓
Garden beets Sugar beets Cauliflower Lettuce Cantaloupe	Currant Lilac Ash Yew Beech Lucaena Maple Poplar Tulip tree		▓▓▓	▓
Spinach Red clovers Peas Cabbage Kentucky blue grass White clovers Carrots	Philibert Juniper Myrtle Elm Apricot		▓▓▓▓	
Cotton Timothy Barley Wheat Fescue (tall and meadow) Corn Soybeans Oats Alsike clover Crimson clover Rice Bermuda grass Tomatoes Vetches Millet Cowpeas Lespedeza Tobacco Rye Buckwheat	Birch Dogwood Fir Magnolia Oaks Cedar Hemlock (Canadian) Cypress Flowering cherry Laurel Andromeda Willow oak Pine oak Red spruce Honey locust	▓▓	▓▓▓	
Red top Potatoes Bent grass (except creeping) Fescue (red and sheep's) Western wheatgrass	American holly Aspen White spruce White Scotch pines Loblolly pine Black locust	▓▓	▓	
Poverty grass Eastern gamagrass Love grass, weeping Redtop grass Cassava Napier grass	Autumn olive Blueberries Cranberries Azalea (native) Rhododendron (native) Tea Hemlock (NC) Blackjack oak Sumac Birch Coffee	▓▓		

FIGURE 9.14 Relation of higher plants to the physiological conditions presented by mineral soils of different reactions. Note that the correlations are very broad and are based on pH ranges. The fertility level will have much to do with the actual relationship in any specific case. Such a chart is of great value in deciding whether or not to apply chemicals such as lime or sulfur to change the soil pH.

Dye Methods

Dye methods take advantage of the fact that certain organic compounds change color as the pH is increased or decreased. Mixtures of such dyes provide significant color changes over a wide pH range (3 to 8). A few drops of the dye solutions are placed in contact with the soil, usually on a white spot plate (see Figure 9.15). After standing a few minutes, the color of the dye is compared to a color chart that indicates the approximate pH.

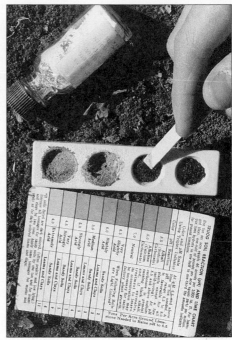

FIGURE 9.15 (Left) Soil pH is measured quickly and inexpensively in the laboratory using an electrometric pH meter. Hundreds of analyses can be made in a day. (Right) In the field, advantage is taken of the fact that pH-sensitive dyes give shades of color that can be calibrated against standard color charts, providing an equally simple and inexpensive means of measuring pH. Although the dye method is not quite as accurate as the electrometric method, it is quite satisfactory for most field management purposes. (Photos courtesy of R. Weil)

9.7 METHODS OF INTENSIFYING SOIL ACIDITY

There are some acid-loving plants such as rhododendrons and azaleas that cannot tolerate even moderately high pH values. To accommodate these plants, it is sometimes necessary to increase the acidity of even acid soils. This is done by adding acid-forming organic and inorganic materials.

Acid Organic Matter

As organic residues that are low in base-forming cations decompose, organic and inorganic acids are formed. Leaf mold from coniferous trees, pine needles, tanbark, pine sawdust and acid moss peat are quite satisfactory organic materials to add around ornamental plants. In contrast to these materials, farm manures, particularly poultry manures, as well as leaf mold of high-base trees such as beech and maple may be alkaline, and may actually increase the soil pH.

Inorganic Chemicals

When the addition of acid organic matter is not feasible, inorganic chemicals such as ferrous sulfate ($FeSO_4$) may be used. This chemical provides available iron (Fe^{2+}) for the plant and, upon hydrolysis, enhances acidity by reactions such as the following:

$$Fe^{2+} + 2H_2O \rightarrow Fe(OH)_2 + 2H^+$$

$$4Fe^{2+} + 6H_2O + O_2 \rightarrow 4Fe(OH)_2^+ + 4H^+$$

The H^+ ions released lower the pH locally and liberate some of the iron already present in the soil.

Other materials that are often used to increase soil acidity are elemental sulfur and, in some irrigation systems, sulfuric acid. Sulfur usually undergoes rapid microbial oxidation in the soil (see Section 12.21), and sulfuric acid is produced.

$$2S + 3O_2 + 2H_2O \rightarrow 2H_2SO_4$$

Under favorable conditions, sulfur is four to five times more effective, kilogram for kilogram, in developing acidity than is ferrous sulfate. Although ferrous sulfate brings about more rapid plant response, sulfur is less expensive, is easy to obtain, and is often used for other purposes. The quantities of ferrous sulfate or sulfur that should be applied will depend upon the buffering capacity of the soil and its original pH level.

9.8 DECREASING SOIL ACIDITY: LIMING MATERIALS

Soil acidity is commonly decreased by adding carbonates, oxides, or hydroxides of calcium and magnesium, compounds that are referred to as **agricultural limes.** Also, wood ashes are used locally to help control soil acidity.

Carbonate Forms

Sources of carbonates include marl, oyster shells, basic slag, and precipitated carbonates, but ground limestone is the most common and is by far the most widely used of all liming materials. The two important minerals occurring in varying proportions in limestones are **calcite,** which is mostly calcium carbonate ($CaCO_3$), and **dolomite,** which is primarily calcium-magnesium carbonate [$CaMg(CO_3)_2$]. When little or no dolomite is present, the limestone is referred to as *calcitic.* As the magnesium increases, this grades into a *dolomitic* limestone.

Oxide and Hydroxide Forms

Two other forms of lime are important, especially for small-scale use such as on home gardens where a very rapid change in pH may be required. One is calcium oxide (CaO), sometimes called *quicklime* or *burned lime* since it is formed by heating limestone. The other, calcium hydroxide ($Ca(OH)_2$), commonly referred to as *hydrated lime,* is formed when water is added to calcium oxide. These limes are more irritating to handle than is limestone and are more expensive, but they continue to find a market where rapid adjustment of soil pH is needed.

Dolomitic limestones and the oxides or hydroxides made from them provide both calcium and magnesium to the soil–plant system. In situations where available magnesium is low, dolomitic limestones would be the products of choice. The nutrient-supplying ability of all liming materials should not be overlooked.

9.9 REACTIONS OF LIME IN THE SOIL

The calcium and magnesium compounds, when applied to an acid soil, react with carbon dioxide and with the acid colloidal complex. These reactions will be considered in order.

Reaction with Carbon Dioxide

All liming materials—whether the oxide, the hydroxide, or the carbonate—react with carbon dioxide and water to yield the bicarbonate form when applied to an acid soil. The carbon dioxide partial pressure in the soil, usually several hundred times greater than that in atmospheric air, is generally high enough to drive reactions such as the following:

$$CaCO_3 + H_2O + CO_2 \rightarrow Ca(HCO_3)_2$$

As we have seen, $Ca(HCO_3)_2$ is very important in buffering the soil.

Reaction with Soil Colloids

These liming materials react directly with acid soils, the calcium and magnesium replacing hydrogen and aluminum on the colloidal complex. The adsorption with respect to calcium may be indicated as follows, using $CaCO_3$ as an example:

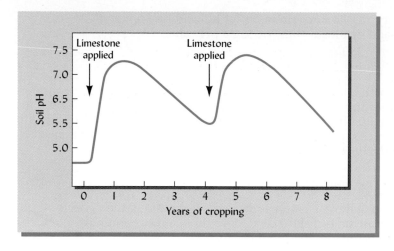

FIGURE 9.16 Diagram to illustrate why repeated applications of limestone are needed to maintain the appropriate chemical balances in the soil. Soil pH likewise increases after 4 to 8 Mg/ha (3.5 to 4.5 tons/acre) of limestone is applied to a temperate region soil. The pH reaches a peak after about one year. Leaching and crop removal deplete the calcium and magnesium and, in time, the pH decreases until a renewed application of limestone is necessary.

$$\boxed{\text{Micelle}}\ {}^{H^+}_{Al^{3+}}\ + 2CaCO_3 + H_2O \longrightarrow \boxed{\text{Micelle}}\ {}^{Ca^{2+}}_{Ca^{2+}}\ + Al(OH)_3 + 2CO_2\uparrow$$

The insolubility of $Al(OH)_3$ and the release of CO_2 to the atmosphere pulls this reactions to the right. In addition, the adsorption of the calcium and magnesium ions raises the percentage base saturation of the colloidal complex, and the pH of the soil solution increases correspondingly.

Depletion of Calcium and Magnesium

As the soluble calcium and magnesium compounds are removed from the soil by the growing plants or by leaching, the percentage base saturation and pH are gradually reduced; eventually, another application of lime is necessary. This type of cyclic activity is typical of much of the calcium and magnesium added to arable soils in humid regions (Figure 9.16).

9.10 LIME REQUIREMENTS: QUANTITIES NEEDED

The amount of liming material required to bring about a desired pH change is determined by several factors, including (1) the change in pH required, (2) the buffer capacity of the soil, (3) the chemical composition of the liming materials to be used and (4) the fineness of the liming materials.

The range of soil pH optimums for various plants, discussed in Section 9.5, suggests the increase in soil pH that may be desired. For example, a higher pH is needed for alfalfa than for corn or soybeans. The limestone requirements of soils with different textures (and therefore likely different buffer capacities) is shown in Figure 9.17. Because of

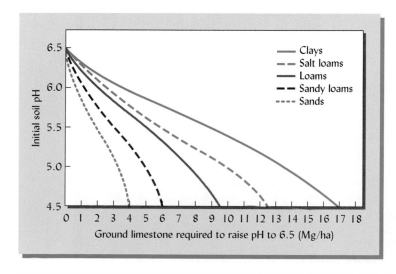

FIGURE 9.17 Effect of soil textural class on the amount of limestone required to raise the pH of soils from their initial state to 6.5. Note the very high quantities needed for the finer-textured soils that have higher clay and organic matter levels and, consequently, higher buffering capacities. The soils are assumed to be typical of cool, humid areas where 2:1-type clays predominate. In warmer climates such as the southeastern United States, where 1:1-type clays and Fe, Al oxides are often dominant, the quantities would likely be no more than one-half or even one-third of those suggested. In all cases, however, no more than 7 to 9 Mg/ha (3 to 4 tons/acre) should be applied at one time. If more is needed to meet the above requirements, subsequent applications after 2 to 3 years could be made.

greater buffering capacity, the lime requirement for an acid fine-textured clay is much higher than that for a sand or a loam with the same pH value.

Soil Buffering and Lime Requirements

Soils in a given region with comparable chemical properties exhibit a fairly common pH–base saturation relationship (see Figure 9.6). Knowing the pH of an unlimed soil, its cation exchange capacity, and the desired pH for the crops to be grown, it is possible to use this relationship to ascertain the approximate amount of liming material needed. Box 9.2 illustrates how the lime needs can be calculated.

Influence of Chemical Composition

The chemical composition of liming materials determines their long-term effects on soil pH, how much of each material is needed to achieve these effects, and how much calcium and magnesium are being supplied. To properly inform users of the chemical contents of liming materials, *guarantees* as to their chemical composition are usually required (see Box 9.3).

Limestone Fineness and Reactivity

The finer a liming material, the more rapidly it will react with the soil. The oxide and hydroxide of lime usually appear on the market as powders, so their fineness is always satisfactory. However, if limestones are not finely ground, they react very slowly with

---

BOX 9.2 CALCULATING LIME NEEDS

Assume that you must provide a recommendation as to how much limestone should be applied for the production of a high-pH-requiring crop—asparagus—to be grown in a loam soil that is currently quite acid (pH = 5.0). The cation exchange capacity (CEC) is about 10 cmol$_c$/kg, and you want to raise the pH to about 7.0. Assuming the pH–base saturation relationship is about the same as that shown in Figure 9.6, the soil is now at 40% base saturation and will need to be brought to about 80% base saturation to reach the goal of pH = 7.0. How much of a dolomitic limestone with a CaCO$_3$ equivalent of 90 will you need to apply to bring about the desired pH changes?

First, we need to know the cmol$_c$ of Ca/kg soil needed to bring about the change. Since the CEC is 10 cmol$_c$/kg and we need a 40% change in base saturation (from 40 to 90%) the Ca^{2+} required is

$$10 \text{ cmol}_c/\text{kg} \times 40/100 = 4 \text{ cmol}_c \text{ Ca/kg soil}$$

Second, since each Ca^{2+} has two charges, 4 cmol$_c$ can be expressed in grams of Ca^{2+} by multiplying by the molecular mass of Ca (40) and then dividing by 2, the Ca^{2+} charge, and by 100 since we are dealing with a *centimol$_c$*.

$$\frac{4 \text{ cmol}_c}{\text{kg soil}} \times \frac{40 \text{ gCa}}{\text{mol Ca}} \times \frac{1 \text{ mol Ca}}{2 \text{ mol}_c} \times \frac{1 \text{ mol}_c}{100 \text{ cmol}_c} = \frac{0.8 \text{ gCa}}{\text{kg soil}}$$

Third, we calculate the amount of CaCO$_3$ needed to provide the 0.8 gCa^{2+} by multiplying by the ratio of the molecular masses of CaCO$_3$ and Ca.

$$0.8 \text{ g/kg soil} \times \frac{100 \text{ (CaCO}_3)}{40 \text{ (Ca)}} = 2 \text{ gCaCO}_3/\text{kg soil}$$

Fourth, to express the 2 gCaCO$_3$/kg soil in terms of the mass needed to change the pH of 1 ha 15 cm deep, we must multiply by 2 × 10^6 kg/ha (see footnote 9, page 138).

$$2 \text{ gCaCO}_3/\text{kg} \times 2 \times 10^6 \text{ kg/ha} = 4 \times 10^6 \text{ g or 4 MgCaCO}_3/\text{ha}$$

Fifth, since the CaCO$_3$ equivalency of our limestone is only 90, some 100 kg would be required to match 90 kg of pure CaCO$_3$. Consequently, the amount of limestone needed would be adjusted upward by a factor of 100/90.

$$4 \times \frac{100}{90} = 4.4 \text{ Mg/ha or about 2 tons/acre.}$$

BOX 9.3 CHEMICAL GUARANTEES AND EQUIVALENCES

Laws require that the chemical composition of liming materials be guaranteed. The means of expressing this composition varies from one governing body to another, but will usually include guarantees of the contents as expressed by one or more of the following:

1. As elemental Ca and Mg
2. As oxides of these cations (CaO, MgO)
3. As CaO equivalent (neutralizing ability of all compounds expressed as CaO)
4. As total carbonates (CaCO₃ plus MgCO₃)
5. As CaCO₃ equivalent (neutralizing ability of all compounds expressed in terms of CaCO₃)

There are advantages and disadvantages to each of these means of expressing the liming effectiveness. However, it is relatively easy to convert one means of expression to another using the concept of *chemical equivalency*. In other words, one atom (or molecule) of Ca, Mg, CaO, MgO, CaCO₃, or MgCO₃ will neutralize the same amount of acidity as another. Consequently, to make comparisons among different liming materials we need merely multiply by the ratios of their molecular masses. For example, to calculate the CaCO₃ equivalent of a pure burned lime (CaO) we need simply multiply by the molecular ratio of CaCO₃ to CaO:

$$\frac{CaCO_3}{CaO} = \frac{100}{56} = 1.786$$

Thus, 1 Mg (1000 kg) of this pure burned lime will neutralize as much acidity as 1.786 Mg of a pure limestone. Some adjustment will need to be made, of course, for the percentage purity of the respective liming materials.

Using chemical equivalencies it is possible to compare a different means of expressing the liming abilities of four materials: a burned lime (oxide), a hydrated lime (hydroxide), and two limestones (one dolomitic, the other calcitic). Each is considered to be 95% pure.

		Means of expressing composition						
Material	Actual chemicals, %	Element, %		Conventional oxide, %		CaO equivalent, %	Total carbonates, %	CaCO₃[a] equiv., %
		Ca	Mg	CaO	MgO			
Burned lime	77 CaO	55.0	—	77.0	—	102.2	—	182.5
	18 MgO	—	10.9	—	18.0			
Hydrated lime	75 Ca(OH)₂	40.5	—	58.6	—	76.0	—	135.8
	20 Mg(OH)₂	—	8.3	—	13.8			
Calcitic limestone	95 CaCO₃	38.0	—	53.2	—	53.2	95	95
Dolomitic limestone	35 CaCO₃	(14.0)	—	(19.6)	—	(19.6)	35	(35)
	60 CaMg(CO₃)₂	(13.0)	(7.9)	(18.2)	(13.1)	(36.4)	60	(65)
		27.0	7.9	37.8	13.1	56.0	95	100

[a] Sometimes referred to as total neutralizing power.

the soil, and give little crop response. For these reasons a **fineness guarantee** is commonly required for limestones. Materials with at least 50% of the particles small enough to pass through a 60-mesh screen (smaller than 0.25 mm in diameter) are quite satisfactory for most liming purposes.

9.11 PRACTICAL CONSIDERATIONS

The principles we have discussed lead to a number of practical considerations, including the following:

1. Since limestone is bulky and is commonly transported long distances (Figure 9.18) the rate of application should be no higher than to meet soil and crop needs, commonly 7 to 9 Mg/ha (3 to 4 tons/acre) or less.

FIGURE 9.18 Bulk application of limestone by specially equipped trucks is the most widespread method of applying liming materials. Because of the weight of such machinery, much limestone is applied to sod land and plowed under. In many cases this same method is used to spread commercial fertilizers.

2. In choosing limestone, give attention to calcium and magnesium as plant nutrients, the gains and losses of which help determine soil quality (Figure 9.19).

3. Special situations call for innovative methods of applying liming materials:

   • Aerial application of limestone on forested areas with acid sandy soils may be needed to supply calcium for the trees.

   • Since no-till farming acidifies the upper few centimeters of soil, limestone should be deeply incorporated before commencing the no-till system, and later applications with minimum tillage could combat the surface soil acidity.

   • Add limestone if needed at seeding time for lawns and golf greens, and properly time future liming applications with annual tillage operations that leave openings into the soil into which the lime could move.

   • Wherever practical, choose the plant or variety that suits the soil rather than try to change the soil to suit the plant. Plant breeders are creating plant varieties that are more tolerant of acidity.

   • Grow selected green manure crops (crops grown specifically to provide organic residues) that have the ability to bind toxic aluminum ions in unavailable forms as they decompose. This lowers lime needs and is most helpful for farmers in low-income countries where limestone availability and costs constrain attempts to adjust the soil pH.

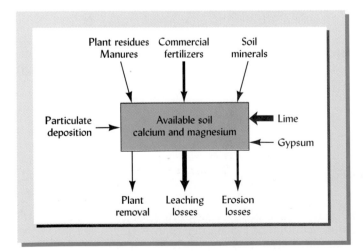

FIGURE 9.19 Important ways by which *available* calcium and magnesium are supplied to and removed from soils. The major losses are through leaching and erosion. These losses may be largely replaced by lime and fertilizer applications. Fertilizer additions are much higher than is generally realized, for some phosphate fertilizers contain large quantities of calcium. Calcium-rich dust particles in arid regions and in industrially polluted areas play an important role in uncultivated ecosystems where soil amendments are not used.

4. Last, avoid the dangers of overliming, which are most prevalent on sandy soils with low buffering capacities.

9.12 AMELIORATING ACIDITY IN SUBSOILS[6]

The subsoils of limed areas commonly remain acid since the Ca^{2+} and Mg^{2+} ions provided by limestone do not move readily down the profile. Root growth and, in turn, crop yields are often constrained by the excessively acid subsoils. Aluminum toxicity and/or calcium deficiency have been found in some of these acid subsoils.

Amelioration by Gypsum

Researchers have found that although gypsum ($CaSO_4 \cdot 2H_2O$) does not change the soil pH, it can ameliorate aluminum toxicity in subsoils and thereby increase crop yields. Applied on the soil surface, the gypsum slowly dissolves and is leached into the subsoil. After about one year increased yields have been obtained with common crops such as corn, soybeans, alfalfa, beans, wheat, apples, and cotton, and the penetration of crop roots into the subsoil is increased (Figure 9.20).

9.13 ALKALINE AND SALT-AFFECTED SOILS

Soils of arid and semiarid areas are mostly alkaline because water from rain or snow is insufficient to leach the base-forming cations (Ca^{2+}, Mg^{2+}, K^+, Na^+, etc.) that are slowly released as the rocks and minerals weather. In some areas the leaching is insufficient to remove even soluble salts, such as NaCl, $CaCl_2$, $MgCl_2$, and KCl, leading to *saline*, as well as alkaline, conditions. Salt buildup is of special concern in some irrigated soils of arid regions.

Salt-affected soils are widely distributed throughout the world, the largest areas being found in Australia, Africa, Latin America, and the Near and Middle East. They are most often found in areas with precipitation-to-evaporation ratios of 0.75 or less, and in low, flat areas with high water tables that may be subject to seepage from higher elevations. Figure 9.21 illustrates the distribution of these soils in the contiguous United States.

---

[6] For a review of this subject, see Sumner (1993).

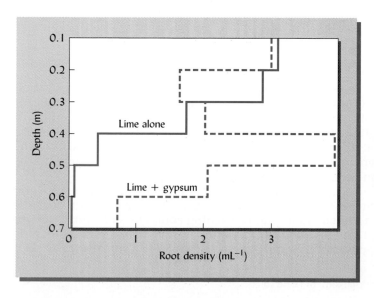

FIGURE 9.20 Surface-applied gypsum moved down the profile to help ameliorate the extreme acidity of the subsoil of this Ultisol in South Africa and helped to permit an increase in the density of corn roots. This extension of the roots into the subsoil can increase the availability of both water and nutrients. [Redrawn from Farina and Channon (1988); used with permission of the Soil Science Society of America]

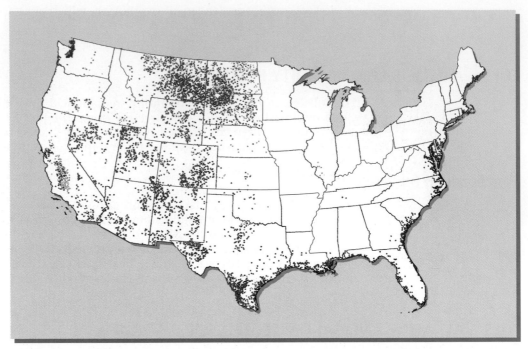

FIGURE 9.21 Distribution of salt-affected soils in the continental United States in 1992. Small areas are found near the ocean shore lines, but note the much larger areas located in vast regions of the West. Each dot represents about 4050 ha. [From the USDA Natural Resources Conservation Service National Resources Inventory, USDA (1996)]

9.14 DEVELOPMENT OF SALT-AFFECTED SOILS[7]

Natural Salt Accumulation

Salts accumulate naturally in some surface soils of arid and semiarid regions because there is insufficient rainfall to flush them from the upper soil layers. In the United States, about one-third of the soils in these regions suffer from some degree of salinity. The salts are primarily chlorides and sulfates of calcium, magnesium, sodium, and potassium.

Other localized but important sources are fossil deposits of salts laid down during geological time in the bottom of now-extinct lakes or oceans or in underground saline water pools. These fossil salts can be dissolved in underground waters that move horizontally over underlying impervious geological layers and ultimately rise to the surface of the soil in the low-lying parts of the landscape, often forming **saline seeps** (Figure 9.22). The water then evaporates, leaving salts in place at or near the soil's surface and creating a saline soil.

Irrigation-Induced Salinity and Alkalinity

Irrigation-induced salinity has become of increasing concern in recent years. Irrigation waters, whether from upstream watersheds or underground pumping, carry significant quantities of soluble salts. In areas where the irrigated land is well drained, sufficient water can percolate through the soil to leach the excess soluble salts from the upper soil layers. If the soil is not well drained, however, downward movement of the salts to the groundwater is impaired, and the salts are left in the soil and are later brought up to the surface as the irrigation water evaporates.

During the past three decades, low-income countries in the dry regions of the world have greatly expanded their areas of irrigated lands. As a result, the proportion of arable land that is irrigated has increased dramatically, reaching about 45% in China, 25% in India, 72% in Pakistan, and 28% in Indonesia. Unfortunately, in many project areas the need for good soil drainage was overlooked and the process of **salinization** has been

---

[7] See Bressler et. al. (1982) for a discussion of these soils.

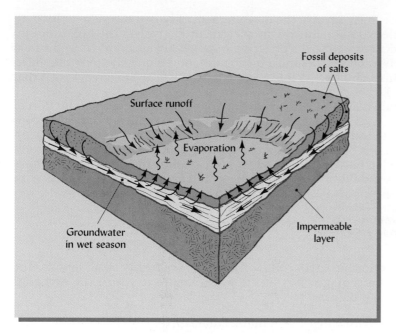

FIGURE 9.22 Field cut showing conditions that encourage salt buildup in soils. Mineral-containing waters from fossil deposits of salt or from soil weathering in upland areas move down the slope toward lower basins, either on the surface during torrential storms or as groundwater on top of impervious geological layers. When the water in the low-lying areas evaporates, it carries the salts up to or near the surface, where the salts accumulate.

accelerated. As a consequence, salts have accumulated to levels that are already adversely affecting crop production.

These events remind the world of the flat wastelands of southeastern Iraq that have been barren since the 12th century. In biblical times, the soils of these areas were fertile and productive. The Euphrates and Tigris rivers provided irrigation water that stimulated such high crop productivity that the overall region was called the Fertile Crescent. Unfortunately, in broad areas the soils were not well drained and the process of salinization proceeded. The salts accumulated to such a high level that crop production declined and the area had to be abandoned. Today, this same process is being repeated at different locations around the world, and some irrigated farming areas are having to be abandoned. Truly, the world is justified in giving serious attention to salt-affected soils.

9.15 MEASURING SALINITY AND ALKALINITY

Plants are detrimentally affected by excess salts in some soils and by high levels of exchangeable sodium in others (sodic soils). Techniques have been developed to measure three primary properties that, along with pH, can be used to characterize salt-affected and sodic soils: (1) soil **salinity**, (2) **exchangeable sodium percentage** (ESP), and (3) the **sodium adsorption ratio** (SAR). Each will be discussed briefly.

Salinity[8]

Pure water is a poor conductor of electricity, but conductivity increases as more and more salt is dissolved in the water. Thus, the **electrical conductivity** (EC) of the soil solution gives us an indirect measurement of the salt content. The EC is measured by both laboratory and field methods and is expressed in terms of decisiemens per meter (dS/m).[9]

The **saturation paste extract** method is the most commonly used laboratory procedure. The soil sample is saturated with distilled water and mixed to a paste consistency. After standing overnight to dissolve the salts, the electrical conductivity of the water extracted from the paste is measured (EC_e). A variant of this method involves the EC of the solution extracted from a 1:2 soil–water mixture after 0.5 hours of shaking (EC_w). The latter method takes less time but is often is not as well related to the salinity of soil solution as is the saturation paste extract method.

---

[8] For an informative discussion of these methods, see Rhoades (1993).

[9] Formerly expressed as millimhos per centimeter (mmho/cm). Since 1 S = 1 mho, 1 dS/m = 1 mmho/cm.

Field methods in use today involve the measurement of bulk soil conductivity that, in turn, is directly related to soil salinity. One such method involves the insertion into the soil of four electrode sensors and the direct measurement of apparent EC in the field (EC_a). The sensors can be mounted behind a tractor and moved quickly from one site in the field to another. This technique is rapid, simple, and practical, and gives values that can be correlated with EC_e.

A second field method is even more rapid. It employs **electromagnetic induction** of electrical current in the body of the soil, the level of which is related to electrical conductivity and, in turn, to soil salinity. The instrument can measure ground EC (designated E^*_a) without the use of probes, and is proving to be useful in surveying saline soils.

Since electrical conductivity is influenced by both salinity and soil water content, the field measurements should be made when the soil is near the field capacity. With irrigated soils, measurements are best made soon after the soil is irrigated to assure reasonably high soil moisture levels.

Sodium Status

Two expressions are used to characterize the sodium status of highly alkaline soils. The **exchangeable sodium percentage** (ESP) identifies the degree to which the exchange complex is saturated with sodium.

$$ESP = \frac{\text{Exchangeable sodium, cmol}_c/\text{kg}}{\text{Cation exchange capacity, cmol}_c/\text{kg}} \times 100$$

ESP levels of 15 are associated with pH values of 8.5 and above. Higher levels may bring the pH to above 10.

The **sodium adsorption ratio** (SAR) is a second more easily measured property that is becoming even more widely used than ESP. The SAR gives information on the comparative concentrations of Na^+, Ca^{2+}, and Mg^{2+} in soil solutions. It is calculated as follows:

$$SAR = \frac{[Na^+]}{\sqrt{\frac{1}{2}([Ca^{2+}] + [Mg^{2+}])}}$$

where $[Na^+]$, $[Ca^{2+}]$, and $[Mg^{2+}]$ are the concentrations (in $mmol_c/L$) of the sodium, calcium, and magnesium ions in the soil solution. The SAR of a soil extract takes into consideration that the adverse effect of sodium is moderated by the presence of calcium and magnesium ions. The SAR also is used to characterize the irrigation water added to these soils.

9.16 CLASSES OF SALT-AFFECTED SOILS

Using EC, ESP, SAR characteristics, and soil pH, salt-affected soils are classified as **saline**, **saline-sodic** and **sodic** (Figure 9.23). These classes will be considered in order.

Saline Soils

If the concentration of neutral soluble salts (mainly chlorides and sulfates of sodium, calcium, magnesium, and potassium) in the soil solution is sufficiently high to give an electrical conductivity EC_e reading of 4 ds/m,[10] the soil is classed as **saline**. The growth of most plants on these soils is adversely affected by the salt levels. Salts commonly accumulate on the surface of these soils, creating a white crust which accounts for the name *white alkali* that is sometimes used to designate them (see Figure 9.24).

The exchange complex of saline soils is dominated by calcium and magnesium. As a consequence, the exchangeable sodium percentage (ESP) is less than about 15, the sodium adsorption ratio (SAR) is 13 or less, and the pH usually is less than 8.5.

Plant growth on saline soils is not generally constrained by poor soil physical conditions. The soluble salts help prevent dispersion of the soil colloids that otherwise would be encouraged by high sodium levels (see Figure 9.25).

---

[10] Some fruit and vegetable crops are adversely affected when the EC is lower than 4 dS/m, leading some scientists to recommend that 2 dS/m be the level above which a soil could be classed as saline.

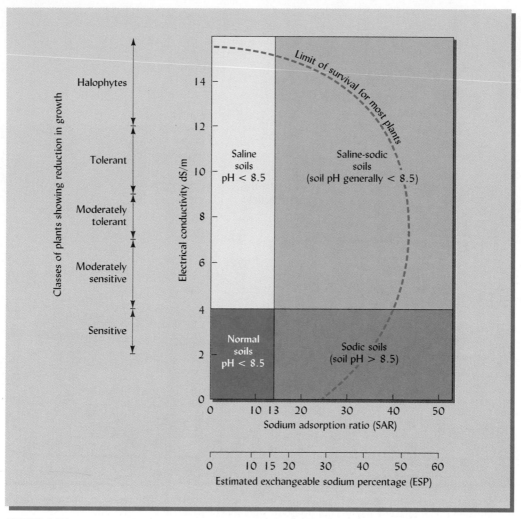

FIGURE 9.23 Diagram illustrating the classification of normal, saline, saline-sodic, and sodic soils in relation to soil pH, electrical conductivity, sodium adsorption ratio (SAR), and exchangeable sodium percentage (ESP). Also shown are the ranges for different degrees of sensitivity of plants to salinity.

FIGURE 9.24 (Left) White alkali spot in a field of alfalfa under irrigation. Because of upward capillarity and evaporation, salts have been brought to the surface where they have accumulated. (Right) White crust on a saline soil from Colorado. The white salts are in contrast with the darker-colored soil underneath. (Scale is shown in inches and centimeters.) [Lower photo courtesy USDA Agricultural Research Service]

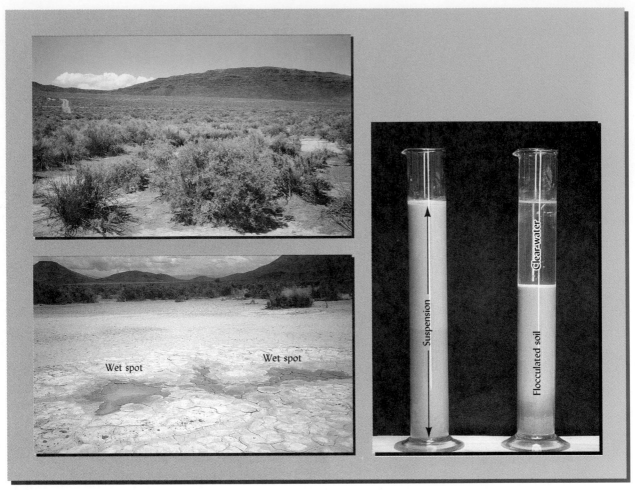

FIGURE 9.25 Illustration of the dispersion of colloids in a sodic soil from southern Colorado. The sodic soil (lower left) is bare of vegetation compared to the adjacent alkaline but nonsodic soil (upper left). Because of the complete dispersion of the soil colloids, water moves through the sodic soil very slowly, as the wet spot in the center of the lower left photograph suggests. To verify the dispersion of the clay, samples of the sodic soil were placed in cylinders and shaken thoroughly with water. The clay was highly dispersed, no observable settling occurring over a period of three weeks (left cylinder). To stimulate the flocculation (joining together) of the clay particles, a teaspoon of table salt (NaCl) was added to the cylinder on the right, and the mixture was again thoroughly shaken with water. Flocculation began to occur and within 24 hours was quite noticeable (right cylinder). The salt stimulated the flocculation, which in the soil could be the first step toward aggregation, improved drainage, and good aeration. This demonstration illustrates the deplorable physical condition of sodic soils. It also shows how the salts in saline-sodic soils can prevent the undesirable dispersion of the soil colloids that is responsible for much of the ill effects of sodic soils.

Saline-Sodic Soils

Saline-sodic soils have characteristics intermediate between those of saline and sodic soils (see Figure 9.23). Like saline soils, they contain appreciable levels of neutral soluble salts, as shown by EC_e levels of more than 4 dS/m. But they have higher ESP levels (greater than 15) and higher SAR values (at least 13). Crop growth can be adversely affected by both excess salts and sodium levels. Because of the moderating effects of the neutral salts the physical and chemical conditions of saline-sodic soils are similar to those of saline soils.

Unfortunately, this situation is subject to rather rapid change if the soluble salts are leached from the soil, especially if the leaching waters are high in Na^+ ions—that is, if the SAR of the water is high. In such a case, the exchangeable sodium level will increase, as will the pH (to above 8.5). Since Na^+ ions are not attracted very closely to the soil colloids, the effective electronegativity of the colloidal particles causes them to repel each other or to disperse (Figure 9.25), thereby breaking up the soil aggregates. The dispersed colloids clog the soil pores as they move down the profile. Water infiltration is thus greatly reduced (Figure 9.26) and the puddled condition, so characteristic of sodic soils, pertains.

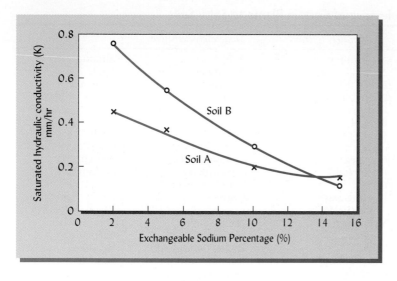

FIGURE 9.26 Effect of increasing exchangeable sodium percentage (ESP) on the saturated hydraulic conductivity of two soils in Italy. Note the steady decline in conductivity with increased ESP and the very low conductivity at ESP = 15%. [Redrawn from Crescimanno, et al. (1995)]

Sodic Soils

The preceding brief description of how sodic soils form suggests that they are the most troublesome of the salt-affected soils. While their levels of neutral soluble salts are low (EC_e less than 4.0 dS/m), their ESP values are above 15 and their SARs are above 13, suggesting a comparatively high level of sodium on the exchange complex. The pH values of sodic soils exceed 8.5, rising to 10 or higher in some soils.

Few plants can tolerate these conditions. Plant growth on these soils is constrained by toxicities of Na^+, OH^-, and HCO_3^- ions, as well as by their very poor soil physical conditions and slow permeability to water.

Because of the extreme alkalinity resulting from the high sodium content, the surface of sodic soils usually is discolored by the dispersed humus that can be carried upward by the capillary water and deposited when it evaporates. Hence, the name **black alkali** has been used to describe these soils. Sometimes located in small areas called **slick spots**, sodic soils may be surrounded by soils that are relatively productive.

9.17 SELECTIVE TOLERANCE OF HIGHER PLANTS TO SALINE AND SODIC SOILS

Plant Sensitivity

Numerous tests have made it possible to classify many domestic plants into four general groups based on their salinity tolerance. Figure 9.27 shows the relative productivity of these general groups influenced by soil salinity as measured by electrical conductivity

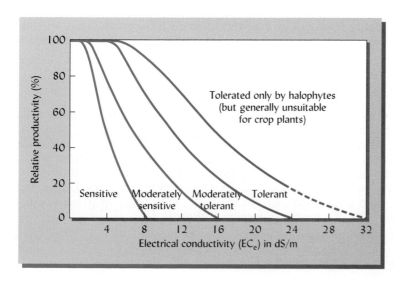

FIGURE 9.27 Relative productivity of four groups of plants classified by their sensitivity to salinity as measured by electrical conductivity. The plant groupings are shown in Table 9.1. [Generalized from data of Carter (1981)]

TABLE 9.1   Relative Tolerance of Certain Plants to Salty Soils

Tolerant	Moderately tolerant	Moderately sensitive	Sensitive
Barley (grain)	Ash (white)	Alfalfa	Almond
Bermuda grass	Aspen	Arborvitae	Apple
Black cherry	Barley (forage)	Boxwood	Apricot
Bougainvillea	Birch (black)	Broad bean	Azalea
Cedar (red)	Beet (garden)	Cauliflower	Beech
Cotton	Broccoli	Cabbage	Bean
Date	Brome grass	Celery	Birch
Elm	Catlaw acacia	Clover (alsike,	Blackberry
Locust	Cow pea	ladino, red,	Boysenberry
Natal plum	Fescue (tall)	strawberry,	Burford holly
Nutall alkali grass	Fig	and berseem)	Carrot
Oak (red and white)	Harding grass	Corn	Celery
Olive	Honeysuckle	Cucumber	Dogwood
Prostrate kochia	Hydrangea	Dallas grass	Elm (American)
Rescue grass	Juniper	Grape	Grapefruit
Rosemary	Kale	Hickory (shagbark)	Hemlock
Rugosa	Locust (black)	Juniper	Hibiscus
Salt grass	Mandarin	Lettuce	Larch
Tamarix	Orchard grass	Maple (red)	Lemon
Wheat grass (crested)	Oats	Pea	Linden
Wheat grass (fairway)	Pomegranate	Peanut	Maple (sugar)
Wheat grass (tall)	Privet	Radish	Onion
Wild rye (altai)	Rye (hay)	Rice (paddy)	Orange
Wild rye (Russian)	Ryegrass (perennial)	Squash	Peach
Willow	Safflower	Sugar cane	Pear
	Sorghum	Sweet clover	Pine (red and white)
	Soybean	Sweet potato	Pineapple
	Squash (zucchini)	Timothy	Plum (prune)
	Sudan grass	Turnip	Potato
	Trefoil (birdsfoot)	Vetch	Raspberry
	Wheat	Viburnum	Rose
	Wheat grass (western)		Star jasmine
			Strawberry
			Tomato

Selected from Carter (1981) and Maas (1984). See Figure 9.27 for tolerance ranges.

EC_e. Table 9.1 lists many plants according to this classification. Note that trees, shrubs, fruits, vegetables, and field crops are included in the different categories.

To this list of domestic species should be added two other potential sources of plants to grow on salty soils: (1) wild *halophytes* (salt-loving plants), and (2) salt-tolerant cultivars developed by plant breeders. A number of wild halophytes have been found that are quite tolerant to salts and that possess qualities that could make them useful for human and/or animal consumption.[11]

Plant breeders have been able to develop new strains with salt tolerance greater than that possessed by conventional varieties. Plant selection and improvement should be one of the goals of the future to maintain or even increase food production on salt-affected soils. However, improved plant tolerance must not be viewed as a substitute for proper salinity control.

Salt Problems Not Related to Arid Climates

In cities where deicing salts are used in abundance during winter months, these salts may impact roadside soils and plants. Repeated application of the deicing salts can result in salinity levels sufficiently high as to adversely affect plants growing alongside a treated highway or sidewalk. Figure 9.28 illustrates tree damage from such treatments. To avoid the specific chemical and physical problems associated with sodium salts, many municipalities have switched from NaCl to KCl for deicing purposes. Unlike soil salinity in arid regions, such salt contamination is usually temporary as the abundant rainfall will leach out the salts in a matter of weeks or months.

---

[11] For a review of such halophytes, see NAS (1990).

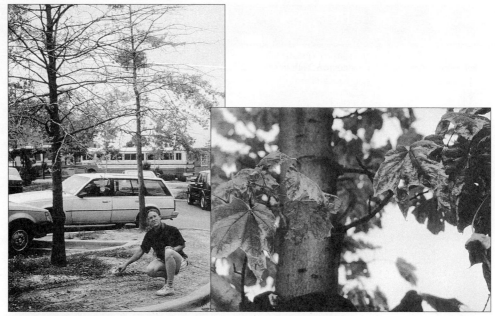

FIGURE 9.28 During the winter months in some temperate region areas, sodium chloride and other salts are used to melt the snow and ice from the streets and sidewalks. Heavy applications of these chemicals can result in salinity damage to plants similar to that experienced on saline soils of arid regions. (Left) Localized area of salt concentration resulting from the piling of salt-treated snow alongside a street. (Right) Leaf damage resulting from the excess salt uptake by the plants. (Photos courtesy of R. Weil)

Salinity can also be a problem for potted plants, particularly perennials that remain in the same pot for long periods of time. Salts in the water as well as those applied in fertilizers can build up if care is not taken to flush them out occasionally.

9.18   MANAGEMENT OF SALINE AND SODIC SOILS

Water Quality Considerations

The chemical quality of water added to salt-affected soils is a prime management tool. If the water is too low in salts, such as in rainwater, it can hasten the change from a saline-sodic soil to a sodic soil. The rainwater not only leaches soluble salts from the upper few centimeters of soil, but the impact of the raindrops encourages dispersion of the soil colloids, an initial step in the process of sodic soil development.

If the salt and SAR levels of irrigation water are too high, the process of forming sodic soils will accelerate. Also, the presence of bicarbonates in the irrigation water can reduce concentrations of Ca^{2+} and Mg^{2+} in the soil solution by precipitating these ions as insoluble carbonates. Such chemical characteristics of the irrigation water can increase the SAR of the soil solution and move the soil toward the sodic class.

Wastewater Management

The quality and disposition of *waste irrigation waters* must also be carefully monitored because of the potential harm to downstream users and habitats. Drainage water from one irrigation system that may be high in salt concentration or in the concentration of one or more specific elements may reenter a stream and be used again at lower elevations. At some locations in the western United States, for example, toxic levels of trace elements such as selenium,[12] molybdenum, and boron have accumulated in downstream wetlands or evaporative ponds. Box 9.4 illustrates how such concentrations can occur. Plants growing on these affected areas can accumulate levels of these elements that are unsafe for livestock and/or wildlife. Indeed, selenium toxicity is known to be damaging to waterfowl in parts of California.

---

[12] For a recent review of selenium in irrigation water of the western United States, see Nolan and Clark (1997).

BOX 9.4 SELENIUM IN IRRIGATED SOILS

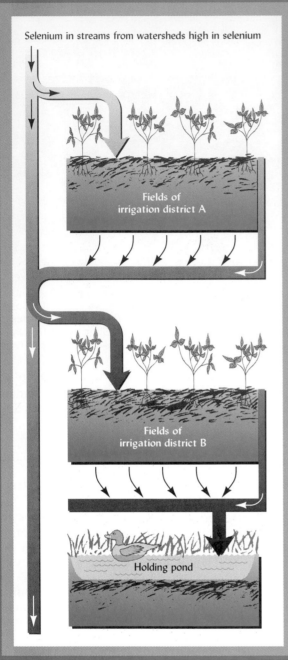

Selenium in streams from watersheds high in selenium

Fields of irrigation district A

Fields of irrigation district B

Holding pond

The use of selenium-laden irrigation waters can result in the buildup of selenium to toxic levels in soils and in irrigation wastewaters coming from these soils. Selenium accumulates in plants growing on these soils and in nearby wetland areas where irrigation wastewaters are disposed.

In some areas of the western United States are found certain marine sediments and shales of the Cretaceous geological period that are high in selenium. Under alkaline and well-drained conditions, the selenate (SeO_4^{2-}) form of Se prevails. Much like sulfates, selenates are reasonably soluble in water and are readily available for plant uptake. They move from watersheds that contain high selenium into streams and rivers from which water is diverted for irrigation.

The drawing illustrates how reuse of high-selenium irrigation waters can cause serious environmental problems. Streams from high-selenium watersheds provide water for irrigation district A. The selenium level in the water (shown by the darkness of the arrow) is high but not sufficiently so as to cause environmental problems.

As evapotranspiration of the irrigation water takes place, selenium accumulates in the soil and in the waste and drainage water coming from the soil. This drainage water moves (or is pumped) back into a stream that later supplies water for irrigation district B downstream. The process is repeated, Se in the soil is increased, and the levels of Se in the drainage water is increased (darker color), being sufficiently high as to make the water unusable for further irrigation.

The water is then diverted or pumped into a nearby wetland area or a shallow holding pond. In either case, much of the water evaporates, leaving the selenium in the water and soil of the holding area. Plants living in the wetland or pond area absorb the selenium, accumulating it to levels that are toxic to farm animals and wildlife. Deformities and deaths of migratory birds have been caused by such high levels of selenium. Similarly, livestock may suffer toxic effects by eating high-selenium forages.

Steps are being taken to reduce or eliminate the release of high-selenium waters into wetlands. Researchers are also seeking plants that are very high selenium accumulators that can remove the selenium from the holding area[a] and can then be destroyed. Other research has shown that under waterlogged, reducing conditions some microorganisms can stimulate the formation of volatile organic selenium compounds, which can be slowly released into the atmosphere. The environmental effects of selenium remind us of the need to manage soil and water with attention to the impacts on all creatures.

[a] For example, see Banuelos, et al. (1997).

The soluble salt level of different water sources varies greatly. Data for a number of water sources, varying from high-quality irrigation water to that in the Pacific Ocean, is shown in Table 9.2. Note that the Colorado River has sufficiently high salt content to alert users to the need for wise irrigation management if soil quality is to be maintained.

Management Objectives

The primary objective in managing salt-affected soils is to lessen the major constraints from excess soluble salts and exchangeable sodium. The removal of these constraints causes several chemical and physical changes that bring the soils back to a more productive state. Consequently, the overall processes are referred to as the **reclamation** of these soils, which will now be considered.

TABLE 9.2 The Comparative Salinity of Six Different Water Sources as Measured by Electrical Conductivity and Dissolved Solids Content

Salinity measurement	Irrigation water quality		Colorado River	Alamo River	Negev (Middle East) groundwater	Pacific Ocean
	Good	Marginal				
Electrical conductivity, dS/m	0–1	1–3	1.3	4.0	4.0–7.0	46
Dissolved solids, ppm	0–500	500–1,500	850	3,000	3,000–4,500	35,000

Adapted from a number of sources by the U.S. National Academy of Sciences (NAS; 1990).

9.19 RECLAMATION OF SALINE SOILS

The removal of excess salts from saline soils requires high-quality irrigation water and good soil drainage. If the natural soil drainage does not permit leaching of the salts, an artificial drainage network must be installed.

The amount of water needed to remove the excess salts from saline soils, called the **leaching requirement LR,** is determined by the sensitivity of the crop to be grown, the quality of the irrigation water, and the soil. An approximation of the LR is given by the ratio of the salinity of the irrigation water (expressed as its EC_w) to the maximum permissible salinity of the soil solution for the crop to be grown (expressed as EC_{dw}, the EC of the drainage water).

$$LR = \frac{EC_w}{EC_{dw}}$$

The LR is water added in excess of the moisture needed to thoroughly wet the soil, and to meet evapotranspiration needs.

Note that if the EC_w is high and the EC_{dw} is low, a very large leaching requirement LR would result. An example of how the LR values are calculated is given in Box 9.5.

BOX 9.5   LEACHING REQUIREMENT LR

A farmer should know how much leaching water is required to prevent the buildup of salts in a soil or, if the salts are already high, to reduce their levels in the soil. The concept of *leaching requirement LR* has been developed to help farmers make this assessment.

The LR is the irrigation water needed (in excess of that required to saturate the soil) to sufficiently leach the soil so as to assure a proper salt balance for the crop being grown. It is approximated by the ratio of the electrical conductivities (ECs) of the incoming irrigation water EC_w and of the outgoing drainage water EC_{dw} that has an acceptable EC level for the crop being grown.

$$LR = \frac{EC_w}{EC_{dw}}$$

As an example, consider the situation where the EC of the irrigation water EC_w is 2.5 dS/m and that of the acceptable draining soil solution is 5.5 dS/m. Then,

$$LR = \frac{2.5 \text{ dS/m}}{5.5 \text{ dS/m}} = 0.45$$

If this ratio (0.45) is multiplied by the amount of water needed to completely saturate the soil—perhaps 8 cm of water—the water to be leached can be calculated as follows:

$$8 \text{ cm} \times 0.45 = 3.6 \text{ cm water}$$

This is the minimum amount of water that must be leached through a water-saturated soil to maintain proper salt balance. In some cases, additional leaching may be needed to reduce the excess concentration of specific elements such as boron.

The modern means of measuring bulk soil conductivity EC_a using the four-electrode probe or remote-sensing electromagnetic induction devices (see Section 9.15) can be used to readily monitor soil salinity changes resulting from leaching practices.

Attention must be paid to the disposal of the drainage water once the soil leaching has taken place (see Box 9.4). If this water cannot be returned to a river or stream without overburdening the stream with salts or other contaminants, then other means must be sought to dispose of the water.

Tillage and planting practices can influence crop production on saline soils. Any tillage or surface residue management practice (such as conservation tillage) that reduces evaporation from the soil surface should reduce the upward transport of soluble salts. Likewise, specific irrigation schemes such as those utilizing sprinkling and trickle systems (see Section 6.11), as well as planting techniques for row crops, can reduce the salt concentration immediately around the young plant roots (Figure 9.29).

FIGURE 9.29 Effect of irrigation on salt movement and plant growth in saline soil. (*a*) With irrigation water applied to furrows on both sides of the row, salts move to the center and damage plant root systems. (*b*) Placing plants on edges of the bed rather than the center helps avoid the damage seen in (*a*). (*c*) Application of water to every other furrow and placement of plants on the side of the bed nearest the water also helps plants avoid the highest salt concentrations. Sprinkle irrigation (*d*) or uniform flood irrigation moves salts downward and the problem is alleviated. (*e*) Drip or trickle irrigation results in salt removal or accumulation, depending on the placement of the trickle tube. (Courtesy of Wesley M. Jarrell)

9.20 RECLAMATION OF SALINE-SODIC AND SODIC SOILS

For both saline-sodic and sodic soils, attention must first be given to reducing the level of exchangeable Na^+ ions. This is accomplished by adding materials that will replace Na^+ ions with either Ca^{2+} or H^+ ions, and then by leaching the displaced Na^+ ions from the soil.

Gypsum

Providing Ca^{2+} ions in the form of gypsum ($CaSO_4 \cdot 2H_2O$) is the most practical way to bring about this exchange. When gypsum is added, reactions such as the following take place:

$$2NaHCO_3 + CaSO_4 \longrightarrow CaCO_3 + Na_2SO_4 + CO_2\uparrow + H_2O$$

$$Na_2CO_3 + CaSO_4 \rightleftharpoons CaCO_3 + \underset{\text{(leachable)}}{Na_2SO_4}$$

$$\begin{matrix} Na^+ \\ Na^+ \end{matrix} \boxed{Micelle} + CaSO_4 \rightleftharpoons Ca^{2+} \boxed{Micelle} + \underset{\text{(leachable)}}{Na_2SO_4}$$

Note that in each case the soluble salt Na_2SO_4 is formed, which can be easily leached from the soil as was done in the case of the saline soils.

Several tons of gypsum per hectare are usually necessary. In Box 9.6, calculations are made to approximate the amount of gypsum that is theoretically needed to remove an acceptable portion of the Na^+ ion from the exchange complex.

Sulfur and Sulfuric Acid

Elemental sulfur and sulfuric acid can be used to help reclaim sodic and saline-sodic soils. The sulfur, upon biological oxidation (see Sections 9.4 and 12.21), yields sulfuric acid, which not only changes the sodium bicarbonate to the less harmful sodium sulfate but also decreases the alkalinity. The reactions of sulfuric acid with the compounds containing sodium may be shown as follows:

$$2NaHCO_3 + H_2SO_4 \longrightarrow 2CO_2 + 2H_2O + Na_2SO_4$$

$$Na_2CO_3 + H_2SO_4 \longrightarrow CO_2 + H_2O + \underset{\text{(leachable)}}{Na_2SO_4}$$

$$\begin{matrix} Na^+ \\ Na^+ \end{matrix} \boxed{Micelle} + H_2SO_4 \rightleftharpoons \begin{matrix} H^+ \\ H^+ \end{matrix} \boxed{Micelle} + \underset{\text{(leachable)}}{Na_2SO_4}$$

Not only are the sodium carbonate and bicarbonate changed to sodium sulfate, a mild neutral salt, but the carbonate radical is removed from the system. When gypsum is used, however, a portion of the carbonate may remain as a calcium compound ($CaCO_3$).

In research trials, both sulfur and sulfuric acid have proven to be very effective in the reclamation of sodic soils, especially if large amounts of $CaCO_3$ are present. In practice, however, gypsum is much more widely used since it is less expensive, is widely available, and is more easily handled.

Physical Condition

The effects of gypsum and sulfur on the physical condition of sodic soils are perhaps more spectacular than are the chemical effects. When the exchangeable Na^+ ions are replaced by Ca^{2+} or H^+, soil aggregation and improved water infiltration results (Figure 9.30).

The reclamation effects of gypsum or sulfur are greatly accelerated by plants growing on the soil. Crops such as sugar beets, cotton, barley, sorghum, berseem clover, or rye can be grown initially. Their roots help provide channels through which gypsum or $H_2SO_4^-$ can move downward into the soil. Deep-rooted crops, such as alfalfa, are especially effective in improving the water conductivity of gypsum-treated sodic soils. Figure 9.30 illustrates the ameliorating effects of the combination of gypsum and deep-rooted crops.

Irrigation Timing

The timing of irrigation is extremely important on salt-affected soils, particularly during the spring planting season. Because young seedlings are especially sensitive to salts, irrigation often precedes or follows planting to move the salts downward and away from the seedling roots (see Figure 9.29). After the plants are well established, their salt tolerance is somewhat greater.

Ammonia Effects

The practice in recent years of adding nitrogen as anhydrous ammonia (NH$_3$) to irrigation water as it is applied to a field has created some soil problems. The NH$_3$ reacts with

FIGURE 9.30 The influence of gypsum and growing crops on the hydraulic conductivity (readiness of water movement) of the upper 20 cm of a saline-sodic soil (Natrustalf) in Pakistan. The use of gypsum to increase water conductivity in all plots was more effective when deep-rooted alfalfa and rotation of sesbenia-wheat were grown. [Drawn from selected data from Ilyas, et al. (1993); used with permission of the Soil Science Society of America]

the irrigation water to form NH_4OH, which, in turn, reacts with CO_2 and soluble calcium and magnesium compounds to form insoluble calcium and magnesium carbonates. This removal of the Ca^{2+} and Mg^{2+} ions from the irrigation water raises the sodium adsorption ratio (SAR) and the hazard of increased exchangeable sodium percentage (ESP). To counteract these difficulties, sulfuric acid is sometimes added to the irrigation water to reduce its pH as well as that of the soil.

9.21 CONCLUSION

No other single chemical soil characteristic is more important in determining the chemical environment of higher plants and microbes than the soil pH. Deficiencies and toxicities of mineral elements are found in soils with extremes of acidity or alkalinity. Few chemical or biological reactions involving soil components are not sensitive to soil pH. This sensitivity must be recognized in any soil-management system.

Soil pH is largely controlled by soil colloids and associated cations. Aluminum and hydrogen enhance soil acidity, whereas calcium and other base-forming cations (especially sodium) encourage soil alkalinity. The exchange complex also helps buffer the soil, which resists rapid changes in soil reaction, giving stability to most plant–soil systems.

The maintenance of soil pH levels suitable for the plants being grown may require the use of lime or limestone materials to combat acidity and of acid-forming materials such as sulfur to reduce alkalinity. When the application of these materials is not practical, it may be necessary to choose a plant that will tolerate the existing soil pH level rather than go to the expense of trying to change the soil pH. Also, the incorporation into the soil of some green manures may alleviate toxicity from excess aluminum concentrations that are common in acid soils.

Alkaline and salt-affected soils dominate arid and semiarid areas where half the world's arable soils are found. Plant growth is constrained in many of the salt-affected soils, the worldwide area of which is expanding because salts are being brought into these areas in irrigation waters. Plants growing on the salt-affected soils may be adversely affected by excess salts (saline soils), by high sodium saturation of the soil colloids (sodic soils), or by a combination of excess salts and high sodium saturation (saline-sodic soils). Also, the colloids in sodic soils are largely dispersed, the soil is puddled, water infiltration is low, and plant growth is severely stressed.

The reclamation of salt-affected soils requires two major actions: (1) the removal of excess salts from the upper profile levels, and (2) removal of excess exchangeable sodium from the soil. Leaching with low-sodium water can remove the salts. The exchangeable sodium can be replaced with calcium from gypsum, or with hydrogen from oxidized sulfur (H_2SO_4), and the sodium can then be leached from the soil solution.

STUDY QUESTIONS

1. Describe the role of aluminum and its associated ions in enhancing soil acidity, identifying the ionic species involved and the effect of these species on the CEC of soils.

2. If you could somehow extract the soil solution from the upper 16 cm of 1 ha of acid soil (pH = 5), only a few kilograms of limestone would be needed to neutralize the soil solution. Yet under field conditions up to 6 Mg of limestone may be required to bring the pH of this soil layer to a pH of 6.5. How do you explain this difference?

3. What is meant by *buffering?* Why is it so important in soils, and what are the mechanisms by which it occurs?

4. What is *acid rain,* and why does it seem to have greater impact on forests than on commercial agriculture?

5. How much of a limestone with a $CaCO_3$ equivalent of 90 would you need to apply to raise the percent base saturation of a silt loam soil (CEC = 20 $cmol_c$/kg) from 60 to 90%?

6. A neighbor complained when his azaleas were adversely affected by a generous application of limestone to the lawn immediately surrounding the azaleas. To what do you ascribe this difficulty? How would you remedy it?

7. Compare the availability of the following essential elements in alkaline soils with that in acid soils: (1) iron, (2) nitrogen, (3) molybdenum, (4) phosphorus, and (5) manganese.

8. The pH of an arid-region soil, when it was first cleared for cropping, was about 8.0. After several years of irrigation, the crop yield began to decline, the soil aggregation tended to break down, and the pH had risen to 10. What is the likely explanation for this situation?

9. What physical and chemical treatments would you suggest to bring the soil described in question 5 back to its original state of productivity?

10. What are some of the adverse consequences of using wetlands as recipients of irrigation wastewaters?

11. What are the advantages of using gypsum ($CaSO_4 \cdot 2H_2O$) in the reclamation of a sodic soil? Show the chemical reactions that take place.

12. Calculate the quantity of gypsum needed to reclaim a sodic soil (ESP = 30%) when CEC = 25 $cmol_c$/kg and pH is 10.2. Assume you want an ESP no higher than 4%.

REFERENCES

Banuelos, G. S., H. A. Ajwa, B. Mackey, L. Wu, C. Cook, S. Akohoue, and S. Zambruzuski. 1997. "Evaluation of different plant species used for Phytoremediation of high selenium," *J. Envir. Qual.,* **26**:639–648.

Blevens, R. L., G. W. Thomas, M. S. Smith, W. W. Frye, and P. L. Cornelius. 1983. "Changes in soil properties after 10 years continuous non-tilled and conventionally tilled corn," *Soil Tillage Res.,* **3**:135–146.

Bresler, E., B. L. McNeal, and D. L. Carter. 1982. *Saline and Sodic Soils, Principles—Dynamics—Modeling.* (Berlin: Springer-Verlag).

Carter, D. L. 1981. "Salinity and plant productivity," in *CRC Handbook Series in Nutrition and Food.* (Boca Raton, Fla: CRC Press).

Crescimanno, G., M. Iovino, and G. Provenzano. 1995. "Influence of salinity and sodicity on soil structural and hydraulic characteristics," *Soil Sci. Soc. Am. J.,* **59**:1701–1708.

Farina, M. P. W., and P. Channon. 1988. "Acid subsoil amelioration: II. Gypsum effects on growth and subsoil chemical properties," *Soil Sci. Soc. Amer. J.,* **52**:175–180.

Helling, C. S., G. Chesters, and H. B. Corey. 1964. "Contributions of organic matter and clay to soil cation exchange capacity as affected by pH of the saturating solution," *Soil Sci. Soc. Amer. Proc.,* **28**:517–520.

Hinsinger, P., and R. J. Gilkes. 1996. "Mobilization of phosphate from phosphate rock and alumina-sorbed phosphate by the roots of ryegrass and clover as related to rhizosphere pH," *European J. Soil Sci.,* **47**:533–554.

Ilyas, M., R. W. Miller, and R. H. Qureski. 1993. "Hydraulic conductivity of saline-sodic soil after gypsum application," *Soil Sci. Soc. Amer. J.,* **57**:1580–1585.

Jenny, H., et al. 1968. "Interplay of soil organic matter and soil fertility with state factors and soil properties," in *Organic Matter and Soil Fertility.* Pontificiae Academia Scientiarum Scripta Varia 32. (New York: Wiley).

Joslin, J. D., J. M. Kelly, and H. van Miegroet. 1992. "Soil chemistry and nutrition of North American spruce-fir stands: Evidence for recent change," *J. Environ. Qual.,* **21**:12–30.

Likens, G. E., C. T. Driscoll, and D. C. Buso. 1996. "Longterm effects of acid rain: Response and recovery of a forest ecosystem," *Science,* **272**:244–246.

Maas, E. U. 1984. "Crop tolerance," *Calif. Agric.,* **38**:20–21.

Magdoff, F. R., and R. J. Bartlett. 1985. "Soil pH buffering revisted," *Soil Sci. Soc. Amer. J.,* **49**:145–148.

National Academy of Sciences. 1990. *Saline Agriculture: Salt Tolerant Plants for Developing Countries.* (Washington, D.C.: National Academy Press).

National Research Council. 1983. *Acid deposition: Atmospheric Processes in Eastern North America.* (Washington, D.C.: National Research Council).

Nolan, B. T., and M. L. Clark. 1997. "Selenium in irrigated agricultural areas of the western United States," *J. Envir. Qual.,* **26**:849–857.

Rhoades, J. D. 1993. "Electrical conductivity methods for measuring and mapping soil salinity," *Advances in Agronomy,* **49**:201–251.

Robarge, W. P., and D. W. Johnson. 1992. "The effects of acidic deposition on forested soils," *Advances in Agronomy,* **47**:1–84.

Ronse, A., L. De Temmerman, M. Guns, and R. De Borger. 1988. "Evaluation of acidity, organic matter content and CEC in uncultivated soils of Northern Belgium during the past 25 years," *Soil Sci.,* **146**:453–460.

Rowell, D. L. 1988. "Soil acidity and alkalinity," in A. Wild (ed.), *Russell's Soil Condition and Plant Growth,* 11th edition. (New York: Longman Scientific and Technical/Wiley).

Sumner, M. E. 1993. "Gypsum and acid soils: The world scene," *Advances in Agronomy,* **51**:1–32.

U.S./Canada Work Group No. 2. 1982. *Atmospheric Science and Analysis.* Final report; H. L. Ferguson and L. Machita (Co-chairmen). (Washington, D.C. U.S. Environmental Protection Agency).

USDA. 1996. *America's Private Land: A Geography of Hope.* (Washington, D.C.: USDA Natural Resources Conservation Service).

10

ORGANISMS AND ECOLOGY OF THE SOIL[1]

Under the silent, relentless chemical jaws of the fungi, the debris of the forest floor quickly disappears. . . .
—*A. FORSYTH AND K. MIYATA,* TROPICAL NATURE

The terms *ecosystem* and *ecology* usually call to mind scenes of lions stalking vast herds of wildebeest on the grassy savannas of East Africa, or the interplay of phytoplankton, fish, and fishermen in some great estuary. Like a savanna or an estuary, a soil is an ecosystem in which thousands of different creatures interact and contribute to the global cycles that make all life possible. This chapter will introduce some of the organisms in the living drama that goes on largely unseen in the soil beneath our feet. If our bodies were small enough to enter the tiny passages in the soil, we would discover a world populated by a wild array of creatures all fiercely competing for every leaf that falls to the forest floor, every root that sloughs off the growing grass, every fecal pellet, every dead body that reaches the soil. We would also find predators of every description lurking in the dark, some with fearsome jaws to snatch unwary victims, others whose jellylike bodies simply engulf and digest their hapless prey.

Most of the work of the soil community is carried out by creatures whose "jaws" are chemical enzymes that eat away at all manner of organic substances left in the soil by their coinhabitants. The diversity of substrates and environmental conditions found in every handful of soil spawns a diversity of adapted organisms that staggers the imagination. The collective vitality, diversity, and balance among these organisms engender a healthy ecosystem and make possible the functions of a high-quality soil.

We will learn how these organisms, both flora and fauna, interact with one another, what they eat, how they affect the soil, and how soil conditions affect them. The central theme will be how this community of organisms assimilates plant and animal residues and waste products, creating soil humus, recycling carbon and mineral nutrients, and supporting plant growth. We will emphasize the activities rather than the scientific classification of these organisms. Consequently, we will consider only very broad, simple taxonomic categories (Figure 10.1).

10.1 THE DIVERSITY OF ORGANISMS IN THE SOIL

Soil organisms are creatures that spend all or part of their lives in the soil environment. Every handful of soil is likely to contain billions of organisms, with representatives of nearly every division and phylum of living things.

---

[1] For a review of soil biology see Paul and Clark (1996) or Sylvia, et al. (1997).

290

Animals (fauna), all heterotrophs

Macrofauna: Largely herbivores and detritivores

Vertebrates	Gophers, mice, squirrels
Arthropods	Ants, beetles and their larvae, maggots, termites, grubs, spiders, millipedes, woodlice
Annelida	Earthworms
Mollusca	Snails, slugs

Largely predators

Vertebrates	Moles, snakes
Arthropods	Beetles, ants, centipedes

Mesofauna: Largely detritivores

Arthropods	Mites, collembola (springtails)
Annelida	Enchytraeid (pot) worms

Largely predators

Arthropods	Mites, protura

Microfauna: Detritivores, predators, fungivores, bacterivores

Nematoda	Nematodes
Roterifera*	Rotifers, water bears
Protozoa*	Amoebae, ciliata

Plants (flora)

Macroflora: Largely autotrophs

Vascular plants	Feeder roots
Bryophytes	Mosses

Microflora: Largely autotrophs

Vascular plants	Root hairs
Algae	Greens, yellow-greens, diatoms

Largely heterotrophs, aerobic

Fungi	Yeasts, mildews, molds, rusts, mushrooms
Actinomycetes[†]	Many kinds of actinomycetes

Autotrophs and heterotrophs

Bacteria[†]	Aerobes, anaerobes
Cyanobacteria[†]	Blue-green algae

*In some classification schemes these are placed in the kingdom Protista, rather than Animalia.
[†] Often classified together in the kingdom Monera.

FIGURE 10.1   General classification of some important groups of soil organisms. Some organisms subsist on living plants (herbivores), others on dead plant debris (detritivores). Some consume animals (predators); some devour fungi (fungivore) or bacteria (bacterivores); and some live off of, but do not consume, other organisms (parasites). Heterotrophs rely on organic compounds for their carbon and energy needs while autotrophs obtain their carbon mainly from carbon dioxide and their energy from photosynthesis or oxidation of various elements. The classification shown here is highly generalized. Most biologists classify living organisms into five kingdoms rather than two, since it has proven difficult to fit many microorganisms into either the plant or animal kingdom. More recently, classifications based on similarities in genetic material place all living things into three primary groups: *Eucarya* (which includes all plants, animals and fungi), *Bacteria,* and *Archaea.*

SIZES OF ORGANISMS. The animals (fauna) of the soil range in size from **macrofauna** (such as moles, prairie dogs, earthworms, and millipedes), through **mesofauna** (such as tiny springtails and mites), to **microfauna** (such as nematodes and single-celled protozoans). Plants (**flora**) include the roots of higher plants, as well as microscopic algae and diatoms. Other microorganisms (too small to be seen without the aid of a microscope) include fungi, bacteria, and actinomycetes, which tend to predominate in terms of numbers, mass, and metabolic capacity.

Note that we have classified the organisms according to their size (*macro* being larger than 2 mm in width, *meso* being between 0.2 and 2 mm, and *micro* being less than 0.2 mm), as well as by their ecological functions (what they eat). A typical, healthy soil might contain several species of vertebrate animals (mice, gophers, snakes, etc.), a half dozen species of earthworms, 20 to 30 species of mites, 50 to 100 species of insects (collembola, beetles, ants, etc.), dozens of species of nematodes, hundreds of species of fungi, and perhaps thousands of species of bacteria and actinomycetes.

TYPES OF DIVERSITY. Great diversity is possible because of the nearly limitless variety of foods and the wide range of habitat conditions found in soils. Within a handful of soil there may be areas of good and poor aeration, high and low acidity, cool and warm temperatures, moist and dry conditions, and localized concentrations of dissolved nutrients, organic substrates, and competing organisms. The populations of soil organisms tend to be concentrated in zones of favorable conditions, rather than evenly distributed throughout the soil.

Soil scientists use the concept of biological diversity as an indicator of soil quality. A high **species diversity** indicates that the organisms present are fairly evenly distributed among a large number of species. Such complexity and species diversity is paralleled by a high degree of **functional diversity**—the capacity to utilize a wide variety of substrates and carry out a wide array of processes.

ECOSYSTEM DYNAMICS. In most healthy soil ecosystems there are several—and in some cases many—different species capable of carrying out each of the thousands of different enzymatic or physical processes that proceed every day. This **functional redundancy**—the presence of several organisms to carry out each task—leads both to ecosystem **stability** and **resilience**. *Stability* describes the ability of soils, even in the face of wide variations in environmental conditions and inputs, to continue to perform such functions as the cycling of nutrients, assimilation of organic wastes, and maintenance of soil structure. *Resilience* describes the ability of the soil to "bounce back" to functional health after a severe disturbance has disrupted normal processes. Given a high degree of diversity, no single organism is likely to become completely dominant. By the same token, the loss of any one species is unlikely to cripple the entire system. Nonetheless, for certain soil processes, such as ammonium oxidation (see Section 12.7) or the creation of aeration macropores (see Section 4.6) primary responsibility may fall to only one or two species. The activity and abundance of these **keystone species** (for example, certain nitrifying bacteria or burrowing earthworms) merit special attention, for their populations may indicate the health of the entire soil ecosystem.

GENETIC RESOURCES. The diversity of organisms in a soil is important for reasons in addition to the safeguarding of ecological functions—soils also make an enormous contribution to **global biodiversity.** Many scientists believe that there are more species in existence below the surface of the earth than above it. The soil is therefore a major storehouse of the genetic innovations that nature has written into the DNA code over hundreds of millions of years. Humans have always found ways to make use of some of the genetic material in soil organisms (beer, yogurt, and antibiotics are examples). However, with recent developments in biological engineering that allow the transfer of genetic material from one type of organism to another, the genes from soil organisms may now be used to produce plants and animals of superior utility to the human community.

10.2 ORGANISMS IN ACTION

The activities of soil flora and fauna are intimately related in what ecologists call a *food chain* or, more accurately, a *food web*. Some of these relationships are shown in Figure 10.2, which illustrates how various soil organisms are involved in the degradation of residues from higher plants.

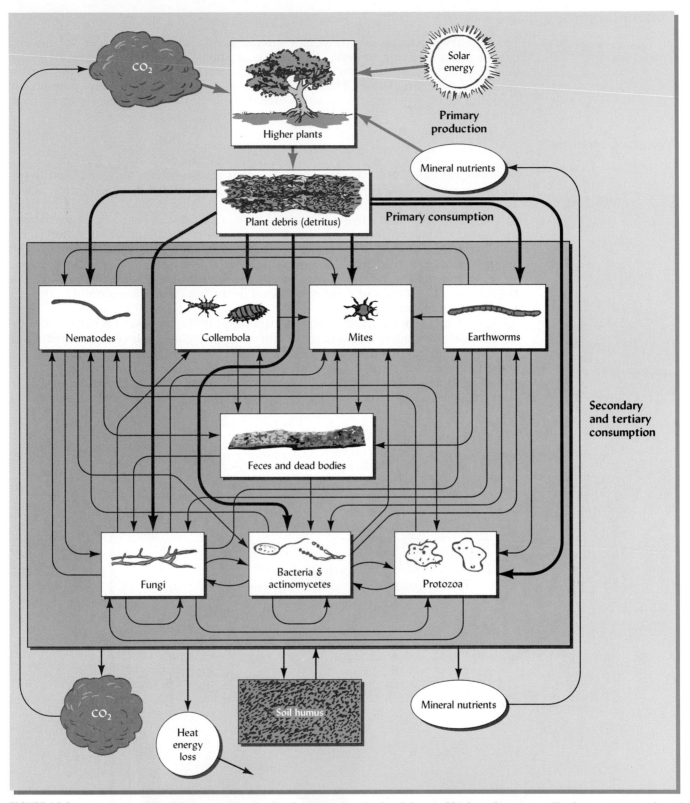

FIGURE 10.2 Greatly simplified diagram of the food web involved in the breakdown of higher-plant tissue. The boxes represent broad groups of organisms and pools of organic material, while the arrows represent transfers of carbon, energy, and nutrients between these pools. Because they capture carbon dioxide and energy, the higher plants are known as *primary producers*. Heavy arrows from the plant debris (detritus) to various organism groups represent *primary consumption*. The arrows within the large box represent *secondary* and *tertiary consumption*. Although all the groups shown play important roles in the process, the microorganisms represented by the lower three boxes account for 80 to 90% of the total metabolic activity. As a result of this metabolism, soil humus is synthesized, and carbon dioxide, heat energy, and mineral nutrients are released into the soil environment.

Primary Producers

The **primary producers** in soils include vascular plants, single-celled algae, lichens, and certain photosynthetic bacteria. As in all ecosystems, primary producers use water, energy from the sun, and carbon from atmospheric carbon dioxide to make organic molecules and living tissues. These organic materials, and the energy they contain, are then utilized by other organisms, either directly or after having been passed on through intermediaries.

Primary Consumers

HERBIVORES AND DETRITIVORES.   Certain soil organisms eat live plants and are called **herbivores.** Examples are parasitic nematodes and insect larvae that attack plant roots, as well as termites, ants, beetle larvae, woodchucks, and mice that devour aboveground plant parts. Because they attack living plants that may be of value to humans, many of these soil herbivores are considered pests. For the vast majority of soil organisms, however, the principal source of food is the debris of dead tissues left by plants and animals. This debris is called **detritus**, and the animals that directly feed on it are called **detritivores.**

Plant and animal detritus is continually being deposited in and on the soil. The assimilation of this mass of plant and animal debris is a critical function of both large and small soil organisms.

As soon as a leaf, a stalk, or a piece of bark drops to the ground, it is subject to coordinated attack by microflora and by detritivores (see Figure 10.2). The detritivore animals, which include mites, tiny insects called collembola (Figure 10.3), and earthworms, chew or tear holes in the tissue, opening it up to more rapid attack by the microflora. The animals and microflora that use the energy stored in the plant residues are termed *primary consumers.*

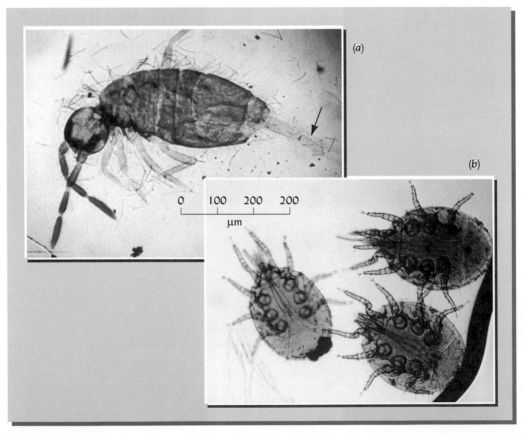

FIGURE 10.3   Springtails (collembala) (*a*) and mites (*b*) are mesofauna detritivores commonly found in soils. The springtail gets its name from the springlike furcula or "tail" that it uses to hop about (see arrow). (Photos courtesy of R. Weil)

ACTION OF THE FAUNA.   While the actions of the microflora are mostly biochemical, those of the fauna are both physical and chemical. The mesofauna and macrofauna chew the plant residues and move them from one place to another on the soil surface and into the soil. The physical incorporation of animal dung (manure) into the soil greatly reduces the loss of nutrients by volatilization and erosion. *Dung beetles* greatly enhance nutrient cycling by burying animal dung in tiny tunnels that they excavate into the upper soil horizons. Earthworms literally eat their way through the soil as they incorporate plant residues into the mineral soil by passing them through their bodies along with mineral soil particles. Larger animals, such as gophers, moles, prairie dogs, and rats, also burrow into the soil and bring about considerable soil mixing and granulation.

The actions of these animals enhance the activity of the microflora in several ways. First, the chewing action fragments the litter, cutting through the resistant waxy coatings on many leaves to expose the more easily decomposed cell contents for microbial digestion. Second, the chewed plant tissues are thoroughly mixed with microorganisms in the animal gut, where conditions are ideal for microbial action. Third, the mobile animals carry microorganisms with them and help the latter to disperse and find new food sources to decompose. The ability of the soil fauna to enhance plant residue decomposition is illustrated in Figure 10.4.

Secondary Consumers

The bodies (cells) of primary consumers soon become food sources for predators and parasites in the soil. These **secondary consumers** include microflora, such as bacteria, fungi, and actinomycetes, as well as **carnivores**, which consume other animals. Examples of carnivores include centipedes and mites that attack small insects; spiders; nematodes; snails; and certain moles that feed primarily on earthworms. Examples of **microphytic feeders**, organisms that use microflora as their source of food, are certain collembola (springtail insects), mites (Figure 10.3), termites, nematodes, and protozoa. These microphytic feeders exert considerable influence over the activity and growth of fungal and bacterial populations. The grazing by these meso- and microfauna on microbial colonies may stimulate faster growth and activity among the microbes, in much the same way that grazing animals can stimulate the growth of pasture grasses. In other cases, the attack of the microphytic feeders may kill off so much of a microbial colony as to inhibit the work of the microorganisms.

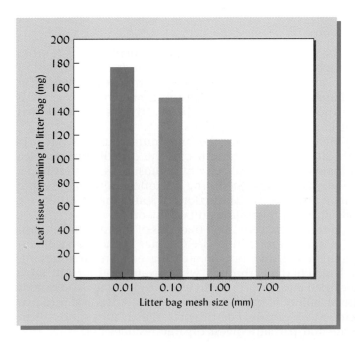

FIGURE 10.4   Influence of various sizes of soil organisms on the decomposition of leaf tissue buried in soil. Small bags made of nylon material with four different-size openings (mesh size) were filled with 558 mg (dry weight) of leaf tissue and buried in the soil for 10 weeks. The amount of leaf tissue remaining in the bags was considerably greater (less decomposition had taken place) where the meso- and macrofauna were excluded by the smaller mesh sizes. [Data from Weil and Kroontje (1979)]

Tertiary Consumers

Moving farther up the food chain, the secondary consumers are prey for still other carnivores, called *tertiary consumers*. For example, ants consume centipedes, spiders, mites, and scorpions—all of which can themselves prey on primary or secondary consumers.

The microflora are intimately involved in every level of the process. In addition to their direct attack on plant tissue (as primary consumers), microflora are active within the digestive tracts of many soil animals, helping these animals digest more resistant organic materials. The microflora also attack the finely shredded organic material in animal feces, and they decompose the bodies of dead animals. For this reason they are referred to as the ultimate decomposers.

As indicated in Figure 10.2, broad groups of organisms, such as the fungi or mites, include some species that are primary consumers, some that are secondary consumers, and others that are tertiary consumers. For example, some mites attack detritus directly, while others eat mainly fungi or bacteria that grow on the detritus. Still others attack and devour the mites that eat fungi. Also note that there are two-way interactions between most groups. For example, some nematodes eat fungi, and some fungi attack nematodes.

10.3 ORGANISM ABUNDANCE, BIOMASS, AND METABOLIC ACTIVITY

Soil organism numbers are influenced primarily by the amount and quality of food available. Other factors affecting their numbers include physical factors (e.g., moisture and temperature), biotic factors (e.g., predation and competition) and chemical characteristics of the soil (e.g., acidity, dissolved nutrients, and salinity). The species that inhabit the soil in a desert will certainly be different from those in a humid forest, which, in turn, will be quite different from those in a cultivated field. Acid soils are populated by species different from those in alkaline soils. Likewise, species diversification and abundance in a tropical rain forest are different from those in a cool temperate area.

Despite these variations, there are a few generalizations that can be made. For example, forested areas support a more diverse soil fauna and a more fungi-dominated microflora than do grasslands, although the total faunal mass per hectare and level of activity are generally higher in grasslands. Cultivated fields are generally lower than undisturbed native lands in numbers and biomass of soil organisms, especially the fauna.

Total **soil biomass**, the living fraction of the soil, is generally related to the amount of organic matter present. On a dry-weight basis, the living portion is usually between 1 and 8% of the total soil organic matter. Also, scientists commonly observe that the ratios of soil organic matter to detritus to microbial biomass to faunal biomass are approximately 1000:100:10:1.

Comparative Organism Activity

The activities of specific groups of soil organisms are commonly identified by (1) their numbers in the soil, (2) their weight (biomass) per unit volume or area of soil, and (3) their metabolic activity (often measured as the amount of carbon dioxide given off in respiration). The numbers and biomass of groups of organisms that commonly occur in soils are shown in Table 10.1. Although the relative metabolic activities are not shown, they are generally related to the biomass of the organisms.

As might be expected, the microorganisms are the most numerous and have the highest biomass. Together with earthworms (or termites, in the case of some tropical soils), the microflora also dominate the biological activity in most soils. It is estimated that about 80% of the total soil metabolism is due to the microflora, although, as previously mentioned, their activity is enhanced by the actions of soil fauna. For these reasons, major attention will be given to the microflora and earthworms.

Even those animals that account for only very small fractions of the total metabolism in the soil can play important roles in soil formation (see Chapter 2) and management. Rodents pulverize, mix, and granulate soil, and incorporate surface organic residues into lower horizons. They provide large channels through which water and air can move freely. Ants are important in localized areas, especially in warm-region grasslands and in boreal forests, for their exceptional ability to break down woody materials and turn over soil materials as they build their nests. Mesofauna detritivores (mostly mites and collembola; see Figure 10.3) translocate and partially digest organic residues and leave their

TABLE 10.1 Relative Numbers and Biomass of Fauna and Flora Commonly Found in the Surface 15 cm of Soil[a]

Since metabolic activity is generally related to biomass, microflora and earthworms dominate the life of most soils.

Organisms	Number		Biomass[b]	
	Per m²	Per gram	kg/ha	g/m²
Microflora				
Bacteria	10^{13}–10^{14}	10^8–10^9	400–5000	40–500
Actinomycetes	10^{12}–10^{13}	10^7–10^8	400–5000	40–500
Fungi	10^{10}–10^{11}	10^5–10^6	1,000–15,000	100–1500
Algae	10^9–10^{10}	10^4–10^5	10–500	1–50
Fauna				
Protozoa	10^9–10^{10}	10^4–10^5	20–200	2–20
Nematodes	10^6–10^7	10–10^2	10–150	1–15
Mites	10^3–10^6	1–10	5–150	.5–1.5
Collembola	10^3–10^6	1–10	5–150	.5–1.5
Earthworms	10–10^3		100–1500	10–150
Other fauna	10^2–10^4		10–100	1–10

[a] A greater depth is used for earthworms.
[b] Biomass values are on a liveweight basis. Dry weights are about 20 to 25% of these values.

excrement for microfloral degradation. By living in the soil, many animals rearrange soil particles to form biopores, thus favorably affecting the soil's physical condition. Others use the soil as a habitat for destructive action against higher plants.

Source of Energy and Carbon

Soil organisms may be classified as either **autotrophic** or **heterotrophic** on the basis of where they obtain the *carbon* needed to build their cell constituents (Table 10.2). The heterotrophic soil organisms obtain their carbon from the breakdown of organic materials previously produced by other organisms. Nearly all heterotrophs also obtain their *energy* from the oxidation of the carbon in organic compounds. These organisms, which include the soil fauna, the fungi, actinomycetes, and most bacteria, are far more numerous than the autotrophs. They are responsible for organic decay.

The autotrophs obtain their carbon from simple carbon dioxide gas (CO_2) or carbonate minerals, rather than from carbon already fixed in organic materials. Autotrophs can be further classified on the basis of how they obtain energy (Table 10.2). While the autotrophs are in the distinct minority, their carbon fixation and inorganic oxidation reactions allow them to play crucial roles in the soil system.

10.4 EARTHWORMS AND TERMITES[2]

Earthworms are probably the most important macroanimals in soils (Figure 10.5). They (along with their much smaller mesofauna cousins, the **enchytraeid worms**) are egg-laying *hermaphrodites* (organisms without separate male and female genders) that eat detritus, soil organic matter, and microorganisms found on these materials. They do not eat living plants or their roots, and so do not act as pests to crops.[3]

Of the up to 3000 species of earthworms reported worldwide, *Lumbricus terrestris*, a deep-boring, reddish organism, and *Allolobophora caliginosa*, a shallow-boring, pale pink organism, are the two most common in Europe and in the eastern and central United States. *Lumbricus terrestris*—commonly called *nightcrawlers*—live in the soil head-up. They come to the surface to pull detritus into their burrows, often creating **middens** of

---

[2] For a readable overview of earthworm biology and roles in the soil, see Lee (1985). For a more technical but wide-ranging review of earthworm ecology, see Hendrix (1995). Edwards (1998) provides up-to-date information on earthworm distribution and ecology.

[3] Earthworms should not be confused, in this regard, with wireworms, rootworms, or other soil-dwelling insect larvae that are commonly referred to as "worms," and which may be plant pests.

TABLE 10.2 Metabolic Grouping of Soil Organisms According to Their Source of Metabolic Energy and Their Source of Carbon for Biochemical Synthesis

Source of carbon	Source of energy	
	Biochemical oxidation	Solar radiation
Combined organic carbon	**Chemoheterotrophs:** All animals, fungi, actinomycetes, and most bacteria Examples: Earthworms *Aspergillus* *Pseudomonas*	**Photoheterotrophs:** A few algae
Carbon dioxide	**Chemoautotrophs:** Mainly bacteria Examples: Ammonia oxidizers—*Nitrosomonas* Sulfur oxidizers—*Thiobacillus denitrificans*	**Photoautotrophs:** Algae and cyanobacteria Examples: *Chorella* *Nostoc*

gathered detritus near the burrow entrance (see Figure 10.5). *Allolobophora caliginosa*—commonly called *red worms*—live head-down in the soil and expel their casts onto the soil surface.

In the tropics and semitropics, other types of earthworms are prevalent and dominate the soil fauna in regions with at least 800 mm of annual rainfall. Some tropical species are small, and others are surprisingly large. In Australia, aboriginal peoples have been known to hunt earthworms that grow up to 1 m long. Enchytraeid worms, on the other hand, range from about 1 mm to no more than a few centimeters in length.

Influence on Soil Fertility and Productivity

BURROWS.   Earthworms literally eat their way through the soil. They may ingest a weight of soil equal to 2 to 30 times their own weight in a single day. In a year, the earthworm population of 1 ha of land may ingest between 50 and 1000 Mg of soil, the higher figure occurring in moist, tropical climates. In so doing, they create extensive systems of burrows. In 1 m³ these burrows may range from 5 to over 100 m in length, contributing from 0.1 to over 9 L of biopore volume. The burrows may greatly improve soil drainage, aeration, and ease of root penetration.

CASTS.   Earthworms also eject enormous quantities of partially digested soil and organic materials. The earthworms' wastes are called *casts,* and usually take the form of globular soil aggregates, some of which can be seen in Figure 10.5. The casting behavior of

FIGURE 10.5 Earthworms are perhaps the most significant macroorganism in soils of humid temperate regions, particularly in relation to their effects on the physical conditions of soils. Surface feeding species such as *Lumbricus terrestris* incorporate large amounts of plant litter into the soil and often gather plant debris into piles called *middens* near their burrow entrances. (Photo courtesy of R. Weil)

earthworms generally enhances the aggregate stability of the soil (Table 10.3). The casts are deposited within the soil profile or on the soil surface, depending on the species of earthworm. Both the piles of casts on the soil surface and the circular holes of burrow entrances are important signs of earthworm activity. In many cases, a *midden* of plant litter may have to be moved aside to observe the burrow entrances.

NUTRIENTS. Earthworms increase the availability of mineral nutrients to plants in two ways. First, as soil and organic materials pass through an earthworm's body, they are ground up physically as well as attacked chemically by the digestive enzymes of the earthworm and its gut microflora. Compared to the bulk soil, the casts are significantly higher in bacteria and organic matter and available plant nutrients (see Table 10.3). Second, as earthworms ingest detritus and soil organic matter of relatively low nitrogen, phosphorus, and sulphur concentrations, they assimilate part of this material into their own body tissues. Consequently, earthworm tissues contain high concentrations of these nutrients, which are then readily released into plant-available form when the earthworms die and decay.

One aspect of concern is that water rapidly percolating down vertical earthworm burrows may carry potential pollutants toward the groundwater (see Figure 6.19). However, the organic-matter-enriched material lining earthworm burrows has been shown to have two to five times as great a capacity to adsorb certain herbicides as the bulk soil. Therefore, transport of such pollutants through earthworm burrows may be much less than is suggested by the mass flow of water through these large biopores. Clearly, on balance, earthworms are very beneficial and their activity is generally to be encouraged by soil managers.

Factors Affecting Earthworm Activity

Earthworms are found most abundantly in medium-textured upland soils where the moisture capacity is high rather than in droughty sands or poorly drained lowlands. Earthworms must have organic matter as a source of food. Generally they grow best on fresh, undecomposed organic matter. Consequently, they thrive where manure or plant residues are plentiful. A few species are reasonably tolerant to low pH, but most earthworms thrive best where the soil is not too acid (pH 5.5 to 8.5) and has an abundant supply of calcium (which is an important component of their slime excretions). Most earthworms are quite sensitive to excess salinity. Enchytraeid worms are much more tolerant of acid conditions and are therefore more active than earthworms in some forested Spodosols.

A soil temperature of about 10°C appears optimum for *Lumbricus terrestris*. This temperature sensitivity, together with their preference for moist soil, probably accounts for the maximum earthworm activity noted in spring and autumn in temperate regions. Some earthworms burrow as deep as 1 to 2 m into the profile, thereby avoiding unfavorable moisture and temperature conditions. However, in soils not insulated by

TABLE 10.3 **Comparative Characteristics of Earthworm Casts and Soils**

Average of six Nigerian soils.

Characteristic	Earthworm casts	Soils
Silt and clay, %	38.8	22.2
Bulk density, Mg/m^3	1.11	1.28
Structural stability[a]	849	65
Cation exchange capacity, cmol/kg	13.8	3.5
Exchangeable Ca^{2+}, cmol/kg	8.9	2.0
Exchangeable K$^+$, cmol/kg	0.6	0.2
Soluble P, ppm	17.8	6.1
Total N, %	0.33	0.12

[a] Numbers of raindrops required to destroy structural aggregates.
From de Vleeschauwer and Lal (1981).

residue mulches, a sudden heavy frost in the fall may decimate earthworm populations before they can move deeper into the profile.

Other factors that depress earthworm populations are predators (moles, mice, and certain mites and millipedes), very sandy soils (partly because of the abrasive effect of sharp sand grains), direct contact with ammonia fertilizer, application of certain insecticides (especially carbamates), and tillage. The last factor is often the overriding deterrent to earthworm populations in agricultural soils. Minimum tillage, with plenty of crop residues left as a mulch on the soil surface, is ideal for encouraging earthworms.

Termites[4]

Termites, sometimes called *"white ants,"* are major contributors to the breakdown of organic material in or at the surface of soils. There are about 2000 species of termites, including several that commonly build protective soil tunnels to enable them to invade (and subsequently destroy) wooden structures built without protective metal termite shields.

Termites are found in about two-thirds of the land areas of the world but are most prominent in the grasslands (savannas) and forests of tropical and subtropical areas (both humid and semiarid). In the drier tropics (less than 800 mm annual rainfall) termites surpass earthworms in dominating the soil fauna.

MOUND BUILDING. Termites are social animals that live in very complex labyrinths of nests, passages, and chambers that they build both below and above the soil surface. Termite mounds built of cemented soil particles are characteristic features of many landscapes in Africa, Latin America, Australia, and Asia (Figure 10.6). Although a few species eat living woody plants and sound deadwood, most termites eat rotting woody materials and plant residues. Several species, such as *Macrotermes spp.* in Africa, use plant residues to grow fungi in their mounds as a source of food.

In building their mounds, termites transport soil from lower layers to the surface, thereby extensively mixing the soil and incorporating into it the plant residues they use as food. Scavenging a large area around each mound, these insects remove up to 4000 kg/ha of leaf and woody material annually, a substantial portion of the plant litter produced in many tropical ecosystems. They also annually move 300 to 1200 kg/ha of soil in their mound-building activities. These activities have significant impacts on soil formation, as well as on current soil fertility and productivity. Attempts to level an occupied mound are usually frustrating, as the termites rebuild very rapidly unless the queen termite is destroyed.

---

[4] For an excellent discussion of termites and ants in the tropics, see Lal (1987).

(a)

(b)

(c)

FIGURE 10.6 Termite mounds in cultivated fields in Africa. (*a*) A termite mound in a cornfield. (*b*) Dissected profile of another mound showing the size and depth of the underground chambers. (*c*) Individual termites dragging cut pieces of leaves into their underground nest. [Photos (*a*) and (*c*) courtesy of R. Weil, photo (*b*) courtesy of R. Lal, Ohio State University]

EFFECTS ON SOIL PRODUCTIVITY.   Unlike earthworms, termites do not generally have a beneficial effect on soil productivity. This is because the digestive processes of termites, aided by microorganisms in their gut, are generally more efficient than those of earthworms. Also, termites mix plant residues into the soil only in the localized areas of their nests, while denuding the remaining land area of surface residues. This termite behavior can make it extremely difficult to provide cropland with the benefits of a protective crop residue mulch.

In digesting woody materials, termites not only destroy wooden structures, but also produce a substantial fraction of the global production of methane (CH_4), an important greenhouse gas (see Chapter 11).

10.5   SOIL MICROANIMALS

From the viewpoint of microscopic animals, soils present many habitats that are essentially aquatic, at least intermittently so. For this reason the soil microfauna are closely related to the microfauna found in lakes and streams. The two groups exerting the greatest influence on soil processes are the nematodes and protozoa.

Nematodes

Nematodes—commonly called *threadworms* or *eelworms*—are found in almost all soils, often in surprisingly large numbers (see Table 10.1). They are unsegmented roundworms, about 4 to 100 µm in cross section and up to several millimeters in length (Figure 10.7). Therefore, they can seldom be seen with the naked eye. Moist, sandy soils typically have especially high nematode populations, as they typically contain abundant pores large enough to accommodate the swimming activities of these highly mobile creatures. When the large soil pores dry up, the nematodes survive by coiling up into a **cryptobiotic** or resting state, in which they seem to be nearly impervious to environmental conditions and use no detectable oxygen for respiration. In semiarid rangelands, nematode activity has been shown to be largely restricted to the first few days after each rainfall that awakens the nematodes from their cryptobiotic state.

PREDATION.   The nematodes comprise a highly diverse group. Most nematodes are predatory on other nematodes, fungi, bacteria, algae, protozoa, and insect larvae. As a result, certain predatory nematodes are sold for use as biological control agents for soilborne insect pests, such as the corn rootworm (a far more environmentally friendly method than the use of toxic pesticides). A bacteria-eating nematode is illustrated in Figure 10.8, along with other soil microorganisms of various sizes. Nematode grazing can have a marked effect on the growth and activities of fungal and bacterial populations. Since bacterial cells contain more nitrogen than the nematodes can use, nematode activity often stimulates the cycling and release of plant-available nitrogen in the soil, accounting for 30 to 40% of the nitrogen released in some ecosystems.

PLANT PARASITES.   Some nematodes, especially those of the genus *Heterodera,* can infest the roots of practically all plant species by piercing the plant cells with a sharp spearlike mouth part. These wounds often allow infection by secondary pathogens and cause the formation of knotlike growths on the roots. Minor nematode infestations are nearly ubiquitous and often have little observable effect on the host plant. However, infestations beyond a certain threshold level result in serious stunting of the plant. Cyst- (egg-sac-) forming nematodes are major pests of soybeans, while root-knot-forming nematodes cause widespread damage to fruit trees and solanaceous crops.

FIGURE 10.7   A nematode commonly found in soil (magnified about 120 times). More than 1000 species of soil nematodes are known. Most nematodes feed on bacteria and fungi, but plant parasitic species can be very detrimental to plant growth. (Courtesy William F. Mai, Cornell University)

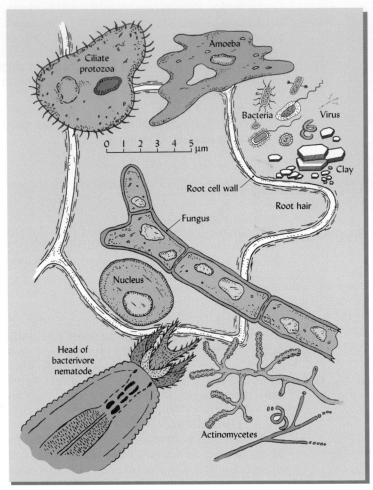

FIGURE 10.8 A depiction of representative groups of soil microorganisms, showing their relative sizes. (Drawing courtesy of R. Weil)

NEMATODE CONTROL. Until recently the principal methods of controlling plant parasitic nematodes were long rotations with nonhost crops (often five years is required for the parasitic nematode populations to sufficiently dwindle), use of genetically resistant crop varieties, and soil fumigation with toxic chemicals (**nematicides**). The use of soil nematicides, such as methyl bromide, has been sharply restricted because of undesirable environmental effects (see Section 10.14).

New approaches to nematode control include the use of hardwood bark for containerized plants and interplanting susceptible crops with marigolds or canola (rapeseed), both of which produce root exudates with nematicidal properties. Often a series of different resistant varieties will be planted in succeeding years to prevent the buildup of parasitic nematode populations.

Protozoa

Protozoa are mobile, single-celled creatures that capture and engulf their food. They are the most varied and numerous of the soil microfauna (though often they are classified as Protista rather than as animals; see Figure 10.9 and Table 10.1). They are considerably larger than bacteria (see Figure 10.8), having a diameter range of 6 to 100 μm. Their cells do not have true cell walls, and have a distinctly more complex organization than bacterial cells. Soil protozoa include amoebas (which move by extending and contracting pseudopodia), ciliates (which move by waving hairlike structures; see Figure 10.12), and flagellates (which move by waving a whiplike appendage called a *flagellum*). Like nematodes, they swim about in the water-filled pores and water films in the soil and can form resistant resting stages (called *cysts*) when the soil dries out or food becomes scarce.

(a)   (b)

FIGURE 10.9 Two microanimals typical of those found in soils. (*a*) Scanning electron micrograph of a ciliated protozoan, *Glaucoma scintillaus*. (*b*) Photomicrograph of two species of rotifer, *Rotaria rotatoria* (stubby one at left) and *Philodina acuticornus* (slender one at right). [(*a*) Courtesy of J. O. Corliss, University of Maryland; (*b*) courtesy of F. A. Lewis and M. A. Stirewalt, Biomedical Research Institute, Rockville, Md.]

More than 350 species of protozoa have been isolated in soils; sometimes as many as 40 or 50 of such groups occur in a single sample of soil. A considerable number of serious animal and human diseases are attributed to infection by protozoa, but mainly by those that are waterborne, rather than soilborne. Most soil-inhabiting protozoa prey upon soil bacteria, exerting a significant influence on the populations of these microflora in soils.

Protozoa generally thrive best in moist, well-drained soils and are most numerous in surface horizons. In pursuit of their bacterial prey, some soil-dwelling protozoa are adapted to squeezing into soil pores with openings as small as 10 μm. Protozoa are especially active in the area immediately around plant roots. Their main influence on organic matter decay and nutrient release is through their effects on bacterial populations.

10.6   ROOTS OF HIGHER PLANTS

Higher plants store the sun's energy and are the primary producers of organic matter (see Figure 10.2). Their roots grow and die in the soil and are classified as soil organisms in this text. They typically occupy about 1% of the soil volume and may be responsible for a quarter to a third of the respiration occurring in a soil. Roots usually compete for oxygen, but they also supply much of the carbon and energy needed by the soil community of fauna and microflora.

Depending on their size, roots may be considered to be either meso- or microorganisms. Fine feeder roots range in diameter from 100 to 400 μm, while root hairs are only 10 to 50 μm in diameter—similar in size to the strands of microscopic fungi (see Figure 10.8). Root hairs are elongated protuberances of single cells of the outer (epidermal) layer (Figure 10.10). One function of root hairs is to anchor the root as it pushes its way through the soil. Another function is to increase the amount of root surface area available to absorb water and nutrients from the soil solution.

FIGURE 10.10 (*a*) Photograph of a root tip illustrating how roots penetrate soil and emphasizing the root cells through which nutrients and water move into and up the plant. (*b*) Diagram of a root showing the origins of organic materials in the rhizosphere. [(*a*) From Chino (1976), used with permission of Japanese Society of Soil Science and Plant Nutrition, Tokyo; (*b*) redrawn from Rovira, et al. (1979), used with permission of Academic Press, London]

Rhizosphere

The importance of root residues in helping to maintain soil organic matter is often overlooked. In grasslands, fires may remove most of the aboveground biomass, so that the deep, dense root systems are the main source of organic matter added to these soils. In plantation and natural forests, more than half of the total biomass production may be in the form of tree roots. In arable soils, the mass of roots remaining in the soil after crop harvest is commonly 15 to 40% that of the aboveground crop. The mass of organic compounds contributed by roots is seen to be even greater when the rhizosphere effects discussed in the following subsections are considered.

The zone of soil significantly influenced by living roots is termed the **rhizosphere** and usually extends about 2 mm out from the root surface. Soil acidity may be 10 times higher (or lower) in the rhizosphere than in the bulk soil. Roots greatly affect the nutrient supply in this zone by withdrawing dissolved nutrients on one hand and by solubilizing nutrients from soil minerals on the other. By these and other means, roots affect the mineral nutrition of soil microbes, just as the microbes affect the nutrients available to the plant roots. Microbial densities in the rhizosphere are commonly 2 to 10 times as great as in the bulk soil.

Rhizodeposition

Significant quantities of at least three broad types of organic compounds are released at the surface of young roots (see Figure 10.10). First, low-molecular-weight organic compounds are exuded by root cells, including organic acids, sugars, amino acids, and phenolic compounds. Second, high-molecular-weight mucilages secreted by root-cap cells

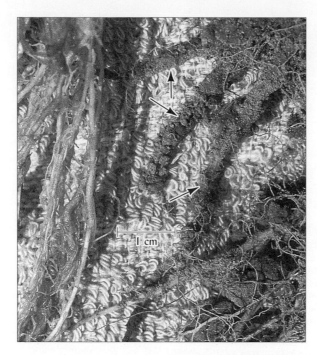

FIGURE 10.11 The zone of soil within 1 to 2 mm of living plant roots is termed the *rhizosphere*. This zone is greatly enriched in organic compounds excreted by the roots. These exudates and the microorganisms they support have caused the rhizosphere soil to adhere to some of these wheat roots as a sheath (roots on the right, see arrows). The roots on the left have been washed in water and the soil of the rhizosphere soil has been removed. The background is a terrycloth towel. (Photo courtesy of R. Weil)

and epidermal cells near apical zones form a substance called **mucigel** when mixed with microbial cells and clay particles. This mucigel lubricates the root's movement through the soil, improves root–soil contact, and provides an ideal environment for the growth of the rhizosphere microorganisms. Third, cells from the root cap and epidermis continually slough off as the root grows and enrich the rhizosphere with a wide variety of cell contents.

Taken together, these types of **rhizodeposition** typically account for 2 to 30% of total dry-matter production in young plants. The roots of common grain and vegetable plants have been observed to rhizodeposit 5 to 40% of the organic substances translocated to them from the plant shoot. Sometimes when a plant is carefully uprooted from a loose soil, these root exudates cause a thin layer of soil, approximating the rhizosphere, to form a sheath around the actively growing root tips (Figure 10.11). Rhizodeposition decreases with plant age but increases with soil stresses, such as compaction and low nutrient supply.

10.7 SOIL ALGAE

Like higher plants, algae consist of eukaryotic cells, those with nuclei organized inside a nuclear membrane. (Organisms formerly called *blue-green algae* are prokaryotes and therefore will be considered with the bacteria.) Also like higher plants, algae are equipped with chlorophyll, enabling them to carry out photosynthesis. As photoautotrophs, algae need light, and are therefore mostly found very near the surface of the soil (Figure 10.12).

Most soil algae range in size from 2 to 20 μm. Many algal species are motile and swim about in soil pore water, some by means of flagella (whiplike "tails"). Most grow best under moist to wet conditions, but some are also very important in hot or cold desert environments. Sometimes the growth of algae may be so great that the soil surface is covered with a green or orange algal mat. Some algae (as well as certain cyanobacteria) form *lichens*, symbiotic associations with fungi. These are important in colonizing bare rock and other low-organic-matter environments. Unvegetated patches in deserts commonly are covered with *algal crusts* that reduce water evaporation and soil erosion but are very sensitive to disruption by trampling or off-road vehicles.

In addition to producing a substantial amount of organic matter in some fertile soils, certain algae excrete polysaccharides that have very favorable effects on soil aggregation (see Section 4.8).

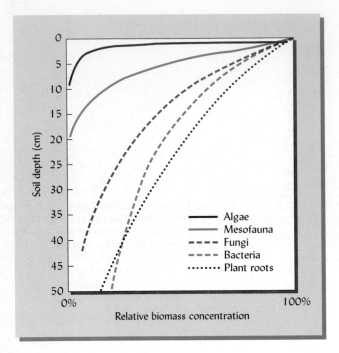

FIGURE 10.12 Relative vertical distribution of various groups of soil organisms in a representative grassland soil. Concentrations shown for a group are relative to the highest biomass concentration for that group. The biomass concentration for all of the organisms is near 100% of the maximum at the soil surface, but, the absolute biomass amounts (not shown here) differ greatly among the groups. For example, the actual biomass per gram of soil would be much higher for the fungi than for the mesofauna.

10.8   SOIL FUNGI

Soil fungi comprise an extremely diverse group of microorganisms. Tens of thousands of species have been identified in soils, representing some 170 genera. As many as 2500 species have been reported to occur in a single location. Scientists believe there are at least 1 million fungal species in the soil still awaiting discovery. Although their numbers are usually somewhat smaller than those of bacteria, their relatively large size results in their dominating the biomass and metabolic activity in many soils (see Table 10.1).

Fungi are eukaryotes with a nuclear membrane and cell walls. As heterotrophs, they depend on living or dead organic materials for both their carbon and their energy. Fungi are aerobic organisms, although some can tolerate the rather low oxygen concentrations and high levels of carbon dioxide found in wet or compacted soils. Strictly speaking, fungi are not entirely microscopic, since some of these organisms, such as mushrooms, form macroscopic structures that can easily be seen without magnification.

For convenience of discussion, fungi may be divided into three groups: (1) yeasts, (2) molds, and (3) mushroom fungi. *Yeasts,* which are single-celled organisms, live principally in waterlogged, anaerobic soils. *Molds* and *mushrooms* are both considered to be filamentous fungi, because they are characterized by long, threadlike, branching chains of cells. Individual fungal filaments, called **hyphae** (Figure 10.13*a*), are often twisted together to form **mycelia** that appear somewhat like woven ropes. Fungal mycelia are often visible as thin, white or colored strands running through decaying plant litter. The filamentous fungi reproduce by means of spores, often formed on fruiting bodies, which may be microscopic (i.e., molds) or macroscopic (i.e., mushrooms).

Molds

The molds are distinctly filamentous, microscopic or semimacroscopic fungi that play a much more important role in soil organic matter breakdown than the mushroom fungi. Molds develop vigorously in acid, neutral, or alkaline soils. Some are favored, rather than harmed, by lowered pH (see Figure 9.13). Consequently, they may dominate the microflora in acid surface soils, where bacteria and actinomycetes offer only mild competition. The ability of molds to tolerate low pH is especially important in decomposing the organic residues in acid forest soils. Many genera of molds are found in soils. Four of the most common are *Penicillium, Mucor, Fusarium,* and *Aspergillus.* The complexity of the organic compounds being attacked seems to determine which particular mold (or molds) prevail.

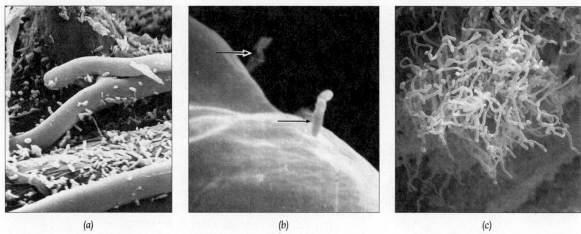

(a) (b) (c)

FIGURE 10.13 Scanning electron micrographs of fungi, bacteria, and actinomycetes. (*a*) Fungal hyphae associated with much smaller rod-shaped bacteria. (*b*) Rod-shaped bacteria attached to a plant root hair. (*c*) Actinomycete threads. [(*a*) Courtesy of R. Campbell, University of Bristol, used with permission of American Phytopathological Society; (*b, c*) courtesy of Maureen Petersen, University of Florida]

Mushroom Fungi

These fungi are associated with forest and grass vegetation where soil moisture and organic residues are ample. Although the mushrooms of many species are extremely poisonous to humans, some are edible—and a few have been domesticated. The aboveground fruiting body of most mushrooms is only a small part of the total organism. An extensive network of hyphae permeate the underlying soil or organic residue. While mushrooms are not as widely distributed as the molds, these fungi are very important, especially in the breakdown of woody tissue, and because some species form a symbiotic relationship with plant roots (see "Mycorrhizae," following).

Activities of Fungi

DECOMPOSITION.   As decomposers of organic materials in soil, fungi are the most versatile and persistent of any group. Cellulose, starch, gums and lignin, as well as the more easily metabolized proteins and sugars, succumb to their attack. Fungi play major roles in the processes of humus formation (see Section 11.4) and aggregate stabilization (see Section 4.7). Fungi function more efficiently than bacteria in that they assimilate into their tissues a larger proportion of the organic materials they metabolize. About 50% of the substances decomposed by fungi may become fungal tissue, compared to about 20% for bacteria. Soil *fertility* depends in no small degree on nutrient cycling by fungi, since they continue to decompose complex organic materials after bacteria and actinomycetes have essentially ceased to function. Soil *tilth* also benefits from fungi as their hyphae stabilize soil structure (see Figure 4.24). Some of the nutrient cycling and ecological activities of soil fungi are made easily visible in the case of "fairy rings," commonly seen on lawns in early spring (Figure 10.14).

ANTAGONISM.   In addition to the breakdown of organic residues and the formation of humus, numerous other fungal activities have significant impact on soil ecology. Some fungi are predators of soil animals. For example, certain species even trap nematodes (see Figure 10.15). Soil fungi can synthesize a wide range of complex organic compounds in addition to those associated with soil humus. It was from a soil fungus, a *Penicillium* species, that the first modern antibiotic drug, penicillin, was obtained. By killing bacteria, such compounds probably help fungi to outcompete rival microorganisms in the soil.

TOXINS AND DISEASE.   Unfortunately, not all the compounds produced by soil fungi benefit humans or higher plants. A few fungi produce chemicals (**mycotoxins**) that are highly toxic to plants or animals (including humans). An important example of the latter is the production of highly carcinogenic aflatoxin by the fungus *Aspergillus flavus* growing on grains such as corn or peanuts, especially when grain is exposed to soil and moisture. Other

FIGURE 10.14 A "fairy ring" of fungal growth and the fungi's fruiting bodies (mushrooms, inset). As the fairy ring fungi (most commonly *Marasmius spp.*) metabolize accumulated grass thatch and residues, they release excess nitrogen that stimulates the lush green growth of the grass. Later, bacteria decompose the aging and dead fungi, producing a second release of nitrogen. The fungus produces a chemical (thought to be hydrogen cyanide) that is toxic to itself. Therefore, each generation must grow into uncolonized soil, producing an ever-expanding circle of fungi and decay marked by an ever-larger ring of dark green grass. The grass in the center of the ring is often brown, stunted and water-stressed, most likely because the fungi render the upper soil layers somewhat hydrophobic. (Photos courtesy of R. Weil)

fungi produce compounds that allow them to invade the tissues of higher plants, causing such serious plant diseases as wilts (e.g., *Verticillium*) and root rots (e.g., *Rhizoctonia*).

BIOLOGICAL CONTROL. On the other hand, efforts are now under way to develop the potential of certain fungi (such as *Beauvaria*) as biological control agents against some insects and mites that damage higher plants. These examples merely hint at the impact of the complex array of fungal activities in the soil.

Mycorrhizae[5]

One of the most ecologically and economically important activities of soil fungi is the mutually beneficial association (**symbiosis**) between certain fungi and the roots of higher plants. This association is called **mycorrhizae**, a term meaning "fungus root." In

---

[5] For two excellent reviews of mycorrhizae and their effects on the soil–plant system, see Bethlenfalvay and Linderman (1992) and Read, et al. (1992).

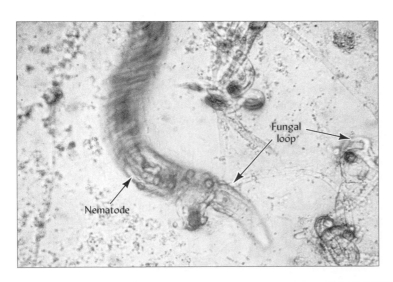

Nematode

Fungal loop

FIGURE 10.15 Several species of fungi prey on soil nematodes, often on those nematodes that parasitize higher plants. Some species of nematode-killing fungi attach themselves to and slowly digest the nematodes. Others make a loop with their hyphae and wait for a nematode to swim through this lassolike trap. The loop is then constricted and the nematode is trapped. The nematode shown here is thrashing about in a vain effort to escape from such a trap. (Photo courtesy of the American Society of Agronomy) natural ecosystems many plants are quite dependent on mycorrhizal relationships and cannot survive without them. Mycorrhizae are the rule, not the exception, for most plant species, including the majority of economically important plants.

Mycorrhizal fungi derive an enormous survival advantage from teaming up with plants. Instead of having to compete with all the other soil heterotrophs for decaying organic matter, the mycorrhizal fungi obtain sugars directly from the plant's root cells. In return, plants receive some extremely valuable benefits from the fungi. The fungal hyphae grow out into the soil some 5 to 15 cm from the infected root, reaching farther and into smaller pores than could the plant's own root hairs. This extension of the plant root system increases its efficiency, providing perhaps 10 times as much absorptive surface as the root system of an uninfected plant.

Mycorrhizae greatly enhance the ability of plants to take up phosphorus and, sometimes, other nutrients that are relatively immobile and present in low concentrations in the soil solution (Table 10.4). In contrast, mycorrhizae prevent the uptake of excessive amounts of salts and toxic metals in saline, acid, or contaminated soils. Water uptake may also be improved by mycorrhizae, making plants more resistant to drought. There is evidence that mycorrhizae protect plants from certain soilborne diseases and parasitic nematodes by producing antibiotics, altering the root epidermis, and competing with fungal pathogens for infection sites.

ECTOMYCORRHIZA. Two types of mycorrhizal associations are of considerable practical importance: **ectomycorrhiza** and **endomycorrhiza.** The ectomycorrhiza group includes hundreds of different fungal species associated primarily with temperate- or semiarid-region trees and shrubs, such as pine, birch, hemlock, beech, oak, spruce, and fir. These fungi, stimulated by root exudates, cover the surface of feeder roots with a fungal mantle. Their hyphae penetrate the roots and develop in the free space around the cells of the cortex but do not penetrate the cortex cell walls (hence the term *ecto,* meaning outside). Ectomycorrhizae cause the infected root system to consist primarily of stubby, white rootlets with a characteristic Y shape (Figure 10.16). These Y-shaped rootlets provide visible evidence of mycorrhizal infection.

For some tree species, satisfactory growth and survival are dependent on proper inoculation with mycorrhizal fungi before planting if the proper fungi are not already in soils at the site. On infertile soils, it is not uncommon for ectomycorrhizal inoculation to increase tree growth by 50 to 500 percent. Sometimes soil from a mature forest containing the desired tree species is used as inoculum to bring the necessary mycorrhizal fungi to forest plantings in such highly disturbed sites as reclaimed mines and landslides. Many ectomycorrhizal fungi are facultative symbionts (they can also live independently in the soil) and therefore can be cultured in large quantities on artificial media. *Pisolithus tinctorus* is a commercial ectomycorrhizal fungal inoculant widely used in tree nurseries.

ENDOMYCORRHIZA. The most important members of the endomycorrhiza group are called **arbuscular mycorrhizae (AM)**, formerly known as *vesicular-arbuscular mycorrhizae (VAM)*. When forming AM, fungal hyphae actually penetrate the cortical root cell walls and, once

TABLE 10.4 **Effect of Inoculation with Mycorrhiza and of Added Phosphorus on the Content of Different Elements in the Shoots of Corn**

Expressed in micrograms per plant.

Element in plant	No phosphorus		25 mg/kg phosphorus added	
	No mycorrhiza	Mycorrhiza	No mycorrhiza	Mycorrhiza
P	750	1,340	2,970	5,910
K	6,000	9,700	17,500	19,900
Ca	1,200	1,600	2,700	3,500
Mg	430	630	990	1,750
Zn	28	95	48	169
Cu	7	14	12	30
Mn	72	101	159	238
Fe	80	147	161	277

From Lambert, et al. (1979).

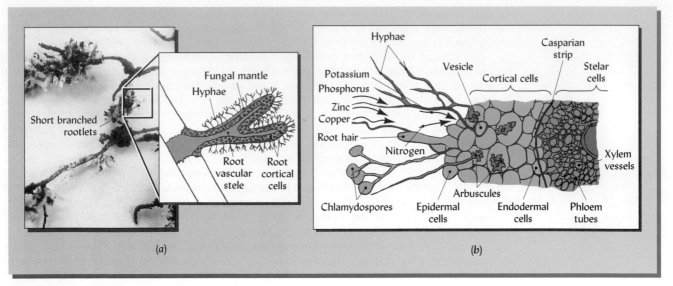

(a)　　　　　　　　　　　　　　　　　　　　　(b)

FIGURE 10.16 Diagram of ectomycorrhiza and arbuscular mycorrhiza (AM) association with plant roots. (*a*) The ectomycorrhiza association produces short branched rootlets that are covered with a fungal mantle, the hyphae of which extend out into the soil and between the plant cells but do not penetrate the cells. (*b*) In contrast, the AM fungi penetrate not only between cells but into certain cells as well. Within these cells, the fungi form structures known as *arbuscules* and *vesicles*. The former transfer nutrients to the plant, and the latter store these nutrients. In both types of association, the host plant provides sugars and other food for the fungi and receives in return essential mineral nutrients that the fungi absorb from the soil. [Redrawn from Menge (1981); Photo courtesy of R. Weil]

inside the plant cell, form small, highly branched structures known as **arbuscules.** These structures serve to transfer mineral nutrients from the fungi to the host plants and sugars from the plant to the fungus. Other structures, called **vesicles,** are sometimes also formed and serve as storage organs for the mycorrhizae (see Figure 10.16).

The roots of most agronomic crops, including corn, cotton, wheat, potatoes, soybeans, alfalfa, sugarcane, cassava, and dryland rice, form AM, as do most vegetables and fruits, such as apples, grapes, and citrus. Many forest trees, including maple, yellow poplar, and redwood, as well as such important tree crops as cacao, coffee, and rubber, also form AM. Two important groups of crop plants that do *not* form mycorrhizae are the *Cruciferae* (cabbage, mustard, canola, and broccoli) and the *Chenopodiaceae* (sugar beet, red beet, and spinach).

Research on AM continues to reveal the ecological and practical significance of this symbiosis. The importance of mycorrhizal hyphae in stabilizing soil aggregate structure is becoming increasingly clear (see Figure 4.24). The presence of mycorrhizae is known to enhance nodulation and N fixation by legumes. Also, AM have been observed to transfer nutrients from one plant to another via hyphal connections, sometimes resulting in a complex four-way symbiotic relationship (Figure 10.17). Interplant transfer of nutrients has been observed in forest, grassland, and pasture ecosystems, but researchers are still trying to determine the extent and ecological significance of this AM-mediated plant interaction.

Although definite differences exist in mycorrhizal efficiency among various species and strains of AM fungi, the ubiquitous distribution of native strains ensures natural infection in plants under most circumstances. Thus, gains from inoculation with efficient mycorrhizal fungi are usually quite small. Exceptions occur where native populations of fungi are low due to soil fumigation or elimination of host plants, such as by surface mining or forest clear-cutting. Practical AM technology has been limited by difficulties in culturing these obligate symbionts on artificial media, a process that would allow production of large quantities of disease-free inoculum.

10.9 SOIL ACTINOMYCETES

Like the fungi, **actinomycetes** are filamentous and often profusely branched (see Figure 10.13*c*), though their mycelial threads are much smaller than those of fungi. Actinomycetes are similar to bacteria in that they are unicellular and of about the same diam-

FIGURE 10.17 Mycorrhizal fungi, rhizobia bacteria, legumes, and nonlegume plants can all interact in a four-way, mutually beneficial relationship. Both the fungi and the bacteria obtain their energy from sugars supplied through photosynthesis by the plants. The rhizobia form nodules on the legume roots and enzymatically capture atmospheric nitrogen, providing the legume with nitrogen to make amino acids and proteins. The mycorrhizal fungi infect both types of plants and form hyphal interconnections between them. The mycorrhizae then not only assist in the uptake of phosphorus from the soil, but can also directly transfer nutrients from one plant to the other. Isotope tracer studies have shown that, by this mechanism, nitrogen is transferred from the nitrogen-fixing legume to the nonlegume (e.g., grass) plant, and phosphorus is mostly transferred to the legume from the nonlegume. The nonlegume grass plant has a fibrous root system and an extensive mycorrhizal network, which is relatively more efficient in extracting P from soils than the root system of the legume. Research indicates that some direct transfer of nutrients via mycorrhizal connections occurs in many mixed plant communities, such as in forest understories, grass–legume pastures, and mixed cropping systems.

eter. Although historically they have often been classified with the fungi, actinomycetes are bacterialike in that they have no nuclear membrane (prokaryote), have about the same diameter, and often break up into spores that closely resemble bacterial cells. They are now commonly classified with bacteria as *Monera*.

Generally aerobic heterotrophs, actinomycetes live on decaying organic matter in the soil, or on compounds supplied by plants with which certain species form parasitic or symbiotic relationships. Many actinomycete species produce antibiotic compounds that kill other microorganisms. Actinomycin, neomycin, and streptomycin are examples of familiar antibiotic drugs produced by growing soil actinomycetes in pure culture. In forest ecosystems, much of the nitrogen supply depends on certain actinomycetes that are capable of fixing atmospheric nitrogen gas into ammonium nitrogen that is then available to higher plants (see Section 12.12).

Actinomycetes develop best in moist, warm, well-aerated soil. But in times of drought, they remain active to a degree not usually exhibited by either bacteria or fungi. Actinomycetes are tolerant of low osmotic potential, and are important in arid-region, salt-affected soils. They are generally rather sensitive to acid soil conditions, with optimum development occurring at pH values between 6.0 and 7.5. Some actinomycete species tolerate relatively high temperatures.

NUMBERS AND ACTIVITIES OF ACTINOMYCETES. Actinomycete numbers in soil exceed those of all other organisms except bacteria (see Table 10.1). With their biomass as high as 5000 kg/ha, they often exceed the bacteria in actual liveweight. These organisms are especially numerous in soils high in humus, such as old meadows or pastures, where the acidity is not too great. The earthy aroma of organic-rich soils and freshly plowed land is mainly due to actinomycetes products called *geosmins* that are volatile derivatives of terpene.

Actinomycetes undoubtedly are of great importance in the decomposition of soil organic matter and the liberation of its nutrients. Apparently, they reduce even the more resistant compounds, such as cellulose, chitin, and phospholipids, to simpler forms. They often become dominant in the later stages of decay when the easily metabolized substrates have been used up.

CHARACTERISTICS.   Bacteria are very small, single-celled prokaryotic organisms. They range in size from 0.5 to 5 μm, the smaller ones approaching the size of the average clay particle (see Figure 10.8). Bacteria are found in various shapes: nearly round (coccus), rod-like (bacillus), or spiral (spirillum). In the soil, the rod-shaped organisms seem to predominate (see Figure 10.13*b*). Many bacteria are motile, swimming about in the soil water by means of hairlike cilia or whiplike flagella.

Bacteria are perhaps the most diverse group of soil organisms; a gram of soil typically contains 20,000 different *species*. They have evolved mechanisms to adapt to life in the most extreme of environments, from the Antarctic to the Amazon, from anaerobic wetlands to desiccated salt flats, from forest litter to deep groundwaters, from sodic soils at pH 10 to acid sulfate soils at pH 2. Most soil bacteria are heterotrophs, but both photoautotrophs and chemautotrophs also exist.

BACTERIAL POPULATIONS IN SOILS.   As with other soil organisms, the numbers of bacteria are extremely variable, but high, ranging from a few billion to more than a trillion in each gram of soil. A biomass of a 400 to 5000 kg/ha liveweight is commonly found in the upper 15 cm of fertile soils (see Table 10.1).

Their ability to form extremely resistant resting stages that survive dispersal by winds, sediments, ocean currents and animal digestive tracts has allowed bacteria to spread to almost all soil environments. Their extremely rapid reproductive potential (generation times of only a few hours) enables bacteria to increase their populations quickly in response to favorable changes in soil environment and food availability. As the early soil microbiologist Beijerinck expressed it, "everything is everywhere" when it comes to bacteria. Change the soil environment, add a new substrate—even an industrial waste—and soon you'll find new populations of bacteria.

IMPORTANCE OF BACTERIA.   Bacteria participate vigorously in virtually all of the organic transactions that characterize a healthy soil system. Because soil bacteria as a group possess such a broad range of enzymatic capabilities, scientists are now finding ways of harnessing, even improving, the metabolic activities of bacteria to help with the remediation of soils polluted by crude oil, pesticides, and various other organic toxins. Bacteria are usually the most important group in the breakdown of hydrocarbon compounds, such as gasoline and diesel fuel.

In addition, bacteria hold near monopolies in several basic enzyme-mediated transformations. One such process is the oxidation or reduction of selected chemical elements in soils (see Sections 7.3 and 7.5). Certain autotrophic bacteria obtain their energy from such inorganic oxidations, while anaerobic and facultative bacteria reduce a number of substances other than oxygen gas. Many of these biochemical oxidation and reduction reactions have significant implications for environmental quality (see Box 10.1) as well as for plant nutrition.

A second critical process in which bacteria are prominent is nitrogen fixation—the biochemical combining of atmospheric nitrogen with hydrogen to form organic nitrogen compounds usable by higher plants (see Section 12.10).

CYANOBACTERIA.   Previously classified as blue-green algae, **cyanobacteria** contain chlorophyll, allowing them to photosynthesize like plants. Cyanobacteria are especially numerous in rice paddies and other wetland soils, and fix appreciable amounts of atmospheric nitrogen when such lands are flooded. Certain cyanobacteria, growing within the leaves of the aquatic fern *azolla,* are able to fix nitrogen in quantities great enough to be of practical significance for rice production (see Section 12.12). These organisms also exhibit considerable tolerance to saline environments.

10.11 CONDITIONS AFFECTING THE GROWTH OF SOIL MICROORGANISMS

ORGANIC MATTER REQUIREMENTS.   The addition of almost any energy-rich organic substance, including the compounds excreted by plant roots, stimulates microbial growth and

BOX 10.1 HARNESSING SOIL BACTERIA TO FIGHT POLLUTION

Organic pollutants from industrial processes, leaks, and spills have caused *acutely* toxic contamination of soils, producing barren areas sometimes referred to as *brownfields*. Industry and government are spending billions of dollars annually to remediate (clean up) these contaminated soils. Very expensive, but widespread methods of soil remediation involve physical and/or chemical treatment of the soil, either in situ (in place) or by excavating and moving the soil to treatment facilities. The polluted soil is incinerated, washed with solvents, or heated and subjected to vacuum extraction to remove or destroy toxic contaminants.

For many contaminated soils there is a more economical *biological* alternative to incineration, soil washing, or landfilling—namely *bioremediation*. Simply put, this technology uses microbes—ususally soil bacteria—to degrade organic toxins into harmless metabolic products. Certain soil bacteria can even break down such toxic petroleum constituents as polyacrylic aromatic hydrocarbons (PAHs), as well as such synthetic toxins as pentachlorophenol and trichloroethylene. The job may be carried out by organisms currently in the soil, or microbes specifically selected for their ability to destroy the contaminants may be introduced into the soil. For example, bacteria isolated from sites with a long history of the specific contamination (or grown in laboratory culture on a diet rich in the pollutant) tend to evolve an unusually high efficiency at metabolizing the target chemical.

Inoculating a polluted soil with such specialized exotic organisms is called *bioaugmentation*. For example, a bacterium has been identified that can detoxify perchloroethene (PCE), a common, highly toxic groundwater pollutant suspected of causing cancer. The bacteria's enzymes remove the four chlorines from the PCE, producing ehtylene, a naturally occurring and relatively harmless gas.

In some cases indigenous microorganisms are capable of degrading the contaminant, but their growth and metabolic rates may be far too slow because of insufficient mineral nutrients, especially nitrogen and phosphorus (see Section 11.3 for a discussion of the C/N ratio in organic decomposition). To alleviate such nutritional limitations, special fertilizers have been formulated and used with innovative application techniques to stimulate and greatly speed up microbial degradation processes in soils. This *biostimulation* approach was pioneered in 1989 when an oil-soluble fertilizer was used to stimulate the breakdown of oil spilled on Alaska's pristine beaches (Entisols) by the tanker *Exxon Valdez*.

Some situations call for the use of bioremediation techniquews deep in the soil where low soil porosity causes oxygen deficiency that limits microbial activity. For example, deep-soil biostimulation can be used to clean up toxic organic solvents in the subsoil and groundwater under leaking storage tanks. Typically, special horizontal drilling rigs are used to pipe in a mixture of air (for oxygen), methane (to act as a carbon source to stimulate specific methanotrophic bacteria), and phosphorus (a nutrient needed for bacteria growth). Soil scintists in the young field of bioremediation are constantly developing new ways to harness the abilities of soil microorganisms to clean up our environment.

activity. Certain bacteria and fungi are stimulated by specific amino acids and other growth factors found in the rhizosphere or produced by other organisms.

Bacteria tend to respond most rapidly to additions of simple compounds such as starch and sugars, while fungi and actinomycetes overshadow the bacteria if cellulose and more resistant compounds dominate the added organic materials. Also, if organic materials are left on the soil surface (as in coniferous forest litter), fungi dominate the microbial activity. Bacteria commonly play a larger role if the substrates are mixed into the soil, as by earthworms, root distribution, or tillage.

OXYGEN REQUIREMENTS.   While most microorganisms are *aerobic* and use O_2 as the electron acceptor in their metabolism, some bacteria are *anaerobic* and use substances other than O_2 (e.g., NO_3^-, SO_4^{2-}, Fe^{3+}, or other electron acceptors). *Facultative* bacteria can use either aerobic or anaerobic forms of metabolism. All three of the above types of metabolism are usually carried out simultaneously in different habitats within a soil.

MOISTURE AND TEMPERATURE.   Optimum moisture level for higher plants (moisture potential of −10 to −70 kPa) is usually best for most aerobic microbes. Too high a water content will limit the oxygen supply. Microbial activity is generally greatest when temperatures are 20 to 40°C. The warmer end of this range tends to favor actinomycetes. Ordinary soil temperature extremes seldom kill bacteria; however, except for certain **cryophilic** species, most microorganisms cease metabolic activity below about 5°C, a temperature sometimes referred to as *biological zero* (see Section 7.8).

EXCHANGEABLE CALCIUM AND pH.   Levels of exchangeable calcium and pH help determine which specific organisms thrive in a particular soil. Although in any chemical condition found in soils some bacterial species will thrive, high calcium and near-neutral pH generally result in the largest, most diverse bacterial populations. Low pH allows fungi to become dominant. The effect of pH and calcium helps explain why fungi tend to dominate in forested soils, while bacterial biomass generally exceeds fungal biomass in most subhumid to semiarid prairie and rangeland soils.

10.12   BENEFICIAL EFFECTS OF SOIL ORGANISMS

The soil fauna and flora are indispensable to the plant productivity and the ecological functioning of soils. Of their many beneficial effects, only the most important can be emphasized here.

ORGANIC MATERIAL DECOMPOSITION.   Soil organisms assimilate organic residues from plants and animals (including human sewage) and other organic materials added to soils. As a by-product of their metabolism, microbes release mineral nutrients and synthesize new compounds, some of which help to stabilize soil structure, or contribute to humus formation.

BREAKDOWN OF TOXIC COMPOUNDS.   Most biologically produced toxins do not remain long in the soil, for soil ecosystems include organisms that not only are unharmed by these compounds, but can produce enzymes that allow them to use these toxins as food.

Some toxins are **xenobiotic** (artificial) compounds foreign to biological systems, and these may resist attack by commonly occurring microbial enzymes (Box 10.1). The detoxifying activity of microorganisms is by far the greatest in the surface layers of soil, where microbial numbers are concentrated in response to the greater availability of organic matter and oxygen (Figure 10.18).

INORGANIC TRANSFORMATIONS.   Organically bound forms of nitrogen, sulfur, and phosphorus are converted by the microbes into plant-available forms. The availability of the other essential elements, such as iron and manganese, is determined largely by microbial action. In well-drained soils, these elements are oxidized by autotrophic organisms to their higher valence states, in which forms they are quite insoluble. Microbial oxidation also controls the potential for toxicity in soil contaminated with selenium or chromium (see Figure 7.6).

NITROGEN FIXATION.   While actinomycetes in the genus *Frankia* fix major amounts of nitrogen in forest ecosystems, and cyanobacteria are important in flooded rice paddies, rhizobia bacteria are the most important group for the capture of gaseous nitrogen in agricultural soils.

PLANT PROTECTION.   Certain soil organisms attack higher plants, but others act to protect plant roots from invasion by soil parasites and pathogens (see following section).

FIGURE 10.18 (*a*) Populations of aerobic bacteria and fungi at various depths in a Dundee soil (Aqualfs). (*b*) Breakdown of the herbicide metribuzin in the same soil at various depths. Soil was sampled from each depth and incubated with the herbicide. Note that the fungi and bacteria were concentrated in the upper 10 cm, and that the breakdown of the herbicide was far more rapid in this layer than in the deeper layers. [From Moorman and Harper (1989)]

10.13 SOIL ORGANISMS AND DAMAGE TO HIGHER PLANTS

Although most of the activities of soil organisms are vital to a healthy soil ecosystem and economic plant production, some soil organisms affect plants in detrimental ways that cannot be overlooked.

SOIL FAUNA.   It already has been suggested that certain of the soil fauna are injurious to higher plants. For instance, some rodents may severely damage crops. Snails and slugs in some climates are dreaded pests, especially of vegetables. Ants transfer aphids onto certain plants, and so contribute to plant damage. Also, most plant roots are infested with nematodes, sometimes so seriously that the successful growth of certain crops is both difficult and expensive.

MICROFLORA AND PLANT DISEASE.   Disease infestations occur in great variety and are induced by many different organisms. Among microorganisms, bacteria and actinomycetes contribute their quota of plant diseases, but fungi are responsible for most of the common soilborne diseases of crop plants. Included are wilts, damping-off, root rots, clubroot of cabbage, and blight of potatoes.

Soils are easily infested with disease organisms, which are transferred from soil to soil by many means, including tillage or planting implements, transplant material, manure from animals that were fed infected plants, soil erosion, and wind-borne fungal spores and bacteria. Once a soil is infested, it is apt to remain so for a long time.

Disease Control by Soil Management

Prevention is the best defense against soilborne diseases. Strict quarantine systems will restrict the transfer of soilborne pathogens from one area to another. Elimination of the host crop from the infested field will help in cases where no alternate hosts are available. Crop rotation and tillage practices can be used to help control a disease by growing nonsusceptible plants for several years between susceptible crops, and by burying plant residues on which fungal spores might overwinter. By preventing the splashing of soil onto foliage during rainstorms, mulches can reduce the transmission of soilborne diseases.

SOIL pH.   Regulation of soil pH is effective in controlling some diseases. For example, keeping the pH low (<5.2) can control both the actinomycete-caused *potato scab* and the fungal disease of turf grass known as *spring dead spot*. Raising soil pH to about 7.2 can control *clubroot* disease in the cabbage family, because the spores of the fungal pathogen germinate poorly, if at all, under alkaline conditions.

AIR AND TEMPERATURE.   Wet, cold soils favor some seed rots and seedling diseases, known as *damping-off*. Good drainage and planting on ridges can help control these diseases. Steam or chemical sterilization is a practical method of treating greenhouse soils for a number of pathogens. It should be remembered, however, that sterilization kills beneficial microorganisms, such as mycorrhizal fungi, as well as pathogens, and so may do more harm than good.

DISEASE-SUPPRESSIVE SOILS.   Research on plant diseases ranging from *take-all* of wheat in Washington state to *fusarium wilt* of bananas in Central America has documented the existence of **disease-suppressive soils** in which a particular disease fails to develop even though both the virulent pathogen and a susceptible host are present. The reason that certain soils become disease suppressive is not entirely understood, but much evidence points to a role for **antagonism** by certain bacteria and fungi that inhibit the pathogenic organisms (Figure 10.19).

Horticulturalists have been able to control *Fusarium* diseases by replacing traditional potting mixes with growing media made from tree bark **compost**. Apparently, large numbers of antagonistic organisms colonize the organic material during the final stages of composting (see Section 11.5).

COMPETITION FOR NUTRIENTS AND OXYGEN.   Soil organisms can quickly absorb essential nutrients into their own cells, leaving the more slowly growing higher plants to use only what is left. Nitrogen is the element for which competition is usually greatest (Section 11.3), although similar competition occurs for phosphorus, potassium, calcium, and even the micronutrients.

Under conditions of poor drainage, active soil microflora may deplete the already limited soil oxygen supply. This oxygen depletion may affect plants adversely in two ways. First, the plant roots require a certain minimum amount of O_2 for normal growth and nutrient uptake. Second, oxidized forms of several elements, including nitrogen, sulfur, iron, and manganese, will be chemically reduced by microbial action.

FIGURE 10.19   The biological basis of disease-suppressive soils that control take-all disease of wheat. (Left) in the center of the petri dish grows a colony of certain *Pseudomonas* bacteria isolated from the rhizosphere of wheat that grew in soils suppressive of take-all disease. The *Pseudomonas* colony is producing an antibiotic that is toxic to *Gaeumannomyces graminis* (the fungal pathogen that causes take-all disease), preventing the pathogen colonies from growing to the center of the plate. (Right) These plots in eastern Washington have grown monoculture wheat for 15 years and have developed high populations of organisms antagonistic to the take-all pathogen. In the 15th year of the study, the entire field was inoculated with *G. graminis* for experimental purposes, but the disease developed (seen as light-colored, prematurely ripened plots) only where the soil was fumigated prior to the inoculation. The fumigation killed most of the antagonistic organisms, leaving the pathogenic fungi free to infect the wheat plants. (Photos by R. J. Cook; courtesy of the American Phytopathological Society)

10.14   ECOLOGICAL RELATIONSHIPS AMONG SOIL ORGANISMS

Mutualistic Associations

We have already mentioned a number of mutually beneficial associations between plant roots and other soil organisms (e.g., mycorrhiza and nitrogen fixing nodules) and between several microorganisms (e.g., lichens). Other examples of such associations abound in soils. For example, photosynthetic algae reside within the cells of certain protozoans. Several types of associations, among them algal-fungal associations on or in soils and rocks, are very important cyclers of nutrients and producers of biomass in desert ecosystems.

Competitive Interactions

COMPETITION FOR FOOD.   The population of soil microbes is generally limited by the food supply, increasing rapidly whenever more organic materials become available. As a result, the soil presents an intensely competitive environment for its microscopic inhabitants. When fresh organic matter is added, the vigorous heterotrophic soil organisms (bacteria, fungi, and actinomycetes) compete with each other for this source of food. If sugars, starches, and amino acids are available in the organic material, the bacteria dominate initially because they reproduce most rapidly and prefer such simple compounds. As these simple compounds are broken down, the fungi, and particularly the actinomycetes, become more competitive.

ANTIBIOTICS AND OTHER COMPETITIVE MECHANISMS.   The soil microbes use many different tactics in their battle for resources. Certain organisms alter the acidity level in their vicinity to the disadvantage of competitors. Others produce substances (**siderophores**) that so strongly bind to iron that other organisms in their immediate vicinity cannot grow for lack of this essential element. Certain soil bacteria, fungi, and actinomycetes produce **antibiotics**, compounds that kill specific types of other organisms on contact. When these and other antagonistic activities of soil microorganisms are turned against potentially plant-pathogenic organisms, significant suppression of plant diseases can occur. Microbial ecologists are attempting to harness some of these competitive mechanisms to favor desirable microorganisms in soils (Box 10.2).

Effects of Management Practices on Soil Organisms

While agricultural practices have different effects on different organisms, a few generalizations can be made (Table 10.5). For example, some agricultural practices (e.g., extensive tillage and monoculture) generally reduce the diversity of species as well as

TABLE 10.5   The Generalized Effects of Major Soil-Management Practices on the Overall Diversity and Abundance of Soil Organisms

Note that the practices that tend to enhance biological diversity and activity in soils are also those associated with efforts to make agricultural systems more sustainable.

Decreases biodiversity and populations	Increases biodiversity and populations
Fumigants	Balanced fertilizer use
Nematicides	Lime on acid soils
Some insecticides	Proper irrigation
Compaction	Improved drainage and aeration
Soil erosion	Animal manures and composts
Industrial wastes and heavy metals	Domestic (clean) sewage sludge
Moldboard plow–harrow tillage	Reduced or zero tillage
Monocropping	Crop rotations
Row crops	Grass–legume pastures
Bare fallows	Cover crops or mulch fallows
Residue burning or removal	Residue return to soil surface
Plastic mulches	Organic mulches

BOX 10.2   CHOOSING SIDES IN THE WAR AMONG THE MICROBES

Recent advances in our understanding of microbial genetics, combined with new techniques for transferring genetic material among different species of microorganisms, have opened many interesting and promising opportunities for scientists to join in the microbial wars in the soil.

Scientists are developing innovative approaches to harness the antagonisms long observed among microorganisms. Two examples are (1) the development of a bacterial seed treatment that helps ward off pathogenic fungi, and (2) a chemically armed *Rhizobium* bacteria that may enable farmers to efficiently use improved strains of nitrogen fixing bacteria on leguminous crops.

The first biological-control seed treatments are now commercially available. These seed treatments use antagonistic soil bacteria instead of chemical fungicide to protect seed from rot-causing soil pathogens. The protective bacteria are specially selected and enhanced strains of *Pseudomonas fluorescens*, a species known to inhabit rhizosphere soil and help protect plant roots from soilborne diseases. When a preparation of these bacteria was applied in the furrow as wheat seeds were sown, the seeds germinated safely in soil known to be infested with seed-rotting fungi. Seeds sown without the treatment decayed before they could germinate (Figure 10.20). Widespread use of such biologically based seed protection in the place of conventional antifungal chemicals should reduce the loading of toxic compounds into the environment.

The second example involves the use of genetic engineering to improve rhizobial inoculum effectiveness. Scientists have been quite successful in finding strains of rhizobia and bradyrhizobia bacteria that have superior nitrogen-fixing ability. Unfortunately, there has been little practical success in using these improved strains to enhance nitrogen fixation in the field. The problem is that most soils contain large populations of indigenous rhizobia strains that, while not as efficient in fixing nitrogen, are very com-

FIGURE 10.20   (Right) The wheat seed protected with antagonistic bacteria remain healthy, while unprotected seed (left) becomes heavily infected with pathogenic fungi. (Photo courtesy of Ecogen, Inc.)

petitive in forming nodules at infection sites on legume roots. Because indigenous rhizobia are successfully competing for survival in the soil and for nodule-formation sites on legume roots, improved strains of rhizobia bacteria generally have little success. Even when high levels of bacterial inoculum are applied to the legume seed or to the soils in which the seed is sown, the vast majority (generally 80 to 95%) of nodules found on the legume roots are formed by indigenous strains, not by the improved, introduced strains. In order to take advantage of the improved strains greater nitrogen-fixing ability, it is necessary to give the improved strains a competitive edge over their indigenous cousins. Dr. Eric Triplett and coworkers in Wisconsin discovered that the strain, *Rhizobia leguminosarium* bv. *trifolii*, produced a substance they called *trifolitoxin* that inhibits other rhizobial strains but did not inhibit the strain that produced it [Fitzmaurice, (1995)]. After working out the sequence of the cluster of genes controlling the production and reaction to trifolitoxin, the scientists were able to transfer the relevant genes to other strains of rhizobia. Their hope is that, armed with these genes, improved rhizobia strains that are highly efficient in nitrogen fixation will be able to outcompete the less efficient indigenous strains for nodule formation sites on legume roots. If successful, this use of microbial antagonism should lead to enhanced nitrogen fixation by legumes, with consequent increased yields of the legumes and increased potential to substitute biologically fixed nitrogen for fertilizer nitrogen in the production of nonlegume crops grown together or in rotation with the legumes.

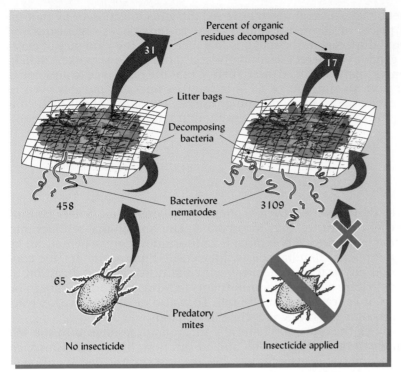

FIGURE 10.21 The indirect effects of insecticide treatment on the decomposition of creosote bush litter in desert ecosystems. Litter bags filled with creosote bush leaves and twigs were buried in desert soils in Arizona, Nevada, and California, either with or without an insecticide (chlordane) treatment. The insecticide killed virtually all the insects and mites. Without predatory mites to hold them in check, bacterivore nematodes multiplied rapidly and devoured a large portion of the bacterial colonies responsible for litter decomposition and nutrient cycling. Thus the insecticide reduced the rate of litter decomposition nearly in half, not by any direct effect on the bacteria, but by the indirect effect of killing the predators of their predators. [Data calculated from Whitford, et al. (1982)]

the abundance (number) of individuals. However, monoculture may *increase* the population of a few species.

Adding lime and fertilizers (either organic or inorganic) to an infertile soil generally will increase microbial and fauna activity, largely due to the increase in the plant biomass that is likely to be returned to the soil as roots, root exudates, and shoot residues. Tillage, on the other hand, is a drastic disturbance of the soil ecosystem, disrupting fungal hyphae networks and earthworm burrows, as well as speeding the loss of organic matter. Reduced tillage therefore tends to increase the role of fungi at the expense of the bacteria, and usually increases overall organism numbers as organic matter accumulates. Addition of animal manure stimulates even higher microbial and faunal (especially earthworm) activity.

Pesticides are highly variable in their effects on soil ecology. Soil fumigants and nematicides can sharply reduce organism numbers, especially for fauna, at least on a temporary basis. On the other hand, application of a particular pesticide often stimulates the population of a specific microorganism, either because the organism can use the pesticide as food, or, more likely, because the predators of that organism have been killed. Figure 10.21 illustrates how a practice that affects one group of organisms will likely affect other groups as well, and will eventually impact on the productivity and functioning of the soil in the ecosystem. It is wise to remember that the interrelationships among soil organisms are intricate and the effects of any perturbation of the system are difficult to predict.

Genetically Engineered Microorganisms

Scientists have developed techniques for transferring specific genetic material to a not-necessarily-related other organism. These technologies have enabled the engineering of new genetic combinations—**recombinant organisms**—not seen in nature. There are many potential uses for **genetically engineered microorganisms** (**GEMs**) in the soil environment. Already, soils contaminated by chemical spills are being cleaned by soil bacteria that have been endowed with an enhanced ability to degrade certain organic pollutants and other GEMs produce toxins that kill plant pathogens or pests. Other potential uses of GEMs in soil are increased nitrogen fixation in roots or rhizospheres, improved mycorrhizal efficiency, and enhanced antagonism against plant pathogens.

The use of GEMs in soils brings the risk of undesirable side effects, which are difficult to predict. Nontarget organisms may be killed. Of broader concern is the possibility that the new genetic material might be transferred (by virus infection or by transfer between bacteria of pieces of chromosome-free DNA) to indigenous organisms well adapted to survival and dispersial in the soil, possibly increasing the virulence of existing pathogens. Also, genetic engineers use antibiotic-resistance genes as markers to help identify and track genetic material. The concern here is that the resistance to antibiotics might be transferred from the introduced GEM to another microbe, possibly even to human pathogens, rendering our life-saving drugs useless. Questions about such concerns continue to require extensive testing and caution in releasing GEMs into the soil environment.

10.15 CONCLUSION

The soil is a complex ecosystem with a diverse community of organisms. Soil organisms are vital to the cycle of life on earth. They incorporate plant and animal residues into the soil and digest them, returning carbon dioxide to the atmosphere, where it can be recycled through higher plants. Simultaneously, they create humus, the organic constituent so vital to good physical and chemical soil conditions. During digestion of organic substrates, they release essential plant nutrients in inorganic forms that can be absorbed by plant roots or be leached from the soil. They also mediate the redox reactions that influence soil color and nutrient cycling in wetlands.

Animals, particularly earthworms, mechanically incorporate residues into the soil and leave open channels through which water and air can flow. Microorganisms such as fungi, actinomycetes, and bacteria are responsible for most organic decay, although their activity is greatly influenced by the soil fauna. Certain of the microorganisms form symbiotic associations with higher plants, playing special roles in plant nutrition and nutrient cycling. Competition among soil microbes, and between these organisms and higher plants, for mineral nutrients can result in plant nutrient deficiencies. Microbial requirements are factors in determining the success of most soil-management systems. Genetically engineered microorganisms offer the promise of enhanced beneficial microbial action, but must be used with caution as they present the potential for troubling unintended side effects in the soil system.

The organisms of the soil must have energy and nutrients if they are to function efficiently. To obtain these, soil organisms break down organic matter, aid in the production of humus, and leave behind compounds that are useful to higher plants. These decay processes and their practical significance are considered in the next chapter, which deals with soil organic matter.

STUDY QUESTIONS

1. What is *functional redundancy,* and how does it help soil ecosystems continue to function in the face of environmental shocks such as fire, clear-cutting or tillage?

2. Give two specific examples of soil organisms that play the roles of *primary producers, primary consumers, secondary consumers,* and *tertiary consumers.*

3. Describe some of the ways in which mesofauna play significant roles in soil metabolism even though their biomass and respiratory activity is only a small fraction of the total in the soil.

4. What are the four main types of metabolism carried out by soil organisms relative to their sources of energy and carbon?

5. What role does O_2 play in aerobic metabolism? What elements take its place under anaerobic conditions?

6. A *mycorrhizae* is said to be a symbiotic association. What are the two parties in this symbiosis, and what are the benefits derived by each party?

7. In what ways is soil improved as a result of earthworm activity? Are there possible detrimental effects as well?

8. What is the *rhizosphere,* and in what ways does the soil in the rhizosphere differ from the rest of the soil?

9. Explain and compare the effects of tillage and manure application on the abundance and diversity of soil organisms.

10. What are some of the potential benefits and potential risks involved with the release of *genetically engineered organisms* (GEMs) into the soil?

REFERENCES

Bethlenfalvay, G. J., and R. G. Linderman (eds.). 1992. *Mycorrhizae in Sustainable Agriculture.* ASA Special Publication no. 54. (Madison, Wis.: Amer. Soc. of Agron.).

Chino, M. 1976. "Electron microprobe analysis of zinc and other elements within and around rice root growth in flooded soils," *Soil Sci. and Plant Nut. J.,* **22:**449.

de Vleeschauwer, D., and R. Lal. 1981. "Properties of worm casts under secondary tropical forest regrowth, *Soil Sci.,* **132:**175–181.

Edwards, C. (ed.). 1998. *Earthworm Ecology.* (Delray Beach, Fla: St. Lucie Press).

Fitzmaurice, L. 1995. "Giving *Rhizobium* the competitive edge," *Wisconsin Bioissues,* **6:**17.

Hendrix, P. F. (ed.). 1995. *Earthworm Ecology and Biogeography in North America.* (Boca Raton, Fla.: Lewis Publishers).

Hendrix, P. F., D. A. Crossley, Jr., J. M. Blair, and D. C. Coleman. 1990. "Soil biota as components of sustainable agricultural agroecosystems," in C. Edwards and R. Lal, (eds.), *Sustainable Agricultural Systems.* (Akeny, Iowa: Soil Water Conserv. Soc. Am.), pp. 637–654.

Ingham, E. R., and W. G. Thies. 1996. "Responses of soil foodweb organisms in the first year following clearcutting and application of chloropicrin to control laminated root rot," *Applied Soil Ecology,* **3:**35–47.

Killham, K. 1994. *Soil Ecology.* (New York: Cambridge University Press).

Lal, R. 1987. *Tropical Ecology and Physical Edaphology.* (New York: Wiley).

Lambert, D. H., D. E. Baker, and H. Cole, Jr. 1979. "The role of Mycorrhizae in the interactions of phosphorus with zinc, copper, and other elements," *Soil Sci. Soc. Amer. J.,* **43:**976–980.

Lee, K. E. 1985. *Earthworms, Their Ecology and Relationships with Soils and Land Use.* (New York: Academic Press).

Menge, J. A. 1981. "Mycorrhizae agriculture technologies," in *Background Papers for Innovative Biological Technologies for Lesser Developed Countries,* Paper no. 9., Office of Technology Assessment Workshop, Nov. 24–25, 1980. (Washington, D.C.: U.S. Government Printing Office), pp. 383–424.

Mooreman, T. B., and S. S. Harper. 1989. "Transformation and mineralization of Metribuzin in surface and subsurface horizons of a Mississippi Delta soil," *J. Environ. Qual.,* **18:**302–306.

Paul, E. A., and F. E. Clark. 1996. *Soil Microbiology and Biochemistry,* 2nd ed. (New York: Academic Press).

Read, D. J., D. Lewis, A. Fitter, and I. Alexander. 1992. *Mycorrhizas in Ecosystems.* (Wallingford, U.K.: CAB International).

Rovira, A. D., R. C. Foster, and J. K. Martin. 1979. "Origin, nature and nomenclature of the organic materials in the rhizosphere," in J. L. Harley and R. S. Russell (eds.), *The Soil–Root Interface.* (New York: Academic Press).

Sylvia, D., J. Fuhrmann, P. Hartel, and D. Zuberer. 1997. *Principles and Applications of Soil Microbiology.* (Upper Saddle River, N.J.: Prentice Hall).

Weil, R. R., and W. Kroontje. 1979. "Organic matter decomposition in a soil heavily amended with poultry manure," *J. Environ. Qual.,* **8:**584–588.

Whitford, W. G., D. W. Freckman, P. F. Santos, N. Z. Elkins, and L. W. Parker. 1982. "The role of nematodes in decomposition in desert ecosystems," in Diana Freckman, (ed.), *Nematodes in Soil Ecosystems.* (Austin, Tex.: University of Texas Press), pp. 98–116.

11

SOIL ORGANIC MATTER

The earth lay rich and dark and fell apart lightly
under the points of their hoes.
—*P. S. BUCK, THE GOOD EARTH*

All organic substances, by definition, contain carbon. Organic matter in the world's soils contains about three times as much carbon as is found in all the world's vegetation. Soil organic matter,[1] therefore, plays a critical role in the global carbon balance that is thought to be the major factor affecting global warming, or the **greenhouse effect.** Although organic matter comprises only a small fraction of the total mass of most soils, this dynamic soil component exerts a dominant influence on many soil physical, chemical, and biological properties.

Soil organic matter is a complex and varied mixture of organic substances. It provides much of the cation exchange and water-holding capacities of surface soils. Certain components of soil organic matter are largely responsible for the formation and stabilization of soil aggregates. Soil organic matter also contains large quantities of plant nutrients and acts as a slow-release nutrient storehouse, especially for nitrogen. Furthermore, organic matter supplies energy and body-building constituents for most of the microorganisms whose general activities were discussed in Chapter 10. In addition to enhancing plant growth through the just-mentioned effects, certain organic compounds found in soils have direct growth-stimulating effects on plants. For all these reasons, the quantity and quality of soil organic matter is a central factor in determining **soil quality.**

We will first examine the role of soil organic matter in the **global carbon cycle** and the process of decomposition of organic residues. Next, we will focus on inputs and losses with regard to soil carbon in specific ecosystems. Finally, we will study the processes and consequences involved in soil organic matter management.

11.1 THE GLOBAL CARBON CYCLE

The element *carbon* is the foundation of all life. From cellulose to chlorophyll, the compounds that comprise living tissues are made of carbon atoms arranged in chains or rings and associated with many other elements. The cycle of carbon on earth is the story of life on this planet. The carbon cycle is all-inclusive because it involves the soil, higher plants of every description, and all animal life, including humans. Disruption of the carbon cycle would mean disaster for all living organisms.

---

[1] For a detailed review of organic matter in soils see Tate (1987). For a practical review of organic matter management in agricultural soils, see Magdoff (1992).

The basic pathways involved in the global carbon cycle are shown in Figure 11.1. Plants take in carbon dioxide from the atmosphere. Then, through the process of photosynthesis, the energy of sunlight is trapped in the carbon-to-carbon bonds of organic molecules (such as those described in Section 11.2). Some of these organic molecules are used as a source of energy (via respiration) by the plants themselves (especially by the plant roots), with the carbon being returned to the atmosphere as carbon dioxide. The remaining organic materials are stored temporarily as constituents of the standing vegetation, most of which is eventually added to the soil as plant litter (including crop residues) or root deposition (see Section 10.6). Some plant material may be eaten by animals (including humans), in which case about half of the carbon eaten is exhaled into the atmosphere as carbon dioxide. The carbon not returned to the atmosphere is eventually returned to the soil as bodily wastes or body tissues. Once deposited on or in the soil, these plant or animal tissues are metabolized (digested) by soil organisms, which gradually return this carbon to the atmosphere as carbon dioxide.

Much smaller amounts of carbon dioxide react in the soil to produce carbonic acid (H_2CO_3) and the carbonates and bicarbonates of calcium, potassium, magnesium, and other base-forming cations. The bicarbonates are readily soluble and may be removed in drainage. Eventually, much of the carbon in the carbonates and bicarbonates is also returned to the atmosphere as carbon dioxide.

Microbial metabolism in the soil produces some organic compounds of such stability that decades or even centuries may pass before the carbon in them is returned to the atmosphere as carbon dioxide. Such resistance to decay allows organic matter to accumulate in soils.

FIGURE 11.1 A simplified representation of the global carbon cycle emphasizing those pools of carbon which interact with the atmosphere. The numbers in boxes indicate the petagrams (Pg = 10^{15}/g) of carbon stored in the major pools. The numbers by the arrows show the amount of carbon annually flowing (Pg/yr) by various pathways between the pools. Note that the soil contains almost twice as much carbon as the vegetation and the atmosphere combined. Imbalances caused by human activities can be seen in the flow of carbon to the atmosphere from fossil fuel burning (5.5) and in the fact that more carbon is leaving (62 + 0.5) than entering (60) the soil. These imbalances are only partially offset by increased absorption of carbon by the oceans. The end result is that a total of 219.5 Pg/yr enters the atmosphere while only 215 Pg/yr of carbon is removed. It is easy to see why carbon dioxide levels in the atmosphere are rising. [Data from several sources; soil C estimate from Batjes (1996)]

Carbon Sources

The original source of soil organic matter is plant tissue. Animals are secondary sources of organic matter. As they eat the original plant tissues, they contribute waste products, and they leave their own bodies when they die (review Figure 10.2). Certain forms of animal life, especially earthworms, termites, ants, and dung beetles, also play an important role in the incorporation and translocation of organic residues.

Globally, at any one time, approximately 2400 petagrams (Pg, 10^{15}) of carbon are stored in soil profiles as soil organic matter (excluding surface litter), about one-third of that at depths below 1 m. An additional 700 Pg are stored as soil carbonates, which can release CO_2 upon weathering. Altogether, about twice as much carbon is stored in the soil than in the world's vegetation and atmosphere combined (see Figure 11.1). Of course, this carbon is not equally distributed among all types of soils (Table 11.1). About 45% of the total is contained in soils of just three orders, Histosols, Inceptisols, and Gelisols. Histosols (and Histels in the order Gelisols) are of limited extent but contain very large amounts of organic matter per unit land area. Inceptisols (and nonhistic Gelisols) contain only moderate concentrations of carbon, but cover vast areas of the globe. The reasons for the varying amounts of organic carbon in different soils will be detailed in Section 11.10.

11.2 THE PROCESS OF DECOMPOSITION IN SOILS

Because plant residues are the principal material undergoing decomposition in soils and, hence, are the primary source of soil organic matter, we will begin by considering the makeup of these materials.

Composition of Plant Residues

Green plant tissues contain from 60 to 90% water (Figure 11.2). On a weight basis, the dry matter consists mostly (at least 90 to 95%) of carbon, oxygen, and hydrogen.

During photosynthesis plants obtain these elements from carbon dioxide and water. If plant dry matter is burned (oxidized), these elements become carbon dioxide and water once more. In the ash and smoke can be found the many nutrient elements originally taken up by the plants from the soil. These essential nutrient elements,

TABLE 11.1 **Mass of Organic Carbon in the World's Soils**

Values for the upper 1 m represent most of the carbon in the soil profile. The upper 15 cm is most readily influenced by land use and soil management.

Soil order	Global area, 10^3 km^2	Organic carbon[a] in upper 100 cm			Organic carbon[a] in upper 15 cm	
		Mg/ha	Global Pg[b]	% of global	Range,[c] %	Typical,[c] %
Entisols	14,921	99	148	9	0.06–6.0	—[d]
Inceptisols[e]	21,580	163	352	22	0.06–6.0	—[d]
Histosols[e]	1,745	2,045	357	23	12–57	47
Andisols	2,552	306	78	5	1.2–10	6
Vertisols	3,287	58	19	1	0.5–1.8	0.9
Aridisols	31,743	35	110	7	0.1–1.0	0.6
Mollisols	5,480	131	73	5	0.9–4.0	2.4
Spodosols	4,878	146	71	5	1.5–5.0	2.0
Alfisols	18,283	69	127	8	0.5–3.8	1.4
Ultisols	11,330	93	105	7	0.9–3.3	1.4
Oxisols	11,772	101	119	8	0.9–3.0	2.0
Misc. land	7,644	24	18	1	—	—
Total	135,215		1576	100		

[a] Organic matter may be roughly estimated as 1.7 to 2.0 times this value. Organic nitrogen may also be estimated from organic carbon values by dividing by 12 for most soils (see Section 11.3).
[b] Petagram = 10^{15} g.
[c] Percent on mass basis (i.e., g/100 g).
[d] These soils are too variable to suggest a typical value.
[e] Carbon stored in Gelisols is included with these soils.
Data calculated from Eswaran, et al. (1993) and Brady (1990).

such as nitrogen, sulfur, phosphorus, potassium, and micronutrients, will be given detailed consideration in later chapters.

ORGANIC COMPOUNDS IN PLANT RESIDUES. The organic compounds in plant tissue can be grouped into broad classes. Although representative percentages of these classes are shown in Figure 11.2, tissues from different plant species, as well as from different parts (leaves, roots, stems, etc.) of a given plant differ considerably in their makeup. Carbohydrates, which range in complexity from simple sugars and starches to cellulose, are usually the most plentiful of plant organic compounds.

Lignins, which are complex compounds with multiple ring-type or *phenol* structures, are components of plant cell walls. The content of lignin increases as plants mature and is especially high in woody tissues. Other **polyphenols,** such as tannins, may comprise as much as 6 or 7% of the leaves and bark of certain plants (for example, the brown color of steeped tea is due to tannins). Certain plant parts, especially seed and leaf coatings, contain significant amounts of fats, waxes, and oils, which are more complex than carbohydrates but less so than lignins. **Proteins** contain about 16% nitrogen and smaller amounts of other essential elements, such as sulfur, manganese, copper, and iron.

RATE OF DECOMPOSITION. Organic compounds may be listed in terms of ease of decomposition as follows:

1. Sugars, starches, and simple proteins   Rapid decomposition
2. Crude proteins
3. Hemicellulose
4. Cellulose
5. Fats, waxes, and so forth
6. Lignins and phenolic compounds   Very slow decomposition

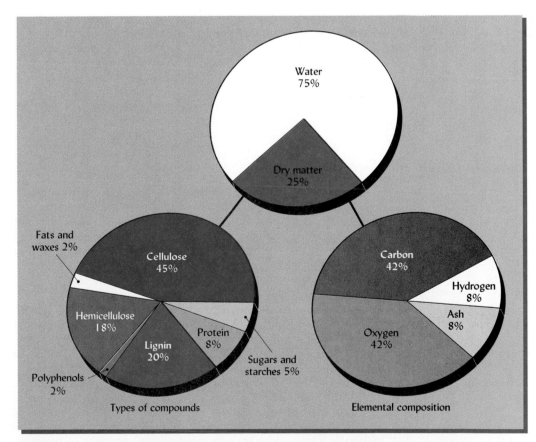

FIGURE 11.2 Typical composition of representative green-plant materials. The major types of organic compounds are indicated at left and the elemental composition at right. The *ash* is considered to include all the constituent elements other than carbon, oxygen, and hydrogen (nitrogen, sulfur, calcium, etc.).

Decomposition of Organic Compounds in Aerobic Soils

When organic tissue is added to an aerobic soil, three general reactions take place:

1. Carbon compounds are enzymatically oxidized to produce carbon dioxide, water, energy, and decomposer biomass.
2. The essential nutrient elements, such as nitrogen, phosphorus, and sulfur, are released and/or immobilized by a series of specific reactions that are relatively unique for each element.
3. Compounds very resistant to microbial action are formed, either through modification of compounds in the original tissue or by microbial synthesis.

DECOMPOSITION: AN OXIDATION PROCESS.   In a well-aerated soil, all of the organic compounds found in plant residues are subject to oxidation.

$$\underset{\substack{\text{Carbon- and} \\ \text{hydrogen-containing} \\ \text{compounds}}}{R-(C, 4H)} + 2O_2 \xrightarrow[\text{oxidation}]{\text{Enzymatic}} CO_2\uparrow + 2H_2O + \text{energy (478 kJ mol}^{-1}\text{ C)}$$

Many intermediate steps are involved in this overall reaction, and it is accompanied by important side reactions that involve elements other than carbon and hydrogen. Even so, this basic reaction accounts for most of the organic matter decomposition in the soil, as well as for the oxygen consumption and CO_2 release.

When plant proteins decay, they yield not only carbon dioxide and water, but amino acids such as glycine (CH_2NH_2COOH) and cysteine ($CH_2HSCHNH_2COOH$). In turn, these nitrogen and sulfur compounds further break down, eventually yielding such simple inorganic ions as ammonium (NH_4^+), nitrate (NO_3^-), and sulfate (SO_4^{2-}), forms available for plant nutrition.

BREAKDOWN OF LIGNIN.   Lignin molecules are very large and complex, consisting of hundreds of interlinked phenolic ring subunits. Because the linkages among these structures are so varied and strong, only a few microorganisms (mainly *white rot fungi*) can break them down. Decomposition proceeds very slowly at first, and is generally assisted by the physical activities of soil fauna. Once the lignin subunits are separated, many types of microorganisms participate in their breakdown. It is thought that microorganisms use some of the ring structures from lignin in the synthesis of stable soil organic matter.

Example of Organic Decay

The process of organic decay in time sequence is illustrated in Figure 11.3. Assume the soil has not been disturbed or amended with plant residues for some time. Initially, no readily decomposable materials are present. The principal microorganisms actively metabolizing are small populations of **autochthonous organisms**, which survive by slowly and steadily digesting the very resistant, stable soil organic matter. Competition for food is severe and microbial activity is relatively low, as reflected in the low **soil respiration** rate or level of CO_2 production in the soil. The supply of soil carbon is slowly but steadily being depleted.

Then, suddenly, an abundance of fresh, decomposable tissue is added to the soil. Maybe deciduous trees are losing their leaves in fall, or a farmer has plowed in the residue of a harvested crop. In any case, the appearance of easily decomposable and often water-soluble compounds, such as sugars, starches, and amino acids, stimulates an almost immediate increase in metabolic activity among the soil microbes. Soon the slower-acting autochthonous populations are overtaken by rapidly multiplying populations of **zymogenous** (opportunist) organisms that have been awakened from their dormant state by the presence of new food supplies. Cellulose-digesting organisms rapidly join in the attack. Microbial numbers and carbon dioxide evolution from microbial respiration both increase exponentially in response to the new food resource.

Soon microbial activity is at its peak, energy is being rapidly liberated, and carbon dioxide is being formed in large quantities. As they multiply and increase their biomass,

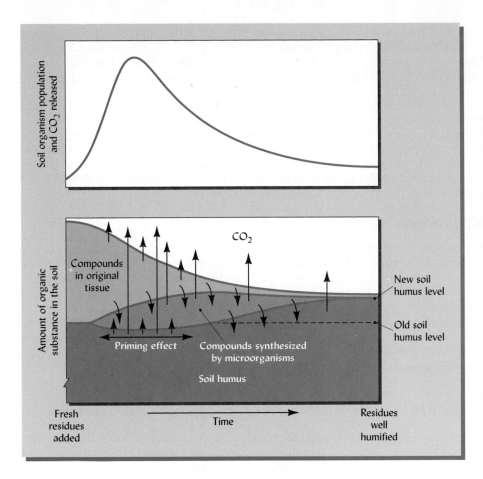

FIGURE 11.3 Diagrammatic illustration of the general changes that take place when fresh plant residues are added to a soil. The arrows indicate transfers of carbon among compartments. The time required for the process will depend on the nature of the residues and the soil. Most of the carbon released during the initial rapid breakdown of the residues is converted to carbon dioxide, but the smaller amounts of carbon converted into microbially synthesized compounds (biomass) and, eventually, into soil humus should not be overlooked. Note that although the peak of microbial activity appears to accelerate the decay of the original humus, a phenomenon known as the *priming effect*, the humus level is increased by the end of the process. Where vegetation, environment, and management remain stable for a long time, the soil humus content will reach an *equilibrium level* in which, the carbon added to the humus pool through the decomposition of plant residues each year is balanced by carbon lost through the decomposition of existing soil humus.

the microbes are also synthesizing new organic compounds. The microbial biomass at this point may account for as much as one-sixth of the organic matter in a soil. The intense microbial activity may even stimulate the breakdown of some resistant soil organic matter, a phenomenon known as the **priming effect.**

With all this frenetic microbial activity, the easily decomposed compounds are soon exhausted. While the cellulose and lignin decomposers continue their slow work, most of the zymogenous microorganisms begin to die of starvation. As microbial populations plummet, the dead cells provide a readily digestible food source for the survivors, which continue to evolve carbon dioxide and water. The decomposition of the dead microbial cells is also associated with the **mineralization** or release of simple inorganic products, such as nitrates and sulfates.

As food supplies are further reduced, microbial activity continues to decline, and the general-purpose zymogenous soil organisms again sink back into comparative quiescence. Very little of the original residue material persists, mainly tiny particles that have become **physically protected** from decay by lodging inside soil pores too tight to allow access by most organisms. Some of the remaining carbon has been **chemically protected** by conversion into **soil humus,** the dark-colored, heterogeneous, mostly colloidal mixture of modified lignin and newly synthesized organic compounds that strongly resists further decay. Some of the fine humus is further protected by binding strongly to clay particles. Thus, a small percentage of the carbon in the added residues has been retained, increasing slightly the pool of stable soil organic matter. In a mature ecosystem this increase will likely be offset during each annual cycle by slow authochthonous decomposition, resulting in little net change in the level of soil organic matter from year to year.

Decomposition in Anaerobic Soils

Microbial decomposition proceeds most rapidly in the presence of plentiful supplies of O_2, which acts as the electron acceptor during aerobic oxidation of organic compounds. Oxygen supplies may become depleted when soil pores filled with water prevent the diffusion of O_2 into the soil from the atmosphere. Without sufficient oxygen present, aerobic organisms cannot function, so anaerobic or facultative organisms become dominant. Under low-oxygen or anaerobic conditions, decomposition takes place much more slowly than when oxygen is plentiful. Hence, wet, anaerobic soils tend to accumulate large amounts of organic matter in a partially decomposed condition.

The products of anaerobic decomposition include a wide variety of partially oxidized organic compounds, such as organic acids, alcohols, and methane gas. Anaerobic decomposition releases relatively little energy for the organisms involved; therefore, the end products still contain much energy. (For this reason, alcohol and methane can serve as fuel.) Some of the products of anaerobic decomposition are of concern because they produce foul odors or inhibit plant growth. The methane gas produced in wet soils is a major contributor to the greenhouse effect (Section 11.11).

Production of Simple Inorganic Products

As proteins are attacked by microbes, the amide (—R—NH$_2$) and sulfide (—R—S) groups of the amino acids are broken off to produce, first, ammonium (NH$_4^+$) and sulfide (S^{2-}) compounds and, finally, nitrates (NO$_3^-$) and sulfates (SO$_4^{2-}$). Similar decomposition of other organic compounds releases these and other inorganic nutrient ions. The term **mineralization** applies to the overall process that releases elements from organic compounds to produce inorganic (mineral) forms. Most of the inorganic ions released by mineralization are readily available to higher plants and to microorganisms. The decay of organic tissues is an important source of nitrogen, sulfur, phosphorus, and other essential elements for plants.

11.3 FACTORS CONTROLLING RATES OF DECOMPOSITION AND MINERALIZATION[2]

The time needed to complete the processes of decomposition and mineralization may range from days to years, depending mainly on two broad factors: (1) the environmental conditions in the soil, and (2) the quality of the added residues as a food source for soil organisms.

The environmental conditions conducive to rapid decomposition and mineralization include a near-neutral pH, sufficient soil moisture, and good aeration (about 60% of the soil pore space filled with water), and warm temperatures (25 to 35°C). These conditions were discussed in Section 10.11 in relation to microbial activity and will be considered again in Section 11.10 as they affect the levels of organic matter accumulating in soils.

Physical Factors Influencing Residue Quality

The location of residues in or on the soil is a physical factor with a critical impact on decomposition rates. Surface placement of plant residues, as in forest litter or conservation tillage mulch, usually results in slower, more variable rates of decomposition than where similar residues are incorporated into the soil by root deposition, faunal action, or tillage. The incorporated residues are in intimate contact with soil organisms and generally are kept uniformly moist. Surface residues, on the other hand, are physically out of reach of most soil organisms, save fungal mycelia and the larger fauna, such as earthworms. Surface residues are subject to drying, as well as extremes of temperature. Nutrient elements mineralized from surface-applied residues are also more susceptible to loss than those from incorporated residues.

Residue particle size is another important physical factor—the smaller the particles, the more rapid the decomposition. Small particle size may result from the nature of the residues (e.g., twigs versus branches), from mechanical treatment (grinding, chopping, tillage, etc.), or from the chewing action of soil fauna. Diminution of residues into smaller particles physically exposes more surface area to decomposition, and also breaks up lignacious cell walls and waxy outer coatings on leaves so as to expose the more readily decomposed tissues and cell contents.

---

[2] For an excellent collection of papers dealing with litter decomposition in soils, see Cadisch and Giller (1997). Many of our current ideas about decomposition were first put forward in classic book by Swift, et al. (1979).

Carbon/Nitrogen Ratio of Organic Materials and Soils

The carbon content of typical plant dry matter is about 42% (see Figure 11.2); that of soil organic matter ranges from 40 to 58%.[3] In contrast, the nitrogen content of plant residues is much lower and varies widely (from <1 to >6%). The ratio of carbon to nitrogen C/N in organic residues applied to soils is important for two reasons: (1) intense competition among microorganisms for available soil nitrogen occurs when residues having a high C/N ratio are added to soils, and (2) the C/N ratio in residues helps determine their rate of decay and the rate at which nitrogen is made available to plants.

C/N RATIO IN PLANTS AND MICROBES. The C/N ratio in plant residues ranges from between 10:1 to 30:1 in legumes and young green leaves to as high as 600:1 in some kinds of sawdust (Table 11.2). Generally, as plants mature, the proportion of protein in their tissues declines, while the proportion of lignin and cellulose, and the C/N ratio, increase. These differences in composition have pronounced effects on the rate of decay when plant residues are added to the soil.

In the bodies and cells of microorganisms, the C/N ratio is not only less variable than in plant tissues, but also much lower, ordinarily falling between 5:1 and 10:1. Among microorganisms, bacteria are generally somewhat richer in protein than fungi, and consequently have a lower C/N ratio.

C/N RATIO IN SOILS. The C/N ratio in the organic matter of arable (cultivated) surface (Ap) horizons commonly ranges from 8:1 to 15:1, the median being near 12:1. The ratio is generally lower for subsoils than for surface layers in a soil profile. In a given climatic region, little variation occurs in the C/N ratio for similarly managed soils. For instance, in calcium-rich soils of semiarid grasslands (e.g., Mollisols and tropical Alfisols), the C/N ratio is relatively narrow. In more severely leached and acidic soils in humid regions, the C/N is relatively wide; C/N ratios as high as 30:1 are not uncommon. When such soils

---

[3] The traditional value used in converting between values of soil organic matter and values of soil carbon is 58% C in the organic matter. However, this value probably applies only to highly stabilized humus, especially in subsoils. For most situations, a value of 50% is more accurate.

TABLE 11.2 Typical Carbon and Nitrogen Contents and C/N Ratios of Some Organic Materials Commonly Associated with Soils

Organic material	% C	% N	C/N
Spruce sawdust	50	0.05	600
Hardwood sawdust	46	0.1	400
Wheat straw	38	0.5	80
Papermill sludge	54	0.9	61
Corn stover	40	0.7	57
Sugarcane trash	40	0.8	50
Rye cover crop, anthesis	40	1.1	37
Bluegrass from fertilized lawn	37	1.2	31
Rye cover crop, vegetative stage	40	1.5	26
Mature alfalfa hay	40	1.8	25
Rotted barnyard manure	41	2.1	20
Finished household compost	30	2.0	15
Young alfalfa hay	40	3.0	13
Hairy vetch cover crop	40	3.5	11
Digested municipal sewage sludge	31	4.5	7
Soil microorganisms			
Bacteria	50	10.0	5
Actinomycetes	50	8.5	6
Fungi	50	5.0	10
Soil organic matter			
Spodosol O horizon	50	0.5	90
Tropical evergreen litter	50	2.0	25
Mollisol Ap horizon	56	4.9	11
Ultisol A1 horizon	52	2.3	23
Average B horizon	46	5.1	9

Data calculated from many sources.

---

are brought under cultivation and limed to increase their pH and calcium content, the enhanced decomposition tends to lower the C/N ratio to near 12:1.

Influence of Carbon/Nitrogen Ratio on Decomposition

Soil microbes, like other organisms, require a balance of nutrients from which to build their cells and extract energy. The majority of soil organisms metabolize carbonaceous materials both in order to obtain carbon for building essential organic compounds and to obtain energy for life processes. However, no creature can multiply and grow on carbon alone. Organisms must also obtain sufficient nitrogen to synthesize nitrogen-containing cellular components, such as amino acids, enzymes, and DNA.

On the average, soil microbes must incorporate into their cells about eight parts of carbon for every one part of nitrogen (i.e., assuming the microbes have an average C/N ratio of 8:1). Because only about one-third of the carbon metabolized by microbes is incorporated into their cells (the remainder is respired and lost as CO_2), the microbes need to find about 24 parts of carbon for every part of nitrogen assimilated into their bodies (i.e., a C/N ratio of 24:1 in their "food").

This requirement results in two extremely important practical consequences. First, if the C/N ratio of organic material added to soil exceeds about 25:1, the soil microbes will have to scavenge the soil solution to obtain enough nitrogen. Thus, the incorporation of high C/N residues will deplete the soil's supply of soluble nitrogen, causing higher plants to suffer from nitrogen deficiency. Second, the decay of organic materials can be delayed if sufficient nitrogen to support microbial growth is neither present in the material undergoing decomposition nor available in the soil solution. These concepts are illustrated by the example in Figure 11.4.

Examples of Inorganic Nitrogen Release During Decay

The practical significance of the C/N ratio becomes apparent if we compare the changes that take place in the soil when residues of either high or low C/N ratio are added (Figure 11.5). Consider a soil with a moderate level of soluble nitrogen (mostly nitrates). General-purpose decay organisms are at a low level of activity in this soil, as evidenced by low carbon dioxide production. If no nitrogen were lost or taken up by plants, the level of nitrates would very slowly increase as the native soil organic matter decays.

LOW NITROGEN MATERIAL.   Now consider what happens when a large quantity of readily decomposable organic material is added to this soil. If this material has a C/N ratio greater than 25, changes will occur according to the pattern shown in Figure 11.5a. In the example shown, the initial C/N ratio of the residues is about 60, typical for cornstalks or many kinds of leaf litter. As soon as the residues contact the soil, the microbial community responds to the new food supply (see Section 11.2). Heterotrophic zymogenous microorganisms become active, multiply rapidly, and yield carbon dioxide in large quantities. Because of the microbial demand for nitrogen, little or no mineral nitrogen (NH_4^+ or NO_3^-) is available to higher plants during this period.

NITRATE DEPRESSION.   This condition, often called the **nitrate depression period**, persists until the activities of the decay organisms gradually subside due to lack of easily oxidizable carbon. As their numbers decrease, carbon dioxide formation drops off, and nitrogen demand by microbes becomes less acute. As decay proceeds, the C/N ratio of the remaining plant material decreases because carbon is being lost (by respiration) and nitrogen is being conserved (by incorporation into microbial cells). Generally, one can expect mineral nitrogen to begin to be released when the C/N ratio of the remaining material drops below about 20. Then, nitrates appear again in quantity, and the original conditions prevail, except that the soil is somewhat richer in both nitrogen and humus.

The nitrate depression period may last for a few days, a few weeks, or even several months. A longer, more severe period of nitrate depression is typical when added residues are easily decomposed and have a higher C/N ratio, and when a larger quantity of residues is added. To avoid producing seedlings that are stunted, chlorotic, and nitrogen-starved, planting should be delayed until after the nitrate depression period, or additional sources of nitrogen can be applied to satisfy the nutritional requirements of both the microbes and the plants.

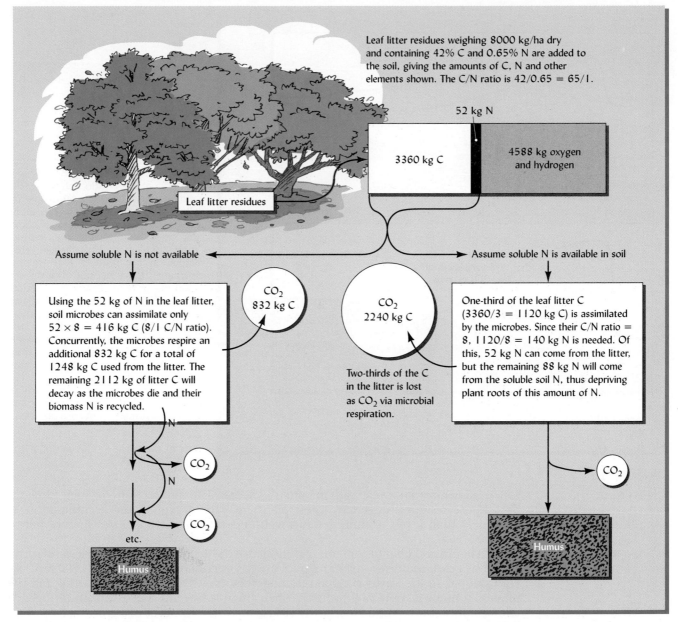

Leaf litter residues weighing 8000 kg/ha dry and containing 42% C and 0.65% N are added to the soil, giving the amounts of C, N and other elements shown. The C/N ratio is 42/0.65 = 65/1.

52 kg N

3360 kg C

4588 kg oxygen and hydrogen

Assume soluble N is not available

Assume soluble N is available in soil

Using the 52 kg of N in the leaf litter, soil microbes can assimilate only 52 × 8 = 416 kg C (8/1 C/N ratio). Concurrently, the microbes respire an additional 832 kg C for a total of 1248 kg C used from the litter. The remaining 2112 kg of litter C will decay as the microbes die and their biomass N is recycled.

CO_2 832 kg C

CO_2 2240 kg C

Two-thirds of the C in the litter is lost as CO_2 via microbial respiration.

One-third of the leaf litter C (3360/3 = 1120 kg C) is assimilated by the microbes. Since their C/N ratio = 8, 1120/8 = 140 kg N is needed. Of this, 52 kg N can come from the litter, but the remaining 88 kg N will come from the soluble soil N, thus depriving plant roots of this amount of N.

N

CO_2

N

CO_2

etc.

Humus

CO_2

Humus

FIGURE 11.4   A simplified, quantitative example of plant residue decay illustrating the fates of carbon and nitrogen and the consequences for decomposition and soil nitrogen availability. Note that if an adequate supply of nitrogen is available, the potential for humus creation is increased.

HIGH-NITROGEN MATERIAL.   The effects on soil nitrate level will be quite different if the residues added have a C/N ratio lower than 20, as in the case represented by Figure 11.5*b*. With organic materials of low C/N ratio, more than enough nitrogen is present to meet the needs of the decomposing organisms. Therefore, soon after decomposition begins, some of the nitrogen from organic compounds is released into the soil solution, augmenting the level of soluble nitrogen available for plant uptake. Generally, nitrogen-rich materials decompose quite rapidly, resulting in a period of intense microbial growth and activity, but no nitrate depression period.

Influence of Lignin and Polyphenol Content of Organic Materials

LITTER QUALITY.   Because they support only low levels of microbial activity and biomass, residues high in phenols and/or lignin are considered to be *poor quality resources* for the

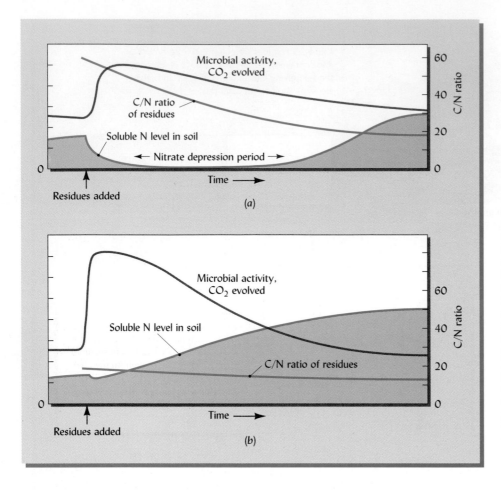

FIGURE 11.5 Changes in microbial activity, in soluble nitrogen level, and in residual C/N ratio following the addition of either high (*a*) or low (*b*) C/N ratio organic materials. Where the C/N ratio of added residues is above 25, microbes digesting the residues must supplement the nitrogen contained in the residues with soluble nitrogen from the soil. During the resulting nitrate depression period, competition between higher plants and microbes would be severe enough to cause nitrogen deficiency in the plants. Note that in both cases soluble N in the soil ultimately increases from its original level once the decomposition process has run its course. The trends shown are for soils without growing plants, which, if present, would continually remove a portion of the soluble nitrogen.

soil organisms that cycle carbon and nutrients. The production of such slow-to-decompose residues by certain forest plants may help explain the accumulation of extremely high levels of humified nitrogen and carbon in the soils of mature boreal forests.

The lignin and phenol contents also influence the decomposition and release of nitrogen from **green manures**—plant residues used as to enrich agricultural soils (Table 11.3). For example, in the leaves of certain legume trees, the C/N ratio is quite narrow, but the phenol content is quite high, so that when these leaves are added to soil, nitrogen is released only slowly—often too slowly to keep up with the needs of a growing crop. Similarly, residues with a lignin content of more than 20 to 25% will decompose too slowly

TABLE 11.3 Litter Quality in Relation to the Lignin Content, Polyphenol Content and C/N Ratio of Several Types of Plant Residues

Prunings (leaves and small twigs) of three common agroforestry tree species and afterharvest residues of two cereal crops were applied at a rate of 5 Mg/ha to an Oxic Paleudult in a humid tropical region of Nigeria. Low values of C/N, lignin, and polyphenols all contribute to high litter quality and high speed of decomposition. The inhibitory effect of polyphenol content can be seen by comparing Gliricidia to Leucaena.

Plant species	Plant parts	Lignin, %	Polyphenols, %	C/N	Decomposition constant,[a] k/week	Litter quality
Gliricidia sepium	Prunings	12	1.6	13	0.255	High
Leucaena leucocephala	Prunings	13	5.0	13	0.166	Medium–high
Oryza sativa	Straw	5	0.6	42	0.124	Medium
Zea mays	Stover	7	0.6	43	0.118	Medium
Dactyladenia barteri	Prunings	47	4.1	28	0.011	Low

[a] As each type of residue decomposed during a 14-week season, researchers periodically determined the proportion Y of the original residue dry matter remaining. The decomposition rate k was determined from the equation $Y = e^{-kt}$, in which e is the base of natural logarithms, and t is time in weeks. Therefore, the larger the decomposition constant k, the faster the decomposition.
Data selected from Tian, et al. (1992 and 1995).

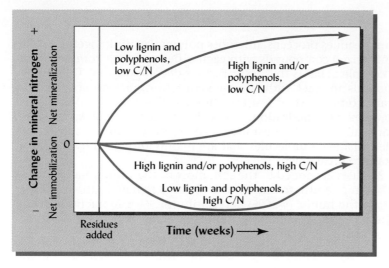

FIGURE 11.6 Temporal patterns of nitrogen release from organic residues differing in quality based on their C/N ratios and contents of lignin and polyphenols. Lignin contents greater than 20%, polyphenol contents greater than 5%, and C/N ratios greater than 30 would all be considered high in the context of this diagram, the combination of these properties characterizing litter of poor quality—that is, litter that has a limited potential for microbial decomposition and mineralization of plant nutrients.

to be effective as green manure for rapidly growing annual crops. For perennial crops or forest systems, however, the slow release of nitrogen from such residues may be advantageous in the long run, as the nitrogen may be less subject to losses. By the same token, the slow decomposition of phenol- or lignin-rich materials means that even if their C/N ratio is very high, the nitrate depression will not be pronounced (Figure 11.6).

11.4 HUMUS: GENESIS AND NATURE[4]

The general term **soil organic matter** (SOM) encompasses all the organic components of a soil: (1) living biomass (intact plant and animal tissues and microorganisms), (2) dead roots and other recognizable plant residues, as well as (3) a largely amorphous and colloidal mixture of complex organic substances no longer identifiable as tissues. Only the third category of organic material is properly referred to as **soil humus** (Figure 11.7).[5]

---

[4] For discussions of humus formation, composition, and reactions, see Frimmel and Christman (1988) and Stevenson (1994).

[5] Although many soil scientists use the term *soil organic matter* synonymously with *humus,* we exclude soil biomass and dead but recognizable tissues from the latter, while including them in the former. As a practical matter, particles of plant residue that do not pass a 2-mm sieve opening are often excluded from consideration as soil organic matter.

FIGURE 11.7 Classification of soil organic matter components separable by chemical and physical criteria. Although surface residues (litter) are not universally considered to be part of the soil organic matter, we include them because they are the principal component of the O horizons in soil profiles. Solubility in alkali and acid is a widely used criterion for grouping different fractions of soil humic substances. The classical scheme dividing soil humic substances into humin, fulvic acids, and humic acids fractions (not shown) is based on their insolubility in NaOH (humin) and their subsequent solubility (fulvic acids) and insolubility (humic acids) in acid solutions (pH = 1).

Microbial Transformations

As decomposition of plant residues proceeds, microbes polymerize (link together) some of the simpler newly synthesized compounds with each other and with complex residual products into long, complex chains that resist further decomposition. These high-molecular-weight compounds interact with nitrogen-containing amino compounds, giving rise to a significant component of resistant humus. The presence of colloidal clays stimulates the complex polymerization. These ill-defined, complex, resistant, polymeric compounds are called **humic substances.** The term **nonhumic substances** refers to the group of identifiable biomolecules that are mainly produced by microbial action and are less resistant to breakdown.

One year after plant residues are added to the soil, most of the carbon has returned to the atmosphere as CO_2, but one-fifth to one-third is likely to remain in the soil either as live biomass ($\approx 5\%$) or as the humic ($\approx 20\%$) and nonhumic ($\approx 5\%$) fractions of soil humus (Figure 11.8). The proportion remaining from root residues tends to be somewhat higher than that remaining from incorporated leaf litter.

Humic Substances

Humic substances comprise about 60 to 80% of the soil organic matter. They are comprised of huge molecules with variable, complex structures characterized by many aromatic rings. Humic substances generally are dark-colored, amorphous substances with molecular weights varying from 2000 to 300,000 g/mol. Because of their complexity, they are the organic materials most resistant to microbial attack.

Depending on the environment, the half-life (the time required to destroy half the amount of a substance) of humic substances may be from decades to centuries.

Nonhumic Substances

About 20 to 30% of the humus in soils consists of nonhumic substances. These substances are less complex and less resistant to microbial attack than those of the humic

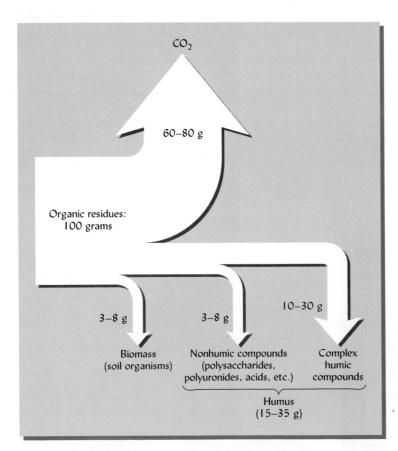

FIGURE 11.8 Disposition of 100 g of organic residues one year after they were incorporated into the soil. More than two-thirds of the carbon has been oxidized to CO_2, and less than one-third remains in the soil—some in the cells of soil organisms, but a larger component as soil humus. The amount converted to CO_2 is generally greater for aboveground residues than for belowground (root) residues. (Estimates from many sources)

group. Unlike humic substances, they are comprised of specific biomolecules with definite physical and chemical properties. Some of these nonhumic substances are microbially modified plant compounds, while others are compounds synthesized by soil microbes as by-products of decomposition.

Included among the nonhumic substances are polysaccharides, polymers which have sugarlike structures and a general formula of $C_n(H_2O)_m$, where n and m are variable. Polysaccharides are especially important in enhancing soil aggregate stability (see Section 4.7).

Some even simpler compounds (such as low-molecular-weight organic acids and some proteinlike materials) are part of the nonhumic group. Although none of these simpler materials are present in large quantities, they may influence the availability of plant nutrients, such as nitrogen and iron, and may also directly affect plant growth.

Clay-Humus Combinations[6]

Interaction with clay minerals provides an important means of stabilizing soil nitrogen and organic matter. Certain clays are known to attract and hold such substances as amino acids, peptides, and proteins, forming complexes that protect the nitrogen-containing compounds from microbial degradation. Organic matter that is entrapped in the very small (<1 μm) pores formed by clay particles is physically inaccessible to decomposing organisms. It is also possible that layer silicate clays, such as vermiculite, may bind organic matter in their interlayers in forms that strongly resist decomposition.

Thus, clay, along with humic and polysaccharide polymers, can protect relatively simple nitrogen compounds from microbial attack. In many soils more than half the organic matter is associated with clay and other inorganic constituents. Although the extent and mechanisms are not yet fully understood, clay–humus interactions partially account for the high organic matter content of clay soils.

Colloid Characteristics of Humus

The colloidal nature of humus was emphasized in Section 8.1. The surface area of humus colloids per unit mass is very high, generally exceeding that of silicate clays. The colloidal surfaces of humus are negatively charged, as a result of H^+ dissociation from carboxylic (—COOH) or phenolic (—OH) groups. The extent of the negative charge is pH-dependent. At high pH values, the cation exchange capacity of humus on a mass basis (150 to 300 cmol/kg) far exceeds that of most silicate clays. However, because humus is much less dense than clay, the cation exchange capacity of humus on a volume basis (40 to 80 cmol/L) is comparable to that of silicate clays.

The water-holding capacity of humus on a mass basis (but not on a volume basis) is four to five times that of the silicate clays. Humus plays a role in aggregate formation and stability. The highly complex humus molecules contain chemical structures that absorb nearly all wavelengths of visible light, giving the substance a characteristic black color (see, for example, the humus-rich A horizons in Plates 2 and 8).

11.5 COMPOSTS AND COMPOSTING

Composting is the practice of creating humuslike organic materials by aerobic decomposition outside of the soil. With composting, decay occurs in such a concentrated fashion as to generate considerable heat. The finished product, **compost**, is popular as a mulch, as an ingredient for potting mixes, and as an organic soil conditioner and slow-release fertilizer.

COMPOSTING PROCESS.   When a sufficient mass of suitable organic materials is kept in a moist, well-aerated, state, it goes through a three-stage process of decomposition (Figure 11.9). (1) During a brief initial *mesophilic* stage sugars and readily available microbial food sources are rapidly metabolized, causing the temperature in the compost pile to

---

[6] For a detailed account of clay–humus complexes, see Huang and Schnitzer (1986) and Loll and Bollag (1983).

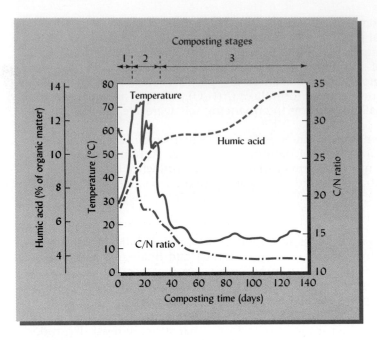

FIGURE 11.9 Changes in the temperature, C/N ratio, and humic acid contents of the organic matter during the production of compost from municipal solid wastes (MSW). The compost was turned (mixed for aeration) once a week for the first 49 days and once in two weeks thereafter. The stages of the compost process are: (1) initial mesophilic stage, (2) thermophilic stage, and (3) final mesophilic or curing stage. See text for explanation of the composting process. [Data redrawn from Chefetz, et al. (1996)]

gradually rise from ambient to over 40°C. (2) A *thermophilic* stage occurs during the next week or two, during which temperatures rise to 50 to 75°C while thermophilic organisms decompose cellulose and other more resistant materials. Frequent mixing during this stage is essential to maintain oxygen supplies and assure even heating of all the material. The easily decomposed compounds are used up, and humuslike compounds are formed during this stage. (3) A second mesophilic or *curing stage* follows, during which the temperature falls back to near ambient, and the material is recolonized by mesophilic organisms, including certain beneficial organisms that produce plant-growth-stimulating compounds or are antagonistic to plant pathogenic fungi. This stage may last from several weeks to several months.

NATURE OF THE COMPOST PRODUCED. As raw organic materials are humified in a compost pile, the content of nonhumic substances declines, and the content of humic acid increases markedly (see Figure 11.9). During the composting process, the C/N ratio of organic materials in the pile decreases until a fairly stable ratio, in the range of 14:1 to 20:1, is achieved. The CEC of the organic matter increases to about 50 to 70 cmol$_c$/kg of compost.

Although 50% or more of the carbon in the initial material is typically lost during composting, mineral nutrients are mostly conserved. Finished compost is therefore generally more concentrated in nutrients than the initial combination of raw materials used and can serve as an effective means of building soil fertility. Finished compost should be free of viable weed seeds and pathogenic organisms, as these are generally destroyed during the thermophilic phase. However, inorganic contaminants such as heavy metals are *not* destroyed by composting.

Proper management of compost is essential if the finished product is to be desirable for use as a potting media or soil amendment (see Box 11.1).

BENEFITS OF COMPOSTING. Although making compost may involve more work and expense than applying uncomposted organic materials directly to the soil, the process offers at least seven distinct advantages:

1. *Safe storage.* Composting provides a means of effectively and safely storing organic materials until it is convenient to apply them to soils.

2. *Easier handling.* As a result of CO_2 losses and settling, the volume of composted organic materials decreases by about 30 to 50% during the composting process. The smaller volume and greater uniformity of the resulting material may greatly ease the handling and eventual use of the organic matter as a soil amendment or potting medium.

BOX 11.1   MANAGEMENT OF A COMPOST PILE

MATERIALS TO USE.   For homeowners, good materials for composting include leaves (preferably shredded), grass clippings, weeds, kitchen scraps, wood shavings, gutter cleanings, pine needles, spoiled hay, and straw. Commercial compost is often made from garbage, paper, sewage sludge, wood chips, animal manures, municipal leaves, and food processing wastes.

MATERIALS TO AVOID.   Avoid using meat scraps (attract rodents), cat droppings (carry pathogens), wood wastes from pressure-treated lumber and plywood (contain heavy metals and arsenic), and plastics and glass (nuisance or dangerous in final product).

BALANCING NUTRIENTS.   Although highly carbonaceous materials can be made to compost satisfactorily if they are turned frequently and kept moist, best results are obtained if high-C/N-ratio materials (such as brown leaves, straw, or paper) are mixed with low-C/N-ratio materials (such as green grass clippings, legume hay, blood meal, sewage sludge, or livestock manure) so as to achieve an overall C/N ratio of 30 or less.

Other materials often added to compost to improve nutrient balance and content are fertilizers (N, P, and K), wood ashes (K, Ca, and Mg), bone meal or phosphate rock powder (P and Ca), and seaweed (K, Mg, Ca, and micronutrients). Some of these materials contain enough soluble salts to necessitate leaching of the finished compost before using it on salt-sensitive plants.

COMPOSTING METHODS.   Allow for good aeration throughout the pile while still providing sufficient mass to generate heat and prevent excessive drying. Backyard compost piles should measure at least 1 m square and 1 m high. Compost bins are available to make turning the compost easier. Large-scale composting is usually carried out in windrows, about 2 to 3 m wide, 2 m tall, and many meters long. In dry climates, composting is often done in pits dug about 1 m deep to protect the material from drying. The various materials to be composted can be mixed together or applied in thin layers. Often, a small amount of garden soil or finished compost is added to ensure that plenty of decomposer organisms will be immediately available. Microbial compost activators are commercially available, but while some may speed the initial heating of the pile, scientific tests rarely show any other advantage to using these preparations.

(a)

(b)

FIGURE 11.10   An efficient and easily managed method of composting suitable for homeowners is the three-bin method (a), in which materials are moved from one bin to the next when they are turned. Perforated white plastic pipes may be inserted for enhanced aeration. The bin on the left contains relatively fresh materials, while the one at the right contains finished compost. These bins at the Rodale Research Institute in Pennsylvania are being used to compost mainly lawn cuttings and plant residues from a vegetable garden. Most large-scale composting operations use long windrows that can be turned by special machines or aerated by pulling air through the pile and into perforated pipes (b). In the case shown here, sewage sludge mixed with wood chips is being composted by the USDA Beltsville aerated pile method. Air is being drawn through the large composting windrow and expelled through the smaller piles of finished compost, which absorb odors from the airstream. (Photos courtesy of R. Weil)

OXYGEN AND MOISTURE CONTROL.   Low oxygen levels, usually due to inadequate turning combined with excessive moisture, can produce putrid odors as anaerobic decomposition takes over. Monitoring temperature and oxygen levels in the pile can help avoid this situation. Good aeration can be promoted by mixing in a bulking agent such as wood chips, by avoiding excessive packing, and by either turning the pile or pulling a stream of air through it (Figure 11.10). The pile should be turned during the thermophilic stage whenever the temperature begins to drop.

The compost should be kept moist, but not wet. When squeezed, compost containing the proper amount of moisture (50 to 70%) will feel damp, but not dripping wet. Turning the pile during dry weather can help reduce excess moisture, while turning it during rain can help moisten a too-dry pile.

3. *Nitrogen competition avoidance.*   For residues with a high initial C/N ratio, proper composting ensures that any nitrate depression period will occur in the compost pile, not in the soil, thereby avoiding induced plant nitrogen deficiency.

4. *Nitrogen stabilization.*   When applied to the soil, composted materials generally decompose and mineralize much more slowly than uncomposted organic materials. **Cocomposting** low-C/N-ratio materials (such as livestock manure and sewage sludge) with high-C/N-ratio materials (such as sawdust, wood chips, senescent tree leaves, or municipal solid waste) provides sufficient carbon for microbes to immobilize the excess nitrogen and minimize any nitrate leaching hazard from the low-C/N materials. It also provides sufficient nitrogen to speed the decomposition of the high-C/N materials.

5. *Partial sterilization.*   High temperatures during the thermophilic stage in well-managed compost piles kill most weed seeds and pathogenic organisms in a matter of a few days. Under less ideal conditions, temperatures in parts of the pile may not exceed 40 to 50°C, so weeks or months may be required to achieve the same results.

6. *Detoxification.*   Most toxic compounds that may be in organic wastes (pesticides, natural phytotoxic chemicals, etc.) are destroyed by the time the compost is considered mature and ready to use. Composting is therefore often used as a method of biological treatment of polluted soils and wastes (see Box 10.1).

7. *Disease suppression.*   Some composts can effectively suppress soilborne plant diseases by encouraging microbial antagonisms (see Section 10.14). Most success in disease suppression has occurred when well-cured compost is used as a main component of potting mixes for greenhouse-grown plants. Some disease suppression has also been observed with field applications of compost.

11.6   DIRECT INFLUENCES OF ORGANIC MATTER ON PLANT GROWTH

Long ago, the observation that plants generally grow better on organic-matter-rich soils led people to think that plants derive much of their nutrition by absorbing humus from the soil. We now know that higher plants derive their carbon from carbon dioxide and that most of their nutrients come from inorganic ions dissolved in the soil solution. In fact, plants can complete their life cycles growing totally without humus, or even without soil (as in soilless or **hydroponic** production systems using only aerated nutrient solutions). This is not to say that soil organic matter is less important to plants than was once supposed, but rather that most of the benefits accrue to plants indirectly through the many influences of organic matter on soil properties. These will be discussed in Section 11.7, after we consider two types of direct organic matter effects on plants.

Direct Influence of Humus on Plant Growth

It is well established that certain organic compounds are absorbed by higher plants. Plants can absorb a very small portion of their nitrogen and phosphorus needs as soluble organic compounds. Various growth-promoting compounds such as vitamins, amino acids, auxins, and gibberellins, are formed as organic matter decays. These substances may at times stimulate growth in both higher plants and microorganisms.

Small quantities of humic substances in the soil solution are known to enhance certain aspects of plant growth (Table 11.4). Components of these humic substances probably act as regulators of specific plant-growth functions, such as cell elongation and lateral root initiation. The concentrations of humic substances commonly present in the soil solution in humid regions (50 to 100 mg/L or ppm) are effective in stimulating plant growth. Commercial humate products have been marketed with claims that small amounts enhance plant growth, but scientific tests of many of these products have failed to show any benefit from their use. Perhaps this is because effective levels of humic substances are naturally present in most soils.

ALLELOCHEMICAL EFFECTS.[7]   **Allelopathy** is the process by which one plant infuses the soil with a chemical that affects the growth of other plants. The plant may do this by directly exuding **allelochemicals**, or the compounds may be leached out of the plant

---

[7] For a comprehensive review of allelopathy, see Inderjit, et al. (1995).

TABLE 11.4   Some Direct Effects of Humic Substances on Plant Growth

Effect on plant growth	Humic substance	Concentration range, mg/L
Accelerated water uptake and enhanced germination of seeds	Humic acid	1–100
Stimulated root initiation and elongation	Humic and fulvic acids	50–300
Enhanced root cell elongation	Humic acid	5–25
Enhanced growth of plant shoots and roots	Humic and fulvic acids	50–300

From Chen and Aviad (1990).

foliage by through-fall rainwater. In other cases, microbial metabolism of dead plant tissues (residues) forms the allelochemicals (Figure 11.11). Occasionally, the term *allelochemical* is also applied to plant chemicals that inhibit microorganisms. In principal, the interactions are much like the antagonistic relationships among certain microorganisms discussed in Section 10.14.

Allelochemicals present in the soil are apparently responsible for many of the effects observed when various plants grow in association with one another. Because they produce such chemicals, certain weeds (e.g., johnsongrass and giant foxtail) damage crops far out of proportion to the size and number of weeds present. Crop residues left on the soil surface may inhibit the germination and growth of the next crop planted (e.g., wheat residues often inhibit sorghum plants). Other allelopathic interactions influence the succession of species as forest and grassland ecosystems mature.

Allelopathic interactions are usually very specific, involving only certain species, or even varieties, on both the producing and receiving ends. The effects of allelopathic chemicals are many and varied. Although the term *allelopathy* most commonly refers to negative effects, allelochemical effects can also be positive (as in certain **companion plantings**). While they vary in chemical composition, most allelochemicals are relatively simple phenolic or organic acid compounds that could be included among the nonhumic substances found in soils. Because most of these compounds can be rapidly destroyed by soil microorganisms or easily leached out of the root zone, effects are usually relatively short-lived once the source is removed.

11.7   INFLUENCE OF ORGANIC MATTER ON SOIL PROPERTIES AND THE ENVIRONMENT

Soil organic matter affects so many soil properties and processes that a complete discussion of the topic is beyond the scope of this chapter. Indeed, in almost every chapter in this book there is mention of the roles of soil organic matter. Figure 11.12 summarizes some of the more important effects of organic matter on soil properties and on soil–environment interactions. Often one effect leads to another, so that a complex chain of multiple benefits results from the addition of organic matter to soils. For example (beginning at the upper left in Figure 11.12), adding organic mulch to the soil surface encourages earthworm activity, which in turn leads to the production of burrows and other biopores, which in turn increases the infiltration of water and decreases its loss as runoff, a result that finally leads to less pollution of streams and lakes.

FIGURE 11.11 Positive and negative allelopathic effects of winged beans on grain amaranth plants. In the pot on the left (T4) amaranth is growing in fresh soil (no association with winged beans). In the center pot (T20) amaranth is growing in soil previously used to grow winged beans (positive effect). In the pot on the right (T28) the amaranth is growing in fresh soil, but the plant was watered three times with a water extract of winged bean tissue (negative effect). The average dry weight of the amaranth plants for each treatment is shown. All pots were watered with a complete nutrient solution. [From Weil and Belmont (1987)]

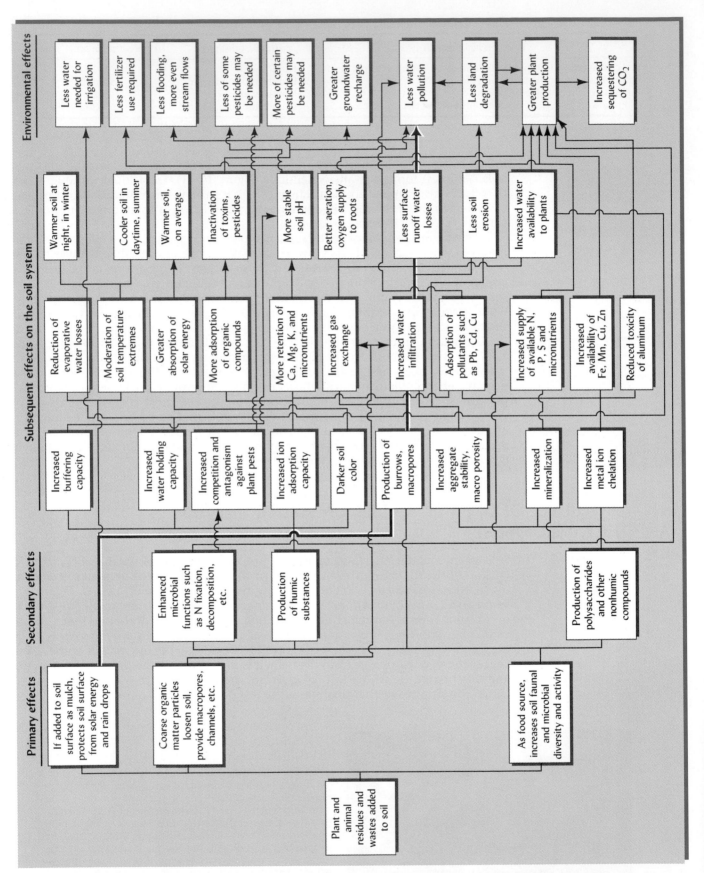

FIGURE 11.12 Some of the ways in which soil organic matter influences soil properties, plant productivity, and environmental quality. Many of the effects are indirect, the arrows indicating the cause-and-effect relationships. It can readily be seen that the influences of soil organic matter are far out of proportion to the relatively small amounts present in most soils. Many of these influences are discussed in this and other chapters in this book. The thicker line shows the sequence of effects referred to in the text in Section 11.7.

INFLUENCE ON SOIL PHYSICAL PROPERTIES. Humus tends to give surface horizons dark brown to black colors. Granulation and aggregate stability are encouraged, especially by the non-humic substances produced during decomposition. The humic fractions help reduce the plasticity, cohesion, and stickiness of clayey soils, making these soils easier to manipulate. Soil water retention is also improved, since organic matter increases both infiltration rate and water-holding capacity.

INFLUENCE ON SOIL CHEMICAL PROPERTIES. Through its high cation exchange capacity and acid and base functional groups, organic matter also provides much of the cation-adsorption and pH buffering capacities of soils (see Section 9.3). In addition, nitrogen, phosphorus, sulfur, and micronutrients are stored as constituents of soil organic matter, from which they are slowly released by mineralization.

Humic acids also attack soil minerals and accelerate their decomposition, thereby releasing essential nutrients as exchangeable cations. Organic acids, polysaccharides, and fulvic acids all can attract such cations as Fe^{3+}, Cu^{2+}, Zn^{2+}, and Mn^{2+} from the edges of mineral structures and **chelate** or bind them in stable organomineral complexes. Some of these metals are made more available to plants as micronutrients because they are kept in soluble, chelated form (see Chapter 13). In very acid soils, organic matter alleviates aluminum toxicity by binding the aluminum ions in nontoxic complexes (see Section 9.1).

BIOLOGICAL EFFECTS. Soil organic matter greatly influences the biology of the soil, because it provides most of the food for the community of heterotrophic soil organisms described in Chapter 10. The quality of soil organic matter markedly affects decomposition rates and, therefore, the amount of organic matter accumulating in soils.

11.8 MANAGEMENT OF AMOUNT AND QUALITY OF SOIL ORGANIC MATTER

Perhaps the most useful approach to defining soil organic matter quality is to recognize different pools of organic carbon that vary in their susceptibility to microbial metabolism. As discussed in Section 11.2, plant residues contain some components (*metabolic carbon*), such as sugars, proteins, and starches, that are quite readily metabolized by soil microbes. Other components (*structural carbon*) exist mostly in the structure of the plant cell walls, including lignin, polyphenols, cellulose, and waxes, and are resistant to decomposition. The model in Figure 11.13 denotes these groups as metabolic and structural pools within the plant residues.

The total organic matter content of a soil is the sum of several different pools of soil organic matter, namely, the **active**, **slow**, and **passive** fractions.

Active Fraction of Soil Organic Matter

The **active fraction** of soil organic matter consists of materials with relatively high C/N ratios (about 15 to 30) and short half-lives (half of these materials can be metabolized in a matter of a few months to a few years). Components probably include the living biomass, some of the fine particulate detritus (referred to as **particulate organic matter**, or POM), most of the polysaccharides, and other nonhumic substances described in Section 11.4, as well as some of the more labile (easily decomposed) fulvic acids. This active fraction provides most of the readily accessible food for the soil organisms and most of the readily mineralizable nitrogen. It is responsible for most of the beneficial effects on structural stability that lead to enhanced infiltration of water, resistance to erosion, and ease of tillage. The active fraction can be readily increased by the addition of fresh plant and animal residues, but it is also very readily lost when such additions are reduced or tillage is intensified. This fraction rarely comprises more than 10 to 20% of the total soil organic matter.

Passive and Slow Fractions of Organic Matter

The **passive fraction** of soil organic matter consists of very stable materials remaining in the soil for hundreds or even thousands of years. This fraction includes most of the humus physically protected in clay–humus complexes, most of the humin, and much of the humic acids. The passive fraction accounts for 60 to 90% of the organic matter in

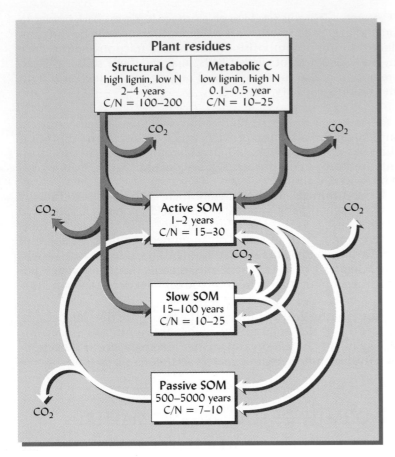

FIGURE 11.13 A conceptual model that recognizes various pools of soil organic matter (SOM) differing by their susceptibility to microbial metabolism. Models that incorporate *active, slow,* and *passive* fractions of soil organic matter have proven very useful in explaining and predicting real changes in soil organic matter levels and in attendant soil properties. Note that microbial action can transfer organic carbon from one pool to another. For example, when the nonhumic substances and other components of the active fraction are rapidly broken down, some resistant, complex by-products may be formed, adding to the slow and passive fractions. Note that all these metabolic changes result in some loss of carbon from the soil as CO_2. [Adapted from Paustian, et al. (1992)]

most soils, and its quantity is increased or diminished only slowly. The passive fraction is most closely associated with the colloidal properties of soil humus, and it is responsible for most of the CEC and water-holding capacity contributed to the soil by organic matter.

Intermediate in properties between the active and passive fractions is the *slow fraction* of soil organic matter. This fraction probably includes very finely divided plant tissues, high in lignin and other slowly decomposable and chemically resistant components. The half-lives of these materials are typically measured in decades. The slow fraction is an important source of mineralizable nitrogen and other plant nutrients, and it provides the underlying food source for the steady metabolism of the autochthonous soil microbes (see Section 11.2). The slow fraction also probably makes some contribution to the effects associated primarily with the active and passive fractions.

Changes in Active and Passive Fractions with Soil Management

The existence of a passive fraction as well as an active fraction of soil organic matter explains why conversion of native forests or grassland into cultivated cropland results in a very rapid decline in soil organic matter during the first few years, followed by a much slower decline thereafter (see, for example, Figure 11.14). The active fraction also explains why soil-management practices that cause only very small changes in total soil organic matter often cause rather pronounced alterations in aggregate stability, nitrogen mineralization rate, or other soil properties attributed to organic matter.

The accumulated plant residues and the active fraction of the soil organic matter are the first to be affected by changes in land management, accounting for most of the early losses in soil organic matter when cultivation of virgin soil begins. In contrast, losses from the passive fraction are very gradual. As a result, the soil organic matter remaining after some years is far less effective in promoting structural stability and nutrient cycling than the original organic matter in the virgin soil. If a favorable change in environmental conditions or management regime occurs, the plant litter and active fractions of soil organic matter are also the first to positively respond (see right-hand portion of Figure 11.14).

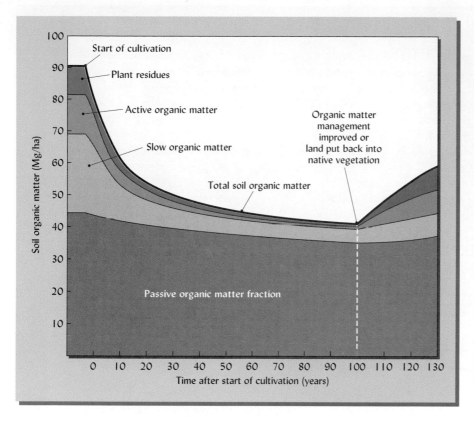

FIGURE 11.14 Changes in various fractions of organic matter in the upper 25 cm of a representative soil after bringing virgin land under cultivation. Initially, under natural vegetation, this soil contained about 91 Mg/ha of total organic matter. The resistant *passive fraction* accounted for about 44 Mg/ha, or about half of the total soil organic matter. The rapidly decomposing *active fraction* accounted for about 14 Mg, or about 16% of the total soil organic matter. After about 40 years of cultivation, the passive fraction had declined by about 11% to about 39 Mg/ha, while the active fraction had lost 90% of its mass, declining to only 1.4 Mg/ha. Note that much of the organic loss due to the change in land management came at the expense of the active fraction. This was also the fraction that most quickly increased when improved organic matter management was adopted after the 100th year. The susceptibility of the active fraction to rapid change explains why even relatively small changes in total soil organic matter can produce dramatic changes in important soil properties, such as aggregate stability and nitrogen mineralization, associated with this soil organic matter fraction.

11.9   CARBON BALANCE IN THE SOIL–PLANT–ATMOSPHERE SYSTEM

Whether the goal is to reduce greenhouse gas emissions or to enhance soil quality and plant production, proper management of soil organic matter requires an understanding of the factors and processes influencing the cycling and balance of carbon in an ecosystem. Although each type of ecosystem, whether a deciduous forest, a prairie, or a wheat field, will emphasize particular compartments and pathways in the carbon cycle, consideration of a specific example, such as that described in Box 11.2, can help us develop a general model that can be applied to many different situations.

Agroecosystems

The rate at which soil organic matter either increases or decreases is determined by the balance between *gains* and *losses* of carbon. The gains come primarily from plant residues grown in place and from applied organic materials. The losses are due mainly to respiration (CO_2 losses), plant removals, and erosion (Table 11.5).

TABLE 11.5   Factors Affecting the Balance between Gains and Losses of Organic Matter in Soils

Factors promoting gains	*Factors promoting losses*
Green manures or cover crops	Erosion
Conservation tillage	Intensive tillage
Return of plant residues	Whole plant removal
Low temperatures and shading	High temperatures and exposure to sun
Controlled grazing	Overgrazing
High soil moisture	Low soil moisture
Surface mulches	Fire
Application of compost and manures	Application of only inorganic materials
Appropriate nitrogen levels	Excessive mineral nitrogen
High plant productivity	Low plant productivity
High plant root:shoot ratio	Low plant root:shoot ratio

BOX 11.2   CARBON BALANCE—AN AGROECOSYSTEM EXAMPLE

The principal carbon pools and annual flows in a terrestrial ecosystem are illustrated in Figure 11.15 using a hypothetical cornfield in a warm temperate region. During a growing season the corn plants produce (by photosynthesis) 17,500 kg/ha of dry matter containing 7500 kg/ha of carbon (C). This C is equally distributed (2500 kg/ha each) among the roots, grain, and unharvested aboveground residues. In this example, the harvested grain is fed to cattle, which oxidize and release as CO_2 about 50% of this C (1250 kg/ha), assimilate a small portion as weight gain, and void the remainder (1100 kg/ha) as manure. The corn stover and roots are left in the field and, along with the manure from the cattle, are incorporated into the soil by tillage or by earthworms.

The soil microbes decompose the crop residues (including the roots) and manure, releasing as CO_2 some 75% of the manure C, 67% of the root C, and 85% of the C in the surface residues. The remaining C in these pools is assimilated into the soil as humus. Thus, during the course of one year, some 1475 kg/ha of C enters the humus pool (825 kg from roots, plus 375 from stover, plus 275 from manure). These values are in general agreement with Figure 11.18, but they will vary widely among different soil conditions and ecosystems.

At the beginning of the year, the upper 30 cm of soil in our example contained 65,000 kg/ha organic C in humus. Such a soil cultivated for row crops would typically lose about 2.5% of its organic C by soil respiration each year. In our example this loss amounts to some 1625 kg/ha of C. Smaller losses of soil organic C occur by soil erosion (160 kg/ha), leaching (10 kg/ha), and formation of carbonates and bicarbonates (10 kg/ha).

Comparing total losses (1805 kg/ha) with the total gains (1475 kg/ha) for the pool of soil humus, we see that the soil in our example suffered a *net annual loss* of 330 kg/ha of C, or 0.5% of the total C stored in the soil humus. If this rate of loss were to continue, degradation of soil quality and productivity would surely result.

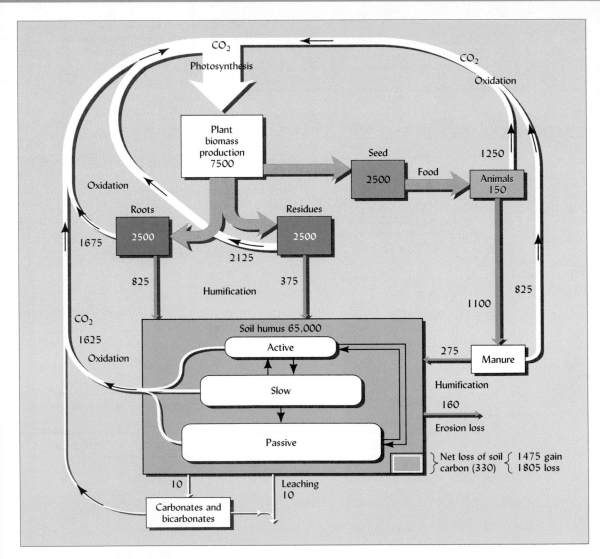

FIGURE 11.15   Carbon cycling in an agroecosystem.

In order to halt or reverse the net carbon loss shown in Figure 11.15, management practices would have to be implemented that would either *increase the additions* of carbon to the soil or *decrease the losses* of carbon from the soil. Since all crop residues and animal manures in the example are already being returned to the soil, additional carbon inputs could most practically be achieved by growing more plant material (i.e., increasing crop production or growing cover crops during the winter).

Specific practices to reduce carbon losses would include better control of soil erosion and the use of conservation tillage. Using a no-till production system would leave crop residues as mulch on the soil surface where they would decompose much more slowly. Refraining from tillage might also reduce the annual respiration losses from the original 2.5% to perhaps 1.5%. A combination of these changes in management would convert the system in our example from one in which soil organic matter is degrading (declining) to one in which it is aggrading (increasing).

Natural Ecosystems

Those interested in natural ecosystems may want to compare the carbon cycle of a natural forest with that of the cornfield in Figure 11.15. If the forest soil fertility were not too low, the total annual biomass production would probably be similar to that of the cornfield. The standing biomass, on the other hand, would be much greater in the forest since the tree crop is not removed each year. While some litter would fall to the soil surface, much of the annual biomass production would remain stored in the trees.

The rate of humus oxidation in the undisturbed forest would be considerably lower than in the tilled field because the litter would not be incorporated into the soil through tillage and the absence of physical disturbance would result in slower soil respiration. The litter from certain tree species may also be rich in phenolics and lignin, factors that greatly slow decomposition and C losses (see Figure 11.6). Furthermore, losses of organic matter through soil erosion would be much smaller on the forested site. Taken together, these factors allow annual net gains in soil organic matter in a young forest and maintenance of high soil organic matter levels in mature forests.

GRASSLANDS. Similar trends occur in natural grasslands, although the total biomass production is likely to be considerably less, depending mainly on the annual rainfall. Among the principles illustrated in Box 11.2, and applicable to most ecosystems, is the dominant role that plant root biomass plays in maintaining soil organic matter levels. In a grassland, the contribution from the plant roots is relatively more important than in a forest, and so a greater proportion of the total biomass produced tends to accumulate as soil organic matter, and this soil organic C is distributed more uniformly with depth.

11.10 FACTORS AND PRACTICES INFLUENCING SOIL ORGANIC MATTER LEVELS

Because of differences in C gains and losses, the amount of organic matter in soils varies widely; mineral surface soils contain from a mere trace (sandy, desert soils) to as high as 20 or 30% (some forested or poorly drained A horizons). Some soils contain even more organic matter, but those that do are considered to be organic—not mineral—soils and will be discussed in Section 11.12. In practice, soil organic matter content is usually estimated from analysis of soil organic carbon content, because the latter can be determined more precisely. Therefore, for quantitative discussions scientists usually refer to soil organic carbon. The ranges and representative organic carbon contents of different soil orders are given in Table 11.1 and illustrated in Figure 11.16.

Influence of Climate

The processes of organic matter production (plant growth) and organic matter destruction (microbial decomposition) respond to quite differently to **temperature**. At low temperatures plant growth outstrips decomposition, but the opposite is true above approximately 25°C. Therefore, as one moves from a warmer to a cooler climate, the organic matter and associated nitrogen content of comparable soils tend to increase (Figure 11.17). Within zones of uniform moisture conditions and comparable vegetation, the average total amounts of organic matter and nitrogen in soils increase from two to three times for each 10°C decline in mean annual temperature.

FIGURE 11.16   Vertical distribution of organic carbon in well-drained soils of four soil orders. Note the higher content and deeper distribution of organic carbon in the soils formed under grassland (Mollisols) compared to the Alfisol and Spodosol, which formed under forests. Also note the bulge of organic carbon in the Spodosol subsoil due to illuvial humus in the spodic horizon (see Chapter 3). The Aridisol has very little organic carbon in the profile, as is typical of dry-region soils.

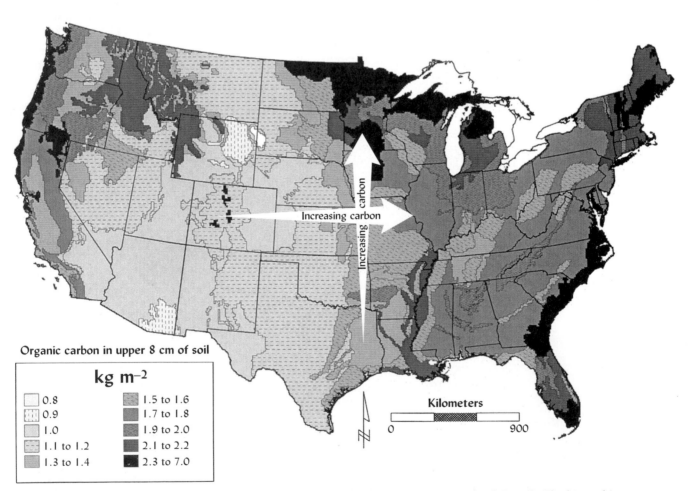

FIGURE 11.17   Influence of mean annual temperature and precipitation on organic matter levels in soils. The large white arrows on the map indicate that in the North American Great Plains region, soil organic matter increases with cooler temperatures going north, and with higher rainfall going east, provided that the soils compared are similar in texture, type of vegetation, drainage, and all other aspects except temperature and rainfall. These trends can be further generalized for global environments. [Kern (1994); Map courtesy of J. Kern, U.S. Environmental Protection Agency.]

Soil **moisture** also exerts a major influence on the accumulation of organic matter and nitrogen in soils. Under comparable conditions, the nitrogen and organic matter content of soils increase as the effective moisture becomes greater. The explanation lies mostly in the sparser vegetation of the drier regions. In determining this rainfall correlation, however, it must be remembered that the level of organic matter in any one soil is influenced by both temperature and precipitation, as well as by other factors.

The lowest natural levels of soil organic matter and the greatest difficulty in maintaining those levels are found where annual mean temperature is high and rainfall is low (Figure 11.17). The relationship between climate and soil C accumulation is extremely important to the productivity and conservation of soils and to the relative difficulty of sustainable natural resource management.

Influence of Natural Vegetation

Climate and vegetation usually act together to influence the soil contents of organic carbon and nitrogen. The greater plant productivity engendered by a well-watered environment generally leads to greater additions to the pool of soil organic matter. Grasslands generally dominate the subhumid and semiarid areas, while trees are dominant in humid regions. In climatic zones where the natural vegetation includes both forests and grasslands, the total organic matter is higher in soils developed under grasslands than under forests (see Figure 11.16). With grassland vegetation, a relatively high proportion of the plant residues consist of root matter, which decomposes more slowly and contributes more efficiently to soil humus formation than does forest leaf litter.

Effects of Texture and Drainage

While climate and natural vegetation affect soil organic matter over broad geographic areas, soil texture and drainage are often responsible for marked differences in soil organic matter within a local landscape. All else being equal, soils high in clay and silt are generally higher in organic matter than are sandy soils. The finer-textured soils accumulate more organic matter because (1) they produce more plant biomass, (2) they are less well aerated, and (3) clay–humus complexes protect some of the carbon.

In poorly drained soils, the high moisture supply promotes plant dry-matter production and relatively poor aeration inhibits organic matter decomposition. Poorly drained soils therefore generally accumulate much higher levels of organic matter and nitrogen than similar but better-aerated soils (Figure 11.18).

Influence of Agricultural Management and Tillage

It is safe to generalize that cultivated land contains much lower levels of both soil nitrogen and organic matter than do comparable areas under natural vegetation. This is not surprising; under natural conditions all the organic matter produced by the vegetation is returned to the soil, and the soil is not disturbed by tillage. In contrast, in cultivated areas much of the plant material is removed for human or animal food and relatively less finds its way back to the land. Also, soil tillage aerates the soil and breaks up the organic residues, making them more accessible to microbial decomposition. The rapid decline in organic matter upon conversion of natural vegetation to cultivation was illustrated by Figure 11.14.

Results from many long-term experimental plots suggest that cultivated soils kept highly productive by supplemental applications of nutrients, lime, and manure and by the choice of high-yielding crop varieties are likely to have more organic matter than comparable, less productive soils.

Recommendations for Managing Soil Organic Matter

In terms of productivity and other ecological functions, achieving a particular level of total soil organic matter is far less important than maintaining a substantial proportion in the active fraction, thus allowing biological metabolism to constantly enhance soil tilth and nutrient cycling.

FIGURE 11.18 Distribution of organic carbon in four soil profiles, two well drained and two poorly drained. Poor drainage results in higher organic carbon content, particularly in the surface horizon.

The following general principles apply to managing soil organic matter in many situations:

1. A continuous supply of organic materials must be added to the soil to maintain an appropriate level of soil organic matter, especially in the active fraction. Plant residues (roots and tops), animal manures, composts, and organic wastes are the primary sources of these organic materials. Cover crops can provide protective cover and additional organic material for the soil. It is almost always preferable to keep the soil vegetated than to keep it in bare fallow.

2. It is generally not practical to try to maintain higher soil organic matter levels than the soil–plant–climate control mechanisms dictate. For example, 1.5% organic matter might be an excellent level for a sandy soil in a warm climate, but would be indicative of a very poor condition for a finer-textured soil in a cool climate.

3. Because of the linkage between soil nitrogen and organic matter, adequate nitrogen inputs are requisite for adequate organic matter levels. Accordingly, the inclusion of legumes in the crop rotation and the judicious use of nitrogen-containing fertilizers to enhance high soil productivity are two desirable practices.

4. Maximum plant growth will increase the amount of organic matter added to soil from crop residues. Even if some plant parts are removed in harvest, vigorously growing plants provide below- and aboveground residues as major sources of organic matter for the soil. Moderate applications of lime and nutrients may be needed to help free plant growth from the constraints imposed by chemical toxicities and nutrient deficiencies.

5. Because tillage accelerates organic matter losses both by increased oxidation of soil organic matter and by erosion, it should be limited to that needed to control weeds and to maintain adequate soil aeration. Conservation practices that minimize tillage leave much of the plant residues on or near the soil surface, and thereby slow down the rate of residue decay and reduce erosion losses (see Sections 4.8 and 15.5). In time, conservation tillage can lead to higher organic matter levels (see Figure 11.19).

6. Perennial vegetation, especially natural ecosystems, should be encouraged and maintained wherever feasible. Improved agricultural production on existing farmlands should be pursued to allow land currently supporting natural ecosystems to be left relatively undisturbed. In addition, there should be no hesitation about taking land out of cultivation and encouraging its return to natural vegetation where such a move is appropriate. In the United States, the Conservation

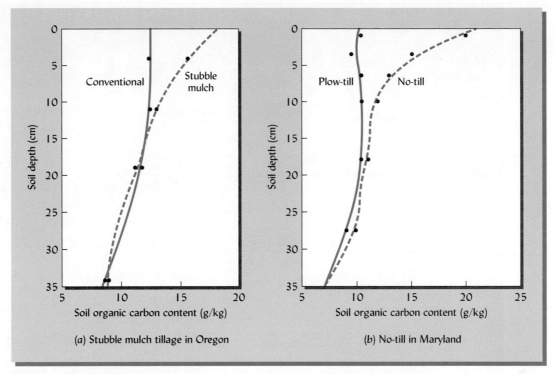

FIGURE 11.19 Effects of two conservation tillage systems on soil organic carbon content. The stubble mulch system was compared to conventional clean tillage on a Mollisol (Typic Haploxeroll) after 44 years of these practices for dryland wheat grown in a semiarid region of Oregon [data from Rasmussen and Rhode (1988)]. The no-till system was compared to conventional plow tillage after eight years of growing corn on an Ultisol (Typic Hapludult) in the humid coastal plain of Maryland [data from Weil, et al. (1988)]. Both conservation tillage systems leave much crop residue on or near the soil surface. The no-till system also leaves the soil almost completely unstirred, thus further slowing decomposition. In both cases, most of the increase in soil organic carbon occurred very near the soil surface.

Reserve Program provides incentives for such action (see Section 15.10). The fact is that large areas of land under cultivation today never should have been cleared.

11.11 SOILS AND THE GREENHOUSE EFFECT

CARBON DIOXIDE.   Through decomposition and accumulation of organic matter, soils have a major effect on the composition of the Earth's atmosphere. A case in point is the carbon in soil organic matter—carbon that originated from the CO_2 in the atmosphere. When soil organic matter is allowed to decline, atmospheric CO_2 is likely to increase, with serious global consequences. Box 11.3 illustrates the importance of soil organic matter in regulating atmospheric CO_2 levels. Today's atmosphere contains about 370 ppm CO_2, as compared to about 280 ppm before the industrial revolution. Levels are increasing at about 0.5% per year. Much of the increase in atmospheric carbon dioxide levels comes from a net loss of organic matter from the soil system.

Carbon dioxide is one of the **greenhouse gases** in the earth's atmosphere that cause the earth to be much warmer than it would otherwise be. Like the glass panes of a greenhouse, these gases allow short-wavelength solar radiation in, but trap much of the outgoing long-wavelength radiation. This heat-trapping action of the atmosphere is a major determinant of global temperature and, hence, global climates. Many scientists believe that the average global temperature has increased by 1.0°C during the past century, and if the trend continues, major changes in rainfall distribution, growing season length, and sea level may occur during the next century.

Much effort and expense are currently being directed at reducing the anthropogenic (human-caused) contributions to climate change. Soil science has a major role to play in dealing with global warming. Gases produced by biological processes, such as those occurring in the soil, account for approximately half of the problem (Figure 11.20). Of

The Biosphere 2 structure, a huge, sealed, ecological labora-tory. (Photo by C. Allen Morgan, © 1995 by Decisions Invest-ments Corp. Reprinted with permission.)

Biospherians at work growing their own food supply in the intensive agriculture biome with compost-amended soils rich in organic matter. (Photo by Pascale Maslin, © 1995 by Deci-sions Investments Corp. Reprinted with permission.)

An interesting experience with the global impact of soil organic matter dynamics was had by the eight biospherians who lived in Biosphere 2, a 1.3-ha sealed glass building in the Arizona desert. This structure was designed to hold a self-contained, self-supporting ecosystem in which scientists could live and study as part of the ecosystem. Costing over $200 million to build, Biosphere 2 contains a minia-ture ocean, coral reef, marsh, forests, and farms designed to maintain the ecological balance. How-ever, not only did the biospherians have a very hard time growing enough food for themselves, their atmosphere soon ran so low on oxygen and became so enriched in carbon dioxide that engineers had to pump in oxygen and install a scrubber to cleanse carbon dioxide from the air. It turned out that the ecosystem was thrown out of kilter by the organic-matter-rich soil hauled in for the Biosphere farms. The designers had underestimated the rate at which soil organic matter would decompose when placed in the warm environment and aerated by garden tillage (see Section 11.2). The decomposition of the soil organic matter was carried out by aerobic soil microbes, which use up oxygen and give off carbon dioxide as they respire.

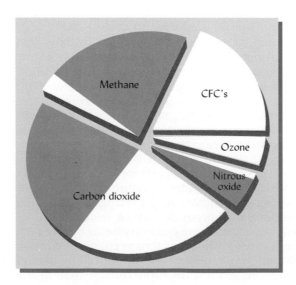

FIGURE 11.20 Relative contribution of dif-ferent gases to the global greenhouse effect. The shaded portions indicate the emissions related to biological systems (in which soils play a role), the white portions being indus-trial contributions. [Modified from Dale, et al. (1993)]

the five primary greenhouse gases, only chlorofluorocarbons (CFCs) are exclusively of industrial origin.

METHANE. Methane (CH_4) makes a substantial contribution to the greenhouse effect, for although it is found in the atmosphere in far smaller amounts than carbon dioxide, each molecule of CH_4 is about 30 times as effective as carbon dioxide in trapping outgoing radiation. In 1990 there was about 1.7 ppm CH_4 in the atmosphere, more than double the preindustrial levels. The level of CH_4 is rising at about 0.6% per year. Soils serve as both a source and a sink for CH_4—that is, they both add CH_4 to and remove it from the atmosphere.

When they are poorly aerated, as is the case in rice paddies and wetlands, soils produce CH_4, rather than CO_2, as organisms decompose organic matter (see Sections 11.2 and 7.5). The rate of CH_4 production in wet soils is very dependent on the availability of oxidizable carbon, either in the soil organic matter or in plant residues returned to the soil (Figure 11.21). The CH_4 contribution to the atmosphere from such soils may also be regulated by the nature and management of the plants growing on them. For example, about 70 to 80% of the methane released from some flooded soils escapes to the atmosphere through the stems of rice plants or natural marsh grasses. In well-aerated soils, soil-dwelling termites may produce large quantities of methane (see Section 10.4). These biological processes account for much of the methane emitted into the atmosphere.

Certain **methanotrophic bacteria** produce the enzyme *methane monooxygenase*, which allows them to oxidize methane as an energy source:

$$CH_4 + \tfrac{1}{2}O_2 \longrightarrow CH_3OH$$

Methane   Methanol

This reaction, which is largely carried out in well-aerated soils, reduces the global greenhouse gas burden by about 1 billion Mg of methane annually. Unfortunately, the long-term use of inorganic (especially ammonium) nitrogen fertilizer on cropland, pastures,

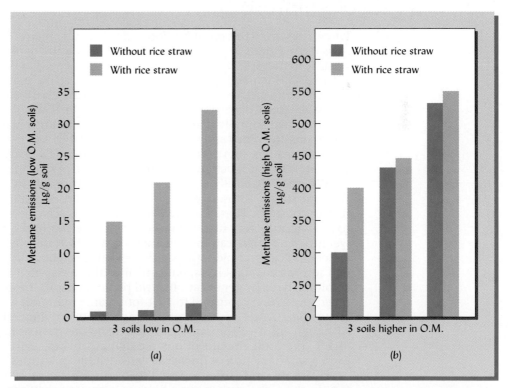

FIGURE 11.21 Global warming is enhanced by the release of methane from poorly drained soils. Methane is a greenhouse gas with atmospheric warming potential about 30 times that of CO_2. Oxygen-deficient, flooded rice soils supply food for over 2 billion people, but also account for up to one-fifth of the global emissions of CH_4. Methane release is increased by organic matter, either in the soil (compare left to right) or in crop residues (compare dark bars to light bars). Note the different scales for the low (left) and high (right) organic matter soils. [Data selected from Neue, et al. (1994)]

TABLE 11.6 Effect of Nitrogen Fertility Management Systems on Methane Oxidation by an Arable Soil (Mollisol) in Germany

The four nitrogen treatments were applied annually for 92 years to a rotation of sugar beets, spring barley, potatoes, and winter wheat. The measurements were made on soil sampled in spring, just before the annual nitrogen applications were made. Note that farmyard manure increased methane oxidation, while inorganic nitrogen fertilizer (NH_4NO_3) reduced methane oxidation from the levels in the control and the manure-only treatment.

Soil treatment	Soil pH	Soil NO_3^--N, kg/ha	Soil NH_4^+-N, kg/ha	Methane oxidation rate, nL CH_4 L^{-1}/hr^{-1}
1. Control—no N added in any form	6.8	0.83	0.20	4.60
2. Fertilizer N—40 to 130 kg/ha N as NH_4NO_3 to meet crop needs	6.9	15.36	3.1	1.34
3. Farmyard manure applied at 20 Mg/ha	7.0	1.98	0.22	11.2
4. Farmyard manure plus N fertilizer as in #3.	7.2	5.01	0.71	3.76

Data from the Static Fertilization Experiment begun in 1902 at Bad Lauchstädt, Germany and reported in Willison, et al. (1996).

and forests has been shown to reduce the capacity of the soil to oxidize methane. The evidence suggests that the rapid availability of ammonium from fertilizer stimulates ammonium-oxidizing bacteria at the expense of the methane-oxidizing bacteria. Long-term experiments in Germany and England indicate that supplying nitrogen in organic form (as manure) actually enhances the soil's capacity for methane oxidation (Table 11.6).

Nitrous oxide (N_2O) is another greenhouse gas produced by microorganisms in poorly aerated soils, but since it is not directly involved in the carbon cycle it will be discussed in the next chapter (see Section 12.9).

Because the soil can act as a major source or sink for carbon dioxide, methane, and nitrous oxide, it is clear that, together with steps to modify industrial outputs, soil management can play a major role in controlling the atmospheric levels of greenhouse gases.

11.12 ORGANIC SOILS (HISTOSOLS)

Discussion up to this point has focused on organic matter in mineral soils. Under waterlogged conditions in bogs and marshes, the oxidation of plant residues is so retarded that organic matter may accumulate to the point that soils are dominated by organic matter. If these soils contain greater than 20 to 30% organic matter by weight, depending on clay content, they are termed *organic soils* and are classified in the Histosols order in *Soil Taxonomy* (see Chapter 3). The organic matter in Histosols may be either peat or muck. *Peat* is brownish, only partially decomposed, fibrous remains of plant tissues. Some of these soils are mined and sold as peat, a material widely used in containerized plant production (see Box 11.4). *Muck,* on the other hand, is a black, powdery material in which decomposition is much more complete and the organic matter is highly humified.

Whether artificially drained for cultivation or left in their natural water-saturated wetland state, Histosols possess unique properties resulting from their high organic matter contents. These include dark brown to black colors, very low bulk densities (0.2 to 0.4 Mg/m^3), very high water-holding capacities (often two to three times their dry weight), and high cation exchange capacities (typically 150 to 200 $cmol_c/kg$). The water- and cation-holding capacities are much higher than those of even clay-rich mineral soils on a weight basis, but similar to those clay-rich mineral soils if considered on a volume basis (water or cations held per liter of soil). Despite relatively high C/N ratios (average of 20:1, compared to 12:1 for organic matter in a mineral soil), Histosols often show high levels of nitrification and accompanying nitrate release when drained. Although most of the carbon in peat is in the passive fraction that resists microbial attack, the total amount of organic matter in these soils is so great that the decay of even a small fraction of it releases considerable quantities of mineral nitrogen.

Histosols (and Histels—permafrost soils with organic surface horizons) are important in the global carbon cycle because, although they cover little more than 1% of the world's land area, they hold about 20% of the global soil carbon. While drainage can make these soils extremely valuable for vegetable and floriculture crops, it also speeds up oxidation of organic matter, which over time leads to Histosol destruction and increased release of CO_2 to the atmosphere (see Figure 11.22). To preserve these soils, the water table should be kept as near the surface as practical.

FIGURE 11.22 Soil subsidence due to rapid organic matter decomposition after artificial drainage of Histosols in the Florida Everglades. The house was built at ground level, with the septic tank buried about 1 m below the soil surface. Over a period of about 60 years, more than 1.2 m of the organic soil has "disappeared." The loss has been especially rapid because of Florida's warm climate, but artificial drainage that lowers the water table and continually dries out the upper horizons is an unsustainable practice on any Histosol. (Photo courtesy of George H. Snyder, Everglades Research and Education Center, Belle Glade, Fla.)

Histosols may help regulate long-term global climate change. As increased atmospheric CO_2 causes the global temperature to rise, seawater expands and the polar ice caps may melt more rapidly, both processes leading to a rise in sea level. The rise in sea level would have two effects that would help to counteract the impact of increased CO_2 content in the atmosphere. First, the greater volume of water in the world's oceans would absorb more CO_2. Second, the rising ocean waters would flood more coastal land area, creating conditions conducive to the formation of more Histosols in tidal marshes. The organic matter accumulation in these new Histosols would represent a significant sequestering of CO_2. This removal of CO_2 from the atmosphere may help to mitigate the global warming trend.

11.13 CONCLUSION

Organic matter is a complex and dynamic soil component that exerts a major influence on soil behavior, properties, and functions in the ecosystem. Because of the enormous amount of carbon stored in soil organic matter and the dynamic nature of this soil component, soil management may be an important tool for moderating the global greenhouse effect.

Organic residue decay, nutrient release and humus formation are controlled by environmental factors and by the quality of the organic materials. High contents of lignin and polyphenols, along with high C/N ratios, markedly slow the decomposition process, causing organic matter to accumulate while reducing the availability of nutrients.

For some purposes it is advantageous to manage the decomposition of organic matter outside of the soil in a process known as *composting*. Composting transforms various organic waste materials into a humuslike product that can be used as a soil amendment or a component of potting mixes. The aerobic decomposition in a compost pile can conserve nutrients while avoiding certain problems, such as noxious odors and the presence of either excessive or deficient quantities of soluble nitrogen, which can occur if fresh organic wastes are applied directly to soils.

Soil organic matter comprises three major pools of organic compounds. The *active pool* consists of microbial biomass and relatively easily decomposed compounds, such as polysaccharides and other nonhumic substances. Although only a small percentage of the total carbon, the active fraction plays a major role in nutrient cycling, micronutrient chelation, maintenance of structural stability and soil tilth, and as a food source underpinning biological diversity and activity in soils.

Most of the organic matter is in the *passive pool,* which contains very stable materials that resist microbial attack and may persist in the soil for centuries. This fraction provides cation exchange and water-holding capacities, but is relatively inert biologically. The so-called *slow fraction* is intermediate in stability and resistance to decomposition. It provides sources of food and energy for the steady metabolism of the authochthonous soil organisms that exist in the soil between times of residue additions. When soil is cleared of natural vegetation and brought under cultivation, the initial decline in soil organic matter is principally at the expense of the active fraction. The passive fraction is depleted only slowly and over very long periods of time.

The carbon-to-nitrogen (C/N) ratio of most soils is relatively constant, generally near 12:1. This means that the level of organic matter will be partially determined by the level of nitrogen available for assimilation into humus. Soil management for enhancing organic matter levels must, therefore, include some means of supplying nitrogen, for example, by the inclusion of leguminous plants.

The level of soil organic matter is influenced by climate (being higher in cool, moist regions), drainage (being higher in poorly drained soils), and by vegetation type (being generally higher where root biomass is greatest, as under grasses).

The maintenance of soil organic matter, especially the active fractions, in mineral soils is one of the great challenges in natural resource management around the world. By encouraging vigorous growth of crops or other vegetation, abundant residues (which contain both carbon and nitrogen) can be returned to the soil directly or through feed-consuming animals. Also, the rate of destruction of soil organic matter can be minimized by restricting soil tillage, controlling erosion, and keeping most of the plant residues at or near the soil surface.

Organic soils (Histosols) contain a large share of the world's soil organic carbon. Histosols are far more permeable to air and water, and generally can hold somewhat more water and cations per hectare, than most mineral soils. While drainage of these soils allows them to be used for production of high-value vegetable and floriculture crops, it also leads to accelerated oxidation of the organic matter they contain and increased CO_2 release.

The decay and mineralization of soil organic matter is one of the main processes governing the economy of nitrogen and sulfur in soils—the subject of the next chapter.

STUDY QUESTIONS

1. Compare the amounts of carbon in Earth's standing vegetation, soils, and atmosphere.

2. If you wanted to apply an organic material that would make a long-lasting mulch on the soil surface, you would choose an organic material with what chemical and physical characteristics?

3. Describe how the addition of certain types of organic materials to soil can cause a nitrate depression period. What are the ramifications of this phenomenon for plant growth?

4. In addition to humic substances, what other categories of organic materials are found in soils?

5. Some scientists include plant litter (surface residues) in their definition of soil organic matter, while others do not. Write two brief paragraphs, one justifying the inclusion of litter as soil organic matter and one justifying its exclusion.

6. What soil properties are mainly influenced by the active fraction of organic matter?

7. In this book and elsewhere, the terms *soil organic carbon* and *soil organic matter* are used to mean almost the same thing. How are these terms related, conceptually and quantitatively? Why is the term *organic carbon* generally more appropriate for quantitative scientific discussions?

8. Explain, in terms of the balance between gains and losses, why agricultural soils generally contain much lower levels of organic carbon than similar soils under natural vegetation.

9. In what ways are soils involved in the greenhouse effect that is thought to be warming up the Earth? What are some common soil-management practices that could be changed to reduce the negative effects and increase the beneficial effects of soils on the greenhouse effect?

10. What causes subsidence of Histosols? How is this phenomenon related to the sustainable use of Histosols?

REFERENCES

Batjes, N. H. 1996. "Total carbon and nitrogen in the soils of the world," *European J. Soil Sci.*, **47**:151–163.

Bouwman, A. F. (ed.). 1990. *Soils and the Greenhouse Effect.* (Chichester, U.K.: Wiley).

Brady, N. C. 1990. *The Nature and Properties of Soils,* 10th ed. (New York: MacMillan).

Cadisch, G. and K. E. Giller (eds.). 1997. *Driven by Nature—Plant Litter Quality and Decomposition.* (Walingford, U.K.: CAB International).

Chefetz, B., P. Hatcher, Y. Hadar, and Y. Chen. 1996. "Chemical and biological characterization of organic matter during composting of municipal solid waste," *J. Environ. Qual.*, **25**:776–785.

Chen, Y., and T. Aviad. 1990. "Effects of humic substances in plant growth," in P. MacCarthy, C. E. Clapp, R. L. Malcolm, and P. R. Bloom (eds.), *Humic Substances in Soil and Crop Sciences: Selected Readings.* (Madison, Wis.: ASA Special Publications), pp. 161–186.

Dale, V. H., R. A. Houghton, A. Grainger, A. E. Lugo, and S. Brown. 1993. "Emissions of greenhouse gases from tropical deforestation and subsequent uses of the land," in National Research Council, *Sustainable Agriculture and the Environment in the Humid Tropics.* (Washington, D.C.: National Academy Press), pp. 215–260.

Eswaran, H., E. Van Den Berg, and P. Reich. 1993. "Organic carbon in soils of the world," *Soil Sci. Soc. Amer. J.,* **57**:192–194.

Frimmel, F. H., and R. F. Christman. 1988. *Humic Substances and Their Role in the Environment.* (New York: Wiley).

Haynes, R. J. 1986. "The decomposition process: Mineralization, immobilization, humus formation and degradation," in R. J. Haynes (ed.), *Mineral Nitrogen in the Plant–Soil System.* (New York: Academic Press).

Huang, P. M., and M. Schnitzer (eds.). 1986. *Interactions of Soil Minerals with Natural Organics and Microbes.* SSSA Special Publication no. 17. (Madison, Wis.: Soil Sci. Soc. Amer.).

Inderjit, K. M., M. Dakshini, and F. A. Einhellig. 1995. *Allelopathy—Organisms, Processes, and Applications.* ACS Symposium Series 582. (Washington, D.C.: American Chemical Society).

Kern, J. S. 1994. "Spatial patterns of soil organic carbon in the contiguous United States," *Soil Sci. Soc. Amer. J.,* **58**:439–455.

Loll, M. J., and J. M. Bollag. 1983. "Protein transformation in soil," *Advances in Agronomy,* **36**:352–382.

Magdoff, F. 1992. *Building Soils for Better Crops.* (Lincoln, Nebr.: University of Nebraska Press).

Neue, H. U., R. Wasserman, R. S. Lantin, M. C. Alberto, and J. B. Aduna. 1994. "Methane production potential of soils," *International Rice Research Notes,* **19**:37–38. (Manila, Philippines: Rice Research Institute).

Paustian, K., W. J. Parton, and J. Persson. 1992. "Modeling soil organic matter-amended and nitrogen-fertilized long-term plots," *Soil Sci. Soc. Amer. J.,* **56**:476–488.

Rasmussen, P. E., and C. R. Rhode. 1988. "Long-term tillage and nitrogen fertilization effects on organic nitrogen and carbon in a semi-arid soil," *Soil Sci. Soc. Amer. J.,* **52**:1114–1117.

Stevenson, F. J. 1986. *Cycles of Soil.* (New York: Wiley).

Stevenson, F. J. 1994. *Humus Chemistry—Genesis, Composition Reactions,* 2nd ed. (New York: Wiley).

Stewart, B. A., R. Lal, and S. A. El-Swaify. 1991. "Sustaining the resource base of an expanding world agriculture," in R. Lal and F. J. Pierce (eds.), *Soil Management for Sustainability.* (Ankeny, Iowa: Soil and Water Conservation Soc.).

Swift, M. J., O. W. Heal, and J. M. Anderson. 1979. *Decomposition in Terrestrial Ecosystems.* Studies in Ecology, vol. 5. (Berkeley: University of California Press).

Tate, R. L. 1987. *Soil Organic Matter: Biological and Ecological Effects.* (New York: Wiley).

Tian, G., B. T. Kang, and L. Brussaard. 1992. "Biological effects of plant residues with contrasting chemical compositions under humid, tropical conditions—decomposition and nutrient release," *Soil Biol. and Biochem.,* **24**:1051–1060.

Tian, G. B., L. Brussaard, and B. T. Kang. 1995. "An index for assessing the quality of plant residues and evaluating their effects on soil and crop in the (sub-) humid tropics," *Applied Soil Ecol.,* **2**:25–32.

Waksman, S. A. 1948. *Humus.* (Baltimore, Md.: Williams & Wilkins).

Weil, R. R., P. W. Benedetto, L. J. Sikora, and V. A. Bandel. 1988. "Influence of tillage practices on phosphorus distribution and forms in three Ultisols," *Agron. J.,* **80**:503–509.

Weil, R. R., and G. S. Belmont. 1987. "Interactions between winged bean and grain amaranth," *Amaranth Newsletter,* **3**(1):3–6.

Willison, T., R. Cook, A. Müller, and D. Powlson. 1996. "CH_4 oxidation in soils fertilized with organic and inorganic N: Differential effects," *Soil Biol. and Biochem.,* **28**:135–136.

12

NITROGEN AND SULFUR ECONOMY OF SOILS

The pulse and body of the soil . . .
—*D. H. LAWRENCE, THE RAINBOW*

Nitrogen and sulfur share some important characteristics as essential plant nutrients. Both are found primarily in organic forms in soils, both move in soils and plants mostly in the anionic form, and both are responsible for serious environmental problems. Consequently, they will be considered together, starting with nitrogen.

More money and effort have been, and are being, spent on the management of nitrogen[1] than any other mineral element. And for good reason: The world's ecosystems are probably influenced more by deficiencies or excesses of nitrogen than by those of any other essential element. The pale yellowish green foliage of nitrogen-starved crops forebodes crop failure, financial ruin, and hunger for people in all corners of the world. Likewise, deficiencies of nitrogen are widespread among plants growing in the wild. Were it not for the biological fixation of nitrogen from the atmosphere by certain microorganisms, and for the recycling back to the soil of much of the nitrogen taken up in natural ecosystems, deficiencies of nitrogen would be even more widespread.

Excesses of some nitrogen compounds in soils can adversely affect human and animal health and can denigrate the quality of the environment. High nitrate levels in soils can lead to sufficiently high nitrates in drinking water as to endanger the health of human infants and ruminant animals. For that reason, nitrate levels are monitored in wells, reservoirs, and other drinking supplies. Likewise, the movement of soluble nitrogen compounds from soils to aquatic systems can disrupt the balance of those systems, leading to eutrophication, decline in oxygen content of the water, and the subsequent death of fish and other aquatic species.

Nitrogen is an essential component of protein. Because of its nutritional importance and relative scarcity, protein is highly sought after by most animals, humans included. Supplying sufficient nitrogen often represents a major expense in agricultural production. Also, the manufacture of nitrogen fertilizer accounts for a large part of the fossil fuel energy used by the agricultural sector. Yet another way in which nitrogen links the soil to the wider environment is the ozone-destroying action of soil-generated nitrous oxide gas. Clearly, for the management of the soil nitrogen cycle, the ecological, financial, and environmental stakes are very high.

---

[1] Nitrogen in soils and in crop production is covered extensively in the reviews edited by Stevenson (1982) and Hauck (1984). A comprehensive treatment of nitrogen in agriculture with an emphasis on environmental quality is given by Addiscott, et al. (1991).

12.1 INFLUENCE OF NITROGEN ON PLANT GROWTH AND DEVELOPMENT

ROLES IN THE PLANT.   Nitrogen is an integral component of many essential plant compounds. It is a major part of all amino acids, which are the building blocks of all proteins—including the enzymes, which control virtually all biological processes. Other critical nitrogenous plant components include the nucleic acids, in which hereditary control is vested, and chlorophyll, which is at the heart of photosynthesis. A good supply of nitrogen stimulates root growth and development, as well as the uptake of other nutrients.

DEFICIENCY.   Plants deficient in nitrogen tend to have a pale yellowish green color (**chlorosis**), have a stunted appearance, and develop thin, spindly stems (see Plates 22 and 33 after page 178). Nitrogen is quite mobile (easily translocated) within the plant, and when plant uptake is inadequate supplies are transferred to the newest foliage, causing the older leaves to show pronounced chlorosis first.

OVERSUPPLY.   When too much nitrogen is applied, excessive vegetative growth occurs, and the top-heavy plants are prone to falling over (lodging) with heavy rain or wind. High nitrogen applications may delay plant maturity and cause the plants to be more susceptible to disease (especially fungal disease) and to insect pests. These problems are especially noticeable if other nutrients, such as potassium, are in relatively low supply.

An oversupply of nitrogen also degrades crop quality. The color and flavor of fruits may be poor, as may the sugar and vitamin contents of certain vegetables and root crops. Flower production on ornamentals is reduced in favor of overabundant foliage. Under certain conditions, an oversupply may result in high-nitrate foliage harmful to livestock in the case of forages, or to human babies in the case of leafy vegetables.

FORMS OF NITROGEN TAKEN UP BY PLANTS.   Plant roots take up nitrogen from the soil solution principally as nitrates (NO_3^-) and NH_4^+ ions.[2] Although certain plants grow best when provided mainly one or the other of these forms, a relatively equal mixture of the two ions gives the best results with most plants. These two ions differ in their effect on the pH of the rhizosphere. Nitrate *anions* (negatively charged ions) move easily to the root with the flow of soil water and exchange at the root surface with HCO_3^- or OH^- ions that, in turn, stimulate an increase in the pH of the soil solution immediately around the root. In contrast, ammonium *cations* (positively charged ions) exchange at the root surface with hydrogen ions, thereby lowering the pH of the solution around the roots. (See Figure 9.12 for an illustration of these pH effects).

12.2 ORIGIN AND DISTRIBUTION OF NITROGEN

Some 300,000 Mg of nitrogen gas is found in the air above 1 ha of soil.[3] But the very strong triple bond between two nitrogen atoms makes this gas quite inert and not directly usable by plants or animals. Were it not for the ability of certain microorganisms to break this triple bond and to form nitrogen compounds (see Section 12.10), vegetation around the world would be rather sparse, and little nitrogen would be found in soils.

The nitrogen content of surface mineral soils normally ranges from 0.02 to 0.5%, a value of about 0.15% being representative for cultivated soils. A hectare of such a soil would contain about 3.5 Mg nitrogen in the A horizon and perhaps an additional 3.5 Mg in the deeper layers. In forest soils the litter layer (O horizons) might contain another 1 to 2 Mg of nitrogen. While these figures are low compared to those for the atmosphere, the soil contains 10 to 20 times as much nitrogen as does the standing vegetation (including roots) of either forested or cultivated areas. Most of the nitrogen in terrestrial systems is found in the soil.

---

[2] Nitrite (NO_2^-) can also be taken up, but this ion is toxic to plants. Fortunately, it rarely occurs in greater than trace quantities in soils.

[3] About 98% of the earth's nitrogen is contained in the igneous rocks deep under the planet's crust, where it is effectively out of contact with the soil–plant–air–water environment in which we live. We will therefore focus our attention on the remaining 2% that cycles in the biosphere.

Most soil nitrogen occurs as part of organic molecules. Soil organic matter typically contains about 5% nitrogen; therefore, the distribution of soil nitrogen closely parallels that of soil organic matter (see Section 11.3). Except where large amounts of chemical fertilizers have been applied, inorganic (i.e., mineral) nitrogen seldom accounts for more than 1 to 2% of the total nitrogen in the soil. Unlike most of the organic nitrogen, the mineral forms of nitrogen are mostly quite soluble in water, and may be easily lost from soils through leaching and volatilization.

12.3 THE NITROGEN CYCLE

As it moves through the **nitrogen cycle**, an atom of nitrogen may appear in many different chemical forms, each with its own properties, behaviors, and consequences for the ecosystem. This cycle explains why vegetation (and, indirectly, animals) can continue to remove nitrogen from a soil for centuries without depleting the soil of this essential nutrient. The biosphere does not run out of nitrogen, because it uses the same nitrogen over and over again. The principal pools and forms of nitrogen, and the processes by which they interact in the cycle, are illustrated in Figure 12.1. This figure deserves careful study; we will refer to it frequently as we discuss each of the major divisions of the nitrogen cycle.

12.4 IMMOBILIZATION AND MINERALIZATION

The great bulk (95 to 99%) of the soil nitrogen is in organic compounds that protect it from loss but leave it largely unavailable to higher plants. Much of this nitrogen is present as amine groups ($R-NH_2$), largely in proteins or as part of humic compounds. When soil microbes attack these compounds, simple amino compounds[4] ($R-NH_2$) are formed. Then the amine groups are hydrolyzed, and the nitrogen is released as ammonium ions (NH_4^+), which can be oxidized to the nitrate form. This enzymatic process, termed *mineralization,* may be indicated as follows, using an amino compound ($R-NH_2$) as an example of the organic nitrogen source:

$$\xrightarrow{\hspace{2cm}} \text{Mineralization} \xrightarrow{\hspace{2cm}}$$

$$R-NH_2 \underset{-2H_2O}{\overset{+2H_2O}{\rightleftharpoons}} OH^- + R-OH + NH_4^+ \underset{-O_2}{\overset{+O_2}{\rightleftharpoons}} 4H^+ + \text{energy} + NO_2^- \underset{-\frac{1}{2}O_2}{\overset{+\frac{1}{2}O_2}{\rightleftharpoons}} \text{energy} + NO_3^-$$

$$\xleftarrow{\hspace{2cm}} \text{Immobilization} \xleftarrow{\hspace{2cm}}$$

Many studies have shown that only about 1.5 to 3.5% of the organic nitrogen of a soil mineralizes annually. Even so, this rate of mineralization provides sufficient mineral nitrogen for normal growth of natural vegetation in most all soils. Research shows that mineralized soil nitrogen constitutes a major part of the nitrogen taken up by a crop. If the organic matter content of a soil is known, one can make a rough estimate of the amount of nitrogen likely to be released by mineralization during a typical growing season (Box 12.1).

The opposite of mineralization is *immobilization,* the conversion of inorganic nitrogen ions (NO_3^- and NH_4^+) into organic forms (see Figure 12.1). As microorganisms decompose carbonaceous organic residues in the soil, they may require more nitrogen than is contained in the residues themselves. The microorganisms then incorporate mineral nitrogen ions into their cellular components, such as proteins, leaving the soil solution temporarily devoid of NH_4^+ and NO_3^- ions. When the organisms die, some of the

---

[4] Amino acids such as lysine (CH_2NH_2COOH) and alanine (CH_3CHNH_2COOH) are examples of these simpler compounds. The R in these formulas represents the part of the organic molecule with which the amino group (NH_2) is associated. For example, for lysine, the R is CH_2COOH.

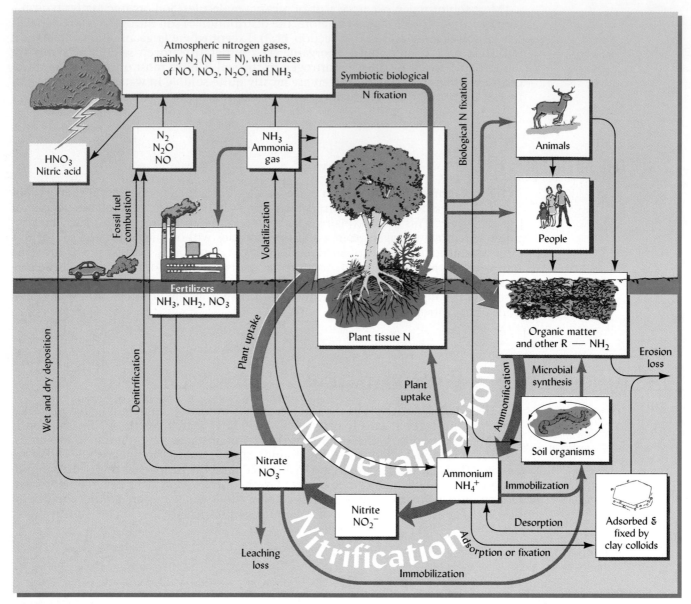

FIGURE 12.1 The nitrogen cycle, emphasizing the primary cycle (heavy, dark arrows) in which organic nitrogen is mineralized, plants take up the mineral nitrogen, and eventually organic nitrogen is returned to the soil as plant residues. Note also the pathways by which nitrogen is lost from the soil and the means by which it is replenished. The compartments represent various forms of nitrogen; the arrows represent processes by which one form is transformed into another.

Ammonium nitrogen is subject to five possible fates in the nitrogen cycle: (1) *immobilization* by microorganisms; (2) removal by *plant uptake;* (3) ammonium ions may be *fixed* in the interlayers of certain 2:1 clay minerals; (4) ammonium ions may be transformed into ammonia gas and lost to the atmosphere by *volatilization;* and (5) ammonium ions may be oxidized to nitrite and subsequently to nitrate by a microbial process called *nitrification.*

Nitrogen in the nitrate form is highly mobile in the soil and in the environment. Whether added as fertilizer or produced in the soil by nitrification, nitrate may take any of four paths in the nitrogen cycle: (1) *immobilization* by microorganisms; (2) removal by *plant uptake;* (3) nitrate ions may be lost by *leaching* in drainage water; or (4) by *volatilization* to the atmosphere as several nitrogen-containing gases.

organic nitrogen in their cells may be converted into forms that make up the humus complex, and some may be released as NO_3^- and NH_4^+ ions. Mineralization and immobilization occur simultaneously in the soil; whether the *net* effect is an increase or a decrease in the mineral nitrogen available depends primarily on the ratio of carbon to nitrogen in the organic residues undergoing decomposition (see Section 11.3).

BOX 12.1 CALCULATION OF NITROGEN MINERALIZATION

Assume you have a silt loam soil from the midwestern part of the United States that has a soil organic matter content (SOM) of about 2.5% (2.5 kg SOM/100 kg soil), and you want to determine the approximate quantity of nitrogen that would be mineralized each year in the upper 15 cm of a hectare of that soil.

First, calculate the mass of SOM in the upper 15 cm of a hectare of the soil, which has a mass of 2×10^6 kg.

$$\left(\frac{2.5 \text{ kg SOM}}{100 \text{ kg soil}}\right)\left(\frac{2 \times 10^6 \text{ kg soil}}{\text{ha 15 cm deep}}\right) = \frac{5 \times 10^4 \text{ kg SOM}}{\text{ha 15 cm deep}} \quad \text{or 50,000 kg SOM/ha}$$

Since the nitrogen content of the SOM is about 5% (5 kg total N/100kg SOM), the total N in the 50,000 kg SOM can be calculated:

$$\left(\frac{5 \text{ kg total N}}{100 \text{ kg SOM}}\right)\left(\frac{50,000 \text{ kg SOM}}{\text{ha 15 cm deep}}\right) = \frac{2500 \text{ kg total N}}{\text{ha 15 cm deep}} \quad \text{or 2500 kg total N/ha}$$

The amount of nitrogen mineralized each year depends upon the soil texture, climate, and management practices. Values of around 2% are common for fine-textured soils, while values of around 3.5% are typical for coarse-textured soils. Slightly higher values are found in warm climates and lower values in cool climates. If we assume a value of 2.5% for our soil (2.5 kg mineralized N/100 kg total N), the mass of mineralized nitrogen released each year can be calculated:

$$\left(\frac{2.5 \text{ kg mineralized N}}{100 \text{ kg total N}}\right)\left(\frac{2500 \text{ kg total N}}{\text{ha 15 cm deep}}\right) = \frac{62.5 \text{ kg mineralized N}}{\text{ha 15 cm deep}} \quad \text{or 62.5 kg mineralized N/ha}$$

Contributions from the deeper layers of this soil might be expected to bring total nitrogen mineralized in the root zone of this soil during a growing season to over 120 kg N/ha. Most nitrogen mineralization occurs during the growing season when the soil is relatively moist and warm.

These calculations estimate the nitrogen mineralized annually from a soil that has not had large amounts of organic residues added to it. Animal manures, legume residues, or other nitrogen-rich organic soil amendments would mineralize much more rapidly than the native soil organic matter, and thus would substantially increase the amount of nitrogen available in the soil.

12.5 AMMONIUM FIXATION BY CLAY MINERALS[5]

Like other positively charged ions, ammonium ions are held in exchangeable form, available for plant uptake. However, because of the particular size of the ammonium ion (and potassium also), it can become entrapped within cavities in the crystal structure of certain clays (see Figure 12.1). Several 2:1-type clay minerals, especially vermiculite, have the capacity to *fix* both ammonium and potassium ions in this manner (see Figure 13.21). Ammonium and potassium ions fixed in the rigid part of a crystal structure are held in a nonexchangeable form, from which they are released only slowly to higher plants and microorganisms. Ammonium fixation by clay minerals is generally greater in subsoil than in topsoil, due to the higher clay content of subsoils.

12.6 AMMONIA VOLATILIZATION

Ammonia gas (NH_3) that is produced in the soil–plant system from the breakdown of organic residues and farm manures, and from chemical fertilizers such as anhydrous ammonia, is in equilibrium with ammonium ions according to the following reversible reaction:

---

[5] This chemical fixation of ammonia is caused by the entrapment or other strong binding of NH_4^+ ions by certain silicate clays. This type of fixation (by which K ions are similarly bound) is not to be confused with the very beneficial biological fixation of atmospheric nitrogen gas into compounds usable by plants (see Section 12.10).

$$NH_4^+ + OH^- \rightleftharpoons H_2O + NH_3\uparrow$$

Dissolved ions   Gas

From this reaction we can draw three conclusions. First, ammonia volatilization will be more pronounced at high pH levels (i.e., OH^- ions drive the reaction to the right); second, ammonia-gas-producing amendments will drive the reaction to the left, raising the pH of the solution in which they are dissolved; and, finally, as a moist soil dries, the water is removed from the right-hand side of the equation, again driving the equation to the right.

Incorporation of manure and fertilizers into the top few centimeters of soil can reduce ammonia losses by 25 to 75%. In natural grasslands and pastures, incorporation of animal wastes by earthworms and dung beetles is critical in maintaining a favorable nitrogen balance and a high animal-carrying capacity in these ecosystems.

VOLATILIZATION FROM WETLANDS. Gaseous ammonia loss from nitrogen fertilizers applied to the surface of fishponds and flooded rice paddies can also be appreciable, even on slightly acid soils. The applied fertilizer stimulates algae growing in the paddy or pond water. As the algae photosynthesize, they extract CO_2 from the water and reduce the amount of carbonic acid formed. As a result, the pH of the paddy water increases markedly, especially during daylight hours, to levels commonly above 9.0. At these pH levels, ammonia is released from ammonium compounds and goes directly into the atmosphere. Natural wetlands lose ammonium by a similar daily cycle.

AMMONIA ABSORPTION. By the reverse of the ammonium loss mechanism just described, both soils and plants can absorb ammonia from the atmosphere. Thus the soil–plant system can help cleanse ammonia from the air, while deriving usable nitrogen for plants and soil microbes. Forests may receive a significant proportion of their nitrogen requirements as ammonia carried by wind from fertilized cropland and cattle feedlots located many kilometers away.

12.7   NITRIFICATION

Ammonium ions in the soil may be enzymatically oxidized by certain soil bacteria, yielding first nitrites and then nitrates. These bacteria are classed as **autotrophs** because they obtain their energy from oxidizing the ammonium ions rather than organic matter. The process termed **nitrification** (see Figure 12.1) consists of two main sequential steps. The first step results in the conversion of ammonium to nitrite by a specific group of autotrophic bacteria (***Nitrosomonas***). The nitrite so formed is then immediately acted upon by a second group of autotrophs, ***Nitrobacter.*** The enzymatic oxidation releases energy, and may be represented very simply as follows:

Step 1

$$NH_4^+ + 1\tfrac{1}{2}O_2 \xrightarrow[\text{bacteria}]{\textit{Nitrosomonas}} NO_2^- + 2H^+ + H_2O + 275 \text{ kJ energy}$$

Ammonium   Nitrite

Step 2

$$NO_2^- + \tfrac{1}{2}O_2 \xrightarrow[\text{bacteria}]{\textit{Nitrobacter}} NO_3^- + 76 \text{ kJ energy}$$

Nitrite   Nitrate

So long as conditions are favorable for both reactions, the second transformation is thought to follow the first closely enough to prevent accumulation of nitrite. This is fortunate, because even at concentrations of just a few parts per million, nitrite is quite toxic to most plants and to mammals.

Regardless of the source of ammonium (i.e., ammonia-forming fertilizer, sewage sludge, animal manure, or any other organic nitrogen source), nitrification will significantly increase soil acidity by producing H^+ ions, as shown in the above reaction. In humid regions, liming materials must be added to counteract acidity (see Chapter 9).

Soil Conditions Affecting Nitrification

The nitrifying bacteria are much more sensitive to environmental conditions than are the broad groups of heterotrophic organisms responsible for the release of ammonium from organic nitrogen compounds (**ammonification**). Nitrification requires a supply of ammonium ions, but excess NH_4^+ is toxic to *Nitrobacter* and must be avoided. The nitrifying organisms, being aerobic, require oxygen to make NO_2^- and NO_3^- ions, and are therefore favored in well-drained soils. The optimum moisture for these organisms is about the same as that for most land plants (about 60% of the pore space filled with water). Since they are autotrophs, their carbon sources are bicarbonates and CO_2. They perform best if the temperature is kept between 20 and 30°C, and perform very slowly if the soil is cold (below 5°C).

Nitrification proceeds most rapidly where there is an abundance of exchangeable base-forming cations, and where nutrient levels are optimum for the growth of higher plants. Nitrifying organisms are quite sensitive to some pesticides, being constrained if these pesticides are applied at high rates. Most studies suggest, however that at ordinary field rates, the majority of the pesticides have only a minimal effect on nitrification.

In recent years chemicals have been found that can inhibit or slow down the nitrification process, thereby reducing the nitrate leaching potential. Such compounds, as well as others that retard the dissolution of urea, are discussed in Box 12.2.

BOX 12.2   REGULATION OF SOLUBLE NITROGEN WITH FERTILIZER TECHNOLOGY

Perhaps the most challenging aspect of nitrogen control is to regulate the levels of soluble nitrogen forms so that at any given time they will not exceed the demands of the plant for this element. One way of doing this is to split the fertilizer nitrogen application into several small doses timed to meet the increasing plant demands as the crop develops. Although expensive in terms of time and labor, such practices prevent the buildup in the soil solution of nitrates that would be subject to losses by leaching and denitrification

NITRIFICATION INHIBITORS.[a]   A second means of controlling nitrate buildup is to employ *nitrification inhibitors* which inhibit the activity of *Nitrosomonas* bacteria that convert ammonia to nitrite in the first step of nitrification. (Note that a chemical that inhibited *Nitrobacter* would *not* be useful, as this would cause toxic nitrite to accumulate.) Three commercially available nitrification inhibitors are dicyandiamide (DCD), nitrapyrin (N-Serve®) and etridiazol (Dwell®). When mixed with nitrogen fertilizers, these compounds can temporarily prevent nitrate formation. The key word here is *temporarily*, for when conditions are favorable for nitrification, the inhibition usually lasts only a few weeks (less if soils are above 20°C). Research has shown that these materials may pay for themselves with improved fertilizer efficiency when conditions are very conducive to nitrogen loss by leaching or denitrification. Thus, nitrification inhibitors have been of value primarily in relatively wet years on sandy soils (reduced leaching) and on imperfectly drained soils (reduced denitrification).

SLOW-RELEASE NITROGEN FERTILIZERS.   The use of slow-release fertilizer materials provides another means of reducing nitrogen losses from fertilized soils. Unlike most inorganic fertilizers, these materials release soluble nitrogen slowly in the soil so that nitrogen availability is better sychronized with plant uptake. Some stabilized organic materials, such as compost and digested sewage sludge (e.g., Milorganite®), are used for this purpose. However, most slow-release nitrogen fertilizers are made by treating urea with materials that slow its dissolution or inhibit its hydrolysis to ammonium. Urea-formaldehyde, isobutylidene diurea (IBDU), resin-coated fertilizers (e.g., Osmocote®), and sulfur-coated urea are all examples of slow-release nitrogen fertilizers. In the case of the latter, the sulfur content (10 to 20%) may be seen as a benefit where sulfur is in low supply (Section 12.17), or as a problem where the extra acidity generated by the sulfur (Section 12.21) would be undesirable. Because of additional manufacturing costs and lower concentrations of nitrogen, slow-release fertilizers cost from 1.5 (for sulfur-coated ureas) to 4 times as much as plain urea per unit of nitrogen. Nonetheless, slow-release materials are practical for high-value plant production where the cost of fertilizers is not critical (certain vegetables, turf, and ornamentals) and for certain situations in which nitrogen loss from urea would be very high (e.g., some rice paddy soils[b]). They are widely applied to turf grass because they are not likely to cause fertilizer burn, and because they provide the convenience of fewer applications. Turf fertilized by a single annual application of soluble nitrogen may lose 10 to 50% of the applied nitrogen by leaching, while 3% or less is lost if slow-release fertilizers are used.

---

[a] For a review of this subject, see Prasad and Power (1995).

[b] Sulfur-coated urea should not be applied in flooded rice paddies, as the reduced iron in the floodwater will combine with the sulfur coating to form insoluble FeS, which locks up the nitrogen in the fertilizer.

12.8 THE NITRATE LEACHING PROBLEM

Since the negatively charged nitrate ions are not adsorbed by the negatively charged colloids that dominate most soils, they move downward freely with drainage water and are thus readily leached from the soil. The loss of nitrogen in this manner is of concern for two basic reasons: (1) such loss represents an impoverishment of the ecosystem whether or not cultivated crops are grown, and (2) leaching of nitrate causes several serious environmental problems.

Productivity Loss

Productivity of the ecosystem suffers not only because of the loss of nitrogen, but also because the leaching of nitrate from acidic sources (nitrification or acid rain) facilitates the loss of calcium and other nutrient cations. Both types of nutrient losses will generally reduce the ecosystem productivity. In the case of managed land, there is also an economic loss equal to the value of the lost nitrogen.

Environmental Impact

Environmental problems caused by nitrogen are mainly associated with the movement of nitrate through drainage waters to the groundwater. It may reach domestic wells, and may also eventually flow underground to surface waters, such as streams, lakes, and estuaries. The nitrate may contaminate drinking water and cause eutrophication and associated problems (Box 12.3).

Nitrate losses are determined by the amount of water leaching through the soil, and the nitrate content of that water. They would be highest in a heavily fertilized sandy soil in a high-rainfall area, or a similar soil under irrigation. Leaching may also be accentuated in conservation tillage systems which increase water infiltration and thereby increase leaching and concomitant losses of nitrates.

Some mature forest ecosystems achieve a very close balance between plant uptake of nitrogen and the return of nitrogen in litter fall and dead trees, the groundwater commonly containing less than 1 mg/L nitrate. However, inputs of nitrogen from the atmosphere (e.g., nitrate in acid rain or from nitrogen fixation) and disturbances of the system (e.g., timber harvest) can overload the forest ecosystem, resulting in annual leaching losses of up to 25 or 30 kg N/ha.

BOX 12.3 HUMAN HEALTH AND ENVIRONMENTAL EFFECTS OF NITRATE LEACHING

If either the groundwater or surface water might be used as drinking water, nitrate has the potential to cause *toxic effects*. **Methemoglobinemia**, also called *blue baby syndrome*, occurs when certain bacteria found in the guts of ruminant animals and human infants convert ingested nitrate into nitrite. The nitrite then interferes with the ability of the blood to carry oxygen to the body cells. Unoxygenated blood is not red; hence, infants with this condition take on a bluish skin color. Although human death from methemoglobinemia is a very rare occurrence, regulatory agencies in most countries limit the amount of nitrate permissible in drinking water to less than half the concentration known to cause this toxicity. In the United States the limit on nitrate is 45 mg/L nitrate (or 10 mg/L N in the nitrate form). In Europe the standard is 50 mg/L nitrate. It should be noted that nitrate in the diet (from water and other sources, such as meat preservatives) is also suspected of leading to the synthesis of colon-cancer-causing nitroso compounds, though the evidence for this relationship is very weak at present.

A much more widespread water quality problem associated with nitrate is the degradation of aquatic ecosystems, especially marine systems which are most often nitrogen-limited. Once it arrives in surface waters, nitrate, along with other forms of nitrogen and phosphorus from numerous sources, contributes to the problem of *eutrophication*. Eutrophication may stimulate the growth of large masses of algae, certain species of which produce flavors and toxins that make the water unfit to drink. The depletion of dissolved oxygen that ensues when algae die and decompose severely degrades the aquatic ecosystem, bringing death to fish and other aquatic organisms. This subject is given further consideration in Section 13.2.

Even greater losses may come from agricultural systems in which inputs of nitrogen regularly exceed the amounts removed by plant uptake and harvest. Heavy nitrogen fertilization (especially common for vegetables, corn, and other cash crops) exceeding what the plants are able to utilize can be a major cause of excessive nitrate leaching (Figure 12.2).

Ineffective management of manure from concentrated livestock production facilities is another common cause of nitrate contamination of ground and surface waters. When combined with high fertilizer nitrogen applications, manure can provide this element in quantities far in excess of plant uptake and can result in the pollution of both water and the atmosphere. Data from three European countries illustrate this point (Table 12.1). It is no wonder that shallow groundwater under some farm conditions contains nitrate in excess of 45 mg/L (10 mg N/L),[6] the legal limit for drinking water in the United States.

MANAGEMENT CAN REDUCE LOSSES. Even in regions of high leaching potential, careful soil management can prevent excessive nitrate losses. The timing of modest fertilizer and manure applications should provide nitrogen when the plants need it, not much before or after the period of active plant uptake. If this is not economically possible, other crops in the rotation (including cover crops) should be planted immediately following the cash crop to take up the unused nitrates (see Figure 12.2). If such guidelines are followed, nitrogen leaching may be kept less than 5 or 10% of the nitrogen applied.

---

[6] Nitrate concentrations are sometimes reported as *N in nitrate form*. To convert mg nitrate (NO_3^-)/L to mg (N in NO_3^- form)/L, divide by 4.4.

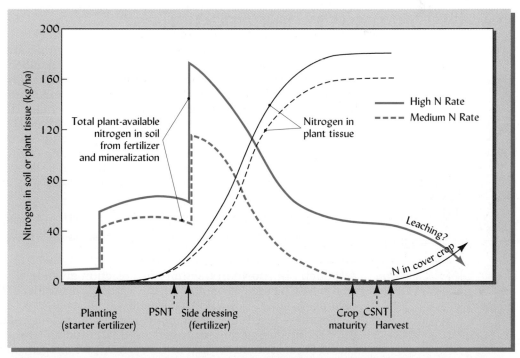

FIGURE 12.2   Two nitrogen fertilizer systems for corn production in the Midwest. The system that has been dominant in the past (solid lines) involves very high nitrogen applications and continuous corn culture. At least 150 kg N/ha is applied, part as starter fertilizer at or before planting time, the remainder as a side dressing just before the most rapid growth stage. Unfortunately when the crop matures and is harvested, much soluble nitrogen remains in the soil, probably in the form of nitrates. If the nitrogen is not captured by a cover crop planted in the corn at harvest time, the nitrogen is subject to leaching during the spring and winter months. This leads to contamination of groundwater and, eventually, surface water. More environmentally sound systems involve crop rotations that include legumes, or if corn is grown continuously, the rate of nitrogen fertilizer is greatly reduced (broken lines). A presidedress nitrate soil test (PSNT) is used to determine the amounts of nitrogen to apply. At the end of the season, a cornstalk nitrate test (CSNT) is used to assess plant nitrogen status. Crop yields and economic returns are about the same, but the nitrogen remaining in the soil at crop harvest is low and nitrate contamination is minimized.

TABLE 12.1 The Input into Soils of Nitrogen from Animal Manures and Fertilizer in Three European Countries

The uptake of nitrogen by crops and residual nitrogen not assimilated are shown. Some of the residual nitrogen is incorporated into soil organic matter, but much moves into the environment through leaching, runoff, and volatilization into the atmosphere.

| | Nitrogen supply, 1000 Mg | | | | Residual nitrogen | |
Country	Manure	Fertilizer	Total	N uptake by crops, 1000 Mg	Total, 1000 Mg	Per hectare, kg
Belgium/Luxembourg	380	199	580	211	369	240
Denmark	434	381	816	287	529	187
The Netherlands	752	504	1,255	285	970	480

Data from Leuch, et al. (1995).

NITRATE RECYCLING IN THE HUMID TROPICS.   Much of the nitrate mineralized in certain highly weathered, acid, tropical Oxisols and Ultisols leaches below the root zone before annual crops such as corn can take it up. Recently it has been found that some of this leached nitrate is not lost to groundwater, but is stored several meters deep in the profile on the highly weathered positively charged clays. Deep-rooted trees are capable of taking up this deep subsoil nitrate and subsequently using it to enrich the surface soil when they shed their leaves. Trees such as *Sesbania*, grown in rotation with annual food crops, can make this pool of leached nitrogen available for food production and prevent its further movement to groundwater. Agroforestry practices such as this have the potential to make a significant contribution to both crop protection and environmental quality in the humid tropics.

12.9 GASEOUS LOSSES BY DENITRIFICATION

Nitrogen may be lost to the atmosphere when nitrate ions are converted to gaseous forms of nitrogen by a series of widely occurring biochemical reduction reactions termed **denitrification**.[7] The organisms that carry out this process are mostly facultative anaerobic bacteria in genera such as *Pseudomonas, Bacillus, Micrococcus,* and *Achromobacter.* These organisms are *heterotrophs*, which obtain their energy and carbon from the oxidation of organic compounds. Other denitrifying bacteria are *autotrophs*, such as *Thiobacillus denitrificans,* which obtain their energy from the oxidation of sulfide. The exact mechanisms for denitrification vary depending on the conditions and organisms involved. In the reaction, NO_3^- [N(V)] is reduced in a series of steps to NO_2^- [N(III)], and then to nitrogen gases that include NO [N(II)], N_2O [N(I)], and eventually N_2 [N(0)]:

$$2NO_3^- \xrightarrow{-2[O]} 2NO_2^- \xrightarrow{-2[O]} 2NO\uparrow \xrightarrow{-[O]} N_2O\uparrow \xrightarrow{-[O]} N_2\uparrow$$

Nitrate ions (+5) Nitrite ions (+3) Nitric oxide gas (+2) Nitrous oxide gas (+1) Dinitrogen gas (0) ⟵ Valence state of nitrogen

Although not shown in the simplified reaction given here, the oxygen released at each step would be used to oxidize organic carbon (or sulfide, if *Thiobacillus* is the nitrifying organism), and CO_2 would be released.

For these reactions to take place, sources of organic residues should be available to provide the energy the denitrifiers need. The soil air in the microsites where denitrification occurs should contain no more than 10% oxygen, and lower levels of oxygen are preferred. Optimum temperatures for denitrification are from 25 to 35°C, but the process will occur between 2 and 50°C. Very strong acidity (pH < 5.0) inhibits rapid denitrification and favors the formation of N_2O).

Generally, when oxygen levels are very low the end product released from the overall denitrification process is dinitrogen gas (N_2). However, NO and N_2O are commonly also released during denitrification under the fluctuating aeration conditions that often occur in the field (Figure 12.3). The proportion of the three main gaseous products seems to be dependent on the prevalent pH, temperature, degree of oxygen depletion,

---

[7] Nitrate can also be reduced to nitrite and to nitrous oxide gas by nonbiological chemical reactions, but these reactions are quite minor in comparison with biological denitrification.

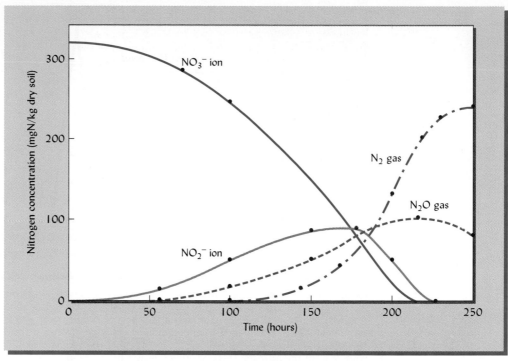

FIGURE 12.3 Changes in various forms of nitrogen during the process of denitrification in a moist soil incubated in the absence of atmospheric oxygen. [From Leffelaar and Wessel (1988)]

and concentration of nitrate and nitrite ions available. For example, the release of nitrous oxide (N_2O) is favored if the concentrations of nitrite and nitrate are high and the supply of oxygen is not too low. Under very acid conditions, almost all of the loss occurs in the form of N_2O. Nitric oxide (NO) loss is generally small and apparently occurs most readily under acid conditions.

Atmospheric Pollution

The question of how much of each nitrogen-containing gas is produced is not merely of academic interest. Dinitrogen gas is quite inert and environmentally harmless, but the oxides of nitrogen are very reactive gases and have the potential to do serious environmental damage in at least four ways. First, NO and N_2O can contribute to the formation of nitric acid, one of the principal components of acid rain. Second, the nitrogen oxide gases can react with volatile organic pollutants to form ground-level ozone, a major air pollutant in the photochemical smog that plagues many urban areas. Third, when NO rises into the upper atmosphere it contributes to the greenhouse effect (as much as 300 times that of an equal amount of CO_2) by absorbing infrared radiation that would otherwise escape into space (see Section 11.11).

Finally, and perhaps most significantly, as N_2O moves up into the stratosphere, it may participate in reactions that result in the destruction of ozone (O_3), a gas that helps shield the earth from harmful ultraviolet solar radiation. In recent years this protective ozone layer has been measurably depleted, in part by reaction with N_2O and other gases. As this protective layer is further degraded, thousands of additional cases of skin cancer are likely to occur annually. While there are other important sources of N_2O, such as automobile exhaust fumes, a major contribution to the problem is being made by denitrification in soils, especially in rice paddies, wetlands, and heavily fertilized or manured agricultural soils.

Quantity of Nitrogen Lost through Denitrification

The exact magnitude of denitrification loss is difficult to predict and will depend on management practices and soil conditions. Studies of forest ecosystems have shown that during periods of adequate soil moisture, denitrification results in a slow, but rel-

FIGURE 12.4 Changes in denitrification and soil conditions in spring on a well-drained, but somewhat fine-textured soil (East Keswick silty clay loam, Udalfs) in Wales, U.K. Two pasture systems were studied: ryegrass receiving 150 kg N/ha annually and a clover–grass mixture. The fertilizer nitrogen applied to the ryegrass system was intended to approximately equal the nitrogen fixed from the atmosphere by the clover in the mixed system (see Section 12.11). Note the sporadic nature of the denitrification process, with most of the nitrogen loss occurring during a brief period when the soil was warm, wet, and high in nitrate. This episodic pattern is typical of disturbed agroecosystems and stands in contrast to the slow, steady pattern of denitrification usually found in undisturbed natural forests. [Data from Colbourn (1993)]

atively steady, loss of nitrogen from these undisturbed natural systems. In contrast, most field losses from agricultural soils are highly variable in both time and space. The greater part of the annual nitrogen loss often occurs during just a few days in summer, when heavy rain has temporarily caused the warm soils to become poorly aerated (Figure 12.4).

Low-lying, organic-rich areas and other hot spots may lose nitrogen 10 times as fast as the average rate for a typical field. Although as much as 10 kg/ha of nitrogen may be lost in a single day from the sudden wetting of a well-drained, humid-region soil, such soils rarely lose more than 5 to 15 kg N/ha annually by denitrification. But where drainage is restricted and where large amounts of nitrogen fertilizer are applied, substantial losses might be expected. Losses of 30 to 60 kg N/ha/yr of nitrogen have been observed in agricultural systems.

Denitrification in Flooded Soils

In flooded soils, such as those found in natural wetlands or rice paddies (Figure 12.5), losses by denitrification may be very high. The soils of rice paddies are commonly subject to alternate periods of wetting and drying. Nitrates that are produced by nitrification during the dry periods are then subject to denitrification when the soils are submerged. Even when submerged, the soil permits both reactions to take place at once, nitrification at the soil–water interface where some oxygen derived from the water is present, and denitrification at lower soil depths (see Figure 12.6). However, nitrogen losses can be dramatically reduced by keeping the soil flooded and by deep placement of the fertilizer into the reduced zone of the soil. In this zone there is insufficient oxygen to allow nitrification to proceed, so the nitrogen remains in the ammonium form and is not susceptible to loss by denitrification (Figure 12.6).

The sequential combination of nitrification and denitrification also operates in natural and artificial wetlands. Tidal wetlands, which become alternately anaerobic and aerated as the water level rises and falls, have particularly high potentials for converting nitrogen to gaseous forms. Often the resulting rapid loss of nitrogen is considered to be

FIGURE 12.5 Denitrification can be very efficient in removing nitrogen in flooded systems that combine aerobic and anaerobic zones and have high concentrations of available organic carbon. Some examples of such systems are (*a*) a flooded rice paddy, (*b*) a tidal wetland, (*c*) a site for treating sewage effluent by overland flow (effluent applied by sprinklers, see arrows), and (*d*) a manure storage lagoon. (Photos courtesy of R. Weil)

a beneficial function of wetlands, in that the process protects estuaries and lakes from the eutrophying effects of too much nitrogen.

Denitrification in Groundwater

Recent studies on the movement of nitrate in groundwater have documented the significance of denitrification taking place in the poorly drained soils under **riparian** vegetation (mainly woodlands adjacent to streams). In most cases studied in humid temperate regions, contaminated groundwater lost most of its nitrate load as it flowed through the riparian zone on its way to the stream. The apparent removal of nitrate may be quite dramatic, whether the nitrate source is septic drainfields or fertilized cropland (Figure 12.7). Most of nitrate is believed to be lost by denitrification, stimulated by organic compounds leached from the decomposing forest litter and by the anaerobic conditions that prevail in the wet riparian zone soils.

We have just discussed a number of biological processes that lead to losses of nitrogen from the soil system. We now turn our attention to the principal biological process by which soil nitrogen is replenished.

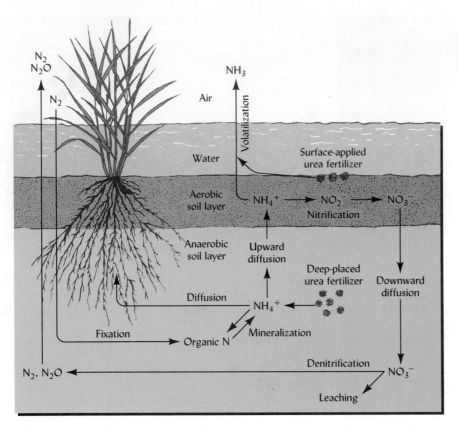

FIGURE 12.6 Nitrification–denitrification reactions and kinetics of the related processes controlling nitrogen loss from the aerobic–anaerobic layers of a flooded soil system. Nitrates, which form in the thin aerobic soil layer just below the soil–water interface, diffuse into the anaerobic (reduced) soil layer below and are denitrified to the N_2 and N_2O gaseous forms, which are lost to the atmosphere. Placing the urea or ammonium-containing fertilizers deep in the anaerobic layer prevents the oxidation of ammonium ions to nitrates, thereby greatly reducing the loss. [Modified from Patrick (1982)]

FIGURE 12.7 Denitrification in riparian wetlands receiving groundwater with high- and low-nitrate contents. The high-nitrate groundwater (left) came from sites with heavy development of houses using septic drainfields for sewage disposal (see Section 6.10). The low-nitrate groundwater (right side) came from an area of undeveloped forest. The riparian zones on both sides of the stream were covered with red maple–dominated forest. Within a few meters after entering the riparian wetland, the nitrate content of the contaminated groundwater was reduced by 75%, from 31 ppm nitrate to less than 7 ppm nitrate. The soils in this study site are very sandy Inceptisols and Entisols. In other regions, riparian zones with finer-textured soils have shown even more complete removal of nitrate from groundwater. [Data shown are from Hanson, et al. (1994)]

Next to plant photosynthesis, **biological nitrogen fixation** is probably the most important biochemical reaction for life on earth. Through this process, certain organisms convert the inert dinitrogen gas of the atmosphere (N_2) to nitrogen-containing organic compounds that become available to all forms of life through the nitrogen cycle. The process is carried out by a limited number of microorganisms, including several species of bacteria, a number of actinomycetes, and certain cyanobacteria (blue-green algae).

Globally, enormous amounts of nitrogen are fixed biologically each year. Terrestrial systems alone fix an estimated 139 million Mg, about twice as much as is industrially fixed in the manufacture of fertilizers (Table 12.2).

The Mechanism.   Regardless of the organisms involved, the key to biological nitrogen fixation is the enzyme *nitrogenase*, which catalyzes the reduction of dinitrogen gas to ammonia.

$$N \equiv N + 8H^+ + 6e^- \xrightarrow[\text{(Fe,Mo)}]{\text{(Nitrogenase)}} 2N{-}H + H_2$$

The ammonia, in turn, is combined with organic acids to form amino acids and, ultimately, proteins.

$$NH_3 + \text{organic acids} \longrightarrow \text{amino acids} \longrightarrow \text{proteins}$$

The site of N_2 reduction is the enzyme **nitrogenase**, a complex consisting of two proteins, the smaller of which contains iron while the larger contains molybdenum and iron. Several salient facts about this enzyme and its function are worth noting, for nitrogenase is unique and its role in the nitrogen cycle is of great importance to humankind.

The reduction of N_2 to NH_3 by nitrogenase requires a great deal of energy to break the triple bond between the nitrogen atoms. Therefore, the process is greatly enhanced by association with higher plants, which can supply this energy from photosynthesis.

The reduction reaction is end-product inhibited—an accumulation of ammonia will inhibit nitrogen fixation. Also, too much nitrate in the soil will inhibit the formation of nodules (see Section 12.11). Nitrogen-fixing organisms have a relatively high requirement for molybdenum, iron, phosphorus, and sulfur, because these nutrients are either part of the nitrogenase molecule or are needed for its synthesis and use.

TABLE 12.2   **Global Nitrogen Fixation from Different Sources**

	Nitrogen fixed per year		
Source of N fixation	Area, 10^6 ha	Rate, kg/ha	Total fixed, 10^6 Mg
Biological fixation			
Legume crops	250	140	35
Nonlegume crops	1,150	8	9
Meadows and grassland	3,000	15	45
Forest and woodland	4,100	10	40
Other vegetated land	4,900	2	10
Ice-covered land	1,500	0	0
Total land	14,900		139
Sea	36,100	1	36
Total biological	51,000		175
Lightning			8
Fertilizer industry			77
Grand Total			260

Calculated from Jenkinson (1990) and Burns and Hardy (1975).

TABLE 12.3   Information on Different Systems of Biological Nitrogen Fixation

N-fixing systems	Organisms involved	Plants involved	Site of fixation
Symbiotic			
Obligatory			
Legumes	Bacteria *Rhizobia* and *Bradyrhizobia*	Legumes	Root nodules
Nonlegumes (angiosperms)	Actinomycetes (*Frankia*)	Nonlegumes (angiosperms)	Root nodules
Associative			
Morphological involvement	Cyanobacteria, bacteria	Various higher plants and microorganisms	Leaf and root nodules, lichens
Nonmorphological involvement	Cyanobacteria, bacteria	Various higher plants and microorganisms	Rhizosphere (root environment) Phyllosphere (leaf environment)
Nonsymbiotic	Cyanobacteria, bacteria	Not involved with plants	Soil, water independent of plants

FIXATION SYSTEMS.   Biological nitrogen fixation occurs through a number of microorganism systems, with or without direct association with higher plants (Table 12.3). Although the legume–bacteria symbiotic systems have received the most attention, recent findings suggest that the other systems involve many more families of plants worldwide, and may even rival the legume-associated systems as suppliers of biological nitrogen to the soil. Each major system will be discussed briefly.

12.11   SYMBIOTIC FIXATION WITH LEGUMES

The **symbiosis** (mutually beneficial relationship) of legumes and bacteria of the genera *Rhizobium* and *Bradyrhizobium* provide the major biological source of fixed nitrogen in agricultural soils. These organisms infect the root hairs and the cortical cells, ultimately inducing the formation of **root nodules** that serve as the site of nitrogen fixation (Figure 12.8). In a mutually beneficial association, the host plant supplies the bacteria with carbohydrates for energy, and the bacteria reciprocate by supplying the plant with fixed-nitrogen compounds.

ORGANISMS INVOLVED.   A given *Rhizobium* or *Bradyrhizobium* species will infect some legumes but not others. For example, *Rhizobium trifolii* inoculates *Trifolium* species

    (a)   (b)   (c)

FIGURE 12.8   Photos illustrating soybean nodules. In (*a*) the nodules are seen on the roots of the soybean plant, and a closeup (*b*) shows a few of the nodules associated with the roots. A scanning electron micrograph (*c*) shows a single plant cell within the nodule stuffed with the bacterium *Bradyrhizobium japonicum*. (Courtesy of W. J. Brill, University of Wisconsin) (most clovers), but not sweet clover, which is in the genus *Melilotus*. Likewise, *Rhizobium phaseoli* inoculates *Phaseolus vulgaris* (dry beans), but not soybeans, which are in the genus *Glycine*. This specificity of interaction is one basis for classifying rhizobia (see Table 12.4). Legumes that can be inoculated by a given *Rhizobium* species are included in the same cross-inoculation group.

In areas where a given legume has been grown for several years, the appropriate species of *Rhizobium* is probably present in the soil. Often, however, the natural *Rhizobium* population in the soil is too low or the strain of the *Rhizobium* species present is not effective. In such circumstances, special mixtures of the appropriate *Rhizobium* and *Bradyrhizobium* inoculant may be applied, either by coating the legume seeds or by applying the inoculant directly to the soil. Effective and competitive strains of *Rhizobium,* which are available commercially, often give significant yield increases, but only if used on the proper crops.

QUANTITY OF NITROGEN FIXED. The rate of biological fixation is greatly dependent on soil and climatic conditions. High levels of available nitrogen, whether from the soil or added in fertilizers, tend to depress biological nitrogen fixation (Figure 12.9). Apparently, plants make the heavy energy investment required for symbiotic nitrogen fixation only when short supplies of mineral nitrogen make nitrogen fixation necessary.

Although quite variable from site to site, the amount of nitrogen biologically fixed can be quite high, especially for those systems involving nodules, which supply energy from photosynthates and protect the nitrogenase enzyme system (Table 12.5). Nonnodulating or nonsymbiotic systems generally fix relatively small amounts of nitrogen. Nonetheless, many natural plant communities and agricultural systems (generally involving legumes) derive the bulk of their nitrogen needs from biological fixation.

EFFECT ON SOIL NITROGEN LEVEL. The data in Table 12.5 show that symbiotic nitrogen fixation is definitely beneficial to both forestry and agriculture. This source of nitrogen is not available to most coniferous and hardwood trees, grasses, cereal crops, and vegetables that lack symbiotic associations, unless they are grown in association with nodulated species. Over time, the presence of nitrogen-fixing species can significantly increase the nitrogen content of the soil and benefit nonfixing species grown in association with fixing species (see Figure 12.10).

Some crops, such as beans and peas, are such weak nitrogen fixers that most of the nitrogen they absorb must come from the soil. Consequently, it should not be assumed that the symbiotic systems always increase soil nitrogen. Only in cases where the soil is low in available nitrogen and vegetation includes strong nitrogen fixers would this be likely to be true.

TABLE 12.4 Classification of Rhizobia Bacteria and Associated Legume Cross-Inoculation Groups

The genus Rhizobium *contains fast-growing, acid-producing bacteria, while those of* Bradyrhizobium *are slow growers that do not produce acid. A third genus,* Azorhizobium, *which is not shown, produces stem nodules on* Sesbania rostrata.

Bacteria		Host legume
Genus	*Species/subgroup*	*Host legume*
Rhizobium	*R. leguminosarum*	
	bv. *viceae*	*Vicia* (vetch), *Pisum* (peas), *Lens* (lentils), *Lathyrus* (sweet pea)
	bv. *trifolii*	*Trifolium* spp. (most clovers)
	bv. *phaseoli*	*Phaseolus* spp. (dry bean, runner bean, etc.)
	R. Meliloti	*Melilotus* (sweet clover, etc.), *Medicago* (alfalfa), *Trigonella,* (fenugreek)
	R. loti	*Lotus* (trefoils), *Lupinus* (lupins), *Cicer* (chickpea), *Anthyllis, Leucaena,* and many other tropical trees
	R. Fredii	*Glycine* spp. (e.g., soybean)
Bradyrhizobium	*B. japonicum*	*Glycine* spp. (e.g., soybean)
	B. sp.	*Vigna* (cowpeas), *Arachis* (peanut), *Cajanus* (pigeon pea), *Pueraria* (kudzu), *Crotolaria* (crotolaria), and many other tropical legumes

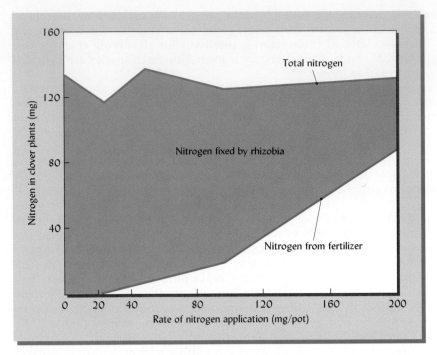

FIGURE 12.9 Influence of added inorganic nitrogen on the total nitrogen in clover plants, the proportion supplied by the fertilizer, and that fixed by the rhizobium organisms associated with the clover roots. Increasing the rate of nitrogen application decreased the amount of nitrogen fixed by the organisms in this greenhouse experiment. [From Walker, et al. (1956)]

In the case of legume crops harvested for seed or hay, most of the nitrogen fixed is removed from the field with the harvest. Nitrogen additions from such crops should be considered as nitrogen *savers* for the soil rather than nitrogen builders. On the other hand, considerable buildup of soil nitrogen can be achieved by perennial hay crops (such as alfalfa), and by annual legumes (such as hairy vetch) whose entire growth is returned to the soil as **green manure.** Such nitrogen buildup should be taken into account when estimating nitrogen fertilizer needs for maximum plant production with minimal environmental pollution (see Section 14.3).

TABLE 12.5   Typical Levels of Nitrogen Fixation from Different Systems

Crop or plant	Associated organism	Typical levels of nitrogen fixation, kg N/ha/yr
Symbiotic		
Legumes (nodulated)		
Ipil Ipil tree (*Leucena leucocephala*)	Bacteria (*Rhizobium*)	100–500
Locust tree (*Robina* spp.)		75–200
Alfalfa (*Medicago sativa*)		150–250
Clover (*Trifolium pratense L.*)		100–150
Lupine (*Lupinus*)		50–100
Vetch (*Vicia vilbosa*)		50–150
Bean (*Phaseolus vulgaris*)		30–50
Cowpea (*Vigna unguiculata*)	Bacteria (*Bradyrhizobium*)	50–100
Peanut (*Arachis*)		40–80
Soybean (*Glycine max L.*)		50–150
Pigeon pea (*Cajunus*)		150–280
Kudzu (*Pueraria*)		100–140
Nonlegumes (nodulated)		
Alders (*Alnus*)	Actinomycetes (*Frankia*)	50–150
Species of Gunnera	Cyanobacteria[a] (*Nostoc*)	10–20
Nonlegumes (nonnodulated)		
Pangola grass (*Degetaria decumbens*)	Bacteria (*Azospirillum*)	5–30
Bahia grass (*Pasalum notatum*)	Bacteria (*Azobactor*)	5–30
Azolla	Cyanobacteria[a] (*Anabena*)	150–300
Nonsymbiotic	Bacteria (*Azobactor, Clostridium*)	5–20
	Cyanobacteria[a] (various)	10–50

[a] Sometimes referred to as *blue-green algae.*

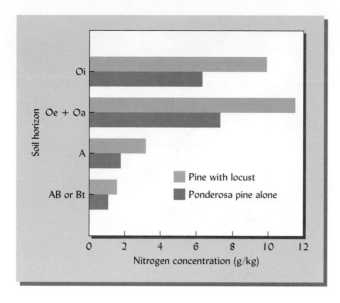

FIGURE 12.10 Nitrogen contents of forest soil horizons showing the effects of New Mexican locust trees (*Robinia neomexicana*) growing in association with ponderosa pine (*Pinus ponderosa*) in a region of Arizona receiving about 670 mm rainfall per year. The data are means from 20 stands of pondersosa pine, half of them with the nitrogen-fixing legume trees (locust) in the understory. The soils are Eutrustalfs and Argiustolls with loam and clay loam textures. [Data from Klemmedson (1994)]

FATE OF NITROGEN FIXED BY LEGUME BACTERIA. Some of the nitrogen fixed in root nodules is used directly by the host plant, which thereby benefits greatly from the symbiosis described in Section 12.10. Second, some fixed nitrogen may become available to nonfixing plants growing in association with nitrogen-fixing plants. The vigorous development of a grass in a legume–grass mixture is evidence of this rapid release, as are the relatively high nitrate concentrations sometimes measured in groundwater under legume crops.

A third pathway for the fixed nitrogen is immobilization by heterotrophic microorganisms and eventual incorporation into the soil organic matter (see Section 12.4).

12.12 SYMBIOTIC FIXATION WITH NONLEGUMES

Nodule-Forming Nonlegumes

Nearly 200 species from more than a dozen genera of nonlegumes are known to develop nodules and to accommodate symbiotic nitrogen fixation. Included are several important groups of angiosperms, listed in Table 12.6. The roots of these plants, which are present in certain forested areas and wetlands, form distinctive nodules when their root hairs are invaded by soil actinomycetes of the genus *Frankia*.

The rates of nitrogen fixation per hectare compare favorably with those of the legume–*Rhizobium* complexes (see Table 12.5). On a worldwide basis, the total nitrogen fixed in this way may even exceed that fixed by agricultural legumes (see Table 12.2). Because of their nitrogen-fixing ability, certain of the tree–actinomycete complexes are able to colonize infertile soils and newly forming soils on disturbed lands, which may have extremely low fertility as well as other constraints on plant growth (Figure 12.11). Once nitrogen-fixing plants become established and begin to build up the soil nitrogen supply through leaf litter and root exudation, the land becomes more hospitable for colonization by other species. *Frankia* thus play a very important role in the nitrogen economy of areas undergoing succession, as well as in established forest marshes.

Certain cyanobacteria are known to develop nitrogen-fixing symbiotic relations with green plants. One involves nodule formation on the stems of *Gunnera*, an angiosperm common in marshy areas of the southern hemisphere. In this association, cyanobacteria of the genus *Nostoc* fix 10 to 20 kg N/ha/yr (see Table 12.5).

Symbiotic Nitrogen Fixation without Nodules

Among the most significant nonnodule nitrogen-fixing systems are those involving cyanobacteria. One system of considerable practical importance is the *Azolla–Anabaena* complex, which flourishes in certain rice paddies of tropical and semitropical areas. The *Anabaena* cyanobacteria inhabit cavities in the leaves of the floating fern *Azolla* and fix

FIGURE 12.11 Soil actinomycetes of the genus *Frankia* can nodulate the roots of certain woody plant species and form a nitrogen-fixing symbiosis that rivals the legume–rhizobia partnership in efficiency. The actinomycete-filled root nodule (*a*) is the site of nitrogen fixation. The red alder tree (*b*) is among the first pioneer tree species to revegetate disturbed or badly eroded sites in high-rainfall areas of the Pacific Northwest in North America. This young alder is thriving despite the nitrogen-poor, eroded condition of the soil, because it is not dependent on soil nitrogen for its needs. (Photos courtesy of R. Weil)

quantities of nitrogen comparable to those of the more efficient *Rhizobium*–legume complexes (see Table 12.5).

A more widespread but less intense nitrogen-fixing phenomenon is that which occurs in the *rhizosphere* of certain grasses and other nonlegume plants. The organisms responsible are bacteria, especially those of the *Spirillum* and *Azotobacter* genera. Plant root exudates supply these microorganisms with energy for their nitrogen-fixing activities.

TABLE 12.6 Number and Distribution of Major Actinomycete-Nodulated Nonlegume Angiosperms

In comparison, there are about 13,000 legume species.

Genus	Family	Species[a] nodulated	Geographic distribution
Alnus	Betulaceae	33/35	Cool regions of the northern hemisphere
Ceanothus	Rhamnaceae	31/35	North America
Myrica	Myricaceae	26/35	Many tropical, subtropical, and temperate regions
Casuarina	Casuarinaceae	24/25	Tropics and subtropics
Elaeagnus	Elaeagnaceae	16/45	Asia, Europe, North America
Coriaria	Coriariaceae	13/15	Mediterranean to Japan, New Zealand, Chile to Mexico

[a] Number of species nodulated/total number of species in genus.
Selected from Torrey (1978).

Scientists have reported a wide range of rates for rhizosphere nitrogen fixation, with the highest values observed in association with certain tropical grasses. Even if typical rates are only 5 to 30 kg N/ha/yr, the vast areas of tropical grasslands suggest that the total quantity of nitrogen fixed by rhizosphere organisms is likely very high (see Table 12.2).

12.13 NONSYMBIOTIC NITROGEN FIXATION

Certain free-living microorganisms present in soils and water are able to fix nitrogen. Because these organisms are not directly associated with higher plants, the transformation is referred to as *nonsymbiotic* or *free-living*.

Fixation by Heterotrophs

Several different groups of bacteria and cyanobacteria are able to fix nitrogen nonsymbiotically. In upland mineral soils, the major fixation is brought about by species of two genera of heterotrophic aerobic bacteria, *Azotobacter* (in temperate zones) and *Beijerinckia* (in tropical soils). Certain anaerobic bacteria of the genus *Clostridium* are also able to fix nitrogen.

The amount of nitrogen fixed by these heterotrophs ranges from 5 to 20 kg N/ha/yr—only a small fraction of the nitrogen needed by crops. However, in some natural ecosystems this level of nitrogen fixation would make a significant contribution toward meeting nitrogen needs.

Fixation by Autotrophs

Among the autotrophs able to fix nitrogen are certain photosynthetic bacteria and cyanobacteria. In the presence of light, these organisms are able to fix carbon dioxide and nitrogen simultaneously. The contribution of the photosynthetic bacteria is uncertain, but that of cyanobacteria is thought to be of some significance, especially in wetland areas and in rice paddies. In some cases, these algae have been found to fix sufficient nitrogen for moderate rice yields, but normal levels may be no more than 20 to 30 kg N/ha/yr. Nitrogen fixation by cyanobacteria in upland soils also occurs, but the level is much lower than is found under wetland conditions.

12.14 ADDITION OF NITROGEN TO SOIL IN PRECIPITATION

The atmosphere contains ammonia gas, nitrates, and other nitrogen compounds derived primarily from the combustion of coal and petroleum products. These atmosphere-borne nitrogen compounds are added to the soil through rain, snow, and dust. Although the rates of addition per hectare are typically small, the total quantity of nitrogen added annually is significant.

The quantity of ammonia and nitrates in precipitation are greater in humid tropical regions than in humid temperate regions, and are larger in the latter than in semi-arid, temperate climates. Rainfall additions of nitrogen are highest near cities and industrial areas and near large animal feedlots (Table 12.7). The environmental impact of acid rain (see Section 9.4) and its influence on vegetation (forests and crops)

TABLE 12.7 Amounts of Nitrogen Brought Down in Precipitation Annually in Different Parts of the United States

| Areas in the United States | Range in annual deposition, kg/ha | | |
	Nitrate nitrogen	Ammonium nitrogen	Total nitrogen, kg/ha
Rural area of Ohio[a]	6.6–10.6	6.2–8.2	12.8–18.8
Industrialized Northeast[b]	4.3–7.4	8.6–14.8	12.9–22.2
Borders of NE industrialized areas[b]	2.8–4.1	5.6–8.2	8.4–12.3
Open areas in West[b]	0.4–0.6	0.8–1.2	1.2–1.8

[a] Owens, et al. (1992).
[b] Ammonium nitrogen is calculated as twice the nitrate deposition, the approximate ratio from numerous measurements, although it may be incorrect at any specific site [U.S./Canada Work Group No. 2 (1982)].

and surface waters tends to overshadow the nutrient benefit of precipitation-supplied nitrates.

Typically, the nitrogen in precipitation is about two-thirds ammonium and one-third nitrate. The range of total nitrogen (nitrate + ammonium) added by precipitation annually is 1 to 25 kg N/ha. A figure of 5 to 8 kg N/ha would be typical for nonindustrial temperate regions. This modest annual acquisition of nitrogen is probably of more significance to the nitrogen budget in natural ecosystems than in agriculture.

Effects on Forest Ecosystems

In a mature forest, net plant uptake and microbial immobilization are in such close balance with release of nitrogen by mineralization that many of these systems have very little ability to retain incoming nitrogen. Nitrogen from the atmosphere enters the forest largely as nitric acid (HNO_3) or as ammonium, which is converted to nitric acid upon microbial nitrification (see Section 12.7). As a result, the H^+ ions from HNO_3 displace base-forming cations on the soil colloids. The displaced cations, especially calcium and magnesium, leach downward with the nitrate (and sulfate; see Section 12.22). This causes imbalances in the nutrition of the trees and a decline in forest productivity. Nitrogen in precipitation might be considered beneficial fertilizer when it falls on farmland, but it can be a serious pollutant when added to a soil supporting a mature forest. Chapter 9 and Section 12.22 should be consulted for a more detailed explanation of these reactions and soil differences.

12.15   REACTIONS OF NITROGEN FERTILIZERS

Most commercial fertilizers supply nitrogen in soluble forms, such as nitrate or ammonium, or as urea, which rapidly hydrolyzes to form ammonium. Ammonium and nitrate ions from fertilizer are taken up by plants and participate in the nitrogen cycle in exactly the same way as ammonium and nitrate derived from organic matter mineralization or other sources. The main difference is that nitrogen from heavy doses of fertilizer is far more concentrated in time (available all at once) and space (applied in narrow bands or in a thin layer in the soil) than is nitrogen from other sources. While judicious application of such concentrations may be very beneficial for plant growth, it may also have serious detrimental effects.

First, high concentrations of ammonia gas from ammonium-releasing fertilizers (such as urea and anhydrous ammonia) favor loss of this gas to the atmosphere, especially in alkaline soils. Also, because nitric acid is one of the products of nitrification, use of ammonium fertilizers increases soil acidity (see Section 12.7). Soil acidity may also be increased by ammonia released from a heavy application of manure.

To save time and money, most land managers in both agriculture and forestry apply most of a year's supply of nitrogen at one time. Unfortunately the plants and microorganisms cannot assimilate the applied nitrogen fast enough to prevent major losses by leaching, surface runoff, denitrification, and ammonia volatilization (See Figure 12.1). The environmental consequences of these losses have already been discussed. As a general rule, it is more environmentally benign to apply nitrogen fertilizer in multiple small doses that coincide with the plant demand.

12.16   PRACTICAL MANAGEMENT OF SOIL NITROGEN IN AGRICULTURE

The goals of nitrogen control are threefold: (1) the maintenance of an adequate nitrogen supply in the soil, (2) the regulation of the soluble forms of nitrogen to ensure that enough is readily available to provide all the nitrogen plants need for optimum growth, and (3) the minimization of environmentally damaging leakage from the soil–plant system.

Major gains and losses of available nitrogen in ecosystems are shown in Figure 12.12. In natural ecosystems essentially all plant residues are returned to the soil, leaching and erosion losses are minimal, and little commercial fertilizer is needed. In most agricultural systems, however, nitrogen removed in crops is high, as are losses from leaching and erosion. These losses must be balanced by applying manures and chemi-

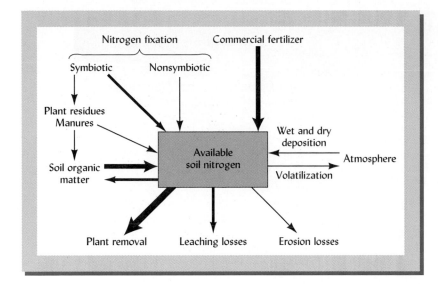

FIGURE 12.12 Major gains and losses of available soil nitrogen. The widths of the arrows roughly indicate the magnitude of the losses and the additions often encountered. It should be emphasized that the diagram represents average conditions only and that much variability is to be expected in the actual and relative quantities of nitrogen involved.

cal fertilizers. Unfortunately, in practice the quantity and timing of such additions stimulate increased leaching and erosion losses.

Several basic strategies are proving useful in achieving a rational reduction of excessive nitrogen inputs while maintaining or improving production levels and profitability in agriculture (Box 12.4). These approaches include: (1) taking into account the nitrogen contribution from *all* sources and reducing the amount of fertilizer applied accordingly; (2) improving the efficiency with which fertilizer, as well as such so-called waste products as animal manure, are used; (3) avoiding overly optimistic yield goals that lead to fertilizer application rates designed to meet crop needs that are much higher than actually occur in most years; and (4) improving crop response knowledge, which identifies the lowest nitrogen application that is likely to produce optimum profit.

12.17 IMPORTANCE OF SULFUR[8]

Sulfur has long been recognized as indispensable for many reactions in living cells. In addition to its vital roles in plant and animal nutrition, sulfur is also responsible for several types of air, water, and soil pollution and is therefore of increasing environmental interest. The environmental problems associated with sulfur include acid precipitation, certain types of forest decline, acid mine drainage, acid sulfate soils, and even some toxic effects in drinking water used by humans and livestock.

Roles of Sulfur in Plants and Animals

Sulfur is a constituent of the amino acids methionine, cysteine, and cystine, deficiencies of which result in serious human malnutrition. The vitamins biotin, thiamine, and B1 contain sulfur, as do many protein enzymes that regulate such activities as photosynthesis and nitrogen fixation. It is believed that sulfur-to-sulfur bonds link certain sites on long chains of amino acids, causing proteins to assume the specific three-dimensional shapes that are the key to their catalytic action. Sulfur is closely associated with nitrogen in the processes of protein and enzyme synthesis. Sulfur is also an essential ingredient of the aromatic oils that give the cabbage and onion families of plants their characteristic odors and flavors. It is not surprising that among the plants, the legume, cabbage, and onion families require especially large amounts of sulfur.

Deficiencies of Sulfur

Healthy plant foliage generally contains 0.15 to 0.45% sulfur, or approximately one-tenth as much sulfur as nitrogen. Plants deficient in sulfur tend to become spindly and to

---

[8] For a discussion of sulfur and agriculture, see Tabatabai (1986).

BOX 12.4 RATIONALIZING NITROGEN INPUTS IN AGRICULTURAL SYSTEMS

When we examine the nitrogen balance for cropland on a national scale, it is clear that more nitrogen is being applied than can be properly used (Table 12.8). In the United States, cropland receives approximately 20.9 million Mg of nitrogen annually, while it exports some 13.5 million Mg of nitrogen as crop harvests and residues. In other words, nitrogen inputs are more than 50% greater than intended outputs, leaving over 7 million Mg of nitrogen to either accumulate in soils (unlikely in most cases) or leak into the environment via leaching, erosion, and runoff, or via gaseous losses to the atmosphere. While some losses from agroecosystems are inevitable, it is becoming widely acknowledged that cropland in the United States and in other industrial countries has been receiving more nitrogen than plants can effectively use, and that this excess will have to be reduced if nitrogen pollution is to be controlled.

Strategies to reduce inputs of commercial fertilizers include the maximization of inputs from organic sources This would involve the recycling of nitrogen in farm manures and organic residues. It would also utilize more legumes in cropping systems, and thereby increase biological nitrogen fixation. Chemical fertilizers would be used primarily to complement the organic sources of nitrogen, including that from the mineralization of soil organic matter. Such strategies should better match the inputs and outputs of soil nitrogen, and reduce the loss of this element in liquid and gaseous forms.

TABLE 12.8 Nitrogen Inputs and Outputs for All Cropland in the United States in 1987 and 1977

Only the major categories are considered. These are aggregate data for the entire country and should not be considered representative of any given farm field. The estimates for the two years were made independently of each other, so the close agreement lends validity to both.

	Millions of Mg	
Nitrogen output or input	*1987*[a]	*1977*[b]
Outputs		
Harvested crops	10.60	8.9
Crop residues	2.89	3.0
Total outputs	13.50	11.9
Inputs to cropland		
As commercial fertilizer	9.39	9.5
As legume N fixation	6.87	7.2
In crop residues returned	2.89	3.0
Recoverable manure	1.73	1.4
Total inputs	20.9	21.1
Balance = inputs – outputs	7.42	9.2

[a] Data from Table 6-3 in National Research Council (1993).
[b] Data abstracted from Power (1981) as cited in NRC (1993).

develop thin stems and petioles. Their growth is slow, and maturity may be delayed. They also have a chlorotic light green or yellow appearance similar to that associated with nitrogen deficiency (see Section 12.1). However, the chlorosis develops first on the youngest leaves as sulfur supplies are depleted. Sulfur-deficient leaves on some plants show interveinal chlorosis or faint striping that distinguishes them from nitrogen-deficient leaves.

As a result of the following three independent trends, sulfur deficiencies in agricultural plants have become increasingly common during the past several decades:

1. Enforcement of clean air standards has led to reduction of sulfur dioxide (SO_2) emissions to the atmosphere from the burning of fossil coal and oil.

2. The sulfur contents of today's highly concentrated N-P-K fertilizers are far lower than those of a generation ago. Thus, less sulfur is being added.

3. As crop harvests have increased, larger amounts of sulfur are being removed from the soils. Consequently, the need for sulfur has increased just as the inputs from all sources have declined.

AREAS OF DEFICIENCY.   Sulfur deficiencies have been reported in most areas of the world but are most prevalent in areas where soil parent materials are low in sulfur, where extreme weathering and leaching has removed this element, or where there is little replenishment of sulfur from the atmosphere. In many tropical countries, one or more of these conditions prevail and sulfur-deficient areas are common.

In the United States, deficiencies of sulfur are most common in the Southeast, the Northwest, California, and the Great Plains. In the Northeast and in other areas with heavy industry and large cities, sulfur deficiencies are not yet widespread.

12.18   NATURAL SOURCES OF SULFUR

The three major natural sources of sulfur that can become available for plant uptake are: (1) *organic matter,* (2) *soil minerals,* and (3) *sulfur gases in the atmosphere.* In natural ecosystems where most of the sulfur taken up by plants is eventually returned to the same soil, these three sources combined are usually sufficient to supply the needs of growing plants. These three natural sources of sulfur will be considered in order.

Organic Matter

In surface soils in temperate, humid regions, 90 to 98% of the sulfur is usually present in organic forms (see Figure 12.13). As is the case for nitrogen, the exact forms of the sulfur in the organic matter are not known. Over time, however, soil microorganisms break down these organic sulfur compounds into soluble inorganic forms, mainly sulfate. This mineralization of organic compounds to release sulfate is analogous to the release of ammonium and nitrate from organic matter discussed in Section 12.4.

In arid and semiarid regions, less organic matter is present in the surface soils. Furthermore, the proportion of organic sulfur is not likely to be as high in arid- and semi-arid-region soils as it is in humid-region soils, especially in the lower soil horizons where gypsum ($CaSO_4 \cdot 2H_2O$) may be prominent.

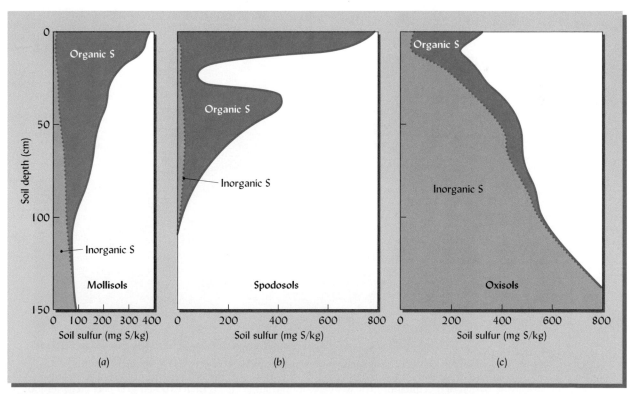

FIGURE 12.13   The distribution of organic and inorganic sulfur in representative soil profiles of the soil orders Mollisols, Spodosols, and Oxisols. In each, soil organic forms dominate the surface horizon. Considerable inorganic sulfur, both as adsorbed sulfate and calcium sulfate minerals, exists in the lower horizons of Mollisols. Relatively little inorganic sulfur exists in Spodosols. However, the bulk of the profile sulfur in the humid tropics (Oxisols) is present as sulfate adsorbed to colloidal surfaces in the subsoil.

The inorganic forms of sulfur are not as plentiful as the organic forms. However, they include sulfates that are reasonably soluble in water and are available for plant uptake. These sulfates are most prominent in the lower horizons of soils of low-rainfall areas. Also included are the sulfide forms found in some poorly drained humid-region areas. If these soils are drained, the sulfides are oxidized to sulfuric acid, and very low pH *acid sulfate* soils result (see Section 9.4).

Another mineral source of sulfur is the clay fraction of some soils high in Fe, Al oxides and kaolinite. These clays are able to strongly adsorb sulfate from soil solution and subsequently release it slowly by anion exchange, especially at low pH. Oxisols and other highly weathered soils of the humid tropics and subtropics may contain large stores of sulfate, especially in their subsoil horizons (see Figure 12.13c). Considerable sulfate may also be bound by the metal oxides in the spodic horizons under certain temperate forests.

Atmospheric Sulfur[9]

The atmosphere contains varying quantities of a number of sulfur-containing gases, including carbonyl sulfide (COS), hydrogen sulfide (H_2S), and sulfur dioxide (SO_2). Some of these materials are oxidized in the atmosphere to sulfates, forming H_2SO_4 and sulfate salts, such as $CaSO_4$ and $MgSO_4$. When these solids and gases return to the earth as dry

---

[9] For a discussion of this topic, see National Academy of Sciences (1983).

FIGURE 12.14 The apparatus shown in the picture collects both wet and dry sulfur deposition. A sensor (*a*) triggers the small roof (*b*) to move over and cover the dry deposition collection chamber (*c*) at the first sign of precipitation. The wet deposition chamber (*d*) is then exposed to collect precipitation. When the precipitation ceases, the sensor triggers the roof to move back over the wet deposition collection chamber so that dry deposition can again be collected. The map shows the typical geographic distribution of sulfur deposition from the atmosphere in eastern North America. The data are based on measurements of sulfur in wet deposition and the assumption that dry deposition contributes approximately an equal share of the total. [Photo courtesy of R. Weil; map modified from Olsen and Slavich (1986)]

particles and gases, it is called *dry deposition;* when they are brought down with precipitation, it is called *wet deposition.* Although the proportion of these two forms of deposition varies from one place to another, each typically supplies about half of the total.

The industrialized northeastern states have the highest deposition of sulfur in the United States, commonly ranging from 30 to 75 kg S/ha/yr downwind from industrial sites. In rural areas of the western United States, away from industrial cities and smelting plants, only 2 to 5 kg S/ha/yr is deposited. In rural Africa, 1 to 4 kg S/ha/yr is deposited. Regional patterns of air quality and sulfur deposition are evident as the less acute effects of industrial emissions are felt hundreds of kilometers downwind (Figure 12.14).

Atmospheric sulfur, whether in the wet or dry form, is absorbed by soils or directly by the plant foliage. The quantity that plants can absorb directly is variable, but in some cases 25 to 35% of the plant sulfur can come from this source even if available soil sulfate is adequate. In sulfur-deficient soils, about half of the plant needs can come from the atmosphere.

Environmental concerns about high sulfur levels in the atmosphere and the resulting acid rain are of great practical significance to both forestry and agriculture. The elevated levels of atmospheric sulfur and associated acid effects are causing serious damage to certain types of forest ecosystems. On the other hand, continued reductions in atmospheric sulfur are resulting in plant deficiencies of this element in other areas, especially for high-yield-potential agricultural crops. In these cases, part of the crop's sulfur requirement may have to be supplied by increasing the sulfur content of the fertilizers used, thereby increasing crop production costs.

12.19 THE SULFUR CYCLE

The major transformations that sulfur undergoes in soils are shown in Figure 12.15. The inner circle shows the relationships among the four major forms of this element: (1) *sulfides,* (2) *sulfates,* (3) *organic sulfur,* and (4) *elemental sulfur.* The outer portions show the most important sources of sulfur and how this element is lost from the system.

Considerable similarity to the nitrogen cycle is evident (compare Figures 12.15 and 12.1). In each case, the atmosphere is an important source of the element in question. Both elements are held largely in the soil organic matter; both are subject to microbial oxidation and reduction; both can enter and leave the soil in gaseous forms; and both are subject to some degree of leaching in the anionic form. Microbial activities are responsible for many of the transformations that determine the fates of both nitrogen and sulfur.

Figure 12.15 should be referred to frequently in conjunction with the following more detailed examination of sulfur in plants and soils.

12.20 BEHAVIOR OF SULFUR COMPOUNDS IN SOILS

Mineralization

Sulfur behaves much like nitrogen as it is absorbed by plants and microorganisms and moves through the sulfur cycle. The organic forms of sulfur must be mineralized by soil organisms if the sulfur is to be used by plants. When conditions are favorable for general microbial activity, sulfur mineralization occurs. Some of the more easily decomposed organic compounds in the soil are sulfate esters, from which microorganisms release sulfate ions directly. However, in much of the soil organic matter, sulfur in the reduced state is bonded to carbon atoms in protein and amino acid compounds. In the latter case the mineralization reaction might be expressed as follows:

$$\text{Organic sulfur} \longrightarrow \text{decay products} \xrightarrow{\text{O}_2} \text{SO}_4^{2-} + 2\text{H}^+$$

| Proteins and other organic combinations | H_2S and other sulfides are simple examples | Sulfates |

Because this release of available sulfate is mainly dependent on microbial processes, the supply of available sulfate in soils fluctuates with seasonal, and sometimes daily, changes in environmental conditions (Figure 12.16). These fluctuations lead to the

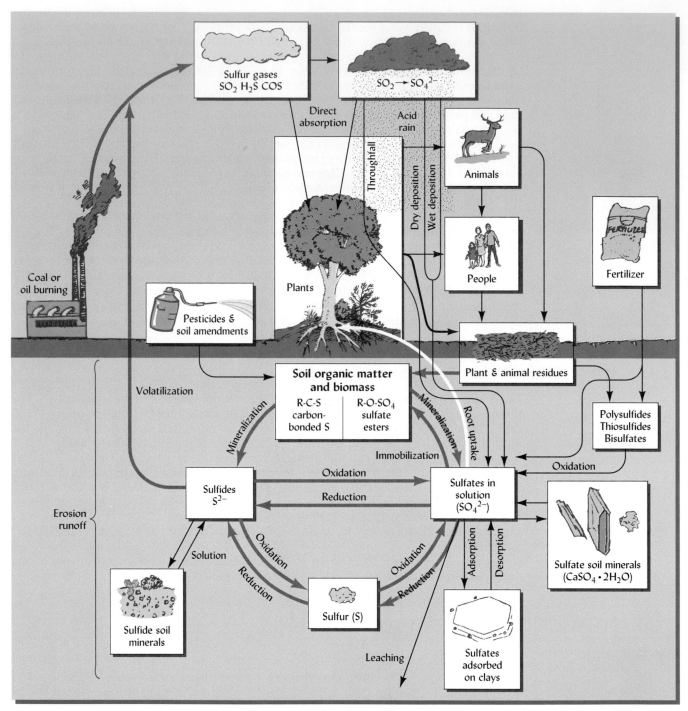

FIGURE 12.15 The sulfur cycle, showing some of the transformations that occur as this element is cycled through the soil–plant–animal–atmosphere system. In the surface horizons of all but a few types of arid-region soils, the great bulk of sulfur is in organic forms. However, in deeper horizons or in excavated soil materials, various inorganic forms may dominate. The oxidation and reduction reactions that transform sulfur from one form to another are mainly mediated by soil microorganisms.

same difficulties in predicting and measuring the amount of sulfur available to plants as were discussed in the case of nitrogen.

Immobilization

Immobilization of inorganic forms of sulfur occurs when low-sulfur, energy-rich organic materials are added to soils. The immobilization mechanism is thought to be the same as for nitrogen—the energy-rich material stimulates microbial growth, and the

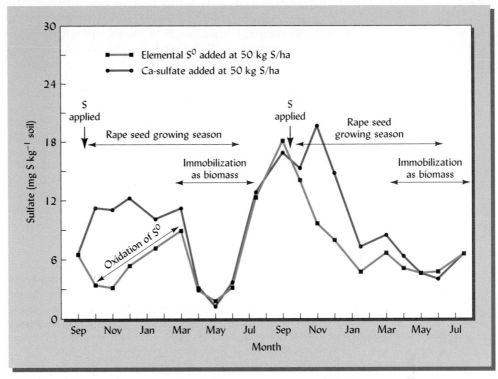

FIGURE 12.16 Seasonal changes in the sulfate form of sulfur available in the surface horizon of a soil (Argixeroll) in Oregon used to grow the oilseed crop, rape. This crop is sown in the fall, grows slowly during the winter, and then grows rapidly during the cool spring months. Data are shown for plots that were fertilized with either elemental S or calcium sulfate. The vertical arrows indicate dates on which these amendments were applied. Note that sulfate concentration was greater in the calcium sulfate–fertilized soils for the first few months after each application while the elemental S was slowly converted to sulfate by microbial oxidation. A distinct depression in sulfate concentration occurred each spring as the soil warmed up, stimulating both immobilization of sulfate into microbial biomass and uptake of sulfate into the rapeseed crop. Sulfate concentrations reached a peak in late summer and early fall, when crop uptake ceased after harvest and microbial mineralization was rapidly occurring. Movement of dissolved sulfate from the lower horizons up into the surface soil may also have occurred during hot, dry weather. [Modified from Castellano and Dick (1991)]

inorganic sulfate is assimilated into microbial tissue. A C/S ratio greater than 400:1 generally leads to such immobilization of sulfur. When the microbial activity subsides, the inorganic sulfate reappears in the soil solution.

The pattern of S immobilization in soils suggests that, like nitrogen, sulfur in soil organic matter may be associated with organic carbon in a reasonably constant ratio. Data from a number of soils in temperate regions suggest that the C/N/S ratio of 100:8:1 is reasonably representative.

12.21 SULFUR OXIDATION AND REDUCTION

The Oxidation Process

During the microbial decomposition of organic carbon-bonded sulfur compounds, sulfides are formed along with other incompletely oxidized substances, such as elemental sulfur (S^0), thiosulfates ($S_2O_3^{2-}$), and polythionates ($S_2O_{3x}^{2-}$). These reduced substances are subject to oxidation, just as are the ammonium compounds formed when nitrogenous materials are decomposed. The oxidation reactions may be illustrated as follows, with hydrogen sulfide and elemental sulfur:

$$H_2S + 2O_2 \rightarrow H_2SO_4 \rightarrow 2H^+ + SO_4^{2-}$$

$$2S + 3O_2 + 2H_2O \rightarrow 2H_2SO_4 \rightarrow 4H^+ + SO_4^{2-}$$

The oxidation of some sulfur compounds, such as sulfites (SO_3^{2-}) and sulfides (S^{2-}), can occur by strictly chemical reactions. However, most sulfur oxidation in soils is *biochemical* in nature, carried out by a number of autotrophic bacteria, which include five species of the genus *Thiobacillus*. Since the environmental requirements and tolerances of these five species vary considerably, the process of sulfur oxidation occurs over a wide range of soil conditions. For example, sulfur oxidation may occur at pH values ranging from <2 to >9. This flexibility is in contrast to the comparable nitrogen oxidation process, nitrification, which requires a rather narrow pH range closer to neutral.

Note that sulfur oxidation is an acidic process which, as we learned in Chapter 9, can be both beneficial and detrimental. The H_2SO_4 produced from sulfur oxidation can be used to help reclaim sodic soils, and to lower the soil pH for acid-loving plants such as rhododendrons. But the process also stimulates the formation of highly toxic acid sulfate soils with pH values as low as 1.5. Likewise, it contirbutes to the problems of acid rain[10] with its detrimental effects, particularly on forests. Last, it increases the quantities of liming materials that must be added to maintain appropriate pH levels in humid-region soils.

The Reduction Process

Like nitrate ions, sulfate ions tend to be unstable in anaerobic environments. They are reduced to sulfide ions by a number of bacteria of two genera, *Desulfovibro* (five species) and *Desulfotomaculum* (three species). The organisms use the oxygen in sulfate to oxidize organic materials. A representative reaction showing the reduction of sulfur coupled with organic matter oxidation is as follows:

$$2R\text{---}CH_2OH + SO_4^{2-} \rightarrow 2R\text{---}COOH + 2H_2O + S^{2-}$$

Organic alcohol   Sulfate   Organic acid   Sulfide

In poorly drained soils, the sulfide ion reacts immediately with iron or manganese, which in anaerobic conditions are typically present in the reduced forms. By tying up the soluble reduced iron, the formation of iron sulfides helps prevent iron toxicity in rice paddies and marshes. This reaction may be expressed as follows:

$$Fe^{2+} + S^{2-} \rightarrow FeS$$

Dissolved   Sulfide   Iron sulfide
ferrous iron   (solid)

The oxidation and reduction reactions of inorganic sulfur compounds play an important role in determining the quantity of sulfate (the plant-available nutrient form of sulfur) present in soils at any one time. Also, the state of sulfur oxidation is an important factor in the acidity of soil and water draining from soils.

12.22   SULFUR RETENTION AND EXCHANGE

The sulfate ion is the form in which plants absorb most of their sulfur from soils. Since many sulfate compounds are quite soluble, the sulfate would be readily leached from the soil, especially in humid regions, were it not for its adsorption by the soil colloids. As was pointed out in Chapter 8, most soils have some anion exchange capacity, which is associated with iron and aluminum oxide coatings and clays and, to a limited extent, with 1:1-type silicate clays. Sulfate ions are attracted by the positive charges that characterize acid soils containing these clays. They also react directly with hydroxy groups exposed on the surfaces of these clays. Figure 12.17 illustrates sulfate adsorption mechanisms on the surface of some Fe, Al oxides and 1:1-type clays. Note that adsorption increases at lower pH values as positive charges that become more prominent on the particle surfaces attract the sulfate ions. Some sulfate reacts with the clay particles, becoming tightly bound, and is only slowly available for plant uptake and leaching.

---

[10] Note that some of this acidity results from nitrogen oxides as they are oxidized in the atmosphere to nitric and nitrous acids. Sulfur and nitrogen are jointly responsible for this problem.

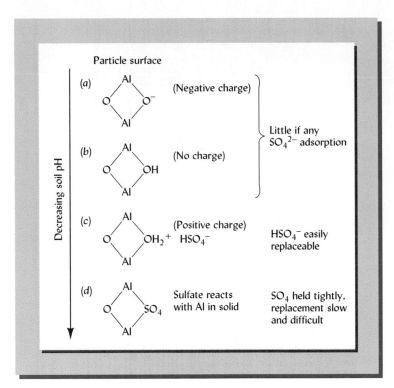

Particle surface

(a) (Negative charge)

(b) (No charge)

(c) (Positive charge) HSO_4^-

(d) Sulfate reacts with Al in solid

Decreasing soil pH

Little if any SO_4^{2-} adsorption

HSO_4^- easily replaceable

SO_4 held tightly, replacement slow and difficult

FIGURE 12.17 Effect of decreasing soil pH on the adsorption of sulfates by 1:1-type silicate clays and oxides of Fe and Al (reaction with a surface-layer Al is illustrated). At high pH levels (*a*), the particles are negatively charged, the cation exchange capacity is high, and base-forming cations are adsorbed. Sulfates are repelled by the negative charges. As acidity is increased (*b*), the H$^+$ ions are attracted to the particle surface, and the negative charge is satisfied but the SO_4^{2-} ions are still not attracted. At still lower pH values (*c*), more H$^+$ ions are attracted to the particle surface, resulting in a positive charge that attracts the SO_4^{2-} ion. This is easily exchanged with other anions. (*d*) At still lower pH levels, the SO_4^{2-} reacts directly with Al and becomes a part of the crystal structure. Such sulfate is tightly bound, and it is removed very slowly, if at all.

In warm, humid regions, surface soils are typically quite low in sulfur. However, much sulfate may be held by the iron and aluminum oxides and 1:1-type silicate clays that tend to accumulate in the subsoil horizons of the Ultisols and Oxisols of these regions (see Figure 12.13). Symptoms of sulfur deficiency commonly occur early in the growing season on Ultisols in the southeastern United States, especially those with sandy, low-organic-matter surface horizons. However, the symptoms may disappear as the crop matures and its roots reach the deeper horizons where sulfate is retained.

Sulfate Adsorption and Leaching of Base-Forming Cations

When the sulfate ion leaches from the soil, it is usually accompanied by equivalent quantities of cations, including Ca and Mg and other base-forming cations. In soils with high sulfate adsorption capacities, sulfate leaching is low and the loss of companion cations is also low. (Figure 12.18). In contrast, sulfate leaching losses from low-sulfate-adsorbing soils are commonly high, and take with them considerable quantities of base-forming cations. Sulfur is thus seen as an indirect conserver of these cations in the soil solution. This is of considerable importance in soils of forested areas that receive acid rain.

12.23   SULFUR AND SOIL FERTILITY MAINTENANCE

Figure 12.19 depicts the major gains and losses of sulfur from soils. The problem of maintaining adequate quantities of sulfur for mineral nutrition of plants is becoming increasingly important. Even though chances for widespread sulfur deficiencies are generally less than for nitrogen, phosphorus, and potassium, increasing crop removal of sulfur makes it essential that farmers be attentive to prevent deficiencies of this element. In some parts of the world (especially in certain semiarid grasslands), sulfur is already the next most limiting nutrient after nitrogen.

Crop residues and farmyard manures can help replenish the sulfur removed in crops, but these sources generally can help to recycle only those sulfur supplies that

FIGURE 12.18 Diagrams illustrating cation leaching losses as influenced by sulfate adsorption capacities of forest soils. When acid rain containing SO_4^{2-} and H^+ ions falls on soils with high SO_4^{2-}-adsorbing capacities (left), the small quantities of SO_4^{2-} available for leaching are accompanied by correspondingly small amounts of cations such as Ca^{2+} and Mg^{2+}. Where acid rain falls on soils with low SO_4^{2-} adsorbing capacity (right), most of the sulfate remains in the soil solution and is leached from the soil along with equivalent quantities of cations, including Ca^{2+} and Mg^{2+}. Al^{3+} ions that commonly replace the Ca^{2+} and Mg^{2+} lost from the soil exchange complex are toxic to many forest species. This likely accounts for at least part of the negative effects of acid rain on some forested areas. [Redrawn from Mitchell, et al. (1992)]

already exist within a farm. In regions with low-sulfur soils, greater dependence must be placed on fertilizer additions. Regular applications of sulfur-containing materials are now necessary for good crop yields in large areas far removed from industrial plants. There will certainly be an increased necessity for the use of sulfur in the future.

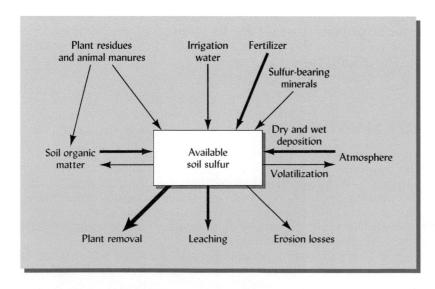

FIGURE 12.19 Major gains and losses of available soil sulfur. The thickness of the arrows indicates the relative amounts of sulfur involved in each process under average conditions. Considerable variation occurs in the field.

12.24 CONCLUSION

Sulfur and nitrogen have much in common concerning the manner in which they cycle in soils. Both are held by soil colloids in slowly available forms. Both are found in proteins and other organic forms as part of the soil organic matter. Their release to inorganic ions (SO_4^{2-}, NH_4^+, and NO_3^-), in which forms they are available to higher plants, is accomplished by soil microorganisms. Anaerobic soil organisms are able to change these elements into gaseous forms, which are then released to the atmosphere to be joined by similar gases released from industrial plants and from vehicle engines. These gases are then deposited on plants, soils, and other objects in forms that are popularly termed *acid rain*. This has serious consequences to forestry, to agriculture, and to society in general.

There are some significant differences between nitrogen and sulfur. Large amounts of both soil nitrogen and soil sulfur are found in organic compounds. However, a higher proportion of soil sulfur is found in inorganic compounds, especially in some drier areas where gypsum ($CaSO_4 \cdot 2H_2O$) is abundant in the subsoil. Some soil organisms have the ability to fix elemental N_2 gas into compounds usable by plants. No analogous process occurs for sulfur. Also, nitrogen, which is removed in much larger quantities by plants, must be replenished regularly by organic residues, manure, or chemical fertilizers.

In rural areas away from cities and industrial plants, sulfur deficiencies in plants are increasingly common. As a result of increased crop removal and reductions in incidental applications of sulfur to soils, this element will likely join nitrogen as a regular component of soil fertility programs.

Several serious environmental problems are caused by excessive amounts of nitrogen or sulfur in certain forms. While water pollution by nitrates is probably the most prominent of the nitrogen-related problems, acid deposition and drainage are the main environmental concerns associated with sulfur cycling in soils. Using our knowledge of the nitrogen and sulfur cycles, much can now be done to alleviate both sets of problems. However, much remains to be learned about the reactions of nitrogen and sulfur in soils and in the environment.

STUDY QUESTIONS

1. A total of 300 kg/ha of nitrogen was supplied through animal manure and chemical fertilizer for a corn crop, but only 200 kg/ha was found in the harvested silage corn, all of which was removed from the land. What happened to the 100 kg/ha N not taken up by the corn crop? Explain.

2. Discuss the benefits and ill effects of sulfur and nitrogen that are added to soils by humans and by natural processes each year.

3. Some 20 kg/ha of nitrogen contained in wheat straw was added to a soil that had a nitrate level of 10 kg/ha. A couple of weeks later a soil test showed no nitrates. How do you explain this situation?

4. Modern domestic animal production systems have some environmental and health implications relating to nitrogen. What are they, and how can they be managed?

5. What differences would you expect in nitrate contents of streams from a forested watershed and one where agricultural crops are grown, and why?

6. What is *acid rain,* what are the sources of this precipitation, and what are its implications for forestry and agriculture?

7. Nitrogen is fixed from the atmosphere and is also fixed by vermiculite clays and humus. Differentiate between these two processes and indicate the role of microbes in each process.

8. Chemical fertilizers and manures are commonly added to agricultural soils, and yet these soils are often lower in nitrogen and other nutrients than are nearby forested soils. Explain why this is the case.

9. Sulfur deficiencies in agricultural crops are more widespread today than 15 years ago. Why is this the case?

10. How do riparian forests help reduce nitrate contamination of streams and rivers?

11. What are *acid sulfate soils*, how do they form, and how can they best be managed?

12. In some tropical regions, agroforestry systems that involve mixed cropping of trees and food crops are used. What advantages in nitrogen management do such systems have over monocroping systems that do not involve trees?

REFERENCES

Addiscott, T. M., A. P. Whitmore, and D. S. Powlson. 1991. *Farming, Fertilizers and the Nitrate Problem.* (Wallingford, U.K.: CAB International).

Burns, R. C., and R. W. F. Hardy. 1975. *Nitrogen Fixation in Bacteria and Higher Plants.* (Berlin: Springer-Verlag).

Castellano, S. D., and R. P. Dick. 1991. "Cropping and sulfur fertilization influence on sulfur transformations in soil," *Soil Sci. Soc. Amer. J.,* **54**:114–121.

Colbourn, P. 1993. "Limits to denitrification in two pasture soils in a maritime temperate climate," *Agriculture, Ecosystems and Environment,* **43**:49–68.

Hanson, G. C., P. M. Groffman, and A. J. Gold. 1994. "Denitrification in riparian wetlands receiving high and low groundwater nitrate inputs," *J. Environ. Qual.,* **23**:917–922.

Hauck, R. D. (ed.). 1984. *Nitrogen in Crop Production.* (Madison, Wis.: Amer. Soc. Agron., Crop Sci. Soc. Amer., Soil Sci. Soc. Amer.).

Jenkinson, D. S. 1990. "An introduction to the global nitrogen cycle," *Soil Use and Management,* **6**:56–61.

Klemmedson, J. O. 1994. "New Mexican locust and parent material: Influence on forest floor and soil macronutrients," *Soil Sci. Soc. Amer. J.,* **58**:974–980.

Leffelaar, P. A., and W. W. Wessel. 1988. "Denitrification in a homogeneous, closed system: Experimental and simulation," *Soil Sci.,* **146**:335–349.

Leuch, D., S. Haley, P. Liapis, and B. McDonald. 1995. *The EU Nitrate Directive and CAP Reform: Effects on Agricultural Production, Trade and Residual Soil Nitrogen.* Report 255. (Washington, D.C.: USDA Economic Research Service).

Mitchell, M. J., et al. 1992. "Sulfur dynamics of forest ecosystems," in Howarth, et al. (eds.), *Sulfur Cycling on the Continents,* 1992 SCOPE. (New York: Wiley).

National Academy of Sciences. 1983. *Acid Deposition: Atmospheric Processes in Eastern North America.* (Washington, D.C.: National Academy Press).

National Research Council. 1993. *Soil and Water Quality: An Agenda for Agriculture.* (Washington, D.C.: National Academy of Sciences).

Olsen, A. R., and A. L. Slavich. 1986. *Acid Precipitation in North America: 1984 Annual Data Summary,* from Acid Deposition System Data Base. Environmental Protection Agency Report EPA/600/4-86/033. (Washington, D.C.: U.S. Environmental Protection Agency).

Owens, L. B., W. M. Edwards, and R. W. Van Keuren. 1992. "Nitrate levels in shallow groundwater under pastures receiving ammonium nitrate or slow-release nitrogen fertilizer," *J. Environ. Qual.,* **21**(4):607–613.

Patrick, W. H., Jr. 1982. "Nitrogen transformations in submerged soils," in F. J. Stevenson (ed.), *Nitrogen in Agricultural Soils.* Agronomy Series no. 27 (Madison, Wis.: Amer. Soc. Agron., Crop Sci. Soc. Amer., Soil Sci. Soc. Amer.).

Power, J. F. 1981. "Nitrogen in the cultivated ecosystem," in F. E. Clark and T. Rosswall (eds.), *Terrestrial Nitrogen Cycles—Processes, Ecosystem Strategies and Management Impacts.* Ecological Bulletin no. 33. (Stockholm, Sweden: Swedish National Research Council), pp. 529–546.

Prasad, R., and J. F. Power. 1995. "Nitrification inhibitors for agriculture, health and the environment," *Advances in Agronomy,* **54**:233–280.

Stevenson, F. J. (ed.). 1982. *Nitrogen in Agricultural Soils.* Agronomy Series no. 22. (Madison, Wis.: Amer. Soc. Agron., Crop Sci. Soc. Amer., Soil Sci Soc. Amer.).

Tabatabai, S. J. 1986. *Sulfur in Agriculture.* Agronomy Series no. 27 (Madison, Wis.: Amer. Soc. Agron., Crop Sci. Soc. Amer., Soil Sci. Soc. Amer.).

Torrey, J. G. 1978. "Nitrogen fixation by actinomycete-induced angiosperms," *BioScience,* **28**:586–592.

U.S./Canada Work Group No. 2. 1982. *Atmospheric Sciences and Analysis.* Final Report; J. L. Ferguson and L. Machata (Cochairmen). (Washington, D.C.: U.S. Environmental Protection Agency).

Walker, T. W., et al. 1956. "Fate of labeled nitrate and ammonium nitrogen when applied to grass and clover grown separately and together," *Soil Sci.,* **81**:339–352.

13

SOIL PHOSPHORUS, POTASSIUM, AND MICRONUTRIENTS

*They are not only roots
beneath the bloodstained stones,
Not only do their demolished bones
definitely till the soil.*
—*Pablo Neruda, CHILEAN POET*

Phosphorus is a critical element in natural and agricultural ecosystems throughout the world. It is second only to nitrogen in its importance in the production of healthy plants and profitable yields. The quantity of phosphorus in native soils is low, and a high proportion of that which is present is not readily available to plants. This low availability leads to major social and environmental problems. Unproductive soils lead to crop failures, especially in tropical and subtropical regions. This forces farmers to clear more land in order to survive. The cleared lands with poor vegetative cover are subject to erosion, the soils are degraded, and downstream ponds, lakes, and reservoirs are polluted with the soil-laden runoff waters. This scenario pertains in much of Africa where inadequate supplies of soil phosphorus have contributed to the decline of per capita food production in recent decades.

Most industrialized countries have overcompensated for the low phosphorus availability in their soils (see Figure 13.1). For many years they have returned to their soils far more phosphorus than has been removed in harvested crops. Large quantitites of phosphate fertilizers and P-containing organic wastes have been applied to the land, and a buildup of phosphorus in the surface soil has resulted. Watershed runoff from such soils carries the excess phosphorus, which leads to nutrient imbalances in downstream rivers, lakes, and estuaries. The added phosphorus inputs contribute to the process of *cultural eutrophication,*[1] which may jeopardize drinking water supplies and restrict the use of these aquatic systems for fisheries, recreation, industry, and aesthetics. Phosphorus control is therefore a high priority of most national and regional water-quality programs.

Potassium does not play as direct a role in environmental quality as does phosphorus, but inadequate supplies of this element commonly limit crop growth and reduce crop quality. Even though most soils have large total supplies of this element, most of that present is tied up in the form of insoluble minerals and is unavailable for plant use. Also, plants require potassium in such large amounts that careful management practices are necessary to make this nutrient available rapidly enough to optimize plant growth.

Of the 18 elements known to be essential for plant growth, 9 are required in such small amounts that they are called *micronutrients* or *trace elements*. The ill effects of deficiencies and even toxicities of these elements are becoming more noted throughout the

---

[1] The word *eutrophication* comes from the Greek *eutrophos,* meaning well-nourished or nourishing. Natural accumulation of nutrients over centuries causes lakes to fill in with plants and lose dissolved oxygen needed by fish—the process of eutrophication. Excessive inputs of nutrients under human influence tremendously speeds this process and is called *cultural eutrophication.*

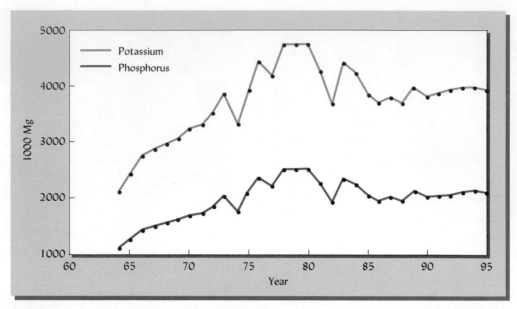

FIGURE 13.1 Estimated annual addition of fertilizer phosphorus and potassium to cropland in the United States since 1964. Note that annual applications of phosphorus increased steadily until about 1980, then declined somewhat to level off at about 2 million Mg per year. The decline and leveling off of phosphorus application rates reflects the reduced need for phosphorus buildup in agricultural soils in which the phosphorus fixation capacity has been largely saturated. The trend toward lower rates of phosphorus application since 1980 was clearly driven by economic and agronomic influences, not environmental concerns, since the trends for potassium, a nonpolluting nutrient, are almost identical. As an example of economic effects, the oil embargo of 1973–1974 increased fertilizer prices dramatically, resulting in the notable dip in application rates in 1974. The sharp dip in 1983 reflected U.S. government policy of "payment in kind" to induce farmers to restrict their corn plantings. In other parts of the world, sustainable crop production will probably require increasing applications of phosphorus and potassium for many years to come. (Based on data from the USDA Economic Research Service and the United Nations Food and Agriculture Organization)

world in terms of stunted growth, low yields, and even plant death. Very small applications of micronutrients can produce dramatic results, while problems of toxicity loom when an oversupply is present.

13.1 ROLE OF PHOSPHORUS IN PLANT NUTRITION AND SOIL FERTILITY[2]

Neither plants nor animals can grow without phosphorus. It is an essential component of the organic compound often called the *energy currency* of the living cell, **adenosine triphosphate** (ATP), which drives most energy-requiring biochemical processes, including the uptake of nutrients and their transport within the plant. Phosphorus is also an essential component of **deoxyribonucleic acid** (DNA), the seat of genetic inheritance, and of **ribonucleic acid** (RNA), which directs protein synthesis in both plants and animals. Phospholipids, which play critical roles in cellular membranes, are another class of universally important phosphorus-containing compounds. For most plant species, the total phosphorus content of healthy leaf tissue is not high, usually only 0.2 and 0.4% of the dry matter.

Phosphorus and Plant Growth

ASPECTS OF PLANT GROWTH ENHANCED. Adequate phosphorus nutrition enhances many aspects of plant physiology, including the fundamental processes of photosynthesis, nitrogen fixation, flowering, fruiting (including seed production), and maturation. Root growth, particularly development of lateral roots and fibrous rootlets, is encouraged by phosphorus, as is straw strength in cereal crops. Improvement of crop quality, especially in forages and vegetables, is another benefit attributed to this nutrient.

---

[2] For a review of the significance of this element, see Khasawneh, et al. (1980).

SYMPTOMS OF PHOSPHORUS DEFICIENCY IN PLANTS. A phosphorus-deficient plant is usually stunted, thin-stemmed, and spindly, but its foliage is often dark, almost bluish, green. In severe cases, phosphorus deficiency can cause yellowing and senescence of leaves. Some plants develop purple colors (see Plate 28, after page 178) in their leaves and stems as a result of phosphorus deficiency. Phosphorus is needed in especially large amounts in meristematic tissues, where cells are rapidly dividing and enlarging.

Phosphorus is very mobile within the plant, so when the supply is short, phosphorus in the older leaves is mobilized and transferred to the newer, rapidly growing leaves. Both the purpling and premature senescence associated with phosphorus deficiency are therefore most prominent on the older leaves. Phosphorus-deficient plants are also characterized by delayed maturity, sparse flowering, and poor seed quality.

The Phosphorus Problem in Soil Fertility

The phosphorus problem in soil fertility is threefold. *First,* the total phosphorus level of soils is low, usually no more than one-tenth to one-fourth that of nitrogen, and one-twentieth that of potassium. The phosphorus content of soils ranges from 200 to 2000 kg phosphorus in the upper 15 cm of 1 ha of soil, with an average of about 1000 kg P. *Second,* the phosphorus compounds commonly found in soils are mostly unavailable for plant uptake, often because they are highly insoluble. *Third,* when soluble sources of phosphorus, such as those in fertilizers and manures, are added to soils, they are fixed[3] (changed to unavailable forms) and, in time, form highly insoluble compounds. These three factors that constrain levels of available phosphorus in soils help account for the low crop productivity in regions where replenishment of soil phosphorus is inadequate. They also help us understand the environmental consequences of human mismanagement of the phosphorus, as discussed in the next section.

13.2 EFFECTS OF PHOSPHORUS ON ENVIRONMENTAL QUALITY

The principal environmental problems related to soil phosphorus are **land degradation** caused by too little available phosphorus and **accelerated eutrophication** caused by too much. Both problems are related to the role of phosphorus as a plant nutrient.

Land Degradation

Many highly weathered soils in the warm, humid, and subhumid regions of the world have very little capacity to supply phosphorus for plant growth. The low phosphorus availability is partly a result of extensive losses of phosphorus during long periods of relatively intense weathering and partly due to the low availability of phosphorus in the aluminum and iron combinations that are the dominant forms of phosphorus in these soils.

Undisturbed natural ecosystems in these regions usually contain enough phosphorus in the biomass and soil organic matter to maintain a substantial standing crop of trees or grasses. Most of the phosphorus taken up by the plants is that released from the decomposing residues of other plants. Very little is lost as long as the system remains undisturbed.

Once the land is cleared for agricultural use (by timber harvest or by forest fires), the losses of phosphorus in eroded soil particles, in runoff water, and in biomass removals (harvests) can be substantial. Within just a few years the system may lose most of the phosphorus that had cycled between the plants and the soils. The remaining inorganic phosphorus in the soil is largely unavailable for plant uptake. In this manner, the phosphorus-supplying capacity of the disturbed soil rapidly becomes so low that regrowth of natural vegetation is sparse and, on land cleared for agricultural use, crops soon fail to produce useful yields (Figure 13.2).

Leguminous plants that might be expected to replenish soil nitrogen supplies are particularly hard hit by phosphorus deficiency, because low phosphorus supply inhibits effective nodulation and retards the biological nitrogen-fixation process. The spindly plants, deficient in both phosphorus and nitrogen, can provide little vegetative cover to prevent heavy rains from washing away the surface soil. The resulting

---

[3] Note that the term *fixation* as applied to phosphorus has the same general meaning as the chemical fixation of potassium or ammonium ions; that is, the chemical being fixed is bound, entrapped, or otherwise held tightly by soil solids in a form that is relatively unavailable to plants. In contrast, the fixation of gaseous nitrogen refers to the biological conversion of N_2 gas to combined forms that plants can use.

FIGURE 13.2 On many highly weathered soils in the tropics and subtropics, low availability of phosphorus, as well as of other nutrients, may limit the regrowth of natural vegetation or the production of crops after natural vegetation is cleared. Phosphorus has been so depleted in this African woman's cornfield that there will be little or no grain to harvest from her severely phosphorus-deficient crop. The poor plant growth leaves the soil susceptible to accelerated erosion and further degradation. (Photo courtesy of R. Weil)

erosion will further reduce soil fertility and water-holding capacity. The increasingly impoverished soils can support less and less vegetative cover, and so the degradation accelerates. Meanwhile, the soil particles lost by erosion become sediment farther down in the watershed, filling in reservoirs and increasing the turbidity of the rivers.

There are probably 1 to 2 billion ha of land in the world where phosphorus deficiency limits growth of both crops and native vegetation. Much of this land lies in poor countries whose farmers have little money for fertilizers. In much of sub-Saharan Africa,[4] additions of phosphorus fertilizers for food crops are a fraction of the rate of removal of this element in the harvested crops. The soils have been mined of phosphorus for years, and crop yields remain low. Without added phosphorus, there can be little hope of restoring these lands to productivity and their inhabitants to prosperity.

Water Quality Degradation

Accelerated or cultural eutrophication (Box 13.1) is caused by phosphorus entering streams from both **point sources** and **nonpoint sources.** Point sources, such as sewage treatment plant outflows, industries, and the like, are relatively easy to identify, regulate, and clean up. Nonpoint sources, which are principally runoff water from soils throughout the affected watershed, are more difficult to identify and control. Consequently, these diffuse sources of phosphorus are major causes of eutrophication in most industrialized countries. In intensively cultivated areas of these countries, phosphorus additions from fertilizers, manures, and municipal wastes far exceed the removal of this element in harvested crops. This leads to a buildup of phosphorus in the soils, to increases in the content of this element in runoff and sediment from these soils, and to eutrophication (see Box 13.1). Figure 13.4 shows how the phosphorus buildup influences the balance of this element in the surface soils of intensively cultivated areas compared to that in natural forested watersheds.

Phosphorus Losses in Runoff

Agricultural management that involves disturbing the soil surface with tillage generally increases the amount of phosphorus carried away on eroded sediment (i.e., **par-**

---

[4] For a recent review of soil phosphorus problems in Africa, see Buresh, et al. (1997).

BOX 13.1   EUTROPHICATION

Abundant growth of plants in terrestrial systems is usually considered beneficial, but in aquatic systems such growth can cause water-quality problems. The unwanted growth of algae (floating single-celled plants) and of aquatic weeds, termed *eutrophication*, can make ponds and lakes unsatisfactory environments for fish and can make the water unsuitable for drinking.

In natural ecosystems, lakes are commonly clear and free of excess plant growth and have highly diverse communities of organisms. This is because some factor (such as low levels of nitrogen or phosphorus) is limiting the growth of algae and other plants. Brackish and salty waters are most often nitrogen-limited, and plant growth is stimulated by high-nitrogen pollutants.

Most freshwater lakes and streams, on the other hand, are usually phosphorus-limited, partly because phosphorous levels in the water are typically very low and partly because nitrogen is being supplied through nitrogen fixation by some cyanobacteria (blue-green algae). Critical levels of phosphorus in water, above which eutrophication is likely to be triggered, are approximately 0.03 mg/L of dissolved phosphorus and 0.1 mg/L of total phosphorus.

When phosphorus is added to a phosphorus-limited lake, it stimulates a burst of algal growth (referred to as an *algal bloom*) and, often, a shift in the dominant algal species. The phosphorus-stimulated algae may cover the surface of the water with mats of algal scum. The lake may also be choked with higher plants that are also stimulated by the added phosphorus. When these aquatic weeds and algal mats die, they sink to the bottom, where their decomposition by microorganisms uses up much of the oxygen dissolved in the water. The decrease in oxygen (anoxic conditions) severely limits the growth of many aquatic organisms, especially fish. Such eutrophic lakes often become turbid, limiting growth of submerged aquatic vegetation and benthic (bottom feeding) organisms that serve as food for much of the fish community. In extreme cases eutrophication can lead to massive fish kills (Figure 13.3).

Eutrophication can transform clear, oxygen-rich, good-tasting water into cloudy, oxygen-poor, foul-smelling, bad-tasting, and possibly toxic water. Eutrophic conditions favor the growth of *Cyanobacter*, blue-green algae that are mostly undesirable food for zooplankton, a major food source for fish. These *Cyanobacter* produce toxins and bad-tasting and -smelling compounds that make the water unsuitable for human or animal consumption. Some filamentous algae can clog water treatment intake filters and thereby increase the cost of water remediation. Dense growth of both algae and aquatic weeds may make the water useless for boating and swimming. Furthermore, eutrophic waters generally have a reduced level of biological diversity (fewer species) and fewer fish, of desirable species.

FIGURE 13.3   In extreme cases of eutrophication, massive fish kills can occur in sensitive lakes and rivers. The kills result from anoxic conditions, which are brought on by the decay of the masses of algae stimulated by elevated inputs of phosphorus (or sometimes of nitrogen). (Photo courtesy of R. Weil)

FIGURE 13.4 Phosphorus balance in surface soils (Ultisols) of adjacent forested and agricultural watersheds. The forest consisted primarily of mature hardwoods that had remained relatively undisturbed for 45 or more years. The agricultural land was producing row crops for more than 100 years. It appears that in the agricultural soil about half of the organic phosphorus has been converted into inorganic forms or lost from the system since cultivation began. At the same time, substantial amounts of inorganic phosphorus accumulated from fertilizer inputs. Compared to the forested soil, mineralization of organic phosphorus was about four times as great in the agricultural soil, and the amount of phosphorus lost to the stream was eight times as great. Flows of phosphorus, represented by arrows, are given as kg/ha/yr. Although not shown in the diagram, it is interesting to note that nearly all (95%) of the phosphorus lost from the agricultural soil was in particulate form, while losses from the forest soil were 33% dissolved and 77% particulate. [Data from Vaithiyanathan and Correll (1992)]

ticulate P). On the other hand, fertilizer or manure that is left unincorporated on the surface of cropland or pastures usually leads to increased losses of phosphorus dissolved in the runoff water (i.e., **dissolved P**). These trends can be seen in Table 13.1 by comparing phosphorus losses from no-till and conventionally tilled wheat. While no-till systems may increase losses of soluble phosphorus, such systems are generally

TABLE 13.1   **Influence of Wheat Production and Tillage on Annual Losses of Phosphorus in Runoff Water and Eroded Sediments Coming from Soils in the Southern Great Plains**

The total phosphorus lost includes the phosphorus dissolved in the runoff water and the phosphorus adsorbed to the eroded particles.[a] Although cattle grazing on the natural grasslands probably increased losses of phosphorus from these watersheds, the losses from the agricultural watersheds were about 10 times as great. The no-till wheat fields tended to lose much less particulate phosphorus, but more soluble phosphorus, than the conventionally tilled wheat fields.

| | | kg P/ha/yr | | |
Location and soil	Management	Soluble P	Particulate P	Total P
El Reno, Okla. Paleustolls, 3% slope	Wheat with conventional plow and disk	0.21	3.51	3.72
	Wheat with no-till	1.04	0.43	1.42
	Native grass, heavily grazed	0.14	0.10	0.24
Woodward, Okla. Ustochrepts, 8% slope	Wheat with conventional sweep plow and disk	0.23	5.44	5.67
	Wheat with no-till	0.49	0.70	1.19
	Native grass, moderately grazed	0.02	0.07	0.09

[a] Wheat was fertilized with up to 23 kg of P each fall.
Data from Smith, et al. (1991).

TABLE 13.2   Effect of Forest Fire on Phosphorus Loss During the Following Year from a Scrub Forest in Northwestern Spain

The site was steeply sloped, with shallow sandy loam soils (lithic Haplumbrepts) relatively high in available P. Note that the more severe the burn, the greater were the losses of sediment and associated P. In all cases, the enrichment ratios were slightly greater than 2, indicating that the sediments eroded from the plots were more than twice as concentrated in phosphorus as were the surface soils themselves. In this, as in most cases, the soil fractions rich in phosphorus were the most susceptible to erosion. The levels of phosphorus lost from this particular forest are higher than levels typically lost from humid-region forests on soils low in available P.

	Sediment lost, Mg/ha	Total P lost, kg/ha	Enrichment ratio[b]
Unburned control	2	1.4	2.6
Moderately burned by prescribed fire	5	4.3	2.1
Severely burned by wildfire	13+[a]	9.1+	2.2

[a] Values for the severely burned plots were measured for only 10 months, while the other plots were measured for 12 months.
[b] Enrichment ratio = (mg P/kg sediment)/(mg P/kg soil).
Data from Saa, et al. (1994).

preferred, not only because they lower particulate losses, but because they constrain soil erosion losses generally.

Disturbances to natural vegetation, such as timber harvest or wildfires, also increase the loss of phosphorus, primarily via eroded sediment. Erosion tends to transport predominantly the clay and organic matter fractions of the soil, which are relatively rich in phosphorus, leaving behind the coarser, lower-phosphorus fractions. Thus, compared to the original soil, eroded sediment is often enriched in phosphorus by a ratio of 2 or more (Table 13.2).

Runoff from animal feedlots and holding areas where hundreds or even thousands of cattle, hogs, or poultry are confined are increasingly significant sources of phosphorus that lead to eutrophication. Manure from a herd of 10,000 beef cattle contains as much phosphorus and nitrogen as is found in human wastes of a city of 100,000. Even if animal manure is spread on nearby fields, heavy rains induce runoff and erosion losses from the treated areas. Significant quantities of soluble and particulate phosphorus are thereby carried downstream into lakes and ponds, where eutrophication occurs.

Such intensive animal production, as well as pollution from domestic and commercial development, are thought to be responsible for algal blooms of ***pfiesteria*** and other such toxic algae that are responsible for the death of millions of fish in estuaries and rivers of the coastal plain area of the eastern United States. Elevated phosphorus in streams is believed to encourage the life stage of *pfiesteria* that releases a toxin 1000 times more deadly than cyanide; it can kill small fish in a matter of minutes and larger ones in a few hours.

The preceding examples make it clear that the control of eutrophication in many rivers and lakes will require improved management of the phosphorus cycle in soils. Specifically, it is critical to reduce the transport of phosphorus from soils to water.

13.3   THE PHOSPHORUS CYCLE[5]

In order to manage phosphorus for economic plant production and for environmental protection, we need to understand the nature of the different forms of phosphorus found in soils and the manner in which these forms of phosphorus interact within the soil and in the larger environment. The cycling of phosphorus within the soil, from the soil to higher plants and back to the soil, is illustrated in Figure 13.5.

PHOSPHORUS IN SOIL SOLUTION.   Compared to other macronutrients, such as sulfur and calcium, the concentration of phosphorus in the soil solution is very low, generally ranging

---

[5] For a review of soil phosphorus, see Olsen and Khasawneh (1980) and Stevenson (1986).

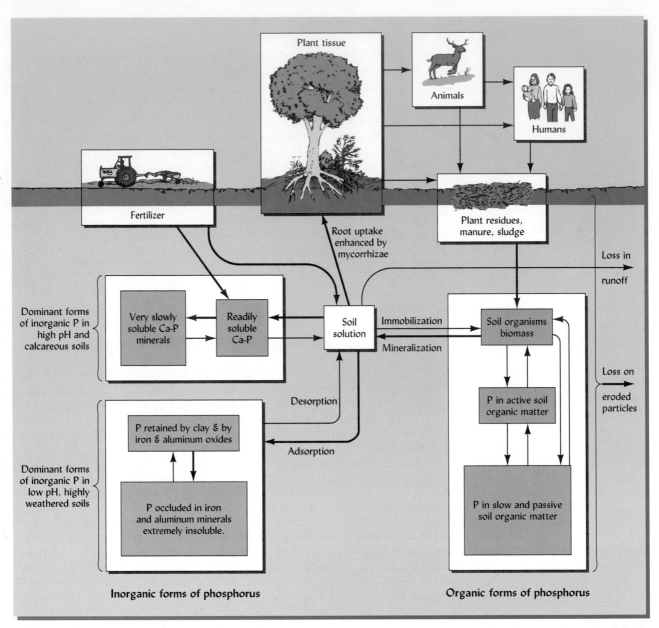

FIGURE 13.5 The phosphorus cycle in soils. The boxes represent pools of the various forms of phosphorus in the cycle, while the arrows represent translocations and transformations among these pools. The three largest white boxes indicate the principal groups of phosphorus-containing compounds found in soils. Within each of these groups, the less soluble, less available forms tend to dominate.

from 0.001 mg/L in very infertile soils to about 1 mg/L in rich, heavily fertilized soils. Plant roots absorb phosphorus dissolved in the soil solution, mainly as phosphate ions (HPO_4^{-2} and $H_2PO_4^{-}$), but some soluble organic phosphorus compounds are also taken up. The chemical species of phosphorus present in the soil solution is determined by the solution pH, as shown in Figure 13.6. In strongly acid soils (pH 4 to 5.5), the monovalent anion $H_2PO_4^{-}$ dominates, while alkaline solutions are characterized by the divalent anion HPO_4^{-2}. Both anions are important in near-neutral soils.

UPTAKE BY ROOTS AND MYCORRHIZAE. Plant uptake of phosphate ions from the soil solution is curtailed by the slow movement of these ions to root surfaces. This is overcome in part by the movement of roots to zones where the ions are held. Also, phosphate ions move to plant roots through symbiosis with mycorrhizal fungi (Figure 13.7). The microscopic, threadlike mycorrhizal hyphae extend out into the soil several centimeters from the root surfaces. The hyphae are able to absorb phosphorus ions as the ions enter the soil solution, and may even be able to access some strongly bound forms of phospho-

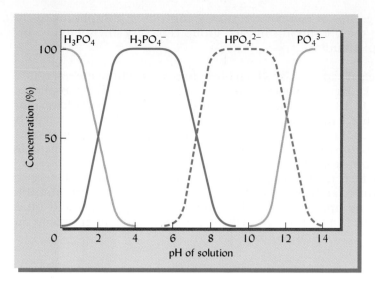

FIGURE 13.6 The effect of pH on the relative concentrations of the three species of phosphate ions. At lower pH values, more H^+ ions are available in the solution, and thus the phosphate ion species containing more hydrogen predominates. In near-neutral soils, HPO_4^{-2} and $H_2PO_4^-$ are found in nearly equal amounts. Both of these species are readily available for plant uptake.

rus. The hyphae then bring the phosphate to the root by transporting it inside the hyphal cells, where soil-retention mechanisms cannot interfere with phosphate movement (see Figure 13.7).

CHEMICAL FORMS IN SOILS. In most soils, the amount of phosphorus available to plants from the soil solution at any one time is very low, seldom exceeding about 0.01% of the total phosphorus in the soil. The bulk of the soil phosphorus exists in three general groups of compounds—namely, *organic phosphorus, calcium-bound inorganic phosphorus,* and *iron- or aluminum-bound inorganic phosphorus* (see Figure 13.5). Some of the organic phosphorus is in the *active, slow,* and *passive* fractions of organic matter (see Section 11.8). Of the inorganic phosphorus, the calcium compounds predominate in most alkaline soils, while the iron and aluminum forms are most important

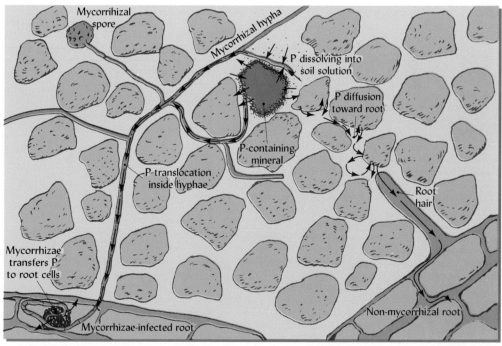

FIGURE 13.7 Roles of diffusion and mycorrhizal hyphae in the movement of phosphate ions to plant roots. In soils with low solution phosphorus concentration and high phosphorus fixation, slow diffusion may seriously limit the ability of roots to obtain sufficient phosphorus. The hyphae of symbiotic mycorrhizal fungi are particularly beneficial to the plant where phosphorus diffusion is slow, because phosphorus is transported inside the hyphae by cytoplasmic streaming, making the plant much less dependent on the diffusion of phosphate ions through the soil.

in acidic soils. All three groups of compounds slowly contribute phosphorus to the soil solution, but most of the phosphorus in each group is of very low solubility and not readily available for plant uptake.

GAINS AND LOSSES.   The principal pathways by which phosphorus is lost from the soil system are plant removal (5 to 50 kg/ha annually in harvested biomass), erosion of phosphorus-carrying soil particles (0.1 to 10 kg/ha annually on organic and mineral particles), and phosphorus dissolved in surface runoff water (0.01 to 3.0 kg/ha annually). For each pathway, the higher figures cited for annual phosphorus loss would most likely apply to cultivated soils. Because phosphorus is so insoluble in soils, the loss of this element by leaching is generally very low except in heavily manured soils.

Little phosphorus is added to soils from the atmosphere, only 0.05 to 0.5 kg/ha being sorbed on dust particles annually. Gains of soil phosphorus levels are from the return of plant residues and applications of fertilizers, manures, sludge, and other wastes. This accounts for the fact that the soils where intensive agricultural systems are practiced commonly have higher phosphorus levels.

13.4   ORGANIC PHOSPHORUS IN SOILS

Both inorganic and organic forms of phosphorus occur in soils, and both are important to plants as sources of this element. The relative amounts in the two forms vary greatly from soil to soil, but the data in Table 13.3 give some idea of their relative proportions in a range of mineral soils. The organic fraction generally constitutes 20 to 80% of the total phosphorus in surface soil horizons (Figure 13.8). The deeper horizons may hold large amounts of inorganic phosphorus, especially in soils from arid and semiarid regions.

Organic Phosphorus Compounds

Three broad groups of organic phosphorus compounds are known to exist in soils: (1) inositol phosphates or phosphate esters of a sugarlike compound, inositol $(C_6H_6(OH)_6)$; (2) nucleic acids; and (3) phospholipids. While other organic phosphorus compounds are present in soils, the identity and amounts present are less well understood.

TABLE 13.3   Total Phosphorus Content of Surface Soils from Different Locations and the Percentage of Total Phosphorus in the Organic Form

Soils	Number of samples	Total P, mg/kg	Organic fraction, %
Western Oregon			
Upland soils	4	357	66
Old valley-filling soils	4	1479	30
Recent valley soils	3	848	26
New York			
Histosols (cultivated)	8	1491	52
Iowa			
Mollisols and Alfisols	6	561	44
Arizona	19	703	36
Australia	3	422	75
Texas			
Ustolls	2	369	34
Hawaii			
Andisol	1	4700	37
Oxisol	1	1414	19
Zimbabwe			
Alfisols	22	899	56
Maryland			
Ultisols (silt loams)	6	650	59
Ultisols (forested sandy loams)	3	472	70
Ultisols (sandy loam, cropland)	4	647	25

Data for Oregon, Iowa, and Arizona from sources quoted by Brady (1974); Australia from Fares, et al. (1974); New York from Cogger and Duxbury (1984); Hawaii from Soltanpour, et al. (1988); Zimbabwe courtesy of R. Weil and F. Folle; Texas from Raven and Hossner (1993); Maryland (silt loams) from Weil, et al. (1988); Maryland (sandy loams) from Vaithiyanathan and Correll (1992).

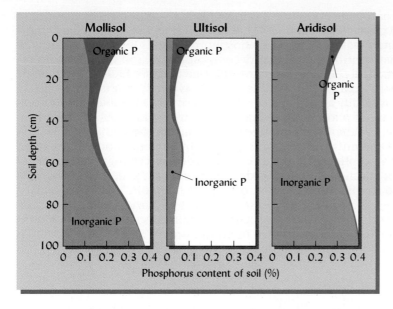

FIGURE 13.8 Phosphorus contents of representative soil profiles from three soil orders. All three soils contain a high proportion of organic phosphorus in their surface horizons. The Aridisol has a high inorganic phosphorus content throughout the profile because rainfall during soil formation was insufficient to leach much of the inorganic phosphorus compounds from the soil. The increased phosphorus in the subsoil of the Ultisol is due to adsorption of inorganic phosphorus by iron and aluminum oxides in the B horizon. In both the Mollisol and Aridisol, most of the subsoil phosphorus is in the form of inorganic calcium-phosphate compounds.

Inositol phosphates are the most abundant of the known organic phosphorus compounds, making up 10 to 50% of the total organic phosphorus. They tend to be quite stable in acid and alkaline conditions, and they interact with the higher-molecular-weight humic compounds. These properties may account for their relative abundance in soils.

Nucleic acids are adsorbed by humic compounds as well as by silicate clays. Adsorption on these soil colloids probably helps protect the phosphorus in nucleic acids from microbial attack. Still, the nucleic acids and phospholipids together probably make up only 1 to 2% of the organic phosphorus in most soils.

The other chemical compounds that contain most of the soil organic phosphorus have not yet been identified, but much of the organic phosphorus appears to be associated with the fulvic acid fraction of the soil organic matter. Our ignorance of the specific compounds involved does not detract from the importance of these compounds as suppliers of phosphorus through microbial breakdown.[6]

Much of the phosphorus in the soil solution and in leachates of areas that have received large quantities of animal wastes is present as **dissolved organic phosphorus** (DOP). DOP is generally more mobile than soluble inorganic phosphates, probably because it is not so readily adsorbed by organic clays and by $CaCO_3$ layers in the soil. In the lower horizons of such soils, the DOP commonly makes up more than 50% of the total soil solution phosphorus. As a consequence, in heavily manured areas with sandy soils DOP can leach downward to nearly 2 m. In fields with high water tables, the phosphorus can move with the groundwater to nearby lakes or streams and thereby contribute significantly to eutrophication.

Mineralization of Organic P

Phosphorus held in organic form can be mineralized and immobilized by the same general processes that release nitrogen and sulfur from soil organic matter (see Chapter 12):

$$\longleftarrow \text{Immobilization} \longleftarrow$$

$$\text{Organic P forms} \underset{\text{Microbes}}{\overset{\text{Microbes}}{\rightleftharpoons}} \underset{\substack{\text{Soluble} \\ \text{phosphate}}}{H_2PO_4^-} \underset{}{\overset{Fe^{3+}, Al^{3+}, Ca^{2+}}{\rightleftharpoons}} \underset{\text{Insoluble fixed P}}{\text{Fe, Al, Ca phosphates}}$$

$$\longrightarrow \text{Mineralization} \longrightarrow$$

---

[6] Although plants are able to absorb some organic phosphorus compounds directly, the level of such absorption is thought to be very low compared to that of inorganic phosphates.

Net immobilization of soluble phosphorus is most likely to occur if residues added to the soil have a C/P ratio greater than about 300:1, while net mineralization is likely if the ratio is below 200:1. Mineralization of organic phosphorus in soils is subject to many of the same influences that control the general decomposition of soil organic matter—namely, temperature, moisture, tillage, and so forth (see Section 11.2). In temperate regions, mineralization of organic phosphorus in soils typically releases 5 to 20 kg P/ha/yr, most of which is readily absorbed by growing plants. These values can be compared to the annual uptake of phosphorus by most crops, trees, and grasses, which generally ranges from 5 to 30 kg P/ha. When forested soils are first brought under cultivation in tropical climates, the amount of phosphorus released by mineralization may exceed 50 kg/ha/yr, but unless phosphorus is added from outside sources these high rates of mineralization will soon decline due to the depletion of readily decomposable soil organic matter. In Florida, rapid mineralization of organic matter in Histosols (Saprists) drained for agricultural use is estimated to release about 80 kg P/ha/yr. Unlike most mineral soils, these organic soils possess little capacity to retain dissolved phosphorus, so water draining from them is quite concentrated in phosphorus (0.5 to 1.5 mg P/L) and is thought to be contributing to the degradation of the Everglades wetland system.

Contribution of Organic Phosphorus to Plant Needs

Recent evidence indicates that the readily decomposable or easily soluble fractions of soil *organic phosphorus* are often the most important factor in supplying phosphorus to plants in *highly weathered soils* (e.g., Ultisols and Oxisols), even though the total organic matter content of these soils may not be especially high. The inorganic phosphorus in the highly weathered soils is far too insoluble to contribute much to plant nutrition. Apparently plant roots and mycorrhizal hyphae are able to obtain some of the phosphorus released from organic forms before it forms inorganic compounds that quickly become insoluble. In contrast, it appears that the more soluble *inorganic forms* of phosphorus play the biggest role in phosphorus fertility of *less weathered soils* (e.g., Mollisols and Vertisols), even though these generally contain relatively high amounts of soil organic matter.

13.5 INORGANIC PHOSPHORUS IN SOILS

As indicated by Figure 13.5, most inorganic phosphorus compounds in soils fall into one of two groups: (1) those containing calcium; and (2) those containing iron and aluminum, and, less frequently, manganese (see Table 13.4). Except for the mono- and dicalcium phosphates supplied by commercial fertilizers, all the inorganic soil phosphorus compounds listed in Table 13.4 are quite insoluble. The calcium phosphates, especially the **apatite** minerals, are stable and extremely insoluble at high pH levels. In contrast, iron- and aluminum-containing phosphates such as **strengite** ($FePO_4 \cdot 2H_2O$) and **variscite** ($AlPO_4 \cdot 2H_2O$) are stable only in very acid soils where they are most insoluble.

TABLE 13.4 Inorganic Phosphorus-Containing Compounds Commonly Found in Soils

In each group, the compounds are listed in order of increasing solubility.

Compound	Formula
Iron and aluminum compounds	
Strengite	$FePO_4 \cdot 2H_2O$
Variscite	$AlPO_4 \cdot 2H_2O$
Calcium compounds	
Fluorapatite	$[3Ca_3(PO_4)_2] \cdot CaF_2$
Carbonate apatite	$[3Ca_3(PO_4)_2] \cdot CaCO_3$
Hydroxy apatite	$[3Ca_3(PO_4)_2] \cdot Ca(OH)_2$
Oxy apatite	$[3Ca_3(PO_4)_2] \cdot CaO$
Tricalcium phosphate	$Ca_3(PO_4)_2$
Octacalcium phosphate	$Ca_8H_2(PO_4)_6 \cdot 5H_2O$
Dicalcium phosphate	$CaHPO_4 \cdot 2H_2O$
Monocalcium phosphate	$Ca(H_2PO_4)_2 \cdot H_2O$

At intermediate pH levels, phosphates are found associated with some 1:1-type silicate clays. Apparently, phosphate ions have reacted with or replaced hydroxyl groups at the edges of the clay particles and are bound in relatively unavailable forms. Once again, soil pH is a major factor in determining the availability of the phosphorus, as we shall see in Section 13.6.

13.6 SOLUBILITY OF INORGANIC PHOSPHORUS IN ACID SOILS

The particular types of reactions that fix phosphorus in relatively unavailable forms are closely related to soil pH (Figure 13.10). In acid soils these reactions involve mostly Al, Fe, or Mn, either as dissolved ions, as oxides, or as hydrous oxides. Many soils contain such hydrous oxides as coatings on soil particles and as interlayer precipitates in silicate clays. At moderate pH values, adsorption on the edges of kaolinite or on the iron oxide coating on kaolinite clays plays an important role.

Precipitation by Iron, Aluminum, and Manganese Ions

Probably the easiest type of phosphorus-fixation reaction to visualize is the simple reaction of phosphate ions with dissolved Fe^{3+}, Al^{3+}, and Mn^{3+} ions to form insoluble hydroxy phosphate precipitates such as those formed in wastewater treatment plants (see Box 13.2). In a moderately acid soil, hydroxy phosphate precipitates (Figure 13.11a) form by similar reactions such as the following (using the Al^{3+} ion as an example):

$$Al^{3+} + H_2PO_4^- + 2H_2O \rightleftharpoons 2H^+ + Al(OH)_2H_2PO_4$$
$$\text{(soluble)} \qquad\qquad\qquad \text{(insoluble)}$$

Freshly precipitated hydroxy phosphates are slightly soluble and, initially at least, somewhat available to plants. Over time, however, as the precipitated hydroxy phosphates age, they become less soluble and the phosphorus in them becomes almost completely unavailable to most plants.

Reaction with Hydrous Oxides and Silicate Clays

Most of the phosphorus fixation in acid soils probably occurs when $H_2PO_4^-$ ions react with, or become adsorbed to, the surfaces of insoluble oxides of iron, aluminum, and manganese, such as gibbsite ($Al_2O_3 \cdot 3H_2O$) and goethite ($Fe_2O_3 \cdot 3H_2O$), and with 1:1-type silicate clays. The large quantities of Fe, Al oxides, and 1:1 clays present in many soils make possible the fixation of extremely large amounts of phosphorus by these reactions.

Although all the exact mechanisms have not been identified, $H_2PO_4^-$ ions are known to react with iron and aluminum mineral surfaces in several different ways, resulting in different degrees of phosphorus fixation. Some of these reactions are shown diagrammatically in Figure 13.11.

The $H_2PO_4^-$ anion may be attracted to positive charges that develop under acid conditions on the surfaces of iron and aluminum oxides and the broken edges of kaolinite clays (see Figure 13.11b). The adsorbed $H_2PO_4^-$ anions are subject to anion exchange with other anions such as OH^-, SO_4^{2-}, and MoO_4^{2-}. Since this type of adsorption of $H_2PO_4^-$ ions is reversible, the phosphorus may slowly become available to plants.

The phosphate ion may also replace a structural hydroxyl to become chemically bound to the oxide (or clay) surface (see Figure 13.11c). This reaction, while reversible, binds the phosphate too tightly to allow its ready replacement by other anions. The availability of phosphate bound in this manner is very low. Over time, a second oxygen of the phosphate ion may replace a second hydroxyl, so that the phosphate becomes chemically bound to two adjacent aluminum (or iron) atoms in the hydrous oxide surface (see Figure 13.11d). With this step, the phosphate becomes an integral part of the oxide mineral and the likelihood of its release back to the soil solution is extremely small.

Finally, as more time passes, the precipitation of additional iron or aluminum hydrous oxide may bury the phosphate deep inside the oxide particle. Such phosphate is termed *occluded* and is the least available form of phosphorus in most acid soils.

BOX 13.2 PHOSPHORUS REMOVAL FROM WASTEWATER

Environmental soil scientists and engineers remove phosphorus from municipal wastes by taking advantage of some of the same reactions that bind phosphorus in soils. After primary and secondary sewage treatment that removes solids and oxidizes most of the organic matter, tertiary treatment in huge, specially designed tanks (Figure 13.9) causes phosphorus to precipitate through reactions with iron and aluminum compounds, such as the following:

$$Al_2(SO_4)_3 \cdot 14H_2O + 2PO_4^{3-} \rightarrow 2AlPO_4 + 3SO_4^{2-} + 14H_2O$$

Alum   Soluble   Insoluble AlP
                 phosphate

$$FeCl_3 + PO_4^{3-} \rightarrow FePO_4 + 3Cl^-$$

Ferric   Soluble   Insoluble FeP
chloride   phosphate

The insoluble aluminum and iron phosphates settle out of solution and are later mixed with other solids from the wastewater to form sewage sludge. The low-phosphorus water, after minor processing, is returned to the river.

FIGURE 13.9 *Increasingly, modern sewage treatment plants are required to include tertiary treatment facilities for phosphorus removal. The chemical reactions encouraged in the sewage treatment process are similar to those that affect phosphorus availability in soils. (Photo courtesy of R. Weil)*

Other less-expensive tertiary treatment approaches involve the spraying of the wastewater on vegetated soils. Natural soil and plant processes clean the phosphorus and other constituents out of the waste water. In some *infiltration systems*, the water percolates through relatively permeable soils. Other systems, termed *overland flow systems*, use finer-textured, less-permeable soils, over which water slowly flows, permitting the upper few centimeters of soil and the vegetation to remove most of the soluble phosphorus and other contaminants. In both systems, advantage is taken of the soil's phosphate-fixing capacity (see Section 13.8).

Soils with significant quantities of iron and aluminum oxides and of 1:1-type clays are best suited for such systems of water purification. Reactions such as those reviewed in this text are responsible for the removal of the excess phosphorus from the wastewater.

Effect of Iron Reduction Under Wet Conditions

It should be noted that phosphorus bound to iron oxides by the mechanisms just discussed is very insoluble under well-aerated conditions. However, prolonged anaerobic conditions can reduce the iron in these complexes from Fe^{3+} to Fe^{2+}, making the iron–phosphate complex much more soluble and causing it to release phosphorus into solution. The release of phosphorus from iron phosphates by means of the reduction and subsequent solubilization of iron improves the phosphorus availability in soils used for paddy rice.

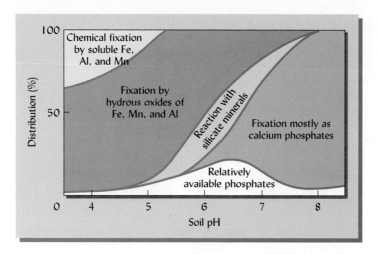

FIGURE 13.10 Inorganic fixation of added phosphates at various soil pH values. Average conditions are postulated, and it is not to be inferred that any particular soil would have exactly this distribution. The actual proportion of the phosphorus remaining in an available form will depend upon contact with the soil, time for reaction, and other factors. It should be kept in mind that some of the added phosphorus may be changed to an organic form in which it would be temporarily unavailable.

These reactions are also of special relevance to concerns about water quality. Phosphorus bound to soil particles may accumulate in river- and lake-bottom sediments, along with organic matter and other debris. As the sediments become anoxic, the reducing environment may cause the gradual release of phosphorus held by hydrous iron oxides. Thus, the phosphorus eroded from soils today may aggravate the problem of eutrophication for years to come, even after the erosion and loss of phosphorus from the land has been brought under control.

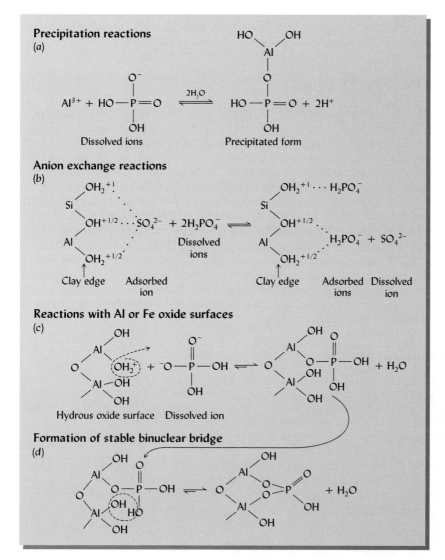

FIGURE 13.11 Several of the reactions by which phosphate ions are removed from soil solution and fixed by reaction with iron and aluminum in various hydrous oxides. Freshly precipitated aluminum, iron, and manganese phosphates (a) are relatively available, though over time they become increasingly unavailable. In (b) the phosphate is reversibly adsorbed by anion exchange. In reactions of the type shown in (c) a phosphate ion replaces an —OH_2 or an —OH group in the surface structure of Al or Fe hydrous oxide minerals. In (d) the phosphate further penetrates the mineral surface by forming a stable binuclear bridge. The adsorption reactions (b, c, d) are shown in order—from those that bind phosphate with the least tenacity (relatively reversible and somewhat more plant-available) to those that bind phosphate most tightly (almost irreversible and least plant-available). It is probable that, over time, phosphate ions added to a soil may undergo an entire sequence of these reactions, becoming increasingly unavailable.

13.7 INORGANIC PHOSPHORUS AVAILABILITY AT HIGH pH VALUES[7]

The availability of phosphorus in alkaline soils is determined principally by the solubility of the various calcium phosphate compounds present. In alkaline soils (e.g., pH = 8) soluble $H_2PO_4^-$ quickly reacts with calcium to form a sequence of products of decreasing solubility. For instance, highly soluble monocalcium phosphate $[Ca(H_2PO_4)_2 \cdot H_2O]$ added as concentrated superphosphate fertilizer rapidly reacts with calcium carbonate in the soil to form first dicalcium phosphate $[CaHPO_4 \cdot 2H_2O)$ and then tricalcium phosphate $[Ca_3(PO_4)_2]$, as follows:

$$Ca(H_2PO_4)_2 \cdot H_2O + 2H_2O \xrightarrow{CaCO_3} 2(CaHPO_4 \cdot 2H_2O) + CO_2\uparrow \xrightarrow{CaCO_3} Ca_3(PO_4)_2 + CO_2\uparrow + 5H_2O$$

| Monocalcium phosphate (soluble) | Dicalcium phosphate (slightly soluble) | Tricalcium phosphate (very low solubility) |

The solubility of these compounds and, in turn, the plant availability of the phosphorus they contain decrease as the phosphorus changes from the $H_2PO_4^-$ ion to tricalcium phosphate $[Ca_3(PO_4)_2]$. Although this compound is quite insoluble, it may undergo further reactions to form even more insoluble compounds, such as the hydroxy-, oxy-, carbonate-, and fluorapatite compounds (apatites) shown in Table 13.4. These compounds are thousands of times less soluble than freshly formed tricalcium phosphates.

Reversion of soluble fertilizer phosphorus to extremely insoluble calcium phosphate forms is most serious in the calcareous soils of low-rainfall regions (e.g., western United States). Iron and aluminum impurities in calcite particles may also adsorb considerable amounts of phosphate in these soils. Because of the various reactions with $CaCO_3$, phosphorus availability tends to be nearly as low in the Aridisols, Inceptisols, and Mollisols of arid regions as in the highly acid Spodosols and Ultisols of humid regions, where iron, aluminum, and manganese limit phosphorus availability.

13.8 PHOSPHORUS-FIXATION CAPACITY OF SOILS

Soils may be characterized by their capacity to fix phosphorus in unavailable, insoluble forms. The phosphorus-fixation capacity of a soil may be conceptualized as the total number of sites on soil particle surfaces capable of reacting with phosphate ions. Phosphorus fixation may also be due to reactive soluble iron, aluminum, or manganese. The different types of fixation mechanisms are illustrated schematically in Figure 13.12.

One way of determining the phosphorus-fixing capacity of a particular soil is to shake a known quantity of the soil in a phosphorus solution of known concentration. After about 24 hours an equilibrium will be approached, and the concentration of phosphorus remaining in the solution [the **equilibrium phosphorus concentration (EPC)**] can be determined. The difference between the initial and final (*equilibrium*) solution phosphorus concentrations represents the amount of phosphorus fixed by the soil. If this procedure is repeated using a series of solutions with different initial phosphorus concentrations, the results can be plotted as a phosphorus-fixation curve (Figure 13.13) and the maximum phosphorus-fixation capacity can be extrapolated from the value at which the curve levels off.

Factors Affecting the Extent of Phosphorus Fixation in Soils

Soils that can fix more than 350 mg P/kg of soil (about 700 kg P/ha) are generally considered to be high phosphorus-fixing soils (see Table 13.5). The phosphorus-fixation capacity of a soil is positively related to certain soil properties, such as soil pH, organic matter content, calcium carbonate content, clay content, and contents of iron, aluminum, and manganese, especially the hydrous oxides of these metals.

AMOUNT OF CLAY PRESENT. Most of the compounds with which phosphorus reacts are in the finer soil fractions. Therefore, if soils with similar pH values and mineralogy are compared, phosphorus fixation tends to be more pronounced and ease of phosphorus release tends to be lowest in those soils with higher clay contents (Table 13.5).

---

[7] See Sample, et al. (1980) for a discussion of this subject.

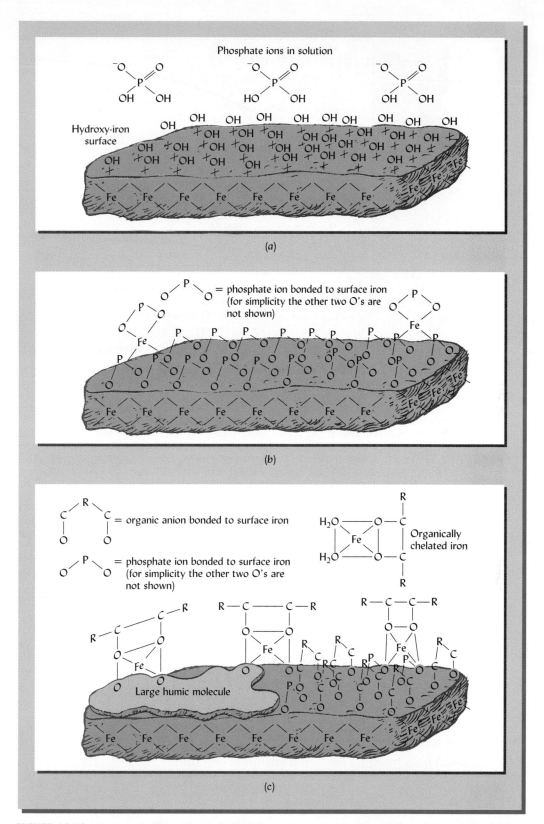

FIGURE 13.12 Schematic illustration of phosphorus-fixation sites on a soil particle surface showing hydrous iron oxide as the primary fixing agent. In part (*a*) the sites are shown as +s, indicating positive charges or hydrous metal oxide sites, each capable of fixing a phosphate ion. In part (*b*) the fixation sites are all occupied by phosphate ions (the soil's fixation capacity is satisfied). Part (*c*) illustrates how organic anions, larger organic molecules, and certain strongly fixed inorganic anions can reduce the sites available for fixing phosphorus. Such mechanisms account for the reduced phosphorus fixation and greater phosphorus availability brought about when organic matter is added to a soil.

FIGURE 13.13 The relationship between phosphorus fixation and phosphorus in solution when two different soils (A and B) are shaken with solutions of various initial phosphorus concentrations. Initially, each soil removes nearly all of the phosphorus from solution, and as more and more concentrated solutions are used, the soil fixes greater amounts of phosphorus. However, eventually solutions are used that contain so much phosphorus that most of the phosphorus-fixation sites are satisfied, and much of the dissolved phosphorus remains in solution. The amount fixed by the soil levels off as the *maximum phosphorus-fixing capacity* of the soil is reached (see horizontal dashed lines: for soil A, 200 mg P/kg; for soil B, 1200 mg P/kg soil). If the initial phosphorus concentration of a solution is equal to the *equilibrium phosphorus concentration (EPC)* for a particular soil, that soil will neither remove phosphorus from nor release phosphorus to the solution (i.e., phosphorus fixation = 0 and EPC = EPC_0). If the solution phosphorus concentration is less than the EPC_0, the soil will release some phosphorus (i.e., the fixation will be negative). In this example, soil B has a much *higher* phosphorus-fixing capacity and a much *lower* EPC than does soil A. It can also be said that soil B is highly *buffered* because much phosphorus must be added to this soil to achieve a small increase in the equilibrium solution phosphorus concentration. On the other hand, if a plant root were to remove a relatively large amount of total phosphorus from the soil, only a small change would occur in the equilibrium solution concentration. [For a recent review of buffering capacity as it relates to the available phosphorus and potassium, as well as other nutrients, see Nair (1996).]

TYPE OF CLAY MINERALS PRESENT. Some clay minerals are much more effective at phosphorus fixation than others. Generally, those clays that possess greater anion exchange capacity (due to positive surface charges) have a greater affinity for phosphate ions. For example, extremely high phosphorus fixation is characteristic of allophane clays typically found in Andisols and other soils associated with volcanic ash. Oxides of

TABLE 13.5 Maximum Phosphorus-Fixation Capacity of Several Soils of Varied Content and Kinds of Clays

Fe, Al oxides (especially amorphous types) fix the largest quantities and silicate clays (especially 2:1-type) fix the least.

Soil Great Group (and series, if known)	Location	Clay Percent	Clay Type	Maximum P fixation, mg P/kg soil
Evesboro (Quartzipsamment)	Maryland	6	Kaolinite, Fe, Al oxides	125
Kandiustalf	Zimbabwe	20	Kaolinite, Fe oxides	394
Kitsap (Xerept)	Washington	12	2:1 clays, allophane	453
Matapeake (Hapludult)	Maryland	15	Chlorite, kaolinite, Fe oxides	465
Rhodustalf	Zimbabwe	53	Kaolinite, Fe oxides	737
Newberg (Haploxeroll)	Washington	38	2:1 clays, Fe oxides	905
Tropohumults	Cameroon	46	Fe, Al oxides, kaolinite	2060

Cameroon data courtesy of V. Ngachie; Washington data from Kuo (1988); Maryland and Zimbabwe data courtesy of R. Weil and F. Folle.

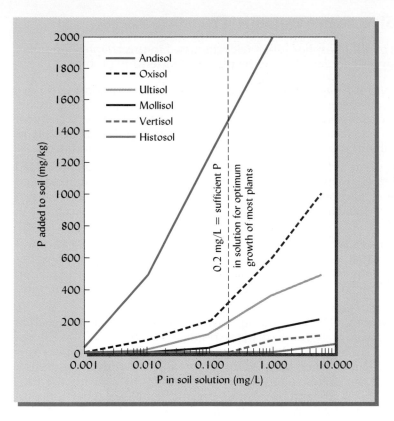

FIGURE 13.14 Phosphorus-fixing tendencies typical of several soil orders. The graph shows the concentrations of phosphorus that would be achieved in the soil solution for each type of soil if varying amounts of soluble phosphorus were added as fertilizer. Although some plants, especially if infected with mycorrhizal fungi, can grow well with lower solution phosphorus concentrations, a concentration of 0.2 mg P/L in the soil solution is adequate for optimum growth of most crops (see Table 13.6). Therefore, the graph indicates that for optimum plant growth, nearly 1500 mg of phosphorus per kg of soil should be added to the Andisol, while somewhat less than 200 mg of phosphorus per kg of soil would be required for the same phosphorus availability in the Ultisol. Within a particular soil order, the phosphorus-fixing tendency of individual soils may vary considerably. The curves given are representative for soils in the taxonomic orders indicated, but it should be borne in mind that high rates of phosphorus application would eventually reduce the fixation capacity (reduce the slope of the curve) as phosphorus-fixation sites were used up. (Curves are representative of data from many sources.)

iron and aluminum, such as gibbsite and goethite, also strongly attract and hold phosphorus ions. Among the layer silicate clays, kaolinite has a greater phosphorus-fixation capacity than most. The 2:1 clays of less-weathered soils have relatively little capacity to bind phosphorus. Differences among clays in their phosphorus-fixing capacities help explain the relatively low phosphorus-fixing capacities of Vertisols and Mollisols (high in 2:1-type clays), and the much higher capacities of Ultisols and Oxisols (high in Fe, Al oxides, and 1:1-types clays) and Andisols (high in allophane; see Figure 13.14).

Effect of Soil pH.   The greatest degree of phosphorus fixation occurs at very low and very high soil pH. As pH increases from below 5.0 to about 6.0, the iron and aluminum phosphates become somewhat more soluble. Also, as pH drops from greater than 8.0 to below 6.0, calcium phosphate compounds increase in solubility. Therefore, as a general rule in mineral soils, phosphate fixation is at its lowest (and plant availability is highest) when soil pH is maintained in the 6.0 to 7.0 range (see Figure 13.10). Even in this range, however, phosphate availability may still be very low, and added soluble phosphates will be readily fixed by soils.

Effect of Organic Matter.   Organic matter generally has little capacity to strongly fix phosphate ions. Soils high in organic matter, especially active fractions of organic matter (see Section 11.8), generally exhibit relatively low levels of phosphorus fixation. Several mechanisms are responsible for the reduced phosphorus fixation associated with high soil organic matter (see Figure 13.12). First, large humic molecules can adhere to the surfaces of clays and metal hydrous oxide particles, masking the phosphorus-fixation sites and preventing them from interacting with phosphorus ions in solution. Second, organic acids produced by plant roots and microbial decay can serve as organic anions, which are attracted to positive charges and hydroxyls on the surfaces of clays and hydrous oxides. These organic anions may compete with phosphorus ions for fixation sites. Third, certain organic acids and similar compounds can entrap reactive Al and Fe in stable organic complexes called *chelates* (see Section 13.20). Once chelated, these metals are unavailable for reaction with phosphorus ions in solution.

13.9 PRACTICAL CONTROL OF PHOSPHORUS AVAILABILITY

Based on the principles of soil phosphorus behavior discussed in this chapter, a number of approaches can be suggested to ameliorate the phosphorus fertility problem.

1. *Saturation of phosphorus-fixing capacity.* By applying ample phosphorus fertilizers, the soil's phosphorus-fixing capacity can be saturated, thereby permitting higher levels of soil solution phosphorus. This is often quite costly, however, especially for low-income countries. It also has the potential for water pollution from large amounts of phosphorus desorbed to runoff water from phosphorus-saturated soils.

2. *Placement of phosphorus fertilizers.* By localized (band) placement of the added phosphorus fertilizers, reactions with the bulk of the soil can be minimized, and the maintenance of phosphorus availability enhanced. Banded applications of starter fertilizers are standard practice, as are the use of pellets and placement in holes around the bases of trees. In untilled systems, a surface band is often created.

3. *Combination of ammonium and phosphorus fertilizers.* The plant uptake of phosphorus can be increased, especially in alkaline soils, if ammonium and phosphorus fertilizers are combined. Apparently, the nitric acid produced in the oxidation of the ammonium ions slows the formation of the more insoluble calcium phosphate compounds.

4. *Cycling of organic matter.* The judicious application of organic materials such as animal manures, green manures, and plant residues can increase phosphorus availability. During the microbial breakdown of these materials, phosphorus is released slowly and can be taken up by plants before it reacts with the soil. Also, organic compounds can protect phosphorus from fixation by masking the fixation sites on the soil colloids, and by forming organic complexes (chelates) with Al, Fe, and Mn ions, thereby limiting the reaction of these ions with phosphorus.

5. *Control of soil pH.* By maintaining pH between 6 and 7, the availability of phosphorus can be optimized in most systems. Proper liming of acid soils and acidifying of alkaline soils (using a material such as elemental sulfur) can help achieve this range.

6. *Enhancement of mycorrhizal symbiosis.* Practices to enhance mycorrhizal symbiosis should be encouraged. In some cases it may be necessary to inoculate tree seedlings with efficient strains of mycorrhizal fungi. For most plants, however, the mycorrhizal symbiosis can best be enhanced by fostering appropriate soil conditions through crop rotations, organic matter additions, and minimum tillage.

7. *Choice of phosphorus-efficient plants.* By choosing plant species known to be efficient in their foraging for phosphorus, the potential for good growth can be enhanced. Table 13.6 suggests that differences in this efficiency exist. While mechanisms for more efficient phosphorus uptake are not always known, advantage should be taken of the differences in efficiency known to exist.

In summary, the small pool of available phosphorus in most soils is depleted and replenished by a number of processes (Figure 13.15). Maintaining sufficient available

TABLE 13.6 Concentration of Phosphorus in Soil Solution Required for Near-Optimal Growth (95% of Maximum Yield) of Various Plants

Plant	Approximate P in soil solution, mg/L
Cassava	0.005
Peanut	0.01
Corn	0.05
Sorghum	0.06
Cabbage	0.04
Soybean	0.20
Tomato	0.20
Head lettuce	0.30

Data from Fox (1981).

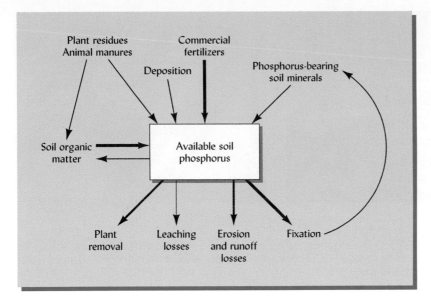

FIGURE 13.15 How the pool of available phosphorus in a soil is depleted and replenished in a typical agroecosystem. The width of the arrows is a relative indication of the quantity of phosphorus involved in each pathway. The two largest fluxes in many arable soils are the addition of phosphorus from soil amendments (usually commercial fertilizer) and the subsequent fixation of much of the added phosphorus in unavailable forms. The return of plant and animal wastes and the subsequent mineralization of soil organic matter can also play major roles in replenishing the available phosphorus. Note that leaching losses are usually negligible.

phosphorus in a soil is basically a twofold program: (1) addition of phosphorus-containing fertilizers or amendments, and (2) regulation to some degree of soil processes that fix both added and native phosphates.

13.10 POTASSIUM: NATURE AND ECOLOGICAL ROLES[8]

Of all the essential elements, potassium is the third most likely, after nitrogen and phosphorus, to limit plant productivity. For this reason it is commonly applied to soils as fertilizer and is a component of most mixed fertilizers.

The potassium story differs in many ways from that of phosphorus. Unlike phosphorus (or sulphur and, to a large extent, nitrogen), potassium is present in the soil solution only as a positively charged cation, K^+. Like phosphorus, potassium does not form any gases that could be lost to the atmosphere. Its behavior in the soil is influenced primarily by soil cation exchange properties (see Chapter 8) and mineral weathering (Chapter 2), rather than by microbiological processes. Unlike nitrogen and phosphorus, potassium causes no off-site environmental problems when it leaves the soil system. It is not toxic and does not cause eutrophication in aquatic systems.

13.11 POTASSIUM IN PLANT AND ANIMAL NUTRITION

Although potassium plays numerous roles in plant and animal nutrition, it is not actually incorporated into the structures of organic compounds. Instead, potassium remains in the ionic form (K^+) in solution in the cell, or acts as an activator for cellular enzymes. Potassium is known to activate over 80 different enzymes responsible for such plant and animal processes as energy metabolism, starch synthesis, nitrate reduction, photosynthesis, and sugar degradation. Certain plants, many of which evolved in sodium-rich semiarid environments, can substitute sodium or other monovalent ions to carry out some, but not all, of the functions of potassium.

Potassium is especially important in helping plants adapt to environmental stresses. Good potassium nutrition is linked to improved drought tolerance, improved winter-hardiness, better resistance to certain fungal diseases, and greater tolerance to insect pests. Potassium also enhances the quality of flowers, fruits, and vegetables by improving flavor and color and strengthening stems (thereby reducing lodging). In many of these respects, potassium seems to counteract some of the detrimental effects of excess nitrogen.

[8] For further information on this topic, see Mengel and Kirkby (1987) and Munson (1985).

(a)   (b)   (c)

FIGURE 13.16 Potassium deficiency often produces easily recognized foliar symptoms, mainly on older leaves: (*a*) chlorotic margins on boxwood leaves; (*b*) chlorotic leaf margins on soybean; and (*c*) small, white necrotic spots on hairy vetch leaflets. [Photos (*a*) and (*c*) courtesy of R. Weil; photo (*b*) courtesy of Potash & Phosphate Institute]

In animals, including humans, potassium plays critical roles in regulating the nervous system and in the maintenance of healthy blood vessels. Diets that include such high-potassium foods as bananas, potatoes, orange juice, and leafy green vegetables have been shown to lower human risk of stroke and heart disease. Maintaining a balance between potassium and sodium is especially important in human diets.

Deficiency Symptoms in Plants

Compared to deficiencies of phosphorus and many other nutrients, a deficiency of potassium is relatively easy to recognize in most plants.

When potassium is deficient the tips and edges of the oldest leaves begin to yellow (chlorosis) and then die (necrosis), so that the leaves appear to have been burned on the edges (Figure 13.16). On some plants the necrotic leaf edges may tear, giving the leaf a ragged appearance. In several important forage and cover-crop legume species, potassium deficiency produces small, white necrotic spots that form a unique pattern along the leaflet margins; this easily recognized symptom is one that people often mistake for insect damage.

13.12 THE POTASSIUM CYCLE

Figure 13.17 shows the major forms in which potassium is held in soils and the changes it undergoes as it is cycled through the soil–plant system. The original sources of potassium are the primary minerals, such as micas (biotite and muscovite) and potassium feldspar (orthoclase and microcline). As these minerals weather, their rigid lattice structures become more pliable. For example, potassium held between the 2:1-type crystal layers of mica is in time made more available, first as nonexchangeable but slowly available forms near the weathered edges of minerals and, eventually, as the readily exchangeable forms and the soil solution forms from which it is absorbed by plant roots.

Potassium is taken up by plants in large quantities. A portion of this potassium is leached from plant foliage by rainwater (through-fall) and returned to the soil, and a portion is returned to the soil with the plant residues. In natural ecosystems, most of the potassium taken up by plants is returned in these ways or as wastes (mainly urine) from animals feeding on the vegetation. Some potassium is lost with eroded soil particles and in runoff water, and some is lost to groundwater by leaching. In agroecosystems, from one-fifth (e.g., in cereal grains) to nearly all (e.g., in hay crops) of the potassium taken up by plants may be exported to distant markets, from which it is unlikely to return.

At any one time, most soil potassium is in primary minerals and nonexchangeable forms. In relatively fertile soils, the release of potassium from these forms to the exchangeable and soil solution forms that plants can use directly may be sufficiently

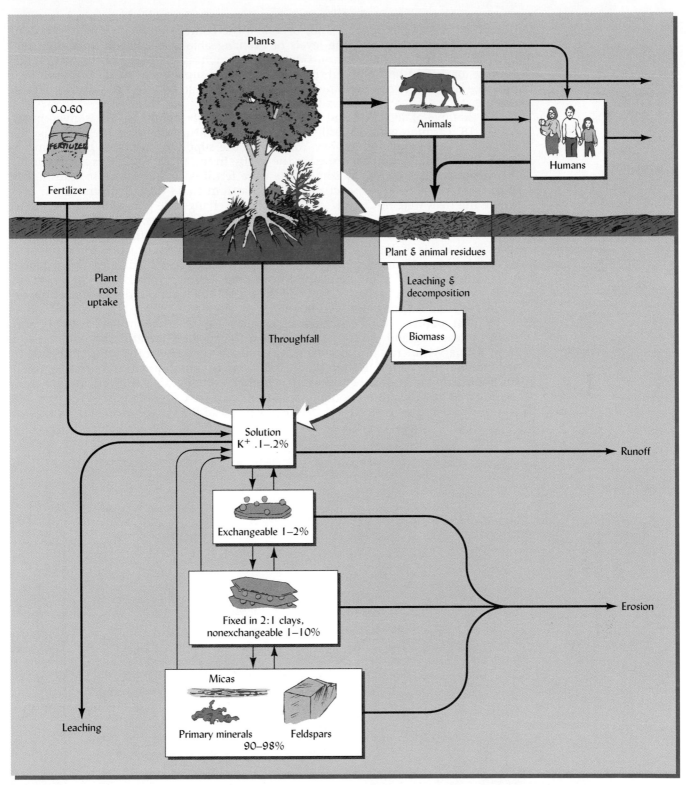

FIGURE 13.17 Major components of the potassium cycle in soils. The inner circle emphasizes the biological cycling of potassium from the soil solution to plants and back to the soil via plant residues or animal wastes. Primary and secondary minerals are the original sources of the element. Exchangeable potassium may include those ions held and released by both clay and humus colloids, but potassium is not a structural component of soil humus. The interactions among solution potassium and exchangeable, nonexchangeable, and structural potassium in primary minerals are shown. The bulk of soil potassium occurs in the primary and secondary minerals, and is released very slowly by weathering processes.

rapid to meet plant needs. On the other hand, where high yields of agricultural crops or timber are removed from the land, or where the content of weatherable potassium-containing minerals is low, the levels of exchangeable and solution potassium may have to be supplemented by outside sources, such as chemical fertilizers, poultry manure, or wood ashes. Without these additions, the supply of available potassium will likely be depleted over a period of years and the productivity of the soil will likewise decline.

An example of depletion and restoration of available soil potassium is given in Figure 13.18. Farming without fertilizers for over a century depleted the exchangeable potassium in a sandy soil in New York. After the supply of available potassium was exhausted, the land was abandoned in the 1920s. In the 1920s and 1930s a red pine forest was planted. The trees in some plots were fertilized with potassium, causing the expected rapid recovery of exchangeable potassium to its preagricultural level. Even where the trees were not fertilized, the level of exchangeable potassium in the surface soil was replenished, but slowly, over a period of about 80 years.

The replenishment of exchangeable potassium in the unfertilized forest plots provides an example of the ability of unharvested perennial plants, such as forest trees and pasture legumes, to ameliorate soil fertility over time. Deep-rooted perennials often act as "nutrient pumps," taking potassium from deep subsoil horizons into their root systems, translocating it to their leaves, and then recycling it back to the surface of the soil via leaf fall and leaching (see Section 14.3).

In most mature natural ecosystems the small (1 to 5 kg/ha) annual losses of potassium by leaching and erosion are more than balanced by weathering of potassium from primary minerals and nonexchangeable forms in the soil profile, followed by vegetative translocation to the surface of the soil. In most agricultural systems, leaching losses are far greater because much higher exchangeable K levels are generally maintained for crop production, but crop roots are active for only part of the year. The sections that follow give greater details on the reactions involved in the potassium cycle.

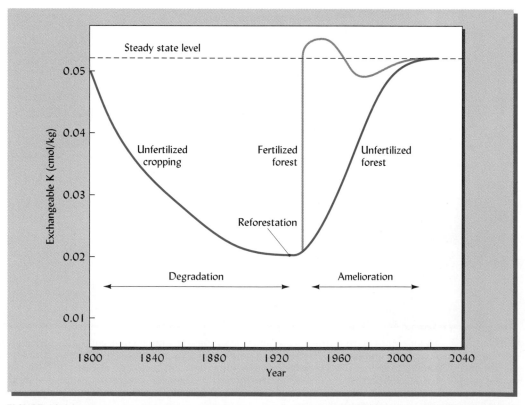

FIGURE 13.18 The general pattern of depletion of A-horizon exchangeable potassium by decades of exploitative farming, followed by its restoration under forest vegetation. The forest consisted of red pine trees planted on a Plainfield loamy sand (Udipsamment) in New York. This soil has a very low cation exchange capacity and low levels of exchangeable K^+. [From Nowak, et al. (1991)]

13.13 THE POTASSIUM PROBLEM IN SOIL FERTILITY

Availability of Potassium

The total quantity of potassiium in soils is generally greater than that of any other macronutrient element. Amounts as great as 30,000 to 50,000 kg potassium in the upper 15 cm of 1 ha of soil are not at all uncommon (see Table 1.2).

Yet the quantity of potassium held in an easily exchangeable condition at any one time often is quite small. Most of this element is held rigidly as part of the primary minerals or is fixed in forms that are, at best, only moderately available to plants. Therefore, the situation with respect to potassium utilization parallels that of phosphorus and nitrogen in at least one way: A very large proportion of all three of these elements in the soil is insoluble and relatively unavailable to growing plants.

Leaching Losses

Potassium is much more readily lost by leaching than is phosphorus. Drainage waters from soils receiving liberal fertilizer applications usually contain considerable quantities of potassium. From representative humid-region soils growing annual crops and receiving only moderate rates of fertilizer, the annual loss of potassium by leaching is usually about 25 to 50 kg/ha, the greater values being typical of acid, sandy soils.

Losses would undoubtedly be much larger were the leaching of potassium not slowed by the attraction of the positively charged potassium ions to the negatively charged cation exchange sites on clay and humus surfaces. Liming an acid soil to raise its pH can reduce the leaching losses of potassium because of the *complementary ion effect* (see Chapter 8 and Figure 13.19). The Ca^{2+} and Mg^{2+} ions on the exchange complex of limed soils are more easily replaced by potassium ions than are the tightly adsorbed trivalent Al^{3+} ions that dominate the complex of acid soils. The potassium so adsorbed in the limed soil resists being lost by leaching.

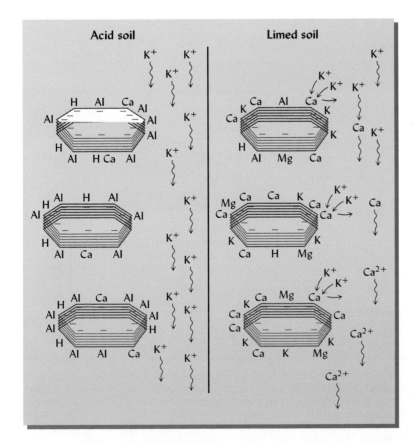

FIGURE 13.19 Diagrammatic illustration of how liming an acid soil can reduce leaching losses of potassium. The fact that the K^+ ions can more easily replace Ca^{2+} ions than they could replace Al^{3+} ions allows more of the K^+ ions to be removed from solution by cation exchange in the limed (high-calcium) soil. The removal of K^+ ions from solution by adsorption on the colloids will reduce their loss by leaching, but they will still be at least moderately available for plant uptake.

Plant Uptake and Removal

Plants take up very large amounts of potassium, often five to ten times as much as for phosphorus and about the same amount as for nitrogen. If most or all of the above-ground plant parts are removed in harvest, the drain on the soil supply of potassium can be very large. For example, a 60-Mg/ha yield of corn silage may remove 160 kg/ha of potassium. Conventional bolewood timber harvest typically removes about 100 kg/ha of potassium. If the entire tree is chipped and removed, as for paper pulp, the removal of potassium may be twice as great. A high-yielding legume hay crop may remove 400 kg/ha of potassium each year.

Luxury Consumption

Moreover, this situation is made even more critical by the tendency of plants to take up soluble potassium far in excess of their needs if sufficiently large quantities are present. This tendency is termed *luxury consumption,* because the excess potassium absorbed does not increase crop yields to any extent.

The principles involved in luxury consumption are shown by the graph in Figure 13.20. For many plants there is a direct relationship between the available potassium (soil plus fertilizer) and the removal of this element by the plants. However, only a certain amount of this element is needed for optimum growth, and this is termed *required potassium.* Potassium taken up by the plant above this critical required level is considered a *luxury.* If plant residues are not returned to the soil, the removal of this excess potassium is decidedly wasteful. In addition, high levels of potassium may depress calcium uptake and cause nutritional imbalances both in the plants and in animals that consume them.

In summary, then, the problem of potassium is at least threefold: (1) a very large proportion is relatively unavailable to higher plants; (2) it is subject to leaching losses; and (3) the removal of potassium by plants is high, especially when luxury quantities of this element are supplied. With these ideas as a background, the various forms and availabilities of potassium in soils will now be considered.

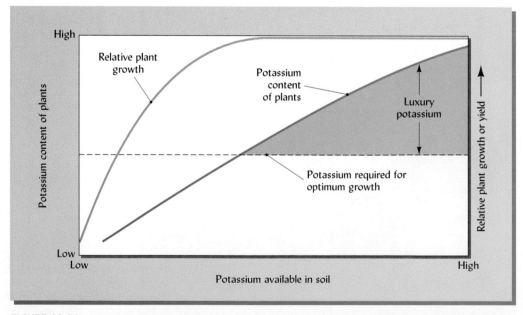

FIGURE 13.20 The general relationship between available potassium level in the soils, plant growth, and plant uptake of potassium. If available soil potassium is raised above the level needed for maximum plant growth, many plants will continue to increase their uptake of potassium without any corresponding increase in growth. The potassium taken up in excess of that needed for optimum growth is termed *luxury consumption.* Such luxury consumption may be wasteful, especially if the plants are completely removed from the soil. It may also cause dietary imbalance in grazing animals.

13.14   FORMS AND AVAILABILITY OF POTASSIUM IN SOILS

Four forms of soil potassium are shown in the potassium-cycle diagram (Figure 13.17, from the bottom upward): (1) K in primary mineral crystal structures, (2) K in nonexchangeable positions in secondary minerals, (3) K in exchangeable form on soil colloid surfaces, and (4) potassium ions soluble in water. The total amount of potassium in a soil and the distribution of potassium among the four major pools shown in Figure 13.17 is largely a function of the kinds of clay minerals present in a soil. Generally, soils dominated by 2:1 clays contain the most potassium; those dominated by kaolinite contain the least (Table 13.7). In terms of availability for plant uptake, the following interpretation applies to the different forms of soil potassium:

K in primary mineral structure	*Unavailable*
Nonexchangeable K in secondary minerals	*Slowly available*
Exchangeable K on soil colloids	*Readily available*
K soluble in water	

Relatively Unavailable Forms

Some 90 to 98% of all soil potassium in a mineral soil is in relatively unavailable forms (see Figure 13.17). The compounds containing most of this form of potassium are the feldspars and the micas. These minerals are quite resistant to weathering and supply relatively small quantities of potassium during a given growing season. However, their cumulative release of potassium over a period of years undoubtedly is of some importance. A few plants adapted to low-fertility sandy soils, such as elephant grass (*Pennisetum purpureum* Schum), have been shown to obtain potassium from even sand-sized primary minerals, a form of potassium usually considered to be unavailable.

Readily Available Forms

Only 1 to 2% of the total soil potassium is readily available. Available potassium exists in soils in two forms: (1) in the soil solution, and (2) exchangeable potassium adsorbed on the soil colloidal surfaces. Although most of this available potassium is in the exchangeable form (approximately 90%), soil solution potassium is most readily absorbed by higher plants but, unfortunately, is subject to considerable leaching loss.

As represented in Figure 13.17, these two forms of readily available potassium are in dynamic equilibrium. This equilibrium is of extreme practical importance. When plants absorb potassium from the soil solution, some of the exchangeable potassium immediately moves into the soil solution until the equilibrium is again established. When water-soluble fertilizers are added, the soil solution becomes potassium enriched and the reverse of the described adjustment occurs—potassium from soil solution moves onto the exchange complex. The exchangeable potassium can be seen as an important buffer mechanism for soil solution potassium.

Slowly Available Forms

In the presence of vermiculite, smectite, and other 2:1-type minerals, the K^+ ions (as well as the similarly sized NH_4^+ ions) in the soil solution (or added as fertilizers) not only

TABLE 13.7   **The Influence of Dominant Clay Minerals on the Amounts of Water-Soluble, Exchangeable, Fixed (Nonexchangeable), and Total Potassium in Soils**

The values given are means for many soils in 10 soil orders sampled in the United States and Puerto Rico.

Potassium pool	Dominant clay mineralogy of soils, mg K/kg soil		
	Kaolinitic (26 soils)	Mixed (53 soils)	Smectitic (23 soils)
Total potassium	3340	8920	15780
Exchangeable potassium	45	224	183
Water-soluble potassium	2	5	4

Data from Sharpley (1990).

Mica
CEC = 0

Fine-grained micas (e.g., illite)
CEC = 15–40

Vermiculite
CEC = 150

Dehydrated K ion •   Hydrated exchangeable cations ⊙ K   ● Ca   ▣ Na   ▲ Mg

FIGURE 13.21 Diagrammatic illustration of the release and fixation of potassium between primary micas, fine-grained mica (illite clay), and vermiculite. In the diagram, the release of potassium proceeds to the right, while the fixation process proceeds to the left. Note that the dehydrated potassium ion is much smaller than the hydrated ions of Na^+, Ca^{2+}, Mg^{2+}, etc. Thus, when potassium is added to a soil containing 2:1-type minerals such as vermiculite, the reaction may go to the left and potassium ions will be tightly held (fixed) in between layers within the crystal, producing a fine-grained mica structure. Ammonium ions (NH_4^+) are of a similar size and charge to potassium ions and may be fixed by similar reactions. [Modified from McLean (1978)]

become adsorbed but also may become definitely fixed by the soil colloids (Figure 13.21). Potassium (and ammonium) ions fit in between layers in the crystals of these normally expanding clays and become an integral part of the crystal. These ions cannot be replaced by ordinary exchange processes and consequently are referred to as *nonexchangeable ions*. As such, these ions are not readily available to most higher plants. This form is in equilibrium, however, with the more available forms and consequently acts as an extremely important reservoir of slowly available nutrients (Figure 13.22).

Release of Fixed Potassium

The quantity of nonexchangeable or fixed potassium in some soils is quite large. The fixed potassium in such soils is continually released to the exchangeable form in amounts large enough to be of great practical importance. The entire equilibrium may be represented for potassium as follows:

$$\text{Nonexchangeable K} \underset{}{\overset{\text{Slow}}{\rightleftharpoons}} \text{Exchangeable K} \underset{}{\overset{\text{Rapid}}{\rightleftharpoons}} \text{Soil solution K}$$

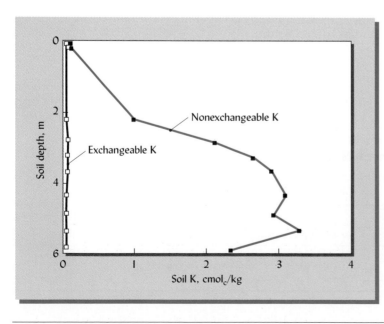

FIGURE 13.22 Exchangeable and nonexchangeable potassium levels in a Ultisol in South Carolina after 30 years' growth of a loblolly pine following 150 years of cultivated crops. Although the exchangeable potassium level was quite low, tree growth was not adversely affected, and large quantities of this element were absorbed by the trees over the 30-year period. This was made possible by the conversion of nonexchangeable potassium to the exchangeable form, which was readily taken up by the trees. The upper horizons may have been depleted somewhat of nonexchangeable K, but the deep tree roots were able to use the potassium released from the nonexchangeable form in the lower horizons. [From Richter, et al. (1994)]

As a result of these relationships, very sandy soils with low exchangeable and nonexchangeable forms of potassium are poorly buffered with respect to this element. In them, the potassium ion concentration may be quite high at the beginning of a growing season or just after fertilization, but the soils have little capacity to maintain the potassium concentration, as plants remove potassium from the soil solution during the growing season. Late-season potassium deficiency may result. In finer-textured soils with higher levels of both exchangeable and nonexchangeable forms, the initial solution concentration of potassium may not be too high, but the soil is capable of maintaining a fairly constant supply of solution potassium ions throughout the growing season.

13.15   FACTORS AFFECTING POTASSIUM FIXATION IN SOILS

The ability of the various soil colloids to fix potassium varies widely. Kaolinite and other 1:1-type clays fix little potassium. On the other hand, clays of the 2:1 type, such as vermiculite, fine-grained mica (illite), and smectite, fix potassium very readily and in large quantities. Even silt-sized fractions of some micaceous minerals fix and subsequently release potassium.

Alternate wetting/drying and freezing/thawing has been shown to enhance both the fixation of potassium in nonexchangeable forms and the release of previously fixed potassium to the soil solution. Although the practical importance of this is recognized, its mechanism is not well understood.

Applications of lime sometimes result in an increase in potassium fixation of soils. This is not surprising, since in strongly acid soils the tightly held H^+ and hydroxy aluminum ions prevent the potassium ions from being closely associated with the colloidal surfaces, which reduces their susceptibility to fixation. As the pH increases, the H^+ and hydroxyl aluminum ions are removed or neutralized, and it is easier for potassium ions to move closer to the colloidal surfaces, where they are more susceptible to fixation in 2:1 clays.

In soils where the negative charge is pH-dependent, liming increases the cation exchange capacity, which results in an increased potassium adsorption by the soil colloids and a decrease in the potassium level in the soil solution (Figure 13.23).

Furthermore, high calcium and magnesium levels in the soil solution may reduce potassium uptake by the plant because cations tend to compete against one another for uptake by roots. Since the absorption of the potassium ion by plant roots is affected by the activity of other ions in the soil solution, some authorities prefer to use the ratio

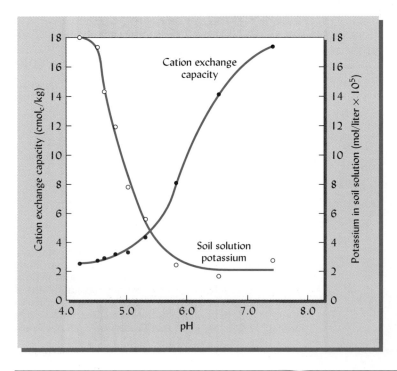

FIGURE 13.23 The influence of increased pH resulting from lime additions on the pH-dependent cation exchange capacity of a soil and the level of potassium in the soil solution. As the cation exchange capacity increases, some of the soil solution potassium is attracted to the adsorbing colloids. [Data from Magdoff and Bartlett (1980)]

$$\frac{[K^+]}{\sqrt{[Ca^{2+}] + [Mg^{2+}]}}$$

rather than the potassium concentration to indicate the available potassium level in solution. Finally, potassium deficiency frequently occurs in calcareous soils even when the amount of exchangeable potassium present would be adequate for plant nutrition on other soils. Potassium fixation as well as cation ratios may be responsible for these adverse effects on calcium carbonate–rich soil.

13.16 PRACTICAL ASPECTS OF POTASSIUM MANAGEMENT

Although there are differences among soils in their potassium-management requirements, a few common suggestions for optimum management of this element are appropriate.

1. Obtain a clear understanding of the sources of gains and losses of available potassium, and of the factors affecting these gains and losses (Figure 13.24). Recognize that the relative importance of each source will vary among soils (e.g., sand versus clays) and among ecosystems (e.g., agricultural versus forests).

2. Take full advantage of the natural potassium-supplying power of soils, remembering that for most soils it may not be necessary to replace all the potassium removed by plants. The natural supplies are commonly adequate for most non-agricultural systems, but generally must be supplemented by outside sources where crops (including trees) are removed from the land.

3. Avoid highly acid soil conditions that may permit extensive leaching losses of potassium.

4. Recycle as much of the plant-absorbed potassium as feasible by returning to the soil plant residues and animal manures, both of which are highly laden with potassium.

5. Where fertilizers are required to meet potassium needs, apply the element in frequent, light (annual) applications rather than in heavy dressings every few years. This will reduce the possibilities of luxury consumption, losses by leaching, and fixation in relatively unavailable forms.

The global use of potassium fertilizers will continue to increase as the pools of available potassium are depleted and increasing crop yields place demands on the soil that cannot be met by mineral weathering alone. This is especially true in cash-crop areas and in regions where sandy soils or highly weathered soils are prominent. In some less developed agricultural regions of the world where current rates of potassium fertilizer use are quite low, such rates will have to increase for many years to come if yields are to be increased or even maintained.

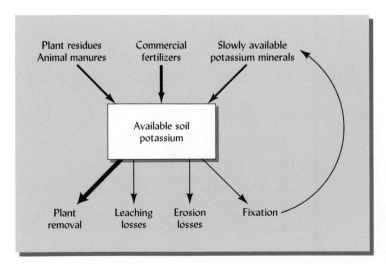

FIGURE 13.24 Gains and losses of *available* soil potassium under average field conditions. The approximate magnitude of the changes is represented by the width of the arrows. For any specific case the actual amounts of potassium added or lost undoubtedly may vary considerably from this representation. As was the case with nitrogen and phosphorus, commercial fertilizers are important in meeting plant demands.

13.17 IMPORTANCE OF MICRONUTRIENT ELEMENTS[9]

The nine so-called *micronutrients* or *trace elements* are no less important to plant growth than are the macronutrients. They are merely needed in much smaller quantities. The trace elements include iron, manganese, zinc, copper, cobalt, nickel, boron, molybdenum, and chlorine. Increased attention is being given to them for the following reasons:

1. Intensive plant production practices have increased crop yields, resulting in greater removal of micronutrients from soils.

2. The trend toward high-analysis fertilizers has reduced the use of impure salts and organic manures, which formerly supplied significant amounts of micronutrients.

3. Increased knowledge of plant nutrition and improved methods of analysis in the laboratory are helping in the diagnosis of micronutrient deficiencies that might formerly have gone unnoticed.

4. Increasing evidence indicates that food grown on soils with low levels of trace elements may provide insufficient human dietary levels of certain elements, even though the crop plants show no signs of deficiency themselves.

Interest in trace elements has also been sparked by problems of toxicity resulting from an oversupply of these elements. Levels of trace elements toxic to plants or to animals may result from natural soil conditions, from pollution, or from soil-management practices.

Micronutrients are required in very small quantities, their concentrations in plant tissue being one or more orders of magnitude lower than for the macronutrients (Figure 13.25). The ranges of plant tissue concentrations considered deficient, adequate, and toxic for several micronutrients are illustrated in Figure 13.26.

---

[9] For review articles on this subject, see Welsh (1995), Mortvedt, et al. (1991), and Stevenson (1986).

FIGURE 13.25 Relative numbers of atoms of the essential elements in alfalfa at bloom stage, expressed logarithmically. Note that there are more than 10 million hydrogen atoms for each molybdenum atom. Even so, normal plant growth would not occur without molybdenum. [Modified from Viets (1965)]

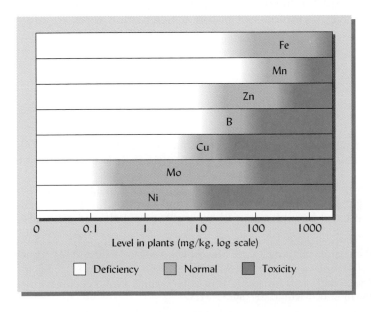

FIGURE 13.26 Deficiency, normal, and toxicity levels in plants for seven micronutrients. Note that the range is shown on a logarithm base and that the upper limit for manganese is about 10,000 times the lower range for molybdenum and nickel. In using this figure, keep in mind the remarkable differences in the ability of different plant species and cultivars to accumulate and tolerate different levels of micronutrients. (Based on data from many sources)

TABLE 13.8　Functions of Several Micronutrients in Higher Plants

Micronutrient	Functions in higher plants
Zinc	Present in several dehydrogenase, proteinase, and peptidase enzymes; promotes growth hormones and starch formation; promotes seed maturation and production.
Iron	Present in several peroxidase, catalase, and cytochrome oxidase enzymes; found in ferredoxin, which participates in oxidation-reduction reactions (e.g., NO_3^- and SO_4^{2-} reduction and N fixation); important in chlorophyll formation.
Copper	Present in laccase and several other oxidase enzymes; important in photosynthesis, protein and carbohydrate metabolism, and probably nitrogen fixation.
Manganese	Activates decarboxylase, dehydrogenase, and oxidase enzymes; important in photosynthesis, nitrogen metabolism, and nitrogen assimilation.
Nickel	Essential for urease, hydrogenases, and methyl reductase; needed for grain filling, seed viability, iron absorption, and urea and ureide metabolism (to avoid toxic levels of these nitrogen-fixation products in legumes).
Boron	Activates certain dehydrogenase enzymes; facilitates sugar translocation and synthesis of nucleic acids and plant hormones; essential for cell division and development.
Molybdenum	Present in nitrogenase (nitrogen fixation) and nitrate reductase enzymes; essential for nitrogen fixation and nitrogen assimilation.
Cobalt	Essential for nitrogen fixation; found in vitamin B_{12}.

13.18　ROLE OF THE MICRONUTRIENTS[10]

Physiological Roles

Micronutrients play many complex roles in plant nutrition. While most of the micronutrients participate in the functioning of a number of enzyme systems (Table 13.8), there is considerable variation in the specific functions of the various micronutrients in plant and microbial growth processes. For example, copper, iron, and molybdenum are capable of acting as electron carriers in the enzyme systems that bring about oxidation-reduction reactions in plants. Such reactions are essential steps in photosynthesis and many other metabolic processes. Zinc and manganese function in many plant enzyme systems as bridges to connect the enzyme with the substrate upon which it is meant to act. The functions of other micronutrients are outlined in Table 13.8.

Deficiency Symptoms

Since most of the micronutrients are relatively immobile in the plant, they are not readily transferred from older leaves to younger ones. Therefore the concentration of the nutrient tends to be lowest, and the symptoms of deficiency most pronounced, in the younger leaves, which develop after the supply of the nutrient has run low.[11] The iron-deficient sorghum in Plate 29 and the zinc-deficient corn and peach leaves in Plates 30 and 31 (following page 178) illustrate the pattern of pronounced deficiency symptoms on the younger leaves. Symptoms of boron deficiency on roses are illustrated in Figure 13.27.

13.19　SOURCE OF MICRONUTRIENTS

Inorganic Forms

Sources of the nine micronutrients vary markedly from area to area. The wide variability in the content of these elements in soils and suggested contents in a representative soil is shown in Table 13.9.

All of the micronutrients have been found in varying quantities in igneous rocks. Two of them, iron and manganese, have prominent structural positions in primary silicate minerals, such as biotite and hornblende. Others, such as cobalt and zinc, also

---

[10] For a review of the role of individual micronutrients, see Mengel and Kirkby (1987) and Marschner (1995). For nickel, see Brown, et al. (1987).

[11] This pattern is in contrast to most of the macronutrients (except sulfur), which are more easily translocated by the plants and so become most deficient in the older leaves.

(a)   (b)

FIGURE 13.27 A misshapen rose, or *bullhead,* on a boron-deficient tea rose (*a*). A well-formed bloom on the same plant after a solution containing boron was applied to the leaves and young buds (*b*). (Photos courtesy of R. Weil)

may occupy structural positions as minor replacements for the major constituents of silicate minerals, including clays.

Anions such as borate and molybdate in soils may undergo adsorption or reactions similar to those of the phosphates. Chlorine, by far the most soluble of the group, is added to soils in considerable quantities each year through rainwater. Its incidental addition to soils in fertilizers and in other ways helps prevent the deficiency of chlorine under field conditions. Chlorine is only very weakly adsorbed to soil colloids.

Organic Forms

Organic matter is an important secondary source of some of the trace elements. Several of them tend to be held as complex combinations by organic (humus) colloids. Copper is especially tightly held by organic matter, so much so that its availability can be very low in organic soils (Histosols). In uncultivated profiles, there is a somewhat greater concentration of micronutrients in the surface soil, much of it presumably in the organic fraction. Correlations between soil organic matter and contents of copper, molybdenum, and zinc have been noted. Although the elements thus held are not always readily available to plants, their release through decomposition is undoubtedly an important fertility factor. Animal manures are a good source of micronutrients, much of it present in organic forms.

TABLE 13.9 Major Sources of Nine Micronutrients, Ranges and Representative Contents of These Nutrients in Soils, Along with Their Representative Contents in Harvested Crops

The ratio of soil to crop contents emphasizes the primary need to increase the efficiency of plants in absorbing these nutrients from the soil.

			Representative contents of		
Element	Major sources	Range, kg/ha/15 cm	Soil, kg/ha/15 cm	Crop, kg/crop	Soil/crop ratio
Fe	Oxides, sulfides, silicates	20,000–220,000	56,000	2	28,000
Mn	Oxides, silicates, carbonates	45–9,000	2200	0.5	4,400
Zn	Sulfides, carbonates, silicates	25–700	110	0.3	366
Cu	Sulfides, hydroxy carbonates, oxides	4–2,000	45	0.1	450
Ni	Silicates (especially serpentine)	10–2,200	45	0.02	2,250
B	Borosilicates, borates	8–200	22	0.2	110
Mo	Sulfides, oxides, molybdates	0.4–10	5	0.02	250
Cl	Chlorides	15–100	22	2.5	0.9
Co	Silicates	2–90	18	0.02	900

TABLE 13.10 Forms of Micronutrients Dominant in the Soil Solution

Micronutrient	Dominant soil solution forms
Iron	Fe^{2+}, $Fe(OH)_2^+$, $Fe(OH)^{2+}$, Fe^{3+}
Manganese	Mn^{2+}
Zinc	Zn^{2+}, $Zn(OH)^+$
Copper	Cu^{2+}, $Cu(OH)^+$
Molybdenum	MoO_4^{2-}, $HMoO_4^-$
Boron	H_3BO_3, $H_2BO_3^-$
Cobalt	Co^{2+}
Chlorine	Cl^-
Nickel	Ni^{2+}, Ni^{3+}

From data in Lindsay (1972).

Forms in Soil Solution

The dominant forms of micronutrients that occur in the soil solution are listed in Table 13.10. The specific forms present are determined largely by the pH and by soil aeration (i.e., redox potential). Note that the cations are present in the form of either simple cations or hydroxy metal cations. The simple cations tend to be dominant under highly acid conditions. The more complex hydroxy metal cations become more prominent as the soil pH is increased.

Molybdenum is present mainly as MoO_4^{2-}, an anionic form that reacts at low pH in ways similar to those of phosphorus. Although boron also may be present in anionic form at high pH levels, research suggests that undissociated boric acid (H_3BO_3) is the form that is dominant in the soil solution and is absorbed by higher plants.

The cycling of micronutrients through the soil–plant–animal system is illustrated in a generalized way by Figure 13.28. Although not every micronutrient will participate in every pathway shown in this figure, it can be seen that organic chelates, soil colloids, soil organic matter, and soil minerals all contribute micronutrients to the soil solution and, in turn, to growing plants. As we turn our attention to micronutrient availability, it will be helpful to refer back to Figure 13.28 to see the relationships among the processes involved.

13.20 FACTORS AFFECTING THE AVAILABILITY OF MICRONUTRIENTS

Micronutrient deficiencies are most apt to occur where the total contents of the nutrients in the soil are low, or where soil conditions constrain the availability of whatever nutrients are present. Highly leached, acid, sandy soils are commonly low in micronutrients, as are most organic soils. Likewise, micronutrient deficiencies are more apt to occur in intensively cropped areas where the harvested crops are removed, and where macronutrients are being replenished through added fertilizers. Also, eroded soils that have lost much of their upper horizons and associated organic matter are more apt to be low in most micronutrients. In all cases, however, the availability of micronutrients is influenced by several general soil conditions which we will now consider.

Micronutrient Cation Availability

The availability of all six micronutrient cations is affected in a similar way by a few soil characteristics.

SOIL pH. The micronutrient cations are most soluble and available under acid conditions. In very acid soils, there is a relative abundance of the ions of iron, manganese, zinc, copper, and nickel. In some cases, concentrations are sufficiently high to be toxic to common plants.

As the pH is increased, the ionic forms of the micronutrient cations are changed first to the hydroxy ions and, finally, to the insoluble hydroxides or oxides of the elements. The following example uses the ferric ion as typical of the group:

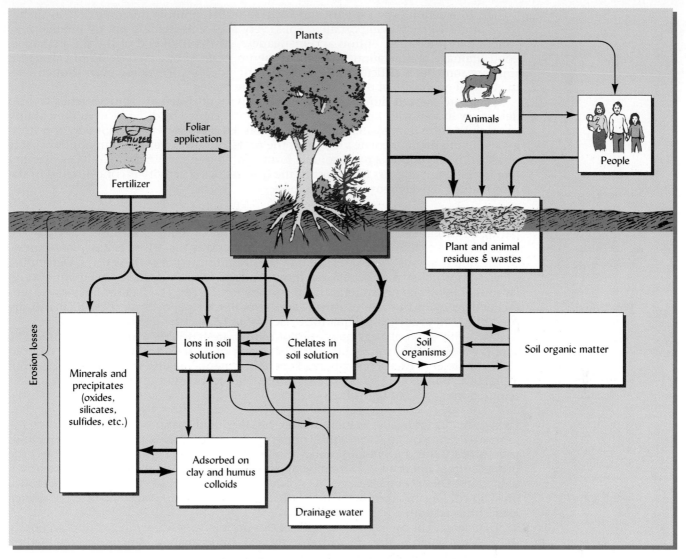

FIGURE 13.28 Cycling and transformations of micronutrients in the soil–plant–animal system. Although all micronutrients may not follow each of the pathways shown, most are involved in the major components of the cycle. The formation of chelates, which keep most of these elements in soluble forms, is a unique feature of this cycle.

$$Fe^{3+} \xrightarrow{OH^-} Fe(OH)^{2+} \xrightarrow{OH^-} Fe(OH)_2^+ \xrightarrow{OH^-} Fe(OH)_3$$

Simple cation   Hydroxy metal cations   Hydroxide
(soluble)   (soluble)   (insoluble)

Thus, as the pH is raised, the solubility (and availability to plants) of micronutrient cations decreases, leading to deficiencies of these elements. Such deficiencies associated with high pH occur naturally in many of the calcareous soils of arid regions. The general desirability of a slightly acid soil (with a pH between 6 and 7) largely stems from the fact that for most plants this pH condition allows micronutrient cations to be soluble enough to satisfy plant needs without becoming so soluble as to be toxic (see Figure 9.13).

OXIDATION STATE AND pH.   Iron, manganese, nickel, and copper occur in soils in more than one valence state. In the lower valence states, the elements are considered *reduced;*[12] in the higher valence state they are *oxidized*. At pH values common in soils, the oxidized states of these micronutrients are generally much less soluble than are the reduced states. For example, the hydroxide of trivalent ferric iron precipitates at pH values of 3.0 to 4.0, whereas ferrous hydroxide does not precipitate until a pH of 6.0 or higher is reached.

---

[12] For a recent review of microbial reduction of certain trace elements, see Lovley (1996).

FACTORS AFFECTING THE AVAILABILITY OF MICRONUTRIENTS   425

The interaction of soil acidity and aeration in determining micronutrient availability is of great practical importance. Thus, very acid soils that are poorly drained may supply toxic quantities of iron and manganese. At the high end of the soil pH range, good drainage and aeration often have the opposite effect. Well-oxidized calcareous soils are sometimes deficient in available iron, zinc, or manganese even though adequate total quantities of these trace elements are present.

There are marked differences in the sensitivity of different plant varieties to iron deficiency in soils with high pH. This is apparently caused by differences in their ability to solubilize iron immediately around the roots. Efficient varieties respond to iron stress by acidifying the immediate vicinity of the roots and by excreting compounds capable of reducing the iron to a more soluble form, with a resultant increase in its availability (Figure 13.29). It appears that most of the iron-reducing activity is concentrated in the root hairs of actively growing young roots.

OTHER INORGANIC REACTIONS.   Micronutrient cations interact with silicate clays by participating in cation exchange reactions, as do calcium or aluminum. Also, they may be tightly bound or fixed to certain silicate clays, especially the 2:1 type. Zinc, manganese, cobalt, and iron ions sometimes occur in the crystal structure of these clays. Depending on conditions, they may be released from the clays or fixed by them in a manner similar to that by which potassium is fixed (see Section 13.14). The fixation may cause serious deficiency in the case of cobalt, and sometimes zinc, because these elements are present in soil in such small quantities (see Table 13.9).

The application of large quantities of phosphate fertilizers can adversely affect the supply of some of the micronutrients. The uptake of both iron and zinc may be reduced in the presence of excess phosphates. For environmental quality reasons, as well as from a practical standpoint, phosphate fertilizers should be used in only those quantities required for good plant growth.

ORGANIC MATTER.   Organic matter, organic residues, and manure applications affect the immediate and potential availability of micronutrient cations. Some organic compounds react with these cations to form water-insoluble complexes (chelates) that can protect the nutrients from interactions with mineral particles that can bind them in even more insoluble forms. Other complexes provide slowly available nutrients as they undergo microbial breakdown. Organic complexes that enhance micronutrient availability are considered later in this section.

Deficiencies of copper and, to a lesser extent, manganese are often found on poorly drained soils high in organic matter (e.g., peats and marshes). Zinc is also retained by organic matter, but deficiencies stemming from this retention are not common.

Micronutrient-enriched organic products have been used as a nutrient source on soils deficient in available trace elements. For example, composts of iron-enriched organic materials, such as forest by-products, peat, animal manures, and plant residues, have been found to be effective on iron-deficient soils.

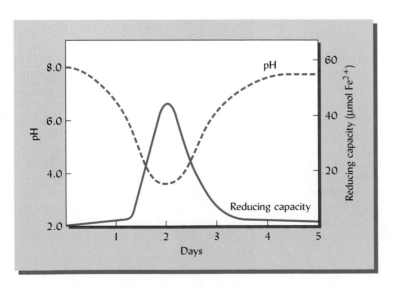

FIGURE 13.29 Response of one variety of sunflowers to iron deficiency. When the plant became stressed owing to iron deficiency, plant exudates lowered the pH and increased the reducing capacity immediately around the roots. Iron is solubilized and taken up by the plant, the stress is alleviated, and conditions return to normal. [From Marschner, et al. (1974) as reported by Olsen, et al. (1981); used with permission of *American Scientist*]

ROLE OF MYCORRHIZAE. A symbiosis between most higher plants and certain soil fungi produces mycorrhizae (fungus roots), which are far more efficient than normal plant roots in several respects. The nature of the mycorrhizal symbiosis and its importance in phosphorus nutrition were described in Sections 10.8 and 13.3, but it is worth mentioning here that mycorrhizae have also been shown to increase plant uptake of micronutrients (Table 10.4). Crop rotations and other practices that encourage a diversity of mycorrhizal fungi may thereby improve micronutrient nutrition of plants.

Surprisingly, mycorrhizae also appear to protect plants from excessive uptake of micronutrients and other trace elements where these elements are present in potentially toxic concentrations. Seedlings of such trees as birch, pine, and spruce are able to grow well on sites contaminated with high levels of zinc, copper, nickel, and aluminum only if their roots are sheathed by ectomycorrhizae. The mycorrhizae apparently help exclude these metallic cations from the root stele and prevent long-distance transport of metal cations within the plant.

Organic Compounds as Chelates

The cationic micronutrients react with certain organic molecules to form the organo-metallic complexes called **chelates** (from the Greek **chele**, claw). In these complexes, a given micronutrient is surrounded by an envelope of C, H, O, and other atoms that protect it from reactions with the soil particles (see Figure 13.30). These organic molecules may be synthesized by plant roots and released to the surrounding soil, may be present in the soil humus, or may be commercial synthetic compounds added to the soil to increase micronutrient availability. Examples of iron chelate ring structures are shown in Figure 13.30.

The effect of chelation can be illustrated with iron. In the absence of chelation, when an inorganic iron salt such as ferric sulfate is added to a calcareous soil, most of the iron is quickly rendered unavailable by reaction with hydroxide, as follows:

$$\underset{\text{(available)}}{Fe^{3+}} + 3OH^- \rightleftharpoons \underset{\text{(unavailable)}}{FeOOH} + H_2O$$

In contrast, if the iron is chelated, it largely remains in the chelate form, which is available for uptake by plants. In this reaction, the available iron chelate reactant is favored:

$$\underset{\text{(available)}}{Fe\ chelate} + 3OH^- \rightleftharpoons \underset{\text{(unavailable)}}{FeOOH} + chelate^{3-} + H_2O$$

The mechanism by which micronutrients from chelates are absorbed by plants is different for different plants. Many dicots appear to remove the metallic cation from the

Fe (III) EDTA

(a)

Fe(III) gluconate

(b)

FIGURE 13.30 Structural formula for two common iron chelates, ferric ethylenedi-aminetetra-acetate (Fe-EDTA) (*a*) and ferric gluconate (*b*). In both chelates, the iron is protected, and yet can be used by plants. [Diagrams from Clemens, et al. (1990); reprinted by permission of Kluwer Academic Publishers] chelate at the root surface, reducing (in the case of iron) and taking up the cation while releasing the organic chelating agent in the soil solution. Roots of certain grasses have been shown to take in the entire chelate–metal complex, reducing and removing the metallic cation inside the root cell, then releasing the organic chelate back to the soil solution (Figure 13.31). In both cases, it appears that the primary role of the chelate is to allow metallic cations to remain in solution so they can diffuse through the soil to the root. Once the micronutrient cations are inside the plant, other organic chelates (such as citrates) may be carriers of these cations to different parts of the plant.

The chelates of different micronutrients vary in their stability and therefore as sources of micronutrients. If a mixture of Fe^{2+}, Mn^{2+}, and Zn^{2+} occurs in the soil solution and the Fe^{2+} forms the most stable chelate, it will be most protected from reaction with the soil and will remain most available for plant uptake. Knowledge of the stabilities of different commercial chelates helps in determining which to use to alleviate the deficiency of a given nutrient. In some cases, mixtures of such chelates are provided since more than one nutrient may be in deficiency status. Commercial chelates are used substantially in industrialized countries, particularly to ameliorate micronutrient deficiencies of fruit trees and ornamentals.

Micronutrient Anion Availability

Unlike the cations needed in trace quantities by plants, the anion micronutrients seem to have relatively little in common with each other. Chlorine, molybdenum, and boron are quite different chemically, so little similarity would be expected in their reaction in soils.

CHLORINE.   Chlorine is absorbed in larger quantities by most crop plants than any of the micronutrients except iron. Most of the chlorine in soils is in the form of the chloride ion, which leaches rather freely from humid-region soils. Except where toxic quantities of chlorine are found in saline soils, there are no common soil conditions that reduce the availability and use of this element. Accretions of chlorine from the atmosphere, along with those from fertilizer salts such as potassium chloride, are sufficient to meet most crop needs.

BORON.   Boron is one of the most commonly deficient of all the micronutrients. The availability of boron is related to the soil pH, this element being most available in acid soils. Soluble boron is present in soils mostly as boric acid [$B(OH)_3$] or as $B(OH)_4^-$. These compounds can exchange with the OH groups on the edges and surfaces of variable charge clays such as kaolinite, and especially with the oxides of iron and aluminum. The boron so adsorbed is quite tightly bound, especially between pH 7 and 9, the range of lowest availability of this element.

FIGURE 13.31   Two ways in which plants utilize micronutrients held in chelated form. (Left) Dicotyledonous plants such as cucumber and peanuts produce strong reducing agents (NADPH) that reduce iron at the outer surface of the root membrane. They then take in only the reduced iron, leaving the organic chelate in the soil solution where it can complex another iron atom. (Right) Grass plants such as wheat or corn apparently take the entire chelate–metal complex into their root cells. They then remove the iron, reduce it, and return the chelate to the soil solution.

Since boron is adsorbed by humus, organic matter serves as a major reservoir of this element in many soils, and exerts considerable control over its availability. Boron availability is impaired by long dry spells that may reduce the rate of mineralization of the organically held boron. Boron deficiencies are also common in calcareous Aridisols and in other soils with high pH.

MOLYBDENUM.   Soil pH is the most important factor influencing the availability and plant uptake of molybdenum. The following equations show the forms of this element present at low and high soil pH:

$$H_2MoO_4 \underset{+H^+}{\overset{+OH^-}{\rightleftharpoons}} HMoO_4^- + H_2O$$

$$HMoO_4^- \underset{+H^+}{\overset{+OH^-}{\rightleftharpoons}} MoO_4^{2-} + H_2O$$

At low pH values, the $HMoO_4^-$ is adsorbed by silicate clays and, more especially, by oxides of iron and aluminum through ligand exchange with hydroxide ions on the surface of the colloidal particles.

The liming of acid soils will usually increase the availability of molybdenum. The effect is so striking that some researchers, especially those in Australia and New Zealand, argue that the primary reason for liming very acid soils is to supply molybdenum. Furthermore, in some instances 30 g or so of molybdenum added to acid soils has given about the same increase in the yield of legumes as has the application of several megagrams of lime.

SELENIUM.   Although not a plant nutrient, selenium is important to animal and human health. Selenium is found in nature in four major solid forms—selenate (SeO_4^{2-}), selenite (SeO_3^{2-}), elemental Se (Se^0), and selenide (Se^{2-}). It also exists in volatile compounds, such as dimethyl selenide and dimethyl diselenide, that can be released as gases to the atmosphere.

Under acid conditions (pH 4.5 to 6.5), *selenite* is usually dominant. This form is only slowly available but may be added to alleviate selenium deficiencies, since it will not likely induce selenium toxicities. At higher pH values (7.5 to 8.5) and in well-aerated soils, the more soluble *selenate* dominates and often gives rise to such toxicities.

Under anaerobic conditions, soil microorganisms can reduce the selenate and selenite to solid selenide and elemental selenium forms, both of which are quite insoluble. Advantage can be taken of such reduction to remove toxic levels of soluble selenium from soil and water, a promising process of bioremediation. This microbial reduction also produces relatively nonphytotoxic volatile selenium compounds (e.g., dimethyl selenide) that can be released to the atmosphere. The production of such volatile compounds is also being considered as a means of reducing toxic levels of selenium in water and soils.

13.21   NEED FOR NUTRIENT BALANCE

Nutrient balance among the trace elements is as essential as, but even more difficult to maintain than, macronutrient balance. Some of the plant enzyme systems that depend on micronutrients require more than one element. For example, both manganese and molybdenum are needed for the assimilation of nitrates by plants. The beneficial effects of combinations of phosphates and molybdenum have already been discussed. Apparently, some plants need zinc and phosphorus for optimum use of manganese. The use of boron and calcium depends on the proper balance between these two nutrients. A similar relationship exists between potassium and copper and between potassium and iron in the production of good-quality potatoes. Copper utilization is favored by adequate manganese, which in some plants is assimilated only if zinc is present in sufficient amounts. These nutrient interactions are examples of *synergism* where one nutrient enhances the utilization of a second nutrient.

Other *antagonistic* effects may be used effectively in reducing toxicities of certain of the micronutrients. For example, copper toxicity of citrus groves caused by residual copper from fungicidal sprays may be reduced by adding iron and phosphate fertilizers. Sulfur additions to calcareous soils containing toxic quantities of soluble molybdenum may reduce the availability, and hence the toxicity, of molybdenum. The hyperaccumulation of phosphorus, manganese, and magnesium in zinc-deficient plants is an example of the complex interactions among essential elements as they influence plant nutrition.

These examples of nutrient interactions, both beneficial and detrimental, emphasize the highly complicated nature of the biological transformations in which micronutrients are involved. The total land area on which unfavorable nutrient balances require special micronutrient treatment is increasing as soils are subjected to more intensive cropping methods.

13.22 SOIL MANAGEMENT AND MICRONUTRIENT NEEDS

In seeking the cause of plant abnormalities, one should keep in mind the conditions under which micronutrient deficiencies or toxicities are likely to occur. Sandy soils, organic soils, and soils having very high or very low pH values are prone to micronutrient deficiencies. Areas of intensive cropping and heavy macronutrient fertilization also may be deficient in the micronutrients.

The optimum management of these soils may differ from one situation to another, but the following suggestions are widely applicable.

1. Recognize and evaluate the major sources of these micronutrients and the general reactions that make them available (Figure 13.32). Give highest priority to those sources, such as organic residues, over which some control can be exerted.

2. Avoid extremes in soil pH to minimize deficiencies and toxicities of the different micronutrients. Generally, pH values between 5.5 and 7 are most appropriate.

3. Avoid extremes in soil drainage status to minimize these same micronutrient constraints. If controlling the drainage status is not practicable, turn to pH modification or choice of plants to alleviate the constraint.

4. Choose plant species or varieties known to tolerate excesses or deficiencies of the micronutrients (see Table 13.11).

5. Utilize organic residues or farm manures to the extent feasible to overcome micronutrient deficiencies. Where these sources are inadequate, turn to commercial fertilizers.

6. Use great care in selecting and applying fertilizer sources of micronutrients to avert risks of overdose and/or upsetting nutrient balances. (See Table 14.8 for a few micronutrient fertilizer products.)

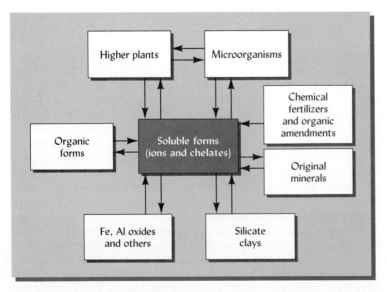

FIGURE 13.32 Diagram of soil sources of soluble forms of micronutrients and their utilization by plants and microorganisms.

TABLE 13.11   Plants Known to Be Especially Susceptible or Tolerant to, and Soil Conditions Conducive to, Micronutrient Deficiencies

Plants which are most susceptible to deficiency of a micronutrient often have a relatively high requirement for that nutrient and may be relatively tolerant to levels of that nutrient that would be high enough to cause toxicity to other plants.

Micronutrient	Common range in rates recommended for soil application[a], kg/ha	Plants most commonly deficient (high requirement or low efficiency of uptake)	Plants rarely deficient (low requirement or high efficiency of uptake)	Soil conditions commonly associated with deficiency
Iron	0.5–10.0	Blueberries, azaleas, roses, holly, grapes, nut trees, maple, bean, sorghum, oaks	Wheat, alfalfa, sunflower, cotton	Calcareous, high pH, waterlogged alkaline soils
Manganese	2–20	Peas, oats, apple, sugar beet, raspberry, citrus	Cotton, soybean, rice, wheat	Calcareous, high pH, drained wetlands, low organic matter, sandy soils
Zinc	0.5–20	Corn, onion, pines, soybeans, beans, pecans, rice, peach, grapes	Carrots, asparagus, safflower, peas, oats, crucifers, grasses	Calcareous soils, acid, sandy soils, high phosphorus
Copper	0.5–15	Wheat, corn, onions, citrus, lettuce, carrots	Beans, potato, peas, pasture grasses, pines	Histosols, very acid, sandy soils
Boron	0.5–5	Alfalfa, cauliflower, celery, grapes, conifers, apples, peanut, beets, rapeseed, pines	Barley, corn, onion, turf grass, blueberry, potato, soybean	Low organic matter, acid, sandy soils, recently limed soils, droughty soils, soils high in 2:1 clays
Molybdenum	0.05–0.5	Alfalfa, sweet clover, crucifers (broccoli, cabbage, etc.), citrus, most legumes	Most grasses	Acid sandy soils, highly weathered soils with amorphous Fe and Al

[a] The lower end of each range is typical for banded application; the higher end is typical for broadcast applications.

Economic responses to micronutrients are becoming more widespread as intensity of plant production increases. For example, responses of fruits, vegetables, and field crops to zinc and iron applications are common in areas with neutral to alkaline soils. Even on acid soils, deficiencies of these elements are increasingly encountered. Molybdenum, which has been used for some time for forage crops and for cauliflower and other vegetables, has received attention in recent years for forest nurseries and soybeans, especially on acid soils. These examples, along with those from muck areas and sandy soils where micronutrients have been used for decades, illustrate the need for these elements if optimum yields are to be maintained.

A soil will often produce a micronutrient deficiency in some plants but not in others. Plant species, and varieties within species, differ widely in their susceptibility to micronutrient deficiency or toxicity. Table 13.11 provides examples of plant species known to be particularly susceptible or tolerant to several micronutrient deficiencies.

Micronutrient Availability

Major sources of micronutrients and the general reactions that make them available to higher plants and microorganisms are summarized in Figure 13.32. Original and secondary minerals are the primary sources of these elements, while the breakdown of organic forms releases ions to the soil solution. The micronutrients are used by higher plants and microorganisms in important life-supporting processes. Removal of nutrients in crop or timber harvest reduces the soluble ion pool, and it may need to be replenished with manures or chemical fertilizers to avoid nutrient deficiencies.

Worldwide Management Problems

Micronutrient deficiencies have been diagnosed in most areas of crop production and some forest areas of the United States and Europe. However, in some developing countries, particularly in the tropics, the extent of these deficiencies is much less well known. Limited research suggests that there may be large areas with deficiencies or toxicities of one or more of these elements. As macronutrient deficiencies are addressed and yields are increased, more micronutrient deficiencies will undoubtedly come to the fore.

13.23 CONCLUSION

Soil phosphorus presents us with a double-edged sword; its management is crucially important from both an environmental standpoint and for soil fertility. On one hand, too little phosphorus commonly limits the productivity of natural and cultivated plants and is the cause of widespread soil and environmental degradation. On the other hand, excess phosphorus from fertilizers and animal wastes has concentrated too much of this element in some soils, which, when eroded, can cause egregious eutrophication in surface waters.

Except in cases of extreme buildup, the availability of phosphorus to plant roots has a double constraint: the low total phosphorus level in soils and the small percentage of this level that is present in available forms. Furthermore, even when soluble phosphates are added to soils, they are quickly fixed into insoluble forms that in time become quite unavailable to growing plants. In acid soils, the phosphorus is fixed primarily by iron, aluminum, and manganese; in alkaline soils, by calcium and magnesium. This fixation greatly reduces the efficiency of phosphate fertilizers, with little of the added phosphorus being taken up by plants.

Potassium is generally abundant in soils, but it, too, is present mostly in forms that are quite unavailable for plant absorption. Fortunately, however, some soils contain considerable nonexchangeable but slowly available forms of this element. Over time this potassium can be released to exchangeable and soil solution forms that can be quickly absorbed by plant roots. This is fortunate since the plant requirements for potassium are high—5 to 10 times that of phosphorus and similar to that of nitrogen.

Micronutrients are becoming increasingly important to world agriculture as crop removal of these essential elements increases. Deficiencies of these elements are due not only to low contents of these elements in soils but more often to their unavailability to growing plants. They are adsorbed by inorganic constituents, such as Fe, Al, oxides, and form complexes with organic matter, some of which are only sparingly available to plants. Other such organic complexes, known as *chelates,* protect some of the micronutrient cations from inorganic adsorption and make them available for plant uptake.

Toxicities of micronutrients are also quite common. They retard both plant and animal growth. Removing these elements from soil and water, or rendering them unavailable for plant uptake, is one of the challenges facing soil and plant scientists. Soil-management practices to overcome deficiencies and toxicities of trace elements are being sought by soil and plant scientists.

STUDY QUESTIONS

1. You have learned that nitrogen, potassium and phosphorus are all "fixed" in the soil. Compare the processes of these fixations and the benefits and constraints they each provide.

2. What is meant by *eutrophication,* and how is it influenced by farm practices involving phosphorus?

3. In the spring a certain surface soil showed the following soil test: soil solution K = 20 kg/ha; exchangeable K = 200 kg/ha. After two crops of alfalfa hay that contained 250 kg/ha of potassium were harvested and removed, a second soil test showed soil solution K = 15 kg/ha and exchangeable K = 150 kg/ha. Explain why there was not a greater reduction in soil solution and exchangeable K levels.

4. What is the effect of soil pH on the availability of phosphorus, and what are the unavailable forms at the different pH levels?

5. What is *luxury consumption* of plant nutrients, and what are its advantages and disadvantages?

6. How does phosphorus that forms relatively insoluble inorganic compounds in soils find its way into streams and other waterways?

7. During a year's time some 250 kg nitrogen and only 30 g molybdenum have been taken up by the trees growing on a hectare of land. Would you therefore conclude that the nitrogen was more essential for the tree growth? Explain.

8. Soybeans growing on a recently limed soil show evidence of a deficiency of a nutrient, thought by some to be molybdenum. Do you agree with this diagnosis? If not, what is your explanation?

9. What are *chelates,* how do they function, and what are their sources?

10. The addition of only 1 kg/ha of a micronutrient to an acid soil on which lime-loving cauliflower was being grown gave considerable growth response. Which of the micronutrients would it likely have been? Explain.

11. Two Aridisols, both at pH 8, were developed from the same parent material, one having restricted drainage, the other being well drained. Plants growing on the well-drained soil showed iron deficiency symptoms while those on the less-well-drained soil did not. What is the likely explanation for this?

REFERENCES

Brady, N. C. 1974. *The Nature and Properties of Soils,* 8th ed. (New York: Macmillan).

Brown, P. H., R. M. Welch, and E. E. Cary. 1987. "Nickel: A micronutrient essential for higher plants," *Plant Physiol.,* **85**:801–803.

Buresh, R. J., P. C. Smithson, and D. T. Hellums. 1997. "Building soil phosphorus capital in Africa," in R. J. Buresh, P. A. Sanchez, and F. Calhoun (eds.), *Replenishing Soil Fertility in Africa.* SSSA Special Publication no. 51. (Madison, Wis.: Soil Sci. Soc. Amer.).

Clemens, D. F., B. M. Whitehurst, and G. B. Whitehurst. 1990. "Chelates in agriculture," *Fertilizer Research,* **25**:127–131.

Cogger, C., and J. M. Duxbury. 1984. "Factors affecting phosphorus loss from cultivated organic soils," *J. Environ. Qual.,* **13**:111–114.

Fares, F., et al. 1974. "Quantitative survey of organic phosphorus in different soil types," *Phosphorus in Agriculture.* (Madison, Wis.: Amer. Soc. Agron.), pp. 25–40.

Fox, R. L. 1981. "External phosphorus requirements of crops," in *Chemistry in the Soil Environment.* ASA Special Publication no. 40. (Madison, Wis.: Amer. Soc. Agron. and Soil Sci. Soc. Amer.), pp. 223–239.

Khasawneh, F. E., et al. (eds.). 1980. *The Role of Phosphorus in Agriculture.* (Madison, Wis.: Amer. Soc. Agron.).

Kuo, S. 1988. "Application of modified Langmuir isotherm to phosphate sorption by some acid soils," *Soil Sci. Soc. Amer. J.,* **52**:97–102.

Lindsay, W. L. 1972. "Inorganic phase equilibria of micronutrients in soils," in J. J. Mortvedt, P. M. Giordano, and W. L. Lindsay (eds.), *Micronutrients in Agriculture* (Madison, Wis.: Soil Sci. Soc. Amer.).

Lovley, D. R. 1996. "Microbial reduction of iron, manganese and other metals," *Advances in Agronomy,* **54**:175–231.

Magdoff, F. R., and R. J. Bartlett. 1980. "Effect of liming acid soils on potassium availability," *Soil Sci.,* **129**:12–14.

Marschner, H. 1995. *Mineral Nutrition of Higher Plants,* 2d ed. (New York: Academic Press).

Marschner, H., A. Kalisch, and V. Romheld. 1974. "Mechanism of iron uptake in different plant species," *Proc. 7th Int. Colloquium on Plant Analysis and Fertilizer Problems,* Hanover, West Germany.

McLean, E. O. 1978. "Influence of clay content and clay composition on potassium availability," in G. S. Sekhon (ed.), *Potassium in Soils and Crops.* (New Delhi, India: Potash Research Institute of India), pp. 1–19.

Mengel, K., and E. A. Kirkby. 1987. *Principles of Plant Nutrition,* 4th ed. (Bern, Switzerland: International Potash Institute).

Mortvedt, J. J., F. R. Cox, L. M. Shuman, and R. M. Welch (eds.). 1991. *Micronutrients in Agriculture.* SSSA Book Series, no. 4. (Madison, Wis.: Soil Sci. Soc. Amer.).

Munson, R. D. (ed.). 1985. *Potassium in Agriculture.* (Madison, Wis.: Amer. Soc. Agron.).

Murphy, L. S., and L. M. Walsh. 1972. "Correction of micronutrient deficiencies with fertilizers," in J. J. Mortvedt, P. M. Giordano, and W. L. Lindsay (eds.), *Micronutrients in Agriculture.* (Madison, Wis.: Soil Sci. Soc. Amer.).

Nair, K. P. P. 1996. "The buffering power of plant nutrients and effects on availability," *Advances in Agronomy,* **57**:237–287.

Nowak, C. A., R. B. Downard, Jr., and E. H. White. 1991. "Potassium trends in red pine plantations at Pack Forest, New York," *Soil Sci. Soc. J.,* **55**:847–850.

Olsen, R. A., R. B. Clark, and J. H. Bennett. 1981. "The enhancement of soil fertility by plant roots," *Amer. Scientist,* **69**:378–384.

Olsen, S. R., and F. E. Khasawneh. 1980. "Use and limitations of physical-chemical criteria for assessing the status of phosphorus in soils," in F. E. Khasawneh, et al. (eds.), *The Role of Phosphorus in Agriculture.* (Madison, Wis.: Amer. Soc. of Agron.).

Raven, K. P., and L. R. Hossner. 1993. "Phosphorus desorption quantity-intensity relationships in soils," *Soil Sci. Soc. Amer. J.,* **57**:1501–1508.

Richter, D. D., et al. 1994. "Soil chemical change during three decades in an old-field Loblolly pine (*Pinus taeda L.*) ecosystem," *Ecology,* **75**:1463–1473.

Saa, A., M. C. Trasar-Cepeda, B. Soto, F. Gil-Sotres, and F. Diaz-Fierros. 1994. "Forms of phosphorus in sediments eroded from burnt soils," *J. Environ. Qual.,* **23**:739–746.

Sample, E. C., et al. 1980. "Reactions of phosphate fertilizers in soils," in F. E. Khasawneh, et al. (eds.), *The Role of Phosphorus in Agriculture.* (Madison, Wis.: Amer. Soc. Agron.).

Sharpley, A. N. 1990. "Reaction of fertilizer potassium in soils of differing mineralogy," *Soil Sci.,* **49**:44–51.

Smith, S. J., A. N. Sharpley, J. W. Naney, W. A. Berg, and O. R. Jones. 1991. "Water quality impacts associated with wheat culture in the Southern Plains," *J. Environ. Qual.,* **20**:244–249.

Soltanpour, P. N., R. L. Fox, and R. C. Jones. 1988. "A quick method to extract organic phosphorus from soils," *Soil Sci. Soc. Amer. J.,* **51**:255–256.

Stevenson, F. J. 1986. *Cycles of Soil Carbon, Nitrogen, Phosphorus, Sulfur, and Micronutrients.* (New York: Wiley).

Vaithiyanathan, P., and D. L. Correll. 1992. "The Rhode River watershed: Phosphorus distribution and export in forest and agricultural soils," *J. Environ. Qual.,* **21**:280–288.

Viets, F. J., Jr. 1965. "The plants' need for and use of nitrogen," in *Soil Nitrogen* (*Agronomy,* no. 10). (Madison, Wis.: Amer. Soc. Agron.).

Weil, R. R., P. W. Benedetto, L. J. Sikora, and V. A. Bandell. 1988. "Influence of tillage practices on phosphorus distribution and forms in three Ultisols," *Agron. J.,* **80**:503–509.

Welch, R. M. 1995. "Micronutrient nutrition of plants," *Critical Reviews in Plant Science,* **14**(1):49–82.

14

PRACTICAL NUTRIENT MANAGEMENT

*Using scientific knowledge and ecological wisdom we
can manage the earth . . .*
—*RENE J. DUBOS,* HUMANIZING THE EARTH

As stewards of the land, soil managers must keep nutrient cycles in balance. By doing so they maintain the soil's capacity to supply the nutritional needs of plants and, indirectly, of us all. While undisturbed ecosystems generally need no intervention from soil managers or from anyone else, few ecosystems are so undisturbed. More often, human hands have directed the output of the ecosystem for human ends. Forests, farms, fairways, and flower gardens are ecosystems modified to provide us with paper, food, recreational opportunities, and aesthetic satisfaction. By their very nature, managed ecosystems need management.

In managed ecosystems, nutrient cycles become unbalanced through increased removals (e.g., harvest of timber and crops), through increased system leakage (e.g., leaching and runoff), through simplification (e.g., monoculture, be it of pine tree or sugarcane), and through increased demands for rapid plant growth (whether the soil is naturally fertile or not). Hence, the land manager is, of necessity, also a nutrient manager.

In this chapter we will discuss the goals of nutrient management and describe the tools that may help achieve those goals. These tools include methods of enhancing nutrient recycling, as well as sources of additional nutrients that can be applied to soils or plants. We will learn how to diagnose nutritional disorders of plants and correct soil fertility problems. Building on the foundation of principles set out in earlier parts of this book, this chapter contains much practical information on the profitable use and conservation of organic and inorganic nutrient resources in producing abundant, high-quality plant products and in maintaining the quality of both the soil and the rest of the environment.

14.1 GOALS OF NUTRIENT MANAGEMENT[1]

Nutrient management aims to achieve four broad, interrelated goals: (1) cost-effective production of high-quality plants, (2) efficient use and conservation of nutrient resources, (3) maintenance or enhancement of soil quality, and (4) protection of the environment beyond the soil. Each of these goals will now be briefly considered.

---

[1] For an overview of issues and advances in nutrient management for agriculture and environmental quality, see Magdoff, et al. (1997). A standard textbook on management of agricultural soil fertility and fertilizers is Tisdale, et al. (1993).

Plant Production

People engage in three primary types of plant production: (1) *agriculture,* (2) *forestry,* and (3) *ornamental landscaping.* The main nutrient-management goal for agriculture is to increase crop and forage yields, thereby helping the subsistent farmers to feed their families and livestock and the commercial farmers to enhance their incomes. Unfortunately, farmers tend to judge the success or failure of their management schemes in only three to six months, the period of growth of the crops they are producing. This is nothing like the span of several human generations that may be needed to fully evaluate the effectiveness of the practices they choose to use.

In forestry, the principal plant product may be measured in terms of volume of lumber or paper produced. In these instances, nutrient management aims to enhance rate of growth so that the time between investment and payoff can be minimized. The time frame in forestry, measured in decades or even centuries, tends to limit the intensity of nutrient-management interventions that can be profitably undertaken.

When soils are used for ornamental landscaping purposes, the principal objective is to produce quality, aesthetically pleasing plants. Whether the plants are produced for sale or not, relatively little attention is paid to yield of biomass produced. In fact, high yield may be undesirable, as in the case of turf grass, for which high yield requires more frequent mowing.

Conservation of Nutrient Resources

Two concepts that are key to the goal of conserving nutrient resources are (1) renewal or reuse of the resources, and (2) nutrient budgeting that reflects a balance between system inputs and outputs.

The first law of thermodynamics suggests that all material resources are ultimately renewable, since the elements are not destroyed by use but are merely recombined and moved about in space. In practical terms, however, once a nutrient has been removed from a plot of land and dispersed into the larger environment, it may be difficult if not impossible to use it again as a nutrient for that particular soil. For instance, phosphorus deposited in a lake bottom with eroded sediment and nitrogen contained in wheat or corn that is exported and consumed thousands of miles away are not available for reuse.

Recycling is a form of reuse in which nutrients are returned to the same land from which they were previously removed. Leaving crop residues in the field and spreading barnyard manure onto the land from which the cattle-feed grains were harvested are both examples of nutrient recycling. The term *renewable resource* best applies to soil nitrogen, which can be replenished from the atmosphere by biological nitrogen fixation (see Sections 12.10 to 12.12). Manufacture of nitrogen fertilizers also fixes atmospheric nitrogen, but at a large cost in nonrenewable fossil fuel energy. Other fertilizer nutrients, such as potassium and phosphorus, are mined or extracted from nonrenewable mineral deposits or from mineral-laden seas and oceans, so their supplies are not renewable.

A very useful step in planning nutrient management is to conceptualize the nutrient flows for the particular system under consideration. Such a flowchart should attempt to account for all the major inputs and outputs of nutrients. Simplified examples of such a conceptual budget are shown in Figure 14.1.

Nutrient Imbalances

In many cases addressing nutrient imbalances, shortages, and surpluses calls for analysis of nutrient flows, not just on a single farm or enterprise, but on a watershed, regional, or even national scale. For example, studies have shown that many countries in Africa are net exporters of nutrients. That is, exports of agricultural and forest products carry with them more nutrients than are imported into the country as fertilizers, food, or animal feed. During the past 30 years, an average of 22 kg/ha N, 2.5 kg/ha P, and 15 kg/ha K have been lost *annually* from about 200 million has of cultivated land in sub-Saharan Africa (excluding South Africa). This net *negative* nutrient balance for African soils[2] appears to be a contributing factor in their impoverishment, the reduction of agricultural productivity, and the stagnation or decline of national economies.

---

[2] For discussions of the negative nutrient balance in African soils, see Sanchez, et al. (1997) and Smaling, et al. (1997).

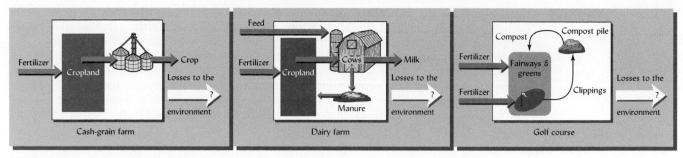

FIGURE 14.1 Representative conceptual nutrient flowcharts for a cash-grain farm, a dairy farm, and a golf course. Only the managed inputs and desired outputs are shown. Unmanaged inputs, such as nutrient deposition in rainfall, are not shown. Outputs that are difficult to manage, such as leaching and runoff losses to the environment, are shown as being variable. Although information on unmanaged inputs and outputs is not always readily available, it must be taken into consideration in developing a complete nutrient-management plan. Such flowcharts are a starting point in identifying imbalances between inputs and outputs that could lead to wasted resources, reduced profitability, and environmental damage.

By contrast, in temperate regions, the cultivated land on average receives nutrients in excess of those removed in crops, runoff, and erosion. During the same 30-year period, the nearly 300 million ha of cultivated temperate-region soils had a net **positive** nutrient balance of at least 60 kg/ha N, 20 kg/ha P, and 30 kg/ha K. While nutrient deficiencies still occur in some fields, the great majority of soils have experienced a nutrient buildup. Some of the excess nutrients move into streams, lakes, or the atmosphere, where they can contribute to environmental damage.

Changes in the structure of agricultural production in some countries have led to *regional* nutrient imbalances and concomitant serious water-pollution problems. The concentration of livestock production in huge feedlots and other confinement facilities using feed imported from other areas is a case in point. The animal manure produced in these production sites contains nutrients far in excess of the amounts that can be used in an efficient and environmentally safe manner by crops in nearby fields. The concentrated poultry industry in Delaware is one example for which the imbalance has been documented (Table 14.1).

Soil Quality and Productivity

The concept of using nutrient management to enhance soil quality goes far beyond simply supplying nutrients for the current year's plant growth. Rather, it includes the long-term nutrient-supplying and -cycling capacity of the soil, improvement of soil physical properties or tilth, maintenance of above- and belowground biological functions and

TABLE 14.1 Statewide Nitrogen and Phosphorus Balance for Poultry-Based Agriculture in Delaware

It is apparent that each year about 10,000 more Mg of N is applied to the soils of Delaware's farms than is needed by the crops grown. This is nearly 50 kg N/ha. Much of this excess nitrogen leaves these farms as nitrates in surface runoff and groundwater. Phosphorus is proportionally even further out of balance, resulting in the buildup of this element in surface soils and increased losses in runoff.

Nitrogen requirement or source	Farmland area, ha	Amount of phosphorus, Mg/yr	Amount of nitrogen, Mg/yr
Requirements			
Corn	69,600	940	9,800
Soybeans	80,600	1,085	0
Wheat	24,300	330	2,200
Barley	11,000	150	1,000
Other crops	32,400	435	3,600
Total	217,900	2,940	16,600
Nutrient Sources			
Poultry manure	—	3,495	7,865
Fertilizer sales	—	2,955	19,275
Other wastes	—	Unknown	Unknown
Total sources of nutrient	—	6,450+	27,140+
Balance (excess N or P)	—	3,510+	10,580+

From Sims and Wolf (1994).

diversity, and the avoidance of chemical toxicities. Likewise, the management tools employed go far beyond the application of various fertilizers (although this may be an important component of nutrient management). Nutrient management requires the integrated management of physical, chemical, and biological processes. The effects of tillage on organic matter accumulation (Chapter 11), the increase in nutrient availability brought about by earthworm activity (Chapter 10), the role of mycorrhizal fungi in phosphorus uptake by plants (Chapter 13), and the impact of fire on soil nutrient and water supplies (Chapter 7) are all examples of components of integrated nutrient management.

14.2 ENVIRONMENTAL QUALITY

Nutrient management impacts the environment most directly in the area of water quality, and that is the aspect which will be emphasized in this chapter. The principal water pollutants to be considered are nitrogen, phosphorus, and sediment. You may therefore wish to consult Chapters 12 and 13 to review the many pathways by which nitrogen and phosphorus can be lost from soils. Clearly, the primary means to avoid damaging surface waters with excess nitrogen or phosphorus is to keep the rate at which these nutrients are supplied in balance with the rate at which they are removed by plants. In addition, a number of strategies can be used to reduce the transport of nutrients (and other pollutants) from soils to groundwater and surface waters. In the United States, practices officially sanctioned to implement these strategies are known as *best management practices* (BMPs). Four general practices will now be briefly considered: (1) buffer strips, (2) cover crops, (3) selective timber cutting, and (4) conservation tillage.

Riparian Buffer Strips[3]

Buffer strips of dense vegetation situated along the bank of a stream or other body of water (the **riparian zone**) are a simple and generally cost-effective method that can be used to protect water from the polluting effects of a nutrient-generating land use. The vegetation in the buffer strips may consist of natural vegetation or planted vegetation, and may include grasses or trees (Figure 14.2).

Water running off the surface of the nutrient-rich land passes through the riparian buffer strip before it reaches the stream. Trees and grass plants and the litter or thatch layer they form reduce the water velocity, thereby permitting the sediment and attached nutrients to settle out. Also, dissolved nutrients are adsorbed by the organic mulch and mineral soil or are taken up by the buffer strip plants. The decreased flow velocity also increases the retention time—the length of time during which microbial action can work to break down pesticides before they reach the stream. Under some circumstances, buffer strips along streams can also reduce the nitrate levels in the groundwater flowing under them (although emission of nitrous oxide to the atmosphere may be a consequence of nitrate reduction; see Figure 12.6).

Cover Crops

A cover crop is not harvested, but rather is allowed to provide vegetative cover for the soil and then is either killed and left on the surface as a mulch, or tilled into the soil as a green manure. These crops, which are commonly grown immediately after the main crop, can reduce the loss of nutrients in both surface runoff and leaching waters. By providing soil cover, they reduce the amount of water running off from the soil and, in turn, the nutrients in that water. Roots of the cover crop also absorb the nitrates and other nutrients remaining from the main cropping period, along with water that might otherwise simply percolate through the soil. Thus, the presence of an actively growing cover crop will slow percolation, and incorporate nutrients into plant tissue that will be available for microbial release during the following year. An ideal cover crop should produce as extensive a root system as possible, as quickly as possible, once the main crop has ceased growth. Grass plants such as winter annual cereals (rye, wheat, oats, etc.) have proven to be more efficient than legumes (vetch, clover, etc.) at mopping up leftover soluble nitrogen (Figure 14.3).

---

[3] For a discussion of buffer strips in forestry, see Belt and O'Laughlin (1994).

FIGURE 14.2 The zone of grassy vegetation maintained along the banks of this stream is an example of a buffer strip installed in an agricultural watershed. A large proportion of the nutrients and other pollutants carried off the crop fields by surface runoff is trapped in these strips before the runoff water can reach the stream. In the forested areas in the background, a similar strip of land along the stream will be left undisturbed to act as a buffer during logging operations. (Photo courtesy of R. Weil)

Selective Timber Cutting

Most forests are characterized by relatively slow rates of nutrient release (mainly by organic matter mineralization) matched rather closely by comparable rates of uptake by the trees. Unfortunately, the most common and economic means of harvesting forest species—clear-cutting, or removing all the trees form large blocks of land—has some serious environmental consequences (Figure 14.4). Not only are the soils left more vulnerable to erosion, but for four primary reasons the nutrient balance is upset.

First, the removal of all the trees leaves the soil with few active roots to take up nutrients dissolved in the soil solution until natural revegetation reestablishes a root network. Second, organic matter decomposition and mineral nutrient release is accelerated by the increased temperature of the unshaded forest floor, by the higher moisture resulting from reduced transpiration from plants, and by the physical mixing of the O and A horizons by log skidding and clearing operations. Third, the slash (branches and leaves) left behind when the logs are removed presents an enormous additional source of nutrients to be released, either by slash burning or by decomposition. Fourth, the major reduction in soil microbial biomass that follows clear cutting results in a large mass of dead, easily decomposed microorganisms. As a result of these factors, streams draining most clear-cut watersheds have been shown to carry elevated loads of nitrate and other nutrients for some time after the timber harvest has occurred.

Selective cutting practices, in contrast, remove only a small percentage of the trees present in any one year. The soil remains permeated with nutrient-scavenging tree roots, and the forest floor suffers only minimum disturbance. Streams are less polluted and the forest soils are less depleted of essential nutrients. A more attractive landscape is another benefit of selective cutting.

Conservation Tillage

The term *conservation tillage* applies to agricultural tillage practices that keep at least 30% of the soil surface covered by plant residues. The effects of conservation tillage on soil properties and on the prevention of soil erosion are discussed elsewhere in this textbook (see Section 15.5). Here, we emphasize the effects on nutrient losses.

Studies in many regions of the world make it clear that, compared to plowed fields with little or no residue cover, conservation tillage reduces the total amount of water

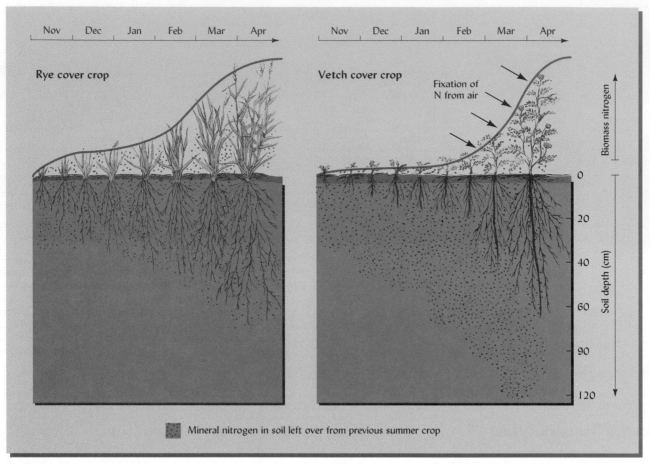

FIGURE 14.3 Relative effectiveness of a nonlegume (rye) and a legume (vetch) winter cover crop in mopping up the nitrate remaining in the soil after the harvest of a heavily fertilized summer crop. Rye grows very rapidly during mild fall weather just after harvest, while the legume grows very little until the soil warms again the following spring. These traits, and the facts that, unlike the legume, rye is dependent on soil nitrogen and has a fibrous root system, combine to make rye an excellent choice of cover crop to reduce winter leaching of soil nitrate.

FIGURE 14.4 Clear-cut harvest of loblolly pine on Coastal Plain soils. A few mature trees were left standing as seed trees. The forest floor is covered with slash of all sizes, but little living vegetation. Decomposition of the high-carbon slash may cause substantial immobilization of nitrogen. (Photo courtesy of R. Weil)

TABLE 14.2 Effect of Land Use and Tillage on the Loss of Nutrients in Runoff from a Silty Clay Loam Hapludalf in Ohio

The forest was a mixed stand of oak and pine, the alfalfa was a three-year-old hayfield, and the cultivated plots were planted to corn after having been in grass for 10 years. Ridge tillage is a form of conservation tillage that leaves the soil covered by residues in winter and disturbs only part of the soil in spring.

Water/nutrient loss	Tillage system			
	Forest	Untilled alfalfa	Ridge-tilled corn	Conventionally tilled corn
Runoff, % of rainfall	5	18	33	40
Nutrient loss[a] kg/ha/yr				
Nitrogen	19	13	49	315
Phosphorus	0.26	0.21	1.12	2.65

[a] Sediment and runoff.
Data from Thomas, et al. (1992).

running off the land surface, and reduces even more the total load of nutrients and sediment carried by that runoff (Tables 14.2 and 13.1). When combined with a cover crop, the effect is greater still.

In the remainder of this chapter we will concentrate on nutrient resources, both those that are internal to the system we are concerned with (farm, watershed, forest, etc.), and those that must be supplied from outside the system. We will examine the advantages of maximizing the use and reuse of internal resources so as to minimize the need for external resources.

14.3 SOIL–PLANT–ATMOSPHERE NUTRIENT CYCLES

Weathering of Parent Materials: An Internal Resource

Depending on the parent materials and climate, weathering of parent materials (see Section 2.3) may release significant quantities of nutrients (Table 14.3). For timber production, most nutrients are released fast enough from either parent materials or decaying organic matter to supply adequate nutrients. In agricultural systems, however, some nutrients generally must be added, since nutrients are removed from the land annually in harvested crops. Negligible amounts of nitrogen are released by weathering of mineral parent materials, so other mechanisms (including biological nitrogen fixation) are needed to resupply this important nutrient. Release of nutrients by mineralization of soil organic matter is important in short-term nutrient cycling, but in the long run, the organic matter and the nutrients it contains must be replenished or soil fertility will be depleted.

Nutrient Economy of Trees[4]

Forests have evolved mechanisms to conserve nutrients in nutrient-poor sites. For instance, forests generally produce more aboveground biomass (100 to 200 kg) per kilogram of nitrogen taken up than other ecosystems (e.g., corn produces about 60 to 70 kg biomass per kg N taken up). Forests of cold climates, in which slow rates of organic mat-

---

[4] For an in-depth treatment of this topic, see Attiwill and Leeper (1987).

TABLE 14.3 Amounts of Selected Nutrients Released by Mineral Weathering in a Representative Humid, Temperate Climate, Compared with Amounts Removed by Silvicultural and Agricultural Harvests

For the forest, weathering release and harvest removal are roughly balanced, but cropping removes much more of some nutrients than can be released by weathering. Leaching losses are not shown.

	Amount released or removed, kg/ha			
	P	K	Ca	Mg
Weathered from igneous parent material over 50 years	5–25	250–1000	150–1500	50–500
Removed in harvest of 50-year-old deciduous bole wood	10–20	60–150	175–250	25–100
Removed by 50 annual harvests of a corn–wheat–soybean rotation	1200	2000	550	500

Estimated from many sources.

TABLE 14.4 Representative Nutrient Distribution in Several Types of Forest Ecosystems[a]

	In vegetation, kg/ha	In forest floor, kg/ha	Residence time,[b] years
	Nitrogen		
Boreal coniferous	300–500	600–1100	200
Temperate deciduous	100–1200	200–1000	6
Tropical rain forest	1000–4000	30–50	0.6
	Phosphorus		
Boreal coniferous	30–60	75–150	300
Temperate deciduous	60–80	20–100	6
Tropical rain forest	200–300	1–5	0.6
	Potassium		
Boreal coniferous	150–350	300–750	100
Temperate deciduous	300–600	50–150	1
Tropical rain forest	2000–3500	20–40	0.2
	Calcium		
Boreal coniferous	200–600	150–500	150
Temperate deciduous	1000–1200	200–400	3
Tropical rain forest	3500–5000	100–200	0.3

[a] Data derived from many sources.
[b] Residence time in forest floor (O horizons).

ter decomposition result in the immobilization of the system's nitrogen in the forest floor, are more efficient in this regard than are temperate or tropical forests (Table 14.4).

In Section 2.14 we saw that trees can act as nutrient pumps, taking up nutrients that occur deep in the profile because of weathering or leaching, and depositing them at the soil surface as litter that will decompose and release the nutrients where they can be of use to relatively shallow-rooted agricultural crops. Overall nutrient-use efficiency can sometimes be increased by combining trees and agricultural crops into what are known as *agroforestry systems* (Figure 14.5). Nitrogen-fixing trees (mostly legumes) can also add nitrogen to the surface soil with their nitrogen-rich leaf litter. Trees may also enhance the fertility of the soil in their vicinity by trapping windblown dust, thus increasing the deposition of such nutrients as calcium, phosphorus, and sulfur.

(a)

(b)

FIGURE 14.5 Two examples of agroforestry systems. (*a*) The deep-rooted *Acacia albida* enriches the soil under its spreading branches (arrow points to a man). Conveniently, these trees leaf out in the dry season and lose their leaves during the rainy season when crops are grown. It is an African tradition to leave these trees standing when land is cleared for crop production. Crops growing under the trees yield more and have higher contents of sulfur, nitrogen, and other nutrients. (*b*) Branches pruned from widely spaced rows of leguminous trees are spread as a mulch on the soil surface in the alleys between the tree rows, thus enriching the alleys with nutrients from the leaves as well as conserving soil moisture. The crops grown in this alley-cropping system may yield better than crops grown alone, but only if competition between trees and crop plants for light and water can be kept to a minimum. (Photos courtesy of R. Weil)

Leguminous Cover Crops to Supply Nitrogen

We have already mentioned many of the benefits of cover crops. Here we will consider the value of legume cover crops in supplying nitrogen for uptake by a subsequent non-legume crop.

Winter annual legumes, such as vetch, clovers, and peas, can be sown in fall after the main crop harvest or, if the growing season is short, they can be seeded while the main crop is still in the field. The objective is to get the legume established before the winter weather becomes too cold for the cover crop to grow. Then, in spring, the cover crop will resume growth and associated microorganisms will fix as much as 3 kg/ha of nitrogen daily during the warmer spring weather. The cover is then killed and left on the surface as a mulch in a no-till system or plowed under in a conventional tillage system. A vigorous winter cover crop can provide a significant amount of nitrogen to the main crop that follows (Figure 14.6).

A cover crop system may be able to replace part or all of the nitrogen fertilizer normally used to grow the main crop. The use of winter legume cover crops to provide nitrogen for nonlegume crops is becoming increasingly popular as more sustainable agricultural systems are developed. Such cover crop systems have been adapted for use in orchards, rice paddies (drained during the dry season), corn fields, vegetable fields, and gardens.

Crop Rotations

Growing the same crop year after year on the same land generally produces lower yields of that crop and engenders more negative impacts on the soil and environment than if that crop is grown in sequence (rotation) with other crops. Crops consistently yield 10 to 20% more in rotations than in continuous culture, especially if legumes are included in the rotation. A legume may substantially reduce the amount of nitrogen needed to grow a subsequent nonleguminous crop, even though much of the biomass

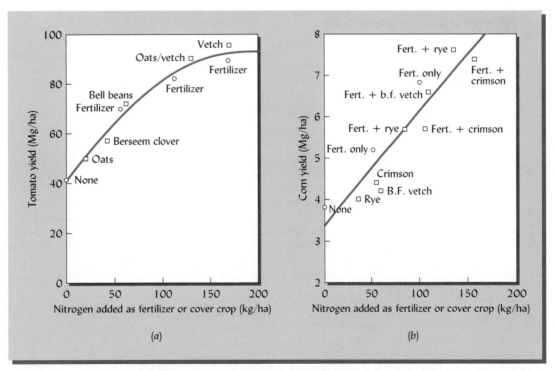

FIGURE 14.6 Comparative effects of winter cover crops and inorganic fertilizer on the yield of the following main crop. (*a*) Yield of processing tomatoes grown on a Xeralf soil in California. (*b*) Yield of corn grown on a Udalf in Kentucky. The amount of nitrogen shown on the *x*-axis was added either as inorganic fertilizer or as aboveground residues of the cover crops, or a combination of both. Note that nitrogen from either source seemed to be equally effective. For the tomato crop, which requires nitrogen over a long period of time, vetch alone or vetch mixed with oats produced enough nitrogen for near-optimal yields. [Data (left) abstracted from Stivers, et al. (1993) © Lewis Publishers, an imprint of CRC Press, Boca Raton, Florida, and (right) Ebelhar, et al. (1984)]

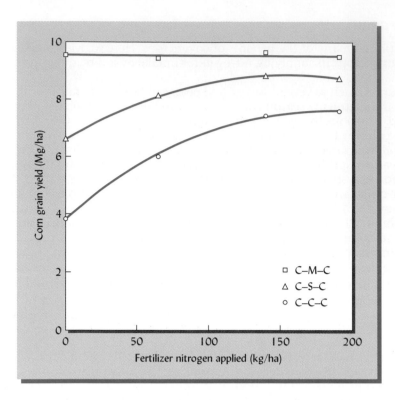

FIGURE 14.7 The nitrogen fertilizer requirements of corn are usually reduced by growing corn in rotation with legumes. This is one reason why the cropping history of a field is a consideration in deciding the optimum amount of fertilizer nitrogen to apply to cereal crops. The cropping sequences represented are: corn-corn-corn (C-C-C), corn-meadow-corn (C-M-C) and corn-soybean-corn (C-S-C). The meadow in the rotations consisted of several years of mixed alfalfa/brome grass. The data represent results on a variety of Iowa soils, but are typical of observations in many parts of the world. Less nitrogen should be applied to corn grown after soybeans than corn grown after corn. Corn following several years of mixed legume meadow is unlikely to respond significantly to nitrogen application. Rotation benefits other than nitrogen supply apparently account for the fact that, regardless of the amount of nitrogen fertilizer applied, corn grown after corn produced less yield than corn grown in rotation. [Based on data of A. M. Blackmer and J. R. Webb as reported in National Research Council (1993)]

and accumulated nitrogen in the legume are removed with the crop harvest (Figure 14.7). Perennial forage legumes, such as alfalfa, tend to produce the greatest effects in this regard.

Protection from Wildfires

Fires convert a great deal of nitrogen and sulfur, and some phosphorus, to gaseous forms in which they are lost from the site. Burned-over land also loses more phosphorus in runoff for several years after a high-intensity burn (see Table 13.2). Ashes, both from prescribed burns and from wildfires, contain high levels of soluble K, Mg, Ca, and P, increasing the short-term availability of these nutrients, but also increasing the rate of loss of these nutrients from the forest ecosystem. Wildfires result in much greater nutrient losses, because the high heat associated with these fires destroys some of the soil organic matter as well as the aboveground biomass. Farmers who burn crop residues after harvest cause similar losses from their fields.

Another nutrient aspect of forest fires is the fact that fire retardants used to fight wildfires are mostly fertilizer-type materials. Diammonium phosphate (DAP), monoammonium phosphate (MAP), and ammonium sulfate (AS) are widely used as fire retardants. The nitrogen and phosphorus in these chemicals may assist rapid regrowth in burned-over areas. However, they may also pose a water-quality threat when lost to streams because of the low rate of plant uptake and increased runoff that occur immediately after most of the vegetation has been destroyed by an intense fire.

Borates are also used as fire suppressants. Levels of boron could be phytotoxic for several years, possibly retarding the growth of plants attempting to revegetate the burned-over, borate-treated area.

14.4 RECYCLING NUTRIENTS THROUGH ANIMAL MANURES

For centuries, the use of farm manure has been synonymous with a successful and stable agriculture. Not only does manure supply organic matter and plant nutrients to the soil, but it is associated with animal agriculture and with forage crops, both of which are soil protecting and conserving.

Huge quantities of farm manure are available each year for possible return to the land. For each kilogram of liveweight of farm animals, about 4 kg dry weight of manure is produced in a year. In the United States the farm animal population voids about 10 times as much manure solids as does the human population.

These nutrient-laden organic manures provide remarkable opportunities for the recycling of essential elements. Unfortunately, there have emerged in the United States and other industrial countries intensified animal production "factories" where manure *disposal* as a problem tends to overshadow manure *utilization* (Figure 14.8). Tens or even hundreds of thousands of animals are concentrated in pens or feedlots, and the disposal of their wastes is a challenge. Furthermore, while the manure is underfoot of the animals and after it is moved to the side in huge piles, much of its nitrogen is lost to the atmosphere as ammonia or through denitrification. Some is also lost through the leaching of nitrates. The groundwater under such feed lots is often polluted with nitrates and pathogens, and is unfit to drink.

To visualize the enormity of the manure disposal problem, consider a 50,000-head beef feedlot that produces some 90,000 Mg annually, after considerable decomposition and loss of urine. If this were to be spread on land at the conventional rate of 15 Mg/ha (7 tons/acre), which would provide about 150 kg N/ha, some 6000 ha (14,815 acres) would be needed. To find this much cropland, the manure would likely have to be hauled 6 to 10 km (3.7 to 6 miles) from the feedlot. To save costs of transport, the manure commonly is applied at higher than needed rates, resulting in salinity damage to crops and soils and the loss of nitrogen and phosphorus to surface and groundwater. Over time, about twice this much land would be needed because manure application rates should be reduced to account for continuing nitrogen release from previous years' applications.

Nutrient Composition of Animal Manures

Since manure as it is applied in the field is a combination of feces and urine, along with bedding (litter) and spilled feed, its composition is quite variable. Therefore, one has to be cautious in interpreting general statements about the value and use of manure.

FIGURE 14.8 Aerial view of a large feedlot in Colorado where 100,000 cattle are fed on grain imported from distant farms. The inset is a closeup of a similar feedlot. It is unlikely that the feedlot managers will be able to recycle the nutrients in the manure back to the land on which the cattle feed was grown. Instead of being a valued resource, the manure in this situation may be considered to be a waste requiring disposal. The challenge is to find ways to structure agriculture in a more ecologically balanced manner. (Large photo courtesy Monfort Feed Lots, Greeley, Colo.; inset courtesy of R. Weil)

Animals excrete large portions of the nutrient elements consumed by them in their feeds. Generally, about three-fourths of the nitrogen, four-fifths of the phosphorus, and nine-tenths of the potassium ingested is voided by the animals and appears in the manures. For this reason, animal manures are valuable sources of both macro- and micronutrients (Table 14.5).

The data in Table 14.5 show that manures have a relatively low nutrient content in comparison with commercial fertilizer, and a nutrient ratio that is considerably lower in phosphorus than in nitrogen and potassium. On a dry-weight basis, animal manures contain from 2 to 5% N, 0.5 to 2% P, and 1 to 3% K. These values are one-half to one-tenth as great as are typical for modern commercial fertilizers.

However, unless it has been specially processed, manure is not spread in the dry form, but contains a great deal of water (see Table 14.5). If the fresh manure is handled as a solid and spread directly on the land (Figure 14.9), the high water content is a nuisance and adds to the expense of hauling. But if the manure is handled and applied in a liquid form or slurry, even more water must be added, making it difficult and expensive to transport the manure to distant fields where it might do the most good.

14.5 STORAGE, TREATMENT, AND MANAGEMENT OF ANIMAL MANURES

Where animal and crop production are integrated on a farm, manure handling is not too much of a problem. The animals themselves spread much of the manure while grazing. The manure from confined animals is produced in small enough quantities to be hauled daily to the fields or be stored under cover during periods when soil conditions are not favorable for spreading manure.

Where animals are concentrated in large confinement systems, the problem of manure disposal takes precedence over its utilization. Four general management systems are used to handle manure from such systems. First, the manure can be collected and *spread daily,* an ideal solution that, unfortunately, is not often used. Second, the manure may be stored and *packed in piles* where it is allowed to partially decompose before spreading. Third, the manure is stored in *aerated ponds* that are either sufficiently

FIGURE 14.9 Spreading solid dairy manure on cropland recycles nutrients efficiently but is labor-intensive and time-consuming for the farmer. Many loads of manure will have to be hauled to fertilize this field. The manure should be incorporated as soon as possible after spreading, and spreading should not be done on frozen soils. Calibration of spreaders is important to prevent unintentional overapplication of nutrients. (Photo courtesy of R. Weil)

TABLE 14.5 Commonly Used Organic Nutrient Sources: Their Nutrient Contents and Other Characteristics

Material	Percent of dry weight							g/Mg of dry weight						
	Water,[a] %	Total N	P	K	Ca	Mg	S	Fe	Mn	Zn	Cu	B	Mo	
Poultry (broiler) manure[b]	35	4.4	2.1	2.6	2.3	1.0	0.6	1,000	413	480	172	40	0.7	May contain high-C bedding, high soluble salts, arsenic, and ammonia.
Dairy cow manure	75	2.4	0.7	2.1	1.4	0.8	0.3	1,800	165	165	30	20	—	May contain high-C bedding.
Swine manure[c]	72	2.1	0.8	1.2	1.6	0.3	0.3	1,100	182	390	150	75	0.6	May contain elevated Cu levels.
Sheep manure	68	3.5	0.6	1.0	0.5	0.2	0.2	—	150	175	30	30	—	
Horse manure	63	1.4	0.4	1.0	1.6	0.6	0.3	—	200	125	25	—	—	May contain high-C bedding.
Feedlot cattle manure[d]	80	1.9	0.7	2.0	1.3	0.7	0.5	5,000	40	8	2	14	1	May contain high (15%) soluble salts.
Young rye green manure	85	2.5	0.2	2.1	0.1	0.05	0.04	100	50	40	5	5	.05	Nutrient content decreases with advanced growth stage.
Spoiled legume hay	40	2.5	0.2	1.8	0.2	0.2	0.2	100	100	50	10	1,500	3	
Municipal solid waste compost[e]	40	1.2	0.3	0.4	3.1	0.3	0.2[f]	14,000	500	650	280	60	7	May have high C/N and contain heavy metals, plastic, and glass.
Sewage sludge	80	4.5	2.0	0.3	1.5[g]	0.2	0.2	16,000[g]	200	700	500	100	15	May contain high soluble salts and toxic heavy metals.
Wood wastes	—	—	0.2	0.2	0.2	1.1	0.2	2,000	8,000	500	50	30	—	Very high C/N ratio; must be supplemented by other N.

[a] Water content given for fresh materials. Processing and storage methods may alter water content to less than 5% (heat-dried) or to more than 93% (slurry).
[b] Broiler and dairy manure composition estimated from means of approximately 800 and 400 samples analyzed by the University of Maryland manure analysis program 1985–1990.
[c] Composition of swine, sheep, and horse manure calculated from North Carolina Cooperative Extension Service Soil Fact Sheets prepared by Zublena, et al. (1993).
[d] Feedlot manure composition is based on average analysis reported in Eghball and Power (1994).
[e] Composition of municipal solid waste compost based on mean values for the products of 10 composting facilities in the United States as reported by He, et al. (1995).
[f] Sulfate-S.
[g] Sludge contents of Ca and Fe may vary 10-fold depending on the wastewater treatment processes used.
Data derived from many sources.

447

TABLE 14.6 Influence of Handling and Storage Methods on Nutrient Losses from Animal Manure

The values are ranges of percent loss of nutrient from the time the manure is excreted until it is applied to the land.

	Percentage lost		
Manure handling and storage method	*N*	*P*	*K*
Solid systems			
Daily scrape and haul	15–35	10–20	20–30
Manure pack or compost	20–40	5–10	5–10
Liquid systems			
Aerobic tank storage	5–25	5–10[a]	0–5
Lagoon (anaerobic) storage	70–80	50–80	50–80

[a] Most of the P and K in a lagoon system settles to the bottom of the lagoon and is recoverable only when the sludge in the lagoon is dredged out.
From Sutton (1994).

shallow to permit fairly ready oxidation of the organic materials, or oxygen is stirred into the slurry. Last is storage in deep *anaerobic lagoons* in which the manure is allowed to ferment in the absence of elemental oxygen. Methane and ammonia are common gaseous products of this fermentation, and considerable denitrification is known to occur. After much organic decomposition has occurred in either the ponds or the lagoons, the remaining materials, which will include some nutrients, such as nitrates, can be applied in liquid form to the soil.

Each of the four methods of handling affects the nutrient management of the manure (Table 14.6). The greatest loss of nutrients, especially nitrogen, occurs either from open-lot storage or from lagoon storage. In open-lot storage much nitrogen is volatilized as ammonia gas (see Section 12.6), and considerable quantities of nutrients may be washed away when it rains. Water and excess nutrients can pass through the lagoon walls to the groundwater and into streams and lakes, giving rise to algal blooms and eutrophication.

Methods of manure handling that both prevent pollution and preserve nutrients in a form that can be easily transported and sold commercially would make a major contribution to ameliorating the manure problem for concentrated animal production enterprises. Several options are currently being developed, including the following:

1. Heat, dry, and compress the manure into **pellets** that are bagged and sold commercially, especially for landscaping and lawn industries.
2. **Compost** the manures, and use the more concentrated products for lawns and gardens.
3. Digest the manure anaerobically to produce **biogas** that can be burned much like commercial natural gas.

In the future such technologies, along with more integrated animal farming systems, may help to redress the serious nutrient imbalances that have developed with regard to manure production in modern agriculture.

14.6 INDUSTRIAL AND MUNICIPAL BY-PRODUCTS

Four major types of organic wastes from industrial and municipal sources are of significance in land application: (1) municipal garbage, (2) sewage effluents and sludges, (3) food-processing wastes, and (4) wastes of the lumber industry. Although the primary reason for adding these wastes to land may be to safely dispose of them, they also serve as valuable soil amendments for agriculture, forestry, and disturbed-land reclamation. Once seen as mere waste products to be flushed into rivers and out to sea, these materials are increasingly seen as sources of nutrients and organic matter that can be used beneficially to promote soil productivity.

Garbage

Municipal garbage, which has traditionally been used extensively to enhance crop production in China and other Asian countries, is becoming more widely used in the United States and Europe. The inorganic glass, metals, and so forth are first removed and municipal organic wastes are composted, sometimes in conjunction with sewage sludge, poultry manure, or other nutrient-rich materials, to produce municipal solid waste (MSW) compost that is then applied to the land. Because of the very low nutrient content of MSW compost (see Table 14.5), and its far distance from the fields to which it can be applied, the transportation of a given quantity of nutrients in MSW compost is even more expensive than that in animal manures. As with manures, however, in some regions attractive products are produced and sold to landscaping and specialty users as soil amendments or planting media.

Food-Processing Wastes

Land application of food-processing wastes is being practiced in selected locations, but the practice is focused almost entirely on pollution abatement and not on crop production. Liquid wastes are commonly applied through sprinkle irrigation to permanent grass. Plant-processing schedules dictate the timing and rate of application, which may not be suitable for optimum crop production.

Sawdust

Sawdust, wood chips, and shredded bark from the lumber industry have long been sources of soil amendments and mulches, especially for home gardeners and landscapers. These wastes are high in lignin and related materials, have very high C/N ratios, and therefore decompose very slowly—desirable properties in a mulch material. However, they do not readily supply plant nutrients. In fact, sawdust incorporated into soils to improve soil physical properties may cause plants to become nitrogen deficient unless an additional source of nitrogen is applied with the sawdust (see Section 11.3).

Recycling by Composting

As the environmental problems associated with the use of landfill sites for organic wastes have mounted, interest in composting has increased significantly. Domestic, garden, community, and even some industrial wastes are combined in compost piles with materials such as sawdust and manures. The number of composting facilities increased more than fourfold between 1989 and 1996, and their products are being used particularly in home gardens.

14.7   SEWAGE EFFLUENT AND SLUDGE

Sewage treatment has evolved over the past century in response to society's desire (and resultant regulations) to avoid polluting our rivers and oceans with pathogens, oxygen-demanding organic debris, and eutrophying nutrients. The amount of material *removed* from the wastewater during the treatment process has increased tremendously. This solid material, known as *sewage sludge* or **biosolids,** must also be disposed of safely. A goal of advanced wastewater treatment is to remove nutrients (mainly phosphorus, but increasingly also nitrogen) from the **effluent** (the treated water that is returned to the stream), a job that can be accomplished safely and economically by allowing the partially treated effluent to interact with a soil–plant system. Therefore, there is a growing interest in using soils to assist with the sewage problem in two ways: (1) as a system of assimilating, recycling, or disposing of the solid sludge; and (2) as a means of carrying out the final removal of nutrients and organics from the liquid effluent.

Sewage Effluent

Sewage effluents have been applied to land for decades in Europe and in selected sites in the United States. A number of cities around the world use nutrient-rich sewage effluent in the irrigation of forest land. Forest irrigation is a cost-effective method of final effluent cleanup and produces enhanced tree growth as a bonus. The rate of wood production is greatly increased as a result of both the additional water and the additional nutrients supplied therewith.

In a carefully planned and managed effluent irrigation system, the combination of (1) nutrient uptake by the plants, (2) adsorption of inorganic and organic constituents by soil colloids, and (3) degradation of organic compounds by soil microorganisms results in the purification of the wastewater.

Sewage Sludge

Sewage sludge (sometimes called **biosolids**) is the solid by-product of domestic and/or industrial wastewater treatment plants. It has been spread on the land for decades, and its use will likely increase in the future. The product Milorganite, a dried sludge sold by the Milwaukee Sewerage Commission, has been used widely in North America since 1927, especially on turf grass. However, the great bulk of sewage sludge used on land is applied as a liquid slurry or as a partially dried cake. Further drying and processing decreases the cost of transportation, an important consideration since large cities have difficulty finding sufficient nearby land to receive their biosolids. One program *daily* ships over 600,000 kg of dried, treated sludge by rail from New York City to western Texas, a distance of over 2000 km.

COMPOSITION OF SEWAGE SLUDGE. Representative values for plant nutrients are given in Table 14.5. Note that, in comparison with organic fertilizers, sludge is generally low in macronutrients. Potassium contents are very low since this element is quite soluble and is easily washed away in the sewage effluent. Because phosphorus is commonly combined with iron and aluminum in sludge, its availability for plant uptake is very low.

Sludges are generally well supplied with micronutrients. In fact, micronutrients and other heavy metals (cadmium, chromium, lead, etc.) can be present at toxic levels in some sludges containing significant industrial waste inputs. Luckily, some reduction programs in the United States and other industrialized countries now require industrial facilities to remove such pollutants *before* sending their sewage to municipal treatment plants.

14.8 PRACTICAL UTILIZATION OF ORGANIC NUTRIENT SOURCES

In Section 11.7 we discussed the many beneficial effects on soil physical and chemical properties that can result from amendment of soils with decomposable organic materials, such as manure or sludge. Here we will focus on the nutrient management aspects of organic waste utilization. Whether the material is sewage sludge, farm manure, or MSW compost, several general principles apply to the ecologically sound application of the material to soils:

1. The rate of application of the organic material is generally governed by its content of nitrogen, the element needed in the largest quantity by most plants. However, since the P/N ratio of most organic sources is higher than in plant tissue, applying sufficient nitrogen to meet plant needs may result in excessive levels of soil phosphorus. Also, contents of heavy metals in some materials may limit the rate of application.

2. Most of the nitrogen in organic sources is not immediately available to plants, as it is with most inorganic commercial fertilizers (see Section 14.9). Table 14.7 lists for a number of materials the percentage of the organic nitrogen that is likely to be released in the first, second, and third years after application. Note that those materials that have been partially decomposed during treatment and handling (e.g., composts and digested sludge) release a lower percentage of the nitrogen. The release rates of other nutrients are less critical because their availability from the organic sources is generally comparable to that from commercial fertilizers.

3. If a field is treated annually with an organic material, the application rate needed will become progressively smaller because, after the first year, the amount of nitrogen released from material applied in previous years must be subtracted from the total to be applied afresh (see Box 14.1, calculations for organic nitrogen application rates).

4. The nutrient and moisture contents of organic amendments vary widely among sources, and even from batch to batch, depending on how the material has been stored and treated. Therefore, general values such as those in Table 14.5 should not be relied upon when calculating rates of material to apply. Instead, represen-

TABLE 14.7 **Rates of Release of Mineral Nitrogen from Various Sources of Organic Nitrogen Applied to Soils**

The values given are percent of the organic nitrogen present in the original material. For example, if 10 Mg of poultry litter initially contains 300 kg (3.0%) nitrogen in organic forms, 50% or 150 kg of nitrogen would be mineralized in the first year. Another 15% (0.15 × 300) or 45 kg of nitrogen would be released in the second year. These values are approximate and may need to be increased for warm climates or sandy soils.

Organic nitrogen source	First year	Second year	Third year	Fourth year
Poultry floor litter	50	15	8	3
Dairy manure (fresh solid)	35	18	9	4
Swine manure lagoon liquid	50	15	8	3
Lime-stabilized, aerobically digested sewage sludge	40	12	5	2
Anaerobically digested sewage sludge	20	8	4	1
Composted sewage sludge	10	5	3	2
Activated, unstabilized sewage sludge	45	15	4	2

tative samples of the material should be analyzed in a laboratory. Analyses are generally required by laws regulating land application of sewage sludge, but other materials are not as widely regulated in this regard.

5. Nutrients from organic sources will produce the highest return and cause the least environmental damage if applied to fields that are relatively poorly supplied with nitrogen and phosphorus. Unfortunately, cost of transport and considerations of convenience provide incentives to apply organic materials on land close to their source, often leading to the nutrient imbalances discussed in Section 14.4.

14.9 INORGANIC COMMERCIAL FERTILIZERS

The worldwide use of fertilizers increased dramatically during the latter half of the 20th century (Figure 14.10), accounting for a significant part of the equally dramatic increases in crop yields during the same period. Improved soil fertility through the application of fertilizer nutrients is an essential factor in enabling the world to feed the billions of people that are being added to its population.

The need to supplement forest soil fertility is increasing as the demands for forest products increase the removal of nutrients, and competing uses of land leave forestry with more infertile, marginal sites. Currently, most fertilization in forestry is concentrated on tree nurseries and on seed trees, where the benefits of fertilization are relatively short term and of high value, and the logistics of fertilizer application are not so difficult and expensive. In the warmer, more humid regions of Japan,[5] about 50% of new forest plantings are fertilized. Forest fertilization is also increasingly practiced in the southeastern United States, the Scandinavian countries, and Australia.

Regional Use of Fertilizers

Worldwide fertilizer statistics provide a perspective on the role of fertilizers in maintaining a global balance between nutrient inputs and outflows. But it is the more specific usage within a region that tells more definitively whether fertilizers are contributing to soil quality enhancement or to environmental degradation. For example, in Europe and East Asia where moisture is abundant and intensified cropping is common, fertilizer nutrient application rates are nearly triple the world average. In contrast, soils in sub-Saharan Africa are literally being mined—far less nutrients are being added from all sources than are being removed in the crops. The fertilizer nutrient rate in this region is only about 10% of the world average. Thus, it is obvious that we must be as site specific

---

[5] In Japan the climate is capable of supporting rapid growth of the more valuable timber species, such as sugi—*Cryptomeria japonica* and *Quercus acustissima*—whose logs are in demand to grow shiitake mushrooms.

BOX 14.1 CALCULATION OF AMOUNT OF ORGANIC NUTRIENT SOURCE NEEDED TO SUPPLY NITROGEN FOR A CROP

This example is based on the principle that the amount of nitrogen made available in any year should meet, but not exceed, the amount of nitrogen that plants can use for optimum growth. Therefore, the rate of release of available nitrogen usually determines the proper amount of manure, sludge, or other organic nutrient source to apply. Our example is a field producing corn two years in a row. The goal is to produce 7000 kg/ha of grain each year. This yield normally requires the application of 120 kg/ha of available nitrogen, which we expect to obtain from a lime-stabilized sewage sludge containing 4.5% total N and 0.2% mineral N (ammonium and nitrate).

Year 1

Calculation of amount of sludge to apply:

% year organic N in sludge = total N − mineral N = 4.5% − 0.2% = 4.3%.

Organic N in 1 Mg of sludge = 0.043×1000 kg = 43 kg N.

Mineral N in 1 Mg of sludge = 0.002×1000 kg = 2 kg N.

Mineralization rate for lime-stabilized sludge in first year (Table 14.7) = 40% of organic N.

Available N mineralized from 1 Mg sludge in first year = 0.40×43 kg N = 17.2 kg N.

Total available N from 1 Mg sludge = mineral N + mineralized N

$$= 2.0 + 17.2 = 19.2 \text{ kg N.}$$

Amount of (dry) sludge needed = 120 kg N/(19.2 kg available N/Mg dry sludge)

$$= 120/19.2 \text{ Mg dry sludge} = 6.25 \text{ Mg dry sludge}$$

Adjust for moisture content of sludge (e.g., assume sludge has 25% solids and 75% water).

Amount of wet sludge to apply: 6.25 Mg dry sludge/(0.25 Mg dry sludge/Mg wet sludge) = 6.25/0.25 = 25 Mg wet sludge.

Year 2

Calculate amount of N mineralized in year 2 from sludge applied in year 1:

Second year mineralization rate (Table 14.7) = 12% of original organic N.

N mineralized from sludge in year 2 = 0.12×43 Kg N/Mg $\times 6.25$ Mg dry sludge

$$= 32.25 \text{ kg N from sludge in second year.}$$

N needed from sludge applied in year 2 = N needed by corn − N from sludge applied in year 1

$$= 120 \text{ kg} - 32.25 \text{ kg} = 87.75 \text{ kg N needed/ha.}$$

Amount of (dry) sludge needed per ha = 87.75 kg N/(19.2 kg available N/Mg dry sludge) = 87.75/19.2 = 4.57 Mg dry sludge/ha.

Adjust for moisture content of sludge (e.g., assume sludge has 25% solids and 75% water).

Mg of wet sludge to apply: 4.57 Mg dry sludge/(0.25 Mg dry sludge/Mg wet sludge) = 4.57/0.25 = 18.3 Mg wet sludge

Note that the 10.82 Mg sludge (6.25 + 4.57) also provided plenty of P: 216 kg P/ha (assuming 2% P; see Table 14.5), an amount that greatly exceeds the crop requirement and will lead to excessive buildup of P.

---

as possible in evaluating the role of fertilizers in meeting humanitarian and environmental goals.

Origin and Processing of Inorganic Fertilizers

Most fertilizers are inorganic salts containing readily available plant nutrient elements. Some are manufactured, but others, such as phosphorus and potassium, are found in natural geological deposits. **Beds of solid salts** located far beneath the earth's surface are the primary sources of potassium. These underground deposits are found in many locations, including sites in Canada, France, Germany, and Russia, as well as in New Mexico. The salts are mined and then purified, yielding such compounds as potassium chloride and potassium sulfate.

Apatite found in **phosphate rock deposits** is the primary source of phosphorus fertilizers. Since apatite is extremely insoluble, however, it is treated with sulfuric, phosphoric, or nitric acid to produce materials, such as triple superphosphate, that yield phosphorus in a readily available form. Phosphate rock deposits are located around the world; among

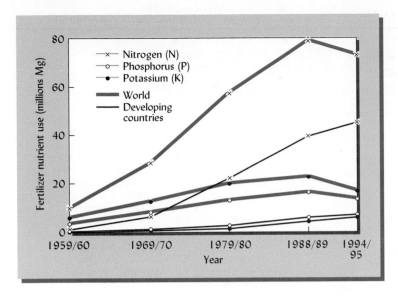

FIGURE 14.10 World and developing country fertilizer use since 1959–1960 by major nutrient element. Note the very high use of nitrogen compared to the other two nutrients. World fertilizer use declined significantly during the 1990s, due primarily to drastic use reductions in nations of the former Soviet Union and to some cutbacks in other industrialized countries. Fertilizer use continued to rise in the developing countries, however, since nutrient removal in crops far exceeds return of nutrients to the soil in many of these countries. [Data from the United Nations Food and Agriculture Organization as reported in Bumb and Baanante (1996)]

the most extensive are found in North Africa and in Florida, North Carolina, Tennessee, and several western states.

Nitrogen in the **atmosphere** is the primary original source of this essential element. Under very high temperatures and pressures, the nitrogen is fixed with hydrogen from natural gas to produce ammonia gas, NH_3. Although this fixation is of vital importance, it consumes tremendous quantities of energy—as critics have pointed out. When the ammonia gas is then put under moderate pressure, it liquifies, forming anhydrous ammonia. Because of its low cost and ease of application, more nitrogen is applied directly to soils in anhydrous ammonia than in any other fertilizer. More important, it is the starting point for the manufacture of most of the other nitrogen carriers, including urea, ammonium nitrate, ammonium sulfate, sodium nitrate, and liquid mixtures known as "nitrogen solutions." The commercialization of the process of nitrogen fixation revolutionized the recycling of nitrogen. Unfortunately, the reduced cost of this element and the variety of materials in which it can be supplied has sometimes encouraged its use in excess of plant needs.

Many commercial fertilizers contain two or more of the primary nutrient elements. In some cases, mixtures of the primary nutrient carriers are used. But in others, a given compound may carry two nutrients, examples being monoammonium phosphate, diammonium phosphate, and potassium nitrate.

Properties and Use of Inorganic Fertilizers

The nutrient element contents and other properties of some of the more commonly used inorganic fertilizers are shown in Table 14.8. In most cases, fertilizers are used to supply plants with the macronutrients nitrogen, phosphorus, and/or potassium—sometimes called the *primary fertilizer elements*. Fertilizers that supply sulfur, magnesium, and the micronutrients are also manufactured.

Note in Table 14.8 that a particular nutrient (say, nitrogen) can be supplied by many different *carriers,* or fertilizer compounds. Table 14.8 also provides information about some of their characteristics, such as salt hazard, acid-forming tendency (see also Section 12.7), tendency to volatilize, ease of solubility, and content of nutrients other than the principal one. Of the nitrogen carriers, anhydrous ammonia, nitrogen solutions, and ureas are the most widely used. Diammonium phosphate and potassium chloride supply the bulk of the phosphorus and potassium used in the United States.

Physical Forms of Marketed Fertilizer

A generation ago, fertilizers were handled primarily in bags. Today less than 10% is handled in this manner, the remainder being marketed as unbagged, dry solids handled in **bulk** form, or as **liquid** or fluid forms. In both cases the cost of the product applied in the field has been reduced.

Bulk spreading, often done at times when trucks or large fertilizer spreaders can get on the land, is the favorite for applying multinutrient fertilizers such as diammonium phosphate.

TABLE 14.8 Commonly Used Inorganic Fertilizer Materials: Their Nutrient Contents and Other Characteristics

Fertilizer	Percent by weight							Salt hazard	Acid-forming tendency[b], kg CaCO$_3$/100 kg	Comments
	N	P	K	S	Ca	Mg	Micro-nutrients			
Primarily sources of nitrogen										
Anhydrous ammonia (NH_3)	82							Low	−148	Pressurized equipment needed; toxic gas.
Urea [$CO(NH_2)_2$]	45							Moderate	−84	Soluble; urease enzyme rapidly hydrolyses urea to ammonium forms.
Ammonium nitrate (NH_4NO_3)	33							High	−59	Absorbs moisture from air; can explode.
Sulfur-coated urea	30–40			13–16				Low	−110	Variable slow rate of release.
UF (ureaformaldehyde)	30–40							Very low	−68	Slowly soluble.
UAN solution	30							Moderate	−52	Most commonly used liquid N.
IBDU (isobutylidene diurea)	30							Very low	—	Slowly soluble.
Ammonium sulfate [$(NH_4)_2SO_4$]	21			24				High	−110	Rapidly lowers soil pH; very easy to handle.
Sodium nitrate ($NaNO_3$)	16						About 0.6% Cl	Very high	+29	Hardens, disperses soil structure.
Potassium nitrate (KNO_3)	13		36	0.2	0.4	0.3	1.2% Cl	Very high	+26	Very rapid plant response.
Primarily sources of phosphorus										
Monoammonium phosphate ($NH_4H_2PO_4$)	11	21–23		1–2				Low	−65	Best as starter.
Diammonium phosphate [$(NH_4)_2HPO_4$]	18–21	20–23		0–1				Moderate	−70	Best as starter.
Triple superphosphate		19–22		1–3	15			Low	0	
Phosphate rock [$Ca_3(PO_4)_2·CaX$]		8–18[a]			30			Very low	Variable	Low to extremely low availability. Best as fine powder on acid soils. Contains some Cd, F, etc.
Single superphosphate		7–9		11	20			Low	0	Nonburning, can place with seed.
Bonemeal	1–3[a]	10[a]	0.4		20			Very low	—	Slow availability of N, P as for phosphate rock.
Colloidal phosphate		8[a]			20			Very low	—	P availability as for phosphate rock.
Primarily sources of potassium										
Potassium chloride (KCl)			50				47% Cl	High	0	Cl may reduce some diseases.
Potassium sulfate (K_2SO_4)			42	17	0.7	1.2		Moderate	0	Use where Cl not desirable.
Wood ashes	0.5–1	1–4	10–20	2–5			0.2% Fe, 0.8% Mn, 0.05% Zn, 0.005% Cu, 0.03% B	Moderate to high	+40	About 2/3 the liming value of limestone; caustic.
Greensand	0.6		6					Very low	0	Very low availability.
Granite dust			4					Very low	0	Very slow availability.

TABLE 14.8 (Cont.)

Fertilizer	N	P	K	S	Ca	Mg	Micro-nutrients	Salt hazard	Acid-forming tendency[b], kg CaCO₃/ 100 kg	Comments
							Primarily sources of other nutrients			
Basic slag		1–7			3–30	3	10% Fe, 2% Mn	Low	+70	Industrial by-product; slow availability; best on acid soils.
Gypsum (CaSO₄·2H₂O)				19	23			Low	0	Stabilizes soil structure; no effect on pH; Ca and S readily available.
Calcitic limestone (CaCO₃)					36			Very low	+95	Slow availability; raises pH.
Dolomitic limestone [CaMg(CO₃)₂]					24	12		Very low	+95	Very slow availability; raises pH.
Epsom salts (MgSO₄·7H₂O)				13	2	10		Moderate	0	No effect on pH; water soluble.
Sulfur, flowers (S)				95				—	−300	Irritates eyes; very acidifying; slow acting; requires microbial oxidation.
Solubor							20.5% B	Moderate	—	Very soluble; compatible with foliar sprays.
EDTA chelate							13% Cu or 10% Fe or 2% Mn or 12% Zn	—	—	One or more micro-nutrients; see label.
Borax (Na₂B₄O₇·10H₂O)							11% B	Low	—	Very soluble; compatible with foliar sprays.
Copper sulfate (CuSO₄·5H₂O)							25% Cu	Low	—	Very soluble; compatible with foliar sprays.
Ferrous sulfate (FeSO₄·7H₂O)							19% Fe	Low	—	Very soluble; compatible with foliar sprays.
Manganese sulfate (MnSO₄·3H₂O)							27% Mn	Low	—	Very soluble; compatible with foliar sprays.
Sodium molybdate (Na₂MoO₄·2H₂O)							39% Mo	Low	—	Very soluble; compatible with foliar sprays.
Zinc Sulfate (ZnSO₄·H₂O)							35% Zn	Low	—	Very soluble; compatible with foliar sprays.

[a] Highly variable contents.
[b] A negative number indicates that acidity is produced; a positive number indicates that alkalinity is produced.
Data derived from many sources.

Liquid fertilizers such as anhydrous ammonia comprise more than half of the single-nutrient carriers sold in the United States, and about 40% of all fertilizers. Labor costs are low for applying either bulk or liquid fertilizers since mechanical means are used to pump or load the fertilizer onto the delivery trucks, transport it to the field, and apply it to the soil. These changes in marketing, which have made it easier and less expensive to apply fertilizer, have encouraged fertilizer usage, sometimes at the expense of the quality of the environment.

Fertilizer Grade

The laws and conventions governing the manufacture and use of chemical fertilizers had their origin in the late 19th and early 20th centuries. One of these conventions that prevails today is that of the *fertilizer grade,* which is expressed as a three-number code, such as 0-15-15 or 6-24-24. These numbers stand for percentages indicating the *total* nitrogen (N) content, the *available* phosphoric acid (P_2O_5) content, and the *soluble* potash (K_2O) content (see Box 14.2). We must remember, however, that plants do not take up phosphorus or potassium in these chemical forms, nor do any fertilizers actually contain P_2O_5 or K_2O.[6]

---

[6] The oxide means of expressing phosphorus (P_2O_5) and potassium (K_2O) is a relic of the days when geochemists reported the contents of rocks and minerals in terms of the oxides formed upon heating. In all scientific work and in this textbook, the simple elemental contents are used (P and K) wherever possible.

BOX 14.2 HOW MUCH NITROGEN, PHOSPHORUS, AND POTASSIUM IS IN A BAG OF 6-24-24?

Conventional labeling of fertilizer products reports percentage N, P_2O_5, and K_2O. Thus, a fertilizer package (Figure 14.11) labeled as 6-24-24 (6% nitrogen, 24% P_2O_5, 24% K_2O) actually contains 6% N, 10.6% P, and 19.9% K (see calculations below).

6-24-24

GUARANTEED ANALYSIS

TOTAL NITROGEN (N) 6.0%
AVAILABLE PHOSPHORIC ACID (P_2O_5) . . 24.0%
SOLUBLE POTASH (K_2O) 24.0%

Potential acidity equivalent to
300 lbs. Calcium Carbonate per ton.

FIGURE 14.11 A typical commercial fertilizer label. Note that a calculation must be performed to determine the percentage of the nutrient elements P and K in the fertilizer since the contents are expressed as if the nutrients were in the forms of P_2O_5 and K_2O. Also note that after interacting with the plant and soil, this material would cause an increase in soil acidity that could be neutralized by 300 units of $CaCO_3$ per 2000 units (1 ton = 2000 lbs) of fertilizer material.

To determine the amount of fertilizer needed to supply the recommended amount of a given nutrient, it is necessary to convert percent P_2O_5 and percent K_2O to percent P and K. This may be done by calculating the proportion of P_2O_5 that is P and the proportion of K_2O that is K. Using the atomic masses listed in Appendix C, we can calculate the proportions:

$$\text{Proportion of P in } P_2O_5 = \frac{2P}{P_2O_5} = \frac{2(30.97)}{2(30.97) + 5(16)} = \frac{61.94}{141.94} = 0.44$$

$$\text{Proportion of K in } K_2O = \frac{2K}{K_2O} = \frac{2(39.1)}{2(39.1) + 16} = \frac{78.2}{94.2} = 0.83$$

We now can convert the oxide percentages to the elemental percentages by multiplying by these proportions.

24% $P_2O_5 \times 0.44 = 10.6\%$ P
24% $K_2O \times 0.83 = 19.9\%$ K

Thus, the elemental nutrient percentage of this fertilizer (N-P-K) is 6-10.6-19.9.

The grade is important from an economic standpoint because it conveys the analysis or concentration of the nutrient in a carrier. The more concentrated carriers are usually the most economical to use, because less weight of fertilizer must be transported to supply the needed quantity of a given nutrient.

Fate of Fertilizer Nutrients

A common *myth* about fertilizers suggests that inorganic fertilizers applied to soil directly feed the plant, and that therefore the biological cycling of nutrients, such as described by Figures 12.1 for nitrogen and 13.5 for phosphorus, is of little consequence where inorganic fertilizers are used. The reality is that nutrients added by normal application of fertilizers, whether organic or inorganic, are incorporated into the complex soil nutrient cycles, and that relatively little of the fertilizer nutrient (from 10 to 60%) actually winds up in the plant being fertilized during the year of application (Table 14.9). Even when the application of fertilizer greatly increases both plant growth and nutrient uptake, the fertilizer stimulates increased cycling of the nutrients, and the nutrient ions taken up by the plant come largely from various pools in the soil and not directly from the fertilizer. Generally, as fertilizer rates are increased, the efficiency of fertilizer nutrient use decreases, leaving behind in the soil an increasing proportion of the added nutrient (Table 14.9).

14.10 THE CONCEPT OF THE LIMITING FACTOR

A famous German chemist named Justus von Liebig is credited with first publishing the concept that *plant production can be no greater than that level allowed by the growth factor present in the lowest amount relative to the optimum amount for that factor.* This growth factor, be it temperature, nitrogen, or water supply, will limit the amount of growth that can occur and is therefore called the *limiting factor* (Figure 14.12).

If a factor is not the limiting one, increasing it may actually reduce plant growth by throwing the system further out of balance. For example, if a plant is limited by lack of phosphorus, adding more nitrogen may only aggravate the phosphorus deficiency.

Looked at another way, applying available phosphorus (the first limiting nutrient in this example) may allow the plant to respond positively to a subsequent addition of nitrogen. Thus, the increased growth obtained by applying two nutrients together often is much greater than the sum of the growth increases obtained by applying each of the two nutrients individually. Such an *interaction* or *synergy* between two nutrients can be seen in the data of Figure 14.13.

TABLE 14.9 **Source of Nitrogen in Corn Plants Grown in North Carolina on an Enon Sandy Loam Soil (Ultic Hapludalf) Fertilized with Three Rates of Nitrogen as Ammonium Nitrate**

The source of the nitrogen in the corn plant was determined by using fertilizer tagged with the isotope ^{15}N. Note that moderate use of nitrogen fertilizer increased the uptake of N already in the soil system as well as that derived from the fertilizer.

Fertilizer nitrogen applied, kg/ha	Corn grain yield, Mg/ha	Total N in corn plant, kg/ha	Fertilizer-derived N in corn, kg/ha	Soil-derived N in corn, kg/ha	Fertilizer-derived N in corn as percent of total N in corn, %	Fertilizer-derived N in corn as percent of N applied, %
50	3.9	85	28	60	33	56
100	4.6	146	55	91	38	55
200	5.5	157	86	71	55	43

Calculated from Reddy and Reddy (1993).

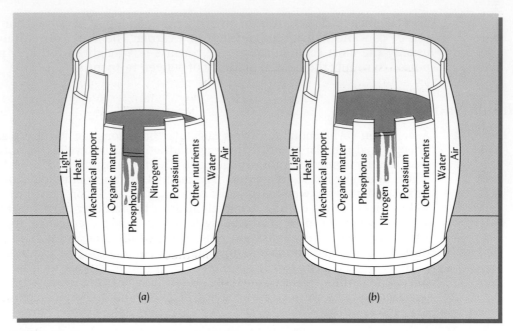

FIGURE 14.12 An illustration of the law of the minimum and the concept of the limiting factor. Plant growth is constrained by the essential element (or other factor) that is most limiting. The level of water in the barrel represents the level of plant production. (*a*) Phosphorus is represented as being the factor that is most limiting. Even though the other elements are present in more than adequate amounts, plant growth can be no greater than that allowed by the level of phosphorus available. (*b*) When phosphorus is added, the level of plant production is raised until another factor becomes most limiting—in this case, nitrogen.

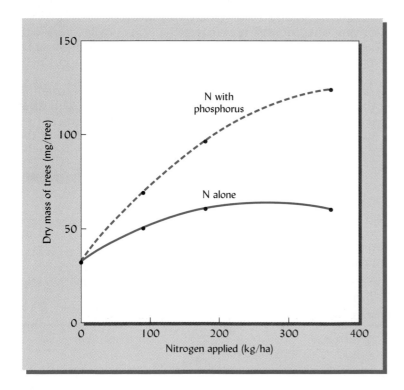

FIGURE 14.13 Increase in biomass of one-year-old white pine seedlings in response to nitrogen fertilizer given with or without phosphorus on a sandy soil in southern Ontario, Canada. Apparently, phosphorus was the most limiting nutrient, and therefore little response was obtained from adding nitrogen until the phosphorus level was raised. This is an example of a nitrogen-by-phosphorus interaction. [Redrawn from Teng and Timmer (1994)]

14.11   FERTILIZER APPLICATION METHODS

Wise, effective fertilizer use involves making correct decisions regarding *which* nutrient element(s) to apply, *how much* of each needed nutrient to apply, *what type* of material or carrier to use (Table 14.8 lists some of the choices), *in what manner* to apply the material, and, finally, *when* to apply it. We will leave information on the first two decisions until Section 14.13. Here we will discuss the alternatives available with regard to the last two decisions.

There are three general approaches to applying fertilizers: (1) *broadcast application,* (2) *localized placement,* (3) and *foliar application.* Each method has some advantages and disadvantages and may be particularly suitable for different situations. Often some combination of the three methods is used.

Broadcasting

In many instances fertilizer is *broadcast* evenly over the entire field or area to be fertilized (Figure 14.14*a, b,* and *c*). The broadcast method is most appropriate when a large amount of fertilizer is being applied with the aim of raising the fertility level of the soil over a long period of time. Broadcasting is the most economical way to spread large amounts of fertilizer over wide areas (Figure 14.15), and is commonly used on close-growing vegetation such as lawns, rangeland, pastures, and small grains.

For phosphorus, zinc, manganese, and other nutrients that tend to be strongly bound by the soil, broadcast applications are usually much less efficient than localized placement. Often 2 to 3 kg of fertilizer must be broadcast to achieve the same response as from 1 kg that is placed in a localized area. While nitrogen does not suffer from reduced availability when broadcast, some of it may be lost by volatilization.

Another disadvantage of surface broadcasting is that the fertilizer is easily washed away with runoff during heavy rains. In fact, many runoff studies have shown that most of the annual loss of nutrients (or herbicides, if surface broadcast) usually occurs during the first one or two heavy-rainfall events after the broadcast application. Finally, it should be pointed out that for crops with wide row spacing or young tree seedlings in forest plantings, broadcasting places the fertilizer where it is just as accessible to the *weeds* as to the target plants.

APPLICATION IN IRRIGATION WATER.   Where irrigation is practiced, liquid fertilizers can be applied in the irrigation water, a practice sometimes called *fertigation.* Liquid ammonia, nitrogen solutions, phosphoric acid, and even complete fertilizers are dissolved in the irrigation stream or the overhead sprinkler system supply. The nutrients are thus broadcast onto the soil in solution and carried down with the infiltrating water. Some care must be taken, however, to prevent ammonia loss by evaporation and to avoid certain fertilizer compounds that can clog the irrigation system with precipitates.

Localized Placement

There are at least two reasons why fertilizer is often more effectively used by plants if it is placed in a localized concentration rather than mixed with soil throughout the root zone. First, localized placement reduces the amount of contact between soil particles and the fertilizer nutrient, thus minimizing the opportunity for adverse fixation reactions. Second, the concentration of the nutrient in the soil solution at the root surface will be very high, resulting in greatly enhanced uptake by the roots.

STARTER APPLICATION.   Localized placement is especially effective for young seedlings, in cool soils in early spring, and for plants that grow rapidly with a big demand for nutrients early in the season. For these reasons, starter fertilizer is often applied in bands on either side of the seed as the crop is planted. The best placement for starter fertilizer is approximately 5 cm below and 5 cm off to the side from the seed row (see Figure 14.14*d* and Figure 14.16).

LIQUID FERTILIZERS.   Liquid fertilizers and slurries of manure and sewage sludge can also be applied in bands rather than broadcast. Bands of these liquids are placed 10 to 30 cm deep in the soil by a process known as *knife injection* (Figures 14.14*e*). Anhydrous ammonia and pressurized nitrogen solutions must be injected into the soil to prevent losses

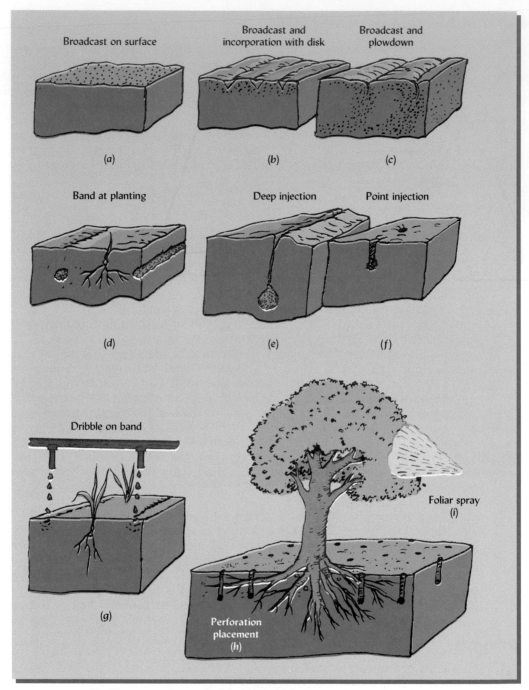

FIGURE 14.14 Fertilizers may be applied by many different methods, depending on the situation. Methods (*a*) to (*c*) represent broadcast fertilizer, with or without incorporation. Methods (*d*) to (*h*) are variations of localized placement. Method (*i*) is foliar application and has special advantages, but also limitations. Commonly, two or three of these methods may be used in sequence. For example, a field may be prepared with (*c*) before planting; (*d*) may be used during the planting operation; (*g*) may be used as a side dressing early in the growing season; and, finally, (*i*) may used to correct a micronutrient deficiency that shows up in the middle of the season.

by volatilization. Injecting bands at depths of 15 and 5 cm, respectively, are considered adequate for these two materials.

DRIBBLE APPLICATION.   Another approach to banding liquids is to dribble a narrow stream of liquid fertilizer alongside the crop row as a side dressing (Figure 14.14*g*). The fertilizer soaks into the soil, thereby reducing volatilization loss of nitrogen.

(a)

(b)

FIGURE 14.15 (a) Broadcasting granular phosphorus and potassium fertilizer on a hayfield in Maryland. (b) Broadcasting a nitrogen solution (mixed with fungicide) on a wheat field in France. (Photos courtesy of R. Weil)

POINT INJECTION. A new system called *point injection* can be used to apply small portions of liquid fertilizer next to every individual plant without significantly disturbing either the plant root or the surface residue cover left by conservation tillage (Figure 14.14*f*).

DRIP IRRIGATION. The use of drip irrigation systems (see Section 6.11) has greatly facilitated the localized application of nutrients in irrigation water. Because drip fertigation is applied at frequent intervals, the plants are essentially spoon-fed, and the efficiency of nutrient use is quite high.

PERFORATION METHOD FOR TREES. Trees in orchards and ornamental plantings can be fertilized individually using the *perforation* method (Figure 14.14*h*). Numerous small holes are bored around each tree within the outer half of the branch-spread zone and extending down into the upper subsoil. Fertilizer is placed into these holes, which are afterward filled up. This method of application places the nutrients within the tree root zone and avoids an undesirable stimulation of the grass or cover that may be growing around the trees.

FIGURE 14.16 Many modern planters are equipped with a fertilizer-banding attachment that places starter fertilizer for row crops slightly below and slightly to the side of the seed. This placement eliminates the danger of fertilizer burn, yet concentrates the nutrients near the seed where the crop roots will encounter them shortly after the seed germinates. (Courtesy National Plant Food Institute, Washington, D.C.)

Foliar Application

Plants are capable of absorbing nutrients through their leaves in limited quantities. Under certain circumstances, the best way to apply a nutrient is *foliar application*—spraying a very dilute nutrient solution directly onto the plant leaves. Diluted NPK fertilizers, micronutrients, or small quantities of urea can be used as foliar sprays, although care must be taken to avoid significant concentrations of Cl^- or NO_3^-, which can be toxic to some plants.

14.12 TIMING OF FERTILIZER APPLICATION

The timing of nutrient applications in the field is governed by two basic considerations: (1) making the nutrient available when the plant needs it, and (2) avoiding excess availability, especially of nitrogen, before and after the principal period of plant uptake.

Availability When the Plants Need It

For mobile nutrients such as nitrogen (and to some degree potassium), the general rule is to make applications as close as possible to the period of rapid plant nutrient uptake. For rapid-growing summer annuals, such as corn, this means making only a small starter application at planting time and applying most of the needed nitrogen as a side dressing just before the plants enter the rapid nutrient accumulation phase, usually about to six weeks after planting. For cool-season plants, such as winter wheat or certain turf grasses, most of the nitrogen should be applied about the time of spring "green-up," when the plants resume a rapid growth rate. For trees, the best time is when new leaves are forming. With slow-release organic sources, some time should be allowed for mineralization to take place prior to the plants' period of maximum uptake.

Environmentally Sensitive Periods

In temperate (**Udic** and **Xeric**) climates, most leaching takes place in the winter and early spring when precipitation is high and evapotranspiration is low. Nitrates left over from the previous growing season have the potential for leaching during this period. With inorganic nitrogen fertilizers, avoiding these leftover nitrates is largely a matter of limiting the amount applied to what the plants are expected to take up. However, for slow-release organic sources applied in late spring or early summer, mineralization is likely to continue to release nitrates after the crop has matured and nutrient uptake has ceased. To the extent that this timing of nitrate release is unavoidable, nonleguminous cover crops should be planted in the fall to absorb the excess nitrate being released.

In high-rainfall conditions and on permeable soils, dividing a large dose of fertilizer into two or more split applications may avoid leaching losses prior to the crop's establishment of a deep root system. Application of nitrogen fertilizers to mature forests is usually carried out when rains can be expected to wash the nutrients into the soil and minimize volatilization losses.

Practical Field Limitations

Sometimes it is simply not possible to apply fertilizers at the ideal time of the year. For example, although a crop may respond to a late-season side dressing, such an application will be difficult if the plants are too tall to drive over without damaging them. Application by airplane may increase the flexibility possible. Early spring applications may be limited by fields too wet to support tractors. Economic costs or the time demands of other activities may also require that compromises be made in the timing of nutrient application.

14.13 DIAGNOSTIC TOOLS AND METHODS[7]

Three basic tools are available for diagnosing soil fertility problems: (1) *field observations,* (2) *plant tissue analysis,* and (3) *soil analysis* (soil testing). To effectively guide the appli-

---

[7] For a detailed discussion of both plant analysis and soil testing, see Westerman (1990).

cation of nutrients, as well as to diagnose problems as they arise in the field, all three approaches should be integrated. There is no substitute for careful observation and *recording* of circumstantial evidence and symptoms in the field. Effective observation and interpretation requires skill and experience, as well as an open mind. It is not uncommon for a supposed soil fertility problem to actually be caused by soil compaction, weather conditions, pest damage, or human error. The task of the diagnostician is to use all the tools available in order to identify the factor that is limiting plant growth, and then devise a course of action to alleviate the limitation.

14.14 PLANT SYMPTOMS AND FIELD OBSERVATIONS

To be an effective soil fertility diagnostician, several general guidelines are helpful. First, develop an organized way to *record* your observations. Second, look for *spatial patterns*—how the problem seems to be distributed in the landscape and in individual plants.

Third, closely examine individual plant leaves to characterize any foliar symptoms. Examples of such symptoms are shown in several figures in Chapters 12 to 14 and in Plates 28 to 33. Determine if the symptoms are most pronounced on the younger leaves (as is the case for most of the micronutrient cations) or on the older leaves (as is the case for nitrogen, potassium, and magnesium).

Fourth, observe and *measure* differences in plant growth and crop yield that may reflect different levels of soil fertility, even though no leaf symptoms are apparent. Check *both* aboveground and belowground growth.

Fifth, obtain records on plant growth or crop yield from previous years, and ascertain the history of management for the site for as many years as possible. It is often useful to sketch a map of the site showing features you have observed and the distribution of symptoms.

14.15 PLANT ANALYSIS AND TISSUE TESTING

The concentration of essential elements in plant tissue is related to plant growth or crop yield, as shown in Figure 14.17. The *sufficiency range* and the *critical range* have been well characterized for many plants, especially for agronomic and major horticultural crops. Less is known about forest trees and ornamentals. Sufficiency ranges for 11 essential elements in a variety of plants are listed in Table 14.10.[8] Plants with tissue concentrations lower than the smaller value given in the sufficiency range for a particular nutrient are likely to respond to additions of that nutrient if no other factor is more limiting.[9]

Tissue analysis can be a powerful tool for identifying plant nutrient problems if several simple precautions are taken. Research has determined for most crops the plant part to be sampled, as well as the appropriate growth stage of the plant at sampling time. Likewise, the importance of nutrient ratios (e.g., Mg/K, S/N, or Fe/Mn) is known and must be taken into consideration. In fact, several elaborate mathematical systems for assessing the ratios or balance among nutrients have proven useful for certain plant species.[10] Nutrient levels in plants from the best and worst areas of the field would help determine the problems of nutrient imbalance.

14.16 SOIL ANALYSIS

Since the total amount of an element in a soil tells us very little about the ability of that soil to supply that element to plants, more meaningful *partial soil analyses* have been developed. *Soil testing* is the routine partial analysis of soils for the purpose of guiding nutrient management.

---

[8]Detailed information on tissue analyses for a large number of plant species can be found in Reuter and Robinson (1986).

[9] One plant test that has received much attention in recent years measures the nitrate level of the lower portion of cornstalks at physiological maturity. It identifies excess nitrate levels in the plant, which are an indication of excess levels in the soil. It should be helpful in diagnosing the current year's management and in fertilizer planning for the coming year.

[10] The most well developed of the multinutrient ratio systems is known as the Diagnostic Recommendation Integrated System (DRIS). For details on the DRIS, see Walworth and Sumner (1987).

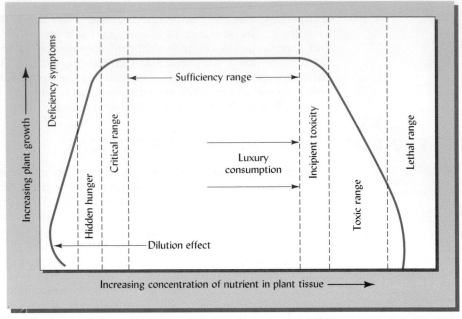

FIGURE 14.17 The relationship between plant growth or yield and the concentration of an essential element in the plant tissue. For most nutrients there is a relatively wide range of values associated with normal, healthy plants (the sufficiency range). Beyond this range, plant growth suffers from either too little or too much of the nutrient. The critical range (CR) is commonly used for the diagnosis of nutrient deficiency. Nutrient concentrations below the CR are likely to reduce plant growth even if no deficiency symptoms are visible. This moderate level of deficiency is sometimes called *hidden hunger*. The odd hook at the lower left of the curve is the result of the so-called dilution effect that is often observed when extremely stunted, deficient plants are given a small dose of the limiting nutrient. The growth response may be so great that even though somewhat more of the element is taken up, it is diluted in a much greater plant mass.

The soil testing process consists of three critical phases: (1) *sampling* the soil, (2) chemically *analyzing* the sample, and (3) *interpreting* the analytical result to make a recommendation on the kind and amounts of nutrients to apply.

Sampling the Soil

Soil sampling is widely acknowledged to be one of the weakest links in the soil testing process. Part of the problem is that about a teaspoonful of soil is eventually used to represent millions of kilograms of soil in the field. Since soils are highly variable, both horizontally and vertically, it is essential to carefully follow the sampling instructions from the soil testing laboratory, which would likely include the following:

1. Take separate samples from each area in the field that is to be treated or used differently, or that you have reason to believe is different from the remainder of the field (e.g., eroded versus noneroded, and flower beds versus lawns).
2. Take 15 to 20 cores (usually 15 to 20 cm deep[11]) from each area to be sampled (Figure 14.18). Composite and thoroughly mix them in a plastic bucket. After air-drying[12] a small subsample of about 0.5 L, place it in a labeled container and send it to the laboratory.
3. Take samples every year or two, always at the same time of year, to monitor changes in nutrient levels.

---

[11] Depth of sampling may vary depending on farming systems. For example, on no-till fields, the upper 1-2 cm should be sampled separately from lower depths.

[12] Samples intended for the analysis of soil nitrates are an exception. These should be dried rapidly under a fan or in an oven at about 65°C.

TABLE 14.10 A Guide to Tissue Analysis for Selected Plant Species

The values given are for sufficiency ranges.

Plant species	Part to sample	Content, %						Content, µg/g				
		N	P	K	Ca	Mg	S	Fe	Mn	Zn	B	Cu
Pine trees (*Pinus* spp.)	Current-year needles near terminal	1.2–1.4	0.10–0.18	0.30–0.45	0.13–0.16	0.05–0.09	0.08–0.12	20–100	50–600	20–50	3–9	2–6
Turf grasses	Clippings	2.75–3.5	0.30–0.55	1.0–2.5	0.50–1.2	0.20–0.60	0.20–0.45	35–1000	25–150	20–55	10–60	5–20
Corn (*Zea mays*)	Ear-leaf at tasseling	2.5–3.5	0.20–0.50	1.5–3.0	0.2–1.0	0.16–0.40	0.16–0.50	25–300	20–200	20–70	6–40	6–40
Soybean (*Glycine max*)	Recently matured trifoliate, flowering stage	4.0–5.0	0.31–0.50	2.0–3.0	0.45–2.0	0.25–0.55	0.25–0.55	50–250	30–200	25–50	25–60	8–20
Apple (*Malus* spp.)	Leaf from base of nonfruiting shoots	1.8–2.4	0.15–0.30	1.2–2.0	1.0–1.5	0.25–0.50	0.13–0.30	50–250	35–100	20–50	20–50	5–20
Wheat (*Triticum* spp.)	First leaf blade from top of plant	2.2–3.3	0.24–0.36	2.0–3.0	0.28–0.42	0.19–0.30	0.20–0.30	35–55	30–50	20–35	5–10	6–10
Tomato (*Lycopersicon esculentum*)	Most recently matured leaf at early bloom	3.2–4.8	0.32–0.48	2.5–4.2	1.7–4.0	0.45–0.70	0.60–1.0	120–200	80–180	30–50	35–55	8–12
Alfalfa (*Medicago sativa*)	Upper third of plant at first flower	3.0–4.5	0.25–0.50	2.5–3.8	1.0–2.5	0.3–0.8	0.3–0.5	50–250	25–100	25–70	6–20	30–80

Data derived from many sources.

465

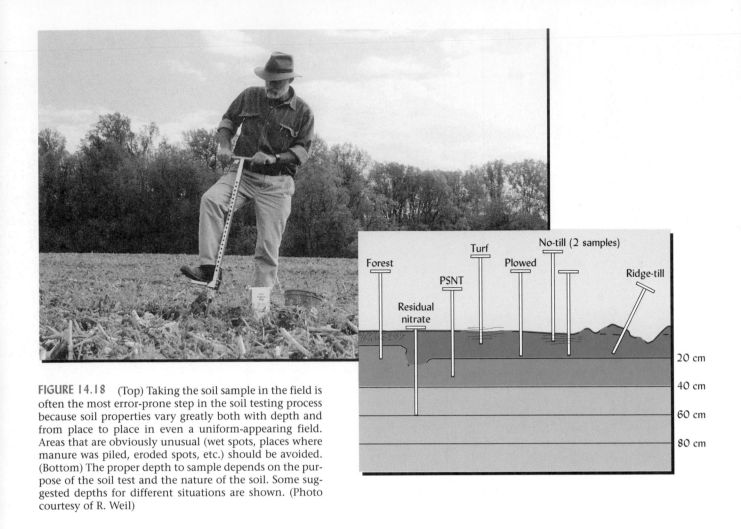

FIGURE 14.18 (Top) Taking the soil sample in the field is often the most error-prone step in the soil testing process because soil properties vary greatly both with depth and from place to place in even a uniform-appearing field. Areas that are obviously unusual (wet spots, places where manure was piled, eroded spots, etc.) should be avoided. (Bottom) The proper depth to sample depends on the purpose of the soil test and the nature of the soil. Some suggested depths for different situations are shown. (Photo courtesy of R. Weil)

4. Some special tests require specific sampling and soil handling procedures different from those for routine analyses.[13] Instructions for these can be obtained from scientists at the testing laboratory.

Chemical Analysis of the Sample

In general, soil tests attempt to extract from the soil amounts of essential elements that are correlated with the nutrients taken up by plants. A small sample of soil is shaken for a specified time with a buffered salt solution such as sodium or ammonium acetate, or mixtures of dilute acids and chelating agents. The amount of the various nutrient elements brought into solution is then determined. The whole process is usually automated so that a modern laboratory can handle hundreds of samples each day.

The most common and reliable tests are those for soil pH, potassium, phosphorus, and magnesium. Micronutrients are sometimes extracted using chelating agents, especially for calcareous soils in the more arid regions. Predicting the availability of nitrogen and sulfur is considerably more difficult because of the many biological factors involved, but the nitrate and sulfate present in the soil at the time of sampling can be measured.

---

[13] For example, two nitrate tests tests are gaining favor to help ascertain nitrogen needs of corn in the Midwest. The *residual nitrate* test is done on 60-cm-deep soil samples obtained at the end of the growing season. Also, to ascertain the needs for sidedressed nitrogen, the *presidedress nitrate test* (PSNT) is made on samples taken from the upper 30 cm of soil when the corn is about 30 cm tall.

Interpreting the Results to Make a Recommendation

This is, perhaps, the most controversial aspect of soil testing. The soil test values themselves are merely *indices* of nutrient-supplying power. They do not indicate the actual amount of nutrient that will be supplied. For this reason, it is best to think of soil test reports as more indicative than quantitative.

Many years of field experimentation at many sites are needed to determine which soil test level indicates a low, medium, or high capacity to supply the nutrient tested. Such categories are used to predict the likelihood of obtaining a profitable response from the application of the nutrient tested (Figure 14.19).

The interpretation of soil test data is best accomplished by experienced and technically trained personnel who fully understand the scientific principles underlying the common field procedures. In modern laboratories, the factors to be considered in making fertilizer recommendations are programmed into a computer, and the interpretation is printed out for the farmer's or gardener's use.

Merits of Soil Testing

It must not be inferred from the preceding discussion that the limitations of soil testing outweigh its advantages. When the precautions already described are observed, soil testing is an invaluable tool in making fertilizer recommendations. These tests are most useful when they are correlated with the results of field fertilizer experiments (Figure 14.20).

Generally, soil testing has been most relied upon in agricultural systems, while foliar analysis has proved more widely useful in forestry. The limited use of soil testing in forestry

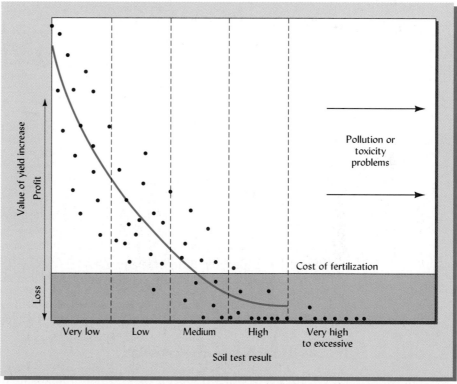

FIGURE 14.19 The relationship between soil test results for a nutrient and the extra yield obtained by fertilizing with that nutrient. Each data point represents the *difference* in plant yield between the fertilized and the unfertilized soil. Because many factors affect yield and because soil tests can only approximately predict nutrient availability, the relationship is not precise, but the data points are scattered about the trend line. If the point falls above the fertilizer cost line, the extra yield was worth more than the cost of the fertilizer and a profit would be made. For a soil testing in the very low and low categories, a profitable response to fertilizer is very likely. For a soil testing medium, a profitable response is a 50:50 proposition. For soils testing in the high category, a profitable response is unlikely.

FIGURE 14.20 (*a*) The yield of corn from plots near North Platte, Nebraska, receiving five different rates of fertilizer as recommended by five soil testing laboratories (A to E). Data averaged over a six-year period. The soil was a Cozad silt loam (Fluventic Haplustoll). (*b*) The nutrient levels recommended by these laboratories and (*c*) the cost of the fertilizer applied each year. Note that the yields from all fertilized plots were about the same, even though rates of fertilizer application differed markedly. The fertilizer rates recommended by laboratory E utilized a sufficiency-level concept that was based on the calibration of soil tests with field yield responses. Those recommended by laboratories A to D were based either on a maintenance concept that required replacement of all nutrients removed by a crop or on the supposed need to maintain set cation ratios (Ca/Mg, Ca/K, and Mg/K). Obviously, the sufficiency-level concept provided more economical results in this trial. As a result of this and many other similar comparison studies, most soil test labs have adopted the sufficiency-level approach. [From Olson, et al. (1982); used with permission of American Society of Agronomy]

is probably a result of the complex stratification in forested soils, which creates a great deal of uncertainty about how to obtain a representative sample of soil for analysis. Also, because of the comparatively long time frame in forestry, little information is available on the correlation of soil test levels with timber yield in the sense that such information is widely available for agronomic crops (see Figure 14.19). Even though the relationship between tree growth and soil test level is not well known for most forest systems, standard agronomic soil testing can still be useful in distinguishing those soils whose ability to supply P or K is adequate from those with very low supplying power for these nutrients.

14.17   SITE-SPECIFIC NUTRIENT MANAGEMENT[14]

Recent developments suggest that nutrient management can be made more site specific when the capabilities of three modern tools are integrated: (1) the map-making capabilities of the computer software called geographic information systems (GIS), (2) the geo-referencing capability of the Global Positioning System (GPS) of earth-orbiting satellites, and (3) satellite-controlled variable-rate fertilizer application equipment. The following example suggests how this integration operates.

A composite soil sample is taken on a grid basis from each hectare in a 20-ha field, and the samples are analyzed. With appropriate GIS and other software, the test data are

---

[14]See Robert, et al. (1996) for a series of papers dealing with site-specific nutrient management.

used to generate a soil test map of the sampled field showing the boundaries between the areas of high-, medium-, and low-fertility status. This map can be overlaid with maps of other spacial variables such as soil classification, past management practices, crop to be planted, and so forth.

The satellite-based GPS is able to plot the exact location of a variable-rate fertilizer drill that is also linked through computers to the soil test map. As the drill moves across the field, its rate of application is automatically adjusted when it crosses a boundary between one soil test level and another. An area showing a high soil test level for a given nutrient would receive little if any of the fertilizer, while one with a low level might receive a higher than average rate.

This site-specific nutrient-management system is an integral part of what is commonly referred to as precision farming.[15] Opportunities for the control of insects and weeds and for modifying plant seeding rates and depths can be utilized on a site-specific basis rather than on a field basis. Such computerized GPS-based systems may not be economically viable for all farms, but their use is spreading, particularly among growers who have access to custom operators with the computer expertise and appropriate equipment. Precision farming may help assure that nutrients will be used only where they are needed to enhance plant production.

14.18 BROADER ASPECTS OF FERTILIZER PRACTICE

Arriving at generalizations on the use of fertilizers is difficult because of the many factors influencing their effectiveness. However, the delicate balance between deficiencies and excesses of mobile nitrogen suggests that initial focus should be on that element in most fertilizer schemes. Applications of phosphorus and potassium are often made to balance the nitrogen supply.

Nitrogen fertilizers should be used to *complement* the supply of this element from the soil, and from manures and crop residues that reach the soil. Research in both the highly intensive farming areas of the United States and in the nutrient-deficient regions of Africa and Asia has shown the synergism between organic and inorganic forms of nitrogen. These studies also suggest that in wide areas of the world, plant production and environmental quality could best be served if somewhat lower rates of inorganic nitrogen were to be used (Figure 14.21).

---

[15]For a review of site-specific management in agriculture, see Pierce and Sadler (1997).

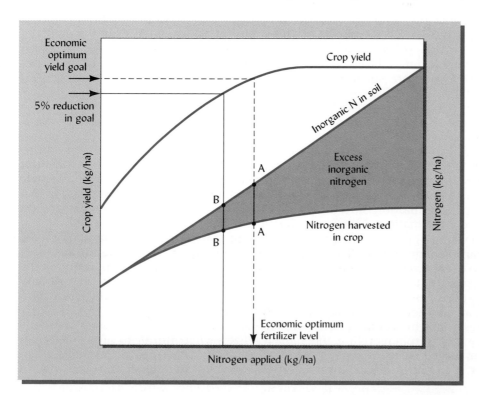

FIGURE 14.21 Influence of rate of nitrogen fertilization on crop yield, nitrogen removed in harvest, and the amount of excess inorganic soil nitrogen potentially available for loss into the environment. The fine lines show that a relatively small decrease (5%) in the yield goal for which nitrogen was applied would result in a relatively large reduction (about 30%) in the potential for nitrogen pollution. Line A-A represents the excess nitrogen present when fertilizer is applied at a rate designed to reach the economic optimum yield goal. Line B-B represents the excess nitrogen if the yield goal is 5% less than the economic optimum. Unfortunately, in many cases a yield reduction of 5% may reduce profits by a much larger percentage. [Adapted from National Research Council (1993)]

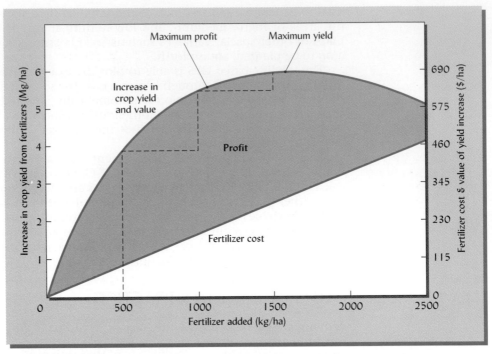

FIGURE 14.22 Relationships among the rate of fertilizer addition, crop yield increase, fertilizer costs, and profit from adding fertilizers. Note that the yield increase (and profit) from the first 500 kg of fertilizer is much greater than from the second and third 500 kg. Also note that the maximum profit is obtained at a lower fertilizer rate than that needed to give a maximum yield. The calculations assume that the yield response to added fertilizer takes the form of a smooth quadratic curve.

Motivational studies have clearly demonstrated that farmers do not use fertilizers just to grow big crops or to increase the nutrient content of their soils. They do so to make a living. Consequently, any fertilizer practice that does not give a fair economic return will not stand the test of time. It has also been shown repeatedly that the most profitable rate of application is commonly less than the rate which produces the highest yield (Figure 14.22). Farmers and those who advise them must always keep this in mind.

While fertilization with nitrogen will continue to receive priority attention, in some regions other nutrients will compete for this attention. For example, even the supply of nitrogen in parts of Africa will depend in part on increased phosphorus levels to support nitrogen-fixing legumes. Likewise, deficiencies of potassium as well as some micronutrients, such as zinc, are noted in some otherwise highly productive rice-growing areas of Asia.

14.19 CONCLUSION

The continuous availability of plant nutrients is critical for the sustainability of most ecosystems. The challenge of nutrient management is threefold: (1) to provide adequate nutrients for plants in the system; (2) to simultaneously ensure that inputs are in balance with plant utilization of nutrients, thereby conserving nutrient resources; and (3) to prevent contamination of the environment with unutilized nutrients.

The recycling of plant nutrients must receive primary attention in any ecologically sound management system. This can be accomplished in part by returning plant residues to the soil. These residues can be supplemented by judicious application of the organic wastes that are produced in abundance by municipal, industrial, and agricultural operations worldwide. The use of cover crops grown specifically to be returned to the soil is an additional organic means of recycling nutrients.

For sites from which crops or forest products are removed, nutrient losses commonly exceed the inputs from recycling. Inorganic fertilizers will continue to supplement natural and managed recycling to replace these losses and to increase the level of soil fertility so as to enable humankind to not only survive, but to flourish on this planet. In extensive areas of the world, fertilizer use will have to be increased above current levels to avoid soil and ecosystem degradation and to enable profitable food production.

The use of fertilizers, both inorganic and organic, should not be done in a simply habitual manner or for so-called insurance purposes. Rather, soil testing and other diagnostic tools should be used to determine the true need for added nutrients. Where soils are low in available nutrients, inorganic fertilizers often return several dollars' worth of improved yield for every dollar invested. However, where the nutrient-supplying power of the soil is already sufficient, adding fertilizers is likely to be damaging both to the bottom line and to the environment.

STUDY QUESTIONS

1. The groundwater under a heavily fertilized field is quite high in nitrates, but by the time it reaches a stream the nitrate level has declined considerably. What is your explanation for this situation?

2. A cornstalk nitrate test at harvest time shows a farmer that the soil likely contains considerable unused nitrates. To minimize nitrate leaching, the farmer wants to grow a cover crop. Would you advise a legume or nonlegume? Why?

3. The nitrate level in waters that are leaching or running off from forested sites is generally quite low. What management practices on forested sites lead to significant nitrate losses, and how can the losses be prevented?

4. What effect do forest fires have on nutrient availabilities and toxicities?

5. What are the main original sources of the materials from which N, P, and K fertilizers are formed? What processes are used to increase their availability for plant absorption?

6. Why is anhydrous ammonia said to be the most important fertilizer nitrogen carrier?

7. Traditionally, animal manures have been considered of major benefit in maintaining soil quality. Today, in many situations they are a liability. Explain, and indicate some means of alleviating the problems.

8. Discuss the concept of the *limiting factor* and indicate its importance in enhancing or constraining plant growth.

9. Why are nutrient cycling problems in agricultural systems more prominent than in those in forested areas?

10. Discuss the value and limitations of soil tests as indicators of plant nutrient needs.

REFERENCES

Attiwill, P. M., and G. W. Leeper. 1987. *Forest Soils and Nutrient Cycles.* (Melbourne, Australia: Melbourne University Press).

Belt, G. H., and J. O'Laughlin. 1994. "Buffer strip design for protecting water quality and fish habitat," *Western J. Applied Forestry,* 9(2):41–45.

Bumb, B. L., and C. A. Baanante. 1996. *The Role of Fertilizer in Sustaining Food Security and Protecting the Environment to 2020.* Food, Agriculture and the Environment Discussion Paper 17. (Washington, D.C.: International Food Policy Research Institute).

Ebelhar, S. A., W. W. Frye, and R. L. Blevins. 1984. "Nitrogen from legume cover crops for no-tillage corn," *Agron. J.,* 76:51–55.

Eghball, B., and J. F. Power. 1994. "Beef cattle feedlot manure management," *J. Soil and Water Conservation,* 49:113–122.

He, X.-T., T. Logan, and S. Traina. 1995. "Physical and chemical characteristics of selected U.S. municipal solid waste composts," *J. Environ. Qual.,* 24:543–552.

Magdoff, F., L. Lanyon, and B. Liebhardt. 1997. "Nutrient cycling, transformations, and flows: Implications for a more sustainable agriculture," *Advances in Agronomy,* 60:2–73.

National Research Council. 1993. *Soil and Water Quality: An Agenda for Agriculture.* (Washington, D.C.: National Academy of Sciences).

Olson, R. A., K. D. Frank, P. H. Graboushi, and G. W. Rehm. 1982. "Economic and agronomic impacts of varied philosophies of soil testing," *Agron. J.,* **74**:492–499.

Pierce, F. J., and E. J. Sadler. 1997. *The state of site-specific management for agriculture.* (Madison, Wis.: Amer. Soc. Agron., Crop Sci. Soc. Amer., and Soil Sci. Soc. Amer.).

Reddy, G. B., and K. R. Reddy. 1993. "Fate of nitrogen-15 enriched ammonium nitrate applied to corn," *Soil Sci. Soc. Amer. J.,* **57**:111–115.

Reuter, D. J., and J. B. Robinson. 1986. *Plant Analysis: An Interpretation Manual.* (Melbourne, Australia: Inkata Press).

Robert, P. C., R. H. Rust, and W. E. Larson (eds.). 1996. *Precision Agriculture.* Proceedings of the Third International Conference on Precision Agriculture. (Madison, Wis.: Amer. Soc. Agron.).

Sanchez, P. A., et al. 1997. "Soil fertility replenishment in Africa: An investment in natural resource capital," in R. J. Buresh, P. A. Sanchez, and F. Calhoun (eds.), *Replenishing Soil Fertility in Africa.* ASA/SSSA Publication. (Madison, Wis.: Soil Sci. Soc. Amer.).

Sims, J. T., and D. C. Wolf. 1994. "Poultry waste management: Agricultural and environmental issues," *Advances in Agronomy,* **52**:1–83.

Smaling, E. M. A., S. M. Nwanda, and B. H. Jensen. 1997. "Soil fertility in Africa is at stake," in R. J. Buresh, P. A. Sanchez, and F. Calhoun (eds.), *Replenishing Soil Fertility in Africa.* ASA/SSSA Publication. (Madison, Wis.: Soil Sci. Soc. Amer.).

Stivers, L. J., C. Shennen, L. E. Jackson, K. Groody, C. J. Griffin, and P. R. Miller. 1993. "Winter cover cropping in vegetable production systems in California," in M. G. Paoletti, W. Foissner, and D. Coleman (eds.), *Soil Biota, Nutrient Cycling and Farming Systems.* (Boca Raton, Fla.: Lewis Publishers).

Sutton, A. L. 1994. "Proper animal manure utilization," in *Nutrient Management,* supplement to *J. Soil Water Conserv.,* **49**(2), pp. 65–70.

Teng, Y., and V. R. Timmer. 1994. "Nitrogen and phosphorus in an intensely managed nursery soil–plant system," *Soil Sci. Soc. Amer. J.,* **58**:232–238.

Thomas, M. L., R. Lal, T. Logan, and N. R. Fausey. 1992. "Land use and management effects on non-point loading from Miamian soil," *Soil Sci. Soc. Amer. J.,* **56**:1871–1875.

Tisdale, S. L., W. L. Nelson, J. D. Beaton, and J. L. Havlin. 1993. *Soil Fertility and Fertilizers,* 5th ed. (New York: Macmillan).

Walworth, J. L., and M. E. Sumner. 1987. "The diagnosis and recommendation integrated system (DRIS)," *Advances in Soil Science,* **6**:149–187.

Westerman, R. L. (ed.). 1990. *Soil Testing and Plant Analaysis,* 3d ed. (Madison, Wis.: Soil Sci. Soc. Amer.).

Wilkinson, D. M., and N. M. Dickinson. 1995. "Metal resistance in trees: The role of mycorrhizae," *Oikos,* **72**:298–299.

Zublena, J. P., J. C. Barker, and T. A. Carter. 1993. "Poultry manure as a fertilizer source," *Soil Facts* (Raleigh, N.C.: North Carolina Cooperative Extension Service, North Carolina State University).

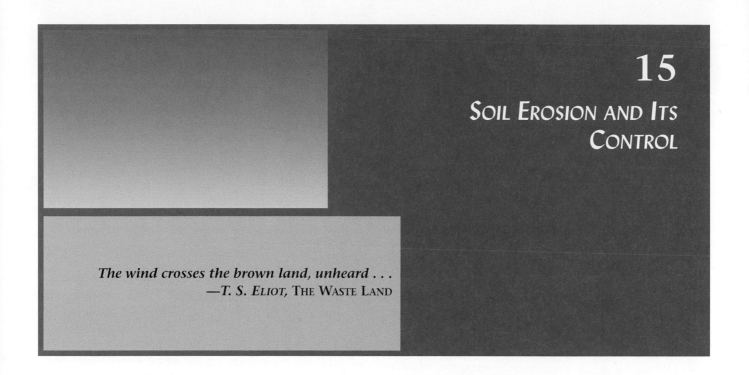

15

SOIL EROSION AND ITS CONTROL

The wind crosses the brown land, unheard . . .
—*T. S. ELIOT, THE WASTE LAND*

No soil phenomenon is more destructive worldwide than the erosion caused by wind and water. Since prehistoric times people have brought the scourge of soil erosion upon themselves, suffering impoverishment and hunger in its wake. Past civilizations have disintegrated as their soils, once deep and productive, washed away, leaving only thin, infertile, rocky relics of the past. Seeing the nearly barren hills in central India, or in parts of Greece, Lebanon and Syria, it is hard to imagine that agricultural communities once flourished in these places.

The current threat of soil erosion is more ominous than at any time in history. In our generation, farmers have had to more than double world food output to feed the unprecedented numbers of people on Earth. In the low-income countries, the ratio of people to available cropland is already very high and rising rapidly. While intensified cultivation of fertile, relatively level lands has helped produce much of the needed food, many nations have had to expand the area of land under cultivation, clearing and burning steep, forested slopes and plowing grasslands. Population pressures have also led to overgrazing rangelands and overexploiting timber resources. All these activities degrade or remove natural vegetation, causing the underlying soils to become much more susceptible to the destructive action of erosion. The result is a vicious downward cycle of deterioration—land degradation leads to poor crops, human poverty, and reduced vegetative protection for the soil, which, in turn, accelerates erosion and drives ever more desperate people to clear, cultivate, and degrade more land.

The degraded productivity of farm, forest, and range lands tells only part of the sad erosion story. Soil particles washed or blown from the eroding areas are subsequently deposited elsewhere—in nearby low-lying landscape sites; in streams and rivers; or in downstream reservoirs, lakes, and harbors. The environmental and economic damages suffered by sites on which the eroded soil materials are deposited may be as great or greater than that incurred on the sites from which the soil material was removed. The displaced soil material (sediment and dust) leads to major water and air pollution problems, bringing enormous economic and social costs to society.

Soil erosion is everybody's business. Fortunately, recent decades have seen much progress in understanding the mechanisms of erosion and in developing techniques that can effectively and economically control soil loss in most situations. This chapter will equip you with some of the concepts and tools you will need to do your part in solving this pressing world problem.

15.1 SIGNIFICANCE OF SOIL EROSION AND LAND DEGRADATION[1]

Land Degradation

During the past half century, human land use and associated activities have degraded some 5 billion ha (about 43% of Earth's vegetated land). Such **land degradation** results in a reduced productive potential and a diminished capacity to provide benefits to humanity. Much of this degradation (on about 2.6 billion ha) is linked to **desertification**, the spreading of desert conditions that disrupt semiarid and arid ecosystems (including agroecosystems). A major cause of desertification is overgrazing by cattle, sheep, and goats, a factor that likely accounts for about a third of all land degradation, mainly that in such dry regions as the Sahel in northern Africa, East Asia, and the rangeland of the American Southwest (see end paper map). Likewise, the indiscriminate felling of rain forest trees has already degraded nearly 0.5 billion ha in the humid tropics. Additionally, inappropriate agricultural practices continue to degrade land in all climatic regions.

Soil–Vegetation Interdependency

Degraded lands may suffer from destruction of native vegetation communities, reduced agricultural yields, lowered animal production, and simplification of once-diverse natural ecosystems with or without accompanying degradation of the soil resource. On about 2 billion of the 5 billion ha of degraded lands in the world, soil degradation is a major part of the problem (see Figure 15.1). In some cases the soil degradation occurs

---

[1] For a very readable account of historical degradation of land and water resources, see Hillel (1991). Overviews of the current extent of soil erosion and other forms of land degradation are given by Oldeman (1994), Daily (1997), and Rosensweig and Hillel (1998).

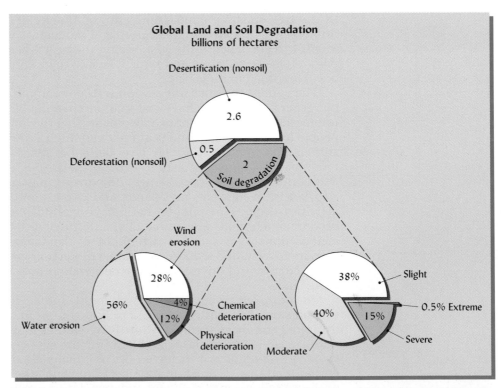

FIGURE 15.1 Soil degradation as a part of global land degradation caused by overgrazing, deforestation, inappropriate agricultural practices, fuel wood overexploitation, and other human activities. About 60% of degraded land has not suffered soil degradation. Of the 2 billion ha of land with degraded soils, most could be restored easily (*slight* degradation) or with considerable financial and technical investments (*moderate* degradation). *Severely* degraded soils are currently useless for agriculture and would require major international assistance for restoration. About 9 million ha (0.5% of degraded soils) are *extremely* degraded and incapable of restoration. About 85% of the soil degradation is caused by erosion by wind and water. [FAO data selected from Oldeman (1994) and Daily (1997)]

---

mainly as deterioration of physical properties by compaction or surface crusting (see Sections 4.5 and 4.8), or as deterioration of chemical properties by acidification (see Section 9.4) or salt accumulation (see Section 9.14). However, most (~85%) soil degradation stems from erosion—the destructive action of wind and water.

The two components of land degradation—damage to soils and to plant communities—are not independent, but interact to cause a downward spiral of accelerating deterioration. Overgrazing, fires, deforestation, and inappropriate farming all diminish the ability of the vegetation to protect the soil. Likewise, soil erosion, crusting, salinization, and nutrient depletion diminish the ability of the soil to support vegetation. Poor farmers, trying to grow food for their families, are often caught up in this cycle of impoverishment as they clear and till ever more land in a desperate attempt to make up for declining soil productivity. Improvements in both soil and vegetation management must go hand-in-hand if the productive potential of the land is to be protected—or even be restored by moving *up* rather than *down* the spiral.

Geological versus Accelerated Erosion

GEOLOGICAL EROSION.   Erosion is a process that transforms soil into **sediment.** Soil erosion that takes place naturally, without the influence of human activities, is termed **geological erosion.** It is a natural leveling process. It inexorably wears down hills and mountains, and through subsequent deposition of the eroded sediments, it fills in valleys, lakes, and bays. Many of the landforms we see around us—canyons, buttes, rounded hills, river valleys, deltas, plains, and pediments—are the result of geological erosion and deposition. The vast deposits that now appear as sedimentary rocks originated in this way.

In most settings, geological erosion wears down the land slowly enough that new soil forms from the underlying rock or regolith faster than the old soil is lost from the surface. The very existence of soil profiles bears witness to the net accumulation of soil and the effectiveness of undisturbed natural vegetation in protecting the land surface from erosion.

The rate of geological soil erosion varies greatly with both rainfall and type of material comprising the regolith. Geological erosion by water tends to be greatest in semiarid regions where rainfall is enough to be damaging, but not enough to support dense, protective vegetation (Figure 15.2).

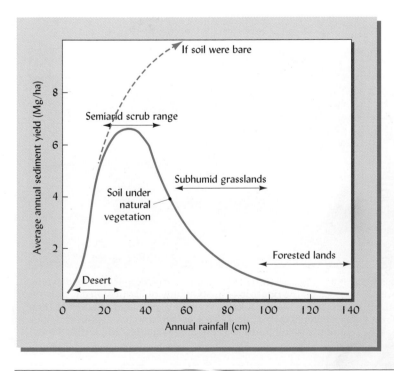

FIGURE 15.2 Generalized relationship between annual rainfall and soil loss from geologic erosion by water. The actual amount of sediment lost annually per hectare will depend on other climatic variables, topography, and the type of soils in the watershed. Note that sediment yields are greatest in semiarid regions. Here a number of severe runoff-generating storms occur in most years, but the total rainfall is too little to support much protective plant cover. By comparison, the very dry deserts have too little rain to cause much erosion, and the well-watered regions support dense forests that effectively protect the soil. Where the natural vegetation is destroyed by plowing, erosion from the bare soils is much higher with increasing rainfall, as indicated by the dashed curve.

ACCELERATED EROSION. **Accelerated erosion** occurs when people disturb the soil or the natural vegetation by grazing livestock, cutting forests, plowing hillsides, or tearing up land for construction of roads and buildings. Accelerated erosion is often 10 to 1000 times as destructive as geological erosion, especially on sloping lands in regions of high rainfall. Rates of erosion by wind and water on agricultural land in Africa, Asia, and South America are thought to average about 30 to 40 Mg/ha annually. In the United States, the average erosion rate on cropland is about 12 Mg/ha—7 Mg by water and 5 Mg by wind. Some cultivated soils are eroding at 10 times these average rates. In comparison, erosion on undisturbed humid-region grasslands and forests generally occurs at rates considerably below 0.1 Mg/ha.

About 4 billion Mg of soil is moved annually by soil erosion in the United States, some two-thirds by water and one-third by wind. More than half of the movement by water and about 60% of the movement by wind takes place on croplands that produce most of the country's food. Much of the remainder comes from semiarid rangelands, from road building and timber harvest on forest lands, and from soils disturbed for highway and building construction. Although progress has been made in reducing erosion (see Section 15.10) current high losses are simply not acceptable for long-term sustainability and must be further reduced.

Under the influence of accelerated erosion, soil is commonly washed or blown away faster than new soil can form by weathering or deposition. As a result, the soil depth suitable for plant roots is often reduced. In severe cases, gently rolling terrain may become scarred by deep gullies, and once-forested hillsides may be stripped down to bare rock.

Accelerated erosion often makes the soils in a landscape more heterogeneous. One example of increased heterogeneity occurs as overgrazed semiarid grasslands degrade into scrub brush ecosystems. During this degradation "islands of fertility" form under shrubs, while the soils between shrubs become increasingly thin and infertile. Another example can be seen in the striking differences in surface soil color that develop as ridgetop soils are truncated, exposing material from the B or C horizon at the land surface, while soils lower in the landscape are buried under organic-matter-enriched sediment (Figure 15.3).

15.2 ON-SITE AND OFF-SITE EFFECTS OF ACCELERATED SOIL EROSION

Erosion damages the site on which it occurs and also has undesirable effects off-site in the larger environment. The off-site costs relate to the effects of excess water, sediment, and associated chemicals on downhill and downstream environments. While the costs

FIGURE 15.3 Erosion and deposition occur simultaneously across a landscape. (Left) The soil on this ridgetop was worn down by erosion during nearly 300 years of cultivation. The surface soil exposed on the ridgetop consists mainly of light-colored C horizon material. At sites lower down the slope the surface horizon shows mainly A and B horizon material, some of which has been deposited after eroding from locations upslope. (Right) Erosion on the sloping wheat field in the background has deposited a thick layer of sediment in the foreground, burying the plants at the foot of the hill. [Photos courtesy of R. Weil (left) and USDA Natural Resources Conservation Service (right)]

associated with either or both of these types of damages may not be immediately apparent, they are real and grow with time. Landowners and society as a whole must eventually foot the bill.

Types of On-Site Damages

The most obviously damaging aspect of erosion is the loss of soil itself. In reality, the damage done to the soil is greater than the amount of soil lost would suggest because the soil material eroded away is almost always more valuable than that left behind. Not only are surface horizons eroded while subsurface horizons (which are usually less useful) remain untouched, but the quality of the remaining topsoil is also impaired. Erosion selectively removes organic matter and fine mineral particles, while leaving behind mainly relatively less active, coarser fractions. Experiments have shown organic matter and nitrogen in the eroded material to be five times as high as in the original topsoil. Comparable **enrichment ratios** for phosphorus (review Table 13.2) and potassium are commonly two and three, respectively. Therefore, erosion can cause a serious loss of nutrients. The soil left behind usually has lower water-holding and cation exchange capacities, less biological activity, and a reduced capacity to supply nutrients for plant growth.

In addition to the just-mentioned reduction in soil-quality factors, soil movement during erosion can spread plant disease organisms from the soil to plant foliage and from a higher- to a lower-lying field. The deterioration of soil structure often leaves a dense crust on the soil surface, which, in turn, greatly reduces water infiltration and increases water runoff. Newly planted seeds and seedlings may be washed downhill, trees may be uprooted, and small plants may be buried in sediment. In the case of wind erosion, fruits and foliage may be damaged by the sandblasting effect of blowing soil particles.

Finally, gullies that carve up badly eroded land may make the use of tractors impossible and may undercut pavements and building foundations, causing unsafe conditions and expensive repairs.

Types of Off-Site Damages

Erosion moves sediment and nutrients off the land, creating the two most widespread water pollution problems in our rivers and lakes. The nutrients impact water quality largely through the process of eutrophication caused by excessive nitrogen and phosphorus, as discussed in Section 13.2. In addition to nutrients, sediment and runoff water may also carry toxic metals and organic compounds, such as pesticides. The sediment itself is a major water pollutant, causing a wide range of environmental damages.

DAMAGES FROM SEDIMENT. Sediment deposited on the land may smother crops and other low-growing vegetation (Figure 15.3). It fills in roadside drainage ditches and creates hazardous driving conditions where mud covers the roadway.

Sediment that washes into streams makes the water cloudy or turbid (Figure 15.4a and Plate 34). High **turbidity** prevents sunlight from penetrating the water and thus reduces photosynthesis and survival of the *submerged aquatic vegetation* (SAV). The demise of the SAV, in turn, degrades the fish habitat and upsets the aquatic food chain. The muddy water also fouls the gills of some fish. Sediment deposited on the stream bottom can have a disastrous effect on many freshwater fish by burying the pebbles and rocks among which they normally spawn. The buildup of bottom sediments can actually raise the level of the river, so that flooding becomes more frequent and more severe. For example, to counter the rising river bottom, flood-control levees along the Mississippi must be constantly enlarged.

A number of major problems occur when the sediment-laden rivers reach a lake, reservoir, or estuary. Here, the water slows down and drops its load of sediment. Eventually reservoirs—even those formed by giant dams—become mere mudflats, completely filled in with sediment (see Figure 15.4b). Prior to that, the capacity of the reservoir to store water for irrigation or municipal water systems is progressively reduced, as is the capacity for floodwater retention or hydroelectric generation. It is estimated that 1.5 billion Mg of sediment is deposited each year in the nation's reservoirs, harbors, and shipping channels.

(a)

(b)

FIGURE 15.4   Off-site damages caused by soil erosion include the effects of sediment on aquatic systems. (*a*) A sediment-laden tributary stream empties into the relatively clear waters of a larger river. The turbid water will foul fish gills, inhibit submerged aquatic vegetation, and clog water-purification systems. Part of the sediment will settle out on the river bottom, covering fish-spawning sites and raising the river bed enough to aggravate the severity of future flooding episodes. (*b*) Expensive dredging and excavation (by the dragline in the foreground) is being undertaken to restore the beauty, recreational value, and flood-control function of this pond in a neighborhood park after accumulated sediment had transformed it into a mere mudflat. The watershed upstream from the pond had undergone a period of rapid suburban development, during which adequate sediment-control practices were not used on the construction sites. [Photos courtesy of USDA Natural Resources Conservation Service (*a*) and R. Weil (*b*)]

WINDBLOWN DUST.   Wind erosion also has its off-site effects. The blowing sands may bury roads and fill in drainage ditches, necessitating expensive maintenance. The sandblasting effect of wind-borne soil particles may damage the fruits and foliage of crops in neighboring fields, as well as the paint on vehicles and buildings many kilometers downwind from the eroding site.

The finest particles blow the farthest, and may present a major human health hazard. While silt-sized particles are generally filtered by nose hairs or trapped in the mucus of the windpipe and bronchial tubes, smaller clay-sized particles often pass through these defenses and lodge in the air sacs (alveoli) of the lungs. Long-term epidemiological studies suggest that the number of deaths resulting from people inhaling this fine *fugitive dust* may exceed the number of deaths from car highway accidents.

ESTIMATED COSTS OF EROSION.   The total on-site costs of erosion in the United States, including costs of replacing nutrients and water lost through accelerated erosion, as well as crop yield reductions due to reduced soil depth, have been estimated at between $4 and $27 billion per year.

The off-site costs of erosion are likely even greater, especially because of the health effects of windblown particles and the reduced recreational (fishing, swimming, and aesthetic) value of muddy waters. The total of these annual off-site costs has been estimated at $5 to $17 billion. The grand total annual cost of erosion in the United States is therefore likely between $9 and $44 billion. Such high costs are a sobering reminder of the burden that society bears as a result of poor land management and would seem to justify increasing the sums allocated to the battle against erosion.

MAINTENANCE OF SOIL PRODUCTIVITY.   Over the long term, accelerated soil erosion that exceeds the rate of soil formation leads to declining productivity on most soils. In the United States, crop yields on severely eroded soils are often 20 to 40% lower than on similar soils with only slight erosion. Ultimately, the rate of decline of soil productivity, or the cost of maintaining constant crop-yield levels, is determined by such soil properties as *depth to a root-restricting layer* and *permeability of the subsoil* (Figure 15.5).

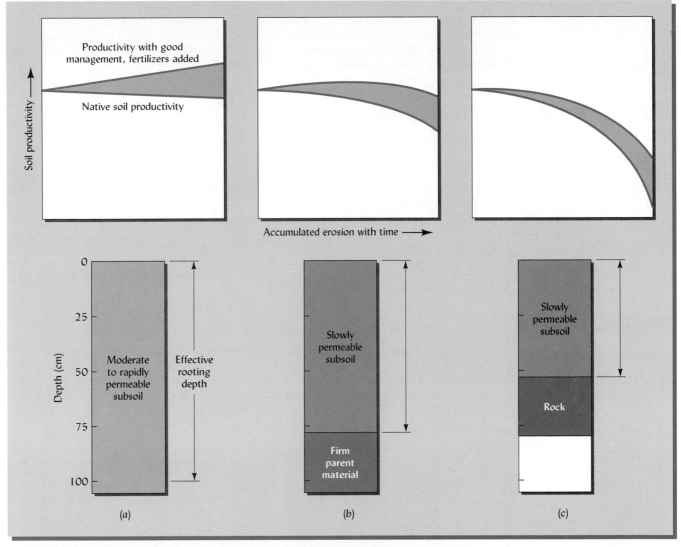

FIGURE 15.5 Effect of erosion over time on the productivity of three soils differing in depth and permeability. Productivity on soil (*a*) actually increases with time because of good management practices and fertilizer additions, even though the native soil productivity declines as a result of erosion. Because soil (*c*) is shallow and has restricted permeability, its productivity declines rapidly as a result of erosion, a decline that good management and fertilizers cannot prevent. Soil (*b*), which is intermediate in both depth and permeability, suffers only a slight decline in productivity due to erosion. Soil characteristics clearly influence the effect of erosion on soil productivity.

Soil-Loss Tolerance

A tolerable soil loss (***T* value**) is the maximum amount of soil that can be lost annually by the combination of water and wind erosion on a particular soil without degrading that soil's long-term productivity. Currently, *T* values are based on the best judgment of informed soil scientists, rather than on rigorous research data. The *T* values for soils in the United States commonly range from 5 to 11 Mg/ha. They depend on a number of soil-quality and -management factors, including soil depth, organic matter content, and the use of water-control practices. Soils with shallow layers of infertile, impermeable, or rocky material are generally assigned *T* values near the low end of the range. In some tropical areas, the low nutrient-supplying power of the subsoil suggests that appropriate *T* values may be considerably lower than those used in the United States.

The majority of agricultural soils in the United States are currently assigned the highest *T* value, 11 Mg/ha. This represents a maximum allowable loss of about 0.9 mm of soil depth annually, a rate at which it would take about 225 years to lose the equivalent of an entire Ap horizon.[2] Under good agricultural management in a deep, perme-

---

[2] This calculation assumes that the Ap horizon is 200 mm thick and the bulk density of the soil is 1.25 Mg/m³.

able soil profile, this may be sufficient time to allow replacement of the lost material as new Ap horizon material forms from underlying subsoil material. Development of horizons in undisturbed soils under natural vegetation is most likely much slower than this, so the *T* value assigned to a particular type of soil probably would not provide sufficient protection for many rangeland or forest sites.

15.3 MECHANICS OF WATER EROSION

Soil erosion by water is fundamentally a three-step process (Figure 15.6):

1. *Detachment* of soil particles from the soil mass
2. *Transportation* of the detached particles downhill by floating, rolling, dragging, and splashing
3. *Deposition* of the transported particles at some place lower in elevation

On comparatively smooth soil surfaces, the beating action of raindrops causes most of the detachment. Where water is concentrated into channels, the cutting action of turbulent flowing water detaches soil particles. In some situations, freezing-thawing action also contributes to soil detachment.

Influence of Raindrops

A raindrop accelerates as it falls until it reaches *terminal velocity*—the speed at which the friction between the drop and the air balances the force of gravity. Larger raindrops fall faster, reaching a terminal velocity of about 30 km/h, or about 2 to 3 times as fast as a person can run. As the speeding raindrops impact the soil with explosive force, they transfer their high kinetic energy to the soil particles (see Figure 15.6).

FIGURE 15.6 The three-step process of soil erosion by water begins with the impact of raindrops on wet soil. (*a*) A raindrop speeding toward the ground. (*b*) The splash that results when the drop strikes a wet, bare soil. Such raindrop impact destroys soil aggregates, encouraging sheet and interrill erosion. Also, considerable soil may be moved by the splashing process itself. The raindrop affects the detachment of soil particles, which are then transported and eventually deposited in locations downhill (*c*).

Raindrop impact exerts three important effects: (1) it detaches soil; (2) it destroys granulation; and (3) its splash, under certain conditions, causes an appreciable transportation of soil. So great is the force exerted by raindrops that they not only loosen and detach soil granules, but may even beat the granules to pieces. As the dispersed material dries it may develop into a hard crust, which will prevent the emergence of seedlings and will encourage runoff from subsequent precipitation (see Section 4.8).

History may someday record that one of the truly significant scientific advances of the 20th century was the realization that most erosion is initiated by the impact of raindrops, rather than the flow of running water. For centuries prior to this realization, soil conservation efforts aimed at controlling the more visible flow of water across the land, rather than protecting the soil surface from the impact of raindrops.

Transportation of Soil

RAINDROP SPLASH EFFECTS.   When raindrops impact on a wet soil surface, they detach soil particles and send them flying in all directions (see Figure 15.6). On a soil subject to easy detachment, a very heavy rain may splash as much as 225 Mg/ha of soil, some of the particles splashing as much as 0.7 m vertically and 2 m horizontally. If the land is sloping or if the wind is blowing, splashing may be greater in one direction, leading to considerable net horizontal movement of soil.

ROLE OF RUNNING WATER.   Runoff water plays the major role in the transportation step of soil erosion. If the rate of rainfall exceeds the soil's infiltration capacity, water will pond on the surface and begin running downslope. The soil particles sent flying by raindrop impact will then land in flowing water, which will carry them down the slope. So long as the water is flowing smoothly in a thin layer (sheet flow), it has little power to detach soil. However, in most cases the water is soon channeled by irregularities in the soil surface and increases in both velocity and turbulence. The channelized flow then not only carries along soil splashed by raindrops, but begins to detach particles as it cuts into the soil mass.

Types of Water Erosion

Three types of water erosion are generally recognized: (1) *sheet,* (2) *rill,* and (3) *gully* (Figure 15.7). In **sheet erosion**, splashed soil is removed more or less uniformly, except that tiny columns of soil often remain where pebbles intercept the raindrops (see Figure 15.7*a*). However, as the sheet flow is concentrated into tiny channels (termed **rills**), **rill erosion** becomes dominant. Rills are especially common on bare land newly planted or in fallow (see Figure 15.7*b*). Rills are channels small enough to be smoothed over by normal tillage, but the damage is already done—the soil is lost. When sheet erosion takes place primarily between irregularly spaced rills, it is called **interrill erosion.**

Where the volume of runoff is further concentrated, the rushing water cuts deeper into the soil, deepening and coalescing the rills into larger channels termed **gullies** (see Figure 15.7*c*). This is **gully erosion.** Gullies on cropland are obstacles to tractors and cannot be removed by ordinary tillage practices. All three types may be serious, but sheet and rill erosion, although less noticeable than gully erosion, are responsible for most of the soil moved.

Deposition of eroded soil

Erosion may send soil particles on a journey of a thousand kilometers—off the hills, into creeks, and down great muddy rivers to the ocean. On the other hand, eroded soil may travel only a meter or two before coming to rest in a slight depression on a hillside or at the foot of a slope (as was shown in Figure 15.3*b*). The amount of soil delivered to a stream divided by the amount eroded is termed the **delivery ratio.** As much as 60% of eroded soil may reach a stream (delivery ratio = 0.60) in certain watersheds where valley slopes are very steep. As little as 1% may reach the streams draining a gently sloping coastal plain. Typically, the delivery ratio is larger for small watersheds than for large ones, because the latter provide many more opportunities for deposition before a major stream is reached. It is estimated that about 5 to 10% of all eroded soil in North America is washed out to sea. The remainder is deposited in reservoirs, river beds, on flood plains, or on relatively level land farther up the watershed.

(a) Sheet erosion

(b) Rill erosion

(c) Gully erosion

FIGURE 15.7 Three major types of soil erosion. *Sheet erosion* is relatively uniform erosion from the entire soil surface. Note that the perched stones and pebbles have protected the soil underneath from sheet erosion. The pencil gives a sense of scale. *Rill erosion* is initiated when the water concentrates in small channels (rills) as it runs off the soil. Subsequent cultivation may erase rills, but it does not replace the lost soil. *Gully erosion* creates deep channels that cannot be erased by cultivation. Although gully erosion looks more catastrophic, far more total soil is lost by the less obvious sheet and rill erosion. [Drawings from FAO (1987); photos courtesy USDA Natural Resources Conservation Service]

15.4 FACTORS AFFECTING EROSION BY WATER

The Universal Soil Loss Equation (USLE)[3]

The major factors that affect erosion by water are quantified in the **universal soil-loss equation (USLE)**:

$$A = RKLSCP$$

A, the predicted soil loss, is the product of

R = rainfall erosivity	} Rain-related factor
K = soil erodibility	
L = slope length	} Soil-related factors
S = slope gradient or steepness	
C = cover and management	
P = erosion-control practices	} Land-management factors

[3] For discussions of the original USLE, see Wischmeier and Smith (1978), and for the RUSLE, see Renard, et al. (1997). In this textbook we use the scientifically acceptable SI units for the *R* and *K* factors in our discussion of these erosion equations. However, since these soil-loss equations were published in the United States for use by landowners and the general public, most maps, tables, and computer programs available supply values for the *R* and *K* factors in customary English units, rather than in SI units. When using English units for the *R* and *K* factors, the soil loss *A* is expressed in tons (2000 lb) per acre, which can be easily converted to Mg/ha by multiplying by 2.24. For details on converting the customary English units to SI units, see Foster, et al. (1981).

The USLE was designed to predict only the amount of soil loss by sheet and rill erosion in an average year for a given location. It cannot predict erosion from a specific year or storm, nor can it predict the extent of gully erosion and sediment delivery to streams. It can, however, show how varying any combination of the soil- and land-management-related factors might be expected to influence soil erosion, and therefore can be used as a decision-making aid in choosing the most effective strategies to conserve soil.

The Revised Universal Soil Loss Equation (RUSLE)

The USLE has been used widely since the 1970s. In the early 1990s, the basic USLE was updated and computerized to create an erosion-prediction tool called the **revised universal soil-loss equation (RUSLE)**. The RUSLE uses the same basic factors of the USLE just shown, although some are better defined and interrelationships among them improve the accuracy of soil-loss prediction. The RUSLE is a computer software package that is constantly being improved and modified as experience is gained from its use around the world.

As we are about to see, the five factors that comprise the USLE provide a useful framework for understanding soil erosion and its control.

Rainfall Erosivity Factor R

The rainfall **erosivity** factor R represents the driving force for sheet and rill erosion. It takes into consideration the total rainfall and, more important, the intensity and seasonal distribution of the rain. Rain intensity is of great importance for two reasons: (1) intense rains have a large drop size, which results in much greater kinetic energy being available to detach soil particles; and (2) the higher the rate of rainfall, the more runoff that occurs, providing the means to transport detached particles. Gentle rains of low intensity may cause little erosion, even if the total annual precipitation is high. In contrast, a few torrential downpours may result in severe damage, even in areas of low annual rainfall. Likewise, soil losses are heavy if the rain falls when the soil is just thawing or is relatively bare because of recent disturbance.

Rainfall index values for locations in the United States are shown in Figure 15.8. Note that they vary from less than 10 in areas of the west to more than 700 along the coasts of Louisiana [the map gives R values in English units that can be converted to SI units of (MJ·mm)/(ha·h·yr) if multiplied by 17.02]. Similar data has been generated in other parts of the world. Generally, rainfall tends to be more intense and more erosive in subtropical and tropical regions than in temperate regions.

Rainfall intensity in most locations is so highly variable that actual erosivity in any one year is commonly 2 to 5 times greater or smaller than the long-term average. In fact, a few unusually intense, heavy storms often account for most of the erosion that takes place during a typical *decade*. Conservation practices based on the predictions of the USLE or RUSLE using long-term average R factors may not be sufficient to limit erosion damages from these relatively rare, but extremely damaging, storms.

Soil Erodibility Factor K

The soil **erodibility** factor K indicates a soil's inherent susceptibility to erosion. The K value assigned to a particular type of soil indicates the amount of soil lost per unit of erosive energy in the rainfall, assuming a standard research plot (22 m long, 9% slope) on which the soil is kept continuously bare by tillage.

The two most significant and closely related soil characteristics influencing erodibility are (1) *infiltration capacity,* and (2) *structural stability.* High infiltration means that less water will be available for runoff, and the surface is less likely to be ponded (which would make it more susceptible to splashing). Stable soil aggregates resist the beating action of rain, and thereby save soil even though runoff may occur. Certain tropical clay soils high in hydrous oxides of iron and aluminum are known for their highly stable aggregates that resist the action of torrential rains. Downpours of a similar magnitude on swelling-type clays would be disastrous.

Basic soil properties that tend to result in high K values include high contents of silt and very fine sand; expansive types of clay minerals; a tendency to form surface crusts; the presence of impervious soil layers; and blocky, platy, or massive soil structure. Soil properties that tend to make the soil more *resistant* to erosion (low K values) include high

FIGURE 15.8 The geographic distribution of *R* values for rainfall erosivity in the continental United States. Note the very high values in the humid, subtropical Southeast, where annual rainfall is high and intense storms are common. Similar amounts of annual rainfall along the coast of Oregon and Washington in the Northwest result in much lower *R* values because there the rain mostly falls gently over long periods. The complex patterns in the West are mainly due to the effects of mountain ranges. Values on map are in units of 100 (ft·ton·in.)/(acre·yr). To convert to Sl units of (MJ·mm)/(ha·h·yr), multiply by 17.02. [Redrawn from Renard, et al. (1997)]

soil organic matter content, nonexpansive types of clays, and strong granular structure. Approximate *K* values for soils at selected locations are shown in Table 15.1. Site-specific *K* values obtained with the RUSLE program will vary somewhat from these values. Note that the *K* factor in Sl units normally varies from near zero to about 0.1. Soils with high rates of water infiltration commonly have *K* values of 0.025 or below, while more easily eroded soils with low infiltration rates have *K* factors of 0.04 or higher.

Topographic Factor LS

The topographic factor *LS* reflects the influence of length and steepness of slope on soil erosion. It is expressed as a unitless ratio with soil loss from the area in question in the numerator, and that from a standard plot (9% slope, 22 m long) in the denominator. The longer the slope, the greater the opportunity for concentration of the runoff water. The *LS* factors in Table 15.2 illustrate the increases that occur as slope length and steepness increase. Three subtables are given for sites with low, moderate, and high ratios of rill to interrill (sheet) erosion.

Cover and Management Factor C

Erosion and runoff are markedly affected by different types of vegetative cover and cropping systems (Table 15.3). Undisturbed forests and dense grass provide the best soil protection and are about equal in their effectiveness. Forage crops (both legumes and grasses) are next in effectiveness because of their relatively dense cover. Small grains, such as wheat and oats, are intermediate and offer considerable obstruction to surface wash. Row crops, such as corn, soybeans, and potatoes, offer relatively little living cover during the early growth stages and thereby leave the soil susceptible to erosion unless residues from previous crops cover the soil surface.

TABLE 15.1   Computed K Values for Soils
at Different Locations

The values listed are in SI units. The computerized RUSLE model may give more accurate values for some locations.

Soil	Location	Computed[a] K
Udalf (Dunkirk silt loam)	Geneva, N.Y.	0.091
Udalf (Keene silt loam)	Zanesville, Ohio	0.063
Udult (Lodi loam)	Blacksburg, Va.	0.051
Udult (Cecil sandy clay loam)	Watkinsville, Ga.	0.048
Udoll (Marshall silt loam)	Clarinda, Iowa	0.044
Udalf (Hagerstown silty clay loam)	State College, Pa.	0.041
Ustoll (Austin silt)	Temple, Tex.	0.038
Aqualf (Mexico silt loam)	McCredie, Mo.	0.034
Udult (Cecil sandy loam)	Clemson, S.C.	0.034
Udult (Cecil sand loam)	Watkinsville, Ga.	0.030
Alfisols	Indonesia	0.018
Alfisols	Benin	0.013
Oxisols	Ivory Coast	0.013
Udult (Tifton loamy sand)	Tifton, Ga.	0.013
Ultisols	Hawaii	0.012
Alfisols	Nigeria	0.008
Udept (Bath flaggy silt loam)	Arnot, N.Y.	0.007
Ultisols	Nigeria	0.005
Oxisols	Puerto Rico	0.001

[a] To convert these *K* values from (Mg · ha · h)/(ha · MJ · mm) to English units of (ton · acre · h)/(100 acres · ft-ton · in.), simply multiply the values in this table by 7.6. From Wischmeier and Smith (1978); data for tropical soils cited by Cassel and Lal (1992).

Cover crops consist of plants that are similar to the forage crops just mentioned. They can provide soil protection during the time of year between the growing seasons for annual crops. For widely spaced perennial plantings, such as orchards and vineyards, cover crops can permanently protect the soil between rows of trees or vines. A mulch of plant residue or applied materials is also effective in protecting soils. Research on all continents has shown that a surface mulch does not have to be thick or cover the

TABLE 15.2   Some Representative Values of the Topographic
Factor LS on Three Types of Sites with Various Slope Gradients
and Lengths

Slope gradient, %	Approximate slope length, m				
	25	50	100	150	200
Sites with low ratios of rill to interrill erosion (e.g., many rangelands)					
2	0.25	0.27	0.30	0.32	0.33
4	0.47	0.55	0.64	0.70	0.75
8	0.93	1.16	1.44	1.65	1.80
12	1.58	2.05	2.65	3.08	3.42
16	2.30	3.05	4.06	4.80	5.40
Sites with moderate ratios of rill to interrill erosion (e.g., most tilled row-crop land)					
2	0.25	0.29	0.35	0.38	0.41
4	0.48	0.62	0.79	0.92	1.02
8	0.94	1.32	1.84	2.24	2.57
12	1.62	2.37	3.48	4.34	5.09
16	2.35	3.54	5.33	6.77	8.03
Sites with high ratios of rill to interrill erosion (e.g., freshly disturbed construction sites and new seedbeds)					
2	0.25	0.33	0.44	0.51	0.58
4	0.49	0.71	1.02	1.27	1.48
8	0.96	1.52	2.38	3.10	3.74
12	1.65	2.70	4.43	5.91	7.24
16	2.39	4.00	6.69	9.03	11.17

Modified from Renard, et al. (1997).

TABLE 15.3   Effect of Plant Cover on Soil Erosion by Water in the Humid Zone of West Africa

The data are averaged over many nearby sites, all with similar slopes, soils, and amounts of rainfall. The erosive influence of cultivation (with various types of tillage), especially bare fallows, is illustrated, as is the protective effect of undisturbed forest vegetation.

Type of cover	Number of sites	Mean rainfall, mm/y	Runoff, % of rainfall	Erosion, Mg/ha
Forest protected from fire	11	1293	0.9	0.10
Forest with light fires	13	1289	1.1	0.27
Natural grass fallow	7	1203	16.6	4.88
Groundnut (peanut)	32	1329	20.7	7.70
Upland rice	17	946	23.3	5.52
Maize (corn)	17	1405	17.7	7.63
Failed crops and bare soil	11	1154	39.5	21.28

Data selected from that cited by Pierre (1992).

soil completely to make a major contribution to soil conservation. Even small increases in surface cover result in large reductions in soil erosion, particularly interrill erosion (Figure 15.9).

Regulation of grazing to maintain a dense vegetative cover on range and pasture land and the inclusion of close-growing hay crops in rotation with row crops on arable land will help control both erosion and runoff. Likewise, the use of conservation tillage systems, which leave most of the plant residues on the surface, greatly decreases erosion hazards.

The *C* factor in the USLE or RUSLE will approach 1.0 where there is little soil cover (e.g., a bare seedbed in the spring or freshly graded bare soil on a construction site). It will be low (e.g., <0.10) where large amounts of plant residues are left on the land or in areas of dense perennial vegetation (Table 15.4).

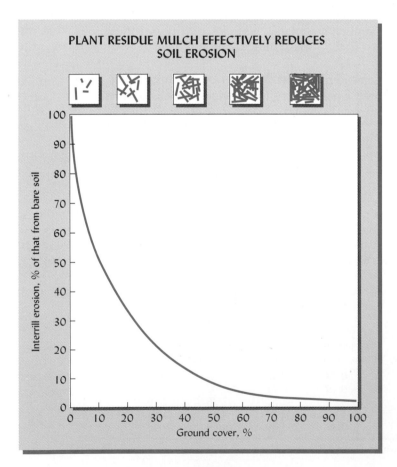

PLANT RESIDUE MULCH EFFECTIVELY REDUCES SOIL EROSION

FIGURE 15.9   Reduction in interrill erosion achieved by increasing ground cover percentage. The diagrams above the graph illustrate 5, 20, 40, 60, and 80% ground cover. Note that even a light covering of mulch has a major effect on soil erosion. The graph applies to interrill erosion. On steep slopes, some rill erosion may occur even if the soil is well covered. [Generalized relationship based on results from many studies]

TABLE 15.4 Examples of C Values for the Cover and Vegetation Management Factor

The C values indicate the ratio of soil eroded from a particular vegetation system to that expected if the soil were kept completely bare. Note the effects of canopy cover, surface litter (residue) cover, tillage, and crop rotation. The C values are site- and situation-specific and must be calculated from local information on plant growth habit, climate, and so on.

Vegetation	Management/condition	C value
Range grasses and low (<1m) shrubs	75% canopy cover, no surface litter	0.17
	75% canopy cover, 60% cover with decaying litter	0.032
Scrub brush about 2 m tall	25% canopy cover, no litter	0.40
	75% canopy cover, no litter	0.28
Trees with no understory, about 4 m drop fall.	75% canopy cover, no litter	0.36
	75% canopy cover, 40% leaf litter cover	0.09
	75% canopy cover, 100% leaf litter cover	0.003
Woodland with understory	90% canopy cover, 100% litter cover	0.001
Permanent pasture	Dense stand of grass sod	0.003
Corn–soybean rotation	Fall plowing, conventional tillage, residues removed	0.53
	Spring chisel plow–plant conservation tillage, 2500 kg/ha surface residues after planting	0.22
	No-till planting, 5000 kg/ha surface residues after planting	0.06
Corn–soybean–wheat–hay rotation	Fall plowing, conventional tillage, residues removed	0.20
	Spring chisel plow–plant conservation tillage, 2500 kg/ha surface residues after planting	0.13
	No-till planting, 5000 kg/ha surface residues after planting	0.05
Corn–oats–hay–hay rotation	Spring conventional plowing before planting	0.05
	No-till planting	0.03

Values typical of midwestern United States. Based on Wischmeier and Smith (1978) and Schwab, et al. (1996).

Values of *C* are specific to each region and type of vegetation or soil management. Estimates based on experiment data and field experience are available from conservation officers and the RUSLE program.

Support Practice Factor P[4]

On some sites with long and/or steep slopes, erosion control achieved by management of vegetative cover, residues, and tillage must be augmented by the construction of physical structures or other steps aimed at guiding and slowing the flow of runoff water. These **support practices** determine the value of the *P* factor in the USLE. The *P* factor is the ratio of soil loss with a given support practice to the corresponding loss if row crops were planted up and down the slope. If there are no support practices, the *P* factor is 1.0. The support practices include tillage on the contour (Figure 15.10), contour strip-cropping, terrace systems, and grassed waterways (Figure 15.11), all of which will tend to reduce the *P* factor.

Narrow strips of permanent vegetation (usually grasses or shrubs) planted on the contour can act as barriers to slow down runoff, trap sediment, and eventually build up "natural" or "living" terraces. In some situations, tropical grasses (e.g., a deep-rooted, drought-tolerant species called *vetiver grass*) have shown considerable promise as an affordable alternative to the construction of terraces (Figure 15.12). The deep-rooted vetiver plants have dense, stiff stems that tend to filter out soil particles from the muddy runoff. This sediment builds up on the upslope side of the grass barrier and, in time, actually creates a terrace that may be more than 1 m above the soil surface on the downslope side of the plants.

Examples of *P* values for contour tillage and strip-cropping at different slope gradients are shown in Table 15.5. Note that *P* values increase with slope and that they are lower for strip-cropping, illustrating the importance of this practice for erosion control. Terracing also reduces the *P* values. Unlike the USLE, the RUSLE takes into account

---

[4] Many of the erosion-control practices or management techniques discussed with regard to the *C* and *P* factors and in later sections of this chapter are considered to be **best management practices (BMPs)** under provisions of the Clean Water Act in the United States. The act defines BMPs as "optimal operating methods and practices for reducing or eliminating water pollution" from land-use activities.

FIGURE 15.10 Contour ridges must be carefully laid out with sufficient height to hold back water from even heavy rainfall. Here, surface retention is fast becoming surface runoff. (Photo courtesy of R. Weil)

interactions between support practices and subfactors such as slope and soil water infiltration.

The five factors of the USLE (*R, K, LS, C,* and *P*) have suggested many approaches to the practical control of soil erosion. A sample calculation is given in Box 15.1 to show how the USLE can help evaluate erosion-control options.

More than half of the soil eroded in the United States comes from agricultural land, with the remainder coming from forests, rangeland, and construction sites. We will now focus on several specific erosion-control technologies appropriate for these various types of land uses.

Grassed
waterway

FIGURE 15.11 An aerial photo of farmland in Kentucky where contour strip-cropping is practiced and grassed waterways (see arrow) control the flow of water off cropland and prevent gully erosion. (Inset) A grassed waterway in action, showing how the permanent grass sod can resist the scouring force of water and safely conduct water off the field. (Photos courtesy of USDA National Resources Conservation Service)

(a)

(b)

FIGURE 15.12   The use of vegetative barriers to create natural terraces. (Photo) A tropical grass (vetiver) has been planted vegetatively on the contour in a cassava field by pushing root sprigs into the soil. (*a*) The root cuttings are planted perpendicular to the slope direction. In a year or so the grass will be well established and its dense root and shoot growth will serve as a barrier to hold soil particles while permitting some water to pass on through. (*b*) Note the buildup of soil above the grass, basically forming a terrace wall. Perennial tall wheatgrass is being used experimentally in the Northern Plains area of the United States to serve as a barrier against snow movement and later against wind erosion. (Photo courtesy of Centro Internacional Agricultura Tropical in Cali, Colombia)

15.5   CONSERVATION TILLAGE[5]

For centuries, conventional agricultural practice around the world encouraged extensive soil tillage that leaves the soil bare and unprotected from the ravages of erosion (Figure 15.13*a*). During the last three decades, the development of herbicides and special planting equipment has allowed many farmers to manage their soils with greatly reduced tillage—or no tillage at all. Farmer interest in reduced tillage heightened as it was shown that these systems produced equal or even higher crop yields in many regions while saving time, fuel, money—and soil. The latter attribute earned these systems the name of **conservation tillage** (Table 15.6).

Conservation tillage systems range from those that merely reduce excess tillage to the no-tillage system, which uses no tillage beyond the slight soil disturbance that occurs as the planter cuts a planting slit through the residues and several cm into the soil

[5] For reviews of conservation tillage technology, see Blevins and Frye (1992) and Carter (1994).

TABLE 15.5   P Factors for Contour and Strip-Cropping at Different Slopes and the Terrace Subfactor at Different Terrace Intervals

The product of the contour or strip-cropping factors and the terrace subfactor gives the P value for terraced fields.

Slope, %	Contour P factor	Strip-cropping P factor	Terrace interval, m	Terrace subfactor Closed outlets	Terrace subfactor Open outlets
1–2	0.60	0.30	33	0.5	0.7
3–8	0.50	0.25	33–44	0.6	0.8
9–12	0.60	0.30	43–54	0.7	0.8
13–16	0.70	0.35	55–68	0.8	0.9
17–20	0.80	0.40	69–60	0.9	0.9
21–25	0.90	0.45	90	1.0	1.0

Contour and strip-cropping factors from Wischmeier and Smith (1978); terrace subfactor from Foster and Highfill (1983).

BOX 15.1 CALCULATIONS OF EXPECTED SOIL LOSS USING USLE

The RUSLE computer software is designed to calculate the expected soil loss from specific cropping systems at a given location.

The principles involved in both USLE and RUSLE can be verified by making calculations using USLE and its associated factors. Note that the factors in the USLE are related to each other in a multiplicative fashion. Therefore, if any one factor can be made to be near zero, the amount of soil loss A will be near zero.

Assume, for example, a location in Iowa on a Marshall silt loam with an average slope of 4% and an average slope length of 50 m. Assume further that the land is clean-tilled and fallowed.

Figure 15.8 shows that the R factor for this location is about 150 in English units or (150×17) 2550 in SI units. The K factor for a Marshall silt loam in central Iowa is 0.044 (Table 15.1) and the topographic factor LS from Table 15.2 is 0.71 (high rill to interrill ratio on soil kept bare). The C factor is 1.0, since there is no cover or other management practice to discourage erosion. If we assume the tillage is up and down the hill, the P value is also 1.0. Thus, the anticipated soil loss can be calculated by the USLE ($A = RKLSCP$):

$$A = (2550)(0.044)(0.71)(1.0)(1.0) = 79.7 \text{ Mg/ha or } 35.6 \text{ tons/acre}$$

If the crop rotation involved corn–soybean–wheat–hay and conservation tillage practices (e.g., spring chisel plow tillage) were used, a reasonable amount of residue would be left on the soil surface. Under these conditions, the C factor may be reduced to about 0.13 (Table 15.4). Likewise, if the tillage and planting were done on the contour, the P value would drop to about 0.5 (Table 15.5). P could be reduced further to 0.4 (0.5×0.8) if terraces with open outlets were installed about 40 m apart. Furthermore, with crops on the land the site would have a moderate rill to interrill erosion ratio, so the LS factor would be only 0.62 (middle of Table 15.2). With these figures the soil loss becomes:

$$A = (2550)(0.044)(0.62)(0.13)(0.4) = 3.6 \text{ Mg/ha or } 1.5 \text{ tons/acre}$$

The units for the calculation are not normallly shown, but they are:

$$\frac{MJ \cdot mm}{ha \cdot h \cdot yr} \times \frac{Mg \cdot ha \cdot h}{ha \cdot MJ \cdot mm} = \frac{Mg}{ha \cdot yr} \text{ (LS, C, and P are unitless ratios)}.$$

The benefits of good cover and management and support practices are obvious. The figures cited were chosen to provide an example of the utility of the universal soil-loss equation, but calculations can be made for any specific location. In the United States, pertinent factor values that can be used for erosion prediction in specific locations are generally available from state offices of the USDA Natural Resource Conservation Service. The necessary factors are also built into the RUSLE software.

(Figure 15.13c). The conventional moldboard plow was designed to leave the field "clean"; that is, free of surface residues. In contrast, conservation tillage systems, such as **chisel plowing** (Figure 15.13b), stir the soil but only partially incorporate surface residues, leaving more than 30% of the soil covered. **Stubble mulching**, whose water-conserving attributes were highlighted in Section 6.5, is another example. **Ridge tillage** is a conservation system in which crops are planted on top of permanent 15- to 20-cm-high ridges. About 30% soil coverage is maintained, even though the ridges are scraped off a bit for planting and then built up again by shallow tillage to control weeds.

With **no-tillage** systems we can expect 50 to 100% of the surface to remain covered (Table 15.7). Well-managed continuous no-till systems in humid regions include cover crops during the winter and high-residue-producing crops in the rotation. Such systems keep the soil completely covered at all times and build up organic surface layers somewhat like those found in forested soils.

Conservation tillage systems generally provide yields equal to or greater than those from conventional tillage, provided the soil is not poorly drained and in a cool region. However, during the transition from conventional tillage to no-tillage, crop yields may decline slightly for several years for reasons associated with some of the effects outlined in the following subsections.

Adaptation by Farmers[6]

In recent years conservation tillage has become increasingly popular, being used on about two-fifths of the nation's cropland. Conservationists project that as much as 75%

---

[6] For the inspiring story of one Chilean farmer's struggle to conquer the forces of erosion and degradation and restore the health of his soil, see Crovetto (1996).

TABLE 15.6   General Classification of Different Conservation Tillage Systems

All systems maintain at least 30% of the crop residues on the surface.

Tillage system	Operation involved
No-till	Soil undisturbed prior to planting, which occurs in narrow seedbed, 2.5 to 7.5 cm wide. Weed control primarily by herbicides.
Ridge till (till, plant)	Soil undisturbed prior to planting, which is done on ridges 10 to 15 cm higher than row middles. Residues moved aside or incorporated on about one-third of soil surface. Herbicides and cultivation to control weeds.
Strip till	Soil undisturbed prior to planting. Narrow and shallow tillage in row using rotary tiller, in-row chisel, and so on. Up to one-third of soil surface is tilled at planting time. Herbicides and cultivation to control weeds.
Mulch till	Soil surface disturbed by tillage prior to planting, but at least 30% of residues left on or near soil surface. Tools such as chisels, field cultivators, disks, and sweeps are used (e.g., stubble mulch). Herbicides and cultivation to control weeds.
Reduced till	Any other tillage and planting system that keeps at least 30% of residues on the surface.

Definitions used by Conservation Technology Information Center, West Lafayette, Ind.

of the cropland in the United States will be managed with some kind of conservation tillage within a decade or so. Already, 80% or more of the cropland in a few areas within the United States is managed with conservation tillage.

Erosion Control by Conservation Tillage

Since conservation tillage systems were initiated, hundreds of field trials have demonstrated that these tillage systems allow much less soil erosion than do conventional tillage methods. Surface runoff is also decreased, although the differences are not as pronounced as with soil erosion. These differences are reflected in the much lower *C* factor values assigned to conservation tillage systems (see Table 15.4).

The erosion-control value of an undisturbed surface residue mulch was discussed in the previous section. Conservation tillage also significantly reduces the loss of nutrients dissolved in runoff water or attached to sediment (review Tables 13.1 and 14.2).

Effect on Soil Properties

When soil management is converted from plow tillage to conservation tillage (especially no-tillage), numerous soil properties are affected, mostly in favorable ways. The

(a)

(b)

(c)

FIGURE 15.13   Conventional and conservation tillage practices. (*a*) In conventional tillage, a moldboard plow inverts the upper soil horizon, burying all plant residues and producing a bare soil surface. (*b*) A chisel plow, one type of conservation tillage implement, stirs the soil but leaves a good deal of the crop residues on the soil surface. (*c*) In no-till systems, one crop is planted directly into a cover crop or the residue of a previous cash crop, with only a narrow band of soil disturbed. No-till systems leave virtually all of the residue on the soil surface, providing up to 100% cover and nearly eliminating erosion losses. Here soybeans are planted into a cover crop that will be killed with a herbicide (weed-killing chemical) to form a surface mulch. (Photos courtesy of R. Weil)

TABLE 15.7 **TABLE 15.7   The Effect of Tillage Systems in Nebraska on the Percentage of Land Surface Covered by Crop Residues**

Note that the conventional moldboard plow system provided essentially no cover, while no-till and planting on a ridge (ridge till) provided best cover.

Tillage system	Number of fields	Percentage of fields with residue cover greater than			
		15%	20%	25%	30%
Moldboard	33	3	0	0	0
Chisel	20	40	15	5	0
Disk	165	40	20	9	4
Field cultivate	13	46	23	0	0
Ridge till (till, plant)	2	100	50	50	0
No-till	3	100	100	100	100
All systems	236	36	18	9	4

From Dickey, et al. (1987).

changes are most pronounced in the upper few centimeters of soil. Generally, the changes are greatest for systems that produce the most plant residue (especially corn and small grains in humid regions), retain the most residue coverage, and cause the least soil disturbance. Many of these changes are illustrated in other chapters of this textbook, so the discussion here will be brief.

PHYSICAL PROPERTIES.   Macroporosity and aggregation (see Sections 4.7 and 4.8) are increased as active organic matter builds up and earthworms and other organisms establish themselves. Infiltration and internal drainage are generally improved, as is soil water-holding capacity. Some of these effects are illustrated in Figure 15.14. The enhanced infiltration capacity of no-till-managed soils is generally very desirable, but in some cases it may lead to more rapid leaching of nitrates and other water-soluble chemicals.

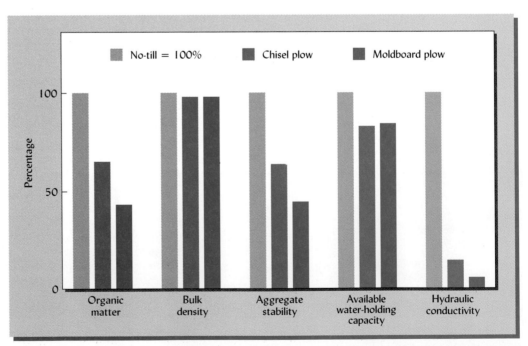

FIGURE 15.14   The comparative effects of 28 years of three tillage systems on soil organic matter content and a number of soil physical properties of an Alfisol in Ohio. Values for the no-till system were taken as 100, and the others are shown in comparison. Bulk density was about the same for each tillage system, but for all other properties the no-till system was decidedly more beneficial than either of the other two systems. The saturated hydraulic conductivity was especially high with no-till. [From Mahboubi, et al. (1993); used with permission of the Soil Science Society of America]

In cool regions, soils with restricted drainage may yield somewhat less using conservation tillage, because soil conditions are wetter and cooler than with conventional tillage. Reduced yields have discouraged the adoption of conservation tillage in these regions. However, limited preplanting tillage over the crop row (see Section 7.11) or ridge tillage are conservation tillage systems that allow at least part of the soil to warm faster and largely overcome these problems.

CHEMICAL PROPERTIES. No-tillage systems significantly increase the organic matter content of the upper few centimeters of soil (see Figures 15.14 and 11.19). During the initial four to six years of no-till management, the buildup of organic matter results in the immobilization of nutrients (see Section 11.3), especially nitrogen. This is in contrast to the mineralization of nutrients that is encouraged by the decline of soil organic matter under conventional tillage. Eventually, when soil organic matter stabilizes at a new higher level, nutrient mineralization rates under no-till increase.

In no-tillage systems, nutrient elements tend to accumulate in the upper few centimeters of soil as they are added to the soil surface in crop residues, animal manures, chemical fertilizers, and lime. However, research indicates that because of the surface mulch, crop roots (like those of trees in the untilled forest soil environment) have no trouble obtaining nutrients from the near-surface soil layers.

Without tillage to mix the soil, the acidifying effects of nitrogen oxidation, residue decomposition, and rainfall are concentrated in the upper few centimeters of soil, the pH of which may drop more rapidly than that of the whole plow layer in conventional systems (see Sections 9.4 and 9.11). In humid regions this acidity must be countered by application of liming materials to the soil.

BIOLOGICAL EFFECTS. The abundance, activity, and diversity of soil organisms tend to be greatest in conservation tillage systems characterized by high levels of surface residue and little physical soil disturbance (see Section 10.14). Earthworms and fungi, both important for soil structure, are especially favored.

15.6 CONTROL OF ACCELERATED EROSION ON RANGE AND FOREST LAND

Rangeland Problems

Many semiarid rangelands lose large amounts of soil under natural conditions (see Figure 15.2), but accelerated erosion can lead to even greater loses if human influences are not carefully managed. Overgrazing, which leads to the deterioration of the vegetative cover on rangelands, is a prime example. Grass cover generally protects the soil better than the scattered shrubs that usually replace it under the influence of poorly managed livestock grazing. In addition, cattle congregating around poorly distributed water sources and salt licks may completely denude the soil. Cattle trails, as well as ruts from off-road vehicles, can channelize runoff water and spawn gullies that eat into the landscape. Because of the prevalence of dry conditions, wind erosion (to be discussed in Section 15.8) also plays a major role in the deterioration of rangeland soils.

Erosion on Forest Lands

In contrast to deserts and rangelands, land under healthy, undisturbed forests loses very small amounts of soil. However, accelerated erosion can be a serious problem on forested land, both because the rates of soil loss may be quite high and because the amount of land involved is often enormous. The main cause of accelerated erosion in forested watersheds is usually the construction of logging roads, timber-harvest operations, and the trampling of trails and off-trail areas by large numbers of recreational users (or cattle, in some areas).

To understand and correct these problems, it is necessary to realize that the secret of low natural erosion from forested land is the undisturbed forest floor, the O horizons that protect the soil from the impact of raindrops and allow such high infiltration rates that surface runoff is very small or absent. Contrary to the common perception, it is the forest floor, rather than the tree canopy or roots, that protects the soil from erosion (Figure 15.15a). In fact, rainwater dripping from the leaves of tall trees often forms very

(a)

(b)

FIGURE 15.15   The leaf mulch on the forest floor, rather than the tree roots or canopy, provides most of the protection against erosion in a wooded ecosystem. (*a*) An undisturbed temperate deciduous forest floor (as seen through a rotten stump). The leafless canopy will do little to intercept rain during winter months. During the summer, rainwater dripping from the foliage of tall trees may impact the forest floor with as much energy as unimpeded rain. (*b*) Severe erosion has taken place under the tree canopy in a wooded area where the protective forest floor has been destroyed by foot traffic. The exposed tree roots indicate that nearly 25 cm of the soil profile has washed away. (Photos courtesy of R. Weil)

large drops that reach terminal velocity and impact the ground with more energy than direct rain from even the most intense of storms. If the forest floor has been disturbed and mineral soil exposed, serious splash erosion can result (see Figure 15.15*b*). Gully erosion can also occur under the forest canopy if water is concentrated, as by poorly designed roads.

Practices to Reduce Soil Loss Caused by Timber Production[7]

The main sources of eroded soil from timber production are *logging roads* (that are built to provide access to the area by trucks), *skid trails* (the paths along which logs are dragged), and *yarding areas* (the areas where collected logs are sized and loaded onto trucks). Relatively little erosion results directly from the mere felling of the trees (except where large tree roots are needed to anchor the soil against landslides). Strategies to control erosion should include consideration of the following: (1) intensity of timber harvest, (2) methods used to remove logs, (3) scheduling of timber harvests, and (4) design and management of roads (Figure 15.16) and trails. Soil disturbance in preparation for tree regeneration (such as tillage to eliminate weed competition or provide better seed-to-soil contact) must also be limited to sites with low susceptibility to erosion.

When forests are harvested, buffer strips as wide as 1.5 times the height of the tallest trees should generally be left untouched along all streams. As discussed in Section 14.2, buffer strips of dense vegetation have a high capacity to remove sediment and nutrients from runoff water. Forested buffers also protect the stream from excessive logging debris. In addition, streamside trees shade the water, protecting it from the undesirable heating that would result from exposure to direct sunlight.

---

[7] For a general introduction to this topic, see Nyland (1996).

FIGURE 15.16 An open-top culvert in a well-designed logging road in Montana. This simple structure, along with proper road bed alignment, can greatly reduce gully erosion caused by water flowing unimpeded along roads and trails in forested areas. Water bars or open-top culverts placed at frequent intervals lead runoff water, a little at a time, off the road and into densely vegetated areas. (Photo courtesy of R. Weil)

15.7 EROSION AND SEDIMENT CONTROL ON CONSTRUCTION SITES

Although active construction sites cover relatively little land in most watersheds, they may still be a major source of eroded sediment because the potential erosion per hectare on drastically disturbed land is commonly 100 times that on agricultural land. Heavy sediment loads are characteristic of rivers draining watersheds in which land use is changing from farm and forest to built-up land. Historically, once urbanization of a watershed is complete (all land being either paved over or covered by well-tended lawns), sedimentation rates return to levels as low (or lower) than before the development took place.

To prevent serious sediment pollution from construction sites, governments in the United States (e.g., through state laws and the federal Clean Water Act of 1992) and in many other industrialized countries require that contractors develop detailed erosion- or sediment-control plans before initiating construction projects that will disturb more than about 1 ha of land. The goals of erosion control on construction sites are (1) to avoid on-site damage, such as undercutting of foundations or finished grades and loss of topsoil needed for eventual landscaping; and (2) to retain eroded sediment on-site so as to avoid all the environmental damages (and liabilities) that would result from deposition of sediment on neighboring land and roads, and in ditches, reservoirs, and streams.

Principles of Erosion Control on Construction Sites

Five basic steps are useful in developing plans to meet the aforementioned goals:

1. When possible, schedule the main excavation activities for low-rainfall periods of the year.
2. Divide the project into as many phases as possible, so that only a few small areas must be cleared of vegetation and graded at any one time.
3. Cover disturbed soils as completely as possible, using vegetation or other materials.
4. Control the flow of runoff to move the water safely off the site without destructive gully formation.
5. Trap the sediment before releasing the runoff water off-site.

The last three steps bear further elaboration. They are best implemented as specific practices integrated into an overall erosion-control plan for the site.

Soils freshly disturbed by excavation or grading operations are characterized by very high erodibility (*K* values). This is especially true for low-organic-matter subsoil materials. Potential erosion can be extremely high (200 to 400 Mg/ha is not uncommon) unless the *C* value is made very low by providing good soil cover. This is best accomplished by allowing the natural vegetation to remain undisturbed for as long as possible, rather than clearing and grading the entire project area at the beginning of construction (see step 1, preceding). Once a section of the site is graded, any sloping areas not directly involved in the construction should be sodded or sown to fast-growing grass species adapted to the soil and climatic conditions.

Seeded areas should be covered with **mulch** or specially manufactured **erosion blankets** (Figure 15.17*b*). Erosion blankets, made of various biodegradable or non-biodegradable materials, provide instant soil cover and protect the seed from being washed away (Table 15.8).

A commonly used technology to protect steep slopes and areas difficult to access, such as road cuts, is the **hydroseeder** (Figure 15.17*a*) that sprays out a mixture of seed, fertilizer, lime (if needed), mulching material, and sticky polymers. Good construction-site management includes removal and stockpiling of the A-horizon material before an area is graded (see Figure 1.15). This soil material is often quite high in fertility and is a potential source of sediment and nutrient pollution. The stockpile should therefore be

FIGURE 15.17 Two methods of establishing vegetative cover on steep, unstable slopes. (Top) A hydroseeder allows vegetative cover to be efficiently established on difficult-to-reach areas. The machine is spraying a mixture of water, chopped straw, grass seed, fertilizer, and sticky polymers that hold the mulch in place until the grass seed can take root. (Bottom) Erosion-control mats made of plastic netting or natural materials like jute are laid down over newly seeded grass to hold the seed and soil in place until the vegetative cover is established. (Photos courtesy of R. Weil)

TABLE 15.8 Effectiveness of Various Erosion-Control Materials in Reducing Runoff and Soil Loss and in Establishing Vegetative Cover on Disturbed Sites

All plots (except sod) were seeded with grass and covered with the materials listed. The sod was by far the most effective material, followed by the straw mulch. In other applications (such as lining of channels) where high velocity of runoff is encountered, the manufactured erosion blankets would have an advantage.

Material	Soil loss,[a] kg/ha	Runoff loss, % of rain	Time to initiation of runoff, s	Ground cover established,[b] %
Bare soil	6650	83	34	50
Jute (woven mesh)	410	68	62	61
Wood shavings in nonwoven polyester netting	810	74	74	69
Coconut fiber mat	1070	76	86	58
Straw (4.5 Mg/ha)	590	60	102	76
Grass sod	100	28	341	NA[c]

[a] Soil loss from a single simulated rain event at a rate of 96 mm/h. Means of two soils: a Sassafras loamy sand with 8% slope and a Matapeake sandy clay loam with 15% slope.
[b] Percent vegetative cover established one year after Kentucky 31 tall fescue grass was seeded and covered by the various materials, in a separate study.
[c] Not applicable since sod provided 100% cover and was not seeded.
Data from Krenisky, et al. (1998)

given a grass cover to protect it from erosion until it is used to provide topsoil for landscaping around the finished structures.

Controlling the Runoff

Freshly exposed and disturbed subsoil material is highly susceptible to the cutting action of flowing water and the gullies so formed may ruin a grading job, undercut pavements and foundations, and produce enormous sediment loads. The flow of runoff water must be controlled by carefully planned grading, terracing, and channel construction. Most construction sites require a perimeter waterway to catch runoff before it leaves the site and to channel it to a retention basin.

The sides and bottoms of such channels must be covered with "armor" to withstand the cutting force of flowing water. Where high water velocities are expected, the soil must be protected with **hard armor** such as **riprap** (large angular rocks, such as are shown in Figure 15.18), **gabions** (rectangular wire-mesh containers filled with hand-sized stone), or interlocking concrete blocks. The soil is first covered with a **geotextile** filter cloth (a tough nonwoven material) to prevent mixing of the soil into the rock or stone.

In smaller channels, and on more gentle slopes where relatively low water velocities will be encountered, **soft armor,** such as grass sod or erosion blankets, can be used. Generally, soft armor is cheaper and more aesthetically appealing than hard armor. Newer approaches to erosion control often involve reinforced vegetation (e.g., trees or grasses planted in openings between concrete block or in tough erosion mats).

The term **bioengineering** describes techniques that use vegetation (locally native, noninvasive species are preferred) and natural biodegradable materials to protect channels subject to rather high water velocities. The so-called **live stake** technique (Figure 15.19) is an example of a bioengineering approach commonly used to stabilize soil along channels subject to high-velocity water.

Trapping the Sediment

For small areas of disturbed soils, several forms of sediment barriers can be used to filter the runoff before it is released. The most commonly used types of silt barriers are straw bales and woven fabric silt fences. If installed properly, both can effectively slow the water flow so that most of the sediment is deposited on the uphill side of the barrier (Figure 15.20), while relatively clear water passes through.

On large construction sites, a system of protected slopes and channels leads storm runoff water to one or more retention and sedimentation ponds located at the lowest elevation of the site. As the flowing water meets the still water in the pond it drops most

FIGURE 15.18   The flow of runoff from large areas of bare soil must be carefully controlled if off-site pollution is to be avoided. Here, a carefully designed channel with a grass sod bottom and sides lined with large rocks (riprap) prevents gully erosion, reduces soil loss, and guides runoff around the perimeter of a construction site. (Photo courtesy of R. Weil)

of its sediment load, allowing the relatively clear water to be skimmed off the top and released to the next pond or off the site. Wetlands (Section 7.7) are often constructed to help purify the overflow from sedimentation ponds before the water is released into the natural stream or river.

Construction site erosion-control measures are commonly designed to retain the runoff from small storms on-site. The retention ponds must also be able to deal with runoff generated from intense rainstorms—the kind that may be expected to occur on a site only once in every 10 to even 100 years. While expensive to construct, well-designed sediment-retention ponds can be incorporated as permanent aesthetic water features that enhance the value of the final project.

FIGURE 15.19   An example of *bioengineering* along a stream bed that had been disturbed during the construction of a commercial airport in Illinois. Living willow branches are pounded into the soft erodible stream bank to anchor the soil and reduce the scouring power of the water during high flows. Eventually the willow branches will take root and sprout (left), providing trees that will permanently stabilize the stream bank and improve wildlife habitat. (Photos courtesy of R. Weil)

FIGURE 15.20 Several types of sediment-control measures used around the periphery of a construction site. (Upper) A line of straw bales pegged to the ground allows water to seep through, but filters out much of the sediment and slows down the flow so that the water drops its sediment load. (Lower) A properly installed silt fence effectively removes sediment from runoff water leaving the edge of a construction site. The silt fencing material is a woven plastic that allows water to flow through at a much reduced rate. Note the light-colored sediment on the inside and the undisturbed forest floor on the outside of the silt fence. Improperly installed silt fencing is useless, as sediment-laden runoff passes underneath. A silt fence must be embedded in the soil to avoid this problem. (Photos courtesy of R. Weil)

15.8 WIND EROSION AND ITS CONTROL

Up to this point we have focused on soil erosion by water, but wind, too, causes much soil loss. Wind erosion is most common in arid and semiarid regions, and it is a problem on some soils in humid climates as well. It occurs when strong winds blow across soils with relatively dry surface layers. All kinds of soils and soil materials are affected. The finer soil particles may be carried to great heights and for thousands of kilometers—even from one continent, across the ocean, to another.

The erosive power of wind is demonstrated in Figure 15.21. The erosive force of the wind tends to be much greater in some regions than in others. For example, the semiarid Great Plains region of the United States is subject to winds with 5 to 10 times the erosive force of the winds common in the humid East. In the Great Plains, the winds are

FIGURE 15.21 Wind erosion in action. One result of wind erosion is dust storms, such as this one moving across the high plains of Texas. The swirling dark cloud consists of fine particles eroded from the soil by high winds sweeping across the flat farmland and range. Apparently, most of the land was not as well covered with vegetation as the wheat field in the foreground. (Photo courtesy of Dr. Chen Weinan, USDA Agricultural Research Service, Warm Springs, Tex.)

most powerful in the winter season. In other regions, high winds occur most commonly during the hot summers.

Wind erosion is a worldwide problem. Large areas of the former Soviet Union and sub-Saharan Africa have been badly damaged by wind erosion. Overgrazing and other misuses of the fragile lands of arid and semiarid areas have allowed wind erosion to depress soil productivity and bring starvation and misery to millions of people in these areas.

Mechanics of Wind Erosion

Like water erosion, wind erosion involves three processes: (1) *detachment,* (2) *transportation,* and (3) *deposition.* The moving air, itself, results in some detachment of tiny soil grains from the granules or clods of which they are a part. However, when the moving air is laden with soil particles, its abrasive power is greatly increased. The impact of these rapidly moving grains dislodges other particles from soil clods and aggregates. These dislodged particles are now ready for one of the three modes of wind-induced transportation, depending mostly on their size.

SALTATION.   The first and most important mode of particle transportation is that of **saltation,** or the movement of soil by a series of short bounces along the ground surface (Figure 15.22). The particles remain fairly close to the ground as they bounce, seldom rising more than 30 cm or so. Depending on conditions, this process may account for 50 to 90% of the total movement of soil.

SOIL CREEP.   Saltation also encourages **soil creep,** or the rolling and sliding along the surface of the larger particles. The bouncing particles carried by saltation strike the large aggregates and speed up their movement along the surface. Soil creep accounts for the movement of particles up to about 1.0 mm in diameter, which may amount to 5 to 25% of the total movement.

SUSPENSION.   The most spectacular method of transporting soil particles is by movement in **suspension.** Here, dust particles of a fine-sand size and smaller are moved parallel to the ground surface and upward. Although some of them are carried at a height no greater than a few meters, the turbulent action of the wind results in others being carried kilometers upward into the atmosphere and many hundreds of kilometers horizontally. These particles return to the earth only when the wind subsides and/or when precipitation washes them down. Although it is the most striking manner of transportation, suspension seldom accounts for more than 40% of the total and is generally no more than about 15%.

Factors Affecting Wind Erosion

SOIL MOISTURE.   Susceptibility to wind erosion is related to the moisture content of soils. Wet soils do not blow because of the adhesion between water and soil particles. Dry

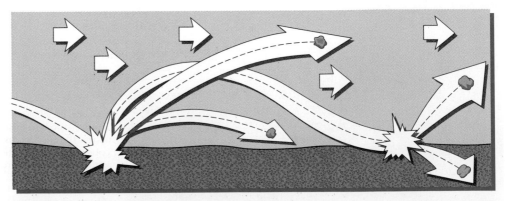

FIGURE 15.22   The process of saltation. Medium-sized particles (0.05 to 0.5 mm in diameter) bounce along the soil surface, striking and dislodging other particles as they move. They are too large to be carried long distances suspended in the air but small enough to be transported by the wind. [From Hughes (1980); used with permission of Deere & Company, Moline, Ill.]

winds generally lower the moisture content to below the wilting point before wind erosion takes place.

WIND VELOCITY AND TURBULENCE. The rate of wind movement, especially gusts having greater than average velocity, will influence erosion. Tests have shown that wind speeds of about 25 km/h are required to initiate soil movement. At higher wind speeds, soil movement is proportional to the cube of the wind velocity. Thus, the quantity of soil carried by wind goes up very rapidly as speeds above 30 km/h are reached. Wind turbulence also influences the capacity of the atmosphere to transport matter. Although the wind itself has some direct influence in picking up fine soil, the impact of wind-carried particles as they strike the soil is probably more important.

SOIL PROPERTIES. Wind erosion is less severe where soil surface roughness is enhanced by proper tillage methods, which leave large clods or ridges on the soil surface. Leaving a stubble mulch (see Section 6.6) is probably an even more effective way of reducing wind-borne soil losses.

Several other soil characteristics influencing wind erosion are (1) mechanical stability of soil clods and aggregates, (2) stability of soil crust, and (3) bulk density and size of erodible soil fractions. Some clods resist the abrasive action of wind-carried particles. If a soil crust resulting from a previous rain is present, it, too, may be able to withstand the wind's erosive power. The presence of clay, organic matter, and other cementing agents is also important in helping clods and aggregates resist abrasion. This is one reason why sandy soils, which are low in such agents, are so easily eroded by wind.

Soil particles or aggregates about 0.1 mm in diameter are more erodible than those larger or smaller in size. Thus, fine sandy soils are quite susceptible to wind erosion. Particles or aggregates about 0.1 mm in size are also partly responsible for the movement of other particles. Saltating particles bounce against larger particles, causing surface creep. The saltating particles also collide with smaller particles, sending them up into suspension.

VEGETATION. Vegetation or a stubble mulch will reduce wind erosion hazards, especially if the rows run perpendicular to the prevailing wind direction. This effectively slows wind movement near the soil surface. In addition, plant roots help bind the soil and make it less susceptible to wind damage.

Predicting and Controlling Wind Erosion

A wind erosion prediction equation (WEQ) has been in use since the late 1960s:

$$E = f(ICKLV)$$

The predicted wind erosion E is a function f of:

I = soil erodibility factor
C = climate factor
K = soil-ridge-roughness factor
L = width of field factor
V = vegetative cover factor

The WEQ involves the major factors that determine the severity of the erosion, but it also considers how these factors interact with each other. Consequently, it is not as simple as is the USLE for water erosion. The **soil erodibility factor** I relates to the properties of the soil and to the degree of slope of the site in question. The **soil-ridge-roughness** factor K takes into consideration the clodiness of the soil surface, vegetative cover V, and ridges on the soil surface. The **climatic factor** C involves wind velocity, soil temperature, and precipitation (which helps control soil moisture). The **width of field factor** L is the width of a field in the downwind direction. Naturally, the width changes as the direction of the wind changes, so the prevailing wind direction is generally used. The **vegetative cover** factor V relates not only to the degree of soil surface covered with residues, but to the nature of the cover—whether it is living or dead, still standing, or flat on the ground.

TABLE 15.9   Some Factors Integrated in the Revised Wind Erosion Equation (RWEQ) Model

The RWEQ program calculates values for each factor for each 15-day period during the year. It also calculates interactions; that is, it uses the value of one factor to modify other factors.

Model factor	Subfactor terms in the model	Comments
Weather factor	Weather factor *WF;* includes terms for wind velocity, direction, air temperature, solar radiation, rainfall, and snow cover.	Modified by other factors such as soil wetness.
Soil factors	Erodible fraction *EF;* fraction smaller than 0.84 mm diameter.	Based on sand, silt, organic matter, and rock cover.
	Soil crust factor *SCF.*	Induced by rainfall, eliminated by tillage.
	Surface roughness *SR;* a random component due to clods and/or an oriented component due to ridges.	Interacts with rainfall and tillage.
	Soil wetness *SW.*	Computed from rainfall minus evapotranspiration.
Tillage factor	Tillage factor *TF;* depends on type of implement, soil conditions, timing, and so forth.	Modifies surface roughness, crust factor, and so forth.
Hill factor	Hill factor HF; slope gradients and length input.	Affects wind speed (high going upslope, lower going down).
Irrigation factor	Irrigation factor *IR.*	Equivalent to added rainfall.
Crops factor	Flat residue SLR_f; factor.	Soil cover by residues lying on the surface is estimated for each crop, including changes over time due to decay, and so forth.
	Standing residue SLR_s; depends on crop, harvest height, plant density, and so forth.	Standing residues reduce the wind speed at the soil surface. Decay is slower than for flat residues.
	Crop canopy factor SLR_c.	Changes daily with crop growth.
Barriers	Barriers (e.g., tree windbreaks); includes orientation, density, height, spacing.	Reduce leeward wind velocity.

Based on information in Fryrear (1998).

A revised, more complex, and more accurate computer-based prediction model has been developed, and is known as the **revised wind erosion equation** (**RWEQ**). It is still an empirical model based on many years of research to characterize the relationship between observable conditions and resulting wind erosion severity. Table 15.9 outlines the factors taken into consideration by RWEQ.

Control of Wind Erosion

SOIL MOISTURE AND VEGETATION.   The factors of the wind erosion equation give clues to methods of reducing wind erosion. For example, since soil moisture increases cohesiveness, the wind speed required to detach soil particles increases dramatically as soil moisture increases. Therefore, where irrigation water is available, it is common to moisten the soil surface when high winds are predicted (Figure 15.23a). Unfortunately, most wind erosion occurs in dry regions without available irrigation. A vegetative cover also discourages soil blowing, especially if the plant roots are well established. In dry-farming areas, however, sound moisture-conserving practices require summer fallow on some of the land, and hot, dry winds reduce the moisture in the soil surface. Consequently, other means must be employed on cultivated lands of these areas.

TILLAGE.   Keeping the soil surface rough and maintaining some vegetative cover is accomplished by using appropriate tillage practices. However, the vegetation should be well anchored into the soil to prevent it from blowing away. Stubble mulch has proven to be effective for this purpose (see Section 6.6).

The effect of tillage depends not only on the type of implement used, but also on the timing of the tillage operation. Tillage can greatly reduce wind erosion if it is done while there is sufficient soil water to cause large clods to form. Tillage on a dry soil may produce a fine, dusty surface that aggravates the erosion problem. Tillage to provide for a cloddy surface condition should be at right angles to the prevailing winds. Likewise, strip-cropping and alternate strips of cropped and fallowed land should be perpendicular to the wind.

(a)

(b)

FIGURE 15.23 Wind breaks to reduce wind erosion. (*a*) When dry, cultivated Histosols are very light and fluffy and susceptible to wind erosion. The rows of trees (mainly willows) were planted perpendicular to the prevailing winds to slow the wind velocity and protect these valuable organic soils from erosion. Wetting the soil surface is another effective means of reducing wind erosion, as seen by the darker-colored field in the background (where the water table was raised) and the darker circles in the foreground where sprinkler irrigation was used. (*b*) The effect of a windbreak on wind velocity. The wind is deflected upward by the trees and is slowed down even before reaching them. On the leeward side further reduction occurs; the effect being felt as far as 20 times the height of the trees. [Photo (*a*) courtesy of USDA Natural Resources Conservation Service, (*b*) from FAO (1987)]

BARRIERS. Barriers such as shelter belts (Figure 15.23) are effective in reducing wind velocities for short distances and for trapping drifting soil. Various devices are used to control blowing of sands, sandy loams, and cultivated peat soils (even in humid regions). Windbreaks and tenacious grasses and shrubs are especially effective. Picket fences and burlap screens, though less efficient as windbreaks than such trees as willows, are often preferred because they can be moved from place to place as crops and cropping practices are varied. Rye, planted in narrow strips across the field, is sometimes used on peat lands and on sandy soils. Narrow rows of such perennial grasses as fall wheatgrass are being evaluated for a combination of wind erosion control and capturing of winter snows in the Northern Plains states.

15.9 LAND CAPABILITY CLASSIFICATION AS A GUIDE TO CONSERVATION

The land capability classification system devised by the U.S. Department of Agriculture uses eight **land capability classes** to indicate the *degree* of limitation imposed on land uses (Figure 15.24), with Class I the least limited and Class VIII the most limited. Each land use class may have four subclasses that indicate the *type* of limitation encountered:

FIGURE 15.24 Several land capability classes in San Mateo County, California. A range is shown from the nearly level land in the foreground (Class I), which can be cropped intensively, to the badly eroded hillsides (Classes VII and VIII). Although topography and erosion hazards are emphasized here, it should be remembered that other factors—drainage, stoniness, droughtiness—also limit soil usage and help determine the land capability class. (Courtesy USDA Natural Resources Conservation Service)

risks of erosion (e); wetness, drainage, or flooding (w); root-zone limitations, such as acidity, density, and shallowness (s); and climatic limitations, such as short growing season (c). The erosion (e) subclasses are the most common, and they will be the focus of our attention here. For example, Class IIe land is slightly susceptible to erosion, while Class VIIIe is extremely susceptible. Figure 15.25 shows the appropriate intensity of use allowable for each of these land capability classes if erosion losses (or problems associated with the other subclasses) are to be avoided.

In the United States, Classes I, V, and VIII each account for less than 3% of the non-federal land. The other classes are much more common, with about 20% of the land in each of Classes II, III, VI, and VII and about 14% in Class IV. This means that about 43% of the land in the United States (242 million ha) is suitable for regular cultivation (Classes I, II, and III). Another 14% (78 million ha) is marginal for growing cultivated crops. The remainder is suited primarily for grasslands and forests, and is used mostly for those purposes. A brief description of the land use capability classes follows.

CLASS I. These soils are deep and well drained, have nearly level topography (see Figure 15.24), and have few limitations on their use. They can be used continuously for row crops, tree nurseries, or for any of the less-intensive uses shown in Figure 15.25. For most purposes, these soils provide the greatest productivity at the least cost.

CLASS II. Soils in Class II have some limitations that reduce the choice of uses or require moderate conservation practices. Tillage and row-crop production should be restricted, or some conservation practices (conservation tillage, grassed waterways, contour strips, etc.) should be installed.

CLASS III. Soils in this class have severe limitations that reduce the choice of plants or require special conservation practices, or both. The same crops may be grown on Class III land as on Classes I and II land, but crops that provide soil cover, such as grasses and legumes, must be more prominent in the rotations used. The amount of land used for row crops is severely restricted. For the w subclass, drainage systems may be needed. Special care and scheduling for timber-harvest operation may be required.

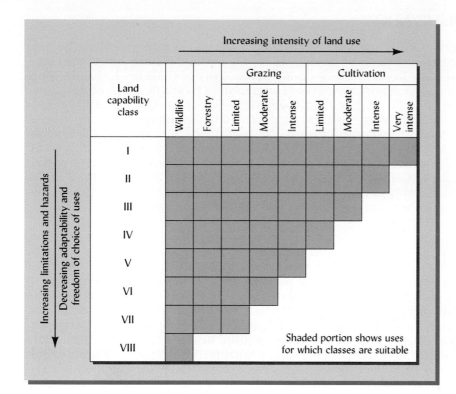

Increasing intensity of land use →

Land capability class	Wildlife	Forestry	Grazing			Cultivation			
			Limited	Moderate	Intense	Limited	Moderate	Intense	Very intense
I									
II									
III									
IV									
V									
VI									
VII									
VIII									

Increasing limitations and hazards
Decreasing adaptability and freedom of choice of uses

Shaded portion shows uses for which classes are suitable

FIGURE 15.25 Intensity with which each land capability class can be used with safety. Note the increasing limitations on the safe uses of the land as one moves from Class I to Class VIII. [Modified from Hockensmith and Steele (1949)]

CLASS IV.   Soils in this class can be used for cultivation, but there are severe limitations on the choice of crops and management practices. If the land is to be cultivated, crops should be mostly close-growing types (e.g., wheat or barley) and sod or hay crops must be used extensively. Row crops cannot be grown safely, except in no-till systems. The choice of crops may be limited by excess moisture (subclass w), as well as by erosion hazards. For tilled cropland, terracing, contour strips, or other special conservation practices are required. For timber harvest on forest land, special techniques may be required.

CLASS V.   These lands are generally not suited to crop production because of factors other than erosion hazards. Such limitations include (1) frequent stream overflow, (2) growing season too short for crop plants, (3) stony or rocky soils, and (4) ponded areas where drainage is not feasible. Often, pastures can be improved on this class of land.

CLASS VI.   Soils in this class have extreme limitations (typically, steep slopes and high erosion hazards) that restrict use largely to pasture, range, woodland, or wildlife. Forest entry should be limited, special care should be used in path construction, and alternative timber-harvest techniques should be used.

CLASS VII.   Severe limitations restrict the use of this land to controlled grazing, woodland, or wildlife. The physical limitations are the same as for Class VI, except they are so strict that pasture improvement is impractical. Cable or balloon timber-harvest methods may be necessary. Only small areas should be harvested, and entry should be strictly limited.

CLASS VIII.   In this land class are soils that should not be used for any kind of commercial plant production. Land use is restricted to recreation, wildlife, water supply, or aesthetic purposes. This land includes sand beaches, river wash, and rock outcrops.
    The eight land capability classes have become the starting point in the development of land-use plans that promote wise use and conservation of the land resource by thousands of farmers, ranchers, and other landowners.

15.10 PROGRESS IN SOIL CONSERVATION

Soil erosion in the United States accelerated when the first European settlers chopped down trees and began to farm the sloping lands of the humid eastern part of the country. Soil erosion was a factor in the declining productivity of these lands that, in time, led to their abandonment and the westward migration of people in search of new farmlands. Now, after some 70 years of soil conservation efforts, soil erosion is still a prominent problem on about half the total cropland area of the country.

Since 1982 rather remarkable progress has been made in reducing soil erosion, largely as a result of two factors: (1) the spread of conservation tillage (see Section 15.5), and (2) the implementation of land-use changes as part of the conservation reserve program. Progress continues to be made on both fronts. However, about one-third of the nation's cultivated cropland is still losing more than 11 Mg/ha/yr, the maximum loss that can be sustained without serious loss of productivity on most soils.

The Conservation Reserve Program

A major part of the reduction in soil erosion experienced in the United States since 1982 is due to government programs that have paid farmers to shift some land from crops to grasses and forests. Establishing grass or trees on these former croplands reduced the sheet and rill erosion from an average of 19.3 to 1.3 Mg/ha and the wind erosion from 24 to 2.9 Mg/ha. Between 1982 and 1992, 13 million ha of cropland were diverted to such noncultivated uses through the **Conservation Reserve Program** (CRP). Only about half (7 million ha) of the CRP land was on what is considered to be **highly erodible land** (HEL).[8] In 1995, the CRP was reoriented to better target these and other environmentally sensitive areas. By 1997, distribution of CRP land (Figure 15.26a) closely reflected the distribution of highly erodible croplands (Figure 15.26b).

The CRP is basically an arrangement by which the nation's taxpayers pay rent to farmers to forego cropping part of their farmland and instead plant grass or trees on it (more rent is paid for trees). The rental agreements (leases) run for 10 to 15 years, during which time the land is undisturbed. The benefits to the nation have been evident, not only in greatly reduced soil losses and sediment pollution, but also in a dramatic

---

[8] Highly erodible land is defined by the USDA Natural Resources Conservation Service as land for which the USLE predicts that soil loss on bare, unprotected soil would be eight times the T value for that soil: $R \times K \times LS \geq 8T$.

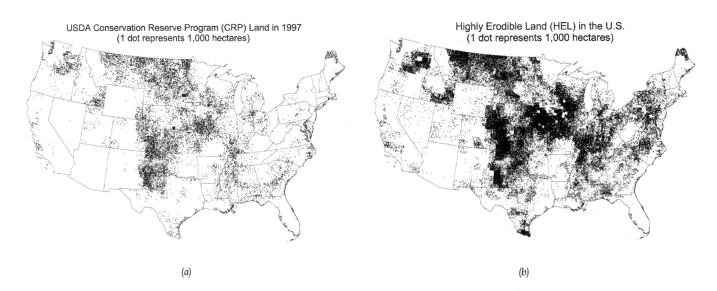

USDA Conservation Reserve Program (CRP) Land in 1997
(1 dot represents 1,000 hectares)

Highly Erodible Land (HEL) in the U.S.
(1 dot represents 1,000 hectares)

(a)   (b)

FIGURE 15.26 Distribution of land in the United States (a) enrolled in the Conservation Reserve Program (CRP) and (b) considered to be highly erodible land (HEL). Note that by 1997 the distributions appear to be quite similar, indicating that the lands most sensitive to erosion are those being taken out of crop production and put into grasses or trees. (Maps courtesy of Scott Brunton, Conservation Technology Information Center, West Lafayette, Ind.)

upswing in wildlife as bird and animal populations take advantage of the newly restored habitat. Where strips of land along streams (riparian zone buffers) have been incorporated into the CRP, water-quality benefits have also been amplified.

As satisfying as is the progress of the past decades, soil losses by erosion are still much too high. Continued efforts must be made to protect the soil and to hold it in place. In the United States, some 30 million ha of highly erodible cropland continues to lose an average of more than 15 Mg/ha of soil each year from water erosion, and an equal amount from wind erosion. In spite of remarkable progress, conservation tillage systems have *not* been adopted for more than half of the nation's cropland. And no one knows what will happen to the CRP lands when the rental leases expire. The battle to bring erosion under control has just begun, not only in the United States but throughout the world.

Conservation Management to Enhance Soil Quality

In a broader sense, conservation management practices are those that improve soil quality in more ways than just by protecting the soil from erosion. Properties that indicate the level of soil quality, especially those associated with soil organic matter, can be enhanced by such conservation measures as minimizing tillage, maximizing residue cover of the soil surface, providing for diversity of plant types, keeping soil under grass sod vegetation for at least part of the time, adding organic amendments where practical, and maintaining balanced soil fertility (Table 15.10). Improved soil quality, in turn, enhances the soil's capacity to support plants, resist erosion, prevent environmental contamination, and conserve water. Conservation management therefore can lead to the upward spiral of soil and environmental improvement referred to in Section 15.1.

Adapting Soil Conservation to the Needs of Resource-Poor Farmers

In much of the world, so little land is available to each farmer for food production that nations cannot afford the luxury of following the land-use capability classification recommendations as outlined in Section 15.9. Many farmers must use *all* land capable of food production simply to stave off starvation and impoverishment. These farmers often realize that farming erodible land jeopardizes their future livelihood and that of their children, but they see no choice. It is imperative to either find nonagricultural employment for these people or to find farming systems that are sustainable on these erodible lands.

Fortunately, some farmers and scientists have developed, through long traditions of adaptation or through innovation and research, farming systems that *can* produce food and profits while conserving such erodible soil resources. Examples include the traditional Kandy Home Gardens of Sri Lanka's humid mountains, in which a rain

TABLE 15.10 The Effect of Conservation Practices on Some Soil-Quality Factors Related to Organic Matter

In each region, soil was analyzed from seven pairs of adjacent fields, one on which conservation practices (reduced tillage, greater crop diversity, more sod crops in rotation, and/or use of organic nutrient sources) were used, while conventional practices (more tillage, less diversity, etc.) were used on the other.

Properties	Coastal plain soils		Piedmont soils	
	Conservation management	Conventional management	Conservation management	Conventional management
Total organic C, g/kg	10	9	19	15
Active organic C,[a] mg/kg	121	75	134	112
Microbial biomass C, % of total organic C	2.5	1.7	4.0	3.4
Nitrogen mineralization rate constant, d^{-1} [b]	44	35	46	34
Aggregate stability, %	72	56	73	66

[a] Mainly sugars extracted from soils after disruption with microwaves.
[b] The rate constant k in the first-order decay equation $N_p = N_o e^{kt}$
Data from Islam and Weil (2000).

forest–like mixed stand of tall fruit and nut trees is combined with an understory of pepper vines, coffee bushes, and spice plants to provide valuable harvests while keeping the soil under perennial vegetative protection. Another example comes from Central America, where farmers have learned to plant thick stands of velvet bean (*Mucuna*) or other viney legumes that can be chopped down by machete to leave a soil-protecting, water-conserving, weed-inhibiting mulch on steep farmlands. In Asia, steep lands have been carefully terraced in ways that allow production of food, even paddy rice, on very steep land without causing significant erosion.

Many examples from the United States and around the world make it clear that when governments cajole, pay, or force farmers into installing soil conservation measures on their land, the results are unlikely to be long-lasting. Usually, farmers will abandon the unwanted practices as soon as the pressure is off. On the other hand, if scientists and conservationists work *with* farmers to help them develop and adapt conservation systems that the farmers feel are of benefit to them and their land, then effective and lasting progress can be made. Experience with conservation tillage systems in the United States, mulch farming systems in Central America, and vegetative contour barriers in Asia have shown that farmers can help develop practices that are good for their land and for their profits: a win-win situation.

15.11 CONCLUSION

Accelerated soil erosion is one of the most critical environmental and social problems facing humanity today. Erosion degrades soils, making them less productive of the plants on which animals and people depend. Equally important, erosion causes great damages downstream in reservoirs, lakes, waterways, harbors, and municipal water supplies. Wind erosion also causes fugitive dust that may be very harmful to human health.

Nearly 4 billion Mg of soil is eroded each year on land in the United States alone. Half of this erosion occurs on the nation's croplands, the remainder on harvested timber areas, rangelands, and construction sites. Some one-third of the cropland still suffers from erosion that exceeds levels thought to be tolerable.

By sheet and rill erosion, water carries most of the sediment in humid areas. Gullies created by infrequent, but violent, storms account for much of the erosion in drier areas. Wind is the primary erosion agent in many drier areas, especially where the soil is bare and low in moisture during the season when strong winds blow.

Protecting soil from the ravages of wind or water is by far the most effective way to constrain erosion. In croplands and forests, such protection is due mainly to the cover of plants and their residues. Conservation tillage practices maintain vegetative cover on at least 30% of the soil surface, and the widening adoption of these practices has contributed to the significant reductions in soil erosion achieved over the past two decades. Crop rotations that include sod and close-growing crops, coupled with such practices as contour tillage, strip cropping, and terracing, also help combat erosion on farmland.

In forested areas, most erosion is associated with timber-harvesting practices and forest road construction. For the sake of future forest productivity and current water quality, foresters must become more selective in their harvest practices and invest more in proper road construction.

Construction sites for roads, buildings, and other engineering projects lay bare many scattered areas of soils that add up to a serious erosion problem. Control of sediment from construction sites requires carefully phased land clearing, along with vegetative and artificial soil covers, and installation of various barriers and sediment-holding ponds. These measures may be expensive to implement, but the costs to society that result when sediment is not controlled are too high to ignore. Once construction is completed, erosion rates on urban areas are commonly as low as those on areas under undisturbed native vegetation.

Erosion-control systems must be developed in collaboration with those who use the land, and especially the poor, for whom immediate needs must overshadow concerns for the future. As put succinctly in *The River*, a classic 1930s documentary film produced during the rebirth of American soil erosion consciousness, "Poor land makes poor people, and poor people make poor land."

STUDY QUESTIONS

1. Explain the distinction between *geologic erosion* and *accelerated erosion*. Is the difference between the two greater in humid or arid regions?

2. When erosion takes place by wind or water, what are three important types of damages that result on the land whose soils are eroding? What are five important types of damages that erosion causes in locations away from the eroding site?

3. What is a common *T* value, and what is meant by this term? Explain why certain soils have been assigned a higher *T* value than other soils.

4. Describe the three main steps in the water erosion process.

5. Many people assume that the amount of soil eroded on the land in a watershed (*A* in the universal soil-loss equation) is the same as the amount of sediment carried away by the stream draining that watershed. What factor is missing that makes this assumption incorrect? Do you think that this means the USLE should be renamed?

6. Why is the total annual rainfall in an area *not* a very good guide to the amount of erosion that will take place on a particular type of bare soil?

7. Contrast the properties you would expect in a soil with either a very high *K* value or a very low *K* value.

8. How much soil is likely to be eroded from a Keene silt loam in central Ohio, on a 12% slope, 100 m long, if it is in dense permanent pasture and has no support practices applied to the land? Use the information available in this chapter to calculate an answer.

9. What type of conservation tillage leaves the greatest amount of soil cover by crop residues? What are the advantages and disadvantages of this system?

10. Why are narrow strips of grass planted on the contour sometimes called a "living terrace"?

11. In most forests, which component of the ecosystem provides the primary protection against soil erosion by water, the *tree canopy, tree roots,* or *leaf litter*?

12. Certain soil properties generally make land susceptible to erosion by wind or erosion by water. List four properties that characterize soils highly susceptible to wind erosion. Indicate which two of these properties should also characterize soils highly susceptible to water erosion, and which two should not.

13. Which three factors in the wind erosion prediction equation (WEQ) can be affected by tillage? Explain.

14. Describe a soil in land capability Class IIw in comparison with one in Class IVe.

15. Why is it important that there be a close relationship between land in the CRP and that considered to be HEL?

REFERENCES

Blevins, R. L., and W. W. Frye. 1992. "Conservation tillage: An ecological approach to soil management," *Advances in Agronomy,* **51:**33–78.

Carter, M. R. 1994. *Conservation Tillage in Temperate Agroecosystems.* (Boca Raton, Fla.: Lewis Publishers).

Cassel, D. K., and R. Lal. 1992. "Soil physical properties of the tropics: Common beliefs and management constraints," in R. Lal and P. A. Sanchez (eds.), *Myths and Science of Soils of the Tropics.* SSA Special Publication no. 29. (Madison, Wis.: Soil Sci. Soc. of Amer.), pp. 61–89.

Crovetto, C. 1996. *Stubble Over the Soil.* (Madison, Wis.: Amer. Soc. Agron.).

Daily, G. 1997. "Restoring value to the world's degraded lands," *Science,* **269:**350–354.

Dickey, E. C., et al. 1987. "Conservation tillage: Perceived and actual use," *J. Soil Water Cons.,* **42:**431–434.

FAO. 1987. *Protect and Produce.* (Rome: U.N. Food and Agriculture Organization).

Follet, R. F., and B. F. Stewart (eds.). 1985. *Soil Erosion and Crop Productivity*. (Madison, Wis.: Amer. Soc. Agron.).

Foster, G. R., D. K. McCool, K. G. Renard, and W. C. Moldenhauer. 1981. "Conversion of the universal soil loss equation to Sl metric units," *J. Soil Water Cons.*, **36**:355–359.

Foster, G. R., and R. E. Highfill. 1983. "Effect of terraces on soil loss: USLEP factor values for terraces," *J. Soil Water Cons.*, **38**:48–51.

Fryrear, D. W. 1998. "RWEQ: Improved wind erosion technology," *J. Soil Water Cons.*, **53**:(in press)

Hillel, D. 1991. *Out of the Earth: Civilization and the Life of the Soil*. (New York: The Free Press).

Hockensmith, R. D., and J. G. Steel. 1949. "Recent trends in the use of the land-capability classification," *Soil Sci. Soc. Amer. Proc.*, **14**:383–388.

Hudson, N. 1995. *Soil Conservation*, 3rd ed. (Ames, Iowa: Iowa State University Press).

Hughes, H. A. 1980. *Conservation Farming*. (Moline, Ill.: John Deere and Company).

Islam, K. R., and R. R. Weil. 2000. Soil quality indicator properties in the mid-Atlantic region as influenced by conservation management. *J. Soil Water Cons.*, **55**:(in press).

IUCN. 1986. *The IUCN Sahel Report*. (Gland, Switzerland: Intl. Union Cons. Nature).

Krenisky, E. C., M. J. Carroll, R. L. Hill, and J. M. Krouse. 1998. "Runoff and sediment losses from natural and man-made erosion control materials," *Crop Sci.*, **38**:1042–1046.

Lattanzi, A. R., L. D. Meyer, and M. F. Baumgardner. 1979. "Influence of mulch rate and slope steepness in interrill erosion," *Soil Sci. Soc. Amer. Proc.*, **38**:946–950.

Mahboubi, A. A., R. Lal, and N. R. Faussey. 1993. "Twenty-eight years of tillage effects on two soils in Ohio," *Soil Sci. Soc. Amer. J.*, **57**:506–512.

Nyland, R. D. 1996. *Silviculture: Concepts and Applications*. (New York: McGraw-Hill).

Oldeman, L. R. 1994. "The global extent of soil degradation," in D. J. Greenland, and I. Szabolcs (eds.), *Soil Resilience and Sustainable Land Use*. (Wallingford, U.K.: CAB International).

Pierre, C. J. M. G. 1992. *Fertility of Soils: A Future for Farming in the West African Savannah*. (Berlin: Springer-Verlag).

Renard, K. G., G. Foster, G. Weesies, D. McCool, and D. Yoder. 1997. *Predicting Soil Erosion by Water: A Guide to Conservation Planning with the Revised Universal Soil Loss Equation (RUSLE)*. Agricultural Handbook no 703. (Washington, D.C.: USDA).

Rosensweig, C., and D. Hillel. 1998. *Climate Change and the Global Harvest: Potential Impacts of the Greenhouse Effect on Agriculture*. (Cary, N.C.: Oxford University Press).

Schwab, G. O., D. D. Fangmeirer, and W. J. Elliot. 1996. *Soil and Water Management Systems*, 4th ed. (New York: Wiley).

Skidmore, E. L., and N. P. Woodruff. 1968. *Wind Erosion Forces in the United States and Their Use in Predicting Soil Loss*. Agricultural Handbook no. 346. (Washington, D.C.: USDA).

USDA. 1994. *Summary Report: 1992 National Resources Inventory*. (Washington, D.C.: USDA Natural Resources Conservation Service).

Wischmeier, W. J., and D. D. Smith. 1978. *Predicting Rainfall Erosion Loss—A Guide to Conservation Planning*. Agricultural Handbook no. 537. (Washington, D.C.: USDA).

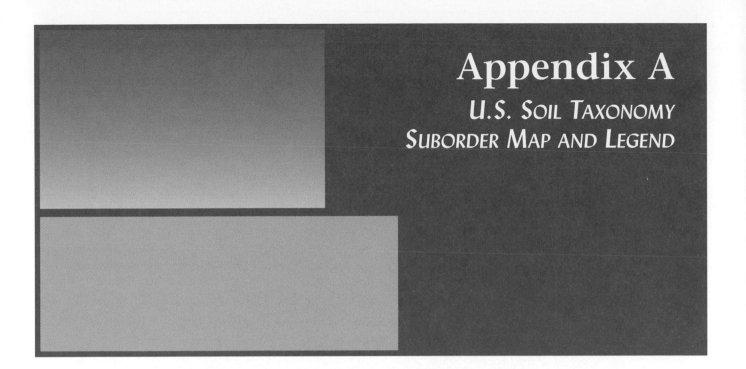

Appendix A
U.S. Soil Taxonomy
Suborder Map and Legend

Soil surveys have been prepared for most counties of the continental United States. The Soil Survey Staff of the U.S. Natural Resources Conservation Service, in cooperation with scientists from other countries, have used information from these surveys and analyses to develop generalized soils maps based on *Soil Taxonomy*. Areas dominated by specific soil orders and suborders are delineated on the map using map symbols (A3a, D1s, etc.). The soil orders and suborders to which these symbols refer are shown on the page facing the map. It should be kept in mind that while the map units represent areas of land dominated by the soil orders and suborders indicated, each unit is also likely to include some soils that differ from those listed in the map legend.

FIGURE A.I General soil map of the United States showing patterns of soil orders and suborders based on *Soil Taxonomy.* Explanations of symbols follow. [Courtesy USDA Natural Resources Conservation Service, National Soil Survey Center (1998)]

ALFISOLS

Aqualfs
A1a Aqualfs with Udalfs, Aquepts, Udolls, Aquolls; gently sloping.

Cryalfs
A2S Cryalfs with Cryolls, Cryepts, Haplocryods, and Rock outcrops; steep.

Udalfs
A2a Udalfs with Udipsamments and Histosols; gently and moderately sloping.

A3a Udalfs with Aqualfs, Aquolls, Rendolls, Udolls, and Udults; gently or moderately sloping.

Ustalfs
A4a Ustalfs with Ustepts, Ustolls, Usterts, Ustipsamments, and Ustorthents; gently or moderately sloping.

Xeralfs
A5S1 Xeralfs with Xerolls, Xerorthents, and Xererts; moderately sloping to steep.

A5S2 Ultic and lithic subgroups of Haploxeralfs with Xerands, Xerults, Xerolls, and Xerepts; steep.

ANDISOLS

Cryands
C1a Cryands, Cryaquepts, Cryaquands, Histosols, and Rock outcrop; gently or moderately sloping.

C1S1 Cryands with Cryepts and Cryods; steep.

Udands
C2a Udands with Udepts, Spodosols, and Aquepts.

Ustands
C3S2 Ustands with Ustepts, Ustolls, Udands, and Folists; moderately sloping to steep.

ARIDISOLS

Argids
D1a Argids with Cambids, Calcids, Orthents, Psamments, and Ustolls; gently and moderately sloping.

D1S Argids with Cambids, gently sloping; and Torriorthents, gently sloping to steep.

Cambids
D2a Cambids with Argids, Orthents, Psamments, and Xerolls; gently or moderately sloping.

D2S Cambids, gently sloping to steep, with Argids, gently sloping; lithic subgroups of Torriorthents and Xerorthents, both steep.

ENTISOLS

Aquents
E1a Aquents with Quartzipsamments, Aquepts, Aquolls, and Aquods; gently sloping.

Orthents
E2a Torriorthents, steep, with xeric and ustic subgroups of Aridisols; Usterts and aridic and vertic subgroups of Ustolls; gently or moderately sloping.

E2b Torriorthents with Torrerts; gently or moderately sloping.

E2c Xerorthents with Xeralfs, Cambids, and Argids; gently sloping.

E2S1 Torriorthents; steep, and Argids, Torrifluvents, and Ustolls; gently sloping

E2S2 Xerorthents with Xeralfs and Xerolls; steep.

E2S3 Cryothents with Cryopsamments and Cryands; gently sloping to steep.

Psamments
E3a Quartizipsamments with Aquults and Udults; gently or moderately sloping.

E3b Udipsamments with Aquolls and Udalfs; gently or moderately sloping.

E3c Ustipsamments with Ustalfs and Aquolls; gently or moderately sloping.

GELISOLS
G1P Gelisols; gently sloping to steep.

G2P Gelisols with Cryorthents, Cryepts, and cryic great groups of Histosols; gently sloping to steep.

HISTOSOLS
H1a Haplohemists with Psammequents and Udipsamments; gently sloping.

H2a Haplohemists and Haplosaprists with Fluvaquents, Hydraquents and Aquepts; gently sloping.

H3a Haplofibrists, Haplohemists, and Haplosaprists with Psammaquents; gently sloping.

INCEPTISOLS

Aquepts
I2a Epi and endo great groups of Aquepts with Aqualfs, Aquerts, Aquolls, Udalfs, and Fluvaquents; gently sloping.

Cryepts
I3a1 Cryepts with Cryaquepts, Histosols, and Cryods; gently or moderately sloping.

I3a2 Cryepts, Cryands with Xerepts, Xerolls, and Cryods; moderately sloping to steep.

Udepts
I3b Eutrudepts with Uderts; gently sloping.

I3c Fragudepts with Fragaquepts; gently or moderately sloping; and Dystrudepts, steep.

I3d Dystrudepts with Udipsamments and Haplorthods; gently sloping.

I3S Dystrudepts, steep, with Udalfs and Udults; gently or moderately sloping.

Xerepts
I3e Xerepts, Xerands with Aquepts and Orthods; gently or moderately sloping.

MOLLISOLS

Aquolls
M1a Aquolls with Udalfs, Fluvents, Udipsamments, Aquepts, and Eutrudepts; gently sloping.

M1b Aquolls and Aquerts with Udolls and Udipsamments; gently sloping.

Udolls
M2a Udolls with Aquolls and Udorthents; gently sloping.

M3a Udolls with Aquolls, Udalfs, Aqualfs, Fluvents, Udipsamments, Udorthents, Aquepts, and Albolls; gently or moderately sloping.

Ustolls
M2b Typic subgroups of Ustolls with Ustipsamments, Ustorthents, and Cryalfs; gently sloping.

M2c Aridic subgroups of Ustolls and Ustic subgroups of Argids, Cambids, and Torriorthents; gently sloping.

M2S Ustolls with Ustalfs, Argids, Torriorthents, and Ustolls; moderately sloping or steep.

M4a Udic subgroups of Ustolls with Ustorthents, Ustepts, Usterts, Aquents, Fluvents, and Udolls; gently or moderately sloping.

M4b Typic subgroups of Ustolls with Ustalfs, Ustipsamments, Ustorthents, Ustepts, Aquolls, and Usterts; gently or moderately sloping.

M4c Aridic subgroups of Ustolls with Ustalfs, Cambids, Ustipsamments, Ustorthents, Ustepts, Torriorthents, and Usterts; moderately sloping.

M4S Ustolls with Argids and Torriorthents; moderately sloping or steep.

Xerolls
M5a Xerolls with Argids, Cambids, Fluvents, Cryalfs, Cryolls, and Xerorthents; gently or moderately sloping.

M5S Xerolls with Cryalfs, Xeralfs, Xerorthents, and Xererts; moderately sloping or steep.

OXISOLS
O1a Ustox with Dystrustepts, Ustolls, Andisols, Udox, and Torrox; gently to steeply sloping.

SPODOSOLS

Aquods
S1a Aquods with Psammaquents, Aquolls, Alorthods, and Aquults; gently sloping.

Orthods
S2a Orthods with Udalfs, Aquents, Udorthents, Udipsamments, Histosols, Aquepts, Fragudepts, and Dystrudepts; gently or moderately sloping.

S2S1 Orthods and Cryods with Histosols, Aquents, and Aquepts; moderately sloping or steep.

Cryods
S2S2 Cryods with Histosols; moderately sloping or steep.

S2S3 Cryods with Cryic great groups of Histosols, Cryands and Cryaquepts; gently sloping to steep.

ULTISOLS

Aquults
U1a Aquults with Aquents, Histosols, Quartzipsamments, and Udults; gently sloping.

Humults
U2Sa Humults and Xeralfs with Xerolls, Aquepts and Xerults; gently sloping to moderately sloping.

U2Sb Humults, Xeralfs, and Xerults with Xerolls and Rock outcrop; gently sloping to steep.

Udults
U3a Udults with Udalfs, Fluvents, Aquents, Quartzipsamments, Aquepts, Dystrudepts, and Aquults; gently or moderately sloping.

U3S Udults with Dystrudepts; moderately sloping or steep.

VERTISOLS

Uderts
V1a Uderts with Aqualfs, Aquerts, Eutrudepts, Aquolls, and Ustolls; gently sloping.

Usterts
V2a Usterts with Aqualfs, Cambids, Aquerts, Fluvents, Aquolls, Ustolls, and Torrerts; gently sloping.

Miscellaneous Areas (with little soil)
X1 Salt flats and Playas.

X2 Rock outcrop (plus permanent snow fields, glaciers, and lava flows).

Slope Classes
Gently sloping—Slopes mainly less than 10%, including nearly level.

Moderately sloping—Slopes mainly between 10 and 25%.

Steep—Slopes mainly steeper than 25%.

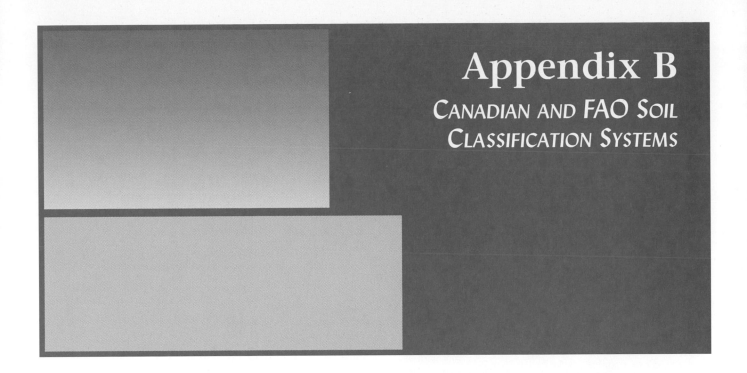

Appendix B
CANADIAN AND FAO SOIL CLASSIFICATION SYSTEMS

The Canadian Soil Classification System is one of many national soil classification systems used in various countries around the world. Of these, it is perhaps the most closely aligned with the U.S. *Soil Taxonomy*. It includes five hierarchical categories: order, great group, subgroup, family, and series. The system is designed to apply principally to the soils of Canada. The soil orders of the Canadian System of Soil Classification are described in Table B.1 and soil orders and some great groups are compared to the U.S. *Soil Taxonomy* in Table B.2.

TABLE B.1 Summary with Brief Descriptions of the Soil Orders in the Soil Classification System of Canada

Brunisolic	Soils with sufficient development to exclude them from the Regosolic order, but lacking the degree or kind of horizon development specified for other soil orders.
Chernozemic	Soils with high base saturation and surface horizons darkened by the accumulation of organic matter from the decomposition of plants from grassland or grassland-forest ecosystems.
Cryosolic	Soils formed in either mineral or organic materials that have permafrost either within 1 m of the surface or within 2 m if more than one-third of the pedon has been strongly cryoturbated, as indicated by disrupted, mixed, or broken horizons.
Gleysolic	Gleysolic soils have features indicative of periodic or prolonged saturation (i.e., gleying, mottling) with water and reducing conditions.
Luvisolic	Soils with light-colored, eluvial horizons that have illuvial B horizons in which silicate clay has accumulated.
Organic	Organic soils developed on well- to undecomposed peat or leaf litter.
Podzolic	Soils with a B horizon in which the dominant accumulation product is amorphous material composed mainly of humified organic matter combined in varying degrees with Al and Fe.
Regosolic	Weakly developed soils that lack development of genetic horizons.
Solonetzic	Soils that occur on saline (often high in sodium) parent materials, which have B horizons that are very hard when dry and swell to a sticky mass of very low permeability when wet. Typically the solonetzic B horizon has prismatic or columnar macrostructure that breaks to hard to extremely hard, blocky peds with dark coatings.
Vertisolic	Soils with high contents of expanding clays that have large cracks during the dry parts of the year and show evidence of swelling, such as gilgae and slickensides.

TABLE B.2 Comparison of U.S. *Soil Taxonomy* and the Canadian Soil Classification System

Note that because the boundary criteria differ between the two systems, certain U.S. Soil Taxonomy soil orders have equivalent members in more than one Canadian Soil Classification System soil order.[a]

U.S. Soil Taxonomy soil order	Canadian system soil order	Canadian system great group	Equivalent lower-level taxa in U.S. Soil Taxonomy
Alfisols	Luvisolic	Gray Brown Luvisols	Hapludalfs
		Gray Luvisols	Haplocryalfs, Eutrocryalfs, Fragudalfs, Glossocryalfs, Palecryalfs, and some subgroups of Ustalfs and Udalfs
	Solonetzic	Solonetz	Natrudalfs and Natrustalfs
		Solod	Glossic subgroups of Natraqualfs, Natrudalfs, and Natrustalfs
Andisols	Components of Brunisolic and Cryosolic		
Aridisols	Solonetzic		Frigid families of Natrargids
Entisols	Regosolic		Cryic great groups and frigid families of Entisols, except Aquents
		Regosol	Cryic great groups and frigid families of Folists, Fluvents, Orthents, and Psamments
Gelisols	Cryosolic	Turbic Cryosol	Turbels
		Organic Cryosol	Histels
		Stagnic Cryosol	Orthels
Histosols	Organic	Fibrisol	Cryofibrists, Sphagnofibrists
		Mesisol	Cryohemists
		Humisol	Cryosaprists
Inceptisols	Brunisolic	Melanic Brunisol	Some Eutrustepts
		Eutric Brunisol	Subgroups of Cryepts; frigid and mesic families of Haplustepts
		Sombric Brunisol	Frigid and mesic families of Udepts, and Ustept and Humic Dystrudepts
		Dystric Brunisol	Frigid families of Dystrudepts and Dystrocryepts
	Gleysolic		Cryic subgroups and frigid families of Aqualfs, Aquolls, Aquepts, Aquents, and Aquods
		Humic Gleysol	Humaquepts
		Gleysol	Cryaquepts and frigid families of Fragaquepts, Epiaquepts, and Endoaquepts
Mollisols	Chernozemic	Brown	Xeric and Ustic subgroups of Argicryolls and Haplocryolls
		Dark Brown	Subgroups of Argicryolls and Haplocryolls
		Black	Typic subgroups of Argicryolls and Haplocryolls
		Dark Gray	Alfic subgroups of Argicryolls
	Solonetzic	Solonetz	Natricryolls and frigid families of Natraquolls and Natralbolls
		Solod	Glossic subgroups of Natricryolls
Oxisols	Not relevant in Canada		
Spodosols	Podzolic	Humic Podzol	Humicryods, Humic Placocryods, Placohumods, and frigid families of other Humods
		Ferro-Humic Podzol	Humic Haplocryods, some Placorthods, and frigid families of humic subgroups of other Orthods
		Humo-Ferric Podzol	Haplorthods, Placorthods, and frigid families of other Orthods and Cryods except humic subgroups
Ultisols	Not relevant in Canada		
Vertisols	Vertisolic	Vertisol	Haplocryerts
		Humic Vertisol	Humicryerts

[a] Based on information in Soil Classification Working Group, 1998, *The Canadian System of Soil Classification,* 3d ed. (Ottawa: Agriculture and Agri-Food Canada). Publication No. A53-1646/1997E.

The FAO Soil Classification System is really a map legend for the world soil map made by FAO/UNESCO. A brief description of the map units is given in Table B.3. The FAO map legend terms are used by many scientists around the world to describe different kinds of soils. Global and regional soil databases are often organized according to this system. The soil characteristics recognized and the terminology used borrow heavily from the Russian soil classification system, as well as from U.S. *Soil Taxonomy*.

TABLE B.3 Soil Map Units for the FAO/UNESCO Soil Map of the World

Acrisols	Low base status soils with argillic horizons
Andosols	Soils formed in volcanic ash that have dark surfaces
Arenosols	Soils formed from sand
Cambisols	Soils with slight color, structure, or consistency change due to weathering
Chernozems	Soils with black surface, high humus under prairie vegetation
Cryosols	Soils of cold climates with permafrost
Ferralsols	Highly weathered soils with sesquioxide-rich clays
Fluvisols	Water-deposited soils with little alteration
Gleysols	Soils with mottled or reduced horizons due to wetness
Greyzems	Soils with dark surface, bleached E horizon, and textural B horizon
Histosols	Organic soils
Kastanozems	Soils with chestnut surface color, steppe vegetation
Lithosols	Shallow soils over hard rock
Luvisols	Medium to high base status soils with argillic horizons
Nitosols	Soils with low CEC clay in argillic horizons
Planosols	Soils with abrupt A–B horizon contact
Phaeozems	Soils with dark surface, more leached than Kastanozem or Chernozem
Podzols	Soils with light-colored eluvial horizon and subsoil accumulation of iron, aluminum, and humus
Podzoluvisols	Soils with leached horizons tonguing into argillic B horizons
Rankers	Thin soils over siliceous material
Regosols	Thin soil over unconsolidated material
Rendzinas	Shallow soils over limestone
Solonchaks	Soils with soluble salt accumulation
Solonetz	Soils with high sodium content
Vertisols	Self-mulching, inverting soils, rich in montmorillinitic clay
Xerosols	Dry soils of semiarid regions
Yermosols	Desert soils

Appendix C

SI Unit Conversion Factors and Periodic Table of the Elements

FACTORS FOR CONVERTING NON-SI UNITS TO SI UNITS

Non-SI Unit	Multiply by[a]	To obtain SI Unit
Length		
inch, in.	2.54	centimeters, cm (10^{-2} m)
foot, ft	0.304	meter, m
mile,	1.609	kilometer, km (10^3 m)
micron, μ	1.0	micrometer, μm (10^{-6} m)
Ångstrom unit, Å	0.1	nanometer, nm (10^{-9} m)
Area		
acre, ac	0.405	hectare, ha (10^4 m^2)
square foot, ft^2	9.29×10^{-2}	square meter, m^2
square inch, in^2	645	square millimeter, mm^2
square mile, mi^2	2.59	square kilometer, km^2
Volume		
bushel, bu	35.24	liter, L
cubic foot, ft^3	2.83×10^{-2}	cubic meter, m^3
cubic inch, in.^{3}	1.64×10^{-5}	cubic meter, m^3
gallon (U.S.), gal	3.78	liter, L
quart, qt	0.946	liter, L
acre-foot, ac-ft	12.33	hectare-centimeter, ha-cm
acre-inch, ac-in.	1.03×10^{-2}	hectare-meters, ha-m
ounce (fluid), oz	2.96×10^{-2}	liter, L
pint, pt	0.473	liter, L
Mass		
ounce (avdp), oz	28.4	gram, g
pound, lb	0.454	kilogram, kg (10^3 g)
ton (2000 lb)	0.907	megagram, Mg (10^6 g)
tonne (metric), t	1000	kilogram, kg
Radioactivity		
curie, Ci	3.7×10^{10}	becquerel, Bq
picocurie per gram, pCi/g	37	becquerel per kilogram, Bq/kg
Yield and Rate		
bushel per acre (60 lb), bu/ac	67.19	kilogram per hectare, kg/ha
bushel per acre (56 lb), bu/ac	62.71	kilogram per hectare, kg/ha
bushel per acre (48 lb), bu/ac	53.75	kilogram per hectare, kg/ha
gallon per acre (U.S.), gal/ac	9.35	liter per hectare, L/ha
ton (2000 lb) per acre	2.24	megagram per hectare, Mg/ha
miles per hour, mph	0.447	meter per second, m/s
gallon per minute (U.S.), gpm	0.227	cubic meter per hour, m^3/h
cubic feet per second, cfs	101.9	cubic meter per hour, m^3/h
Pressure		
atmosphere, atm	0.101	megapascal, MPa (10^6 Pa)
bar	0.1	megapascal, MPa
pound per square foot, lb/ft^2	47.9	pascal, Pa
pound per square inch, lb/in^2	6.9×10^3	pascal, Pa
Temperature		
degrees Fahrenheit (°F − 32)	0.556	degrees, °C
degrees Celsius (°C + 273)	1	Kelvin, K
Energy		
British thermal unit, Btu	1.05×10^3	joule, J
calorie, cal	4.19	joule, J
dyne, dyn	10^{-5}	newton, N
erg	10^{-7}	joule, J
foot-pound, ft-lb	1.36	joule, J
Concentrations		
percent, %	10	gram per kilogram, g/kg
part per million, ppm	1	milligram per kilogram, mg/kg
milliequivalents per 100 grams	1	centimole per kilogram, cmol/kg

[a] To convert from SI to non-SI units, *divide* by the factor given.

Periodic Table of the Elements with Notes Concerning Relevance to Soil Science

Based on atomic mass of $^{12}C = 12.0$. Numbers in parentheses are the mass numbers of the most stable isotopes of radioactive elements.

Group IA	Group IIA	Group IIIB	Group IVB	Group VB	Group VIB	Group VIIB	Group VIIIB			Group IB	Group IIB	Group IIIA	Group IVA	Group VA	Group VIA	Group VIIA	Group VIIIA
1 H 1.01 Hydrogen																	2 He 4.00 Helium
3 Li 6.94 Lithium	4 Be 9.01 Beryllium											5 B 10.81 Boron	6 C 12.01 Carbon	7 N 14.01 Nitrogen	8 O 16.00 Oxygen	9 F 19.00 Fluorine	10 Ne 20.18 Neon
11 Na 22.99 Sodium	12 Mg 24.30 Magnesium											13 Al 26.98 Aluminum	14 Si 28.09 Silicon	15 P 30.97 Phosphorus	16 S 32.07 Sulfur	17 Cl 35.45 Chlorine	18 Ar 39.95 Argon
19 K 39.10 Potassium	20 Ca 40.08 Calcium	21 Sc 44.96 Scandium	22 Ti 47.88 Titanium	23 V 50.94 Vanadium	24 Cr 52.00 Chromium	25 Mn 54.94 Manganese	26 Fe 55.85 Iron	27 Co 58.93 Cobalt	28 Ni 58.69 Nickel	29 Cu 63.55 Copper	30 Zn 65.38 Zinc	31 Ga 69.72 Gallium	32 Ge 72.59 Germanium	33 As 74.92 Arsenic	34 Se 78.96 Selenium	35 Br 79.90 Bromine	36 Kr 83.80 Krypton
37 Rb 85.47 Rubidium	38 Sr 87.62 Strontium	39 Y 88.91 Yttrium	40 Zr 91.22 Zirconium	41 Nb 92.91 Niobium	42 Mo 95.94 Molybdenum	43 Tc (98) Technetium	44 Ru 101.07 Ruthenium	45 Rh 102.91 Rhodium	46 Pd 106.42 Palladium	47 Ag 107.87 Silver	48 Cd 112.41 Cadmium	49 In 114.82 Indium	50 Sn 118.71 Tin	51 Sb 121.75 Antimony	52 Te 127.60 Tellurium	53 I 126.90 Iodine	54 Xe 131.29 Xenon
55 Cs 132.91 Cesium	56 Ba 137.33 Barium	57 La 138.91 Lanthanum	72 Hf 178.49 Hafnium	73 Ta 180.95 Tantalum	74 W 183.85 Tungsten	75 Re 186.21 Rhenium	76 Os 190.2 Osmium	77 Ir 192.22 Iridium	78 Pt 195.08 Platinum	79 Au 196.97 Gold	80 Hg 200.59 Mercury	81 Tl 204.38 Thallium	82 Pb 207.2 Lead	83 Bi 208.98 Bismuth	84 Po (209) Polonium	85 At (210) Astatine	86 Rn (222) Radon
87 Fr (223) Francium	88 Ra (226) Radium	89 Ac (227) Actinium	104 Unq (261) Unniliquadium	105 Unp (262) Unnilpentium	106 Unh (263) Unnilhexium	107 Uns (262) Unnilseptium	108 Uno (265) Unniloctium										

58 Ce 140.12 Cerium	59 Pr 140.91 Praseodymium	60 Nd 144.24 Neodymium	61 Pm (145) Promethium	62 Sm 150.36 Samarium	63 Eu 151.96 Europium	64 Gd 157.25 Gadolinium	65 Tb 158.93 Terbium	66 Dy 162.50 Dysprosium	67 Ho 164.93 Holmium	68 Er 167.26 Erbium	69 Tm 168.93 Thulium	70 Yb 173.04 Ytterbium	71 Lu 174.97 Lutetium
90 Th (232) Thorium	91 Pa (231) Protactinium	92 U (238) Uranium	93 Np (237) Neptunium	94 Pu (244) Plutonium	95 Am (243) Americium	96 Cm (247) Curium	97 Bk (247) Berkelium	98 Cf (251) Californium	99 Es (252) Einsteinium	100 Fm (257) Fermium	101 Md (258) Mendelevium	102 No (259) Nobelium	103 Lr (260) Lawrencium

Metals / Nonmetals

Atomic number
Symbol
Atomic mass

87
Fr
(223)
Francium

Elements known to be nutrients for animals or plants. Some are also toxic in excessive amounts.

Elements toxic to organisms in small amounts, and not known to serve as nutrients.

Other elements commonly studied in Soil Science because of soil-environmental impacts or because of their use as tracers or electrodes. (Br is used to trace anionic solutes such as nitrate. Isotopes of Rb and Sr are used to trace K and Ca in plants and soils. Cs and Ti are used to trace geological processes such as soil erosion. Pt and Ag are used in electrodes for measuring soil redox potential and pH, respectively.)

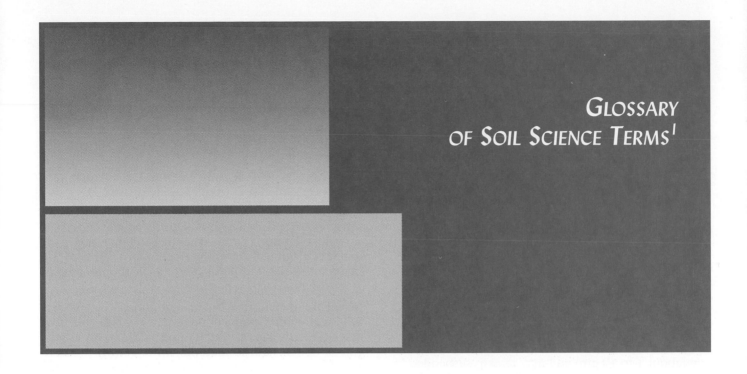

A horizon The surface horizon of a mineral soil having maximum organic matter accumulation, maximum biological activity, and/or eluviation of materials such as iron and aluminum oxides and silicate clays.

abiotic Nonliving basic elements of the environment, such as rainfall, temperature, wind, and minerals.

accelerated erosion Erosion much more rapid than normal, natural, geological erosion; primarily as a result of the activities of humans or, in some cases, of animals.

acid rain Atmospheric precipitation with pH values less than about 5.6, the acidity being due to inorganic acids such as nitric and sulfuric that are formed when oxides of nitrogen and sulfur are emitted into the atmosphere.

acid soil A soil with a pH value <7.0. Usually applied to surface layer or root zone, but may be used to characterize any horizon. *See also* reaction, soil.

acid sulfate soils Soils that are potentially extremely acid (pH < 3.5) because of the presence of large amounts of reduced forms of sulfur that are oxidized to sulfuric acid if the soils are exposed to oxygen when they are drained or excavated. A sulfuric horizon containing the yellow mineral jarosite is often present. *See also* cat clays.

acidity, active The activity of hydrogen ions in the aqueous phase of a soil. It is measured and expressed as a pH value.

acidity, residual Soil acidity that can be neutralized by lime or other alkaline materials but cannot be replaced by an unbuffered salt solution.

acidity, salt replaceable Exchangeable hydrogen and aluminum that can be replaced from an acid soil by an unbuffered salt solution such as KCl or NaCl.

acidity, total The total acidity in a soil. It is approximated by the sum of the salt-replaceable acidity plus the residual acidity.

actinomycetes A group of organisms intermediate between the bacteria and the true fungi that usually produce a characteristic branched mycelium. Includes many, but not all, organisms belonging to the order of Actinomycetales.

activated sludge Sludge that has been aerated and subjected to bacterial action.

active layer The upper portion of a Gellisol that is subject to freezing and thawing and is underlain by permafrost.

active organic matter A portion of the soil organic matter that is relatively easily metabolized by microorganisms and cycles with a half-life in the soil of a few days to a few years.

adhesion Molecular attraction that holds the surfaces of two substances (e.g., water and sand particles) in contact.

---

[1] This glossary was compiled and modified from several sources, including *Glossary of Soil Science Terms* [Madison, Wis.: Soil Sci. Soc. Amer. (1997)], *Resource Conservation Glossary* [Anheny, Iowa: Soil Cons. Soc. Amer. (1982)], and *Soil Taxonomy* (Washington, D.C.: U.S. Department of Agriculture (1998)].

adsorption The attraction of ions or compounds to the surface of a solid. Soil colloids adsorb large amounts of ions and water.

adsorption complex The group of organic and inorganic substances in soil capable of adsorbing ions and molecules.

aerate To impregnate with gas, usually air.

aeration, soil The process by which air in the soil is replaced by air from the atmosphere. In a well-aerated soil, the soil air is similar in composition to the atmosphere above the soil. Poorly aerated soils usually contain more carbon dioxide and correspondingly less oxygen than the atmosphere above the soil.

aerobic (1) Having molecular oxygen as a part of the environment. (2) Growing only in the presence of molecular oxygen, as aerobic organisms. (3) Occurring only in the presence of molecular oxygen (said of certain chemical or biochemical processes, such as aerobic decomposition).

aerosolic dust A type of eolian material that is very fine (about 1 to 10 μm) and may remain suspended in the air over distances of thousands of kilometers. Finer than most *loess*.

aggregate (soil) Many soil particles held in a single mass or cluster, such as a clod, crumb, block, or prism.

agric horizon A diagnostic subsurface horizon in which clay, silt, and humus derived from an overlying cultivated and fertilized layer have accumulated. Wormholes and illuvial clay, silt, and humus occupy at least 5% of the horizon by volume.

agronomy A specialization of agriculture concerned with the theory and practice of field-crop production and soil management. The scientific management of land.

air-dry (1) The state of dryness (of a soil) at equilibrium with the moisture content in the surrounding atmosphere. The actual moisture content will depend upon the relative humidity and the temperature of the surrounding atmosphere. (2) To allow to reach equilibrium in moisture content with the surrounding atmosphere.

air porosity The proportion of the bulk volume of soil that is filled with air at any given time or under a given condition, such as a specified moisture potential; usually the large pores.

albic horizon A diagnostic subsurface horizon from which clay and free iron oxides have been removed or in which the oxides have been segregated to the extent that the color of the horizon is determined primarily by the color of the primary sand and silt particles rather than by coatings on these particles.

Alfisols An order in *Soil Taxonomy*. Soils with gray to brown surface horizons, medium to high supply of bases, and B horizons of illuvial clay accumulation. These soils form mostly under forest or savanna vegetation in climates with slight to pronounced seasonal moisture deficit.

algal bloom A population explosion of algae in surface waters, such as lakes and streams, often resulting in high turbidity and green- or red-colored water, and commonly stimulated by nutrient enrichment with phosphorus and nitrogen.

alkaline soil Any soil that has pH > 7. Usually applied to the surface layer or root zone but may be used to characterize any horizon or a sample thereof. *See also* reaction, soil.

allelochemical An organic chemical by which one plant can influence another. *See* allelopathy.

allelopathy The process by which one plant may affect other plants by biologically active chemicals introduced into the soil, either directly by leaching or exudation from the source plant, or as a result of the decay of the plant residues. The effects, though usually negative, may also be positive.

allophane A poorly defined aluminosilicate mineral whose structural framework consists of short runs of three-dimensional crystals interspersed with amorphous noncrystalline materials. Along with its more weathered companion, it is prevalent in volcanic ash materials.

alluvial fan Fan-shaped alluvium deposited at the mouth of a canyon or ravine where debris-laden waters fan out, slow down, and deposit their burden.

alluvium A general term for all detrital material deposited or in transit by streams, including gravel, sand, silt, clay, and all variations and mixtures of these. Unless otherwise noted, alluvium is unconsolidated.

alpha particle A positively charged particle (consisting of two protons and two neutrons) that is emitted by certain radioactive compounds.

aluminosilicates Compounds containing aluminum, silicon, and oxygen as main constituents. An example is microcline, $KAlSi_3O_8$.

amendment, soil Any substance other than fertilizers, such as lime, sulfur, gypsum, and sawdust, used to alter the chemical or physical properties of a soil, generally to make it more productive.

amino acids Nitrogen-containing organic acids that couple together to form proteins. Each acid molecule contains one or more amino groups (—NH_2) and at least one carboxyl group (—COOH). In addition, some amino acids contain sulfur.

ammonification The biochemical process whereby ammoniacal nitrogen is released from nitrogen-containing organic compounds.

ammonium fixation The entrapment of ammonium ions by the mineral or organic fractions of the soil in forms that are insoluble in water and are at least temporarily nonexchangeable.

amorphous material Noncrystalline constituents of soils.

anaerobic (1) Without molecular oxygen. (2) Living or functioning in the absence of air or free oxygen.

andic properties Soil properties related to volcanic origin of materials, including high organic carbon content, low bulk density, high phosphate retention, and extractable iron and aluminum.

Andisols An order in *Soil Taxonomy*. Soils developed from volcanic ejecta. The colloidal fraction is dominated by allophane and/or Al-humus compounds.

angle of repose The maximum slope steepness at which loose, cohesionless material will come to rest.

anion Negatively charged ion; during electrolysis it is attracted to the positively charged anode.

anion exchange Exchange of anions in the soil solution for anions adsorbed on the surface of clay and humus particles.

anion exchange capacity The sum total of exchangeable anions that a soil can adsorb. Expressed as centimoles of charge per kilogram ($cmol_c/kg$) of soil (or of other adsorbing material, such as clay).

anoxic *See* anaerobic.

anthropic epipedon A diagnostic surface horizon of mineral soil that has the same requirements as the mollic epipedon but that has more than 250 mg/kg of P_2O_5 soluble in 1% citric acid, or is dry more than 10 months (cumulative) during the period when not irrigated. The anthropic epipedon forms under long-continued cultivation and fertilization.

antibiotic A substance produced by one species of organism that, in low concentrations, will kill or inhibit growth of certain other organisms.

Ap The surface layer of a soil disturbed by cultivation or pasturing.

apatite A naturally occurring complex calcium phosphate that is the original source of most of the phosphate fertilizers. Formulas such as $[3Ca_3(PO_4)_2] \cdot CaF_2$ illustrate the complex compounds that make up apatite.

aquic conditions Continuous or periodic saturation (with water) and reduction, commonly indicated by redoximorphic features.

aquiclude A saturated body of rock or sediment that is incapable of transmitting significant quantities of water under ordinary water pressures.

aquifer A saturated, permeable layer of sediment or rock that can transmit significant quantities of water under normal pressure conditions.

arbuscular mycorrhiza A common endomycorrhizal association produced by phycomycetous fungi and characterized by the development, within root cells, of small structures known as *arbuscules*. Some also form, between root cells, storage organs known as *vesicles*. Host range includes many agricultural and horticultural crops. Formerly called vesicular arbuscular mycorrhiza (VAM). *See also* endotrophic mycorrhiza.

arbuscule Specialized branched structure formed within a root cortical cell by endotrophic mycorrhizal fungi.

argillan A thin coating of well-oriented clay particles on the surface of a soil aggregate, particle, or pore. A clay film.

argillic horizon A diagnostic subsurface horizon characterized by the illuvial accumulation of layer-lattice silicate clays.

arid climate Climate in regions that lack sufficient moisture for crop production without irrigation. In cool regions annual precipitation is usually less than 25 cm. It may be as high as 50 cm in tropical regions. Natural vegetation is desert shrubs.

Aridisols An order in *Soil Taxonomy*. Soils of dry climates. They have pedogenic horizons, low in organic matter, that are never moist for as long as three consecutive months. They have an ochric epipedon and one or more of the following diagnostic horizons: argillic, natric, cambic, calcic, petrocalcic, gypsic, petrogypsic, salic, or a duripan.

aspect (of slopes) The direction (e.g., south or north) that a slope faces with respect to the sun.

association, soil *See* soil association.

Atterberg limits Water contents of fine-grained soils at different states of consistency.
> **liquid limit (LL)** The water content corresponding to the arbitrary limit between the liquid and plastic states of consistency of a soil.
> **plastic limit (PL)** The water content corresponding to an arbitrary limit between the plastic and semisolid states of consistency of a soil.

autochthonous organisms Those microorganisms thought to subsist on the more resistant soil organic matter and little affected by the addition of fresh organic materials. *Contrast with* zymogenous organisms.

autotroph An organism capable of utilizing carbon dioxide or carbonates as the sole source of carbon and obtaining energy for life processes from the oxidation of inorganic elements or compounds such as iron, sulfur, ammonium, and nitrites, or from radiant energy. *Contrast with* heterotroph.

available nutrient That portion of any element or compound in the soil that can be readily absorbed and assimilated by growing plants. ("Available" should not be confused with "exchangeable.")

available water The portion of water in a soil that can be readily absorbed by plant roots. The amount of water released between the field capacity and the permanent wilting point.

B horizon A soil horizon, usually beneath the A horizon, that is characterized by one or more of the following: (1) a concentration of silicate clays, iron and aluminum oxides, and humus, alone or in combination; (2) a blocky or prismatic structure; and (3) coatings of iron and aluminum oxides that give darker, stronger, or redder color.

bar A unit of pressure equal to 1 million dynes per square centimeter (10^6 dynes/cm^2). It approximates the pressure of a standard atmosphere.

base-forming cations Those cations that form strong (strongly dissociated) bases by reaction with hydroxyl; e.g., K^+ forms potassium hydroxide ($K^+ + OH^-$).

base saturation percentage The extent to which the adsorption complex of a soil is saturated with exchangeable cations other than hydrogen and aluminum. It is expressed as a percentage of the total cation exchange capacity.

bedding (Engineering) Arranging the surface of fields by plowing and grading into a series of elevated beds separated by shallow depressions or ditches for drainage.

bedrock The solid rock underlying soils and the regolith in depths ranging from zero (where exposed by erosion) to several hundred feet.

bench terrace An embankment constructed across sloping fields with a steep drop on the downslope side.

beta particle A high-speed electron emitted in radioactive decay.

bioaccumulation A buildup within an organism of specific compounds due to biological processes. Commonly applied to heavy metals, pesticides, or metabolites.

biodegradable Subject to degradation by biochemical processes.

biological nitrogen fixation Occurs at ordinary temperatures and pressures. It is commonly carried out by certain bacteria, algae, and actinomycetes, which may or may not be associated with higher plants.

biomass The total mass of living material of a specified type (e.g., microbial biomass) in a given environment (e.g., in a cubic meter of soil).

biopores Soil pores, usually of relatively large diameter, created by plant roots, earthworms, or other soil organisms.

bioremediation The decontamination or restoration of polluted or degraded soils by means of enhancing the chemical degradation or other activities of soil organisms.

biosequence A group of related soils that differ, one from the other, primarily because of differences in kinds and numbers of plants and soil organisms as a soil-forming factor.

biosolids *See* sewage sludge.

blocky soil structure Soil aggregates with blocklike shapes; common in B horizons of soils in humid regions.

broad-base terrace A low embankment with such gentle slopes that it can be farmed, constructed across sloping fields to reduce erosion and runoff.

broadcast Scatter seed or fertilizer on the surface of the soil.

buffering capacity The ability of a soil to resist changes in pH. Commonly determined by presence of clay, humus, and other colloidal materials.

bulk blending Mixing dry individual granulated fertilizer materials to form a mixed fertilizer that is applied promptly to the soil.

bulk density, soil The mass of dry soil per unit of bulk volume, including the air space. The bulk volume is determined before drying to constant weight at 105°C.

buried soil Soil covered by an alluvial, loessal, or other deposit, usually to a depth greater than the thickness of the solum.

by-pass flow *See* preferential flow.

C horizon A mineral horizon, generally beneath the solum, that is relatively unaffected by biological activity and pedogenesis and is lacking properties diagnostic of an A or B horizon. It may or may not be like the material from which the A and B have formed.

calcareous soil Soil containing sufficient calcium carbonate (often with magnesium carbonate) to effervesce visibly when treated with cold 0.1 N hydrochloric acid.

calcic horizon A diagnostic subsurface horizon of secondary carbonate enrichment that is more than 15 cm thick, has a calcium carbonate equivalent of more than 15%, and has at least 5% more calcium carbonate equivalent than the underlying C horizon.

caliche A layer near the surface, more or less cemented by secondary carbonates of calcium or magnesium precipitated from the soil solution. It may occur as a soft, thin soil horizon; as a hard, thick bed just beneath the solum; or as a surface layer exposed by erosion.

cambic horizon A diagnostic subsurface horizon that has a texture of loamy very fine sand or finer, contains some weatherable minerals, and is characterized by the alteration or removal of mineral material. The cambic horizon lacks cementation or induration and has too few evidences of illuviation to meet the requirements of the argillic or spodic horizon.

capillary conductivity (Obsolete) *See* hydraulic conductivity.

capillary fringe A zone in the soil just above the plane of zero water pressure (water table) that remains saturated or almost saturated with water.

capillary water The water held in the capillary or *small* pores of a soil, usually with a tension >60 cm of water. *See also* soil water potential.

carbon cycle The sequence of transformations whereby carbon dioxide is fixed in living organisms by photosynthesis or by chemosynthesis, liberated by respiration and by the death and decomposition of the fixing organism, used by heterotrophic species, and ultimately returned to its original state.

carbon/nitrogen ratio The ratio of the weight of organic carbon (C) to the weight of total nitrogen (N) in a soil or in organic material.

carnivore An organism that feeds on animals.

casts, earthworm Rounded, water-stable aggregates of soil that have passed through the gut of an earthworm.

cat clays Wet clay soils high in reduced forms of sulfur that, upon being drained, become extremely acid because of the oxidation of the sulfur compounds and the formation of sulfuric acid. Usually found in tidal marshes. *See* acid sulfate soils.

catena A sequence of soils of about the same age, derived from similar parent material, and occurring under similar climatic conditions, but having different characteristics because of variation in *relief* and in *drainage*.

cation A positively charged ion; during electrolysis it is attracted to the negatively charged cathode.

cation exchange The interchange between a cation in solution and another cation on the surface of any surface-active material, such as clay or organic matter.

cation exchange capacity The sum total of exchangeable cations that a soil can adsorb. Sometimes called *total-exchange capacity, base-exchange capacity,* or *cation-adsorption capacity.* Expressed in centimoles of charge per kilogram ($cmol_c/kg$) of soil (or of other adsorbing material, such as clay).

cemented Indurated; having a hard, brittle consistency because the particles are held together by cementing substances, such as humus, calcium carbonate, or the oxides of silicon, iron, and aluminum.

channery Thin, flat fragments of limestone, sandstone, or schist up to 15 cm (6 in.) in major diameter.

chelate (Greek, claw) A type of chemical compound in which a metallic ion is firmly combined with an organic molecule by means of multiple chemical bonds.

chert A structureless form of silica, closely related to flint, that breaks into angular fragments.

chisel, subsoil A tillage implement with one or more cultivator-type feet to which are attached strong knife-like units used to shatter or loosen hard, compact layers, usually in the subsoil, to depths below normal plow depth. *See also* subsoiling.

chlorite A 2:1:1-type layer-structured silicate mineral having 2:1 layers alternating with a magnesium-dominated octahedral sheet.

chlorosis A condition in plants relating to the failure of chlorophyll (the green coloring matter) to develop. Chlorotic leaves range from light green through yellow to almost white.

chroma (color) *See* Munsell color system.

chronosequence A sequence of related soils that differ, one from the other, in certain properties primarily as a result of time as a soil-forming factor.

classification, soil *See* soil classification.

clay (1) A soil separate consisting of particles <0.002 mm in equivalent diameter. (2) A soil textural class containing >40% clay, <45% sand, and <40% silt.

clay mineral Naturally occurring inorganic material (usually crystalline) found in soils and other earthy deposits, the particles being of clay size, that is, <0.002 mm in diameter.

claypan A dense, compact, slowly permeable layer in the subsoil having a much higher clay content than the overlying material, from which it is separated by a sharply defined boundary. Claypans are usually hard when dry and plastic and sticky when wet. *See also* hardpan.

clod A compact, coherent mass of soil produced artificially, usually by such human activities as plowing and digging, especially when these operations are performed on soils that are either too wet or too dry for normal tillage operations.

coarse texture The texture exhibited by sands, loamy sands, and sandy loams (except very fine sandy loam).

cobblestone Rounded or partially rounded rock or mineral fragments 7.5 to 25 cm (3 to 10 in.) in diameter.

cohesion Holding together: force holding a solid or liquid together, owing to attraction between like molecules. Decreases with rise in temperature.

collapsible soil Certain soil that may undergo a sudden loss in strength when wetted.

colloid, soil (Greek, gluelike) Organic and inorganic matter with very small particle size and a correspondingly large surface area per unit of mass.

colluvium A deposit of rock fragments and soil material accumulated at the base of steep slopes as a result of gravitational action.

color The property of an object that depends on the wavelength of light it reflects or emits.

columnar soil structure *See* soil structure types.

companion planting The practice of growing certain species of plants in close proximity because one species has the effect of improving the growth of the other, sometimes by positive *allelopathic* effects.

compost Organic residues, or a mixture of organic residues and soil, that have been piled, moistened, and allowed to undergo biological decomposition. Mineral fertilizers are sometimes added. Often called *artificial manure* or *synthetic manure* if produced primarily from plant residues.

concretion A local concentration of a chemical compound, such as calcium carbonate or iron oxide, in the form of grains or nodules of varying size, shape, hardness, and color.

conduction The transfer of heat by physical contact between two or more objects.

conductivity, hydraulic *See* hydraulic conductivity.

conservation tillage *See* tillage, conservation.

consistence The combination of properties of soil material that determine its resistance to crushing and its ability to be molded or changed in shape. Such terms as *loose, friable, firm, soft, plastic,* and *sticky* describe soil consistence.

consistency   The interaction of adhesive and cohesive forces within a soil at various moisture contents as expressed by the relative ease with which the soil can be deformed or ruptured.

consociation   *See* soil consociation.

consolidation test   A laboratory test in which a soil mass is laterally confined within a ring and is compressed with a known force between two porous plates.

constant charge   The net surface charge of mineral particles, the magnitude of which depends only on the chemical and structural composition of the mineral. The charge arises from isomorphous substitution and is not affected by soil pH.

consumptive use   The water used by plants in transpiration and growth, plus water vapor loss from adjacent soil or snow, or from intercepted precipitation in any specified time. Usually expressed as equivalent depth of free water per unit of time.

contour   An imaginary line connecting points of equal elevation on the surface of the soil. A contour terrace is laid out on a sloping soil at right angles to the direction of the slope and nearly level throughout its course.

contour strip-cropping   Layout of crops in comparatively narrow strips in which the farming operations are performed approximately on the contour. Usually strips of grass, close–growing crops, or fallow are alternated with those of cultivated crops.

controlled traffic   A farming system in which all wheeled traffic is confined to fixed paths so that repeated compaction of the soil does not occur outside the selected paths.

convection   The transfer of heat through a gas or solution because of molecular movement.

cover crop   A close-growing crop grown primarily for the purpose of protecting and improving soil between periods of regular crop production or between trees and vines in orchards and vineyards.

creep   Slow mass movement of soil and soil material down relatively steep slopes, primarily under the influence of gravity, but facilitated by saturation with water and by alternate freezing and thawing.

crop rotation   A planned sequence of crops growing in a regularly recurring succession on the same area of land, as contrasted to continuous culture of one crop or growing different crops in haphazard order.

crotovina   A former animal burrow in one soil horizon that has been filled with organic matter or material from another horizon (also spelled *krotovina*).

crumb   A soft, porous, more or less rounded natural unit of structure from 1 to 5 mm in diameter. *See also* soil structure types.

crushing strength   The force required to crush a mass of dry soil or, conversely, the resistance of the dry soil mass to crushing. Expressed in units of force per unit area (pressure).

crust   A surface layer on soils, ranging in thickness from a few millimeters to perhaps as much as 3 cm, that is much more compact, hard, and brittle when dry than the material immediately beneath it.

cryoturbation   Physical disruption and displacement of soil material within the profile by the forces of freezing and thawing. Sometimes called *frost churning*, it results in irregular, broken horizons, involutions, oriented rock fragments, and accumulation of organic matter on the permafrost table.

cryptogam   Assemblage of algae, lichens, liverworts, and mosses that commonly forms an irregular crust on the soil surface, especially on otherwise barren, arid-region soils. Sometimes referred to as *cryptogamic crusts.*

crystal   A homogeneous inorganic substance of definite chemical composition bounded by plane surfaces that form definite angles with each other, thus giving the substance a regular geometrical form.

crystal structure   The orderly arrangement of atoms in a crystalline material.

cultivation   A tillage operation used in preparing land for seeding or transplanting or later for weed control and for loosening the soil.

cutans   A modification of the texture, structure, or fabric at natural surfaces in soil materials due to concentration of particular soil constituents.

cyanobacteria   Chlorophyll-containing bacteria that accommodate both photosynthesis and nitrogen fixation. Formerly called blue-green algae.

deciduous plant   A plant that sheds all its leaves every year at a certain season.

decomposition   Chemical breakdown of a compound (e.g., a mineral or organic compound) into simpler compounds, often accomplished with the aid of microorganisms.

deflocculate   (1) To separate the individual components of compound particles by chemical and/or physical means. (2) To cause the particles of the *disperse phase* of a colloidal system to become suspended in the *dispersion medium.*

delta   An alluvial deposit formed where a stream or river drops its sediment load upon entering a quieter body of water.

denitrification   The biochemical reduction of nitrate or nitrite to gaseous nitrogen, either as molecular nitrogen or as an oxide of nitrogen.

density   *See* particle density; bulk density.

desalinization   Removal of salts from saline soil, usually by leaching.

desert crust A hard layer, containing calcium carbonate, gypsum, or other binding material, exposed at the surface in desert regions.

desert pavement A natural residual concentration of closely packed pebbles, boulders, and other rock fragments on a desert surface where wind and water action has removed all smaller particles.

desorption The removal of sorbed material from surfaces.

detritivore An organism that subsists on detritus.

detritus Debris from dead plants and animals.

diagnostic horizons (As used in *Soil Taxonomy*): Horizons having specific soil characteristics that are indicative of certain classes of soils. Horizons that occur at the soil surface are called *epipedons;* those below the surface, *diagnostic subsurface horizons.*

diatomaceous earth A geologic deposit of fine, grayish, siliceous material composed chiefly or wholly of the remains of diatoms. It may occur as a powder or as a porous, rigid material.

diatoms Algae having siliceous cell walls that persist as a skeleton after death; any of the microscopic unicellular or colonial algae constituting the class Bacillariaceae. They occur abundantly in fresh and salt waters and their remains are widely distributed in soils.

diffusion The movement of atoms in a gaseous mixture or of ions in a solution, primarily as a result of their own random motion.

dioctahedral sheet An octahedral sheet of silicate clays in which the sites for the six-coordinated metallic atoms are mostly filled with trivalent atoms, such as Al^{3+}.

disintegration Physical or mechanical breakup or separation of a substance into its component parts (e.g., a rock breaking into its mineral components).

disperse (1) To break up compound particles, such as aggregates, into the individual component particles. (2) To distribute or suspend fine particles, such as clay, in or throughout a dispersion medium, such as water.

dissolution Process by which molecules of a gas, solid, or another liquid dissolve in a liquid, thereby becoming completely and uniformly dispersed throughout the liquid's volume.

diversion terrace *See* terrace.

drain (1) To provide channels, such as open ditches or drain tile, so that excess water can be removed by surface or by internal flow. (2) To lose water (from the soil) by percolation.

drain field, septic tank An area of soil into which the effluent from a septic tank is piped so that it will drain through the lower part of the soil profile for disposal and purification.

drainage, soil The frequency and duration of periods when the soil is free from saturation with water.

drift Material of any sort deposited by geological processes in one place after having been removed from another. Glacial drift includes material moved by the glaciers and by the streams and lakes associated with them.

drumlin Long, smooth, cigar-shaped low hills of glacial till, with their long axes parallel to the direction of ice movement.

dryland farming The practice of crop production in low-rainfall areas without irrigation.

duff The matted, partly decomposed organic surface layer of forest soils.

duripan A diagnostic subsurface horizon that is cemented by silica, to the point that air-dry fragments will not slake in water or HCl. Hardpan.

dust mulch A loose, finely granular or powdery condition on the surface of the soil, usually produced by shallow cultivation.

E horizon Horizon characterized by maximum illuviation (washing out) of silicate clays and iron and aluminum oxides; commonly occurs above the B horizon and below the A horizon.

earthworms Animals of the Lumbricidae family that burrow into and live in the soil. They mix plant residues into the soil and improve soil aeration.

ectotrophic mycorrhiza (ectomycorrhiza) A symbiotic association of the mycelium of fungi and the roots of certain plants in which the fungal hyphae form a compact mantle on the surface of the roots and extend into the surrounding soil and inward between cortical cells, but not into these cells. Associated primarily with certain trees. *See also* endotrophic mycorrhiza.

edaphology The science that deals with the influence of soils on living things, particularly plants, including human use of land for plant growth.

effective precipitation That portion of the total precipitation that becomes available for plant growth or for the promotion of soil formation.

electrical conductivity (EC) The capacity of a substance to conduct or transmit electrical current. In soils or water, measured in siemens/meter, and related to dissolved solutes.

eluviation The removal of soil material in suspension (or in solution) from a layer or layers of a soil. (Usually, the loss of material in *solution* is described by the term *leaching.*) *See also* leaching.

endotrophic mycorrhiza (endomycorrhiza) A symbiotic association of the mycelium of fungi and roots of a variety of plants in which the fungal hyphae penetrate directly into root hairs, other epidermal cells, and occasionally into cortical cells. Individual hyphae also extend from the root surface outward into the surrounding soil. *See also* arbuscular mycorrhiza.

Entisols An order in *Soil Taxonomy*. Soils have no diagnostic pedogenic horizons. They may be found in virtually any climate on very recent geomorphic surfaces.

eolian soil material Soil material accumulated through wind action. The most extensive areas in the United States are silty deposits (loess), but large areas of sandy deposits also occur.

epipedon A diagnostic surface horizon that includes the upper part of the soil that is darkened by organic matter, or the upper eluvial horizons, or both. *(Soil Taxonomy.)*

equilibrium phosphorus concentration The concentration of phosphorus in a solution in equilibrium with a soil, the EPC_O being the concentration of phosphorus achieved by desorption of phosphorus from a soil to phosphorus-free distilled water.

erosion (1) The wearing away of the land surface by running water, wind, ice, or other geological agents, including such processes as gravitational creep. (2) Detachment and movement of soil or rock by water, wind, ice, or gravity.

esker A narrow ridge of gravelly or sandy glacial material deposited by a stream in an ice-walled valley or tunnel in a receding glacier.

essential element A chemical element required for the normal growth of plants.

eukaryote An organism whose cells each have a visibly evident nucleus.

eutrophic Having concentrations of nutrients optimal (or nearly so) for plant or animal growth. (Said of nutrient solutions or of soil solutions.)

eutrophication Nutrient enrichment of lakes, ponds, and other such waters that stimulates the growth of aquatic organisms, which leads to a deficiency of oxygen in the water body.

evapotranspiration The combined loss of water from a given area, and during a specified period of time, by evaporation from the soil surface and by transpiration from plants.

exchange capacity The total ionic charge of the adsorption complex active in the adsorption of ions. *See also* anion exchange capacity; cation exchange capacity.

exchangeable sodium percentage The extent to which the adsorption complex of a soil is occupied by sodium. It is expressed as follows:

$$ESP = \frac{\text{exchangeable sodium (cmol}_c\text{/kg soil)}}{\text{cation exchange capacity (cmol}_c\text{/kg soil)}} \times 100$$

exfoliation Peeling away of layers of a rock from the surface inward, usually as the result of expansion and contraction that accompany changes in temperature.

expansive soil Soil that undergoes significant volume change upon wetting and drying, usually because of a high content of swelling-type clay minerals.

external surface The area of surface exposed on the top, bottom, and sides of a clay crystal or micelle.

facultative organism An organism capable of both aerobic and anaerobic metabolism.

fallow Cropland left idle in order to restore productivity, mainly through accumulation of nutrients, water, and/or organic matter. Preceding a cereal grain crop in semiarid regions, land is commonly left in *summer fallow* for a period during which weeds are controlled by chemicals or tillage and water is allowed to accumulate in the soil profile. In humid regions, fallow land may be allowed to grow up in natural vegetation for a period ranging from a few months to many years. *Improved fallow* involves the purposeful establishment of plant species capable of restoring soil productivity more rapidly than a natural plant succession.

family, soil In *Soil Taxonomy*, one of the categories intermediate between the great group and the soil series. Families are defined largely on the basis of physical and mineralogical properties of importance to plant growth.

fauna The animal life of a region or ecosystem.

ferrihydrite, $Fe_5HO_8 \cdot 4H_2O$ A dark reddish brown poorly crystalline iron oxide that forms in wet soils.

fertigation The application of fertilizers in irrigation waters, commonly through sprinkler systems.

fertility, soil The quality of a soil that enables it to provide essential chemical elements in quantities and proportions for the growth of specified plants.

fertilizer Any organic or inorganic material of natural or synthetic origin added to a soil to supply certain elements essential to the growth of plants. The major types of fertilizers include:

  bulk blended fertilizers Solid fertilizer materials blended together in small blending plants, delivered to the farm in bulk, and usually spread directly on the fields by truck or other special applicator.
  granulated fertilizers Fertilizers that are present in the form of rather stable granules of uniform size, which facilitate ease of handling the materials and reduce undesirable dusts.
  liquid fertilizers Fluid fertilizers that contain essential elements in liquid forms, either as soluble nutrients, as liquid suspensions, or both.
  mixed fertilizers Two or more fertilizer materials mixed together. May be as dry powders, granules, pellets, bulk blends, or liquids.

fertilizer requirement The quantity of certain plant nutrient elements needed, in addition to the amount supplied by the soil, to increase plant growth to a designated optimum.

fibric materials *See* organic soil materials.

field capacity (field moisture capacity) The percentage of water remaining in a soil two or three days after its having been saturated and after free drainage has practically ceased.

fine texture Consisting of or containing large quantities of the fine fractions, particularly of silt and clay. (Includes clay loam, sandy clay loam, silty clay loam, sandy clay, silty clay, and clay textural classes.)

fine-grained mica A silicate clay having a 2:1-type lattice structure with much of the silicon in the tetrahedral sheet having been replaced by aluminum and with considerable interlayer potassium, which binds the layers together, prevents interlayer expansion and swelling, and limits interlayer cation exchange capacity.

fixation (1) For other than elemental nitrogen: the process or processes in a soil by which certain chemical elements essential for plant growth are converted from a soluble or exchangeable form to a much less soluble or to a nonexchangeable form; for example, potassium, ammonium, and phosphate fixation. (2) For elemental nitrogen: process by which gaseous elemental nitrogen is chemically combined with hydrogen to form ammonia. *See* biological nitrogen fixation.

flagstone A relatively thin rock or mineral fragment 15 to 38 cm in length commonly composed of shale, slate, limestone, or sandstone.

flocculate To aggregate or clump together individual, tiny soil particles, especially fine clay, into small clumps or floccules. Opposite of *deflocculate* or *disperse*.

floodplain The land bordering a stream, built up of sediments from overflow of the stream and subject to inundation when the stream is at flood stage. Sometimes called *bottomland*.

flora The sum total of the kinds of plants in an area at one time. The organisms loosely considered to be of the plant kingdom.

fluorapatite A member of the apatite group of minerals containing fluorine. Most common mineral in phosphate rock.

fluvial deposits Deposits of parent materials laid down by rivers or streams.

fluvioglacial *See* glaciofluvial deposits.

foliar diagnosis An estimation of mineral nutrient deficiencies (excesses) of plants based on examination of the chemical composition of selected plant parts, and the color and growth characteristics of the foliage of the plants.

fragipan Dense and brittle pan or subsurface layer in soils that owes its hardness mainly to extreme density or compactness rather than high clay content or cementation. Removed fragments are friable, but the material in place is so dense that roots penetrate and water moves through it very slowly.

friable A soil consistency term pertaining to soils that crumble with ease.

frigid A soil temperature class with mean annual temperature below 8°C.

fritted micronutrients Sintered silicates having total guaranteed analyses of micronutrients with controlled (relatively slow) release characteristics.

fulvic acid A term of varied usage but usually referring to the mixture of organic substances remaining in solution upon acidification of a dilute alkali extract from the soil.

functional diversity The characteristic of an ecosystem exemplified by the capacity to carry out a large number of biochemical transformations and other functions.

fungi Eukaryote microorganisms with a rigid cell wall. Some form long filaments of cells called *hyphae* that may grow together to form a visible body.

furrow slice The uppermost layer of an arable soil to the depth of primary tillage; the layer of soil sliced away from the rest of the profile and inverted by a moldboard plow.

gamma ray A high-energy ray (photon) emitted during radioactive decay of certain elements.

Gelisols An order in *Soil Taxonomy*. Soils that have permafrost within the upper 1 m, or upper 2 m if cryoturbation is also present. They may have an ochric, histic, mollic, or other epipedon.

gellic materials Mineral or organic soil materials that have *cryoturbation* and/or ice in the form of lenses, veins, or wedges and the like.

genesis, soil The mode of origin of the soil, with special reference to the processes responsible for the development of the solum, or true soil, from the unconsolidated parent material.

genetic horizon Soil layers that resulted from soil-forming (pedogenic) processes, as opposed to sedimentation or other geologic processes.

geological erosion Wearing away of the earth's surface by water, ice, or other natural agents under natural environmental conditions of climate, vegetation, and so on, undisturbed by man. Synonymous with *natural erosion*.

gibbsite, Al(OH)$_3$ An aluminum trihydroxide mineral most common in highly weathered soils, such as oxisols.

gilgai The microrelief of soils produced by expansion and contraction with changes in moisture. Found in soils that contain large amounts of clay that swells and shrinks considerably with wetting and drying. Usually a succession of microbasins and microknolls in nearly level areas or of microvalleys and microridges parallel to the direction of the slope.

glacial drift Rock debris that has been transported by glaciers and deposited, either directly from the ice or from the meltwater. The debris may or may not be heterogeneous.

glacial till *See* till.

glaciofluvial deposits Material moved by glaciers and subsequently sorted and deposited by streams flowing from the melting ice. The deposits are stratified and may occur in the form of outwash plains, deltas, kames, eskers, and kame terraces.

gleyed A soil condition resulting from prolonged saturation with water and reducing conditions that manifest themselves in greenish or bluish colors throughout the soil mass or in mottles.

glomalin A protein-sugar complex substance secreted by certain fungi resulting in a sticky hyphal surface thought to contribute to aggregate stability.

goethite, FeOOH A yellow-brown iron oxide mineral that accounts for the brown color in many soils.

granular structure Soil structure in which the individual grains are grouped into spherical aggregates with indistinct sides. Highly porous granules are commonly called *crumbs*. A well-granulated soil has the best structure for most ordinary crop plants. *See also* soil structure types.

granulation The process of producing granular materials. Commonly used to refer to the formation of soil structural granules, but also used to refer to the processing of powdery fertilizer materials into granules.

grassed waterway A natural or constructed waterway covered with erosion-resistant grasses that permits removal of runoff water without excessive erosion.

gravitational potential That portion of the total *soil water potential* due to differences in elevation of the reference pool of pure water and that of the soil water. Since the soil water elevation is usually chosen to be higher than that of the reference pool, the gravitational potential is usually positive.

gravitational water Water that moves into, through, or out of the soil under the influence of gravity.

great group A category in *Soil Taxonomy*. The classes in this category contain soils that have the same kind of horizons in the same sequence and have similar moisture and temperature regimes.

green manure Plant material incorporated with the soil while green, or soon after maturity, for improving the soil.

greenhouse effect The entrapment of heat by upper atmosphere gases, such as carbon dioxide, water vapor, and methane, just as glass traps heat for a greenhouse. Increases in the quantities of these gases in the atmosphere will likely result in global warming that may have serious consequences for humankind.

groundwater Subsurface water in the zone of saturation that is free to move under the influence of gravity, often horizontally to stream channels.

gully erosion The erosion process whereby water accumulates in narrow channels and, over short periods, removes the soil from this narrow area to considerable depths, ranging from 1 to 2 ft to as much as 23 to 30 m (75 to 100 ft).

gypsic horizon A diagnostic subsurface horizon of secondary calcium sulfate enrichment that is more than 15 cm thick.

gypsum requirement The quantity of gypsum required to reduce the exchangeable sodium percentage in a soil to an acceptable level.

halophyte A plant that requires or tolerates a saline (high salt) environment.

hardpan A hardened soil layer, in the lower A or in the B horizon, caused by cementation of soil particles with organic matter or with such materials as silica, sesquioxides, or calcium carbonate. The hardness does not change appreciably with changes in moisture content and pieces of the hard layer do not slake in water. *See also* caliche; claypan.

harrowing A secondary broadcast tillage operation that pulverizes, smooths, and firms the soil in seedbed preparation, controls weeds, or incorporates material spread on the surface.

heaving The partial lifting of plants, buildings, roadways, fenceposts, etc., out of the ground, as a result of freezing and thawing of the surface soil during the winter.

heavy metals Those metals that have densities of 5.0 Mg/m or greater. Elements in soils include Cd, Co, Cr, Cu, Fe, Hg, Mn, Mo, Pb, and Zn.

heavy soil (Obsolete in scientific use) A soil with a high content of the fine separates, particularly clay, or one with a high drawbar pull, hence difficult to cultivate.

hematite, Fe$_2$O$_3$ A red iron oxide mineral that contributes red color to many soils.

herbicide A chemical that kills plants or inhibits their growth; intended for weed control.

herbivore A plant-eating animal.

heterotroph An organism capable of deriving energy for life processes only from the decomposition of organic compounds and incapable of using inorganic compounds as sole sources of energy or for organic synthesis. *Contrast with* autotroph.

histic epipedon A diagnostic surface horizon consisting of a thin layer of organic soil material that is saturated with water at some period of the year unless artificially drained and that is at or near the surface of a mineral soil.

Histosols An order in *Soil Taxonomy*. Soils formed from materials high in organic matter. Histosols with essentially no clay must have at least 20% organic matter by weight (about 78% by volume). This minimum organic matter content rises with increasing clay content to 30% (85% by volume) in soils with at least 60% clay.

horizon, soil A layer of soil, approximately parallel to the soil surface, differing in properties and characteristics from adjacent layers below or above it. *See also* diagnostic horizons.

horticulture The art and science of growing fruits, vegetables, and ornamental plants.

hue (color) *See* Munsell color system.

humic acid A mixture of variable or indefinite composition of dark organic substances, precipitated upon acidification of a dilute alkali extract from soil.

humic substances A series of complex, relatively high molecular weight, brown- to black-colored organic substances that make up 60 to 80% of the soil organic matter and are generally quite resistant to ready microbial attack.

humid climate Climate in regions where moisture, when distributed normally throughout the year, should not limit crop production. In cool climates annual precipitation may be as little as 25 cm; in hot climates, 150 cm or even more. Natural vegetation in uncultivated areas is forests.

humification The processes involved in the decomposition of organic matter and leading to the formation of humus.

humin The fraction of the soil organic matter that is not dissolved upon extraction of the soil with dilute alkali.

humus That more or less stable fraction of the soil organic matter remaining after the major portions of added plant and animal residues have decomposed. Usually it is dark in color.

hydration Chemical union between an ion or compound and one or more water molecules, the reaction being stimulated by the attraction of the ion or compound for either the hydrogen or the unshared electrons of the oxygen in the water.

hydraulic conductivity An expression of the readiness with which a liquid, such as water, flows through a solid, such as soil, in response to a given potential gradient.

hydric soils Soils that are water-saturated for long enough periods to produce reduced conditions and affect the growth of plants.

hydrogen bonding Relatively low energy bonding exhibited by a hydrogen atom located between two highly electronegative atoms, such as nitrogen or oxygen.

hydrologic cycle The circuit of water movement from the atmosphere to the earth and back to the atmosphere through various stages or processes, as precipitation, interception, runoff, infiltration, percolation, storage, evaporation, and transpiration.

hydrolysis The reaction between water and a compound (commonly a salt). The hydroxyl from the water combines with the cation from the compound undergoing hydrolysis to form a base; the hydrogen ion from the water combines with the anion from the compound to form an acid.

hydronium A hydrated hydrogen ion (H_3O^+), the form of the hydrogen ion usually found in an aqueous system.

hydroponics Plant-production systems that use nutrient solutions and no solid medium to grow plants.

hydrous mica *See* fine-grained mica.

hydroxyapatite A member of the apatite group of minerals rich in hydroxyl groups. A nearly insoluble calcium phosphate.

hygroscopic coefficient The amount of moisture in a dry soil when it is in equilibrium with some standard relative humidity near a saturated atmosphere (about 98%), expressed in terms of percentage on the basis of oven-dry soil.

hyperthermic A soil temperature class with mean annual temperatures >22°C.

hypha (pl. hyphae) Filament of fungal cells. Actinomycetes also produce similar, but thinner, filaments of cells.

hysteresis A relationship between two variables that changes depending on the sequences or starting point. An example is the relationship between soil water content and water potential, for which different curves describe the relationship when a soil is gaining water or losing it.

igneous rock Rock formed from the cooling and solidification of magma that has not been changed appreciably since its formation.

illite *See* fine-grained mica.

illuvial horizon A soil layer or horizon in which material carried from an overlying layer has been precipitated from solution or deposited from suspension. The layer of accumulation.

immature soil A soil with indistinct or only slightly developed horizons because of the relatively short time it has been subjected to the various soil-forming processes. A soil that has not reached equilibrium with its environment.

immobilization The conversion of an element from the inorganic to the organic form in microbial tissues or in plant tissues, thus rendering the element not readily available to other organisms or to plants.

imogolite A poorly crystalline aluminosilicate mineral with an approximate formula $SiO_2Al_2O_3 \cdot 2.5H_2O$; occurs mostly in soils formed from volcanic ash.

impervious Resistant to penetration by fluids or by roots.

improved fallow *See* fallow.

Inceptisols An order in *Soil Taxonomy*. Soils that are usually moist with pedogenic horizons of alteration of parent materials but not of illuviation. Generally, the direction of soil development is not yet evident from the marks left by various soil-forming processes or the marks are too weak to classify in another order.

indurated (soil) Soil material cemented into a hard mass that will not soften on wetting. *See also* consistence; hardpan.

infiltration The downward entry of water into the soil.

infiltration capacity A soil characteristic determining or describing the *maximum* rate at which water *can* enter the soil under specified conditions, including the presence of an excess of water.

inoculation The process of introducing pure or mixed cultures of microorganisms into natural or artificial culture media.

inorganic compounds All chemical compounds in nature except compounds of carbon other than carbon monoxide, carbon dioxide, and carbonates.

insecticide A chemical that kills insects.

intergrade A soil that possesses moderately well-developed distinguishing characteristics of two or more genetically related great soil groups.

interlayer (mineralogy) Materials between layers within a given crystal, including cations, hydrated cations, organic molecules, and hydroxide groups or sheets.

internal surface The area of surface exposed within a clay crystal or micelle between the individual crystal layers. *Compare with* external surface.

interstratification Mixing of silicate layers within the structural framework of a given silicate clay.

ions Atoms, groups of atoms, or compounds that are electrically charged as a result of the loss of electrons (cations) or the gain of electrons (anions).

iron-pan An indurated soil horizon in which iron oxide is the principal cementing agent.

irrigation efficiency The ratio of the water actually consumed by crops on an irrigated area to the amount of water diverted from the source onto the area.

isomorphous substitution The replacement of one atom by another of similar size in a crystal lattice without disrupting or changing the crystal structure of the mineral.

isotopes Two or more atoms of the same element that have different atomic masses because of different numbers of neutrons in the nucleus.

joule The SI energy unit defined as a force of 1 newton applied over a distance of 1 meter; 1 joule = 0.239 calorie.

kame A conical hill or ridge of sand or gravel deposited in contact with glacial ice.

kandic horizon A subsurface diagnostic horizon having a sharp clay increase relative to overlying horizons and having low-activity clays.

kaolinite An aluminosilicate mineral of the 1:1 crystal lattice group; that is, consisting of single silicon tetrahedral sheets alternating with single aluminum octahedral sheets.

labile A substance that is readily transformed by microorganisms or is readily available for uptake by plants.

lacustrine deposit Material deposited in lake water and later exposed either by lowering of the water level or by the elevation of the land.

land A broad term embodying the total natural environment of the areas of the earth not covered by water. In addition to soil, its attributes include other physical conditions, such as mineral deposits and water supply; location in relation to centers of commerce, populations, and other land; the size of the individual tracts or holdings; and existing plant cover, works of improvement, and the like.

land capability classification A grouping of kinds of soil into special units, subclasses, and classes according to their capability for intensive use and the treatments required for sustained use. One such system has been prepared by the USDA Natural Resources Conservation Service.

land classification The arrangement of land units into various categories based upon the properties of the land or its suitability for some particular purpose.

land forming Shaping the surface of the land by scraping off the high spots and filling in the low spots with precision grading machinery to create a uniform, smooth slope, often for irrigation purposes. Also called *land smoothing*.

land-use planning The development of plans for the uses of land that, over long periods, will best serve the general welfare, together with the formulation of ways and means for achieving such uses.

laterite An iron-rich subsoil layer found in some highly weathered humid tropical soils that, when exposed and allowed to dry, becomes very hard and will not soften when rewetted. When erosion removes the overlying layers, the laterite is exposed and a virtual pavement results. *See also* plinthite.

layer (clay mineralogy) A combination in silicate clays of (tetrahedral and octahedral) sheets in a 1:1, 2:1, or 2:1:1 combination.

leaching The removal of materials in solution from the soil by percolating waters. *See also* eluviation.

leaching requirement The leaching fraction of irrigation water necessary to keep soil salinity from exceeding a tolerance level of the crop to be grown.

legume A pod-bearing member of the Leguminosae family, one of the most important and widely distributed plant families. Includes many valuable food and forage species, such as peas, beans, peanuts, clovers, alfalfas, sweet clovers, lespedezas, vetches, and kudzu. Nearly all legumes are associated with nitrogen-fixing organisms.

lichen A symbiotic relationship between fungi and cyanobacteria (blue-green algae) that enhances colonization of bare minerals and rocks. The fungi supply water and nutrients, the cyanobacteria the fixed nitrogen and carbohydrates from photosynthesis.

Liebig's law The growth and reproduction of an organism are determined by the nutrient substance (oxygen, carbon dioxide, calcium, etc.) that is available in minimum quantity with respect to organic needs; the *limiting factor*.

light soil (Obsolete in scientific use) A coarse-textured soil; a soil with a low drawbar pull and hence easy to cultivate. *See also* coarse texture; soil texture.

lignin The complex organic constituent of woody fibers in plant tissue that, along with cellulose, cements the cells together and provides strength. Lignins resist microbial attack and after some modification may become part of the soil organic matter.

lime (agricultural) In strict chemical terms, calcium oxide. In practical terms, a material containing the carbonates, oxides, and/or hydroxides of calcium and/or magnesium used to neutralize soil acidity.

lime requirement The mass of agricultural limestone, or the equivalent of other specified liming material, required to raise the pH of the soil to a desired value under field conditions.

limestone A sedimentary rock composed primarily of calcite ($CaCO_3$). If dolomite ($CaCO_3 \cdot MgCO_3$) is present in appreciable quantities, it is called a *dolomitic limestone*.

limiting factor *See* Liebig's law.

liquid limit (LL) *See* Atterberg limits.

lithosequence A group of related soils that differ, one from the other, in certain properties primarily as a result of parent material as a soil-forming factor.

loam The textural-class name for soil having a moderate amount of sand, silt, and clay. Loam soils contain 7 to 27% clay, 28 to 50% silt, and 23 to 52% sand.

loamy Intermediate in texture and properties between fine-textured and coarse-textured soils. Includes all textural classes with the words *loam* or *loamy* as a part of the class name, such as clay loam or loamy sand. *See also* loam; soil texture.

lodging Falling over of plants, either by uprooting or stem breakage.

loess Material transported and deposited by wind and consisting of predominantly silt-sized particles.

luxury consumption The intake by a plant of an essential nutrient in amounts exceeding what it needs. For example, if potassium is abundant in the soil, alfalfa may take in more than it requires.

lysimeter A device for measuring percolation (leaching) and evapotranspiration losses from a column of soil under controlled conditions.

macronutrient A chemical element necessary in large amounts (usually 50 mg/kg in the plant) for the growth of plants. Includes C, H, O, N, P, K, Ca, Mg, and S. (*Macro* refers to quantity and not to the essentiality of the element.) *See also* micronutrient.

macropores Larger soil pores, generally having a diameter greater than 0.06 mm, from which water drains readily by gravity.

marl Soft and unconsolidated calcium carbonate, usually mixed with varying amounts of clay or other impurities.

marsh Periodically wet or continually flooded area with the surface not deeply submerged. Covered dominantly with sedges, cattails, rushes, or other hydrophytic plants. Subclasses include freshwater and saltwater marshes.

mass flow Movement of nutrients with the flow of water to plant roots.

matric potential That portion of the total *soil water potential* due to the attractive forces between water and soil solids as represented through adsorption and capillarity. It will always be negative.

mature soil A soil with well-developed soil horizons produced by the natural processes of soil formation and essentially in equilibrium with its present environment.

maximum retentive capacity The average moisture content of a disturbed sample of soil, 1 cm high, which is at equilibrium with a water table at its lower surface.

mechanical analysis (Obsolete) *See* particle size analysis; particle size distribution.

medium texture Intermediate between fine-textured and coarse-textured (soils). It includes the following textural classes: very fine sandy loam, loam, silt loam, and silt.

melanic epipedon A diagnostic surface horizon formed in volcanic parent material, that contains more than 6% organic carbon, is dark in color, and has a very low bulk density and high anion adsorption capacity.

mellow soil A very soft, very friable, porous soil without any tendency toward hardness or harshness. *See also* consistence.

mesic A soil temperature class with mean annual temperature 8 to 15°C.

mesofauna Animals of medium size, between approximately 2 and 0.2 mm in diameter.

metamorphic rock A rock that has been greatly altered from its previous condition through the combined action of heat and pressure. For example, marble is a metamorphic rock produced from limestone, gneiss is produced from granite, and slate is produced from shale.

methane, CH_4 An odorless, colorless gas commonly produced under anaerobic conditions. When released to the upper atmosphere, methane contributes to global warming. *See also* greenhouse effect.

micas Primary aluminosilicate minerals in which two silica tetrahedral sheets alternate with one alumina/magnesia octahedral sheet with entrapped potassium atoms fitting between sheets. They separate readily into visible sheets or flakes.

microfauna That part of the animal population which consists of individuals too small to be clearly distinguished without the use of a microscope. Includes protozoans and nematodes.

microflora That part of the plant population which consists of individuals too small to be clearly distinguished without the use of a microscope. Includes actinomycetes, algae, bacteria, and fungi.

micronutrient A chemical element necessary in only extremely small amounts (<50 mg/kg in the plant) for the growth of plants. Examples are B, Cl, Cu, Fe, Mn, and Zn. (*Micro* refers to the amount used rather than to its essentiality.) *See also* macronutrient.

micropores Relatively small soil pores, generally found within structural aggregates and having a diameter less than 0.06 mm. *Contrast to* macropores.

microrelief Small-scale local differences in topography, including mounds, swales, or pits that are only 1 m or so in diameter and with elevation differences of up to 2 m. *See also* gilgai.

mineral soil A soil consisting predominantly of, and having its properties determined predominantly by, mineral matter. Usually contains <20% organic matter, but may contain an organic surface layer up to 30 cm thick.

mineralization The conversion of an element from an organic form to an inorganic state as a result of microbial decomposition.

minimum tillage *See* tillage, conservation.

minor element (Obsolete) *See* micronutrient.

moderately coarse texture Consisting predominantly of coarse particles. In soil textural classification, it includes all the sandy loams except the very fine sandy loam. *See also* coarse texture.

moderately fine texture Consisting predominantly of intermediate-sized (soil) particles or with relatively small amounts of fine or coarse particles. In soil textural classification, it includes clay loam, sandy loam, sandy clay loam, and silty clay loam. *See also* fine texture.

moisture equivalent (Obsolete) The weight percentage of water retained by a previously saturated sample of soil 1 cm in thickness after it has been subjected to a centrifugal force of 1000 times gravity for 30 min.

moisture potential *See* soil water potential.

mole drain Unlined drain formed by pulling a bullet-shaped cylinder through the soil.

mollic epipedon A diagnostic surface horizon of mineral soil that is dark colored and relatively thick, contains at least 0.6% organic carbon, is not massive and hard when dry, has a base saturation of more than 50%, has less than 250 mg/kg P_2O_5 soluble in 1% citric acid, and is dominantly saturated with bivalent cations.

Mollisols An order in *Soil Taxonomy.* Soils with nearly black, organic-rich surface horizons and high supply of bases. They have mollic epipedons and base saturation greater than 50% in any cambic or argillic horizon. They lack the characteristics of Vertisols and must not have oxic or spodic horizons.

montmorillonite An aluminosilicate clay mineral in the smectite group with a 2:1 expanding crystal lattice, with two silicon tetrahedral sheets enclosing an aluminum octahedral sheet. Isomorphous substitution of magnesium for some of the aluminum has occurred in the octahedral sheet. Considerable expansion may be caused by water moving between silica sheets of contiguous layers.

mor Raw humus; type of forest humus layer of unincorporated organic material, usually matted or compacted or both; distinct from the mineral soil, unless the latter has been blackened by washing in organic matter.

moraine An accumulation of drift, with an initial topographic expression of its own, built within a glaciated region chiefly by the direct action of glacial ice. Examples are ground, lateral, recessional, and terminal moraines.

morphology, soil The constitution of the soil, including the texture, structure, consistence, color, and other physical, chemical, and biological properties of the various soil horizons that make up the soil profile.

mottling Spots or blotches of different color or shades of color interspersed with the dominant color.

mucigel The gelatinous material at the surface of roots grown in unsterilized soil.

muck Highly decomposed organic material in which the original plant parts are not recognizable. Contains more mineral matter and is usually darker in color than peat. *See also* muck soil; peat.

muck soil (1) A soil containing 20 to 50% organic matter. (2) An organic soil in which the organic matter is well decomposed.

mulch Any material such as straw, sawdust, leaves, plastic film, and loose soil that is spread upon the surface of the soil to protect the soil and plant roots from the effects of raindrops, soil crusting, freezing, evaporation, etc.

mulch tillage *See* tillage, conservation.

mull A humus-rich layer of forested soils consisting of mixed organic and mineral matter. A mull blends into the upper mineral layers without an abrupt change in soil characteristics.

Munsell color system A color designation system that specifies the relative degrees of the three simple variables of color:

  chroma The relative purity, strength, or saturation of a color.
  hue The chromatic gradation (rainbow) of light that reaches the eye.
  value The degree of lightness or darkness of the color.

mycelium A stringlike mass of individual fungal or actinomycetes hyphae.

mycorrhiza The association, usually symbiotic, of fungi with the roots of seed plants. *See also* ectotrophic mycorrhiza; endotrophic mycorrhiza; arbuscular mycorrhiza.

natric horizon A diagnostic subsurface horizon that satisfies the requirements of an argillic horizon, but that also has prismatic, columnar, or blocky structure and a subhorizon having more than 15% saturation with exchangeable sodium.

necrosis Death associated with discoloration and dehydration of all or parts of plant organs, such as leaves.

nematodes Very small worms abundant in many soils and important because some of them attack and destroy plant roots.

neutral soil A soil in which the surface layer, at least to normal plow depth, is neither acid nor alkaline in reaction. In practice this means the soil is within the pH range of 6.6 to 7.3. *See also* acid soil; alkaline soil; pH; reaction, soil.

nitrate depression period A period of time, beginning shortly after the addition of fresh, highly carbonaceous organic materials to a soil, during which decomposer microorganisms have removed most of the soluble nitrate from the soil solution.

nitrification The biochemical oxidation of ammonium to nitrate, predominantly by autotrophic bacteria.

nitrogen assimilation The incorporation of nitrogen into organic cell substances by living organisms.

nitrogen cycle The sequence of chemical and biological changes undergone by nitrogen as it moves from the atmosphere into water, soil, and living organisms, and upon death of these organisms (plants and animals) is recycled through a part or all of the entire process.

nitrogen fixation The biological conversion of elemental nitrogen (N_2) to organic combinations or to forms readily utilized in biological processes.

nodule bacteria *See* rhizobia.

nonhumic substances The portion of soil organic matter comprised of relatively low molecular weight organic substances; mostly identifiable biomolecules.

no-tillage *See* tillage, conservation.

nucleic acids Complex organic acids found in the nuclei of plant and animal cells; may be combined with proteins as nucleoproteins.

O horizon Organic horizon of mineral soils.

ochric epipedon A diagnostic surface horizon of mineral soil that is too light in color, too high in chroma, too low in organic carbon, or too thin to be a plaggen, mollic, umbric, anthropic, or histic epipedon, or that is both hard and massive when dry.

octahedral sheet Sheet of horizontally linked, octahedral-shaped units that serve as the basic structural components of silicate (clay) minerals. Each unit consists of a central, six-coordinated metallic atom (e.g., Al, Mg, or Fe) surrounded by six hydroxyl groups that, in turn, are linked with other nearby metal atoms, thereby serving as interunit linkages that hold the sheet together.

oligotrophic Environments, such as soils or lakes, which are poor in nutrients.

order, soil The category at the highest level of generalization in *Soil Taxonomy*. The properties selected to distinguish the orders are reflections of the degree of horizon development and the kinds of horizons present.

organic fertilizer By-product from the processing of animal or vegetable substances that contain sufficient plant nutrients to be of value as fertilizers.

organic soil A soil in which more than half of the profile thickness is comprised of organic soil materials.

organic soil materials (As used in *Soil Taxonomy*): (1) Saturated with water for prolonged periods unless artificially drained and having 18% or more organic carbon (by weight) if the mineral fraction is more than 60% clay, more than 12% organic carbon if the mineral fraction has no clay, or between 12 and 18% carbon if the clay content of the mineral fraction is between 0 and 60%. (2) Never saturated with water for more than a few days and having more than 20% organic carbon. Histosols develop on these organic soil materials. There are three kinds of organic materials:

    fibric materials The least decomposed of all the organic soil materials, containing very high amounts of fiber that are well preserved and readily identifiable as to botanical origin.

    hemic materials Intermediate in degree of decomposition of organic materials between the less decomposed fibric and the more decomposed sapric materials.

    sapric materials The most highly decomposed of the organic materials, having the highest bulk density, least amount of plant fiber, and lowest water content at saturation.

ortstein An indurated layer in the B horizon of Spodosols in which the cementing material consists of illuviated sesquioxides (mostly iron) and organic matter.

osmotic potential That portion of the total *soil water potential* due to the presence of solutes in soil water. It will generally be negative.

osmotic pressure Pressure exerted in living bodies as a result of unequal concentrations of salts on both sides of a cell wall or membrane. Water moves from the area having the lower salt concentration through the membrane into the area having the higher salt concentration and, therefore, exerts additional pressure on the side with higher salt concentration.

outwash plain A deposit of coarse-textured materials (e.g., sands and gravels) left by streams of meltwater flowing from receding glaciers.

oven-dry soil Soil that has been dried at 105°C until it reaches constant weight.

oxic horizon A diagnostic subsurface horizon that is at least 30 cm thick and is characterized by the virtual *absence* of weatherable primary minerals or 2:1 lattice clays and the *presence* of 1:1 lattice clays and highly insoluble minerals, such as quartz sand, hydrated oxides of iron and aluminum, low cation exchange capacity, and small amounts of exchangeable bases.

oxidation The loss of electrons by a substance; therefore, a gain in positive valence charge and, in some cases, the chemical combination with oxygen gas.

oxidation ditch An artificial open channel for partial digestion of liquid organic wastes in which the wastes are circulated and aerated by a mechanical device.

Oxisols An order in *Soil Taxonomy*. Soils with residual accumulations of low-activity clays, free oxides, kaolin, and quartz. They are mostly in tropical climates.

pans Horizons or layers in soils that are strongly compacted, indurated, or very high in clay content. *See also* caliche; claypan; fragipan; hardpan.

parent material The unconsolidated and more or less chemically weathered mineral or organic matter from which the solum of soils is developed by pedogenic processes.

particle density The mass per unit volume of the soil particles. In technical work, usually expressed as metric tons per cubic meter (Mg/m^3) or grams per cubic centimeter (g/cm^3).

particle size The effective diameter of a particle measured by sedimentation, sieving, or micrometric methods.

particle size analysis Determination of the various amounts of the different separates in a soil sample, usually by sedimentation, sieving, micrometry, or combinations of these methods.

particle size distribution The amounts of the various soil separates in a soil sample, usually expressed as weight percentages.

partitioning The distribution of organic chemicals (such as pollutants) into a portion that dissolves in the soil organic matter and a portion that remains undissolved in the soil solution.

pascal An SI unit of pressure equal to 1 newton per square meter.

peat Unconsolidated soil material consisting largely of undecomposed, or only slightly decomposed, organic matter accumulated under conditions of excessive moisture. *See also* organic soil materials; peat soil.

peat soil An organic soil containing more than 50% organic matter. Used in the United States to refer to the stage of decomposition of the organic matter, *peat* referring to the slightly decomposed or undecomposed deposits and *muck* to the highly decomposed materials. *See also* muck; muck soil; peat.

ped A unit of soil structure such as an aggregate, crumb, prism, block, or granule, formed by natural processes (in contrast to a *clod*, which is formed artificially).

pedology The science that deals with the formation, morphology, and classification of soil bodies as landscape components.

pedon The smallest volume that can be called *a soil*. It has three dimensions. It extends downward to the depth of plant roots or to the lower limit of the genetic soil horizons. Its lateral cross section is roughly hexagonal and ranges from 1 to 10 m^2 in size, depending on the variability in the horizons.

pedoturbation Physical disturbance and mixing of soil horizons by such forces as burrowing animals (faunal pedoturbation) or frost churning (cryoturbation).

peneplain A once high, rugged area that has been reduced by erosion to a lower, gently rolling surface resembling a plain.

penetrability The ease with which a probe can be pushed into the soil. May be expressed in units of distance, speed, force, or work depending on the type of penetrometer used.

penetrometer An instrument consisting of a rod with a cone-shaped tip and a means of measuring the force required to push the rod into a specified increment of soil.

perc test *See* percolation test.

percolation, soil water The downward movement of water through soil. Especially, the downward flow of water in saturated or nearly saturated soil at hydraulic gradients of the order of 1.0 or less.

percolation test A measurement of the rate of percolation of water in a soil profile, usually to determine the suitability of a soil for use as a septic tank drain field.

perforated plastic pipe Pipe, sometimes flexible, with holes or slits in it that allow the entrance and exit of air and water. Used for soil drainage and for septic effluent spreading into soil.

permafrost (1) Permanently frozen material underlying the solum. (2) A perennially frozen soil horizon.

permanent charge *See* constant charge.

permanent wilting point *See* wilting point.

permeability, soil The ease with which gases, liquids, or plant roots penetrate or pass through a bulk mass of soil or a layer of soil.

petrocalcic horizon A diagnostic subsurface horizon that is a continuous, indurated calcic horizon cemented by calcium carbonate and, in some places, with magnesium carbonate. It cannot be penetrated with a spade or auger when dry; dry fragments do not slake in water; and it is impenetrable to roots.

petrogypsic horizon A diagnostic subsurface horizon that is a continuous, strongly cemented, massive gypsic horizon that is cemented by calcium sulfate. It can be chipped with a spade when dry. Dry fragments do not slake in water and it is impenetrable to roots.

pH, soil The negative logarithm of the hydrogen ion activity (concentration) of a soil. The degree of acidity (or alkalinity) of a soil as determined by means of a glass or other suitable electrode or indicator at a specified moisture content or soil-to-water ratio, and expressed in terms of the pH scale.

phase, soil A subdivision of a soil series or other unit of classification having characteristics that affect the use and management of the soil but do not vary sufficiently to differentiate it as a separate series. Included are such characteristics as degree of slope, degree of erosion, and content of stones.

pH-dependent charge That portion of the total charge of the soil particles that is affected by, and varies with, changes in pH.

photomap A mosaic map made from aerial photographs to which place names, marginal data, and other map information have been added.

phyllosphere The leaf surface.

physical properties (of soils) Those characteristics, processes, or reactions of a soil that are caused by physical forces and that can be described by, or expressed in, physical terms or equations. Examples of physical properties are bulk density, water-holding capacity, hydraulic conductivity, porosity, pore-size distribution, and so on.

phytotoxic substances Chemicals that are toxic to plants.

placic horizon A diagnostic subsurface horizon of a black to dark reddish mineral soil that is usually thin but that may range from 1 to 25 mm in thickness. The placic horizon is commonly cemented with iron and is slowly permeable or impenetrable to water and roots.

plaggen epipedon A diagnostic surface horizon that is human-made and more than 50 cm thick. Formed by long-continued manuring and mixing.

plant nutrients *See* essential element.

plastic limit (PL) *See* Atterberg limits.

plastic soil A soil capable of being molded or deformed continuously and permanently, by relatively moderate pressure, into various shapes. *See also* consistence.

platy Consisting of soil aggregates that are developed predominantly along the horizontal axes; laminated; flaky.

plinthite (brick) A highly weathered mixture of sesquioxides of iron and aluminum with quartz and other diluents that occurs as red mottles and that changes irreversibly to hardpan upon alternate wetting and drying.

plow layer The soil ordinarily moved when land is plowed; equivalent to *surface soil.*

plow pan A subsurface soil layer having a higher bulk density and lower total porosity than layers above or below it, as a result of pressure applied by normal plowing and other tillage operations.

plowing A primary broad-base tillage operation that is performed to shatter soil uniformly with partial to complete inversion.

point of zero charge The pH value of a solution in equilibrium with a particle whose net charge, from all sources, is zero.

polypedon (As used in *Soil Taxonomy*) Two or more contiguous pedons, all of which are within the defined limits of a single soil series; commonly referred to as a *soil individual.*

pore size distribution The volume of the various sizes of pores in a soil. Expressed as percentages of the bulk volume (soil plus pore space).

porosity, soil The volume percentage of the total soil bulk not occupied by solid particles.

preferential flow Nonuniform movement of water and its solutes through a soil along certain pathways, which are often macropores.

primary consumer An organism that subsists on plant material.

primary mineral A mineral that has not been altered chemically since deposition and crystallization from molten lava.

primary producer An organism (usually a photosynthetic plant) that creates organic, energy-rich material from inorganic chemicals, solar energy, and water.

primary tillage *See* tillage, primary.

priming effect The increased decomposition of relatively stable soil humus under the influence of much enhanced, generally biological, activity resulting from the addition of fresh organic materials to a soil.

prismatic soil structure A soil structure type with prismlike aggregates that have a vertical axis much longer than the horizontal axes.

prokaryote An organism whose cells do not have a distinct nucleus.

Proctor test A laboratory procedure that indicates the maximum achievable bulk density for a soil and the optimum water content for compacting a soil.

productivity, soil The capacity of a soil for producing a specified plant or sequence of plants under a specified system of management. Productivity emphasizes the capacity of soil to produce crops and should be expressed in terms of yields.

profile, soil A vertical section of the soil through all its horizons and extending into the parent material.

protein Any of a group of nitrogen-containing organic compounds formed by the polymerization of a large number of amino acid molecules and that, upon hydrolysis, yield these amino acids. They are essential parts of living matter and are one of the essential food substances of animals.

protonation Attachment of protons (H^+ ions) to exposed OH groups on the surface of soil particles, resulting in an overall positive charge on the particle surface.

protozoa One-celled eukaryotic organisms, such as amoeba.

puddled soil Dense, massive soil artificially compacted when wet and having no aggregated structure. The condition commonly results from the tillage of a clayey soil when it is wet.

rain, acid *See* acid rain.

reaction, soil (No longer used in soil science) The degree of acidity or alkalinity of a soil, usually expressed as a pH value or by terms ranging from extremely acid for pH values <4.5 to very strongly alkaline for pH values >9.0.

recharge area A geographic area in which an otherwise confined aquifer is exposed to surficial percolation of water to recharge the groundwater in the aquifer.

redox concentrations Zones of apparent accumulations of Fe-Mn oxides in soils.

redox depletions Zones of low chroma (<2) where Fe-Mn oxides, and in some cases clay, have been stripped from the soil.

redox potential The electrical potential (measured in volts or millivolts) of a system due to the tendency of the substances in it to give up or acquire electrons.

redoximorphic features Soil properties associated with wetness that result from reduction and oxidation of iron and manganese compounds after saturation and desaturation with water. *See also* redox concentrations; redox depletions.

reduction The gain of electrons, and therefore the loss of positive valence charge, by a substance. In some cases, a loss of oxygen or a gain of hydrogen is also involved.

regolith The unconsolidated mantle of weathered rock and soil material on the earth's surface; loose earth materials above solid rock. (Approximately equivalent to the term *soil* as used by many engineers.)

relief The relative differences in elevation between the upland summits and the lowlands or valleys of a given region.

residual material Unconsolidated and partly weathered mineral materials accumulated by disintegration of consolidated rock in place.

rhizobia Bacteria capable of living symbiotically with higher plants, usually in nodules on the roots of legumes, from which they receive their energy, and capable of converting atmospheric nitrogen to combined organic forms; hence the term *symbiotic nitrogen-fixing bacteria*. (Derived from the generic name *Rhizobium*.)

rhizosphere That portion of the soil in the immediate vicinity of plant roots in which the abundance and composition of the microbial population are influenced by the presence of roots.

rill A small, intermittent water course with steep sides; usually only a few centimeters deep and hence no obstacle to tillage operations.

rill erosion An erosion process in which numerous small channels of only several centimeters in depth are formed; occurs mainly on recently cultivated soils. *See also* rill.

riparian zone The area, both above and below the ground surface, that borders a river.

riprap Broken rock, cobbles, or boulders placed on earth surfaces, such as the face of a dam or the bank of a stream, for protection against the action of water (waves); also applied to brush or pole mattresses, or brush and stone, or other similar materials used for soil erosion control.

rock The material that forms the essential part of the earth's solid crust, including loose incoherent masses such as sand and gravel, as well as solid masses of granite and limestone.

root interception Acquisition of nutrients by a root as a result of the root growing into the vicinity of the nutrient source.

root nodules Swollen growths on plant roots. Often in reference to those in which symbiotic microorganisms live.

rotary tillage *See* tillage, rotary.

runoff The portion of the precipitation on an area that is discharged from the area through stream channels. That which is lost without entering the soil is called *surface runoff* and that which enters the soil before reaching the stream is called *groundwater runoff* or *seepage flow* from groundwater. (In soil science *runoff* usually refers to the water lost by surface flow; in geology and hydraulics *runoff* usually includes both surface and subsurface flow.)

salic horizon A diagnostic subsurface horizon of enrichment with secondary salts more soluble in cold water than gypsum. A salic horizon is 15 cm or more in thickness.

saline seep An area of land in which saline water seeps to the surface, leaving a high salt concentration behind as the water evaporates.

saline soil A nonsodic soil containing sufficient soluble salts to impair its productivity. The conductivity of a saturated extract is >4 dS/m, the exchangeable sodium adsorption ratio is less than about 13, and the pH is <8.5.

saline-sodic soil A soil containing sufficient exchangeable sodium to interfere with the growth of most crop plants and containing appreciable quantities of soluble salts. The exchangeable sodium adsorption ratio is >13, the conductivity of the saturation extract is >4 dS/m (at 25°C), and the pH is usually 8.5 or less in the saturated soil.

salinization The process of accumulation of salts in soil.

saltation Particle movement in water or wind where particles skip or bounce along the stream bed or soil surface.

sand A soil particle between 0.05 and 2.0 mm in diameter; a soil textural class.

sapric materials *See* organic soil materials.

saprolite Bedrock that has weathered in place to the point that it is porous and can be dug with a spade.

saturated paste extract The extract from a saturated soil paste, the electrical conductivity E_c of which gives an indirect measure of salt content in a soil.

saturation extract The solution extracted from a saturated soil paste.

saturation percentage The water content of a saturated soil paste, expressed as a dry weight percentage.

savanna (savannah) A grassland with scattered trees, either as individuals or clumps. Often a transitional type between true grassland and forest.

second bottom The first terrace above the normal floodplain of a stream.

secondary mineral A mineral resulting from the decomposition of a primary mineral or from the reprecipitation of the products of decomposition of a primary mineral. *See also* primary mineral.

sediment Transported and deposited particles or aggregates derived from soils, rocks, or biological materials.

sedimentary rock A rock formed from materials deposited from suspension or precipitated from solution and usually being more or less consolidated. The principal sedimentary rocks are sandstones, shales, limestones, and conglomerates.

seedbed The soil prepared to promote the germination of seed and the growth of seedlings.

self-mulching soil A soil in which the surface layer becomes so well aggregated that it does not crust and seal under the impact of rain but instead serves as a surface mulch upon drying.

semiarid Term applied to regions or climates where moisture is more plentiful than in arid regions but still definitely limits the growth of most crop plants. Natural vegetation in uncultivated areas is short grasses.

separate, soil One of the individual-sized groups of mineral soil particles—sand, silt, or clay.

septic tank An underground tank used in the deposition of domestic wastes. Organic matter decomposes in the tank, and the effluent is drained into the surrounding soil.

series, soil The soil series is a subdivision of a family in *Soil Taxonomy* and consists of soils that are similar in all major profile characteristics.

sewage effluent The liquid part of sewage or wastewater; it is usually treated to remove some portion of the dissolved organic compounds and nutrients present from the original sewage.

sewage sludge Settled sewage solids combined with varying amounts of water and dissolved materials, removed from sewage by screening, sedimentation, chemical precipitation, or bacterial digestion. Also called *biosolids*.

shear Force, as of a tillage implement, acting at right angles to the direction of movement.

sheet (mineralogy) A flat array of more than one atomic thickness and composed of one or more levels of linked coordination polyhedra. A sheet is thicker than a plane and thinner than a layer. Examples: tetrahedral sheet, octahedral sheet.

sheet erosion The removal of a fairly uniform layer of soil from the land surface by runoff water.

shelterbelt A wind barrier of living trees and shrubs established and maintained for protection of farm fields. Syn. *windbreak*.

shifting cultivation A farming system in which land is cleared, the debris burned, and crops grown for 2 to 3 years. When the farmer moves on to another plot, the land is then left idle for 5 to 15 years; then the burning and planting process is repeated.

short-range order minerals Minerals, such as allophane, whose structural framework consists of short distances of well-ordered crystalline structure interspersed with distances of noncrystalline amorphous materials.

side-dressing The application of fertilizer alongside row-crop plants, usually on the soil surface. Nitrogen materials are most commonly side-dressed.

silica/alumina ratio The molecules of silicon dioxide (SiO_2) per molecule of aluminum oxide (Al_2O_3) in clay minerals or in soils.

silica/sesquioxide ratio The molecules of silicon dioxide (SiO_2) per molecule of aluminum oxide (Al_2O_3) plus ferric oxide (Fe_2O_3) in clay minerals or in soils.

silt (1) A soil separate consisting of particles between 0.05 and 0.002 mm in equivalent diameter. (2) A soil textural class.

silting The deposition of waterborne sediments in stream channels, lakes, reservoirs, or on floodplains, usually resulting from a decrease in the velocity of the water.

site index A quantitative evaluation of the productivity of a soil for forest growth under the existing or specified environment.

slag A product of smelting, containing mostly silicates; the substances not sought to be produced as matte or metal and having a lower specific gravity.

slash-and-burn *See* shifting cultivation.

slick spots Small areas in a field that are slick when wet because of a high content of alkali or exchangeable sodium.

slickensides Stress surfaces that are polished and striated and are produced by one mass sliding past another.

slope The degree of deviation of a surface from horizontal, measured in a numerical ratio, percent, or degrees.

slow fraction (of soil organic matter) That portion of soil organic matter that can be metabolized with great difficulty by the microorganisms in the soil and therefore has a slow turnover rate with a half-life in the soil ranging from a few years to a few decades. Often this fraction is the product of some previous decomposition.

smectite A group of silicate clays having a 2:1-type lattice structure with sufficient isomorphous substitution in either or both the tetrahedral and octahedral sheets to give a high interlayer negative charge and high cation exchange capacity and to permit significant interlayer expansion and consequent shrinking and swelling of the clay. Montmorillonite, beidellite, and saponite are in the smectite group.

sodic soil A soil that contains sufficient sodium to interfere with the growth of most crop plants, and in which the sodium adsorption ratio is 13 or greater.

sodium adsorption ratio (SAR)

$$SAR = \frac{[Na^+]}{\sqrt{\frac{1}{2}([Ca^{2+}] + [Mg^{2+}])}}$$

where the cation concentrations are in millimoles of charge per liter ($mmol_c/L$).

soil (1) A dynamic natural body composed of mineral and organic materials and living forms in which plants grow. (2) The collection of natural bodies occupying parts of the earth's surface that support plants and that have properties due to the integrated effect of climate and living matter acting upon parent material, as conditioned by relief, over periods of time.

soil air The soil atmosphere; the gaseous phase of the soil, being that volume not occupied by soil or liquid.

soil alkalinity The degree or intensity of alkalinity of a soil, expressed by a value >7.0 on the pH scale.

soil amendment Any material, such as lime, gypsum, sawdust, or synthetic conditioner, that is worked into the soil to make it more amenable to plant growth.

soil association A group of defined and named taxonomic soil units occurring together in an individual and characteristic pattern over a geographic region, comparable to plant associations in many ways.

soil auger A tool used to bore small holes up to several meters deep in soils in order to bring up samples of material from various soil layers. It consists of a long T-handle attached to either a cylinder with twisted teeth or a screwlike bit.

soil classification (*Soil Taxonomy*) The systematic arrangement of soils into groups or categories on the basis of their characteristics. *See* order; suborder; great group; subgroup; family; and series.

soil complex A mapping unit used in detailed soil surveys where two or more defined taxonomic units are so intimately intermixed geographically that it is undesirable or impractical, because of the scale being used, to separate them. A more intimate mixing of smaller areas of individual taxonomic units than that described under *soil association*.

soil compressibility The property of a soil pertaining to its capacity to decrease in bulk volume when subjected to a load.

soil conditioner Any material added to a soil for the purpose of improving its physical condition.

soil conservation A combination of all management and land-use methods that safeguard the soil against depletion or deterioration caused by nature and/or humans.

soil consociation A kind of soil map unit that is named for the dominant soil taxon in the delineation, and in which at least half of the pedons are of the named soil taxon, and most of the remaining pedons are so similar as to not affect most interpretations.

soil correlation The process of defining, mapping, naming, and classifying the kinds of soils in a specific soil survey area, the purpose being to ensure that soils are adequately defined, accurately mapped, and uniformly named.

soil erosion *See* erosion.

soil fertility *See* fertility, soil.

soil genesis *See* genesis, soil.

soil geography A subspecialization of physical geography concerned with the areal distributions of soil types.

soil horizon *See* horizon, soil.

soil loss tolerance (T) The maximum rate of annual soil loss that will permit plant productivity to be maintained economically and indefinitely.

soil management The sum total of all tillage operations, cropping practices, fertilizer, lime, and other treatments conducted on or applied to a soil for the production of plants.

soil map A map showing the distribution of soil types or other soil mapping units in relation to the prominent physical and cultural features of the earth's surface.

soil mechanics and engineering A subspecialization of soil science concerned with the effect of forces on the soil and the application of engineering principles to problems involving the soil.

soil moisture potential *See* soil water potential.

soil monolith A vertical section of a soil profile removed from the soil and mounted for display or study.

soil morphology The physical constitution, particularly the structural properties, of a soil profile as exhibited by the kinds, thicknesses, and arrangement of the horizons in the profile, and by the texture, structure, consistence, and porosity of each horizon.

soil order *See* order, soil.

soil profile   *See* profile, soil.

soil organic matter   The organic fraction of the soil that includes plant and animal residues at various stages of decomposition, cells and tissues of soil organisms, and substances synthesized by the soil population. Commonly determined as the amount of organic material contained in a soil sample passed through a 2-mm sieve.

soil porosity   *See* porosity, soil.

soil productivity   *See* productivity, soil.

soil quality   The capacity of a specific kind of soil to function, within natural or managed ecosystem boundaries, to sustain plant and animal productivity, maintain or enhance water and air quality, and support human health and habitation. Sometimes considered in relation to this capacity in the undisturbed, natural state.

soil reaction   *See* reaction, soil; pH, soil.

soil salinity   The amount of soluble salts in a soil, expressed in terms of percentage, milligrams per kilogram, parts per million (ppm), or other convenient ratios.

soil separates   *See* separate, soil.

soil series   *See* series, soil.

soil solution   The aqueous liquid phase of the soil and its solutes, consisting of ions dissociated from the surfaces of the soil particles and of other soluble materials.

soil strength   A transient soil property related to the soil's solid phase cohesion and adhesion.

soil structure   The combination or arrangement of primary soil particles into secondary particles, units, or peds. These secondary units may be, but usually are not, arranged in the profile in such a manner as to give a distinctive characteristic pattern. The secondary units are characterized and classified on the basis of size, shape, and degree of distinctness into classes, types, and grades, respectively.

soil structure classes   A grouping of soil structural units or peds on the basis of size from the very fine to very coarse.

soil structure grades   A grouping or classification of soil structure on the basis of inter- and intraaggregate adhesion, cohesion, or stability within the profile. Four grades of structure, designated from 0 to 3, are recognized: *structureless, weak, moderate,* and *strong.*

soil structure types   A classification of soil structure based on the shape of the aggregates or peds and their arrangement in the profile, including platy, prismatic, columnar, blocky, subangular blocky, granulated, and crumb.

soil survey   The systematic examination, description, classification, and mapping of soils in an area. Soil surveys are classified according to the kind and intensity of field examination.

soil temperature classes   A criterion used to differentiate soil in *Soil Taxonomy,* mainly at the family level. Classes are based on mean annual soil temperature and on differences between summer and winter temperatures at a depth of 50 cm.

soil textural class   A grouping of soil textural units based on the relative proportions of the various soil separates (sand, silt, and clay). These textural classes, listed from the coarsest to the finest in texture, are sand, loamy sand, sandy loam, loam, silt loam, silt, sandy clay loam, clay loam, silty clay loam, sandy clay, silty clay, and clay. There are several subclasses of the sand, loamy sand, and sandy loam classes based on the dominant particle size of the sand fraction (e.g., loamy fine sand, coarse sandy loam).

soil texture   The relative proportions of the various soil separates in a soil.

soil water potential   (total) A measure of the difference between the free energy state of soil water and that of pure water. Technically it is defined as "that amount of work that must be done per unit quantity of pure water in order to transport reversibly and isothermically an infinitesimal quantity of water from a pool of pure water, at a specified elevation and at atmospheric pressure to the soil water (at the point under consideration)." This *total* potential consists of *gravitational, matric,* and *osmotic* potentials.

solum   (p. **sola**) The upper and most weathered part of the soil profile; the A, E, and B horizons.

sombric horizon   A diagnostic subsurface horizon that contains illuvial humus but has a low cation exchange capacity and low percentage base saturation. Mostly restricted to cool, moist soils of high plateaus and mountainous areas of tropical and subtropical regions.

sorption   The removal from the soil solution of an ion or molecule by adsorption and absorption. This term is often used when the exact mechanism of removal is not known.

species diversity   The variety of different biological species present in an ecosystem. Generally, high diversity is marked by many species with few individuals in each.

species richness   The number of different species present in an ecosystem, without regard to the distribution of individuals among those species.

specific heat capacity   The amount of kinetic (heat) energy required to raise the temperature of 1 g of a substance (usually in reference to soil or soil components).

specific surface   The solid particle surface area per unit mass or volume of the solid particles.

splash erosion   The spattering of small soil particles caused by the impact of raindrops on very wet soils. The loosened and separated particles may or may not be subsequently removed by surface runoff.

spodic horizon   A diagnostic subsurface horizon characterized by the illuvial accumulation of amorphous materials composed of aluminum and organic carbon with or without iron.

Spodosols An order in *Soil Taxonomy*. Soils with subsurface illuvial accumulations of organic matter and compounds of aluminum and usually iron. These soils are formed in acid, mainly coarse-textured materials in humid and mostly cool or temperate climates.

stem flow The process by which rain or irrigation water is directed by a plant canopy toward the plant stem so as to wet the soil unevenly under the plant canopy.

stratified Arranged in or composed of strata or layers.

strip-cropping The practice of growing crops that require different types of tillage, such as row and sod, in alternate strips along contours or across the prevailing direction of wind.

structure, soil *See* soil structure.

stubble mulch The stubble of crops or crop residues left essentially in place on the land as a surface cover before and during the preparation of the seedbed and at least partly during the growing of a succeeding crop.

subgroup, soil The *great groups* in *Soil Taxonomy* are subdivided into central concept subgroups that show the central properties of the great group, intergrade subgroups that show properties of more than one great group, and other subgroups for soils with atypical properties that are not characteristic of any great group.

suborder, soil A category in *Soil Taxonomy* that narrows the ranges in soil moisture and temperature regimes, kinds of horizons, and composition, according to which of these is most important.

subsoil That part of the soil below the plow layer.

subsoiling Breaking of compact subsoils, without inverting them, with a special knifelike instrument (chisel), which is pulled through the soil at depths usually of 30 to 60 cm and at spacings usually of 1 to 2 m.

sulfuric horizon A diagnostic subsurface horizon in either mineral or organic soils that has a pH < 3.5 and fresh straw-colored mottles (called *jarosite mottles*). Forms by oxidation of sulfide-rich materials and is highly toxic to plants.

summer fallow *See* fallow.

surface runoff *See* runoff.

surface seal A thin layer of fine particles deposited on the surface of a soil that greatly reduces the permeability of the soil surface to water.

surface soil The uppermost part of the soil, ordinarily moved in tillage, or its equivalent in uncultivated soils. Ranges in depth from 7 to 25 cm. Frequently designated as the *plow layer*, the *Ap layer*, or the *Ap horizon*.

surface tension The elasticlike phenomenon resulting from the unbalanced attractions among liquid molecules (usually water) and between liquid and gaseous molecules (usually air) at the liquid–gas interface.

symbiosis The living together in intimate association of two dissimilar organisms, the cohabitation being mutually beneficial.

talus Fragments of rock and other soil material accumulated by gravity at the foot of cliffs or steep slopes.

taxonomy, soil The science of classification of soils; laws and principles governing the classifying of soil. Also a specific *soil classification* system developed by the U.S. Department of Agriculture.

tensiometer A device for measuring the negative pressure (or tension) of water in soil in situ; a porous, permeable ceramic cup connected through a tube to a manometer or vacuum gauge.

tension, soil-moisture *See* soil water potential.

terrace (1) A level, usually narrow, plain bordering a river, lake, or the sea. Rivers sometimes are bordered by terraces at different levels. (2) A raised, more or less level or horizontal strip of earth usually constructed on or nearly on a contour and designed to make the land suitable for tillage and to prevent accelerated erosion by diverting water from undesirable channels of concentration; sometimes called *diversion terrace*.

tetrahedral sheet Sheet of horizontally linked, tetrahedron-shaped units that serve as one of the basic structural components of silicate (clay) minerals. Each unit consists of a central four-coordinated atom (e.g., Si, Al, Fe) surrounded by four oxygen atoms that, in turn, are linked with other nearby atoms (e.g., Si, Al, Fe), thereby serving as interunit linkages to hold the sheet together.

texture *See* soil texture.

thermal analysis (differential thermal analysis) A method of analyzing a soil sample for constituents, based on a differential rate of heating of the unknown and standard samples when a uniform source of heat is applied.

thermic A soil temperature class with mean annual temperature 15 to 22°C.

thermophilic organisms Organisms that grow readily at temperatures above 45°C.

thixotrophy The property of certain clay soils of becoming fluid when jarred or agitated and then setting again when at rest. Similar to *quick*, as in quick clays or quicksand.

tile, drain Pipe made of burned clay, concrete, or ceramic material, in short lengths, usually laid with open joints to collect and carry excess water from the soil.

till (1) Unstratified glacial drift deposited directly by the ice and consisting of clay, sand, gravel, and boulders intermingled in any proportion. (2) To plow and prepare for seeding; to seed or cultivate the soil.

tillage The mechanical manipulation of soil for any purpose; but in agriculture it is usually restricted to the modifying of soil conditions for crop production.

tillage, conservation Any tillage sequence that reduces loss of soil or water relative to conventional tillage, which generally leaves at least 30% of the soil surface covered by residues, including the following systems:

minimum tillage The minimum soil manipulation necessary for crop production or meeting tillage requirements under the existing soil and climatic conditions.

mulch tillage Tillage or preparation of the soil in such a way that plant residues or other materials are left to cover the surface; also called *mulch farming, trash farming, stubble mulch tillage,* and *plowless farming.*

no-tillage system A procedure whereby a crop is planted directly into a seedbed not tilled since harvest of the previous crop; also called *zero tillage.*

ridge till Planting on ridges formed by cultivation during the previous growing period.

strip till Planting is done in a narrow strip that has been tilled and mixed, leaving the remainder of the soil surface undisturbed.

tillage, conventional The combined primary and secondary tillage operations normally performed in preparing a seedbed for a given crop grown in a given geographic area.

tillage, primary Tillage that contributes to the major soil manipulation, commonly with a plow.

tillage, rotary An operation using a power-driven rotary tillage tool to loosen and mix soil.

tillage, secondary Any tillage operations following primary tillage designed to prepare a satisfactory seedbed for planting.

tilth The physical condition of soil as related to its ease of tillage, fitness as a seedbed, and its impedance to seedling emergence and root penetration.

topdressing An application of fertilizer to a soil after the crop stand has been established.

toposequence A sequence of related soils that differ, one from the other, primarily because of *topography* as a soil-formation factor.

topsoil (1) The layer of soil moved in cultivation. *See also* surface soil. (2) Presumably fertile soil material used to top-dress roadbanks, gardens, and lawns.

trace element (Obsolete) *See* micronutrient.

trioctahedral An octahedral sheet of silicate clays in which the sites for the six-coordinated metallic atoms are mostly filled with divalent cations, such as Mg^{2+}.

truncated Having lost all or part of the upper soil horizon or horizons.

tuff Volcanic ash usually more or less stratified and in various states of consolidation.

tundra A level or undulating treeless plain characteristic of arctic regions.

Ultisols An order in *Soil Taxonomy.* Soils that are low in bases and have subsurface horizons of illuvial clay accumulations. They are usually moist, but during the warm season of the year some are dry part of the time.

umbric epipedon A diagnostic surface horizon of mineral soil that has the same requirements as the mollic epipedon with respect to color, thickness, organic carbon content, consistence, structure, and P_2O_5 content, but that has a base saturation of less than 50%.

universal soil loss equation (USLE) An equation for predicting the average annual soil loss per unit area per year; $A = RKLSPC$, where R is the climatic erosivity factor (rainfall plus runoff), K is the soil erodibility factor, L is the length of slope, S is the percent slope, P is the soil erosion practice factor, and C is the cropping and management factor.

unsaturated flow The movement of water in a soil that is not filled to capacity with water.

vadose zone The aerated region of soil above the permanent water table.

value (color) *See* Munsell color system.

variable charge *See* pH-dependent charge.

varnish, desert A glossy sheen or coating on stones and gravel in arid regions.

vermiculite A 2:1-type silicate clay, usually formed from mica, that has a high net negative charge stemming mostly from extensive isomorphous substitution of aluminum for silicon in the tetrahedral sheet.

Vertisols An order in *Soil Taxonomy.* Clayey soils with high shrink–swell potential that have wide, deep cracks when dry. Most of these soils have distinct wet and dry periods throughout the year.

vesicles (1) Unconnected voids with smooth walls. (2) Spherical structures formed inside root cortical cells by vesicular arbuscular mycorrhizal fungi.

virgin soil A soil that has not been significantly disturbed from its natural environment.

water potential, soil *See* soil water potential.

water table The upper surface of groundwater or that level below which the soil is saturated with water.

water table, perched The surface of a local zone of saturation held above the main body of groundwater by an impermeable layer of stratum, usually clay, and separated from the main body of groundwater by an unsaturated zone.

water use efficiency Dry matter or harvested portion of crop produced per unit of water consumed.

waterlogged Saturated with water.

watershed All the land and water within the geographical confines of a drainage divide or surrounding ridges that separate the area from neighboring watersheds.

water-stable aggregate A soil aggregate stable to the action of water, such as falling drops or agitation, as in wet-sieving analysis.

weathering All physical and chemical changes produced in rocks, at or near the earth's surface, by atmospheric agents.

wetland An area of land that has hydric soil and hydrophytic vegetation, typically flooded for part of the year, and forming a transition zone between aquatic and terrestrial systems.

wetting front The boundary between the wetted soil and dry soil during infiltration of water.

wilting point (permanent wilting point) The moisture content of soil, on an oven-dry basis, at which plants wilt and fail to recover their turgidity when placed in a dark, humid atmosphere.

windbreak Planting of trees, shrubs, or other vegetation perpendicular, or nearly so, to the principal wind direction to protect soils, crops, homesteads, etc., from wind and snow.

xenobiotic Compounds foreign to biological systems. Often refers to compounds resistant to decomposition.

xerophytes Plants that grow in or on extremely dry soils or soil materials.

zero tillage *See* tillage, conservation.

zymogenous organisms So-called opportunist organisms found in soils in large numbers immediately following addition of readily decomposable organic materials.

Absorption, 23
Absorption fields, 181
Achromobacter, 366
Acidity, 21
  classes of, 255–56
  combating, 255, 268–73, 362
  factors in development of, 35, 47, 253–56, 362, 386
  methods for increasing, 267–68
Acid rain, 252, 260–62, 389
Acids, 341
  as factor in weathering, 33, 34, 54
  sources of, 34, 252, 267, 328
Actinomycetes, 292, 295, 310–12, 320
  nitrogen fixation by, 314, 371, 375
Adenosine triphosphate (ATP), 392
Adhesion, 132, 133, 136
Adsorption, 23, 132
Aerenchyma tissues, 195, 201, 208
Aerosolic dust, 42, 43
Aggregates, 17, 104, 115–21, 307, 310
Agricultural limes, 268. *See also* Lime
Agriculture, *see* Farming
Agroforestry, 442
A horizons, 12, 47, 55
  development of, 51, 52, 53
Albedo, 215
Albic horizons, 63, 82, 83
Alfisols, 81–82
  clays in, 238
  C/N ratio of, 329
  profile of, 80
Algae, 292, 294, 305, 362
  and fungi, 305, 317
  and protozoans, 317
  as soil conditioners, 124
  toxic, 397
Algal bloom, 395, 397, 448

Algal crusts, 305
Alkaline soils, 273–87
Alkalinity, 20, 276, 406
  combating, 257–58
  measuring, 275–76
  sources of, 256–58
Allelochemicals, 338–39
Allelopathy, 338–39
Alley cropping, 442
Allophane, 62, 71, 228, 239
  formation of, 236–37
  phosphorus fixation by, 408
Alluvial fans, 37, 38
Alluvial stream deposits, 37–38, 51
Aluminum, 47, 402
  role in acidity, 253–55
Aluminum oxide, 12, 31, 54, 55, 56
Aluminum oxide clays, *see* Fe, Al oxide clays
Amino acids, 326, 330, 335, 338, 358, 379
Ammonia, 377, 459
  absorption of, 362
  anhydrous, 286–87, 453, 459
  sources of, 448, 453
  volatilization of, 361–62
Ammonification, 363
Ammonium, 326, 328, 359, 378
  fixation by clays, 361, 418
  oxidation of, 362
Anaerobic bacteria, 312, 328
Anaerobic soils, 197
Andic properties, 72
Andisols, 71–72, 236, 409
Angle of repose, 126
Animals:
  effect on soil development, 48–49
  as factor in weathering, 33
Anion exchange, 226, 246–47

Anion exchange capacity (AEC), 249
Anions, 54
Anoxic conditions, 8
Antagonism, 316, 318, 429–30
Anthropic epipedons, 62
Ants, 294, 296, 315, 324
Apatites, 402, 406, 452
Ap horizons, 57
Aquents, 70
Aquic SMR, 64, 208
Aquifers, 173
Aquolls, 65
Arbuscular mycorrhizae (AM), 309–10
Archaeologic digs, 88
Aridic SMR, 64
Aridisols, 65, 67, 75–76, 243
    clays in, 238
    effect of overgrazing on, 76, 77
    phosphorus in, 406
Argiaquolls, 66
Argillans, 63
Argillic horizons, 63, 81
Aspect, 215
Atmosphere, 10
    effect of soils on, 349
    nitrogen in, 453
    sulfur in, 382
Atterberg limits, 128, 129
Autochthonous organisms, 326
Autotrophic organisms, 297, 312, 362, 363, 366, 377
Azolla, 312, 375–76
Azotobacter, 376, 377

Bacillus, 366
Bacteria, 292, 294, 295, 312, 320
    actions of, 118–19, 351, 362, 366
    effect of pH on, 265
    nitrogen fixing by, 312, 371, 377
    as pollution remediator, 312, 313, 315, 319
    symbiosis with plants, 372–77
B horizons, 12, 47, 56
    in epipedons, 62
    time required for formation, 51
Bicarbonates, 268, 323
    role in alkalinity, 256–57
Bioengineering, 497
Biogas, 448
Biomass, 2, 18
Biopores, 113, 114, 144
Bioremediation, 212
Biosequence, 35
Biosolids, 449, 450
Biosphere 2, 350
Biota, 34, 44, 47–50
Biotite, 30, 235, 412, 422
Black alkali, 279
Blocklike soil structure, 104, 105
Blue baby syndrome, 364
Blue-green algae, see Cyanobacteria
Bogs, 73, 74, 207, 352
Border irrigation, 187
Boron, 4, 281, 421, 428–29
    in fire retardants, 444
    forms of, 424

Boulders, 97, 100
Bradyrhizobium, 372, 373
Broadcasting of fertilizers, 459, 461
Bubblers, 189–90
Buffering capacity, 259
Bulk density, 106–13
    factors affecting, 106–7
    management practices and, 110–12
Burned lime, 268
Bypass flow, 176, 177
By-products as nutrient sources, 448–49

Calcareous horizons, 257
Calcic horizons, 63, 75, 256
Calcite, 54, 268
Calcium, 4, 20, 47, 402, 442
    and potassium levels, 419
    requirements for microorganisms, 314
Calcium carbonate, 12, 47, 56, 406
    geographic dispersal of, 43
    use in acid reduction, 255
Cambic horizons, 71
Capillarity, 133, 134–35, 136
Capillary fringes, 174, 175, 183
Capillary rise, 160
Capillary water, 151, 153
Carbon, 322, 323, 324, 326, 341
    accumulation in soil, 7, 19
    amounts in soil, 322, 324, 329
    balance in the environment, 343–44, 345
Carbonates, 323
    role in alkalinity, 256–57
    sources of, 268
Carbonation, 34
Carbon cycle, 322–23, 334, 344, 352
Carbon dioxide, 4, 7, 19, 195, 223, 323
    as greenhouse gas, 349, 351
    reactions of lime with, 268
    in soil air, 22, 197
    sources of, 324, 326, 327
Carbonic acid, 241, 256, 257, 323
Carbon/nitrogen (C/N) ratio, 329–30, 354
Carnivores, 295
Casts, 298–99
Catchments, 160
Catenas, 50
Cation cycling, 47
Cation exchange, 226, 230, 241–46, 250
Cation exchange capacity (CEC), 241–45, 250, 259
    variations between soils, 242–44
Cations, 54, 256
    adsorbed, in soil colloids, 229–30
    effect on behaviors of clays, 116–17
    as nutrients, 245–46
    and pH, 255, 257
    regional variations in, 245
    and sulfur, 387
Cation saturation, 245
Chelates, 409, 427–28, 432
    formation of, 34, 426
    micronutrients in, 424
    uses of, 466
Chelation, 341
Chemical decomposition, 31, 33–34

Chemical protection, 327
Chisel plowing, 490
Chlorine, 421, 423, 428
Chlorites, 233, 235–36
Chlorofluorocarbons (CFCs), 351
Chlorosis, 358, 412
C horizons, 12, 56
Chronosequence, 35, 52
Clay domains, 116
Clayey soil:
  behavior of water in, 140, 141, 153
  versus sandy soil, 109, 113, 145–46, 163, 242, 347
Clay-humus complexes, 335, 341, 347
Claypans, 148, 184
Clays, 17, 23, 55, 98–99. *See also* Silicate clays; Smectite
    clays
  ammonium fixation by, 361
  cations and, 116–17, 230
  effect on soil structure, 54
  erosion resistance of, 483–84
  expansive, 147
  factors in formation of, 35, 237
  flocculation of, 116
  formation of rock from, 31
  geographic distribution of, 237–38
  micronutrients in, 423, 426, 428, 429
  phosphorus fixation by, 403, 406, 408–9
  potassium fixation by, 361, 417–18
  as sealants, 248
  soils formed from, 77, 82
  sources of charges in, 232
  sulfur in, 382, 386–87
  swelling and shrinking of, 9, 117–18, 127, 248, 249
  variations in ion exchange capacities of, 247
  water-holding variations in, 153
Clay skins, 63
Clean Water Act of 1992, 495
Climate, 34, 44
  as determinant of vegetation type, 47
  effect on parent materials, 35, 45–47
  influence on organic matter levels, 344, 346–47
  influence on weathering, 51–52, 67
Climatic factor *C,* 501
Climosequence, 35
Clostridium, 377
Coarse fragment modifiers, 100
Coastal plains, 38–39
Cobalt, 421, 426, 422–23
Cobbles, 97, 100
Cocomposting, 338
Coefficient of Linear Extensibility (COLE), 128
Cohesion, 132, 133, 98
Collapsible soils, 126
Colloids, 99. *See also* Soil colloids
Colluvium, 36, 51
Compaction, 113, 140, 153–54, 163, 204
  in engineering applications, 127
  by farm implements, 110–11, 122
  minimizing, 111–12
Companion plantings, 339
Complexation, 34, 47
Compost, 316, 335–38, 348, 354, 449
  as micronutrient supplier, 426
Compressibility, 8, 127
Conservation Reserve Program (CRP), 348–49, 506–7

Conservation tillage, 221, 284, 344, 349, 439, 441, 489–93
  effect of, 491–93
  erosion protection by, 122–23, 348, 486, 491
  as evaporation controller, 169, 170–71
  extent of use, 490–91, 507
  yields provided by, 490
Consistence, 125
Consistency, 125–26
Consolidation test, 127
Constant charges, 238–39
Controlled traffic systems, 111
Copper, 4, 325, 421, 422, 424
  deficiencies of, 426
  toxicities of, 429
Core cultivation, 206
Cover and management factor *C,* 484–87
Cover crops, 163, 348, 438, 440
  erosion protection by, 485
  for nitrogen supply, 443
Critical range, 463
Crops:
  as mulch, 169, 170–71
  rotation of, 315, 348, 427, 443–44
Crotovinas, 48
Crusting, 123, 481
Cryic temperature regime, 65
Cryoturbation, 73
Cryptobiotic states, 301
Cryptopores, 115
Cultural eutrophication, 391, 394
Cyanobacteria, 305, 312, 314
  nitrogen fixing by, 371, 375, 377
Cysts, 302

Darcy's law, 143, 145
Decomposition, 31, 33–34, 320, 324–33
  in aerobic soils, 326–27
  in anaerobic soils, 327–28
  factors controlling, 328–33
  rate of, 325, 354
  role of fungi in, 307
  role of organisms in, 314
Delivery ratio, 481
Deltas, 37, 38
Denitrification, 366–71
Depth ratio, 141
Desertification, 474
Desert pavement, 76
Desert varnish, 76
Detritivores, 294, 296–97
Detritus, 36, 294, 341
Diagnostic soil horizons, 62–63
Diammonium phosphate (DAP), 444, 453
Diffusion, 25, 195, 196
Dioctahedral sheets, 230, 238
Disease-suppressive soils, 316
Dissolution, 33
Dissolved organic phosphorus (DOP), 401
DNA, 330, 392
Dolomite, 268
Double ring infiltrometers, 146, 147
Drainage, 160, 179–81, 183
  artificial, 177–79, 200
  in building foundations, 181

Drainage (*cont.*):
    design considerations for, 180
    for disease control, 316
    effect on organic matter levels, 347
    enhancing, 176–81
    impediments to, 200
    for salt removal, 283
    in septic systems, 181–85
Drain fields, 181, 183, 191
Dribble application of fertilizers, 460
Drift, 40
Drip irrigation, 189, 190, 461
Drumlins, 41
Dukochaev, V. V., 34
Dune sand, 42
Duripan, 63
Dust, 473, 478

Earthworms, 297–300, 328
    activities of, 49, 162, 295, 320, 324
    casts of, 298–99
    encouraging, 163, 169, 300
    influence on soil productivity, 298, 362
    pores created by, 113, 175
Ecosystems:
    crisis in, 1–2
    interdependency of, 194
    soils as, 7–8, 27, 290–320
Ectomycorrhiza, 309, 310
Eelworms, 301
Effective precipitation, 45, 46, 47
Effluent, 183, 449–50
E horizons, 12, 47, 55
Electrical conductivity (EC), 275
Electrical resistance block, 142–43
Electromagnetic induction, 276
Elements, 4, 20, 203
Eluviation, 55
Enchytraeid worms, 297, 298, 299
Endoaquic soil, 176, 178
Endomycorrhiza, 309–10
Entisols, 66, 70, 116, 244
Enzymes, 330, 358
Enzyme systems, 422, 429
Eolian materials, 42
Epiaquic soil, 176, 178
Epipedons, 62–63
Equilibrium phosphorus concentration (EPC), 406
Erosion, 100, 160, 473–508. *See also* Weathering
    on construction sites, 495
    costs of, 478
    and deposition, 473, 477, 481
    effect of, 476–78, 507–8
    geological versus accelerated, 475–76
    in irrigation, 187, 188
    measures against, 485–86, 491, 494–98, 502–3, 508
    and phosphorus, 394, 397
    reducing, 169, 221
    role in disease spread, 477
    soil losses to, 476, 506, 507, 508
    by water, 480–88
    by wind, 499–503
Erosion blankets, 496, 497
Eskers, 41

Essential elements, 20, 22, 23
Eutrophication, 175, 395
    contributors to, 203, 357, 364, 401, 432, 448, 477
    cultural, 391, 394
Evaporation, 159, 160, 165
    control of, 169–71, 212
Evapotranspiration, 160, 165–67, 169
Exchangeable sodium percentage (ESP), 275, 276
Exchange complex, 241
Exfoliation, 31
Expansive soils, 127

Facultative bacteria, 312, 328
Fairy rings, 307, 308
Fallow land, 169
Farming:
    best practices in, 120, 122–23, 169, 348, 365, 379, 420
    effect on ecosystems, 435
    effect on soil, 110, 113, 119–21, 122, 342, 347, 414
    impact on soil organisms, 317, 319
    land degradation caused by, 474, 475
    nutrient-management goals of, 436
    practices to control disease, 315–16
Fe, Al oxide clays, 228
    formation of, 236
    ion activities in, 254–55
    sources of charges on, 239
Feldspars, 17, 30, 33, 106, 412, 417
Ferrihydrite, 71
Ferrous sulfate, 267–68
Fertigation, 459
Fertilization, 451–57
    benefits of, 348, 362
    combination of ammonium and phosphorus, 410
    effect of, 50, 150, 168, 241, 252, 260, 319
    by fire retardants, 444
    in forestry, 451
    hazards of, 351–52, 365, 378
    and micronutrients, 426, 430
    regional use of, 451–52
    regulation of nitrogen with, 363
Fertilizers, 452–53, 455
    application methods for, 459–62
    fate of nutrients in, 457
    grades of, 455–57
    nutrient content of, 454–55
    proper use of, 462, 469–70, 471
Fibric materials, 44
Field capacity, 151
Field water efficiency, 186
Filter fields, 181
Fine earth fraction, 97
Fire, effect on soil, 214, 444, 475
Flags, 100
Flocculation, 116
Flooding, 179, 208
Floodplains, 37–38
Fluvents, 70
Foliar application of fertilizers, 462
Food web, 292, 293
Forest floor, 55, 221, 493, 494
Forestry:
    erosion and, 493–94, 495
    nutrient-management goals of, 436

Fourier's Law, 217
Fragipans, 63, 88, 112, 148, 184
Fragiudalf, 87
Frankia, 314, 375
Friable soils, 125
Frost churning, 73
Frost heaving, 177, 178, 213, 224
Fugitive dust, 478
Fulvic acids, 229, 341, 401
Functional diversity, 292
Functional redundancy, 292
Fungi, 265, 292, 295, 296, 306–10, 320. *See also*
     Mycorrhizae
  activities of, 307–8
  and algae, 305, 317
  antibiotic compounds produced by, 307
  detrimental effect of, 307–8, 315
  irrigation and, 188
  reducing risk from, 179
  role in aggregation, 118
Furrows, 187
Furrow slice, 13
Fusarium, 306
Fusarium wilt, 316

Gabbro, 31
Gabions, 497
Gelisols, 67, 72–74, 324
Genetically engineered microorganisms (GEM), 319–20
Geographic information system (GIS), 92, 468
Geosmins, 311
Gibbsite, 228, 236
  phosphorus fixation by, 403, 409
Gilgai, 78, 79
Glacial outwash, 40–41
Glacial till, 40
Glaciers, 40
Gleying, 64, 96, 209
Global biodiversity, 292
Global carbon cycle, 322–23
Global Positioning System (GPS), 468, 469
Global warming, 204, 322, 349
Glomalin, 118
Goethite, 228, 236, 403, 409
Granite, 30, 31, 35
Granulation, 113
Gravel, 16, 97, 100, 148
Gravimetric method, 141–42
Gravitational potential, 137, 138
Gravitational water, 150
Gravity, 136, 138
Greenhouse effect, 7, 19, 322, 349–52
  contributors to, 204, 328, 367
Greenhouse gases, 349–52
Green manure, 332, 374, 410, 438
Groundwater, 173–76
  contamination of, 1,2, 174–75, 248, 252, 265, 364, 445
  denitrification in, 369–71
  phosphate movement in, 401
  reducing pollutant levels in, 438
Gully erosion, 161, 481, 482
Gypsic horizons, 63, 75
Gypsum, 12, 33, 54, 381
  as acid neutralizer, 273

as salt reducer, 285, 286
as soil conditioner, 123–24

Halophytes, 280
Harrows, 110, 122, 124
Hematite, 228, 236
Hemic materials, 44
Herbivores, 294
Heterotrophic organisms, 297, 366, 377
Hierarchy of aggregation, 116
Highly erodible land (HEL), 506
Histels, 73, 324, 352
Histic epipedons, 63
Histosols, 74, 202, 243, 324, 352–53, 354–55, 402, 423
Horizons, 10–15, 55–56, 57, 92
  aeration variations in, 200, 201
  diagnostic, 62
  effect of animal burrows on, 48–49, 295
Hornblende, 30, 422
Humans:
  effect on climate, 349
  effect on soil, 49–50, 104, 252, 260–63
  influence on hydrologic cycle, 176
  relation to environment, 158
Humic acids, 34, 341
Humic substances, 334
Humin, 341
Hummocks, 73
Humus, 7, 239, 327, 333–34, 344
  versus clay, 19, 335
  colloid characteristics of, 228, 335
  functions of, 23, 53, 243
  influence on plant growth, 338–39
  influence on soil color, 53, 341
  as soil conditioner, 124
Hydrated lime, 268
Hydration, 33
Hydraulic conductivity, 143–46
Hydraulic gradient, 144
Hydric soils, 208, 209
Hydrogen, 253–55, 324
Hydrogen bonding, 132, 232
Hydrologic cycle, 158–92
Hydrolysis, 33
Hydrophytes, 195, 201, 208, 209
Hydroponics, 6, 338
Hydroseeders, 496
Hygroscopic coefficient, 151, 153
Hyphae, 306, 307, 398–99

Ice, 31–32, 118, 212
Igneous rock, 30–31
Illite, 233, 238, 419
Illuviation, 56, 100
Immobilization:
  of nitrogen, 359–60
  of organic phosphorus, 402
  of sulfur, 384–85
Imogolite, 71, 228, 236–37
Inceptisols, 65, 71, 244, 324, 406
Infiltration, 146, 160, 169
Infiltration capacity, 146, 483
Infiltration systems, 404

Inner-sphere complexes, 246, 247
Inorganic phosphorus, 399, 402–6
Inositol phosphates, 400, 401
Interception, 160
Interlayer expansion, 234
Internal drainage, 181
Interped pores, 113, 114
Interrill erosion, 481
Ion exchange, 23, 25, 226, 250
Ionic double layer, 227
Iron, 4, 47, 55, 209, 325, 371, 372, 402, 421, 422, 424
   chelation of, 427–28
   deficiencies of, 426
   oxidation of, 34, 197, 198
   reduction of, 203, 404
   sources of, 422
Iron oxide, 12, 17, 31, 54, 56
Iron oxide clays, *see* Fe, Al oxide clays
Irrigation, 160–63, 185–91
   effect of, 12, 30, 146, 173, 262
   efficiency of, 186, 191
   as erosion control, 502
   and evapotranspiration, 169
   fertilizer application in, 459
   salinity and alkalinity induced by, 274–75, 281, 284
   timing of, 286
   use of sewage effluent in, 449–50
   water management in, 190–91
Isomorphous substitution, 231–32, 238–39

Jenny, Hans, 34

Kandic horizons, 81
Kaolinite clays, 63, 128, 228, 229, 232–33, 236, 237, 247
   charges on, 239, 246
   phosphorus fixation by, 403, 409
   potassium fixation by, 419
Keystone species, 292

Lacustrine deposits, 41
Land capability classes, 503–5
Land degradation, 474–75
Land forming, 179, 181
Leaching, 54, 175, 179, 246, 462
Leaching requirement, 283
Leaf area index (LAI), 166, 167, 168
Leaf litter, 35, 47, 166, 328, 347
Legumes, 348
   nitrogen fixation by, 372, 443–44
   symbiosis with bacteria, 372–75
Level basin irrigation, 187
Lichens, 294, 305, 317
Liebig, Justus von, 457
Lignins, 325, 341, 342, 344, 354
   breakdown of, 326
   influence on decomposition rate, 331–33
Lime, 348, 362
   amounts needed to change pH, 269–71
   considerations in applying, 271–73
   depletion of, 269
   effect of, 50, 241, 319, 429

   guarantees on, 270, 271
   methods of applying, 272
   and potassium levels, 415, 419
   reactions in soil, 268–69
Limestone, 31, 35, 51
   amounts needed to neutralize acidity, 256
   importance of fineness of, 270
   types of, 268
Limiting factors, 457–58
Liquid limits, 128
Lithosequence, 35
Live stake technique, 497
Loams, 100
Localized placement of fertilizers, 459–62
Loess, 42–43
Lumbricus terrestris, 297–98, 299
Luxury consumption, 416, 420

Macroaggregates, 118
Macrofauna, 292
Macronutrients, 6, 264
Macropores, 113, 144, 151
   and aeration, 195, 200
   chemical movement through, 175–76
Magnesium, 31, 47, 419, 466
Manganese, 179, 203, 325, 402, 421, 422, 424, 429
   application of, 459
   deficiencies of, 426
   oxidation of, 34, 198
Manure, 200, 348
   benefits of, 295, 352, 362, 387–88, 444
   composition of, 397, 420, 445–46, 447
   disposal of, 445, 446
   effect of, 241, 319, 378, 410, 426
   methods of applying, 446, 448
   as micronutrient provider, 421, 423
   as mulch, 169
   and nitrate leaching, 365
Marine sediments, 38–39
Marshes, 74, 207, 352, 386
Mass flow, 25, 195
Mass water content, 141
Master horizons, 55–56
Master variables, 252
Matric force, 136
Matric horizons, 63
Matric potential, 137, 138–39
Matric potential gradient, 144
Maximum retentive capacity, 150
Melanic epipedons, 62, 72
Mesofauna, 292
Metamorphic rock, 30, 31
Methane, 195, 197, 198, 204, 210, 223
   as greenhouse gas, 351
   reducing production of, 179
   sources of, 301, 328, 351, 448
Methanotrophic bacteria, 351
Methemoglobinemia, 364
Micas, 17, 33, 106, 235, 412, 417
   fine-grained, 229, 233, 235, 238
Micelles, 227, 229
Microaggregates, 116
Microbes, 200, 202, 212, 223
   effect of pH on, 264–65

gases produced by, 197, 200
role in reducing elements, 316
Microclines, 33, 412
Micrococcus, 366
Microfauna, 292, 301–3, 312–14
Microflora, 295, 296, 312–14, 315
Microirrigation, 189–90
Micronutrients, 6, 391–92, 421–31, 432
availability of, 264, 424–29
cycling of, 424
deficiencies of, 422, 424, 429, 430, 431, 432
forms of, 424
interactions among, 429–30
versus macronutrients, 421
oxidation of, 425
sources of, 325, 341, 421, 422–23, 424–26, 430
toxicities of, 421, 432
Microphytic feeders, 295
Micropores, 16, 113, 114–15
Middens, 297, 299
Milorganite, 450
Mineralization, 327, 328–33, 341
of nitrogen, 359, 360, 361
of organic phosphorus, 401–2
of sulfur, 383–84
Minerals, 16–18, 30–31, 87, 382, 417, 424. *See also* Primary
minerals; Rocks
expanding, 234
nonexpanding, 235–36
1:1 type, 232–33
secondary, 17, 30, 33
2:1 type, 233–36
weathering of, 29–34, 54
Mites, 292, 294, 295, 296–97
Molds, 306
Mollic epipedons, 62
Mollisols, 59, 66, 79–80, 238, 243
C/N ratio of, 329
phosphorus in, 402, 406, 409
Molybdenum, 281, 371, 372, 421, 422, 423, 424, 429
Monera, 311
Montmorillonite, 229, 234
Moraines, 40
Mucigel, 305
Muck, 44, 352
Mudslides, 126, 131, 214
Mulch, 124, 169–70
decomposition rate of, 328, 449
paper and plastic, 169–70, 222, 223
and soil temperature, 221–23
use in erosion protection, 485–86, 496, 501
Municipal solid waste (MSW), 449
Munsell color charts, 96
Muscovite, 30, 235, 412
Mushroom fungi, 306, 307, 308
Mutualistic associations, 317
Mycelia, 306, 328
Mycorrhizae, 148, 308–10, 317, 410, 427
as aid in phosphate uptake, 398
Mycotoxins, 307

Natric horizons, 75, 81
Necrosis, 412
Nematicides, 302

Nematodes, 292, 295, 296, 301–2
parasitic, 294, 301, 315
Neutron scattering, 142
Nickel, 421, 424
Nightcrawlers, 297–98, 299
Nitrate depression, 330, 333
Nitrates, 179, 198, 358, 377, 378
and composting, 338
hazards of, 357
leaching of, 363, 364–66
origin of, 326, 327, 328
recycling of, 366
reduction of, 200
Nitric oxide, 366, 367
Nitrification, 362–63
Nitrite, 198, 362, 364
Nitrobacter, 362, 363
Nitrogen, 4, 19, 20, 179, 203, 325, 326, 357–79, 389
amounts in soils, 358–59
application of, 459
and climate, 344
competition for, 316
deficiencies of, 330, 338, 357, 358
in eroded material, 477
excesses of, 357, 358
in fertilizers, 378
immobilization of, 359–60
importance of, 330, 348, 357, 469
leaching problems with, 175
losses through fires, 444
mineralization of, 359, 360, 361
versus potassium, 411, 416
in precipitation, 377–78
sources of, 311, 325, 341, 342, 377–78, 442, 453
testing for, 466
as water pollutant, 438
Nitrogenase, 371
Nitrogen cycle, 359
Nitrogen fixation, 312, 314
biological, 371–72, 436
conditions required for, 371–72
nonsymbiotic, 377
and phosphorus, 392, 393
by plant-bacteria complexes, 372–77
regulators of, 379
Nitrogen fixing nodules, 317
Nitrosomonas, 362, 363
Nitrous oxide, 352, 357, 366, 367
Noncohesive soils, 126
Nonhumic substances, 334–35
Nonpoint sources, 394
No-tillage, 170, 344, 396, 490
Nucleic acids, 358, 400, 401
Nutrients:
amounts in soil solution, 23
conservation of, 436
cycling of, 441–44
imbalances among, 436–37
interactions among, 457, 469–70
leaching problems with, 175, 364
management of, 435–71
mineral, 4–6
modes of existence of, 24
necessity for replenishment of, 23, 441

Nutrients (*cont.*):
  organic, 6
  pH and availability of, 264–65
  as pollutant, 477
  reuse of, 7, 436, 444–46, 470
  site-specific management of, 468–69
  sources of, 6, 19, 304–5, 322, 341, 342, 441
  supply systems for, 4–6, 20, 22–23
  uptake by plant roots, 24–26

Ochric epipedons, 62
Octahedral sheets, 230–31
O horizons, 47, 55
1:1-type minerals, 232–33
Organic material, 18–20, 200, 322–55
  as acidifier, 267
  active fraction of, 341, 342
  advice on utilization of, 450–51
  amount in soils, 19, 344–49
  effect of, 96, 119, 252, 409, 426
  in eroded material, 477
  influence on plant growth, 338–39
  influence on soil properties, 339–41
  management of, 341–42, 347–49
  nutrients provided by, 19, 423, 424
  passive fraction of, 341–42
  slow fraction of, 342
  sulfur in, 381
  types of, 44
  and water availability, 153–55
Organic phosphorus, 399, 400–2
Organisms in soil, 292–96
  classes of, 297
  competition for food, 317
  effect of, 314, 315–16
  as nutrient providers, 314
  as plant protectors, 314–16
  relationships among, 317–20
  role in carbon cycle, 323
Organoclays, 248
Orthoclase, 412
Osmosis, 136, 139
Osmotic potential, 137, 139
Outwash plains, 40
Overgrazing, 76, 77, 474, 475, 493
Overland flow systems, 404
Oxic horizons, 63
Oxidation, 34, 197–200, 203–4
  biochemical, 386
  decomposition as, 326
  effect of, 96, 203, 262–63
Oxidizing agents, 198
Oxisols, 59, 66, 84–85, 116
  clays in, 228, 237
  nitrate recycling in, 366
  phosphorus in, 402, 409
  sulfates in, 382, 387
Oxygen, 195. *See also* Soil aeration
  as factor in weathering, 33
  and plants, 4, 23, 324
  requirements for microorganisms, 314
  in soil air, 22, 196–97
  as stimulus to decomposition, 327

Oxygen-reduction potential, 197–200, 424
Ozone layer, depletion of, 1, 2, 131, 357, 367

Packing pores, 114
Pans, 63
Parent materials, 11, 34, 35–44, 51
  residual, 35, 36, 51
Partial pressure, 195
Particle density, determinants of, 106
Particle size, 87, 97–99
  analyzing, 102–4
  and erodibility, 501
Particulate organic matter (POM), 341
Particulate phosphorus, 394, 396
Partitioning, 248
Peat, 43–44, 50, 352
  compressibility of, 127
  types of, 43–44, 353
  uses of, 74, 353
Pebbles, 100
Pedons, 60, 61
Pedoturbation, 49
Peds, 104
Penetrometers, 113
Percentage base saturation, 245, 256
Percentage saturation, 245
Percolation, 145, 146, 171, 172, 173–76, 184
Percolation-evaporation balance, 171–73
Perc tests, 184–85
Periaquic SMR, 208
Permafrost, 73, 74
Permanent wilting percentage, 152
Perudic SMR, 64
Pesticides, 319, 363
Petrocalcic horizons, 75
Pfiesteria, 397
pH, 21–22, 252–73, 287, 328
  and anion exchange, 246
  and CEC, 244–45, 255
  determinants of, 252, 287
  and disease control, 315–16
  effect on ammonia volatilization, 362
  effect on life forms in soils, 264–65
  effect on micronutrients, 424–26, 428, 429, 430
  effect on oxidation, 425
  and environmental quality, 265
  influences of salts on, 257–58
  and phosphorus, 403, 409, 410
  and potassium levels, 419
  ranges in various soils, 21, 254
  and redox potential, 198
  requirements for microorganisms, 314
  and surface charges of colloids, 238, 241, 249, 335
pH-dependent charges, 238, 239–40, 254
Phenols, 325, 344
Phosphorus, 4, 20, 247, 326, 391–432, 442
  application of, 459
  availability of, 391, 393, 399–400, 403, 404, 406, 432
  content in soils, 393, 397–98
  deficiencies of, 391, 393–94
  in eroded material, 477
  fixation of, 372, 403, 406–9, 432
  importance of, 391, 392–93, 429

levels necessary for eutrophication, 395
losses through fires, 444
pathways for gains and losses, 400
versus potassium, 411, 415, 416
runoff losses of, 394, 396–97, 402
sources of, 325, 341, 397, 400, 452, 453
surfeits of, 391, 394
testing for, 466
and water quality, 391, 394, 395, 401, 405, 438
Phosphorus cycle, 397–400
Photosynthesis, 6, 323, 324, 371, 392, 397, 422
Phyllosilicates, 230
Physical disintegration, 31–33
Physical protection, 327
Phytotoxic substances, 4
Placic horizons, 63
Plaggen epipedons, 62
Plant-available water, 152, 153, 155, 156
Plant residues, 324–25, 348
    allelochemical effect of, 339
    effect of burning, 444
    effect on phosphorus availability, 410
    factors influencing quality of, 328
    as sulfur replenisher, 387–88
Plants:
    acid tolerance of, 265, 266
    as aid in salinity reduction, 285, 287
    best times to fertilize, 462
    C/N ratio in, 329
    contribution to soil formation, 53
    damage to, by soil organisms, 315–16
    diagnosing problems with, 463
    effect on aeration, 201, 202
    effect on organic matter levels, 347
    effect on soil development, 47
    effect on water partitioning, 161
    erosion protection by, 475, 484–85, 486, 501, 502
    essential nutrients for, 4–6
    as factor in weathering, 33
    nutrient uptake by, 24–26, 245–46, 358, 416
    optimal growth conditions for, 13, 14, 16, 19, 20, 27, 197
    osmotic potential and, 139
    potted, 151, 156, 204–5, 281
    rain funneling effect of, 161, 162
    respiration process of, 4
    salinity tolerance of, 279–81
    as temperature regulator, 216
    tolerance for poor aeration, 204
    transpiration variations in, 167
    unfavorable growth conditions for, 22, 25, 26, 112, 113, 123, 178, 279
    water uptake mechanisms of, 155
    wilting of, 152, 166
Plasticity, 98, 125
Plasticity index, 128
Plastic limits, 128
Platy soil structure, 104, 105
Plinthite, 82
Plow layer, 13
Plow pans, 110, 122
Plows, 110, 122, 490
Point sources, 394
Polyacrylamides (PAM), 124

Polypedons, 60
Polyphenols, 325, 331–33, 341, 354
Pore size, 113–15
    and capillarity, 134–35
    water movement and, 148–49
Pore space, 113–15
Potassium, 4, 20, 47, 391, 411–20, 432
    availability of, 415, 417–18
    deficiencies of, 391, 412, 420, 470
    depletion and restoration of, 414, 416
    in eroded material, 477
    fixation of, 361, 417–18, 419–20
    leaching of, 414, 415, 417, 420
    luxury consumption of, 416, 420
    versus other nutrients, 411, 415
    release of, 418–19
    sources of, 325, 412, 414, 453
    substitution of sodium for, 411
    testing for, 466
Potassium cycle, 412–14
Potential evapotranspiration rate (PET), 165–66
Potentiometers, 265
Precipitation, 160–63
    and leaching, 176
    as source of nitrogen, 377–78
Preferential flow, 144
Pressure potential, 138–39
Primary consumers, 293, 294–95
Primary minerals, 17, 30, 33
    clay formation from, 236–37
Primary producers, 293, 294
Priming effect, 327
Prismlike soil structure, 104, 105
Proctor test, 127
Proteins, 325, 335, 357, 358
    decay of, 326, 328
Protonation, 265
Protozoans, 292, 295, 302–303
    algae and, 317
Pseudomonas, 316, 366
Pseudosand, 116
Puddled soil, 121, 278

Quartz, 17, 30, 31, 55, 97, 98, 106
Quicklime, 268

Rain:
    effect on infiltration capacity, 146
    erosion by, 480–81, 483, 493–94
    global distribution of, 158
Rainfall erosivity factor *R*, 483, 484
Rain forests, 1, 2, 84
Recombinant organisms, 319
Recycling, 436
Redox potential *(E_h)*, 197–200, 424
Reducing agents, 198
Regolith, 9–11
    effect of living organisms on, 10
    transport from area of origin, 9, 11
    types in U.S., 36
    weathering of, 11, 12
Renewable resources, 436

Resilience, 292
Retention basins, 497–98
Revised universal soil loss equation (RUSLE), 483
Revised wind erosion equation (RWEQ), 502
Rhizobium, 314, 318, 372, 373
Rhizodeposition, 304–5
Rhizosphere, 304, 305, 376
Rice paddies:
    bacteria in, 312, 314, 377
    cover crops in, 443
    gas emissions of, 204, 351, 367
    nutrient losses in, 362, 369
    plant-bacteria complexes in, 375–76
    sulfur reduction in, 386
Rill erosion, 481, 482
Riparian buffer strips, 438
Riparian vegetation, 369, 494, 506
Riprap, 497
R layers, 56
RNA, 392
Rocks:
    types of, 30–31, 475
    weathering of, 29–34
Root interception, 25
Root nodules, 372, 375–77
Root rots, 308, 315
Roots, 303–5
    and aeration, 201
    depth of, 155
    effect of phosphorus on, 392
    effect of soil temperature on, 212
    iron reduction by, 426
    nematode infestation of, 301, 315
    as nutrient providers, 304–5
    pore formation by, 175
    specialized for waterlogged soils, 195, 208
Runoff water, 171, 481, 497–98

Salic horizons, 63, 75
Saline seeps, 274
Saline-sodic soils, 278
    reclamation of, 285–87
    transformation to sodic soils, 278–79, 281
Saline soils, 276
    reclamation of, 283–84
Salinity, measuring, 275–76
Salinization, 274
Salt-affected soils, 273–87
    global distribution of, 252, 273
    management of, 281–82
    reclamation of, 283–87
Saltation, 500
Salts:
    accumulation of, 26–27, 45, 51, 190, 252, 262, 274–75
    composition of, 274
    effect of, 139, 257–58, 275, 279
    as micronutrient provider, 421
    removal of, 173, 178, 179, 186, 283–87
    sources of, 274, 280
Sand, 16, 17, 97, 148
    as factor in weathering, 32
    soil formation potential of, 42
Sandstone, 31, 35, 51

Sandy soil:
    capillarity in, 134–35
    versus clayey soil, 109, 113, 145–46, 163, 242, 347
    saturated conductivity of, 144
    structure of, 107
Sapric materials, 44
Saprolite, 10
Saturated flow, 143–44
Saturated hydraulic conductivity, 143–44, 184
Saturation, 150, 177–78
Saturation paste extract method, 275
Secondary consumers, 293, 295
Secondary minerals, 17, 30, 33, 34
Sediment, 475, 477
    damages caused by, 438, 473, 477
    trapping, 497–98
Sedimentary rock, 30, 31, 475
Selenium, 203, 210, 429
    and irrigation, 281, 282
    toxicities of, 429
Septic systems, 181–85
Settlement, 127
Sheet erosion, 481, 482
Shrinkage limits, 128
Siderophores, 317
Silicate clays, 12, 31, 56, 227–28, 335
    formation of, 17, 33, 34, 54, 236
    mineralogical organization of, 232–36
    structure of, 230–32
Silt, 17, 97–98, 153
Silt fences, 497, 499
Slickenslides, 78
Slick spots, 279
Slope aspect, 50–51
Sludge, 183, 200, 449, 450
    impact on pH, 262
Smectite clays, 17, 128, 233, 234, 238, 246, 247
    potassium fixation by, 417–18, 419
    swelling properties of, 127, 234, 248
Sodic soils, 275, 279
    reclamation of, 285–87
Sodium adsorption ratio (SAR), 275, 276
Soil, 2–9, 26, 27, 35, 292
    analyzing, 12, 463–68
    classification of, 60, 65
    C/N ratio of, 329–30
    cracks in, 77, 78, 117–18, 175, 200, 234
    as ecosystem, 7–8, 27, 290–320
    as engineering medium, 3, 8–9, 125–29
    freezing and thawing of, 212–13
    and hydrologic cycle, 158–92
    as insulator, 4, 219
    as natural body, 9–11, 60
    organic versus mineral, 344, 352
    as recycling system, 3, 7, 18–19
    as regulator of water supplies, 6–7, 158
    sampling, 9, 464, 466
    sterilization of, 316
    thermal properties of, 217–20
    use in water purification, 7, 404, 449
    weight of, 108–9
Soil aeration, 194–210, 223, 328
    characterization of, 196–97
    effect of micronutrients, 204, 424, 426

factors affecting, 194, 200–2
in potted plants, 204–5
seasonal differences in, 202
in trees and lawns, 205–6
Soil air, 22, 196–97
Soil architecture, 95–129
Soil associations, 91
Soil biomass, 296
Soil buffering, 258–59, 270
Soil classification, 59–93
systems of, 59, 129, 503–5
Soil colloids, 17, 19, 99, 226–50
formation of, 236–37
micronutrients in, 424
nucleic acid adsorption of, 401
reactions with lime, 268–69
Soil color:
information provided by, 96, 203
sources of, 96, 203, 215, 341
Soil complexes, 91
Soil conditioners, 123–24
Soil consociations, 91
Soil creep, 500
Soil density, 106–13
Soil depth 87, 154
Soil erodibility factor I, 501
Soil erodibility factor K, 483–84, 485
Soil failure, 126
Soil formation, 10, 20, 29–57
Soil genesis, 52, 57
Soil heterogeneity, 200–2
Soil horizons, see Horizons
Soil individuals, 60
Soil layering:, 148–49, 154
Soil-loss tolerance (T), 479–80
Soil management, 121–25
effect on organic matter, 342
effect on soil organisms, 317, 319
and greenhouse gases, 352
importance of, 252, 354
for micronutrient needs, 430
and nitrate leaching, 365
role in disease control, 315–16
and water, 161, 163
Soil mapping, 89–91
Soil moisture, 166, 195, 328
effect on erosion, 500–1
effect on organic matter levels, 347
requirements for microorganisms, 314
and soil temperature, 217, 223
and thermal conductivity, 219
Soil moisture regimes (SMR), 64–65, 68
Soil orders, 65, 66–85
Soil pits, 11
Soil-plant-atmosphere continuum (SPAC), 163–67
Soil pores, 4, 16, 22. See also Pore size; Pore space
effect on bulk density, 106
types of, 113, 114
Soil productivity:
determinants of, 20
diagnosing problems with, 462–68
nutrient management and, 437–38
role of fungi in, 307
and sulfur, 387–88

Soil profiles, 11–12, 55–57
effect of terrain on, 50
horizons in, 57
Soil properties, 95–129
changeable versus unchangeable, 26
effect of, 3, 163
Soil quality, 8, 26–27
factors in, 322
nutrient management and, 437–38
Soil-ridge-roughness factor K, 501
Soil separates, 97–99, 100
Soil series, 60, 65, 66
Soil solarization, 212
Soil solution, 20–22
Soil stability, 8, 129, 292, 483
Soil strength, 113, 126
Soil structure, 17–18, 95, 104–5
effect of clays on, 54
effect on water content, 140
maintenance of, 115–16
Soil survey reports, 91–92
Soil surveys, 89–92, 93
Soil Taxonomy, 60–89, 93, 208, 352
nomenclature system in, 61, 65–66, 86, 87
Soil temperature, 194, 210–14, 314, 328
cover and, 215–16
effect on plants, 210–12
factors influencing, 214
fluctuations in, 219–20
under plastic mulch, 170
regulating, 169, 211, 221–23, 224
and respiration, 200
and soil color, 215
use in disease control, 212
Soil temperature regimes, 65, 68
Soil testing, 463–64, 467–68
Soil texture, 17, 95, 97–104
effect of, 35, 107, 153, 347
and hydraulic conductivity, 145, 146
and water availability, 153–55
Soil water, 20–22, 131–56
and aeration, 195, 224
availability to plants, 153–55
energy concepts in, 135–39
flow of, 143–46
measuring, 140–43
versus tap water, 20, 131
Soil water potential, 136–38, 140–43
Soil wetness, 150–55
Solar energy, 159, 214–17
Solum, 56
Species diversity, 292
Specific gravity, 106
Specific heat, 217, 218
Specific surface, 234
Spheroidal soil structure, 104, 105
Spirillum, 376
Spitters, 189–90
Spodic horizons, 63, 83
Spodosols, 83–84, 236, 299, 406
profiles of, 80, 85
Sprinkler irrigation, 187–89
Stem flow, 161, 162
Stickiness, 97, 125

Stones, 16, 100
Stubble mulch, 171, 490, 501
Submerged aquatic vegetation (SAV), 477
Submergence potential, 137, 138
Subsoil, 13–15
    ameliorating acidity in, 273
    versus surface soil layers, 16, 200
Subsoiling, 110
Subsurface drainage, 179, 181, 182
Sulfates, 217, 382, 383
    origin of, 326, 327, 328
    reduction of, 386
Sulfides, 328, 382, 383, 386
Sulfites, 386
Sulfur, 4, 325, 326, 372, 379–88, 442
    as acidifier, 267–68
    atmospheric, 382–83
    deficiencies of, 379–81, 383, 387–88, 389
    immobilization of, 384–85
    losses through fires, 444
    mineralization of, 383–84
    versus nitrogen, 357, 389
    oxidation of, 34, 198, 203, 385–86
    versus potassium, 411
    as salt reducer, 285
    sources of, 19, 325, 341, 381–83
    testing for, 466
Sulfur cycle, 383
Sulfuric acid, 285, 382, 452
Support practice factor P, 487–88
Surface area, 98, 99–100
    external, 232–33
    of soil colloids, 227, 249
Surface drainage, 179
Surface runoff, 160
Surface seals, 123
Surface tension, 133
Suspension, 500
Swamps, 74, 173
Symbiosis, 308, 372
Synthetic processes, 10–11

Temperature, 87
    effect of, 45, 47, 73, 344
    as weathering agent, 31, 45
Tensiometers, 142
Terminal velocity, 480
Termites, 294, 295, 300–1, 324, 351
Terraces, 52, 187, 487
Tertiary consumers, 293, 296
Tetrahedral sheets, 230
Textural triangle, 104
Thermal conductivity, 217
Thermal flux, 217
Thiobacillus denitrificans, 366
Thixotropy, 126
Threadworms, 301
Tillage, 122, 491. See also Conservation tillage; Farming
    and aeration, 200
    for disease control, 315
    and earthworms, 300
    effect of, 119, 260, 261, 317, 319, 342, 347
    and phosphorus levels, 394, 396
    on saline soils, 284

and soil management, 121–23, 129
    use in erosion control, 502
Tilth, 121–22
    managing, 124–25
    role of fungi in, 307
Time, and soil formation, 35, 44, 51–52
Time-domain reflectometry (TDR) 142
Tissue testing, 463, 465
Topographic factor LS, 484
Topography, 35, 44
    influence on soil development, 50–51
Toposequences, 35, 50
Topsoil, 13–15
Torric SMR, 64–65
Total porosity, 113
Total soil water potential, 136
Trace elements, see Micronutrients
Traffic pans, 110
Transformation, 52
Translocation, 52
Transpiration, 160, 165
Transpiration ratio, 167
Trees:
    acid tolerance of, 265
    fertilization of, 461, 462
    mycorrhizae on, 309
    nitrogen fixation by, 375
    nutrient economy of, 441–42
    selective cutting of, 439
Trioctahedral sheets, 230, 238–39
Triplett, Eric, 318
Turbels, 73
Turbidity, 477
2:1-type minerals, 233–36
Typic soils, 86

Udic SMR, 64
Ultisols, 66, 82–83, 116, 244
    clays in, 228, 237
    nitrate recycling in, 366
    phosphorus in, 402, 406, 409
    sulfates in, 387
Ultramicropores, 115
Umbric epipedons, 62
Unified System of Classification, 129
Universal soil loss equation (USLE), 482–83, 490
Unsaturated flow, 143, 144–45
Urea, 378, 453, 462
U.S. Army Corps of Engineers, 129, 207
U.S. Bureau of Reclamation, 129
USDA Natural Resource Conservation Service, 91
U.S. Department of Agriculture (USDA), 60, 97, 207, 503
Ustic SMR, 64, 166
Ustolls, 81, 85
Ustox, 84

Vadose zone, 173, 174, 175
Valley fills, 41
Vapor movement, 143, 150
Vapor pressure gradient, 165
Variable charges, 238, 239–40, 254
Variscite, 402

Vegetation, *see* Plants
Vegetative cover factor *V*, 501
Vermiculites, 233, 234–35, 238, 247, 254, 335
    ammonium/potassium fixation by, 361, 417–18, 419
Vertisols, 76–79, 234, 238, 402, 409
Vesicles, 310
Vesicular-arbuscular mycorrhizae (VAM), 309
Volcanic ash, 42, 43
    soils formed from, 71, 228, 236–37
Volumetric water content, 141

Wastewater treatment, 210, 281–82, 403, 404, 449
Water, 20, 31, 131,132, 156, 159
    barriers to, 149, 248
    content in plant tissues, 324
    efficiency in use of, 167–68
    erosion by, 480–88
    as factor in weathering, 31–32, 33, 45
    global distribution of, 158, 159
    management practices, 160
    movement in stratified soils, 148–49
    nutrient management and, 438
    and phosphorus, 391, 394, 395, 401, 402, 405
    soil as purifier of, 7, 404, 449
Water application efficiency, 186
Water balance equation, 160
Water bridges, 126
Water deficit, 166
Waterlogged soils, 195
Water partitioning, 161–63
Water potential gradient, 144
Water release characteristic curves, 140
Watersheds, 160
Water stress, 168, 169, 178
Water tables, 173, 174
    perched, 176, 178

Weathering, 29–34, 54, 100, 412
    effect of, 260, 441
    and ion exchange capacity, 217
    pathways of, 31, 32
    and phosphorus availability, 393
    rates of, 29, 51–52
Weeds, 169, 212, 214, 339
Wetlands, 194, 207–10, 224, 368
    bacteria in, 312, 377
    chemistry of, 209–10
    constructed, 210, 498
    drainage of, 50, 74, 178–79
    gas losses in, 362, 368, 369
    methane emissions of, 204, 351
    soils in, 22, 74, 176, 177, 207
Wetting fronts, 146–48
White alkali, 276, 277
Width of field factor *L*, 501
Wilting coefficient, 151, 152
Wind, 32, 499–503
    parent materials transported by, 42–43
Wind erosion prediction equation (WEQ), 501–2
Winter burn, 212

Xeric SMR, 65, 166
Xerophytes, 150, 156

Yeasts, 306

Zinc, 421–24
    application of, 459
    deficiencies of, 426, 470
    functions of, 422, 429
Zymogenous organisms, 326